PETROLEUM
REFINING
PROCESSES

CHEMICAL INDUSTRIES

A Series of Reference Books and Textbooks

Consulting Editor

HEINZ HEINEMANN
Berkeley, California

ADDITIONAL VOLUMES IN PREPARATION

PETROLEUM REFINING PROCESSES

James G. Speight

CD&W Inc.
Laramie, Wyoming

Baki Özüm

Apex Engineering Inc.
Edmonton, Alberta, Canada

MARCEL DEKKER, INC.   NEW YORK • BASEL

ISBN: 0-8247-0599-8

This book is printed on acid-free paper.

Headquarters
Marcel Dekker, Inc.
270 Madison Avenue, New York, NY 10016
tel: 212-696-9000; fax: 212-685-4540

Eastern Hemisphere Distribution
Marcel Dekker AG
Hutgasse 4, Postfach 812, CH-4001 Basel, Switzerland
tel: 41-61-261-8482; fax: 41-61-261-8896

World Wide Web
http://www.dekker.com

The publisher offers discounts on this book when ordered in bulk quantities. For more information, write to Special Sales/Professional Marketing at the headquarters address above.

Current printing (last digit):
10 9 8 7 6 5 4 3 2 1

PRINTED IN THE UNITED STATES OF AMERICA

Preface

In recent decades, the energy industry has experienced significant changes in oil market dynamics, resource availability, and technological advancements. However, our dependence on fossil fuels as our primary energy source has remained unchanged. Developments in exploration, production, and refining technologies allow utilization of resources that might have been considered unsuitable in the middle decades of the 20th century.

It has been estimated that global energy consumption will grow about 50% by the end of the first quarter of the 21st century, and about 90% of that energy is projected to be supplied by fossil fuels such as oil, natural gas, and coal. In this supply-and-demand scenario, it is expected that the existing peak in conventional oil production will decline within the next two to three decades and the production of oil from residua, heavy oil, and tar sand bitumen will increase significantly.

As in the 1940s and 1950s, the next two decades will see a surge in upgrading technologies to produce marketable products from residua. The need continues for the development of upgrading processes in order to fulfill market demand as well as to satisfy environmental regulations. In one area in particular—the need for residuum conversion—technology has emerged as a result of the declining residual fuel oil market and the necessity to upgrade crude oil residua beyond the capabilities of the visbreaking, coking, and low-severity hydrodesulfurization processes.

The precursor to this book (*Petroleum Processing Handbook*, J. J. McKetta, Editor, Marcel Dekker, New York, 1992) filled the needs of many readers by covering up-to-date processing operations in an easy-to-read, understandable manner. The current book brings the reader further up to date and adds more data as well as processing options that may be the processes of choice in the future.

In the meantime, the refining industry has entered a significant transition period with the arrival of the 21st century and the continued reassessment by various levels of government—and by various governments—of oil-importing and -exporting policies. Therefore it is not surprising that refinery operations have evolved to include a range of *next-generation processes*, as the demand for transportation fuels and fuel oil has steadily grown. These processes differ from one another in method and product slates and will find employment in refineries according to their respective features. Their primary goal is to convert heavy feedstocks, such as residua, to lower-boiling products. Thus, these processes are given some consideration in this volume.

The book is divided into three parts: "Feedstock Terminology, Availability, and Evaluation," "Engineering Aspects of Refining," and "Refining." Each part takes the reader through the steps necessary for crude oil evaluation and refining. Part I (Chapters 1–4) deals with the prerefining steps and outlines how to evaluate a feedstock prior to applying refining processes. Part II (Chapters 5–12) will be of particular interest to the engineer who needs to understand the mathematics of chemical reaction, reaction kinetics, transport phenomena, and reactor engineering. Part III (Chapters 13–20) describes in detail, with relevant process data, the various processes that can be applied to a variety of feedstocks. All processes are described with sufficient detail to explain their operation.

By presenting the evolutionary changes that have occurred to date, this book will satisfy the needs of engineers and scientists at all levels from academia to the refinery, helping them to understand the refining processes and prepare for new changes in the industry.

James G. Speight
Baki Özüm

Contents

PETROLEUM
REFINING
PROCESSES

1
Definitions and Terminology

1. INTRODUCTION

Historically, petroleum and its derivatives have been known and used for millennia (Table 1.1) (Henry, 1873; Abraham, 1945; Forbes, 1958a, 1958b, 1959, 1964; Cobb and Goldwhite, 1995). Ancient workers recognized that certain derivatives of petroleum (e.g., asphalt) could be used for civic and decorative purposes whereas others (e.g., naphtha) could provide certain advantages in warfare.

Scientifically, petroleum is a carbon-based resource and is an extremely complex mixture of hydrocarbon compounds, usually with minor amounts of nitrogen-, oxygen-, and sulfur-containing compounds as well as trace amounts of metal-containing compounds (Chapter 3). Heavy oil is a subcategory of petroleum that contains a greater proportion of the higher boiling constituents and heteroatomic compounds. Tar sand bitumen is different from petroleum and heavy oil insofar as it cannot be recovered by any of the conventional (including enhanced recovery) methods (see Chapter 2).

In their crude state, petroleum, heavy oil, and bitumen have minimal value, but when refined they provide high-value liquid fuels, solvents, lubricants, and many other products. The fuels derived from petroleum account for approximately one-third to one-half of the total world energy supply and are used not only for transportation fuels (gasoline, diesel fuel, and aviation fuel, among others) but also to heat buildings. Petroleum products have a wide variety of uses in forms that vary from gaseous and liquid fuels to near-solid machinery lubricants. In addition, the residue of many refinery processes, asphalt (a once maligned by-product), is now a premium value product for highway surfaces, roofing materials, and miscellaneous waterproofing uses.

The *definition* of petroleum has been varied, unsystematic, diverse, and often archaic. Furthermore, the terminology of petroleum has evolved over a period of many years. Thus the long-established use of an expression, however inadequate, is altered with difficulty, and a new term, however precise, is at best adopted only slowly.

Because of the need for a thorough understanding of petroleum and the associated technologies, it is essential that the definitions and terminology of petroleum science and technology be given careful consideration at the outset. This will aid in gaining a better understanding of petroleum, its constituents, and its various fractions. Of the many forms

Table 1.1   Summary of the Historical Uses of Petroleum Derivatives

3800 B.C.	First documented use of asphalt for caulking reed boats
3500 B.C.	Asphalt used as cement for jewelry and for ornamental applications
3000 B.C.	Documented use of asphalt as a construction cement by Sumerians; also believed to be used as a road material; asphalt used to seal bathing pool or water tank at Mohenjo-Daro
2500 B.C.	Documented use of asphalt and other petroleum liquids (oils) in the embalming process; asphalt believed to be widely used for caulking boats
1500 B.C.	Documented use of asphalt for medicinal purposes and (when mixed with beer) as a sedative for the stomach; continued reference to use of asphalt liquids (oil) as illuminant in lamps
1000 B.C.	Documented use of asphalt as a waterproofing agent by lake dwellers in Switzerland
500 B.C.	Documented use of asphalt mixed with sulfur as an incendiary device in Greek wars; also use of asphalt liquid (oil) in warfare
350 B.C.	Documented occurrence of flammable oils in wells in Persia
300 B.C.	Documented use of asphalt and liquid asphalt as incendiary device (Greek fire) in warfare
300 B.C.– A.D. 250	Documented occurrences of asphalt and oil seepages in several areas of the Fertile Crescent (Mesopotamia); repeated documentation of the use of liquid asphalt (oil) as an illuminant in lamps
A.D. 750	First documented use in Italy of asphalt as a color in paintings
A.D. 950–1000	Report of destructive distillation of asphalt to produce an oil; reference to oil as nafta (naphtha)
A.D. 1100	Documented use of asphalt for covering (lacquering) metalwork
A.D. 1200	Continued use of asphalt and naphthas as an incendiary device in warfare; use of naphtha as an illuminant and incendiary material
A.D. 1500–1600	Documentation of asphalt deposits in the Americas; first attempted documentation of the relationship of asphalts and naphtha (petroleum)
A.D. 1600–1800	Asphalt used for a variety of tasks; relationship of asphalt to coal and wood tar studied; asphalt studied, used for paving; continued documentation of the use of naphtha as an illuminant and the production of naphtha from asphalt; importance of naphtha as fuel realized
A.D. 1859	Discovery of petroleum in North America; birth of modern petroleum science and refining.

of terminology that have been used, not all have survived, but the more common terms are illustrated here. Particularly troublesome, and more confusing, are those terms that are applied to the more viscous materials, for example *bitumen* and *asphalt*. This part of the text attempts to alleviate much of the confusion that exists, but it must be remembered that the terminology of petroleum is still a matter of personal choice and historical usage.

For the purposes of definition and terminology, it is preferable to subdivide petroleum and related materials into three major classes (Table 1.2):

1.   Materials that are of natural origin
2.   Materials that are manufactured

Table 1.2 Subgroups of Petroleum and Related Materials for Definition Purposes

Natural materials	Manufactured materials	Derived materials
Petroleum	Wax	Oils
Heavy oil	Residuum	Resins
Mineral wax	Asphalt	Asphaltenes
Bitumen	Tar	Carbenes
Bituminous rock	Pitch	Carboids
Bituminous sand	Coke	
Kerogen	Synthetic crude oil	
Natural gas		

3. Materials that are integral fractions derived from natural or manufactured products

In the context of this chapter, it is pertinent to note that throughout the millennia in which petroleum has been known and used, it is only in the last four decades that some attempts have been made to standardize petroleum nomenclature and terminology. Confusion may still exist. Therefore it is the purpose of this chapter to impart some semblance of order to the disordered state that exists in the segment of petroleum technology that is known as *terminology*. There is no effort here to define the individual processes, because they will be defined in the relevant chapters. The purpose is to define the various aspects of the feedstocks that are used in a refinery so that the reader can make ready reference to any such word used in the text.

2. NATIVE MATERIALS

2.1 Petroleum

Petroleum is a mixture of gaseous, liquid, and solid hydrocarbon compounds that occur in sedimentary rock deposits throughout the world. It contains small quantities of nitrogen-, oxygen-, and sulfur-containing compounds as well as trace amounts of metallic constituents (Gruse and Stevens, 1960; Speight, 1999 and references therein).

Petroleum is a naturally occurring mixture of hydrocarbons, generally in a liquid state, and may also include compounds of sulfur, nitrogen, oxygen, metals, and other elements (ASTM D-4175). Petroleum has also been defined (ITAA, 1936) as

1. Any naturally occurring hydrocarbon, whether in a liquid, gaseous, or solid state
2. Any naturally occurring mixture of hydrocarbons, whether in a liquid, gaseous, or solid state
3. Any naturally occurring mixture of one or more hydrocarbons, whether in a liquid, gaseous, or solid state, plus hydrogen sulfide and/or helium and/or carbon dioxide

The ITAA definition includes any petroleum as defined in items 1–3 above that has been returned to a natural reservoir.

Crude petroleum is a mixture of compounds that boil at different temperatures and can be separated into a variety of generic fractions by distillation (Table 1.3). The termi-

Table 1.3 General Boiling Ranges of Petroleum Fractions

	Boiling	Range[a]
Fraction	°C	°F
Light naphtha	−1–150	30–300
Gasoline	−1–180	30–355
Heavy naphtha	150–205	300–400
Kerosene	205–260	400–500
Stove oil	205–290	400–550
Light gas oil	260–315	400–600
Heavy gas oil	315–425	600–800
Lubricating oil	> 400	> 750
Vacuum gas oil	425–600	800–1000
Residuum	> 600	> 1000

[a] For convenience, boiling ranges are interconverted to the nearest 5°.

nology of these fractions has been bound by utility and often bears little relationship to their composition.

Because there is wide variation in the properties of crude petroleum (Table 1.4) (Speight, 1999), the proportions in which the different constituents occur vary with origin (Gruse and Stevens, 1960; Koots and Speight, 1975). Thus, some crude oils have higher proportions of the lower boiling components and others (such as heavy oil and bitumen) have higher proportions of higher boiling components (asphaltic components and residuum).

The molecular boundaries of petroleum cover a wide range of boiling points and carbon numbers of hydrocarbon compounds and other compounds containing nitrogen, oxygen, and sulfur as well as metallic (porphyrinic) constituents. However, the actual boundaries of such a *petroleum map* can be only arbitrarily defined in terms of boiling

Table 1.4 Variation in Properties of Different Crude Oils

		Approximate physical composition (%)			
Origin	Specific gravity (water = 1.000)	Gasoline and gas	Kerosene	Gas oil	Residuum (1000°F+)
California	0.858	36.6	4.4	36.0	23.0
Pennsylvania	0.800	47.4	17.0	14.3	1.3
Oklahoma	0.816	47.6	10.8	21.6	20.0
Texas	0.827	33.9	7.3	58.8	
	0.864	44.9	4.2	23.2	27.9
Iraq	0.844	45.3	15.7	15.2	23.8
Iran	0.836	45.1	11.5	22.6	20.8
Kuwait	0.860	39.2	8.3	20.6	31.9
Bahrain	0.861	26.1	13.4	34.1	26.4
Saudi Arabia	0.840	34.5	8.7	29.3	27.5
Venezuela	0.950	15.3	7.4	77.3	

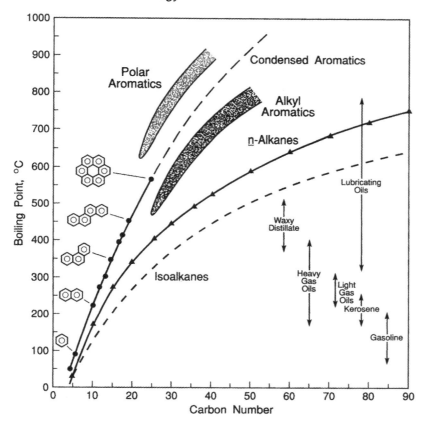

Figure 1.1 Boiling envelope of petroleum.

point and carbon number (Figure 1.1). In fact, petroleum is so diverse that materials from different sources exhibit different boundary limits, and for this reason alone it is not surprising that petroleum has been difficult to map in a precise manner (Speight, 1999).

Thus, *petroleum* and the equivalent term *crude oil* cover a wide assortment of materials that may vary widely in volatility, specific gravity, and viscosity. Metal-containing constituents, notably those compounds that contain vanadium and nickel, usually occur in the more viscous crude oils in amounts of up to several thousand parts per million and can have serious consequences during processing of these feedstocks (Gruse and Stevens, 1960; Speight, 1984). Because petroleum is a mixture of widely varying constituents and proportions, its physical properties also vary widely (Chapter 4); its color ranges from colorless to black. The term *bitumen* also added further complexity to the description of refinery feedstocks.

Petroleum occurs underground at various pressures depending on the depth below sea level. Because of the pressure, it contains a considerable amount of natural gas in solution. Oil is much more fluid underground than it is on the surface and is generally mobile under reservoir conditions because the elevated temperatures in subterranean formations [on average, the temperature rises 1°C for every 100 ft (33 m) of depth] decrease its viscosity.

Petroleum is derived from aquatic plants and animals that lived and died hundreds of millions of years ago. Their remains mixed with mud and sand in layered deposits that, over the millennia, were geologically transformed into sedimentary rock. Gradually the organic matter decomposed and eventually it formed petroleum (or a related precursor), which migrated from the original source beds to more porous and permeable rocks such as *sandstone* and *siltstone*, where it finally became entrapped. Such entrapped accumulations of petroleum are called *reservoirs*. A series of reservoirs within a common rock structure or a series of reservoirs in separate but neighboring formations is commonly referred to as an *oil field*. A group of fields is often found in a single geological environment known as a *sedimentary basin* or *province*.

The major components of petroleum are *hydrocarbons*, compounds of hydrogen and carbon that display great variation in their molecular structure (Speight, 1999 and references therein). The simplest hydrocarbons are a large group of chain-shaped molecules known as the *paraffins*. This broad series extends from methane, which forms natural gas, through liquids that are refined into gasoline, to crystalline waxes. A series of saturated hydrocarbons containing a (usually six-membered) ring, known as the *naphthenes*, ranges from volatile liquids such as *naphtha* to high molecular weight substances isolated as the *asphaltene* fraction. Another group of hydrocarbons, which contains a single or condensed aromatic ring system, are known as the *aromatics*. The chief compound in this series is benzene, a popular raw material for making petrochemicals.

Non-hydrocarbon constituents of petroleum include organic derivatives of nitrogen, oxygen, sulfur, and the metals nickel and vanadium. Most of these impurities are removed during refining.

2.2 Heavy Oil

There are also other types of petroleum that are different from conventional petroleum in that they are much more difficult to recover from the subsurface reservoir. These materials have a higher viscosity (and lower API gravity) than conventional petroleum, and primary recovery of these types of petroleum usually requires thermal stimulation of the reservoir.

When petroleum occurs in a reservoir that allows the crude material to be recovered by pumping operations as a free-flowing dark to light colored liquid, it is often referred to as *conventional petroleum*. Heavy oils are the other types of petroleum that are different from conventional petroleum insofar as they are much more difficult to recover from the subsurface reservoir. The definition of heavy oil is usually based on the API gravity or viscosity and is quite arbitrary, although there have been attempts to rationalize the definition on the basis of viscosity, API gravity, and density (Speight, 1999).

For example, petroleum and heavy oil have been arbitrarily defined in terms of physical properties. Heavy oil was considered to be the type of crude oil that had an API gravity of less than 20°. For example, an API gravity equal to 12° signifies a heavy oil, whereas extraheavy oil, such as tar sand bitumen, usually has an API gravity less than 10° (e.g., Athabasca bitumen = 8° API). Residua vary depending upon the temperature at which distillation was terminated, but usually vacuum residua are in the range of 2–8° API (Speight, 1999, 2000).

The term *heavy oil* has also been used collectively to describe both the heavy oils that require thermal stimulation for recovery from the reservoir and incorrectly the bitumen in bituminous sand formations (see Section 2.3) from which the heavy bituminous material is recovered by a mining operation (Speight, 1990). Convenient as this may be, it is scientifically and technically incorrect.

The term *extra heavy oil* has been used to describe materials that occur in the solid or near-solid state in the deposit or reservoir and are generally incapable of free flow under ambient conditions (bitumen, see next section). Whether or not such a material exists in the near-solid or solid state in the reservoir can be determined from the pour point and the reservoir temperature (Chapter 4).

2.3 Bitumen

The term *bitumen* (also on occasion known as *extra heavy oil* or *native asphalt*, although the latter term is incorrect) includes a wide variety of reddish brown to black materials of near-solid to solid character that exist in nature either with no mineral impurity or with mineral matter contents that exceed 50% by weight. Bitumen is frequently found filling pores and crevices of sandstone, limestone, or argillaceous sediments, in which case the organic and associated mineral matrix is known as *rock asphalt* (Abraham, 1945; Hoiberg, 1964).

Bitumen is also a naturally occurring material that is found in deposits that are incorrectly referred to as *tar sand*, because tar is a product of the thermal processing of coal (Speight, 1994, 1999). The permeability of a tar sand deposit is low, and fluids can pass through the deposit only after fracturing techniques have been applied to it. Alternatively, bitumen recovery can be achieved by converting the bitumen to a product and then recovering the product from the deposit. Tar sand bitumen is a high-boiling material with little if any material boiling below 350°C (660°F), and the boiling range approximates the boiling range of an atmospheric residuum.

There have been many attempts to define tar sand deposits and the bitumen contained therein. In order to define conventional petroleum, heavy oil, and bitumen, the use of a single physical parameter such as viscosity is not sufficient. Other properties such as API gravity, elemental analysis, composition, and, most of all, the properties of the bulk deposit must also be included in any definition of these materials. Only then will it be possible to classify petroleum and its derivatives (Speight, 1999).

In fact, the most appropriate definition of *tar sands* is found in the writings of the United States government (FE-76-4):

> Tar sands are the several rock types that contain an extremely viscous hydrocarbon which is not recoverable in its natural state by conventional oil well production methods including currently used enhanced recovery techniques. The hydrocarbon-bearing rocks are variously known as bitumen-rocks oil, impregnated rocks, oil sands, and rock asphalt.

This definition speaks to the character of the bitumen through the method of recovery. Thus, the bitumen found in tar sand deposits is an extremely viscous material that is *immobile under reservoir conditions* and cannot be recovered through a well by the application of secondary or enhanced recovery techniques. Mining methods match the requirements of this definition (because mining is not one of the specified recovery methods) and the bitumen can be recovered by alteration of its natural state such as by thermal conversion to a product that is then recovered. In this sense, changing the natural state (the chemical composition) such as occurs during several thermal processes (for example, in some in situ combustion processes) also meets the requirements of the definition.

By inference, conventional petroleum and heavy oil are also included in this definition. Petroleum is the material that can be recovered by conventional oil well production methods, whereas heavy oil is the material that can be recovered by enhanced recovery methods. Tar sand is currently recovered by a mining process followed by separation of the bitumen by the hot water process. The bitumen is then used to produce hydrocarbons by a conversion process (Fig. 1.2).

It is incorrect to refer to native bituminous materials as *tar* or *pitch*. Although the word *tar* is descriptive of the black, heavy bituminous material, it is best to avoid its use with respect to natural materials and to restrict its meaning to the volatile or near-volatile products produced by the destructive distillation of such organic substances as coal (Speight, 1994). In the simplest sense, pitch is the distillation residue of the various types of tar.

Thus, alternative names such as *bituminous sand* or *oil sand* are gradually finding use, although the former (bituminous sand) is more technically correct. The term *oil sand* is also used in the same way as the term *tar sand*, and these terms are used interchangeably throughout this text.

2.4 Bituminous Rock and Bituminous Sand

Bituminous rock and *bituminous sand* (see also *bitumen*, Section 2.3) are those formations in which the bituminous material is found filling in veins and fissures in fractured rocks or impregnating relatively shallow sand, sandstone, and limestone strata. The deposits contain as much as 20% bituminous material, and if the organic material in the rock matrix is bitumen, it is usual (although chemically incorrect) to refer to the deposit as *rock asphalt* to distinguish it from bitumen that is relatively mineral-free. A standard test (ASTM D-4) is available for determining the bitumen content of various mixtures with inorganic materials, although the use of the word *bitumen* as applied in this test might be questioned and it might be more appropriate to use the term *organic residues* to include *tar* and *pitch*. If the material is the asphaltite type or asphaltoid type, the corresponding terms should be used: *rock asphaltite* or *rock asphaltoid* (Speight, 1999).

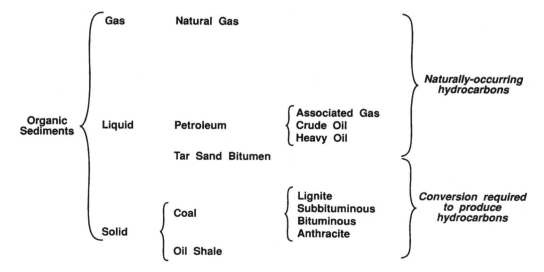

Figure 1.2 Classification of organic sediments.

Bituminous rocks generally have a coarse, porous structure, with the bituminous material filling in the voids. A much more common situation is that in which the organic material is present as an inherent part of the rock composition insofar as it is a diagenetic residue of the organic material detritus that was deposited with the sediment. The organic components of such rocks are usually refractory and are only slightly affected by most organic solvents.

Oil shale is the term applied to a class of rocks that has achieved some importance (Scouten, 1990; Lee, 1991). Oil shale does not contain oil. It is an argillaceous, laminated sediment of generally high organic content (*kerogen*) that can be thermally decomposed to yield appreciable amounts of a hydrocarbon-based oil that is commonly referred to as *shale oil*. Oil shale does not yield shale oil without the application of high temperatures and the ensuing thermal decomposition that is necessary to decompose the organic material (kerogen) in the shale.

2.5 Natural Gas

Just as petroleum was used in antiquity, natural gas was also known in antiquity (Speight, 1999 and references therein). However, the use of petroleum has been relatively well documented because of its use as a mastic for walls and roads as well as in warfare (Table 1.1). Although somewhat less well documented, historical records indicate that the use of natural gas (for other than religious purposes) dates back to about the year 250 A.D., when it was used as a fuel in China. The gas was obtained from shallow wells and was distributed through a piping system constructed from hollow bamboo stems. There is other fragmentary evidence for the use of natural gas in certain old texts, but its use is usually inferred because the gas is not named specifically. However, it is known that natural gas was used on a small scale for heating and lighting in northern Italy during the early seventeenth century. From this it might be conjectured that natural gas found some use from the seventeenth century to the present day, recognizing that gas from coal would be a strong competitor.

The generic term *natural gas* applies to gas commonly associated with petroliferous (petroleum-producing, petroleum-containing) geological formations. Natural gas generally contains high proportions of methane (CH_4), and some of the higher molecular weight paraffins (C_nH_{2n+2}) generally containing up to six carbon atoms may also be present in small quantities. The hydrocarbon constituents of natural gas are combustible, but non-flammable non-hydrocarbon components such as carbon dioxide, nitrogen, and helium are often present in minor proportion and are regarded as contaminants. In addition to reservoirs in which gas is found associated with petroleum, there are also reservoirs in which natural gas may be the sole occupant. And just as petroleum can vary in composition, so can natural gas.

Differences in natural gas composition occur between different reservoirs, and two wells in the same field may also yield gaseous products that are different in composition (Speight, 1990). Indeed, there is no single composition of components that might be termed *typical* natural gas. Methane and ethane constitute the bulk of the combustible components; carbon dioxide (CO_2) and nitrogen (N_2) are the major noncombustible (inert) components. Other constituents such as hydrogen sulfide (H_2S), mercaptans (thiols; $R-SH$), and trace amounts of other sulfur-containing compounds may also be present.

Before the discovery of natural gas, the principal gaseous fuel source was the gas produced by the surface gasification of coal (Speight, 1994). In fact, each town of any size had a plant for the gasification of coal (hence the use of the term *town gas*). Most of the natural gas produced at the petroleum fields was vented to the air or burned in a flare stack; only a small amount of it was pipelined to industrial areas for commercial use. It was only in the years after World War II that natural gas became a popular fuel commodity, leading to the recognition that it has at the present time.

Several general definitions have been applied to natural gas. Thus, *lean gas* is gas in which methane is the major constituent. W*et gas* contains considerable amounts of the higher molecular weight hydrocarbons. S*our gas* contains hydrogen sulfide, whereas *sweet gas* contains very little if any hydrogen sulfide. *Residue gas* is natural gas from which the higher molecular weight hydrocarbons have been extracted, and *casinghead gas* is derived from petroleum but is separated out at the wellhead.

To further define the terms *dry* and *wet* in quantitative measures, the term *dry natural gas* indicates that there is less than 0.1 gallon (1 US gall = 264.2 m^3) of gasoline vapor (higher molecular weight paraffins) per 1000 ft^3 (1 ft^3 = 0.028 m^3). The term *wet natural gas* indicates that there are such paraffins present in the gas, in fact more than 0.1 gal/1000 ft^3. *Associated* or *dissolved natural gas* occurs either as free gas or as gas in solution in the petroleum. Gas that occurs as a solution in the petroleum is *dissolved* gas, whereas the gas that exists in contact with the petroleum (*gas cap*) is *associated* gas.

2.6 Wax

Naturally occurring wax, often referred to as *mineral wax*, occurs as yellow to dark brown solid substances and is composed largely of paraffins. Fusion points vary from 60°C (140°F) to as high as 95°C (203°F). Mineral waxes are usually found associated with considerable mineral matter, as a filling in veins and fissures, or as an interstitial material in porous rocks. The similarity in character of these native products is substantiated by the fact that, with minor exceptions where local names have prevailed, the original term *ozokerite* (*ozocerite*) has served without notable ambiguity for mineral wax deposits (Gruse and Stevens, 1960).

Ozokerite (*ozocerite*), from the Greek for *odoriferous wax*, is a naturally occurring hydrocarbon material composed chiefly of solid paraffins and cycloparaffins (i.e., hydrocarbons) (Wollrab and Streibl, 1969). Ozocerite usually occurs as stringers and veins that fill rock fractures in tectonically disturbed areas. It is predominantly paraffinic material (containing up to 90% nonaromatic hydrocarbons) with a high content (40–50%) of normal or slightly branched paraffins as well as cyclic paraffin derivatives. Ozocerite contains approximately 85% carbon, 14% hydrogen, and 0.3% each of sulfur and nitrogen and is therefore predominantly a mixture of pure hydrocarbons; any non-hydrocarbon constituents are in the minority.

Ozocerite is soluble in solvents that are commonly employed for dissolution of petroleum derivatives, e.g., toluene, benzene, carbon disulfide, chloroform, and ethyl ether.

2.7 Kerogen

Kerogen is the complex carbonaceous macromolecular (organic) material that occurs in sedimentary rocks and shale. It is for the most part insoluble in the common organic solvents. When the kerogen occurs in shale, the material as a whole is often referred to as *oil shale* (see Section 2.4). Kerogen is not the same as the material (bitumen) found in tar sand deposits.

A *synthetic crude oil* (Section 4.5, page 22) is produced from oil shale kerogen by the application of heat, so the kerogen is thermally decomposed (cracked) to produce the lower molecular weight products. Kerogen is also reputed to be a precursor of petroleum, but this concept is still the subject of considerable speculation and debate (Speight, 1999).

For comparison with tar sand, *oil shale* is any fine-grained sedimentary rock containing solid organic matter (kerogen) that yields a hydrocarbon oil when heated (Scouten, 1990). Oil shale varies in mineral composition. For example, clay minerals predominate in true shale, whereas other minerals (e.g., dolomite and calcite) occur in appreciable but subordinate amounts in the carbonates. In all shale types, layers of the constituent mineral alternate with layers of kerogen.

3. DERIVED MATERIALS

Any feedstock or product mentioned in the various sections of this book is capable of being separated into several fractions that are sufficiently distinct in character to warrant the application of individual names (Fig. 1.3) (Pfeiffer, 1950; Speight, 1999 and references therein). For example, distillation fractions that are separated by boiling point (Table 1.4) have been known and named for several decades. There may be some slight variation in the boiling ranges but in general the names are recognized. On the other hand, fractions of feedstocks and product that are separated by other names are less well defined and in many cases can be described only with some difficulty.

For example, treatment of petroleum, residua, heavy oil, or bitumen with a low-boiling liquid hydrocarbon results in the separation of brown to black powdery materials known as *asphaltenes* (Fig. 1.3). The reagents for effecting this separation are usually *n*-

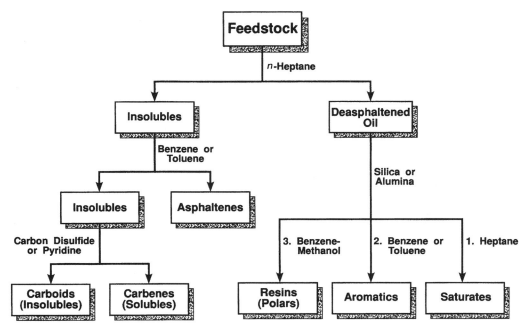

Figure 1.3 Fractionation scheme for various feedstocks.

pentane and *n*-heptane, although other low molecular weight liquid hydrocarbons have also been used (van Nes and van Westen, 1951; Mitchell and Speight, 1973). It must also be recognized that the character and yield of such a fraction varies with the liquid hydrocarbon used for the separation (Fig. 1.4).

Asphaltenes separated from petroleum, residua, heavy oil, and bitumen dissolve readily in benzene, carbon disulfide, chloroform, or other chlorinated hydrocarbon solvents. However, in the case of the higher molecular weight native materials or petroleum residua that have been heated intensively or for prolonged periods, the *n*-pentane-insoluble (or *n*-heptane-insoluble) fraction may not dissolve completely in the aforementioned solvents. Definition of the asphaltene fraction has therefore been restricted to the *n*-pentane- or *n*-heptane-insoluble material that dissolves in such solvents as benzene.

The benzene- or toluene-insoluble materials are collectively referred to as *carbenes* and *carboids*, and the fraction soluble in carbon disulfide (or pyridine) but insoluble in benzene is defined as *carbenes*. *Carboids* are insoluble in carbon disulfide (or pyridine). However, because various solvents might be used in place of benzene, it is advisable to define the carbenes (or carboids) by prefixing them with the name of the solvent used for the separation.

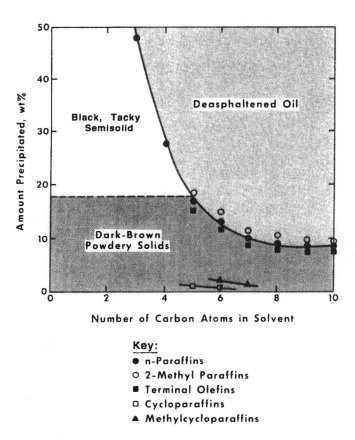

Figure 1.4   Representation of a feedstock as a three-phase system.

The portion of petroleum that is soluble in, for example, pentane or heptane is often referred to as *maltenes* (or *malthenes*). This fraction can be further subdivided by percolation through any surface-active material such as fuller's earth or alumina to yield an *oil* fraction. A more strongly adsorbed red to brown semisolid material known as a *resin* remains on the adsorbent until desorbed by a solvent such as pyridine or a benzene–methanol mixed solvent (Fig. 1.3). The *oil* fraction can be further subdivided into an *aromatics* fraction and a *saturates* fraction. Several other ways have been proposed for separating the resin fraction; for example, a common procedure in the refining industry (Speight, 1999) that can also be used in laboratory practice involves precipitation by liquid propane.

The resin fraction and the oil (maltenes) fraction have also been referred to collectively as *petrolenes*, thereby adding further confusion to this system of nomenclature. However, it has been accepted by many workers in petroleum chemistry that the term *petrolenes* should be applied to that part of the *n*-pentane-soluble (or *n*-heptane-soluble) material that is low boiling [< 300°C (< 570°F) 760 mm] and can be distilled without thermal decomposition. Consequently, the term *maltenes* is now arbitrarily assigned to the pentane-soluble portion of petroleum that is relatively high boiling (> 300°C, 760 mm).

Different feedstocks have different amounts of the asphaltene, resin, and oil fractions (Table 1.5), which can also lead to different yields of thermal coke in the Conradson carbon residue or Ramsbottom carbon residue tests (Chapter 4). Such differences affect the methods chosen for refining the various feedstocks.

4. MANUFACTURED MATERIALS

4.1 Residuum (Residua)

A *residuum* (pl. *residua*, also shortened to *resid*, pl. *resids*) is the residue obtained from petroleum after nondestructive distillation has removed all the volatile materials. The temperature of the distillation is usually maintained below 350°C (660°F), because the rate of thermal decomposition of petroleum constituents is minimal below this temperature but is substantial above it (Fig. 1.5). If the temperature of the distillation unit rises above 350°C (660°F), as happens in certain units where temperatures up to 395°C (740°F) are known to occur, cracking can be controlled by adjustment of the residence time.

Residua are black, viscous materials obtained by the distillation of a crude oil under atmospheric pressure (atmospheric residua) or under reduced pressure (vacuum residua) (Fig. 1.6). They may be liquid at room temperature (generally atmospheric residua) or almost solid (generally vacuum residua) depending upon the nature of the crude oil (Tables 1.5 and 1.6; Fig. 1.7).

When a residuum is obtained from crude oil and thermal decomposition has commenced, it is more usual to refer to this product as *pitch* (Speight, 1999). The differences between conventional petroleum and the related residua are due to the relative amounts of various constituents present, which are removed or remain by virtue of their relative volatility (Fig. 1.8).

The chemical composition of a residuum is complex. Physical methods of fractionation usually indicate high proportions of asphaltenes and resins, up to 50% (or more) of the residuum. In addition, the presence of ash-forming metallic constituents, including such organometallic compounds as those of vanadium and nickel, is also a distinguishing feature of residua and the heavier oils. Furthermore, the deeper the *cut* into the crude oil,

Table 1.5 Miscellaneous Properties of Residua, Heavy Oil, and Bitumen

Crude oil origin	Kuwait[a]	Kuwait		Venezuela (east)	Burzurgan	Boscan	Khafji		California	Cabimas
Residuum type:	Vacuum gas oil	Atmos	Vacuum	Atmos			Atmos	Vacuum	Vacuum	Vacuum
Fraction of crude, vol%	—	42	21	74	52	78	—	—	23	34
Gravity, °API	22.4	13.9	5.5	9.6	3.1	5.0	14.4	6.5	43	6.8
Viscosity										
SUS, 210°F	—	—	—	—	—	—	—	—	—	—
SFS, 122°F	—	553	500,000	27,400	—	—	429	—	—	—
SFS, 210°F	—	—	—	—	—	—	—	—	—	—
cSt, 100°F	—	—	—	—	—	—	—	—	—	—
cSt, 210°F	—	55	1900	—	3355	5250	—	—	15,000	7800
Pour point, °F	—	65	—	95	—	—	—	—	—	162
Sulfur, wt%	2.97	4.4	5.45	2.6	6.2	5.9	4.1	5.3	2.3	3.26
Nitrogen, wt%	0.12	0.26	0.39	0.61	0.45	0.79	—	—	0.98	0.62
Metals, ppm										
Nickel	0.2	14	32	94	76	133	37	53	120	76
Vanadium	0.04	50	102	218	233	1264	89	178	180	614
Asphaltenes, wt%										
Pentane-insoluble	0	—	11.1	—	—	—	—	12.0	19.0	12.9
Hexane-insoluble	0	—	—	—	—	—	—	—	—	—
Heptane-insoluble	0	2.4	7.1	9	18.4	15.3	—	—	—	10.5
Resins, wt%	0	—	39.4	—	—	—	—	—	—	—
Carbon residue, wt%										
Ramsbottom	< 0.1	9.8	—	14.5	—	—	—	—	—	—
Conradson	0.09	12.2	23.1	—	22.5	18.0	—	21.4	24.0	18.7

Crude oil origin: Residuum type:	Arabian light Atmos	Louisiana Vacuum	Saudi Arabia Vacuum	Alaska (North Slope) Atmos	Alaska (North Slope) Vacuum	Boscan	Tar Sand Triangle	P.R. Spring	N.W. Asphalt Ridge
Fraction of crude, vol%	—	13.1	20	58	22	100	100	100	100
Gravity, °API	16.7	11.3	5.0	15.2	8.2	10.3	11.1	10.3	14.4
Viscosity									
SUS, 210°F	—	—	—	1281	—	—	—	—	—
SFS, 122°F	—	—	—	—	—	—	—	—	—
SFS, 210°F	—	—	—	—	—	—	—	—	—
cSt, 100°F	—	—	—	—	—	20,000	7000[b]	200,000[b]	15,000[b]
cSt, 210°F	27	700	2700	42	1950	37	—	—	—
Pour point, °F	—	—	—	75	—	37	—	—	—
Sulfur, wt%	3.00	0.93	5.2	1.6	2.2	5.6	4.38	0.75	0.59
Nitrogen, wt%	—	0.38	0.30	0.36	0.63	—	0.46	1.00	1.02
Metals, ppm									
Nickel	11	20	28	18	47	117	53	98	120
Vanadium	28	—	75	30	82	1220	108	25	25
Asphaltenes, wt%									
Pentane-insoluble	—	6.5	15.0	4.3	8.0	12.6	26.0	16.0	6.3
Hexane-insoluble	—	—	—	—	—	11.4	—	—	—
Heptane-insoluble	2.0	—	—	31.5	—	—	—	—	—
Resins, wt%	—	—	—	—	—	24.1	—	—	—
Carbon residue, wt%									
Ramsbottom	8	12.0	20.0	8.4	17.3	14.0	—	—	—
Conradson	—	—	—	—	—	—	21.6	12.5	3.5

Table 1.5 (Continued)

Crude oil origin:	Athabasca	Lloyd-minster	Cold Lake	Quayarah	Jobo	Bachaquero		Heavy Arabian	West Texas	
Residuum type:	—	—	—	—		Atmos	Vacuum	Vacuum	Vacuum	Atmos
Fraction of crude, vol%	100	100	100	100	100	34	—	27	—	—
Gravity, °API	5.9	14.5	10.0	15.3	8.6	17	2.8	4	9.4	18.4
Viscosity										
SUS, 210°F	513	260	—	—	—	—	—	—	—	—
SFS, 122°F	—	294	—	—	—	—	—	—	313	86
SFS, 210°F	820	—	—	—	247	—	—	—	—	—
cSt, 100°F	—	—	79	—	—	—	—	—	—	—
cSt, 210°F	—	—	—	—	—	—	—	—	—	—
Pour point, °F	58	38	—	—	—	—	—	—	—	—
Sulfur, wt%	4.9	4.3	4.4	8.4	3.9	2.4	3.7	5.3	3.3	2.5
Nitrogen, wt%	0.41	—	0.39	0.7	0.7	0.3	0.6	0.4	0.5	0.6
Metals, ppm										
Nickel	86	40	62	60	103	50	100	230	27	11
Vanadium	167	100	164	130	460	—	900	—	47	20
Asphaltenes, wt%										
Pentane-insoluble	17.0	12.9	15.9	20.4	18	10	—	25	—	—
Hexane-insoluble	13.5	—	—	14.3	—	—	—	—	—	—
Heptane-insoluble	11.4	—	10.8	13.5	—	—	—	—	—	—
Resins, wt%	34.0	38.4	31.2	36.1	—	—	—	—	—	—
Carbon residue, wt%										
Ramsbottom	14.9	—	13.6	—	—	—	—	—	—	—
Conradson	18.5	9.1	13.6	15.6	14	12	27.5	—	16.9	6.6

Crude oil origin: Residuum type:	Tia Juana (light)		Safaniya		P.R. Spring Bitumen	Asphalt Ridge Bitumen	Tar Sand Triangle Bitumen	Sunnyside Bitumen
	Atmos	Vacuum	Atmos	Vacuum				
Fraction of crude, vol%	49	18	40	22	100	100	100	100
Gravity, °API	17.3	7.1	11.1	2.6	10.3	14.4	11.1	—
Viscosity								
SUS, 210°F	165	—	—	—	—	—	—	—
SFS, 122°F	172	—	—	—	—	—	—	—
SFS, 210°F	—	—	—	—	—	—	—	—
cSt, 100°F	890	—	—	—	—	—	—	—
cSt, 210°F	35	7959	—	—	—	—	—	—
Pour point, °F	—	—	—	—	—	—	—	—
Sulfur, wt%	1.8	2.6	4.3	5.3	0.8	0.6	4.4	0.5
Nitrogen, wt%	0.3	0.6	0.4	0.4	1.0	1.0	0.5	0.9
Metals, ppm								
Nickel	25	64	26	46	98	120	53	—
Vanadium	185	450	109	177	25	25	108	—
Asphaltenes, wt%								
Pentane-insoluble	—	—	17.0	30.9	—	—	—	—
Hexane-insoluble	—	—	—	—	—	—	—	—
Heptane-insoluble	—	—	—	—	—	—	—	—
Resins, wt%	—	—	—	—	—	—	—	—
Carbon residue, wt%								
Ramsbottom	—	—	—	—	12.5	3.5	21.6	—
Conradson	9.3	21.6	14.0	25.9	—	—	—	—

[a] Included for comparison.
[b] Estimated

Table 1.6 Properties of Residua from Tijuana Light (Venezuela) Crude Oil

	Whole crude	> 430	> 565	> 650	> 700	> 750	> 850	> 950	> 1050
Boiling range									
°F	Whole crude	> 430	> 565	> 650	> 700	> 750	> 850	> 950	> 1050
°C	Whole crude	> 220	> 295	> 345	> 370	> 400	> 455	> 510	> 565
Yield on crude, vol%	100.0	70.2	57.4	48.9	44.4	39.7	31.2	23.8	17.9
Gravity, °API	31.6	22.5	19.4	17.3	16.3	15.1	12.6	9.9	7.1
Specific gravity	0.8676	0.9188	0.9377	0.9509	0.9574	0.9652	0.9820	1.007	1.0209
Sulfur, wt%	1.08	1.42	1.64	1.78	1.84	1.93	2.12	2.35	2.59
Carbon residue (Conradson), wt%	—	6.8	8.1	9.3	10.2	11.2	13.8	17.2	21.6
Nitrogen, wt%	—	—	—	0.33	0.36	0.39	0.45	0.52	0.60
Pour point, °F	−5	15	30	45	50	60	75	95	120
Viscosity									
Kinematic, cSt									
100°F	10.2	83.0	315	890	1590	3100	—	—	—
210°F	—	9.6	19.6	35.0	50.0	77.0	220	1010	7959
Furol (SFS),s									
122°F	—	—	70.6	172	292	528	—	—	—
210°F	—	—	—	—	25.2	37.6	106	484	3760
Universal (SUS), s at 210°F	—	57.8	96.8	165	234	359	1025	—	—
Metals, ppm									
Vanadium	—	—	—	185	—	—	—	—	450
Nickel	—	—	—	25	—	—	—	—	64
Iron	—	—	—	28	—	—	—	—	48

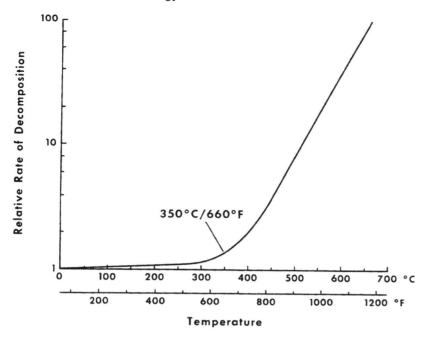

Figure 1.5 Representation of the rate of thermal decomposition of petroleum.

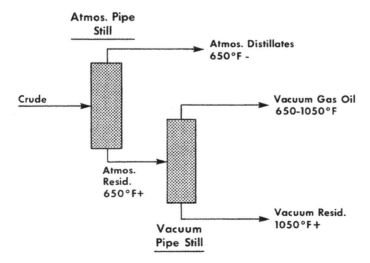

Figure 1.6 Simplified crude oil distillation scheme.

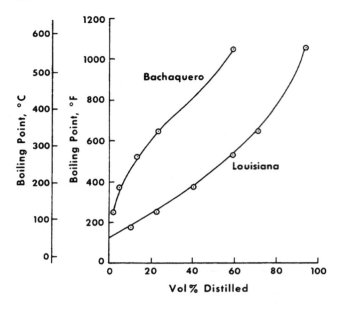

	Louisiana	Bachaquero
Gravity, API	13.1	2.8
Sulfur, wt %	0.9	3.71
Nitrogen, wt %	0.4	0.60
Con. Carbon, wt %	15.8	27.5
Nickel, ppm	20	100
Vanadium, ppm	8	900
Pour Point, °F	-	130

Figure 1.7 Boiling profiles and properties for two different crude oils.

the greater is the concentration of sulfur and metals in the residuum and the greater the deterioration in physical properties (Table 1.6) (Speight, 2000).

4.2 Asphalt

Asphalt is manufactured from petroleum (Fig. 1.9) and is a black or brown material that has a consistency varying from a viscous liquid to a glassy solid (Speight, 1995).

To a point, asphalt can resemble bitumen, hence the tendency to refer to bitumen (incorrectly) as *native asphalt*. It is recommended that asphalt (manufactured) and bitumen (naturally occurring) be differentiated other than by use of the qualifying terms *petroleum* and *native*, because their origins may be reflected in the resulting physicochemical properties of the two types of materials. It is also necessary to distinguish between the asphalt that originates from petroleum by refining and the product in which the source of the asphalt is a material other than petroleum, e.g., *wurtzilite asphalt* (Bland and Davidson, 1967). In the absence of a qualifying word, it is assumed that the term *asphalt* refers to the product manufactured from petroleum.

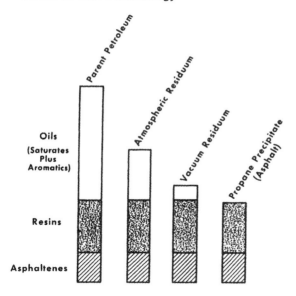

Figure 1.8 Relationship of crude oil to its residua and asphalt.

When the asphalt is produced simply by distillation of an asphaltic crude oil, it can be referred to as *residual asphalt* or *straight run asphalt*. If the asphalt is prepared by solvent extraction of residua or by light hydrocarbon (propane) precipitation, or if blown or otherwise treated, the term should be modified accordingly (e.g., *propane asphalt*, *blown asphalt*).

Asphalt softens when heated and is elastic under certain conditions. The mechanical properties of asphalt are of particular significance when it is used as a binder or adhesive.

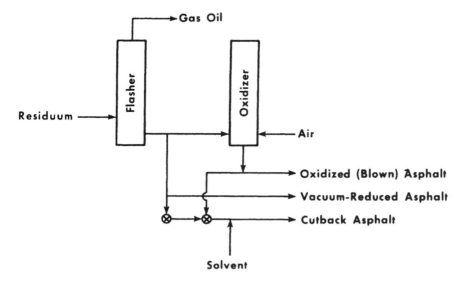

Figure 1.9 Simplified schematic for asphalt production.

The principal application of asphalt is in road surfacing, which may be performed in a variety of ways. Light oil *dust layer* treatments may be built up by repetition to form a hard surface, or a granular aggregate may be added to an asphalt coat, or earth materials from the road surface itself may be mixed with the asphalt.

Other important applications of asphalt include canal and reservoir linings, dam facings, and sea works. The asphalt so used may be a thin, sprayed membrane covered with earth for protection against weathering and mechanical damage, or thicker surfaces, often including riprap (crushed rock). Asphalt is also used for roofing materials, coatings, floor tiles, soundproofing, waterproofing, and other building construction elements and in a number of industrial products, such as batteries. For certain applications an asphaltic emulsion is prepared in which fine globules of asphalt are suspended in water.

4.3 Coke

Coke is the solid carbonaceous material produced from petroleum during thermal processing. It is often characterized as having a high carbon content (95%+ by weight) and a honeycomb type of appearance. The color varies from gray to black, and the material is insoluble in organic solvents.

4.4 Wax

The term *paraffin wax* is restricted to the colorless, translucent, highly crystalline material obtained from the light lubricating fractions of paraffinic crude oils (wax distillates). The commercial products melt in the approximate range of 50–65°C (120–150°F). Dewaxing of heavier fractions leads to semisolid material, generally known as *petrolatum*, and solvent deoiling of the petroleum or of heavy, waxy residua results in dark-colored waxes of a sticky, plastic to hard nature. The waxes are composed of fine crystals and contain, in addition to *n*-paraffins, appreciable amounts of isoparaffins and cyclic hydrocarbon compounds substituted with long-chain alkyl groups. The melting points of the commercial grades are in the 70–90°C (160–195°F) range.

Highly paraffinic waxes are also produced from peat, lignite, or shale oil residua. Paraffin waxes known as *ceresins*, which are quite similar to the waxes from petroleum, can also be prepared from ozokerite.

4.5 Synthetic Crude Oil

Coal, oil shale, bitumen, and heavy oil can be upgraded through thermal decomposition by a variety of processes to produce a marketable and transportable product. These products vary in nature, but the principal one is a hydrocarbon that resembles conventional crude oil—hence the use of the terms *synthetic crude oil* and *syncrude*. The synthetic crude oil, although it may be produced by one of the less conventional conversion processes, can be refined by the usual refining methods.

The unrefined synthetic crude oil from bitumen will generally resemble petroleum more closely than the synthetic crude oil from either coal or oil shale. Unrefined synthetic crude oil from coal can be identified by a high content of phenolic compounds, whereas the unrefined synthetic crude oil from oil shale will contain high proportions of nitrogen-containing compounds.

4.6 Coal Liquids

Coal liquids are products of the thermal decomposition of coal, and the term is often used to indicate the refined (hydrotreated) products (Table 1.7). Their composition can vary from a mixture with a majority of hydrocarbon species to a mixture containing a majority of heteroatomic species. Predominant heteroatomic species contain oxygen, usually in the form of phenolic oxygen (e.g., C_6H_5-OH) or ether oxygen (R_1-O-R_2).

The refined (hydrotreated) coal liquids, in which the heteroatomic content has been reduced to acceptable levels, may also be referred to as *synthetic crude oil* (Section 4.5) and is sent to a refinery for further processing into various products.

4.7 Shale Oil

Oil shale is the term applied to a class of bituminous rocks that contain a complex heteroatomic molecule known as *kerogen* (see Section 2.7) (Scouten, 1990; Lee, 1991). Oil shale does not contain oil. The product commonly referred to as *shale oil* is the hydrocarbon-based oil produced in considerable amounts by the thermal decomposition of kerogen (Table 1.8). Oil shale does not yield shale oil without the application of high temperatures and the ensuing thermal decomposition that is necessary to decompose the organic material (kerogen) in the shale. Shale oil also contains heteroatomic species, predominantly organic nitrogen-containing molecules.

The refined (hydrotreated) shale oil, in which the heteroatomic content has been reduced to acceptable levels, may also be referred to as *synthetic crude oil* and is sent to a refinery for further processing into various products.

Table 1.7 Properties of Coal Liquids from the H-Coal Process

Product (wt%)	Illinois synthetic crude	Wyodak synthetic crude
C_1-C_3 hydrocarbons	10.7	10.2
C_4–200°C distillate	17.2	26.1
200–340°C distillate	28.2	19.8
340–525°C distillate	18.6	6.5
525°C + residual oil	10.2	11.1
Unreacted ash-free coal	5.2	9.8
Gases	15.0	22.7

Table 1.8 Properties of Shale Oil from the TOSCO Process

Component	Vol %	°API	wt% S	wt% N
C_5–400	17	51	0.7	0.4
400–950	60	20	0.8	2.0
950+	23	6.5	0.7	2.9
Total	100	21	0.7	1.9

REFERENCES

Abraham, H. 1945. Asphalts and Allied Substances. Van Nostrand, New York.

ASTM D-4. Test Method for Bitumen Content. Annual Book of Standards. American Society for Testing and Materials, West Conshohocken, PA.

ASTM D-4175. Standard Terminology Relating to Petroleum, Petroleum Products, and Lubricants. Annual Book of Standards. American Society for Testing and Materials, West Conshohocken, PA.

Bland, W. F., and Davidson, R. L. 1967. Petroleum Processing Handbook. McGraw-Hill, New York.

Cobb, C., and Goldwhite, H. 1995. Creations of Fire: Chemistry's Lively History from Alchemy to the Atomic Age. Plenum Press, New York.

Forbes, R. J. 1958a. A History of Technology. Oxford Univ. Press, Oxford, England.

Forbes, R. J. 1958b. Studies in Early Petroleum Chemistry. E. J. Brill, Leiden, The Netherlands.

Forbes, R. J. 1959. More Studies in Early Petroleum Chemistry. E. J. Brill, Leiden, The Netherlands.

Forbes, R. J. 1964. Studies in Ancient Technology. E. J. Brill, Leiden, The Netherlands.

Gruse, W. A., and Stevens, D. R. 1960. The Chemical Technology of Petroleum. McGraw-Hill, New York.

Henry, J. T. 1873. The Early and Later History of Petroleum. Vols I and II. APRP, Philadelphia, PA.

Hoiberg, A. J. 1964. Bituminous Materials: Asphalts, Tars, and Pitches. Wiley, New York.

ITAA. 1936. Income Tax Assessment Act. Government of the Commonwealth of Australia.

Koots, J. A., and Speight, J. G. 1975. Fuel 54:179.

Lee, S. 1991. Oil Shale Technology. CRC Press, Boca Raton, FL.

Mitchell, D. L., and Speight. J. G. 1973. Fuel 52:149.

Pfeiffer, J. H. 1950. The Properties of Asphaltic Bitumen. Elsevier, Amsterdam.

Scouten, C. S. 1990. In Fuel Science and Technology Handbook. J. G. Speight, ed. Marcel Dekker, New York.

Speight, J. G. 1984. In: Characterization of Heavy Crude Oils and Petroleum Residues. S. Kaliaguine and A. Mahay (eds.) Elsevier, Amsterdam, p. 515.

Speight, J. G. ed., 1990. Fuel Science and Technology Handbook. Marcel Dekker, New York.

Speight, J. G. 1994. The Chemistry and Technology of Coal. 2nd ed. Marcel Dekker, New York.

Speight, J. G. 1995. Asphalt. In: Encyclopedia of Energy and the Environment. A Bisio and S. Boots (eds.) Vol. 1. Wiley, New York, p. 321.

Speight, J. G. 1999. The Chemistry and Technology of Petroleum. 3rd ed. Marcel Dekker, New York.

Speight. J. G. 2000. The Desulfurization of Heavy Oils and Residua. 2nd ed. Marcel Dekker, New York.

van Nes, K., and van Westen, H. A. 1951. Aspects of the Constitution of Mineral Oils. Elsevier, Amsterdam.

2
Occurrence and Availability

1. INTRODUCTION

Fossil fuels are those fuels—coal, petroleum (including heavy oil), bitumen, natural gas, and shale oil—produced by the decay of plant remains over geological time (Speight, 1990, 1999). They are carbon-based and represent a vast source of energy. Resources such as bitumen in tar sand formations represent an unrealized potential, with the amounts of liquid fuels from petroleum being only a fraction as great as those that could ultimately be produced from tar sand bitumen.

At the present time, the majority of the energy consumed by humans is produced from fossil fuels, with smaller amounts of energy coming from nuclear and hydroelectric sources (Fig. 2.1). The nuclear power industry (Zebroski and Levenson, 1976; Rahn, 1987) is truly an industry where the future is uncertain or at least at the crossroads. As a result, fossil fuels are projected to be the major sources of energy for the next 50 years.

Petroleum is scattered throughout the earth's crust, which is divided into natural groups or strata, categorized in order of their antiquity (Table 2.1). These divisions are recognized by the distinctive systems of debris (organic material, minerals, and fossils) that form a chronological time chart that indicates the relative ages of the strata.

Petroleum and other carbonaceous materials such as coal and oil shale occur in all of these geological strata from the Precambrian to the recent, and the origin of petroleum within these formations is a question that remains open to conjecture and the basis for much research and speculation (Speight, 1999 and references therein). The answer cannot be given in this text, nor for that matter can it be presented in any advanced treatise. It is more pertinent to the present text that historical data be introduced to illustrate the decreasing quality of crude oil and current prospects for the continuation and evolution of the refining industry.

The modern petroleum industry began in 1859 with the discovery and subsequent commercialization of petroleum in Pennsylvania (Yergin, 1991). After completion of the first well (by Edwin Drake), the surrounding areas were immediately leased and extensive drilling took place. Crude oil output in the United States increased from approximately 2000 barrels [1 barrel (bbl) = 42 U.S. gall = 35 Imperial gall = 5.61 ft^3 = 158.8L] in 1859 to nearly 3,000,000 bbl in 1863 and approximately 10,000,000 bbl in 1874. In 1861 the first

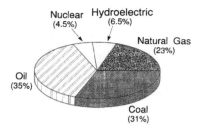

Figure 2.1 Consumption of energy from various sources.

cargo of oil, contained in wooden barrels, was sent across the Atlantic to London, and by the 1870s refineries, tank cars, and pipelines had become characteristic features of the industry, mostly through the leadership of Standard Oil, which was founded by John D. Rockefeller (Yergin, 1991).

At the outbreak of World War I in 1914, the two major petroleum producers were the United States and Russia, with lesser production occurring in Indonesia, Mexico, and Rumania. During the 1920s and 1930s, attention was also focused on other areas for oil production, such as the countries of the Middle East; at this time, African and European countries were not considered major oil-producing areas. In the post-1945 era, oil production in the Middle Eastern countries continued to rise in importance because of new discoveries of petroleum reserves. The United States, although continuing to be the major producer of petroleum, was also the major consumer and thus was not generally

Table 2.1 Nomenclature and Approximate Age of Geological Strata

Era	Period	Epoch	Age (10^6 years)
Cenozoic	Quaternary	Recent	0.01
		Pleistocene	3
	Tertiary	Pliocene	12
		Miocene	25
		Oligocene	38
		Eocene	55
		Paleocene	65
Mesozoic	Cretaceous		135
	Jurassic		180
	Triassic		225
Paleozoic	Permian		275
	Carboniferous		
	Pennsylvanian		350
	Mississippian		
	Devonian		413
	Silurian		430
	Ordovician		500
	Cambrian		600

recognized as an exporter of oil. At this time, oil companies began to roam much farther in the search for oil, and significant deposits were discovered in Europe, Africa, and Canada. Canada, in particular, became a recognized petroleum producer with the discovery and subsequent development of the Leduc (Alberta) field south of Edmonton in 1947. The development of the Cold Lake (Alberta) heavy oil reserves in the 1960s followed by that of the Athabasca (Alberta) tar sand deposits by Suncor in 1967 and Syncrude in 1977 made the Canadian province of Alberta a major producer of liquid fuels. These developments also afforded Canada a degree of self-sufficiency in terms of petroleum and liquid fuel production. The United States is seen to be proceeding toward a degree of self-sufficiency by seeking to increase the number of alternatively fueled vehicles (AFVs) but maturity of the program is still some time in the future (Joyce, 2000) and petroleum remains the fuel source of choice.

In a more general sense, the average quality of crude oil has become worse in recent years. This is reflected in a progressive decrease in API gravity (Fig. 2.2) (i.e., increase in density) and a rise in sulfur content (Fig. 2.3) (Swain, 1991, 1993, 1998). The data show a tendency for the quality of crude oil feedstocks to stabilize, in terms of API gravity and sulfur content, in the mid-1990s, but the downward trend in API gravity and the upward trend in sulfur content resumed.

It is not surprising that the nature of crude oil refining has changed considerably and the decline in the reserves of light conventional crude oil has resulted in an increasing need to develop options to upgrade the abundant supply of known heavy crude oil reserves (Speight, 1999, 2000 and references therein). In addition, there is considerable focus and a renewal of efforts on adapting recovery techniques to the production of heavy oil. In fact, the occurrence of petroleum and its use as the source of much-needed liquid fuels may be likened to the tip of an iceberg (Fig. 2.4) compared to the other fossil fuel resources.

This chapter will deal with the occurrence of petroleum from the perspective of global energy scenarios, recognizing that the more efficient use of petroleum is of paramount

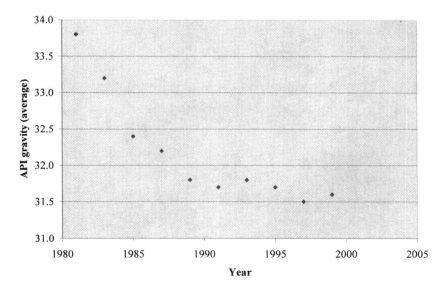

Figure 2.2 Changes in crude oil API gravity during the 1980s and 1990s.

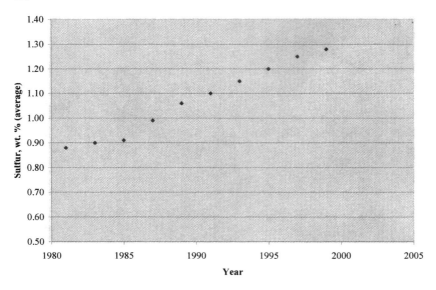

Figure 2.3   Changes in crude oil sulfur content during the 1980s and 1990s.

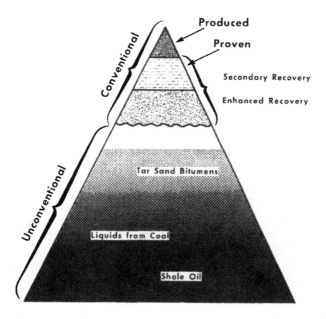

Figure 2.4   Representation of the projected distribution of liquids from fossil fuels.

importance. Furthermore, petroleum technology in one form or another will be with us until suitable alternative forms of energy become readily available. Therefore, a discussion of the importance of a thorough understanding of the limitations on the avail of petroleum and the benefits of refining is now introduced that will be continued throughout the pages of this book.

2. RESERVES

The reservoir rocks that yield crude oil range in age from Precambrian to recent geological time, but rocks deposited during the Tertiary, Cretaceous, Permian, Pennsylvanian, Mississippian, Devonian, and Ordovician periods are particularly productive. In contrast, rocks of Jurassic, Triassic, Silurian, and Cambrian age are less productive, although they do contain some petroleum, and Precambrian rocks yield petroleum only under exceptional circumstances. Most of the crude oil currently recovered and sent to refineries is produced from underground reservoirs, and it is these reserves, which are the predominant source of petroleum, that are discussed here.

The majority of crude oil reserves identified to date are located in a relatively small number of very large fields, known as *giants*. Of the 90 or so oil-producing nations, five Middle Eastern countries contain almost 70% of the known oil reserves in such fields.

The definitions that are used to describe petroleum reserves are often varied and misunderstood, because they are not adequately defined at the time of use. Therefore, as a means of alleviating this problem, it is pertinent at this point to consider the definitions used to describe the amount of petroleum that remains in subterranean reservoirs.

Petroleum is a *resource*. In particular, petroleum is a *fossil fuel resource*. A *resource* is the commodity itself that exists in the sediments and strata, whereas *reserves* are the amount of a commodity that can be recovered economically. However, the use of the term *reserves* as being descriptive of the resource is subject to much speculation. It is also subject to numerous modifications. For example, reserves are classed as *proved reserves*, *unproved reserves*, *probable reserves*, and *possible reserves*, with *undiscovered reserves* or *undiscovered resources* lurking in the background (Figs 2.5 and 2.6).

Proven reserves of petroleum are those reserves that have been identified by drilling operations and are recoverable by means of current technology. Their quantification has a high degree of accuracy because the reservoir or field is often clearly demarcated and estimates of the oil-in-place have a high degree of accuracy. *Proven reserves* are frequently updated as recovery operations proceed by means of reservoir characteristics such as production data and pressure transient analysis.

The term *inferred reserves* is commonly used in addition to, or in place of, *potential reserves*. Quantities quoted for inferred reserves are regarded as of a higher degree of accuracy than those of potential reserves, and the term *inferred* is applied to those reserves that are estimated using an improved understanding of reservoir frameworks. The term also usually includes those reserves that can be recovered by further development of the various recovery technologies.

Probable reserves are those reserves of petroleum about which a slight doubt exists; there is an even greater degree of uncertainty about the probability of their recovery, but some information is known about them. The term *potential reserves* is also used on occasion; these reserves are based upon geological information about the types of sediments where such resources are likely to occur, and they are considered to represent an educated guess. Then there are the so-called *undiscovered reserves*, which can be, and often are, little

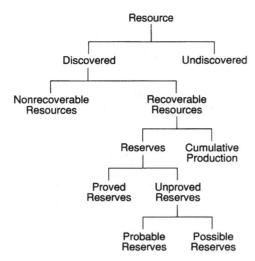

Figure 2.5 Subdivision of resources.

more than figments of the imagination! The terms *undiscovered reserves* and *undiscovered resources* should be used with caution, especially when applied as a means of estimating petroleum reserves. The data are very speculative and are regarded by many energy scientists as having little basis other than unbridled optimism.

The differences between the data obtained from these various estimates can be considerable, but it must be remembered that any data about the reserves of petroleum (and, for that matter, about estimated amounts of any other fuel or mineral resource) will always be open to questions about the degree of certainty (Fig. 2.7). Thus, in reality, and in spite of unjustified word manipulation, *proven reserves* may be a very small part of the total hypothetical and/or speculative amounts of a resource.

At some time in the future, the reserves of certain resources may be increased. This can arise as a result of improvements in recovery techniques that may either make more of the resource accessible or bring about a lowering of the recovery costs and render winning of

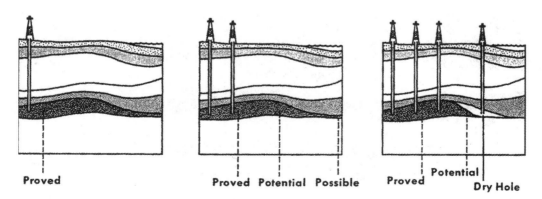

Figure 2.6 Representation of various reserves.

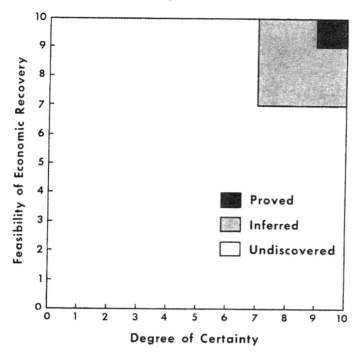

Figure 2.7   Representation of the degree of accuracy of reserve estimates.

the resource an economical proposition. In addition, other uses may be found for a commodity, and the increased demand may result in an increase in probability. Alternatively, a large deposit may become exhausted and unable to produce any more of the resource, thus forcing production to focus on a resource that is of a lower grade and has a higher recovery cost but whose recovery has become more attractive.

Finally, it is rare that petroleum (the exception being tar sand deposits, from which most of the volatile material has disappeared over time) does not occur without an accompanying gas cap, referred to as *natural gas* (Fig. 2.8). It is important, when describing reserves of petroleum, to also acknowledge the occurrence, properties, and character of the natural gas.

2.1   Conventional Petroleum

Several countries are currently recognized as producers of petroleum with available reserves. These available reserves have been defined (Campbell, 1997), but not quite in the manner outlined above. For example, on a worldwide basis the produced conventional crude oil is estimated to be approximately 784 billion (784×10^9 bbl, with approximately 836×10^9 bbl remaining as reserves. It is also estimated that there are 180×10^9 bbl that remain to be discovered with approximately 1 trillion (1×10^{12}) bbl yet to be produced. The annual depletion rate is estimated to be approximately 2.6%.

The United States is one of the largest importers of petroleum (Fig. 2.9). The imports of crude oil into the United States continue to rise, and it is interesting, perhaps even frightening, that projections made in 1990 are quite close (Fig. 2.9). As of the end of the first half of

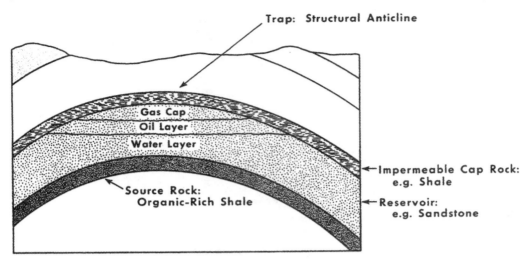

Figure 2.8 Schematic of a petroleum reservoir showing the gas cap.

the calendar year 2000, the United States was importing approximately 65% of its daily crude oil requirements, and demand is continuing to rise (*Oil & Gas Journal*, 2000a; 2000b, 2000c). For example, daily imports for the week of July 7, 2000, were 9,191,000 bbl of petroleum and 1,698,000 bbl of petroleum products, for a total of 10,889,000 bbl. For the same week, daily domestic production of crude oil and lease condensates totaled 5,718,000 bbl. Using the total figure for imports (10,889,000 bbl), crude oil imports represent 65.6% of the daily totals (imports plus domestic production) of the United States. Using the data

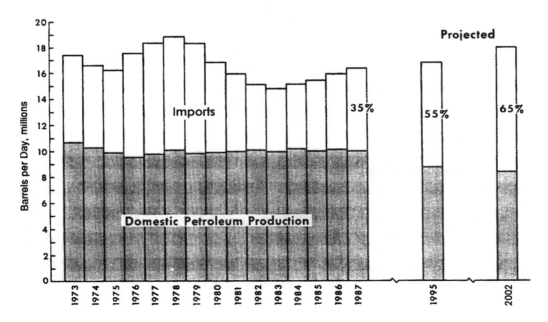

Figure 2.9 Comparison of U.S. production and imports of crude oil.

for the imports of crude oil only (9,191,000 bbl) and excluding crude oil products, crude oil imports represent 61.6% of the daily totals of the United States.

Such data point to policies that offer little direction in terms of stability of supply or any measure of self-sufficiency in liquid fuel precursors, other than resorting to military action. This is particularly important for U.S. refineries because a disruption in supply could cause major shortfalls in feedstock availability. Indeed, fluctuation of gasoline prices upward during the summer of 2000 and the need to appeal to oil-producing countries to increase production to lower gasoline prices (*Oil & Gas Journal*, 2000b) only serve as an indicator of what became reality.

In addition, the crude oils available for refining today are quite different in composition and properties from those available some 30 years ago (Fig. 2.10, see also pp. 27 and 28). The current crude oils are somewhat heavier, because they have higher proportions of nonvolatile (asphaltic) constituents. In fact, by the standards of yesteryear, many of the crude oils currently in use would have been classified as heavy feedstocks, though they may not approach the definitions used today for heavy crude oils. Changes in feedstock character, such as this tendency to heavier materials, require adjustments to refinery operations to reduce the amount of coke formed during processing and to balance the overall product slate.

2.2 Natural Gas

Natural gas is defined as a gaseous mixture that is predominantly methane but also contains other combustible hydrocarbon compounds as well as non-hydrocarbon compounds (Table 2.2) (Speight, 1993). In fact, associated natural gas is believed to be the most economical form of ethane (Farry, 1998).

In addition to composition and thermal content (Btu/scf, Btu/ft^3), natural gas can also be characterized on the basis of the mode of the natural gas found in reservoirs where there is no, or at best only minimal amounts of, crude oil.

Natural gas occurs in the porous rock of the earth's crust either alone (*nonassociated natural gas*) or with accumulations of petroleum (*associated natural gas*). In the latter case, the gas forms a gas cap, a mass of gas trapped between the liquid petroleum and the impervious cap rock of the petroleum reservoir. When the pressure in the reservoir is

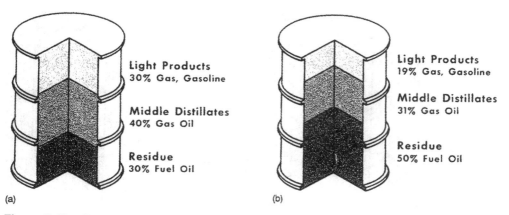

Figure 2.10 Composition of (a) a light crude oil (44° API) compared to (b) a heavier crude oil (31° API)

Table 2.2 Composition of Natural Gas

Category	Component	Amount (%)
Paraffinic	Methane (CH_4)	70–98
	Ethane (C_2H_6)	1–10
	Propane (C_3H_6)	Trace–5
	Butane (C_4H_{10})	Trace–2
	Pentane (C_5H_{12})	Trace–1
	Hexane (C_6H_{14})	Trace–0.5
	Heptane and higher (C_{7+})	None–trace
Cyclic	Cyclopropane (C_3H_6)	Traces
	Cyclohexane (C_6H_{12})	Traces
Aromatic	Benzene (C_6H_6), others	Traces
Nonhydrocarbon	Nitrogen (N_2)	Trace–15
	Carbon dioxide (CO_2)	Trace–1
	Hydrogen sulfide (H_2S)	Trace occasionally
	Helium (He)	Trace–5
	Other sulfur and nitrogen compounds	Trace occasionally
	Water (H_2O)	Trace–5

sufficiently high, the natural gas may be dissolved in the petroleum, in which case it is released upon penetration of the reservoir as a result of drilling operations.

The proven reserves of natural gas are of the order of in excess of 3600 trillion cubic feet (1 Tcf = 1×10^{12} ft^3). Approximately 300 Tcf exist in the United States and Canada (*BP Statistical Review of World Energy*, 1992; *BP Review of World Gas*, 1992), with demand in the United States continuing to increase (*Oil & Gas Journal*, 2000a). It should also be remembered that the total gas resource base (like any fossil fuel or mineral resource base) is dictated by economics. Therefore, when resource data are quoted, some attention must be given to the cost of recovering those resources. Most important, the economics must also include a cost factor that reflects the willingness to secure total, or a specific degree of, energy independence.

In addition to defining natural gas as *associated* and *nonassociated*, the types of natural gas vary according to composition. There is *dry gas* or *lean gas*, which is mostly methane, and *wet gas*, which contains considerable amounts of higher molecular weight and higher boiling hydrocarbons (Table 2.3). *Sour gas* contains high proportions of hydrogen sulfide, whereas *sweet gas* contains little or no hydrogen sulfide. *Residue gas* is the gas remaining (mostly methane) after the higher molecular weight paraffins have been extracted. *Casinghead gas* is the gas derived from an oil well by extraction at the surface. Natural gas has no distinct odor, and its main use is for fuel, but it can also be used to make chemicals and liquefied petroleum gas.

Some natural gas wells also produce helium, which can occur in commercial quantities; nitrogen and carbon dioxide are also found in some natural gases. Gas is usually separated at as high a pressure as possible, reducing compression costs when the gas is to be used for gas lift or delivered to a pipeline. After gas removal, lighter hydrocarbons and hydrogen sulfide are removed as necessary to obtain a crude oil of suitable vapor pressure for transport that retains most of the natural gasoline constituents.

The non-hydrocarbon constituents of natural gas can be classified into two types of materials:

Table 2.3   Range of Composition of Wet and Dry Natural Gas

Constituents	Composition (vol%)		
	Wet	Dry	Range
Hydrocarbons			
Methane	84.6	96.0	
Ethane	6.4	2.0	
Propane	5.3	0.6	
Isobutane	1.2	0.18	
n-Butane	1.4	0.12	
Isopentane	0.4	0.14	
n-Pentane	0.2	0.06	
Hexanes	0.4	0.10	
Heptanes	0.1	0.80	
Nonhydrocarbons			
Carbon dioxide			0–5
Helium			0–0.5
Hydrogen sulfide			0–5
Nitrogen			0–10
Argon			0–0.05
Radon, krypton, xenon			Traces

1. Diluents, such as nitrogen, carbon dioxide, and water vapors
2. Contaminants, such as hydrogen sulfide and/or other sulfur compounds

The diluents are noncombustible gases that reduce the heating value of the gas and are on occasion used as *fillers* when that property is desirable. On the other hand, the contaminants are detrimental to production and transportation equipment in addition to being obnoxious pollutants. Thus, the primary reason for gas processing is to remove the unwanted constituents of natural gas.

The major diluents or contaminants of natural gas are

1. Acid gas, which is predominantly hydrogen sulfide although carbon dioxide does occur to a lesser extent
2. Water, which includes all entrained free water or water in condensed forms
3. Liquids in the gas, such as higher boiling hydrocarbons as well as pump lubricating oil, scrubber oil, and, on occasion, methanol
4. Any solid matter that may be present, such as fine silica (sand) and scaling from the pipe

As with petroleum, natural gas from different wells varies widely in composition and analysis (Table 2.4), and the proportion of non-hydrocarbon constituents can vary over a very wide range. Thus a particular natural gas field could require production, processing, and handling protocols different from those used for another field.

Another product is *gas condensate*, which contains relatively high amounts of the higher molecular weight liquid hydrocarbons. These hydrocarbons may occur in the gas phase in the reservoir.

Liquefied petroleum gas (LPG) is composed of propane (C_3H_8), butanes (C_4H_{10}), and/or mixtures thereof, and small amounts of ethane and pentane may also be present as

Table 2.4 Variation in Natural Gas Composition with Source

Component	Type of gas field			Natural gas separated from crude oil, Ventura[a]		
	Dry gas, Los Medanos[a] (mol%)	Sour gas, Jumping Pound[b] (mol %)	Gas condensate, Paloma[a] (mol%)	400 lb (mol%)	50 lb (mol%)	Vapor (mol%)
Hydrogen sulfide	0	3.3	0	0	0	0
Carbon dioxide	0	6.7	0.7	0.3	0.7	0.8
Nitrogen and air	0.8	0	0	0	—	2.2
Methane	95.8	84.0	74.5	89.6	81.8	69.1
Ethane	2.9	3.6	8.3	4.7	5.8	5.1
Propane	0.4	1.0	4.7	3.6	6.5	8.8
Isobutane	0.1	0.3	0.9	0.5	0.9	2.1
n-Butane	Trace	0.4	1.9	0.9	2.3	5.0
Isopentane	0		0.8	0.2	0.5	1.4
n-Pentane	0		0.6	0.1	0.5	1.4
Hexane	0	0.7	1.3			
Heptane	0			0.1	1.0	4.1
Octane	0		6.3			
Nonane	0					
	100.0	100.0	100.0	100.0	100.0	100.0

[a] In California.
[b] In Canada.

impurities. On the other hand, natural gasoline (like refinery gasoline) consists mostly of pentane (C_5H_{12}) and higher molecular weight hydrocarbons.

The term *natural gasoline* has also on occasion been applied in the gas industry to mixtures of liquefied petroleum gas, pentanes, and higher molecular weight hydrocarbons. Caution should be taken not to confuse natural gasoline with *straight-run gasoline* (often also incorrectly referred to as natural gasoline), which is the gasoline distilled unchanged from petroleum.

2.3 Heavy Oil

When petroleum occurs in a reservoir that allows the crude material to be recovered by pumping operations as a free-flowing dark to light colored liquid, it is often referred to as *conventional petroleum*. *Heavy crude oil* (often shortened to *heavy oil*) is another type of petroleum that is different from conventional petroleum in that it is much more difficult to recover from the subsurface reservoir. Heavy oil has a much higher viscosity (and lower API gravity) than conventional petroleum, and recovery of heavy oil usually requires thermal stimulation of the reservoir.

The generic term *heavy oil* is often applied to a petroleum that has an API gravity of less than 20° and usually, but not always, a sulfur content higher than 2% by weight. Furthermore, in comparison to conventional crude oils, heavy oils are darker in color and may even be black. The term *heavy oil* has also been arbitrarily used to describe both the heavy oils that require thermal stimulation of recovery from the reservoir and incorrectly the bitumen in bituminous sand (tar sand) formations from which the heavy bituminous material is recovered by a mining operation.

2.4 Bitumen (Extra Heavy Oil)

In addition to conventional petroleum and heavy crude oil, there remains an even more viscous material that offers some relief to the potential shortfalls in supply. This is the *bitumen* (*extra heavy oil*) found in *tar sand* (*oil sand*) deposits. However, many of these reserves can be recovered only with some difficulty, and optional refinery scenarios will be necessary for their conversion to liquid products (Speight, 1999, 2000 and references therein) because of the substantial differences in character between conventional petroleum and tar sand bitumen (Table 2.5).

Tar sands, also variously called *oil sands* or *bituminous sands* (Chapter 1), are a loose-to-consolidated sandstone or a porous carbonate rock, impregnated with bitumen, an

Table 2.5 Comparison of Crude Oil and Bitumen Properties

Property	Bitumen	Conventional
Gravity, °API	8.6	25–37
Distillation, °F		
Vol%		
IBP	—	
5	430	
10	560	—
30	820	
50	1,010	< 650
Viscosity		
SUS, 100°F (38°C)	35,000	< 30
SUS, 210°F (99°C)	513	
Pour point, °F	+50	
Elemental analysis, wt%		
Carbon	83.1	86
Hydrogen	10.6	13.5
Sulfur	4.8	0.1–2.0
Nitrogen	0.4	0.2
Oxygen	1.1	
Hydrocarbon type, wt%		
Asphaltenes	19	< 10
Resins	32	
Oils	49	> 60
Metals, ppm		
Vanadium	250	
Nickel	100	
Iron	75	2–10
Copper	5	
Ash, wt%	0.75	0
Conradson carbon, wt%	13.5	1–2
Net heating value, Btu/lb	17,500	~ 19,500

asphaltic oil with an extremely high viscosity under reservoir conditions.

On an international scale, the bitumen in tar sand deposits represents a potentially large supply of energy (Table 2.6).

Because of the diversity of available information and the continuing attempts to delineate the various tar sand deposits throughout the world, it is virtually impossible to present accurate numbers that reflect the extent of the reserves in terms of barrels. Indeed, investigations into the extent of many of the world's deposits are continuing at such a rate that the numbers vary from one year to the next. Accordingly, the data quoted here must be recognized as approximate; actual figures may be quite different at the time of publication.

The only commercial operations for the recovery of bitumen from tar sand and its subsequent conversion to liquid fuels exist in the Canadian province of Alberta, where Suncor (formerly Great Canadian Oil Sands) went on-stream in 1967 and Syncrude (a consortium of several companies) went on-stream in 1977. Thus, throughout this text, frequent reference is made to tar sand bitumen, but because commercial operations have been in place for over 30 years (Spragins, 1978) it is not surprising that more is known about the Alberta tar sand reserves than about any other reserves in the world. Therefore, when tar sand deposits are discussed, reference is made to the relevant deposit, but when the information is not available the Alberta material is used for the purposes of the discussion.

Tar sand deposits are widely distributed throughout the world (Phizackerley and Scott, 1967; Demaison, 1977; Meyer and Dietzman, 1981; Speight, 1997). The various deposits have been described as belonging to one of two types:

1. Materials that are found in stratigraphic traps
2. Deposits that are located in structural traps

There are, inevitably, gradations and combinations of these two types of deposits, and a broad pattern of deposit entrapment is believed to exist. In general terms, the entrapment characteristics for the very large tar sand deposits all involve a combination of stratigraphic and structural traps, and there are no very large ($> 4 \times 10^9$ bbl) oil sand accumulations either in purely structural or in purely stratigraphic traps.

The potential reserves of bitumen that occur in tar sand deposits have been variously estimated on a world basis as being in excess of 3 trillion ($> 3 \times 10^{12}$) barrels of petroleum equivalent (Table 2.7). The reserves that have been estimated for the United States have been estimated to be in excess of 52 million ($> 52 \times 10^6$) barrels (Table 2.7). That com-

Table 2.6 Estimates of Oil Available in Crude Oil Reservoirs and Tar Sand Deposits (Billions of Barrels)

Commodity	Africa	Canada	Europe	Venezuela	Russia	United States	Total
Tar sand						
Resource	1.75	2000.00	0.4	1000.00	168.00	36.2	3206.35
Reserve	0.175	333.00	0.06	100.00	16.00	2.90	452.135
Crude oil						
Resource	475.8	50	198.3	140.0	558.3	260.0	1542.4
Reserve	57.1	6.8	23.8	17.9	67.0	26.5	199.1

Table 2.7 Estimates of Tar Sand Bitumen in Place (10^6 bbl) in the United States

Deposits	Measured	Speculative	Total
Major (> 100×10^6 barrels)			
Alabama	1.8	4.6	6.4
Alaska	—	10.0	10.0
California	1.9	2.6	4.5
Kentucky	1.7	1.7	3.4
New Mexico	0.1	0.2	0.3
S. Oklahoma	—	0.8	0.8
Texas	3.9	0.9	4.8
Tri-state (KS, MO, OK)	0.2	2.7	2.9
Utah	11.9	7.5	19.4
Wyoming	0.1	0.1	0.2
Subtotal	21.6	31.1	52.7
Minor (between 10×10^6 and 100×10^6 barrels)			
Alabama	—	0.1	0.1
California	—	0.2	0.2
Utah	—	0.7	0.7
Total	21.6	32.1	53.7

mercialization has taken place in Canada does not mean that commercialization is imminent for other tar sand deposits. There are considerable differences between the Canadian and U.S. deposits that could preclude across-the-board application of the Canadian principles to the U.S. sands (Speight, 1990).

As discussed earlier, various definitions have been applied to energy reserves, but the crux of the matter is the amount of a resource that is recoverable using current technology. And although tar sands are not a principal energy reserve, they certainly are significant with regard to projected energy consumption over the next several generations.

Thus, in spite of the high estimates of the reserves of bitumen, the two conditions of vital concern for the economic development of tar sand deposits are the concentration of the resource, or the percent bitumen saturation, and its accessibility, usually measured by the overburden thickness. Recovery methods are based either on mining combined with some further processing or a future operation on the oil sands in situ. The mining methods are applicable to shallow deposits, characterized by an overburden ratio (i.e., overburden depth to thickness of tar sand deposit). For example, indications are that for the Athabasca deposit, no more than 10% of the in-place bitumen can be recovered by mining within current concepts of the economics and technology of open-pit mining; this 10% portion may be considered the *proven reserves* of bitumen in the deposit.

REFERENCES

BP Review of World Gas 1992. British Petroleum Company, London, September.
BP Statistical Review of World Energy. 1992. British Petroleum Company, London, June.
Campbell, C. J. 1997. Oil Gas J. 95(52):33.

Demaison, G. J. 1977. In: The Oil Sands of Canada-Venezuela. Can. Inst. Mining Metall. Spec. Vol. 17. D. A. Redford and A. G. Winestock, eds. Canadian Institute of Mining and Metallurgy, Ottawa, p. 9.

Farry, M. 1998. Oil Gas J. 96(23):115.

Joyce, M. 2000. Oil Gas J. 98(28):64.

Meyer, R. F., and Dietzman, W. D. 1981. In: The Future of Heat Crude and Tar Sands. R. F. Meyer and C. T. Steele, eds. McGraw-Hill, New York, p. 16.

Oil & Gas Journal. 2000a. 98(5):22.

Oil & Gas Journal. 2000b. 98(28):22.

Oil & Gas Journal. 2000c. 98(30):74.

Phizackerley, P.H., and Scott, L.O. 1967. Proc. Seventh World Petroleum Congress. 3:551.

Rahn, F.J. 1987. In: McGraw-Hill Encyclopedia of Science and Technology. S. P. Parker, ed. McGraw-Hill, New York, Vol. 12, p. 171.

Speight, J. G., ed. 1990. Fuel Science and Technology Handbook. Marcel Dekker, New York.

Speight, J. G. 1993. Gas Processing: Environmental Aspects and Methods. Butterworth-Heinemann, Oxford, England.

Speight, J. G. 1997. In: Kirk-Othmer Encyclopedia of Chemical Technology. 4th ed. Wiley-Interscience, New York, Vol. 23, p. 717.

Speight. J. G. 1999. The Chemistry and Technology of Petroleum. 3rd ed. Marcel Dekker, New York.

Speight. JG. 2000. The Desulfurization of Heavy Oils and Residua. 2nd ed. Marcel Dekker, New York.

Spragins, F. K. 1978. Development in Petroleum Science. Vol. 7, Bitumens, Asphalts and Tar Sands. T. F. Yen and G. V. Chilingarian, eds. Elsevier, New York, p. 92.

Swain, E. J. 1991. Oil Gas J. 89(36):59.

Swain, E. J. 1993. Oil Gas J. 91(9):62.

Swain, E. J. 1998. Oil Gas J. 96 (40):43.

Yergin, D. 1991. The Prize: The Epic Quest for Oil, Money, and Power. Simon & Schuster, New York.

Zebroski, E., and Levenson, M. 1976. Ann. Rev. Energy 1:101.

3
Composition

1. INTRODUCTION

Petroleum is not a uniform material, and its chemical and physical (fractional) composition can vary not only with the location and age of the oil field but also with the depth of the individual well. On a molecular basis, petroleum is a complex mixture of hydrocarbons with small amounts of organic compounds containing sulfur, oxygen, and nitrogen as well as compounds containing metallic constituents, particularly vanadium, nickel, iron, and copper. The hydrocarbon content may be as high as 97wt%, as in the lighter paraffinic crude oil, or as low as 50wt% in heavy crude oil. However, crude oil with as little as 50wt% hydrocarbon components is still classified as a mixture of naturally occurring hydrocarbons. It will retain most of the essential characteristics of the hydrocarbons even though the non-hydrocarbon portion may actually consist of molecules containing one or perhaps two atoms of elements other than carbon and hydrogen (Gruse and Stevens, 1960; Speight, 1999 and references therein).

Bitumen, and residua (as a result of the concentration effect of distillation) contain more heteroatomic species and less hydrocarbon constituents than crude oil. Thus, to obtain more gasoline and other liquid fuels, there have been different approaches to refining the heavier feedstocks as well as the recognition that knowledge of the constituents of these higher boiling feedstocks is also of some importance. The problems encountered in processing the heavier feedstocks can be attributed to the *chemical character* and the *amount* of complex, higher boiling constituents in the feedstock. Refining these materials is not just a matter of applying know-how derived from refining conventional crude oils but requires knowledge of the *chemical structure* and *chemical behavior* of these more complex constituents.

It is the purpose of this chapter to present a brief overview of the types of constituents that are found in petroleum, heavy oil, and bitumen through the application of a variety of standard tests (ASTM, 1998; IP, 1999).

2. ULTIMATE (ELEMENTAL) COMPOSITION

The analysis of feedstocks to determine the percentages by weight of carbon, hydrogen, nitrogen, oxygen, and sulfur is perhaps the first method used to examine the general nature of and evaluate a feedstock. The atomic ratios of the various elements to carbon (i.e., H/C, N/C, O/C, and S/C) are frequently used for indications of the overall character of the feedstock. It is also of value to determine the amounts of trace elements, such as vanadium and nickel, in a feedstock because these materials can have serious deleterious effects on catalyst performance during refining by catalytic processes.

For example, *carbon content* can be determined by the method designated for coal and coke (ASTM D-3178) or by the method designated for municipal solid waste (ASTM E-777). There are also methods designated for

1. *Hydrogen content* (ASTM D-1018, ASTM D-3178, ASTM D-3343, ASTM D-3701, and ASTM E-777)
2. *Nitrogen content* (ASTM D-3179, ASTM D-3228, ASTM D-3431, ASTM E-148, ASTM E-258, and ASTM E-778)
3. *Oxygen content* (ASTM E-385)
4. *Sulfur content* (ASTM D-124, ASTM D-1266, ASTM D-1552, ASTM D-1757, ASTM D-2662, ASTM D-3177, ASTM D-4045 and ASTM D-4294)

For any feedstock, the higher the atomic hydrogen/carbon ratio, the higher is its value as refinery feedstock because of the lower hydrogen requirements for upgrading. Similarly, the lower the heteroatomic content, the lower the hydrogen requirements for upgrading. Thus, inspection of the elemental composition of feedstocks provides an initial indication of the quality of the feedstock and, with the molecular weight, indicates the molar hydrogen requirements for upgrading (Fig. 3.1).

However, it has become apparent, with the introduction of the heavier feedstocks into refinery operations, that these ratios are not the only requirement for predicting feedstock character before refining. The use of feedstocks of more complex chemical composition has added a new dimension to refinery operations. Thus, although atomic ratios, as determined by elemental analyses, may be used to compare feedstocks, there is no guarantee that a particular feedstock will behave as predicted from these data. Product slates cannot be predicted accurately, if at all, from these ratios. Additional knowledge such as that of the various chemical reactions of the constituents and the reactions of these constituents with each other also plays a role in determining the processability of a feedstock.

In summary, petroleum contains carbon, hydrogen, nitrogen, oxygen, sulfur, and metals (particularly nickel and vanadium) and the amounts of these elements (in percent by weight) in a whole series of crude oils vary over fairly narrow limits:

Carbon	83.0–87.0%
Hydrogen	10.0–14.0%
Nitrogen	0.1–2.0%
Oxygen	0.05–1.5%
Sulfur	0.05–6.0%
Metals (Ni and V)	< 1000 ppm

These narrow ranges are contradictory to the wide variation in physical properties from the lighter, more mobile crude oils at one extreme to the heavier asphaltic crude oils at the

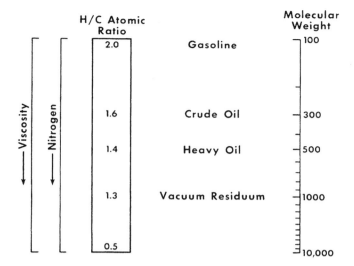

Figure 3.1 Atomic hydrogen/carbon ratio and molecular weight of feedstocks.

other extreme (see also Charbonnier et al., 1969; Draper et al., 1977). And, because of the narrow ranges of carbon and hydrogen content, it is not possible to classify petroleum on the basis of carbon content as coal is classified. The carbon content of coal can vary from as low as 75wt% in lignite to 95wt% in anthracite (Speight, 1994a). Of course, other subdivisions are possible within the various carbon ranges of the coals, but petroleum is restricted to a much narrower range of elemental composition.

The elemental analysis of oil sand bitumen has also been widely reported (Camp, 1976; Bunger et al., 1979; Meyer and Steele, 1981; Speight, 1990), but the data suffer from the disadvantage that identification of the source is too general (i.e., Athabasca bitumen, which covers several deposits) and is often not site specific. In addition, the analysis is quoted for separated bitumen, which may have been obtained by any one of several procedures and may therefore not be representative of the total bitumen in the sand. However, recent efforts have focused on a program to produce sound, reproducible data from samples for which the origin is carefully identified (Wallace et al., 1988).

Like conventional petroleum, from the data that are available the elemental composition of oil sand bitumen is generally constant and, like the data for petroleum, fall into narrow ranges (percent by weight):

Carbon	$83.4 \pm 0.5\%$
Hydrogen	$10.4 \pm 0.2\%$
Nitrogen	$0.4 \pm 0.2\%$
Oxygen	$1.0 \pm 0.2\%$
Sulfur	$5.0 \pm 0.5\%$
Metals (Ni and V)	> 1000 ppm

The major exception to these narrow limits is the oxygen content, which can vary from as little as 0.2wt% to as high as 4.5wt%. This is not surprising, because when oxygen is estimated by difference the analysis is subject to the accumulation of all the errors in the other elemental data. In addition, bitumen is susceptible to aerial oxygen, and the oxygen content is very dependent on sample history.

Heteroatoms in feedstocks affect every aspect of refining. The occurrence of *sulfur* in feedstocks as organic or elemental sulfur or in produced gas as compounds of oxygen (SO_x) and hydrogen (H_2S) is an expensive aspect of refining. It must be removed at some point in the upgrading and refining process. Sulfur contents of many crude oils are on the order of 1% by weight, whereas the sulfur content of tar sand bitumen can exceed 5% or even 10% by weight. Of all of the heteroelements, sulfur is usually the easiest to remove, and many commercial catalysts are available that routinely remove 90% of the sulfur from a feedstock (Speight, 2000).

The *nitrogen* content of petroleum is usually less than 1% by weight, but the nitrogen content of tar sand bitumen can be as high as 1.5% by weight. The presence of nitrogen complicates refining by poisoning the catalysts employed in the various processes. Nitrogen is more difficult to remove than sulfur, and there are fewer catalysts that are specific for nitrogen. If the nitrogen is not removed, the potential for the production of nitrogen oxides (NO_x) during processing and use becomes real.

Metals (particularly *vanadium* and *nickel*) are found in almost every crude oil. Heavy oils and residua contain relatively high proportions of metals either in the form of salts or as organometallic constituents (such as the metalloporphyrins), which are extremely difficult to remove from the feedstock. Indeed, the nature of the process by which residua are produced virtually dictates that all the metals in the original crude oil be concentrated in the residuum (Speight, 1999, 2000). The metallic constituents that may actually *volatilize* under the distillation conditions and appear in the higher boiling distillates are the exceptions here.

Metal constituents of feedstocks cause problems by poisoning the catalysts used for removing sulfur and nitrogen as well as the catalysts used in other processes such as catalytic cracking. Thus, serious attempts are being made to develop catalysts that can tolerate a high concentration of metals without serious loss of catalyst activity or catalyst life.

A variety of tests (ASTM D-1026, D-1262, D-1318, D-1368, D-1548, D-1549, D-2547, D-2599, D-2788, D-3340, D-3341, and D-3605) have been designated for the determination of metals in petroleum products. Determination of metals in whole feeds can be accomplished by combustion of the sample so that only inorganic ash remains. The ash can then be digested with an acid and the solution examined for metal species by atomic absorption (AA) spectroscopy or by inductively coupled argon plasma (ICP) spectrometry.

3. CHEMICAL COMPOSITION

Processability is not only a matter of knowing the elemental composition of a feedstock; it is also a matter of understanding the bulk properties as they relate to the chemical or physical composition of the material. Understanding of the chemical types (or composition) of any feedstock can lead to a better understanding of the chemical aspects of processing.

For example, it is difficult to understand, a priori, the process chemistry of various feedstocks from the elemental composition alone. From such data, it might be surmised that the major difference between a heavy crude oil and a more conventional material is, for example, the H/C atomic ratio. This property indicates that a heavy crude oil (having a lower H/C atomic ratio and being more aromatic in character) would require more hydrogen for upgrading to liquid fuels. This is, indeed, true, but much more information is necessary to understand the *processability* of the feedstock.

Hydrocarbons may constitute as much as 97wt% of a light paraffinic crude oil, and it is convenient to divide the hydrocarbon components of petroleum into the following three classes:

1. *Paraffins*, which are saturated hydrocarbons with straight or branched chains but without any ring structure
2. *Naphthenes*, which are saturated hydrocarbons containing one or more rings, each of which may have one or more paraffinic side chain (more correctly known as *alicyclic hydrocarbons*)
3. *Aromatics*, which are hydrocarbons containing one or more aromatic nuclei, such as benzene, naphthalene, and phenanthrene ring systems, which may be linked up with (substituted) naphthene rings and/or paraffinic side chains

Considerable progress has been made in the isolation and/or identification of the lower molecular weight hydrocarbons as well as accurate estimations of the overall proportions of the various types of hydrocarbons present in petroleum. It has been established that as the boiling point of the petroleum fraction increases, not only the number of the constituents but also their molecular complexity increase (Fig. 3.2).

Nevertheless, it is the non-hydrocarbon (sulfur, oxygen, nitrogen, and metal) constituents that play a large part in determining the processability of the crude oil, and their

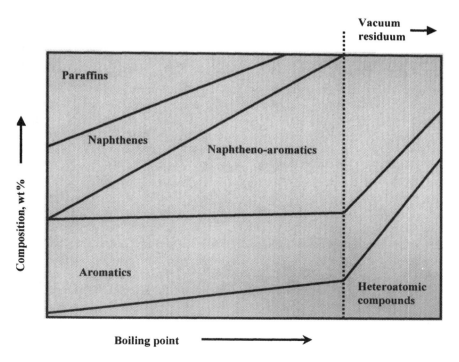

Figure 3.2 Distribution of compound types in petroleum.

influence on the processability of the petroleum is important irrespective of their molecular size (Rossini et al., 1953; Brooks et al., 1954; Lochte and Littmann, 1955; Schwartz and Brasseaux, 1958; Brandenburg and Latham, 1968; Rall et al., 1972; Speight, 2000 and references therein). The occurrence of organic compounds of sulfur, nitrogen, and oxygen serves only to present crude oil as an even more complex mixture, and the appearance of appreciable amounts of these non-hydrocarbon compounds causes some concern in crude oil refining. The non-hydrocarbon constituents (i.e., those organic compounds that contain one or more sulfur, oxygen, and/or nitrogen atoms) tend to concentrate in the higher boiling fractions of petroleum (Speight, 2000)

Although the concentration of heteroatomic constituents in certain fractions may be quite small, their influence is important. For example, the decomposition of inorganic salts suspended in the crude can cause serious breakdowns in refinery operations; the thermal decomposition of deposited inorganic chlorides with evolution of free hydrochloric acid can give rise to serious corrosion problems in the distillation equipment. The presence of organic acid components, such as mercaptans and acids, can also promote metallic corrosion. In catalytic operations, passivation and/or poisoning of the catalyst can be caused by deposition of traces of metals (vanadium and nickel) or by chemisorption of nitrogen-containing compounds on the catalyst, thus necessitating the frequent regeneration of the catalyst or its expensive replacement.

The presence of traces of non-hydrocarbons may impart objectionable characteristics to finished products, such as discoloration, lack of stability in storage, or a reduction in the effectiveness of organic lead antiknock additives. It is thus obvious that a more extensive knowledge of these compounds and their characteristics could result in improved refining methods and even in finished products of better quality. The non-hydrocarbon compounds, particularly the porphyrins and related compounds, are also of fundamental interest in the elucidation of the origin and nature of crude oils. Furthermore, knowledge of their surface-active characteristics is of help in understanding problems related to the migration of oil from the source rocks to the reservoirs.

Sulfur compounds are among the most important heteroatomic constituents of petroleum, and although there are many varieties of sulfur compounds (Speight, 1999, 2000 and references therein), the prevailing conditions during the formation, maturation, and even in situ alteration may dictate that only preferred types exist in any particular crude oil. Nevertheless, sulfur compounds of one type or another are present in all crude oils; in general, the higher the density of the crude oil (or the lower the API gravity of the crude oil), the higher the sulfur content (Speight, 2000 and references therein). The total sulfur in the crude can vary from 0.04% by weight for a light crude oil to about 5.0% for a heavy crude oil. However, the sulfur content of crude oils produced from broad geographic regions varies with time, depending on the composition of newly discovered fields, particularly those in different geological environments.

The presence of sulfur compounds in finished petroleum products often has harmful effects. For example, in gasoline, sulfur compounds are believed to promote corrosion of engine parts, especially under winter conditions, when water containing sulfur dioxide from the combustion may accumulate in the crankcase. In addition, mercaptans in hydrocarbon solution cause the corrosion of copper and brass in the presence of air and also affect lead susceptibility and color stability. Free sulfur is also corrosive, as are sulfides, disulfides, and thiophenes. However, gasoline with a sulfur content of 0.2–0.5 wt% has been used without obvious harmful effect. In diesel fuels, sulfur compounds increase wear and can contribute to the formation of engine deposits. Although a high sulfur content can sometimes be tolerated in industrial fuel oils, a high content of sulfur compounds in lubricating oils seems to lower resistance to oxidation and increase the deposition of solids.

The distribution of sulfur in the various crude oil fractions has been studied many times, beginning in 1891 (Thompson et al., 1976). Although it is generally true that the proportion of sulfur increases with the boiling point during distillation (Speight, 2000 and references therein), the middle fractions may actually contain more sulfur than higher boiling fractions as a result of the decomposition of the higher molecular weight compounds during the distillation. A high sulfur content is generally considered harmful in most petroleum products, and the removal of sulfur compounds or their conversion to less deleterious types is an important part of refinery practice. The distribution of the various types of sulfur compounds varies markedly among crude oils of diverse origin, but fortunately some of the sulfur compounds in petroleum undergo thermal reactions at relatively low temperatures. If elemental sulfur is present in the oil, a reaction, with the evolution of hydrogen sulfide, begins at about 150°C (300°F) and is very rapid at 220°C (430°F), but organically bound sulfur compounds do not yield hydrogen sulfide until higher temperatures are reached. Hydrogen sulfide is, however, a common constituent of many crude oils, and crude oil with > 1% sulfur is often accompanied by gas having substantial properties of hydrogen sulfide.

Oxygen in organic compounds can occur in a variety of forms (Speight, 1999, 2000 and references therein), and it is not surprising that the more common oxygen-containing compounds occur in petroleum. The total oxygen content of crude oil is usually less than 2wt%, although larger amounts have been reported, but when the oxygen content is phenomenally high it may be that the oil has suffered prolonged exposure to the atmosphere either during or after production. However, the oxygen content of petroleum increases with the boiling point of the fractions examined; in fact, the nonvolatile residua may have oxygen contents of up to 8wt%. Although these high molecular weight compounds contain most of the oxygen in petroleum, little is known concerning their structure. Those of lower molecular weight have been investigated with considerably more success and have been shown to contain carboxylic acids and phenols.

It has generally been concluded that the carboxylic acids in petroleum with fewer than eight carbon atoms per molecule are almost entirely aliphatic; monocyclic acids begin at C_6 and predominate above C_{14}. This indicates that the structures of the carboxylic acids correspond with those of the hydrocarbons with which they are associated in the crude oil. In the range in which paraffins are the prevailing type of hydrocarbon, the aliphatic acids may be expected to predominate. Similarly, in the ranges in which monocycloparaffins and dicycloparaffins prevail, one may expect to find principally monocyclic and dicyclic acids, respectively. Carboxylic acids may be less detrimental than other heteroatomic constituents because there is the high potential for decarboxylation to a hydrocarbon and carbon dioxide at the temperatures [> 340°C (> 645°F)] used during distillation or flashing (Speight and Francisco, 1990):

$$R\text{–}CO_2H \longrightarrow R\text{–}H + CO_2$$
carboxylic acid   hydrocarbon

In addition to the carboxylic acids and phenolic compounds (Ar–OH, where Ar is an aromatic moiety), the presence of ketones (>C=O), esters [>C(=O)–OR], ethers (R–O–R), and anhydrides >C(=O)–O–(O=)C<] has been claimed for a variety of crude oils. However, the precise identification of these compounds is difficult, because most of them occur in the higher molecular weight nonvolatile residua. They are claimed to be products of the air blowing of the residua, and their existence in virgin crude oil, heavy oil, or bitumen may yet need to be substantiated.

Nitrogen in petroleum may be classified arbitrarily as basic and nonbasic. The basic nitrogen compounds (Speight, 1999, 2000 and references therein), which are composed mainly of pyridine homologs and occur throughout the boiling ranges, have a decided tendency to exist in the higher boiling fractions and residua. The nonbasic nitrogen compounds, which are usually of the pyrrole, indole, and carbazole types, also occur in the higher boiling fractions and residua.

In general, the nitrogen content of crude oil is low, within the range 0.1–0.9% by weight, although early work indicates that some crude oils may contain up to 2% nitrogen. Crude oils with no detectable nitrogen or with trace amounts are not uncommon, but in general the more asphaltic the oil, the higher its nitrogen content. Insofar as an approximate correlation exists between the sulfur content and API gravity of crude oil (Speight, 2000), there also exists a correlation between nitrogen content and the API gravity of crude oil (Speight, 2000). It also follows that there is an approximate correlation between the nitrogen content and the carbon residue. The higher the carbon residue, the higher the nitrogen content. The presence of nitrogen in petroleum is of much greater significance in refinery operations than might be expected from the small amounts present. Nitrogen compounds can be responsible for the poisoning of cracking catalysts, and they also contribute to gum formation in such products as domestic fuel oil. The trend in recent years to cut deeper into the crude to obtain stocks for catalytic cracking has accentuated the harmful effects of the nitrogen compounds, which are concentrated largely in the higher boiling portions.

Basic nitrogen compounds with a relatively low molecular weight can be extracted with dilute mineral acids; equally strong bases of higher molecular weight remain unextracted because of unfavorable partitioning between the oil and aqueous phases. A method has been developed in which the nitrogen compounds are classified as basic or nonbasic depending on whether they can be titrated with perchloric acid in a 50:50 solution of glacial acetic acid and benzene. Application of this method has shown that the ratio of basic to total nitrogen is approximately constant (0.3 ± 0.05) irrespective of the source of the crude. Indeed, the ratio of basic to total nitrogen was found to be approximately constant throughout the entire range of distillate and residual fractions. Nitrogen compounds extractable with dilute mineral acids from petroleum distillates were found to consist of pyridines, quinolines, and isoquinolines carrying alkyl substituents, as well as a few pyridines in which the substituent was a cyclopentyl or cyclohexyl group. The compounds that cannot be extracted with dilute mineral acids contain the greater part of the nitrogen in petroleum and are generally carbazoles, indoles, and pyrroles.

Porphyrins are a naturally occurring chemical species whose existence in petroleum has been known for more than 50 years. They are given separate consideration in this section because of their uniqueness as separate and distinct chemical entities. They are not usually considered among the usual nitrogen-containing constituents of petroleum, nor are they considered a metalloorganic material that also occurs in some crude oils. As a result of early investigations there arose the concept of porphyrins as biomarkers that could establish a link between compounds found in the geosphere and their corresponding biological precursors (Treibs, 1934; Glebovskaya and Volkenshtein, 1948).

Porphyrins usually occur in the nonbasic portion of the nitrogen-containing concentrate. The simplest porphyrin is porphine, which consists of four pyrrole molecules joined by methine ($-CH=$) bridges (Speight, 1999, 2000 and references therein). The methine bridges establish conjugated linkages between the component pyrrole nuclei, forming a more extended resonance system. Although the resulting structure retains much of the inherent character of the pyrrole components, the larger conjugated system increases the aromatic character of the porphine molecule (Falk, 1964; Smith, 1975).

Almost all crude oil, heavy oil, and bitumen contain detectable amounts of vanadyl and nickel porphyrins. More mature, lighter crude oil usually contains only small amounts of these compounds. Heavy oils may contain large amounts of vanadyl and nickel porphyrins. Vanadium concentrations of over 1000 ppm are known for some crude oils, and a substantial amount of the vanadium in these crude oils is chelated with porphyrins. In high-sulfur crude oil of marine origin, vanadyl porphyrins are more abundant than nickel porphyrins. Low-sulfur crude oils of lacustrine origin usually contain more nickel porphyrins than vanadyl porphyrins.

Of all the metals in the periodic table, only vanadium and nickel have been proven to exist as chelates in significant amounts in a large number of crude oils and tar sand bitumen. The existence of iron porphyrins in some crude oils has been claimed (Franceskin et al., 1986). Geochemical reasons for the absence of substantial quantities of porphyrins chelated with metals other than nickel and vanadium from most crude oils and tar sand bitumen have been advanced (Hodgson et al., 1967; Baker, 1969; Baker and Palmer, 1978; Baker and Louda, 1986; Filby and Van Berkel, 1987; Quirke, 1987).

If the vanadium and nickel contents of crude oils are measured and compared with porphyrin concentrations, it is usually found that not all the metal content can be accounted for as porphyrins (Dunning et al., 1960; Reynolds, 1998). In some crude oils, as little as 10% by weight of total metals appears to be chelated with porphyrins. Only rarely can all measured nickel and vanadium in a crude oil be accounted for as porphyrinic (Erdman and Harju, 1963). Currently some investigators believe that part of the vanadium and nickel in crude oils is chelated with ligands that are not porphyrins. These metal chelates are referred to as nonporphyrin metal chelates or complexes (Crouch et al., 1983; Fish et al., 1984; Reynolds et al., 1987).

On the other hand, the existence of nonporphyrin metal in crude oils has been questioned (Goulon et al., 1984). It is possible that in such systems as the heavy crude oils, in which intermolecular associations are important, measurements of the porphyrin concentrations are unreliable and there is a tendency to understate the actual values. However, for the purposes of this chapter it is assumed that nonporphyrin chelates exist in fossil fuels but the relative amount is as yet unknown.

Finally, during the fractionation of petroleum (Chapter 9) the metallic constituents (metalloporphyrins and nonporphyrin metal chelates) are concentrated in the asphaltene fraction. The deasphalted oils (petrolenes and maltenes) (Chapter 1) contain smaller concentrations of porphyrins than the parent materials and usually very small concentrations of nonporphyrin metals.

Metallic constituents are found in every crude oil, and the concentrations have to be reduced to convert the oil to transportation fuel. Metals affect many upgrading processes and cause particular problems because they poison catalysts used for sulfur and nitrogen removal as well as other processes such as catalytic cracking.

Thus, the occurrence of metallic constituents in crude oil is of considerably greater interest to the petroleum industry than might be expected from the very small amounts present. Even minute amounts of iron, copper, nickel, and vanadium (particularly the latter two) in the charging stocks for catalytic cracking affect the activity of the catalyst and result in increased gas and coke formation and reduced yields of gasoline. In high temperature power generators such as oil-fired gas turbines, the presence of metallic constituents, particularly vanadium, in the fuel may lead to ash deposits on the turbine rotors, thus reducing clearances and disturbing their balance. More particularly, damage by corrosion may be very severe. The ash resulting from the combustion of fuels containing sodium and, especially, vanadium reacts with refractory furnace linings to lower their fusion points and so cause their deterioration.

The ash residue left after combusting a feedstock is due to the presence of these metallic constituents, part of which occur as inorganic water-soluble salts (mainly chlorides and sulfates of sodium, potassium, magnesium, and calcium) in the water phase of crude oil emulsions. These are removed in the desalting operations, either by evaporation of the water and subsequent water washing or by breaking the emulsion, thereby causing the original mineral content of the crude to be substantially reduced. Other metals are present in the form of oil-soluble organometallic compounds as complexes or metallic soaps or in the form of colloidal suspensions, and the total ash from desalted crude oils is of the order of 0.1–100 mg/L. Metals are generally found only in the nonvolatile portion of crude oil (Altgelt and Boduszynski, 1994; Reynolds, 1998).

Two groups of elements appear in significant concentrations in the original crude oil associated with well-defined types of compounds. Zinc, titanium, calcium, and magnesium appear in the form of organometallic soaps with surface-active properties adsorbed in the water/oil interfaces and act as emulsion stabilizers. However, vanadium, copper, nickel, and part of the iron found in crude oils seem to be in a different class and are present as oil-soluble compounds. These metals are capable of complexing with pyrrole pigment compounds derived from chlorophyll and hemoglobin and are almost certain to have been present in plant and animal source materials. It is easy to surmise that the metals in question are present in such form, ending in the ash content. Evidence for the presence of several other metals in oil-soluble form has been produced, and thus zinc, titanium, calcium, and magnesium compounds have been identified in addition to vanadium, nickel, iron, and copper. Analyses of a number of crude oils for iron, nickel, vanadium, and copper indicate a relatively high vanadium content, usually exceeding that of nickel, although the reverse can also occur.

Distillation concentrates the metallic constituents in the residues, although some can appear in the higher boiling distillates, but the latter may be due in part to entrainment (Reynolds, 1998; Speight, 2000). Nevertheless, there is evidence that a portion of the metallic constituents may occur in the distillates due to volatilization of the organometallic compounds present in the petroleum. In fact, as the percentage of overhead obtained by vacuum distillation of a reduced crude is increased, the amount of metallic constituents in the overhead oil is also increased. The majority of the vanadium, nickel, iron, and copper in residual stocks may be precipitated along with the asphaltenes by hydrocarbon solvents. Thus, removal of the asphaltenes with *n*-pentane reduces the vanadium content of the oil by up to 95%, with substantial reductions in the amounts of iron and nickel.

4. BULK (FRACTIONAL) COMPOSITION

In petroleum refining the feedstock is subjected to a series of physical and chemical processes (Fig. 3.3) that generate a variety of products. In some of the processes, e.g., distillation, the constituents of the feedstock are isolated unchanged, whereas in other processes, e.g., cracking, considerable changes are brought about to the constituents. Feedstocks can be defined (on a *relative* or *standard* basis) in terms of three or four general

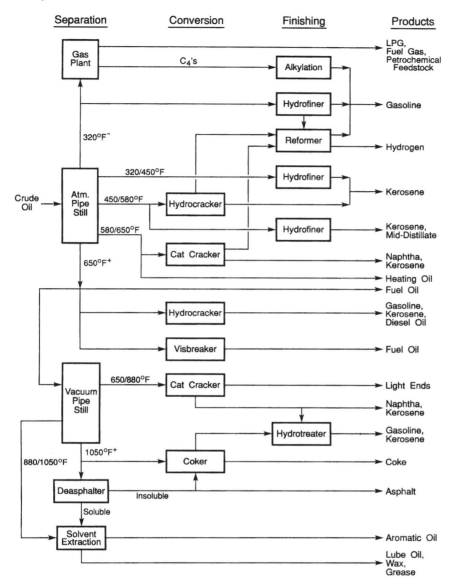

Figure 3.3 Representation of a refinery.

fractions: asphaltenes, resins, saturates, and aromatics (Figs 3.4 and 3.5). Thus, it is possible to compare interlaboratory investigations and then apply the concept of predictability to refining sequences and potential products. Recognition that refinery behavior is related to the composition of the feedstock has led to a multiplicity of attempts to establish petroleum and its fractions as compositions of matter. As a result, various analytical techniques have been developed for the identification and quantification of *every molecule* in the lower boiling fractions of petroleum. It is now generally recognized that the name *petroleum* does not describe a composition of matter but rather a mixture of various

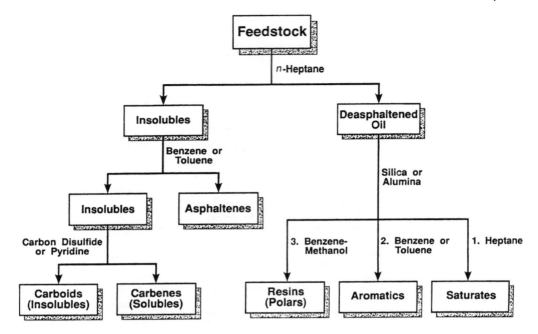

Figure 3.4 Separation scheme for various feedstocks.

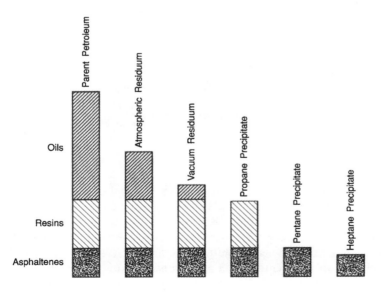

Figure 3.5 Relationship of feedstocks to the parent petroleum.

organic compounds that includes a wide range of molecular weights and molecular types that exist in balance with each other (Speight, 1994b; Long and Speight, 1998). There must also be some questions of the advisability (perhaps *futility* is a better word) of attempting to describe *every molecule* in petroleum. The true focus should be on the end uses to which these molecules can be put.

The fractionation methods available to the petroleum industry allow a reasonably effective degree of separation of hydrocarbon mixtures (Speight, 1999 and references therein). However, there are problems with separating the petroleum constituents without alteration of their molecular structure and obtaining these constituents in a substantially pure state. Thus, the general procedure is to employ techniques that segregate the constituents according to molecular size and molecular type.

It is more generally true, however, that the success of any attempted fractionation procedure depends on the utilization of several integrated techniques, especially those techniques that make use of chemical and physical properties to differentiate the various constituents. For example, the standard processes of physical fractionation used in the petroleum industry are those of distillation and solvent treatment as well as adsorption by surface-active materials. Chemical procedures depend on specific reactions, such as the interaction of olefins with sulfuric acid or the various classes of adduct formation. Chemical fractionation is often but not always successful and, because of the complex nature of crude oil, may result in unprovoked chemical reactions that have an adverse effect on the fractionation and the resulting data. Indeed, caution is advised when using methods that require chemical separation of the constituents.

The order in which the several fractionation methods are used is determined not only by the nature and/or composition of the crude oil but also by the effectiveness of a particular process and its compatibility with the other separation procedures to be employed. Thus, although there are wide variations in the nature of refinery feedstocks, there have been many attempts to devise standard methods of petroleum fractionation. However, the various laboratories are inclined to adhere firmly to, and promote, their own particular methods. Recognition that no one particular method may satisfy all the requirements of petroleum fractionation is the first step in any fractionation study. This is due, in the main part, to the complexity of petroleum not only with respect to the distribution of hydrocarbon species but also with respect to the distribution of heteroatomic (nitrogen, oxygen, and sulfur) species.

4.1 Solvent Treatment

Fractionation of petroleum by distillation is an excellent means by which the volatile constituents can be isolated and studied. However, the nonvolatile residuum, which may actually constitute from 1 to 60wt% of the petroleum, cannot be fractionated by distillation without the possibility of thermal decomposition, and as a result alternative methods of fractionation have been developed.

The distillation process separates *light* (lower molecular weight) and *heavy* (higher molecular weight) constituents by virtue of their volatility and involves the participation of a vapor phase and a liquid phase. These are, however, physical processes that involve the use of two liquid phases, usually a solvent phase and an oil phase.

Solvent methods have also been applied to petroleum fractionation on the basis of molecular weight. The major molecular weight separation process used in the laboratory as well as in the refinery is solvent precipitation. Solvent precipitation occurs in a refinery in a deasphalting unit (Fig. 3.6) and is essentially an extension of the procedure for separation by molecular weight, although some separation by polarity might also be operative. The deasphalting process is usually applied to the higher molecular weight fractions of petroleum such as atmospheric and vacuum residua.

These fractionation techniques can also be applied to cracked residua, asphalt, bitumen, and even virgin petroleum. However, in the last case the possibility of losses of the lower boiling constituents is apparent; hence the recommended procedure for virgin petroleum is distillation first, followed by fractionation of the residua.

The simplest application of solvent extraction consists of mixing petroleum with another liquid, which results in the formation of two phases. This causes the petroleum constituents to be distributed between the two phases; the dissolved portion is referred to as the *extract*, and the nondissolved part of the petroleum is referred to as the *raffinate*.

The ratio of the concentration of any particular component in the two phases is known as the distribution coefficient K:

$$K = C_1/C_2$$

where C_1 and C_2 are the concentrations in the phases 1 and 2, respectively. The distribution coefficient is usually constant, although it may vary slightly, with the concentrations of the other components. In fact, the distribution coefficients may differ for the various components of the mixture to such an extent that the ratios of the concentrations of the

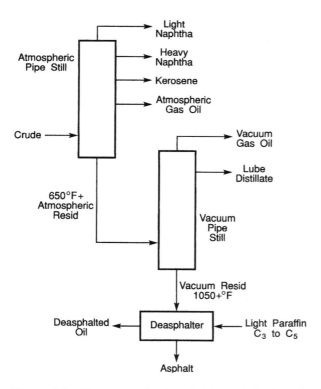

Figure 3.6 Placement of a deasphalting unit in the refinery (see also Fig. 3.3).

various components in the solvent phase differ from those in the original petroleum; this is the basis for solvent extraction procedures.

It is generally molecular type, not molecular size, that is responsible for the solubility of species in various solvents. Thus, solvent extraction separates petroleum fractions according to type, although within any particular series there is a separation according to molecular size. Lower molecular weight hydrocarbons of a series (the light fraction) may well be separated from their higher molecular weight homologs (the heavy fraction) by solvent extraction procedures.

In general, it is advisable to employ selective extraction with fairly narrow boiling range fractions. However, the separation achieved after one treatment with the solvent is rarely complete, and several repetitions of the treatment are required. Such repetitious treatments are normally carried out by moving the liquids countercurrently through the extraction equipment (*countercurrent extraction*), which affords better yields of the extractable materials.

The list of compounds that have been suggested as selective solvents for the preferential extraction fractionation of petroleum contains a large selection of different functional types (Table 3.1). However, before any extraction process is attempted, it is necessary to consider the following criteria:

Table 3.1 Compound Types Used for the Selective Extraction of Petroleum

Esters	$R-C\overset{\displaystyle O}{\underset{\displaystyle OR'}{\big<}}$
Alcohols	$R-OH$
Aldehydes	$R-C\overset{\displaystyle O}{\underset{\displaystyle H}{\big<}}$
Acids	$R-C\overset{\displaystyle O}{\underset{\displaystyle OH}{\big<}}$
Ketones	$R-\overset{\displaystyle O}{\overset{\|}{C}}-R'$
Amines	$R-NH_2$
Amides	$R-C\overset{\displaystyle O}{\underset{\displaystyle NH_2}{\big<}}$
Nitrocompounds	$R-NO_2$
Nitriles	$R-CN$

R and R′ are alkyl or aromatic radicals, and if both are alkyl or both are aromatic they may or may not be the same.

1. The differences in the solubility of the petroleum constituents in the solvent should be substantial.
2. The solvent should be significantly less or more dense than the petroleum (product) to be separated to allow easier countercurrent flow of the two phases.
3. Separation of the solvent from the extracted material should be relatively easy.

It may also be advantageous to consider other properties, such as viscosity, surface tension, and the like, as well as the optimal temperature for the extraction process. Thus, aromatics can be separated from naphthene and paraffinic hydrocarbons by the use of selective solvents. Furthermore, aromatics with differing numbers of aromatic rings that may exist in various narrow boiling fractions can also be effectively separated by solvent treatment.

The separation of crude oil into two fractions, asphaltenes and maltenes, is conveniently brought about by means of low molecular weight paraffinic hydrocarbons, which were recognized to have selective solvency for hydrocarbons, and simple relatively low molecular weight hydrocarbon derivatives. The more complex, higher molecular weight compounds are precipitated particularly well by the addition of 40 volumes of *n*-pentane or *n*-heptane in the methods generally preferred at present (Speight et al., 1984; Speight, 1994b), although hexane is used on occasion (Yan et al., 1997). This method separates the chemical components with the most complex structures from the mixture, and this fraction, which should correctly be called *n-pentane asphaltenes* or *n-heptane asphaltenes*, is qualitatively and quantitatively reproducible (Fig. 3.4).

Variation in the solvent type also causes significant changes in asphaltene yield. For example, branched-chain paraffins or terminal olefins do not precipitate the same amount of asphaltenes as the corresponding normal paraffins (Mitchell and Speight, 1973). Cycloparaffins (naphthenes) have a remarkable effect on asphaltene yield and give results totally unrelated to those from any other nonaromatic solvent. For example, when cyclopentane, cyclohexane, or their methyl derivatives are employed as precipitating media, only about 1% of the material remains insoluble.

To explain those differences it was necessary to consider the solvent power of the precipitating liquid, which can be related to molecular properties (Hildebrand et al., 1970). The solvent power of nonpolar solvents has been expressed as a solubility parameter, δ, and equated to the internal pressure of the solvent, that is, the ratio between the surface tension γ and the cubic root of the molar volume V:

$$\delta_1 = V^{1/3}$$

Alternatively, the solubility parameter of non-polar solvents can be related to the energy of vaporization ΔE^V and the molar volume,

$$\delta_2 = (\Delta E^v / V)^{1/2}$$

or

$$\delta_2 = \left(\Delta H^v - \frac{RT}{V} \right)^{1/2}$$

where ΔH^v is the heat of vaporization, R is the gas constant, and T is the absolute temperature.

The introduction of a polar group (heteroatomic function) into the molecule of the solvent has significant effects on the quantity of precipitate. Treatment of a residuum with a variety of ethers or treatment of asphaltenes with a variety of solvents illustrates this point (Speight, 1979). In the latter instance it was not possible to obtain data from addition of the solvent to the whole feedstock per se, because the majority of the non-hydrocarbon materials were not miscible with the feedstock. It is nevertheless interesting that, as with the hydrocarbons, the amount of precipitate or degree of asphaltene solubility can be related to the solubility parameter.

The solubility parameter allows an explanation of certain apparent anomalies, such as the insolubility of asphaltenes in pentane and their almost complete solubility in cyclopentane. Moreover, the solvent power of various solvents is in agreement with the derivation of the solubility parameter; for any one series of solvents the relationship between amount of precipitate (or asphaltene solubility) and the solubility parameter δ is quite regular.

In any method used to isolate asphaltenes as a separate fraction, standardization of the technique is essential. For many years, the method of asphaltene separation was not standardized, and even now it remains subject to the preferences of the standards organizations of different countries. The use of both *n*-pentane and *n*-heptane has been widely advocated, and although *n*-heptane is becoming the deasphalting liquid of choice, this is by no means a hard and fast rule. And it must be recognized that large volumes of solvent may be required to effect a reproducible separation, similar to the amounts required for consistent asphaltene separation. It is also preferable that the solvents be of sufficiently low boiling point that complete removal of the solvent from the fraction can be effected and, most important, the solvent must not react with the feedstock. Hence the preference for hydrocarbon liquids, although the several standard methods that have been used are not unanimous in the ratio of hydrocarbon liquid to feedstock.

Method	Deasphalting liquid	Volume (mL/g)
ASTM D-893	*n*-Pentane	10
ASTM D-2006	*n*-Pentane	50
ASTM D-2007	*n*-Pentane	10
IP 143	*n*-Heptane	30
ASTM D-3279	*n*-Heptane	100
ASTM D-4124	*n*-Heptane	100

However, it must be recognized that some of these methods were developed for use with feedstocks other than heavy oil and adjustments are necessary.

Although *n*-pentane and *n*-heptane are the solvents of choice in the laboratory, other solvents can be used (Speight, 1979) and cause the separation of asphaltenes as brown to black powdery materials. In the refinery, supercritical low molecular weight hydrocarbons (e.g., liquid propane, liquid butane, or mixtures of the two) are the solvents of choice and the product is a semisolid (tacky) to solid asphalt. The amount of asphalt that settles out of the paraffin–residuum mixture depends on the molecular size of the paraffin, the temperature, and the paraffin-to-feedstock ratio (Girdler, 1965; Corbett and Petrossi, 1978; Speight et al., 1984; Speight, 1999).

Because industrial solvents are very rarely a single compound, it is also of interest to note that the physical characteristics of two different solvent types, in this case benzene and *n*-pentane, are additive on a mole fraction basis (Mitchell and Speight, 1973) and also explain the variation of solubility with temperature. The data also show the effects of blending a solvent with the bitumen itself and allowing the resulting solvent–heavy oil blend to control the degree of bitumen solubility. Varying proportions of the hydrocarbon alter the physical characteristics of the oil to such an extent that the amount of precipitate (asphaltenes) can be varied accordingly within a certain range.

At constant temperature, the quantity of precipitate first increases with increasing ratio of solvent to feedstock and then reaches a maximum (Speight et al., 1984). In fact, there are indications that when the proportion of solvent in the mix is < 35%, little or no asphaltenes are precipitated. In addition, when pentane and the lower molecular weight hydrocarbon solvents are used in large excess, the quantity of precipitate and the composition of the precipitate change with increasing temperature (Mitchell and Speight, 1973; Andersen, 1994).

Contact time between the hydrocarbon and the feedstock also plays an important role in asphaltene separation (Speight et al., 1984). Yields of the asphaltenes reach a maximum after approximately 8 h, which may be ascribed to the time required for the asphaltene particles to agglomerate into particles of a *filterable size* as well as the diffusion-controlled nature of the process. Heavier feedstocks also need time for the hydrocarbon to penetrate their mass.

After removal of the asphaltene fraction, further fractionation of petroleum is also possible by variation of the hydrocarbon solvent. For example, liquefied gases, such as propane and butane, precipitate as much as 50% by weight of the residuum or bitumen. The precipitate is a black, tacky, semisolid material, in contrast to the pentane-precipitated asphaltenes, which are usually brown amorphous solids. Treatment of the propane precipitate with pentane then yields the insoluble brown amorphous asphaltenes and soluble near-black semisolid resins, which are, as near as can be determined, equivalent to the resins isolated by adsorption techniques.

4.2 Adsorption Methods

Separation by adsorption chromatography essentially commences with the preparation of a porous bed of finely divided solid, the adsorbent. The adsorbent is usually contained in an open tube (column chromatography); the sample is introduced at one end of the adsorbent bed and induced to flow through the bed by means of a suitable solvent. As the sample moves through the bed the various components are held (adsorbed) to a greater or lesser extent depending on their chemical nature. Thus, those molecules that are strongly adsorbed spend considerable time on the adsorbent surface rather than in the moving (solvent) phase, but components that are slightly adsorbed move through the bed comparatively rapidly.

It is essential that the asphaltenes be completely removed before application of the adsorption technique to the petroleum; this can be accomplished, for example, by any of the methods outlined in the previous section. The prior removal of the asphaltenes is essential because as they are usually difficult to remove from the earth or clay and may be irreversibly adsorbed on the adsorbent.

By definition, the *saturate fraction* consists of paraffins and cycloparaffins (naphthenes). The single-ring *naphthenes*, or *cycloparaffins*, present in petroleum are primarily alkyl-substituted cyclopentane and cyclohexane. The alkyl groups are usually quite short, with methyl, ethyl, and isopropyl groups the predominant substituents. As the molecular weight of the naphthenes increases, the naphthene fraction contains more condensed rings. with six-membered rings predominating. However, five-membered rings are still present in the complex higher molecular weight molecules.

The *aromatic fraction* consists of those compounds that contain an aromatic ring, varying from *monoaromatics* (one benzene ring in a molecule) to *diaromatics* (substituted naphthalene) to *triaromatics* (substituted phenanthrene). Higher condensed ring systems (*tetra-aromatics*, *penta-aromatics*) are also known but are somewhat less prevalent than the lower ring systems, and each aromatic type will have increasing amounts of condensed ring naphthene attached to the aromatic ring as molecular weight is increased.

However, depending upon the adsorbent employed for the separation, a compound having an aromatic ring (i.e., six aromatic carbon atoms) carrying side chains consisting *in toto* of more than six carbon atoms (i.e., more than six nonaromatic carbon atoms) will appear in the aromatic fraction.

Careful monitoring of the experimental procedures and the nature of the adsorbent has been responsible for the successes achieved with this particular technique. Early procedures consisted of warming solutions of the petroleum fraction with the adsorbent and subsequent filtration. This procedure has continued to the present day, and separation by adsorption is used commercially in plant operations in the form of clay treatment of crude oil fractions and products. In addition, the proportions of each fraction depend on the ratio of adsorbent to deasphaltened oil.

It is also advisable, once a procedure using a specific adsorbent has been established, that the same type of adsorbent be employed for future fractionation, because the ratio of the product fractions varies from adsorbent to adsorbent. It is also necessary that the procedure be used with caution and that the method not only be reproducible but quantitative recoveries be guaranteed. Reproducibility with only, say, 85% of the material recoverable is not considered a criterion of success.

Two procedures have received considerable attention over the years: (1) the United States Bureau of Mines–American Petroleum Institute (USBM-API) method and (2) the saturates-aromatics-resins-asphaltenes (SARA) method. This latter method is often also called the saturates-aromatics-polars-asphaltenes (SAPA) method. These two methods are used to represent the standard methods of petroleum fractionation. Other methods are also noted, especially when they have added further meaningful knowledge to compositional studies.

The USBM-API method (Fig. 3.7) employs ion-exchange chromatography and coordination chromatography with adsorption chromatography to separate heavy oils and residua into seven broad fractions: acids, bases, neutral nitrogen compounds, saturates, and mono-, di-, and polynuclear aromatic compounds. The acid and base fractions are isolated by ion-exchange chromatography, the neutral nitrogen compounds by complexation chromatography using ferric chloride, and the saturates and aromatics by adsorption chromatography on activated alumina (Jewell et al., 1972a) or on a combined alumina-silica column (Hirsch et al., 1972; Jewell et al., 1972b).

The SARA method (Jewell et al., 1974) is essentially an extension of the API method that allows more rapid separations by placing the two ion-exchange resins and the $FeCl_3$–clay–anion-exchange resin packing into a single column. The adsorption chromatography of the nonpolar part of the petroleum fractions is still performed in a separation operation. Becausee the asphaltene content of petroleum (and synthetic fuel) feedstocks is often an important aspect of processability, an important feature of the SARA method is that the asphaltenes are separated as a group. Perhaps more important is that the method is

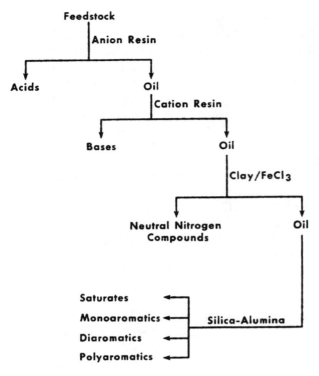

Figure 3.7   The USBM-API fractionation method.

reproducible and is applicable to a large variety of the most difficult feedstocks, such as residual tar sand bitumen, shale oil, and coal liquids.

Both the USBM-API and SARA methods require some caution if the asphaltenes are first isolated as a separate fraction. For example, the asphaltene yield varies with the hydrocarbon used for the separation and with other factors (Girdler, 1965; Mitchell and Speight, 1973; Speight et al., 1984). An inconsistent separation technique can give rise to problems resulting from residual asphaltenes in the deasphaltened oil undergoing irreversible adsorption on the solid adsorbent. The USBM-API and SARA methods are widely used separation schemes for studying the composition of heavy petroleum fractions and other fossil fuels, but several other schemes have also been used successfully and have found common use in investigations of feedstock composition (Speight, 1999, 2000 and references therein).

There are three ASTM methods that provide for the separation of a feedstock into four or five constituent fractions (Figs 3.8, 3.9, and 3.10). It is interesting to note that as the methods have evolved there has been a change from the use of pentane (ASTM D-2006 and D-2007) to heptane (ASTM D-4124) to separate asphaltenes. This is, in fact, in keeping with the production of a more consistent fraction that represents these higher molecular weight complex constituents of petroleum (Girdler, 1965; Speight et al., 1984).

Two of the methods (ASTM D-2007 and D-4124) use adsorbents to fractionate the deasphaltened oil, but the third method (ASTM D-2006) advocates the use of various grades of sulfuric acid to separate the material into compound types. Caution is advised in

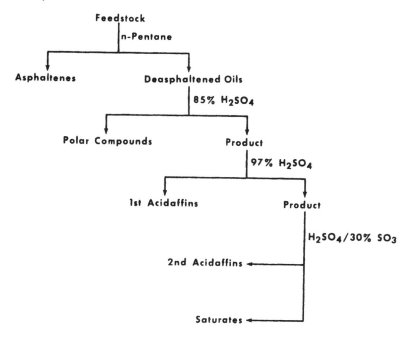

Figure 3.8 The ASTM D-2006 fractionation method.

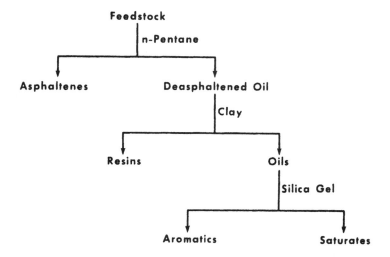

Figure 3.9 The ASTM D-2007 fractionation method.

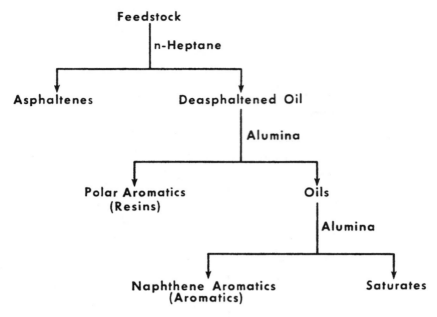

Figure 3.10   The ASTM D-4124 fractionation method.

the application of this method, because it does not work well with all feedstocks. For example, when the *sulfuric acid* method (ASTM D-2006) is applied to the separation of heavy feedstocks, complex emulsions can be produced.

Obviously, precautions must be taken when attempting to separate heavy feedstocks or polar feedstocks into constituent fractions. The disadvantages in using ill-defined adsorbents are that adsorbent performance differs with the same feed and, in certain instances, may even cause chemical and physical modification of the feed constituents. The use of a chemical reactant such as sulfuric acid should be advocated only with caution, because feeds react differently and may even cause irreversible chemical changes and/or emulsification. These advantages may be of little consequence when it is not, for various reasons, the intention to recover the various product fractions in toto or in the original state, but in terms of the compositional evaluation of different feedstocks the disadvantages are very real.

In summary, the terminology used for the identification of the various methods might differ. However, in general terms, group-type analysis of petroleum is often identified by the acronyms for the names: PONA (paraffins, olefins, naphthenes, and aromatics), PIONA (paraffins, isoparaffins, olefins, naphthenes, and aromatics), PNA (paraffins, naphthenes, and aromatics), PINA (paraffins, isoparaffins, naphthenes, and aromatics), or SARA (saturates, aromatics, resins, and asphaltenes). However, it must be recognized that the fractions produced by the use of different adsorbents will differ in content and will also be different from fractions produced by solvent separation techniques.

The variety of fractions isolated by these methods and the potential for differences in composition of the fractions makes it even more important that the method be described accurately and that it be reproducible not only in any one laboratory but also between various laboratories.

4.3 Chemical Methods

Methods of fractionation using chemical reactants are entirely different in nature from the methods described in the preceding sections. Although several methods using chemical reactants have been applied to fractionation—for example, adsorption, solvent treatment, and treatment with alkali (Kalichevsky and Stagner, 1942; Kalichevsky and Kobe, 1956; Nelson, 1958; Gruse and Stevens, 1960; Bland and Davidson, 1967; Speight, 1999)—these methods are often applied to product purification as well as separation. Thus, with the exception of treatment with alkali, they have already been described. The method of chemical separation commonly applied to separate crude oil into various fractions is treatment with sulfuric acid, and because this method has also been applied in the refinery it will be the focus of this section.

The early method for the treatment of deasphaltened crude oil with sulfuric acid usually produced a precipitate (Fig. 3.11) that consisted of material that had been converted to a polar fraction by interaction of the asphaltic constituents with the sulfuric acid. It is nevertheless possible that some of the acid-precipitated material originated as asphaltenes that were incompletely precipitated by the naphtha, because the addition of only 20 volumes of solvent to heavy feedstocks is not sufficient to completely precipitate asphaltene material (see Section 4.1).

However, this method has served to demonstrate the type of separation that can be obtained with sulfuric acid. A later refinement of this principle led to the development of a technique that proposes a resolution of crude oils, crude oil residua, and asphaltic or bituminous materials into five broad fractions (Fig. 3.12) (Rostler, 1965). These fractions

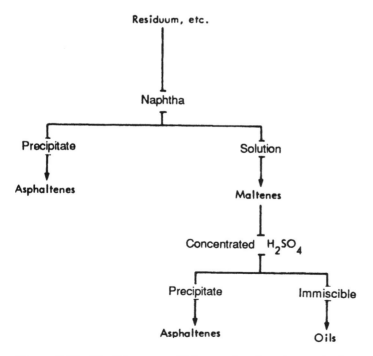

Figure 3.11 The Marcusson–Eickmann fractionation method.

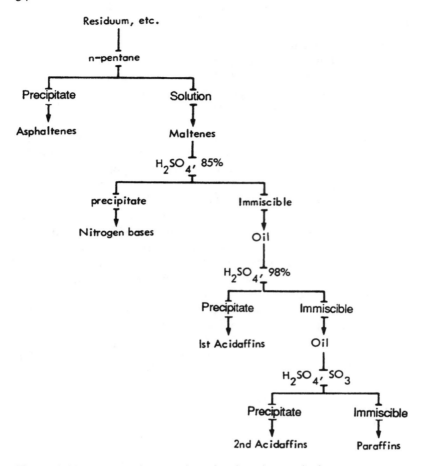

Figure 3.12   The Rostler–Sternberg fractionation method.

are distinctly different in chemical reactivity as measured by their response to cold sulfuric acid of increasing strength (sulfur trioxide concentration). The fractionation involves, like most other methods, initial separation of the sample into asphaltenes and the fraction referred to as maltenes. The fractionation of maltenes into resins and oils is often too vague for identification purposes, and the nature, or composition, of these two fractions may differ from one crude oil to another.

The chemical precipitation method is claimed to provide the needed subdivision of the resins and the oils to yield chemically related fractions. The names given to the individual fractions are descriptive of the steps used in the procedure. For example, the name *paraffin* is used for the saturated nonreactive fraction, whereas the next two groups in ascending reactivity are *second acidaffins* and *first acidaffins*. The term *second acidaffins* denotes the group of hydrocarbons having affinity for strong acid as represented by fuming sulfuric acid. The term *first acidaffins* denotes constituents that have an affinity for ordinary concentrated sulfuric acid. The next fraction, nitrogen bases, separated by sulfuric acid of 85% concentration, includes the most reactive components and may contain, among other components, substantially all the nitrogen-containing compounds.

The fractionation accomplished by the sulfuric acid method is presumed to be the subdivision of crude oils or asphaltic materials into five groups of components by virtue of their chemical makeup. The method was accepted by the American Society for Testing and Materials as a standard method of testing for "Characteristic Groups in Rubber Extender and Processing Oils by the Precipitation Method" (ASTM D-2006). However, there is still some doubt regarding the applicability of the method to a wide variety of petroleum, petroleum residuum, and bituminous materials.

Indeed, the sulfonation index of asphalt (ASTM D-872) requires that the sulfuric acid react with the heavy feedstocks, and, in our experience, application of the sulfuric acid method of separation to conventional crude oil, heavy oil, and bitumen has led, in some instances, to emulsification and hence to difficulties in the separation procedure. It must also be remembered that the use of the strong grades of sulfuric acid leads to sulfonation of the constituents, and it may be difficult if not impossible to regenerate these constituents to their natural state. It therefore appears that a number of refinements of the sulfuric acid method are desirable, because it is apparent that not all carbonaceous liquids react in the same manner with sulfuric acid.

Obviously, the success of this fractionation method is feedstock-dependent. In conclusion, it would appear that the test should be left more as a method of product cleaning, for which it was originally designed, rather than a method of separating the various fractions.

5. USE OF THE DATA

In the simplest sense, crude oil is a composite of four major fractions that are defined by the method of separation (Fig. 3.4) (Speight, 1999 and references therein) but, more important, the behavior and properties of any feedstock are dictated by composition (Speight, 1999 and references therein). Although early studies were primarily focused on the composition and behavior of asphalt, the techniques developed for those investigations have provided an excellent means of studying heavy feedstocks (Tissot, 1984). Later studies focused not only on the composition of petroleum and its major operational fractions but on further fractionation that allows different feedstocks to be compared on a relative basis and provides a very simple but convenient feedstock *map*.

Such a map does not give any indication of the complex interrelationships of the various fractions (Koots and Speight, 1975), although predictions of feedstock behavior are possible using such data. It is necessary to take the composition studies one step further using subfractionation of the major fractions to obtain a more representative indication of petroleum composition.

Thus by careful selection of an appropriate technique it is possible to obtain an overview of petroleum composition that can be used for behavioral predictions. By taking the approach one step further and by assiduous collection of various subfractions it becomes possible to develop the petroleum map and add an extra dimension to compositional studies (Figs 3.4 and 3.13). Petroleum and heavy feedstocks then appear more as a continuum than as four specific fractions. Such a concept has also been applied to the asphaltene fraction of petroleum in which asphaltenes are considered a complex state of matter based on molecular weight and polarity (Long, 1979, 1981; Speight, 1994).

Furthermore, petroleum can be viewed as consisting of two continuous distributions, one of molecular weight and the other of molecular type. Using data from molecular weight studies and elemental analyses, the numbers of nitrogen and sulfur atoms in the

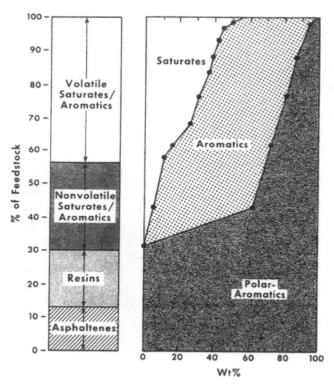

Figure 3.13   A petroleum map based on simple fractionation.

aromatic fraction and in the polar aromatic fraction can also be also exhibited. These data showed that not only can every molecule in the resins and asphaltenes have more than one sulfur atom or more than one nitrogen atom but also some molecules probably contain both sulfur and nitrogen. As the molecular weight of the aromatic fraction decreases, the sulfur and nitrogen contents of the fractions also decrease. In contrast to the sulfur-containing molecules, which appear in both the naphthene aromatics and the polar aromatic fractions, the oxygen compounds present in the heavy fractions of petroleum are normally found in the polar aromatic fraction.

In more recent work, a different type of composition map was developed (Long and Speight, 1989) using the molecular weight distribution and the molecular type distribution as coordinates. The separation involved the use of an adsorbent such as clay, and the fractions were characterized by *solubility parameter* as a measure of the polarity of the molecular types. The molecular weight distribution can be determined by gel permeation chromatography. Using these two distributions, a map of composition can be prepared using molecular weight and solubility parameter as the coordinates for plotting the two distributions. Such a composition map can provide insights into many separation and conversion processes used in petroleum refining.

The molecular type was characterized by the polarity of the molecules, as measured by the increasing adsorption strength on an adsorbent. At the time of the original concept, it was unclear how to characterize the continuum in molecular type or polarity. For this reason, the molecular type coordinate of the first maps was the yield, with the molecular types ranked in order of increasing polarity. However, this type of map can be somewhat misleading because the areas are not related to the amounts of material in a given type. The horizontal distance on the plot is a measure of the yield, and there is not a continuous variation in polarity for the horizontal coordinate. It was suggested that the solubility parameter of the different fractions could be used to characterize both polarity and adsorption strength.

In an attempt to remove some of these potential ambiguities, more recent developments of this concept have focused on the solubility parameter, estimated by the values for the eluting solvents that remove the fractions from the adsorbent. The simplest map that can be derived using the solubility parameter is produced with the solubility parameters of the solvents used in solvent separation procedures, equating these parameters to the various fractions (Fig. 3.14).

Thus, a composition map can be used to show where a particular physical or chemical property tends to concentrate on the map. For example, the *coke-forming propensity*, i.e., the amount of carbon residue, is shown for various regions on the map for a sample of atmospheric residuum (Fig. 3.15) (Long and Speight, 1998; Speight, 1999). The figure shows molecular weight plotted against weight percent yield in order of increasing polarity. The dashed line is the envelope of composition of the total sample. The slanted lines show the boundaries of solvent-precipitated fractions, and the vertical lines show the boundaries of the fractions obtained by clay adsorption of the pentane-deasphaltemed oil.

A composition map can be very useful for predicting the effectiveness of various types of separations or conversions of petroleum (Fig. 3.16) (Long and Speight, 1998; Speight,

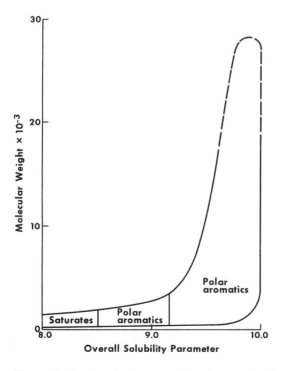

Figure 3.14 A petroleum map based on molecular weight and solubility parameter.

Fractional Composition

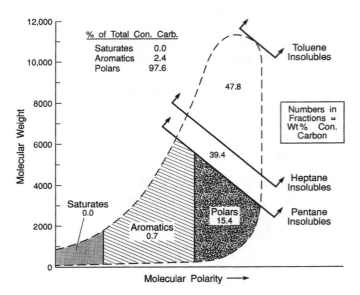

Figure 3.15 Property prediction using a petroleum map.

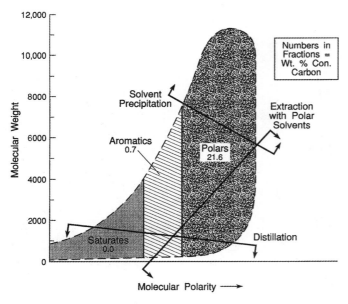

Figure 3.16 Separation efficiency using a petroleum map.

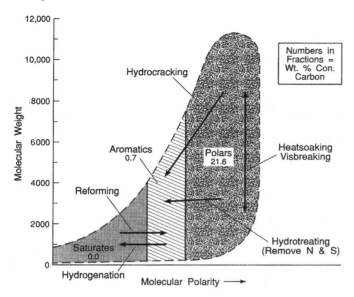

Figure 3.17   Processability from a petroleum map

1999). These processes are adsorption, distillation, solvent precipitation with relatively nonpolar solvents, and solvent extraction with polar solvents. The vertical lines show the cut points between saturates aromatics and polar aromatics as determined by clay chromatography. The slanted lines show how distillation, extraction, and solvent precipitation can divide the composition map. The line for distillation divides the map into distillate, which lies below the dividing line, and bottoms, which lies above the line. As the boiling point of the distillate is raised, the line moves upward, including higher molecular weight materials and more of the polar species in the distillate and rejecting lower molecular weight materials from the bottoms. As more of the polar species are included in the distillate, the carbon residue of the distillate rises. In contrast to the cut lines generated by separation processes, conversion processes move materials in the composition from one molecular type to another (Fig. 3.17) (Long and Speight, 1998; Speight, 1999).

The ultimate decision in the choice of any particular fractionation technique must be influenced by the need for the data. For example, some applications require only that the crude oil be separated into four bulk fractions. On the other hand, there may be the need to separate the crude oil into many subfractions in order to define specific compound types (Green et al., 1988; Vogh and Reynolds, 1988). Neither method is incorrect; each method is merely being used to answer the relevant questions about the character of the crude oil.

REFERENCES

Altgelt, K. H., and Boduszynski, M. M. 1994. Compositional Analysis of Heavy Petroleum Fractions. Marcel Dekker, New York.

Andersen, S. I. 1994. Fuel Sci. Tech. Int. 12:51.

ASTM. 1998. Annual Book of Standards. American Society for Testing and Materials, West Conshohocken, PA.

Baker, E. W. 1969. In: Organic Geochemistry. G. Eglinton and M. T. J. Murphy, eds. Springer-Verlag, New York.

Baker, E. W., and Palmer, S. E. 1978. The Porphyrins. Vol. I. Structure and Synthesis. A. D. Dolphin, ed. Academic Press, New York.

Baker, E. W., and Louda, J. W. 1986. In: Biological Markers in the Sedimentary Record. R. B. Johns, ed. Elsevier, Amsterdam.

Bland, W. F., and Davidson, R. L. 1967. Petroleum Processing Handbook. McGraw-Hill, New York.

Brandenburg, C. R., and Latham, D. R. 1968. J. Chem. Eng. Data 13:391.

Brooks, B. T., Kurtz, S. S., Jr., Boord, C. E., and Schmerling, L. 1954. The Chemistry of Petroleum Hydrocarbons. Reinhold, New York.

Bunger, J. W., Thomas, K. P., and Dorrence, S. M. 1979. Fuel 58:183.

Camp, F. W. 1976. The Tar Sands of Alberta. Cameron Engineers, Denver, CO.

Charbonnier, R. P., Draper, R. G., Harper, W. H., and Yates, A. 1969. Analyses and Characteristics of Oil Samples from Alberta. Information Circular IC 232. Department of Energy Mines and Resources, Mines Branch, Ottawa, ON, Canada.

Corbett, L. W., and Petrossi, U. 1978. Ind. Eng. Chem. Prod. Res. Dev. 17:342.

Crouch, F. W., Sommer, C. S., Galobardes, J. F., Kraus, S., Schmauch, E. M., Galobardes, M., Fatmi, A., Pearsall, K., and Rogers, L. B. 1983. Separ. Sci. Tech. 18:603.

Draper, R. G., Kowalchuk, E., and Noel, G. 1977. Analyses and Characteristics of Crude Oil Samples Performed Between 1969 and 1976. Report ERP/ERL 77-59 (TR). Energy, Mines, and Resources, Ottawa, ON, Canada.

Dunning. H. N., Moore, J. W., Bieber, H., and Williams, R. B. 1960. J. Chem. Eng. Data 5:547.

Erdman, J. G., and Harju, P. H. 1963. J. Chem. Eng. Data 8:252.

Falk, J. E. 1964. Porphyrins and Metalloporphyrins. Elsevier, New York.

Filby, R. H., and Van Berkel, G. J. 1987. In: Metal Complexes in Fossil Fuels. R. H. Filby and J. F. Branthaver, eds. ACS Symp. Ser. No. 344. American Chemical Society, Washington, DC, p. 2.

Fish, R. H., Konlenic, J. J., and Wines, B. K. 1984. Anal. Chem. 56:2452.

Franceskin, P. J., Gonzalez-Jiminez, M. G., DaRosa, F., Adams, O., and Katan, L. (1986). Hyperfine Interactions 28:825.

Girdler, R. B. 1965. Proc. Assoc. Asphalt Paving Technologists 34:45.

Glebovskaya, F. A., and Volkenshtein, M. V. 1948. J. Gen. Chem. USSR 18:1440.

Goulon, J., Retournard, A., Frient, P., Goulon-Ginet, C., Berthe, C. K., Muller, J. R., Poncet, J. L., Guilard, R., Escalier, J. C., and Neff, B. 1984. J. Chem. Soc. Dalton Trans. 1984:1095.

Green, J. B., Grizzle, P. L., Thomson, P. S., Shay, J. Y., Diehl, B. H., Hornung, K. W., and Sanchez, V. 1988. Report No. DE88 001235, Contract FC22-83F460149. U.S. Department of Energy, Washington, DC.

Gruse, W. A., and Stevens, D. R. 1960. The Chemical Technology of Petroleum. McGraw-Hill, New York.

Hildebrand, J. H., Prausnitz, J. M., and Scott, R. L. 1970. Regular Solutions. Van Nostrand Reinhold, New York.

Hirsch, D. E.. Hopkins, R. L., Coleman, H. J., Cotton, F. O., and Thompson, D. J. 1972. Anal. Chem. 44:915.

Hodgson, G. W., Baker, B. L., and Peake, E. 1967. In: Fundamental Aspects of Petroleum Geochemistry. B. Nagy and U. Columbo, eds. Elsevier, Amsterdam, Chapter 5.

IP. 1999. Standard Methods for Analysis and Testing of Petroleum and Related Products. Institute of Petroleum, London, England.

Jewell, D. M., Weber, J. H., Bunger, J. W., Plancher, H., and Latham, D. R. 1972a. Anal. Chem. 44:1391.

Jewell, D. M., Ruberto, R. G., and Davis, B. E. 1972b. Anal. Chem. 44:2318.

Jewell, D. M., Albaugh, E. W., Davis, B. E., and Ruberto, R. G. 1974. Ind. Eng. Chem. Fundam. 13:278.

Kalichevsky, V. A., and Stagner, B. A. 1942. Chemical Refining of Petroleum. Reinhold, New York.

Kalichevsky, V. A., and Kobe, K. A. 1942. Petroleum Refining with Chemicals. Elsevier, Amsterdam, The Netherlands.

Koots, J. A., and Speight, J. G. 1975. Fuel 54:179.

Lochte, H. L., and Littmann, E. R. 1955. The Petroleum Acids and Bases. Chemical Pub. Co., New York.

Long, R. B. 1979. Preprints. Div. Petrol. Chem. Am. Chem. Soc. 24(4):891.

Long, R. B. 1981. In: The Chemistry of Asphaltenes. J. W. Bunger and N. Li, eds. Adv. Chem. Ser. No. 195. American Chemical Society, Washington, DC.

Long, R. B., and Speight, J. G. 1989. Rev. l'Inst. Français Pétrole. 44:205.

Long, R. B., and Speight, J.G. 1998. In Petroleum Chemistry and Refining. J. G. Speight, ed. Taylor & Francis, Washington, DC, Chapter 1.

Meyer, R. F., and Steele, C. F, eds. 1981. The Future of Heavy Crude and Tar Sands. McGraw-Hill, New York.

Mitchell, D. L., and Speight, J. G. 1973. Fuel 52:149.

Nelson, W. L. 1958. Petroleum Refinery Engineering. McGraw-Hill, New York.

Quirke, J. M. E. 1987. In: Metal Complexes in Fossil Fuels. R. H. Filby and J. F. Branthaver, eds. ACS Symp. Ser. No. 344. American Chemical Society, Washington, DC. p. 74.

Rall, H. T., Thompson, C. J., Coleman, H. J., and Hopkins, R. L. 1972. Bulletin 659. Bureau of Mines, U.S. Department of the Interior, Washington, DC.

Reynolds, J. G. 1998. In: Petroleum Chemistry and Refining. J. G. Speight, ed. Taylor & Francis, Washington, DC, Chapter 3.

Reynolds, J. G., Biggs, W. E., and Bezman, S. A. 1987. Metal Complexes in Fossil Fuels. R. H. Filby and J. F. Branthaver, eds. ACS Symp. Ser. No. 344. American Chemical Society, Washington, DC, p. 205.

Rossini, F. D., Mair, B. J., and Streif, A. J. 1953. Hydrocarbons from Petroleum. Reinhold, New York.

Rostler, F. S. 1965. In: Bituminous Materials: Asphalts, Tars, and Pitches. Vol. II, Part I. A. J. Hoiberg, ed. Interscience, New York, p. 151.

Schwartz, R. D., and Brasseaux, D. J. 1958. Anal. Chem. 30:1999.

Smith, K. M. 1975. Porphyrins and Metalloporphyrins. Elsevier, New York.

Speight, J. G. 1979. Information Series No. 84. Alberta Research Council, Edmonton, AB, Canada.

Speight, J. G. 1990. In: Fuel Science and Technology Handbook. J. G. Speight (ed.) Marcel Dekker, New York, Chapter 12.

Speight, J. G. 1994a. The Chemistry and Technology of Coal. 2nd ed. Marcel Dekker, New York.

Speight, J. G. 1994b. In: Asphaltenes and Asphalts. Vol. I. Developments in Petroleum Science. T. F. Yen and G. V. Chilingarian, eds. Elsevier, Amsterdam, The Netherlands, Chapter 2.

Speight, J. G. 1999. The Chemistry and Technology of Petroleum. 3rd ed. Marcel Dekker, New York.

Speight. J. G. 2000. The Desulfurization of Heavy Oils and Residua. 2nd ed. Marcel Dekker, New York.

Speight, J. , and Francisco, M. A. 1990. Rev. l'Inst. Français du Pétrole 45:733.

Speight, J. G., Long, R. B., and Trowbridge, T. D. 1984. Fuel 63:616.

Thompson, C. J., Ward, C. C., and Ball, J. S. 1976. Characteristics of World's Crude Oils and Results of API Research Project 60. Report BERC/RI-76/8. Bartlesville Energy Technology Center, Bartlesville, OK.

Tissot, B.P., ed. 1984. Characterization of Heavy Crude Oils and Petroleum Residues. Editions Technip, Paris.

Treibs, A. 1934. Analen 509:103.

Vogh, J. W., and Reynolds, J. W. 1988. Report No. DE88 001242. Contract FC22-83FE60149. U.S. Department of Energy, Washington, DC.

Wallace, D., Starr, J., Thomas, K. P., and Dorrence, S. M. 1988. Characterization of Oil Sands Resources. Alberta Oil Sands Technology and Research Authority, Edmonton, AB, Canada.

Yan, J., Plancher, H., and Morrow, N. R. 1997. Paper No. SPE 37232. SPE Int. Symp. Oilfield Chemistry, Houston, TX.

4
Testing and Evaluation

1. INTRODUCTION

The physical and chemical characteristics of crude oil and the yields and properties of products or fractions prepared from it vary considerably and depend on the concentration of the various types of hydrocarbons as well as on the amounts of heteroatoms (nitrogen, oxygen, sulfur, and metals) in it. Some types of petroleum have economic advantages as sources of fuels and lubricants with highly restrictive characteristics because they require less specialized processing than that needed for the production of the same products from many types of crude oil. Others may contain unusually low concentrations of components that are desirable constituents of fuel or lubricants, and the production of these products from such crude oils may not be economically feasible.

Because petroleum exhibits a wide range of physical properties, it is not surprising that the behavior of various feedstocks in refinery operations is not simple. The atomic ratios yielded by ultimate analysis (Fig. 4.1) can give an indication of the nature of a feedstock and the generic hydrogen requirements to satisfy the refining chemistry, but it is not possible to predict with any degree of certainty how the feedstock will behave during refining. Any deductions made from such data are pure speculation and are open to much doubt.

The chemical composition of a feedstock (Chapter 3) is also an indicator of refining behavior. Whether the composition is represented in terms of compound types or in terms of generic compound classes, knowledge of it can enable the refiner to determine the nature of the reactions. Hence, chemical composition can play a large part in determining the nature of the products that arise from the refining operations. It can also play a role in determining the means by which a particular feedstock should be processed (Nelson, 1958; Bland and Davidson, 1967; Gary and Handwerk, 1984; Ali et al., 1985; Wallace et al., 1988; Speight, 1999, 2000 and references therein).

Therefore, the judicious choice of a crude oil to produce any given product is just as important as the selection of the product for any given purpose. Thus, initial inspection of the nature of the petroleum will enable deductions about the most logical means of refining. Indeed, careful evaluation of petroleum from physical property data is a major part of the initial study of any petroleum destined to be a refinery feedstock. Proper interpretation

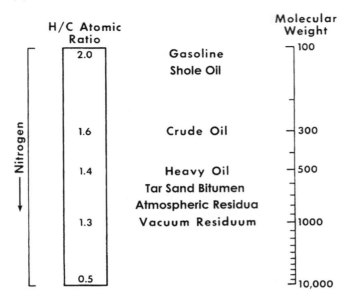

Figure 4.1 Atomic hydrogen/carbon ratio of various feedstocks.

of the data resulting from the inspection of crude oil requires an understanding of their significance.

Evaluation of petroleum for use as a feedstock usually involves an examination of one or more of the physical properties of the material. By this means, a set of basic characteristics can be obtained that can be correlated with utility. The physical properties of petroleum and petroleum products are often equated with those of the hydrocarbons, because although petroleum is indeed a very complex mixture, there is gasoline produced by nondestructive distillation in which fewer than a dozen hydrocarbons make up at least 50% of the material (Speight, 1999).

To satisfy specific needs with regard to the type of petroleum to be processed as well as to the nature of the product, most refiners have, through time, developed their own methods of petroleum analysis and evaluation. However, such methods are considered proprietary and are not normally available. Consequently, various standards organizations, such as the American Society for Testing and Materials (ASTM, 1998) in North America and the Institute of Petroleum in Britain (IP, 1999), have devoted considerable time and effort to the correlation and standardization of methods for the inspection and evaluation of petroleum and petroleum products. A complete discussion of the large number of routine tests available for petroleum fills an entire book (ASTM, 1998). However, it seems appropriate that in any discussion of the physical properties of petroleum and petroleum products reference be made to the corresponding test, and accordingly the various test numbers are included in this text.

It is the purpose of this chapter to present an outline of the tests that can be applied to petroleum or petroleum products, and the resulting physical properties on the basis of which a feedstock can be evaluated. For this purpose, data relating to various physical properties have been included as illustrative examples, but theoretical discussions of the physical properties of hydrocarbons were deemed irrelevant and are omitted.

2. PHYSICAL PROPERTIES

Before any volatility tests are carried out it must be recognized that the presence of more than 0.5% water in test samples of crude oil can cause several problems during various test procedures and produce erroneous results. For example, during various thermal tests, water (which has a high heat of vaporization) requires the application of additional thermal energy to the distillation flask. In addition, water is relatively easily superheated, and therefore excessive *bumping* can occur, leading to erroneous readings and the potential for destruction of the glass equipment. Steam formed during distillation can act as a carrier gas, and high boiling point components may end up in the distillate (often referred to as *steam distillation*).

Removal of water (and sediment) can be achieved by centrifugation if the sample is not a tight emulsion. Other methods that are used to remove water include

1. Heating in a pressure vessel to control loss of light ends
2. Addition of calcium chloride as recommended in ASTM D-1160
3. Addition of an azeotroping agent such as isopropanol or *n*-butanol
4. Removal of water in a preliminary low efficiency or flash distillation followed by reblending of the hydrocarbon that codistills with the water into the sample (see also IP 74)
5. Separation of the water from the hydrocarbon distillate by freezing

Thus, the tests described in the following sections are based on the assumption that water has been reduced to an acceptable level, as usually defined by each standard test.

2.1 Elemental (Ultimate) Analysis

The analysis of petroleum for the percent by weight of carbon, hydrogen, nitrogen, oxygen, and sulfur is perhaps the first method used to examine the general nature of a feedstock and evaluate it. The atomic ratios of the various elements to carbon (i.e., H/C, N/C, O/C, and S/C) are frequently used for indications of the overall character of the feedstock. It is also of value to determine the amounts of trace elements, such as vanadium and nickel, in a feedstock, because these materials can have serious deleterious effects on catalyst performance during refining by catalytic processes.

However, with the introduction of the heavier feedstocks into refinery operations, it has become apparent that these ratios are not the only requirement for predicting feedstock character before refining. The use of more complex feedstocks (in terms of chemical composition) has added a new dimension to refining operations. Thus, although atomic ratios, as determined by elemental analyses, can be used in a comparison of feedstocks, there is no guarantee that a particular feedstock will behave as predicted from these data. Product slates cannot be predicted accurately, if at all, from these ratios.

The ultimate analysis (elemental composition) of petroleum is not reported to the same extent as that of coal (Speight, 1994). Nevertheless, there are ASTM procedures (ASTM, 1995) for the ultimate analysis of petroleum and petroleum products, although many such methods may have been designed for other materials.

For example, *carbon content* can be determined by the method designated for coal and coke (ASTM D-3178) or by the method designated for municipal solid waste (ASTM E-777). There are also methods designated for

1. *Carbon* and *hydrogen content* (ASTM D-1018, D-3178, D-3343, D-3701, D-5291, E-777; IP 338)
2. *Nitrogen content* (ASTM D-3179, D-3228, D-3431, E-148, E-258, D-5291, and E-778)
3. *Oxygen content* (ASTM E-385)
4. *Sulfur content* (ASTM D-124, D-129, D-139, D-1266, D-1552, D-1757, D-2622, D-2785, D-3120, D-3177, D-4045 and D-4294, E-443; IP 30, IP 61, IP 103, IP 104, IP 107, IP 154, IP 243)
5. *Metals content* (ASTM C-1109, C-1111, D-482, D-1026, D-1262, D-1318, D-1368, D-1548, D-1549, D-2547, D-2599, D-2788, D-3340, D-3341, D-3605; IP 288, IP 285)

Of the ultimate analytical data, more has been made of the sulfur content than of any other property. For example, the sulfur content (ASTM D-124, D-1552, and D-4294) and the API gravity represent the two properties that have, in the past, had the greatest influence on determining the value of petroleum as a feedstock.

The sulfur content varies from about 0.1 to about 3 wt% for the more conventional crude oils to as much as 5–6wt% for heavy oil and bitumen. Residua, depending on the sulfur content of the crude oil feedstock, may be of the same order or even have a substantially higher sulfur content. Indeed, the very nature of the distillation process by which residua are produced, that is, removal of distillate without thermal decomposition, dictates that the majority of the sulfur, which is located predominantly in the higher molecular weight fractions, will be concentrated in the residuum.

Metals cause problems during petroleum refining because they poison catalysts used for sulfur and nitrogen removal as well as catalysts for other processes such as catalytic cracking. Heavy oils and residua contain relatively high proportions of metals either in the form of salts or as organometallic constituents (such as the metalloporphyrins), which are extremely difficult to remove from the feedstock. Indeed, the nature of the process by which residua are produced virtually dictates that all the metals in the original crude oil will be concentrated in the residuum (Speight, 2000). Those metallic constituents that may actually *volatilize* under the distillation conditions and appear in the higher boiling distillates are the exceptions here. The deleterious effect of metallic constituents on the catalyst is known, and serious attempts have been made to develop catalysts that can tolerate a high concentration of metals without serious loss of catalytic activity or catalyst life.

2.2 Density and Specific Gravity

The *density* and *specific gravity* of crude oil are two properties that have found wide use in the industry for preliminary assessment of the character of the crude oil.

Density is the mass of a unit volume of material at a specified temperature and has the dimensions of grams per cubic centimeter or grams per milliliter. *Specific gravity* is the ratio of the mass of a volume of the substance to the mass of the same volume of water and is dependent on two temperatures, those at which the masses of the sample and the water are measured. When the water temperature is 4°C (39°F), the specific gravity is equal to the density in the centimeter-gram-second (cgs) system, because the volume of 1 g of water at that temperature is, by definition, $1 \, cm^3$. Thus the density of water, for example, varies with temperature, and its specific gravity at equal temperatures is always unity. The standard temperatures for specific gravity in the petroleum industry in North America are 60/60°F (15.6/15.6°C).

The density or specific gravity of petroleum, petroleum products, heavy oil, and bitumen can be measured by means of a hydrometer (ASTM D-287, D-1298, D-1657, and IP 160) or a pycnometer (ASTM D-70, D-941, D-1217, D-1480, and D-1481), by the displacement method (ASTM D-712), or by means of a digital density meter (ASTM D-4052, IP 365) and a digital density analyzer (ASTM D-5002). Not all of these methods are suitable for measuring the density (or specific gravity) of heavy oil and bitumen although some lend themselves to adaptation. The API gravity of a feedstock (ASTM D-287) is calculated directly from the specific gravity.

Although density and specific gravity are used extensively, the API (American Petroleum Institute) gravity is the preferred property. This property was derived from the Baumé scale:

$$\text{Degrees Baum} = \frac{140}{\text{sp gr at } 60/60°F} - 130$$

However, a considerable number of hydrometers calibrated according to the Baumé scale were found at an early period to be in error by a consistent amount, and this led to the adoption of the equation

$$\text{Degrees API} = \frac{141.5}{\text{sp gr at } 60/60°F} - 131.5$$

The specific gravity of petroleum usually ranges from about 0.8 (45.3° API) for the lighter crude oils to over 1.0 (10° API) for heavy crude oils and bitumen (Speight, 1999, 2000).

Specific gravity is influenced by the chemical composition of petroleum, but quantitative correlation is difficult to establish. Nevertheless, it is generally recognized that increased amounts of aromatic compounds result in an increase in density, whereas an increase in saturated compounds results in a decrease in density. Indeed, it is also possible to recognize certain preferred trends between the density of petroleum and one or more of the physical properties. For example, correlations exist between the density (API gravity) and sulfur content (Fig. 4.2), Conradson carbon residue (Fig. 4.3), and viscosity (Fig. 4.4).

The variation of density with temperature, effectively the coefficient of expansion, is a property of great technical importance, because most petroleum products are sold by volume and specific gravity is usually determined at the prevailing temperature (21°C, 70°F) rather than at the standard temperature (60°F; 15.6°C). The tables of gravity corrections (ASTM D-1555) are based on an assumption that the coefficient of expansion of all petroleum products is a function (at fixed temperatures) of density only. Work in the past decade has focused on the calculation and predictability of density using new mathematical relationships (Gomez, 1989, 1992).

The success of the hot water process used to separate Athabasca bitumen from the associated sand depends on the variation in density (specific gravity) of the bitumen with temperature (Speight, 1990). Over the temperature range 30–130°C (85–265°F) the bitumen is lighter than water, and flotation of the bitumen (with aeration) is facilitated (Berkowitz and Speight, 1975; Spragins, 1978; Speight, 1990).

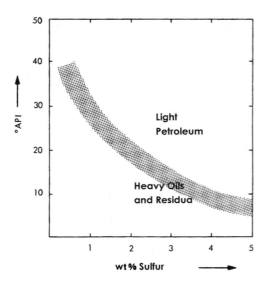

Figure 4.2 Relationship between API gravity and sulfur content.

2.3 Viscosity

Viscosity is the force in dynes required to move a plane of 1 cm^2 area at a distance of 1 cm from another plane of 1 cm^2 area through a distance of 1 cm in 1 s and is actually a measure of the internal resistance of a fluid to motion by reason of the forces of cohesion between molecules or molecular groupings.

In the centimeter-gram-second (cgs) system the unit of viscosity is the poise (P) or centipoise (cP). Two other terms in common use are *kinematic viscosity* and *fluidity*. The kinematic viscosity is the viscosity in centipoises divided by the specific gravity, and the

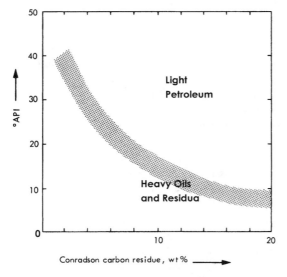

Figure 4.3 Relationship between API gravity and carbon residue (Conradson).

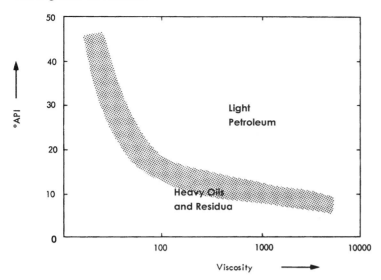

Figure 4.4   Relationship between API gravity and viscosity.

unit is the stokes ($1\,St = 1\,cm^2/s$), although the centistokes (cSt) is in more common use; fluidity is simply the reciprocal of viscosity.

The viscosity (ASTM D-88, D-341, D-445, D-1270, D-1665, D-2161, D-2170, D-2171, D-2270, D-3205, D-3570, E-102; IP 71, IP 212, IP 222, IP 226, IP 319) of crude oils varies markedly over a very wide range. Values vary from less than 10 cP at room temperature to many thousands of centipoises at the same temperature. In the present context, oil sand bitumen is at the higher end of this scale, where a relationship between viscosity and density between various crude oils has been noted (Fig. 4.5).

In the early days of the petroleum industry viscosity was regarded as the *body* of an oil, a significant number for lubricants or for any liquid pumped or handled in quantity (Speight, 1990, 1999, 2000). The changes in viscosity with temperature, pressure, and rate of shear are pertinent not only in lubrication but also for such engineering concepts as heat transfer. The viscosity and relative viscosity of different phases, such as gas, liquid oil, and water, are determining influences in producing the flow of reservoir fluids through porous oil-bearing formations. The rate and amount of oil production from a reservoir are often governed by these properties.

Many types of instruments have been proposed for the determination of viscosity. The simplest and most widely used are capillary types (ASTM D-445), and the viscosity is derived from the equation

$$\mu = \pi r^4 P/8nl$$

where r is the tube radius, l the tube length, P the pressure difference between the ends of a capillary, n the *coefficient of viscosity*, and μ the quantity discharged in unit time. Not only are such capillary instruments the simplest, but when designed in accordance with known principles and used with known necessary correction factors, they are probably the most accurate viscometers available. It is usually more convenient, however, to use relative

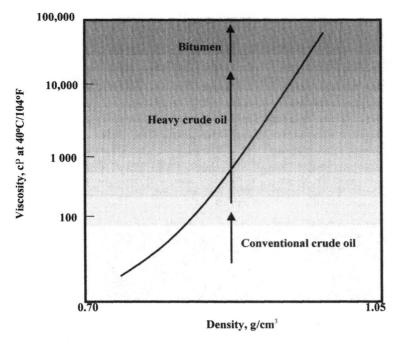

Figure 4.5   Relationship between density and viscosity.

measurements, and for this purpose the instrument is calibrated with an appropriate standard liquid of known viscosity.

Batch flow times are generally used; in other words, the time required for a fixed amount of sample to flow from a reservoir through a capillary is the datum actually observed. Any technique that contributes to longer flow times is usually desirable. Some of the principal capillary viscometers in use are those of Cannon-Fenske, Ubbelohde, Fitzsimmons, and Zeitfuchs.

The Saybolt universal viscosity (SUS) (ASTM D-88) is the time in seconds required for 60 mL of petroleum to flow from a container, at constant temperature, through a calibrated orifice. The Saybolt furol viscosity (SFS) (ASTM D-88) is determined in a similar manner except that a larger orifice is employed.

As a result of the various methods for viscosity determination, it is not surprising that much effort has been spent on interconversion of the several scales, especially converting Saybolt to kinematic viscosity (ASTM D-2161):

$$\text{Kinematic viscosity} = a \times \text{Saybolt seconds} + \frac{b}{\text{Saybolt seconds}}$$

where a and b are constants.

The Saybolt universal viscosity equivalent to a given kinematic viscosity varies slightly with the temperature at which the determination is made because the temperature of the calibrated receiving flask used in the Saybolt method is not the same as that of the oil. Conversion factors are used to convert kinematic viscosities of 2–70 cSt at 38°C (100°F) and 99°C (210°F) to equivalent Saybolt universal viscosities in seconds (ASTM D-2161).

Appropriate multipliers are listed to convert kinematic viscosities over 70 cSt. For a kinematic viscosity determined at any other temperature, the equivalent Saybolt universal value is calculated by use of the Saybolt equivalent at 38°C (100°F) and a multiplier that varies with the temperature:

Saybolt s at 100°F (38°C) = cSt × 4.635

Saybolt s at 210°F (99°C) = cSt × 4.667

Various studies have also been made on the effect of temperature on viscosity because the viscosity of petroleum or a petroleum product decreases as the temperature increases. The rate of change appears to depend primarily on the nature or composition of the petroleum, but other factors, such as volatility, may also have a minor effect. The effect of temperature on viscosity is generally represented by the equation

$$\log\log(n + c) = A + B\log T$$

where n is absolute viscosity, T is temperature, and A and B are constants. This equation has been sufficient for most purposes and has come into very general use. The constants A and B vary widely with different oils, but c remains fixed at 0.6 for all oils having a viscosity over 1.5 cSt; it increases only slightly at lower viscosities (0.75 at 0.5 cSt). The viscosity–temperature characteristics of any oil, so plotted, thus create a straight line, and the parameters A and B are equivalent to the intercept and slope of the line. To express the viscosity and viscosity–temperature characteristics of an oil, the slope and the viscosity at one temperature must be known; the usual practice is to select 38°C (100°F) and 99°C (210°F) as the observation temperatures.

Suitable conversion tables are available (ASTM D-341), and each table or chart is constructed in such a way that for any given petroleum or petroleum product the viscosity–temperature points result in a straight line over the applicable temperature range. Thus, only two viscosity measurements need be made at temperatures far enough apart to determine a line on the appropriate chart from which the approximate viscosity at any other temperature can be read.

The charts can be applied only to measurements made in the temperature range in which a given petroleum oil is a Newtonian liquid. The oil may cease to be a simple liquid near the cloud point because of the formation of wax particles or near the boiling point because of vaporization. Thus the charts do not give accurate results when either the cloud point or boiling point is approached. However, they are useful over the Newtonian range for estimating the temperature at which an oil attains a desired viscosity. The charts are also convenient for estimating the viscosity of a blend of petroleum liquids at a given temperature when the viscosities of the component liquids at the given temperature are known (Fig. 4.6).

Because the viscosity–temperature coefficient of lubricating oil is an important expression of its suitability, a convenient number to express this property is very useful, and therefore a viscosity index (ASTM D-2270, ASTM D-2270, IP 226) was derived. It is established that naphthenic oils have higher viscosity–temperature coefficients than those of paraffinic oils at equal viscosity and temperature. The Dean and Davis scale was based on the assignment of a value of zero to a typical naphthenic crude oil and a value of 100 to a typical paraffinic crude oil; intermediate oils were rated by the formula

$$\text{Viscosity index} = \frac{L - U}{L - H} \times 100$$

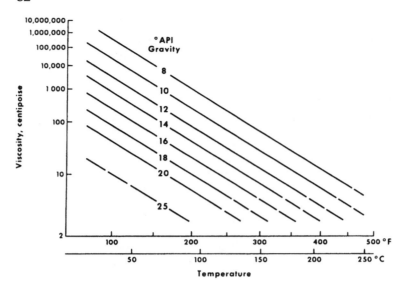

Figure 4.6   Relationship between API and viscosity and temperature.

where L and H are the viscosities of the 0 and 100 index reference oils, both having the same viscosity at 99°C (210°F), and U is that of the unknown, all at 38°C (100°F). Originally the viscosity index was calculated from Saybolt viscosity data, but subsequently figures were provided for kinematic viscosity.

The viscosity of petroleum fractions increases when pressure is applied, and this increase may be very large. The pressure coefficient of viscosity correlates with the temperature coefficient even when oils of widely different types are compared. A plot of the logarithm of the kinematic viscosity against pressure for several oils has given reasonably linear results up to about 20,000 psi, and the slopes of the isotherms are such that extrapolated values for a given oil intersect. At higher pressures the viscosity decreases with increasing temperature, as at atmospheric pressure; in fact, viscosity changes of small magnitude are usually proportional to density changes, whether the latter are caused by pressure or by temperature.

Because of the importance of viscosity in determining the transport properties of petroleum, recent work has focused on the development of an empirical equation for predicting the dynamic viscosity of low molecular weight and high molecular weight hydrocarbon vapors at atmospheric pressure (Gomez, 1995). The equation uses molar mass and specific temperature as the input parameters and offers a means of estimating the viscosity of a wide range of petroleum fractions. Other work has focused on the prediction of the viscosity of blends of lubricating oils as a means of accurately predicting the viscosity of the blend from the viscosity of the base oil components (Al-Besharah et al., 1989)

3.   THERMAL PROPERTIES

Although several important physical properties were described in Section 2, in the context of this chapter the tests that produce data relating to the various thermal properties of petroleum, heavy oil, and bitumen are described in this section. It is these properties that provide valuable information relevant to the movement of feedstocks around a refinery as well as indications of the yields of various products.

3.1 Volatility

The volatility of a liquid fuel (or liquefied gas) is the tendency of the liquid to vaporize, that is, to change from the liquid to the vapor or gaseous state. Because one of the three essentials for combustion in a flame is that the fuel be in the gaseous state, volatility is a primary characteristic of liquid fuels. Another purpose for determining the volatility of petroleum serves to indicate the comparative ease with which the material can be refined. In all senses, investigation of the volatility of petroleum is, like other test methods, carried out under standard conditions that allow comparisons to be made between data obtained from various laboratories.

The vaporizing tendencies of petroleum and petroleum products are the basis for the general characterization of liquid petroleum fuels, such as liquefied petroleum gas, natural gasoline, motor and aviation gasoline, naphtha, kerosene, gas oil, diesel fuel, and fuel oil (ASTM D-2715). A test (ASTM D-6, IP 45) also exists for determining the loss of material when crude oil and asphaltic compounds are heated, and another test (ASTM D-20) is applicable to the distillation of road tars and can also be used for estimating the volatility of high molecular weight residua. For other uses it is important to know the tendency of a product to partially vaporize or to completely vaporize, and in some cases to know if small quantities of high-boiling components are present. For such purposes, chief reliance is placed on the distillation methods. Such tests include determination of the *flash point*, the *fire point*, *vapor pressure*, and *boiling range*.

The *flash point* of petroleum or a petroleum product is the temperature to which the product must be heated under specified conditions to give off sufficient vapor to form a mixture with air that can be ignited momentarily by a specified flame (ASTM D-56, D-92, D-93, D-1310, D-3828; IP 34, IP 36, IP 170, IP 303, IP 304, IP 403, IP 404). The Pensky–Marten apparatus using a closed or open system (ASTM D-93, IP 34, IP 35) is the standard instrument for flash points above 50°C, and the Abel apparatus (IP 170) is used for more volatile oils with flash points below 50°C. The Cleveland open-cup method (ASTM D-92, IP 36) is also used for the determination of the *fire point* (the temperature at which the sample will ignite and burn for at least 5 s).

The flash point of a petroleum product is also used to detect contamination. A substantially lower flash point than expected for a product is a reliable indicator that the product has become contaminated with a more volatile product such as gasoline. The flash point is also an aid in establishing the identity of a petroleum product.

The *fire point* is the temperature to which the product must be heated under the prescribed conditions of the method to burn continuously when the mixture of vapor and air is ignited by a specified flame (ASTM D-92, IP 36).

Information about the flash point is of most significance at or slightly above the maximum temperatures (30–60°C, 86–140°F) that may be encountered in the storage, transportation, and use of liquid petroleum products in either closed or open containers. In this temperature range the relative fire and explosion hazard can be estimated from the flash point. For products with flash points below 40°C (104°F) special precautions are necessary for safe handling. Flash points above 60°C (140°F) gradually lose their safety significance until they become indirect measures of some other quality.

A further aspect of volatility that receives considerable attention is the vapor pressure of petroleum and its constituent fractions. *Vapor pressure* is the force exerted on the walls of a closed container by the vaporized portion of a liquid. Conversely, it is the force that must be exerted on the liquid to prevent it from vaporizing further (ASTM D-323, IP 69). The *Reid vapor pressure* (ASTM D-323, IP 69, IP 402) is a measure of the vapor pressure of petroleum or a petroleum product such as an oil at 37.8°C (100°F) expressed as milli-meters of mercury (mmHs). The vapor pressure increases with temperature for any given product. The temperature at which the vapor pressure of a liquid, either a pure compound of a mixture of many compounds, equals 1 atm (14.7 psi, absolute) is designated as the *boiling point* (or *initial boiling point*) of the liquid.

In each homologous series of hydrocarbons, the boiling points increase with molecular weight (Table 4.1). Structure also has a marked influence; as a general rule, branched paraffin isomers have lower boiling points than the corresponding *n*-alkanes (Table 4.2). However, the most dramatic illustration of the variation in boiling point with carbon num-ber is an actual plot for different hydrocarbons (Fig. 4.7). In any given series, steric effects notwithstanding, there is an increase in boiling point with an increase in carbon number of the alkyl side chain (Speight, 1999, p. 326). This particularly applies to alkyl aromatic compounds where alkyl-substituted aromatic compounds can have higher boiling points than polycondensed aromatic systems. And this fact is very meaningful when attempts are made to develop hypothetical structures for asphaltene constituents (see Chapter 10).

The boiling points of petroleum fractions are rarely, if ever, distinct temperatures; it is, in fact, more correct to refer to the boiling ranges of the various fractions. To determine these ranges, the petroleum is tested in various methods of distillation, either at atmo-spheric pressure or at reduced pressure. In general, the limiting molecular weight range for distillation at atmospheric pressure without thermal degradation is 200–250, whereas the limiting molecular weight range for conventional vacuum distillation is 500–600.

Thus, petroleum can be subdivided by distillation into a variety of fractions of dif-ferent *boiling ranges* (*cut points*) using a variety of standard methods specifically designed for this task:

ASTM D-86 (IP 123)—Distillation of Petroleum Products
ASTM D-216 (IP 191)—Distillation of Natural Gasoline

Table 4.1 Boiling Points of *n*-Hydrocarbons (Straight-Chain Hydrocarbons)

	Boiling point	
Hydrocarbon (normal)	°C	°F
Butane	−0.5	18
Pentane	36	97
Hexane	68	154
Heptane	98	208
Octane	126	259
Nonane	151	304
Decane	174	345
Pentadecane	271	520
Eicosane	343	649

Table 4.2 Boiling Points of *n*-Octane and Branched-Chain
Octanes

Hydrocarbon	Boiling point	
	°C	°F
n-Octane	126	259
2-Methylheptane	117	243
2,4-Dimethylhexane	109	228
2,2,3,3-Tetramethylbutane	106	223
2,2,4-Trimethylpentane	99	210

ASTM D-285—Distillation of Crude Petroleum
ASTM D-1160—Distillation of Petroleum Products at Reduced Pressure
ASTM D-2887—Test Method for Boiling Range Distribution of Petroleum Fractions
  by Gas Chromatography
ASTM D 2892—Distillation of Crude Petroleum (15 Theoretical Plate Column)

Petroleum can be subdivided by distillation into a variety of fractions of different cut points (Table 4.3). In fact, distillation is the method by which petroleum feedstocks are determined to be suitable for various refinery options.

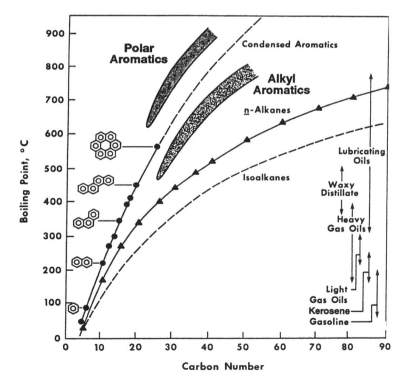

Figure 4.7 Relationship of boiling point to carbon number.

Table 4.3 General Boiling Fractions of Petroleum

Fraction	Boiling range[a]	
	°C	°F
Light naphtha	−1–150	30–300
Gasoline	−1–180	30–355
Heavy naphtha	150–205	300–400
Kerosene	205–260	400–500
Stove oil	205–290	400–550
Light gas oil	260–315	400–600
Heavy gas oil	315–425	600–800
Lubricating oil	> 400	> 750
Vacuum gas oil	425–600	800–1100
Residuum	> 600	> 1100

[a] For convenience, boiling ranges are interconverted to the nearest 5°F.

As an early part of characterization studies, a correlation was observed between the quality of petroleum products and their hydrogen content, because gasoline, kerosene, diesel fuel, and lubricating oil are made up of hydrocarbon constituents containing high proportions of hydrogen. Thus, it is not surprising that tests to determine the volatility of petroleum and petroleum products were among the first to be defined. Indeed, volatility is one of the major tests for petroleum products, and it is inevitable that all products will, at some stage of their history, be tested for volatility characteristics.

Distillation involves the general procedure of vaporizing the petroleum liquid in a suitable flask either at *atmospheric pressure* (ASTM D-86, D-216, D-285, D447, D-2892; IP 24, IP 123, IP 191) or at *reduced pressure* (ASTM D-1160). There are also test methods for the distillation of pitch (ASTM D-2569) and cutback asphalt (ASTM D-402) that can be applied to heavy oil and bitumen. However, most of the methods specify an upper atmospheric equivalent temperature (AET) limit of 360°C (680°F) and therefore are too limited to be of value in the analysis of tar sand bitumen.

The five distillation procedures that are commonly used by laboratories, either singly or in combination, to determine the distillation curve of, or produce fractions from, heavy oil and bitumen are

1. ASTM D-2892—Distillation of Crude Petroleum (15 Theoretical Plate Column)
2. ASTM D-1160—Distillation of Petroleum Products at Reduced Pressures
3. ASTM D-5236—Distillation of Heavy Hydrocarbon Mixtures (Vacuum Potstill Method)
4. Spinning band distillation
5. Distillation using a flash still

The distillation results are described as a series of boiling ranges commencing with the *initial boiling point* and terminating with the *end point* or *dry point*.

Two of the methods (ASTM D-1160, ASTM D-2892) are used to determine the boiling ranges of petroleum products to a maximum liquid temperature of 400°C (752°F) atmospheric equivalent temperature (AET) at pressures as low as 1 mmHg. The data produced by such methods are often referred to as the *true boiling point distillation* data.

The *initial boiling point* is the thermometer reading in the neck of the distillation flask when the first drop of distillate leaves the tip of the condenser tube. This reading is materially affected by a number of test conditions, namely, room temperature, rate of heating, and condenser temperature. *Distillation temperatures* are usually observed when the level of the distillate reaches each 10% mark on the graduated receiver, with the temperatures for the 5% and 95% marks often included. Conversely, the volume of the distillate in the receiver—that is, the percentage recovered—is often observed at specified thermometer readings.

End-point or *maximum temperature* is the highest thermometer reading observed during distillation. In most cases it is reached when all of the sample has been vaporized. If a liquid residue remains in the flask after the maximum permissible adjustments are made in heating rate, this is recorded as indicative of the presence of very high boiling compounds. The *dry point* is the thermometer reading at the instant the flask becomes dry and is used for special purposes such as for solvents and for relatively pure hydrocarbons. For these purposes the dry point is considered more indicative of the final boiling point than the end point or maximum temperature.

Recovery is the total volume of distillate recovered in the graduated receiver, and *residue* is the liquid material, mostly condensed vapors, left in the flask after it has been allowed to cool at the end of distillation. The residue is measured by transferring it to an appropriate small graduated cylinder. Low or abnormally high residues indicate the absence or presence, respectively, of high-boiling components. *Total recovery* is the sum of the liquid recovery and residue. *Distillation loss* is determined by subtracting the total recovery from 100%; it is, of course, the measure of the portion of the vaporized sample that does not condense under the conditions of the test. Like the initial boiling point, distillation loss is affected materially by a number of test conditions, namely, condenser temperature, sampling and receiving temperatures, barometric pressure, heating rate in the early part of the distillation, and others. Provisions are made for correcting high distillation losses for the effect of low barometric pressure because of the practice of including distillation loss as one of the items in some specifications for motor gasoline.

Percentage evaporated is the percentage recovered at a specific thermometer reading or other distillation temperature. The amounts that have been evaporated are usually obtained by plotting observed thermometer readings against the corresponding observed recoveries plus, in each case, the distillation loss. The initial boiling point is plotted with the distillation loss as the percentage evaporated. Distillation data (Tables 4.4 and 4.5) are considerably reproducible, particularly for the more volatile products.

Nondestructive distillation data (U.S. Bureau of Mines method) show that, not surprisingly, bitumen is a higher boiling material than the more conventional crude oils (Fig. 4.8). There is usually little, or no, gasoline (naphtha) fraction in bitumen, and the majority of the distillate falls in the gas oil–lubrication distillate range [> 260°C (> 500°F)]. In excess of 50% of each bitumen is nondistillable under the conditions of the test, and the nonvolatile material corresponds very closely to the asphaltic content (asphaltenes plus resins) of each feedstock.

Detailed fractionation of the sample might be of secondary importance. Thus, it must be recognized that the general shape of a one-plate distillation curve is often adequate for making engineering calculations, correlating with other physical properties, and predicting the product slate (Nelson, 1958).

There is also another method that is increasing in popularity for application to a variety of feedstocks—the method commonly known as *simulated distillation* (ASTM D-

Table 4.4 Distillation Characteristics of a Conventional Crude Oil

Fraction number	Cut at °C	Cut at °F	%	Sum %	Sp.gr, 60°F	°API, 60°F	CI	Aniline point (°C)	SUV, 100°F	Cloud test (°F)
Distillation (U.S. Bureau of Mines routine method)										

Distillation (U.S. Bureau of Mines routine method)

Fraction number	Cut at °C	Cut at °F	%	Sum %	Sp.gr, 60°F	°API, 60°F	CI	Aniline point (°C)	SUV, 100°F	Cloud test (°F)
Stage I: Distillation at atmospheric pressure, 758 mmHg; first drop, 26°C (79°F)										
1	50	122	2.8	2.8	0.656	84.2	—	—		
2	75	167	3.3	6.1	0.677	77.5	11	56.3		
3	100	212	5.1	11.2	0.713	67.0	18	53.0		
4	125	257	6.8	18.0	0.741	59.5	22	50.6		
5	150	302	6.2	24.2	0.761	54.4	24	50.5		
6	175	347	5.9	30.1	0.779	50.1	26	51.1		
7	200	392	5.0	35.1	0.794	46.7	27	53.7		
8	225	437	5.0	40.1	0.809	43.4	28	58.5		
9	250	482	5.1	45.2	0.821	40.9	29	63.4		
10	275	527	6.1	51.3	0.833	38.4	30	67.7		
Stage 2: Distillation continued at 40 mmHg pressure										
11	200	392	5.2	56.5	0.849	35.2	33	71.3	40	10
12	225	437	4.8	61.3	0.858	33.4	34	76.2	46	30
13	250	482	5.0	66.3	0.870	31.1	36	80.1	57	50
14	275	527	4.7	71.0	0.882	28.9	39	84.0	85	70
15	300	572	4.8	75.8	0.891	27.3	40	88.8	155	90
Residuum			20.2	96.0	0.946	18.1				

Carbon residue of residuum: 6.0%

	Approximate summary			
	Vol%	Sp. gr.	°API	SUV, 100°F
Light gasoline	11.2	0.688	74.2	
Total gasoline and naphtha	35.1	0.742	59.2	
Kerosene distillate	10.1	0.815	42.1	
Gas oil	15.5	0.845	36.0	
Nonviscous lubricating distillate	9.0	0.862–0.884	32.7–28.6	50–100
Medium lubricating distillate	6.1	0.884–0.895	28.6–26.6	100–200
Viscous lubricating distillate	—	—	—	> 200
Residuum	20.2	0.946	18.1	
Distillation loss	4.0			

Field: Leduc. Specific gravity at 60°F: 0.819. Sulfur, 0.22 wt%. Saybolt university viscosity (SUV) at 100°F, 39 s. API gravity at 60°F, 41.3. Pour point, 20°F. Color: dark green. Carbon residue, 1.4 wt% (Conradson).
Source: Information Circular IC 232, June 1969, p. 166. Copyright © Queen's Printer for Canada, 1970, by permission of the publisher, Canada Centre for Mineral and Energy Technology.

Table 4.5 Distillation Characteristics of Tar Sand Bitumen

Fraction number	Cut at °C	°F	%	Sum %	Sp.gr, 60°F	°API, 60°F	CI	Aniline point (°C)	SUV, 100°F	Cloud test (°F)
	Distillation (U.S. Bureau of Mines routine method)									

Distillation (U.S. Bureau of Mines routine method)

Fraction number	Cut at °C	°F	%	Sum %	Sp.gr, 60°F	°API, 60°F	CI	Aniline point (°C)	SUV, 100°F	Cloud test (°F)
Stage I: Distillation at atmospheric pressure, 762 mmHg										
1	50	122								
2	75	167								
3	100	212								
4	125	257								
5	150	302	0.9	0.9						
6	175	347	0.8	1.7	0.809	43.4	—			
7	200	392	1.1	2.8	0.823	40.4	41			
8	225	437	1.1	3.9	0.848	35.4	47			
9	250	482	4.1	8.0	0.866	31.9	50			
10	275	527	11.9	19.9	0.867	31.7	46			
Stage 2: Distillation continued at 40 mmHg pressure										
11	200	392	1.6	21.5	0.878	29.7	47		36	< 0
12	225	437	3.2	24.7	0.929	20.8	67		66	< 0
13	250	482	6.1	30.8	0.947	17.9	73		118	< 0
14	275	527	6.4	37.2	0.958	16.2	75		178	<0
15	300	572	10.6	47.8	0.972	14.1	78		508	< 0
Residuum			49.5	97.3						

Carbon residue of residuum: 39.6%

Carbon residue of crude: 19.6%

	Approximate summary			
	Vol%	Sp. gr.	°API	SUV, 100°F
Light gasoline	—	—	—	
Total gasoline and naphtha	2.8	0.818	41.5	
Kerosene distillate	—	—	—	
Gas oil	19.0	0.867	31.7	
Nonviscous lubricating distillate	4.3	0.901–0.940	25.6–19.0	50–100
Medium lubricating distillate	8.5	0.940–0.959	19.0–16.1	100–200
Viscous lubricating distillate	13.2	0.959–0.981	16.1–12.7	> 200
Residuum	49.5	—	—	
Distillation loss	2.7			

Field: McMurray Specific gravity at 60°F: 1.030. Sulfur, 5.84 wt%. Saybolt furol viscosity at 210°F, 820 s at 250°F, 236 s. API gravity at 60°F, 5.9. Pour point, °F : −. Color: black. Carbon residue, 19.6 wt% (Conradson).
Source: Information Circular IC 232, June 1969, p. 166. Copyright © Queen's Printer for Canada, 1970, by permission of the publisher, Canada Centre for Mineral and Energy Technology.

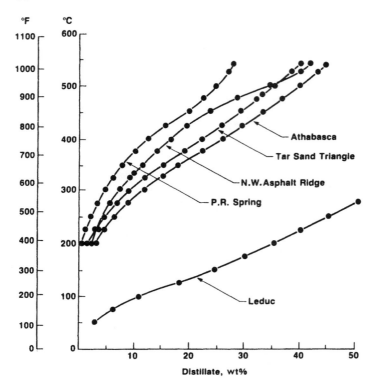

Figure 4.8 Simulated distillation profiles for conventional petroleum and tar sand bitumen.

2887). This method has been well researched in terms of development and application (Hickerson, 1975; Green, 1976; Stuckey, 1978; Vercier and Mouton, 1979; Thomas et al., 1983; Romanowski and Thomas, 1985; MacAllister and DeRuiter, 1987; Schwartz et al., 1987; Thomas et al., 1987). The benefits of the technique include good comparisons with other ASTM distillation data as well as the application to higher boiling fractions of petroleum. In fact, output data include the provision of the corresponding Engler profile (ASTM D-86) as well as the prediction of other properties such as vapor pressure and flash point (DeBruine and Ellison, 1973). When it is necessary to monitor product properties, as is often the case during refining operations, such data provide a valuable aid to process control and on-line product testing.

Simulated distillation by gas chromatography is often applied in the petroleum industry to obtain true boiling point data for distillates and crude oils (Butler, 1979). Two standardized methods (ASTM D-2887 and D-3710) are available for the boiling point determination of petroleum fractions and gasoline, respectively. The regularity of the elution order of the hydrocarbon components allows the retention times to be equated to distillation temperatures (Green et al., 1964). The ASTM D-2887 method uses non-polar, packed gas chromatographic columns in conjunction with flame ionization detection. The upper limit of the boiling range covered by this method is approximately 540°C (1000°F) atmospheric equivalent boiling point. Recent efforts in which high temperature gas chromatography was used have focused on extending the scope of the ASTM D-2887 method for higher boiling petroleum materials to 800°C (1470°F) atmospheric equivalent boiling point.

3.2 Liquefaction and Solidification

Although petroleum and the majority of petroleum products are liquids at ambient temperature, some feedstocks and products are semisolid to solid. Determination of the *melting point* is a test (ASTM D-87, ASTM D-127, IP 55, IP 133) that is widely used by suppliers and consumers of wax; it is particularly applied to the highly paraffinic or crystalline waxes. Quantitative prediction of the melting point of pure hydrocarbons is difficult, but the melting point tends to increase qualitatively with the molecular weight and with symmetry of the molecule. The test can also be applied to solid feedstocks such as atmospheric residua, vacuum residua, and tar sand bitumen.

The *softening point* (ASTM D-36, ASTM D-2398, IP 58) is the temperature at which a disk of the material softens and sags downward a distance of 25 mm under the weight of a steel ball under strictly specified conditions. It finds wider use for residua, asphalt, and bitumen. The *dropping point* (ASTM D-566, IP 132) is the near-equivalent test that is used for lubricating greases.

The reverse process, i.e. *solidification*, has also received considerable attention as it applies to petroleum refining. In fact, solidification of petroleum and petroleum products has been differentiated into four categories: *freezing point, congealing point, pour point*, and *cloud point* (ASTM D-2500, ASTM D-3117, IP 219, IP 444, IP 445, IP 446).

Petroleum becomes more or less a plastic solid when cooled to sufficiently low temperatures. This is due to the congealing of the various hydrocarbons that constitute the oil. The *cloud point* of a petroleum oil is the temperature at which paraffin wax or other solidifiable compounds present in the oil appear as a haze when the oil is chilled under certain prescribed conditions (ASTM D-2500, ASTM D-3117, IP 219, IP 444, IP 445, IP 446). Related to the *cloud point*, the wax appearance temperature or wax appearance point is also determined (ASTM D-3117, IP 389).

As cooling is continued, all petroleum oils become more and more viscous and flow becomes slower and slower. The *pour point* of a petroleum oil is the lowest temperature at which the oil pours or flows under certain prescribed conditions when it is chilled without disturbance at a standard rate (ASTM D-97, D-5327, D-5853, D-5949, D-5950, D-5985; IP 15, IP 219, IP 441).

The pour point of petroleum and petroleum products was originally developed to be of use in determining the fluid properties of waxy crude oil and wax-containing products. More recently, the pour point has also found use as an indicator of the temperature at which crude oil or heavy oil will flow during in situ recovery operations (Wallace, 1988, p. 183). Under similar conditions, bitumen is immobile. For example, for asphaltic crude oils where paraffin precipitation will not occur, if 21°C (70°F) is the pour point of a crude oil in a reservoir where the temperature is 38°C (100°F), the oil is liquid under reservoir conditions and will be mobile and flow under those conditions. On the other hand, tar sand bitumen [pour point 60°C (140°F)] in a deposit at a temperature of 10°C (50°F) will be solid and immobile. This state of the oil in the reservoir can also have consequences for the ability of gases and liquids (e.g., steam, hot water) used for recovery operations to penetrate the reservoir or deposit. Although pressure can have some influence on the pour point, the effect is not great and is unlikely to have any general results. Indeed, for specific types of petroleum (usually asphaltic and nonwaxy) there is a relationship between API gravity and pour point. Thus, any increase in pour point due to an increase in pressure (surface pressure compared to reservoir or deposit pressure) will most likely be negated as the API gravity increases with increase in temperature (60°F compared to reservoir temperature).

The solidification characteristics of a petroleum product depend on its grade or kind. For greases, the temperature of interest is that at which fluidity occurs, commonly known as the dropping point. The *dropping point* of a grease is the temperature at which the grease passes from a plastic solid to a liquid state and begins to flow under the conditions of the test (ASTM D-566, ASTM D-2265, IP 132). For another type of plastic solid, including petrolatum and microcrystalline wax, both melting point and congealing point are of interest.

The melting point of a wax is the temperature at which the wax becomes sufficiently fluid to drop from the thermometer. The congealing point is the temperature at which melted petrolatum ceases to flow when allowed to cool under prescribed conditions (ASTM D-938).

For another type of solid, paraffin wax, the solidification temperature is of interest. In this case, the *melting point* is also the temperature at which the melted paraffin wax begins to solidify, as shown by the minimum rate of temperature change, when cooled under prescribed conditions. For pure or essentially pure hydrocarbons, the solidification temperature is the *freezing point*, the temperature at which a hydrocarbon passes from the liquid to the solid state (ASTM D-910, D-1015, D-1016, D-2386; IP 16, IP 434, IP 435).

The relationship of *cloud point*, *pour point*, *melting point*, and *freezing point* to one another varies widely from one petroleum product to another. Hence, their significance for different types of products also varies. In general, the cloud point, the melting point, and the freezing point are of more limited value, and each has a narrower range of application than the pour point. The cloud point and pour point are useful for predicting the temperature at which the observed viscosity of an oil deviates from the true (Newtonian) viscosity in the low temperature range. They are also useful for identification of oils or when planning the storage of oil supplies, because low temperatures may cause handling difficulties with some oils.

3.3 Carbon Residue

Petroleum products are mixtures of many compounds that differ widely in their physical and chemical properties. Some of them may be vaporized in the absence of air at atmospheric pressure without leaving an appreciable residue. Other nonvolatile compounds leave a *carbonaceous residue* when destructively distilled under such conditions. This residue is known as *carbon residue* when determined in accordance with prescribed procedure.

There are two older methods for determining the carbon residue of a petroleum or petroleum product: the *Conradson method* (ASTM D-189, IP 13) and the *Ramsbottom method* (ASTM D-524, IP 14). Both are applicable to the relatively nonvolatile portion of petroleum and petroleum products, which partially decompose when distilled at a pressure of 1 atm. However, oils that contain ash-forming constituents are assigned erroneously high carbon residue values by either method unless the ash is first removed from the oil; the degree of error is proportional to the amount of ash.

Although there is no exact correlation between the two methods, it is possible to interconnect the data. However, caution is advised when using the portion of the curve below 0.1wt% Conradson carbon residue.

Recently, a newer method (Noel, 1984) has also been accepted (ASTM D-4530, IP 398) that requires smaller sample amounts and was originally developed as a *thermogravimetric method*. The carbon residue produced by this method is often referred to as the *microcarbon residue* (MCR). Agreement between the data from the three methods is good, making it possible to interrelate all of the data from carbon residue tests (Long and Speight, 1989).

Even though the three methods have their relative merits, there is a tendency to advocate the use of the more expedient microcarbon method rather than the Conradson and Ramsbottom methods because of the lesser amounts required in the *microcarbon* method, although it is somewhat less precise in practice.

The carbon residue is a property that can be correlated with several other properties of petroleum (Fig. 4.9); hence it also presents indications of the volatility of the crude oil and its coke-forming (or gasoline-producing) propensity. However, tests for carbon residue are sometimes used to evaluate the carbon-depositing characteristics of fuels used in certain types of oil-burning equipment and internal combustion engines.

The mechanical design and operating conditions of such equipment have such a profound influence on carbon deposition during service that comparison of carbon residues between oils should be considered as giving only a rough approximation of relative deposit-forming tendencies. Recent work has focused on the carbon residue of the different fractions of crude oils, especially the asphaltenes (Fig. 4.10). A more precise relationship between carbon residue and hydrogen content, the H/C atomic ratio, nitrogen

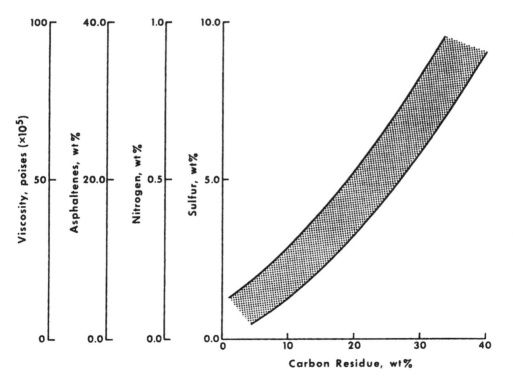

Figure 4.9 Relationship of carbon residue to other physical properties.

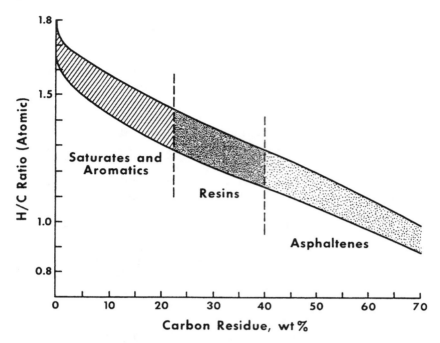

Figure 4.10   Relationship of thermal coke yields to feedstock fractions.

content, and sulfur content has been shown to exist (see also Nelson, 1974). These data can provide more precise information about the anticipated behavior of a variety of feedstocks in thermal processes (Roberts, 1989). Thus, there is a fairly universal linear correlation between the Conradson carbon residue (CR) and the H/C ratio:

$$H/C = 171 - 0.0115\,CR \text{ (Conradson)}$$

This equation holds within two limits; at an H/C values of 1.71, where the carbon residue is zero (no coke formation) and at $H/C = 0.5$, where the carbon residue is 100 (all the material converts to coke under test conditions). There is a relationship between the carbon residue (Conradson) and the nitrogen content.

Because of the extremely small values of carbon residue obtained by the Conradson and Ramsbottom methods when applied to the lighter distillate fuel oils, it is customary to distill such products to 10% residual oil and determine the carbon residue thereof. Such values can be used directly in comparing fuel oils as long as it is kept in mind that the values are carbon residues on 10% residual oil and are not to be compared with straight carbon residues.

3.4   Aniline Point

The *aniline point* of a liquid was originally defined as the consolute or critical solution temperature of the two liquids, that is, the minimum temperature at which they are miscible in all proportions. The term is now most generally applied to the temperature at which exactly equal parts of the two are miscible. This value is more conveniently measured than the original value and is only a few tenths of a degree lower for most substances.

Although it is an arbitrary index (ASTM D-611, IP 2), the aniline point is of considerable value in the characterization of petroleum products. For oils of a given type it increases slightly with molecular weight; for those of a given molecular weight it increases rapidly with increasing paraffinic character. As a consequence, it was one of the first properties proposed for the group analysis of petroleum products with respect to aromatic and naphthene content. It is used alternatively even in one of the more recent methods. The simplicity of the determination makes it attractive for the rough estimation of aromatic content when that value is important for functional requirements, as in the case of the solvent power of naphtha and the combustion characteristics of gasoline and diesel fuel. The aniline point is used in conjunction with the API gravity to determine the *diesel index*.

3.5 Heat of Combustion

The *heat of combustion* (ASTM D-240, IP 12) is a direct measure of fuel energy content. It is determined as the quantity of heat liberated by the combustion of a quantity of fuel with oxygen in a standard bomb calorimeter.

Chemically, the heat of combustion is the energy (heat) released when an organic compound is burned to produce water [$H_2O(l)$], carbon dioxide [$CO_2(g)$], sulfuric acid [$H_2SO_4(l)$], and nitric acid [$HNO_3(l)$]. The value can be calculated using a theoretical equation based upon the elemental composition of the feedstock:

$$H_g/4.187 = 8400C + 27,765H + 1500N + 2500S - 2650O$$

where H_g is given in kilojoules per kilogram ($1.0\,kJ/kg = 0.43\,Btu/lb$), and C, H, N, S, and O are the normalized weight fractions for the corresponding elements in the sample.

The gross heats of combustion of crude oil and its products are given with fair accuracy by the equation

$$Q = 12,400 - 2100d^2$$

where d is the 60/60°F specific gravity. Deviation is generally less than 1%, although many highly aromatic crude oils show considerably higher values; the range for crude oil is 10,000–11,600 cal/g. (Speight, 1999, p. 340). For gasoline, the heat of combustion is 11,000–11,500 cal/g, and for kerosene (and diesel fuel) it falls in the range 10,500–11,200 cal/g. Finally, the heat of combustion for fuel oil is on the order of 9500–11,200 cal/g. Heats of combustion of petroleum gases can be calculated from the analytical data for the pure compounds. Experimental values for gaseous fuels can be obtained by measurement in a water flow calorimeter, and heats of combustion of liquids are usually measured in a bomb calorimeter.

An alternative criterion of energy content is the aniline gravity product (AGP) (ASTM D-1405, IP 193), which is in reasonable agreement with the calorific value. It is the product of the API gravity and the aniline point (ASTM D-611, IP 2) of the sample.

3.6 Pressure–Volume–Temperature Relationships

Hydrocarbon vapors, like other gases, follow the ideal gas law (i.e., $PV = RT$) only at relatively low pressures and high temperatures, that is, far from the critical state. Several empirical equations, such as the well-known van der Waals equation, have been proposed to represent the gas laws more accurately, but they are either inconvenient for calculation or require the experimental determination of several constants. A more useful device is to use the simple gas law and induce a correction, termed the *compressibility factor*, so that the equation takes the form

$$PV = \mu RT$$

For hydrocarbons the compressibility factor μ is very nearly a function only of the reduced variables of state, that is, of the pressure and temperature divided by the respective critical values. The compressibility factor method functions excellently for pure compounds but may become ambiguous for mixtures because the critical constants have a slightly different significance. However, the use of pseudocritical temperature and pressure values, which are generally lower than the true values, permit the compressibility factor to be employed in such cases.

3.7 Critical Properties

A study of the pressure–volume–temperature relationships of a pure component reveals a particular unique state where the properties of a liquid and vapor become indistinguishable from each other. At that state the latent heat of vaporization becomes zero and no volume change occurs when the liquid is vaporized. This state is called the critical state, and the appropriate parameters of state are termed the critical pressure (P_C), critical volume (V_C), and critical temperature (T_C). It is an important characteristic of the critical state for a pure component that with values of P or T greater than either P_C or T_C the vapor and liquid states cannot coexist at equilibrium, and thus P_C and T_C represent the maximum values of P_C and T_C at which phase separation can occur.

Because the critical state of a component is unique it is perhaps not surprising that knowledge of P_C, T_C and V_C allows many predictions to be made concerning the physical properties of substances. These predictions are based on the law of corresponding states, which states that substances behave in the same way when they are in the same state with reference to the critical state. The particular corresponding state is characterized by its reduced properties, i.e. $T_r = T/T_C$, $P_r = P/P_C$, $V_r = V/V_C$.

The use of this concept permits generalized plots in terms of reduced properties to be drawn, which are then applicable to all substances that obey the law and can be of great value in determining thermodynamic relationships. It is rare in petroleum engineering to have to deal with pure substances, and unfortunately the application of the law of corresponding states to mixtures is complicated by the fact that the use of the true critical point for a mixture does not yield correct values of reduced properties for accurate prediction from generalized charts. For a mixture the critical state no longer represents the maximum temperature and pressure at which a liquid and vapor phase can coexist, and phase separation can occur under retrograde conditions.

For engineering purposes this difficulty is resolved by the use of pseudocritical conditions, which are based on the molal average critical temperatures and pressures of the compounds of the mixture. Although the use of pseudo-reduced conditions for mixtures of hydrocarbons is generally satisfactory, this is not true for states near the true critical state or, in general, for mixtures of vapor and liquid.

The temperature, pressure, and volume at the critical state are of considerable interest in petroleum physics, particularly in connection with modern high pressure, high temperature refinery operations and in correlating pressure–temperature–volume relationships for other states. Critical data are known for most of the lower molecular weight pure hydrocarbons, and standard methods are generally used for such determinations.

The *critical point* of a pure compound is the equilibrium state in which its gaseous and liquid phases are indistinguishable and coexistent—they have the same intensive properties. However, localized variations in these phase properties may be evident experimentally. The definition of the critical point of a mixture is the same. However, mixtures generally have a maximum temperature or pressure at other than the true critical point; *maximum* here denotes the greatest value at which two phases can coexist in equilibrium.

Thus, when a pure compound is heated at atmospheric pressure, it eventually reaches its boiling point and is completely vaporized at a constant temperature unless the pressure is increased. If the pressure is increased, the compound is completely condensed and cannot be vaporized again unless the temperature is also increased. This mechanism, alternately increasing the pressure and temperature, functions until at some high temperature and pressure it is found that the material cannot be condensed regardless of the amount of pressure applied. This point is called the *critical point*, and the temperature and pressure at the critical point are called the *critical temperature* and *critical pressure*, respectively.

The liquid phase and vapor phase merge at the critical point so that one phase cannot be distinguished from the other. No volume change occurs when a liquid is vaporized at the critical point, and no heat is required for vaporization, but the coefficient of expansion has become large.

Limited information concerning the behavior of complex mixtures has required that the pseudocritical temperature and pseudocritical pressure be used for many petroleum fractions and products. The *pseudocritical point* is defined as the molal average critical temperature and pressure of the several constituents that make up a mixture. It may be used as the critical point of a mixture in computing reduced temperatures and pressures. However, in computing the pressure–volume–temperature relations of mixtures by use of the pseudocritical point, it must be recognized that the values are not accurate in the region of the critical point and that it cannot be applied to mixtures of gas and liquid.

In the correlation of many properties, reduced properties are useful. A *reduced property* is defined as the ratio of the actual value of the property to its critical value. Thus, for volume the relationship is

$$\text{Reduced volume } V_r = V/V_c$$

where V is the volume at specified conditions and V_c is the volume at the critical point. Similarly,

$$\text{Reduced temperature } T_r = T/T_c$$

and

$$\text{Reduced pressure } P_r = P/P_c$$

where T and P are the temperature and pressure, respectively, at specified conditions and T_c and V_c are the temperature and pressure, respectively, at the critical point.

4. USE OF THE DATA

The data derived from any one or more of the evaluation techniques described here can be used to give an indication of the feedstock type and can also be employed to give the refiner an indication of the means by which the crude feedstock should be processed to produce a specific product slate as well as for the prediction of product properties (Dolbear et al., 1987; Adler and Hall, 1988; Wallace and Carrigy, 1988; Al-Besharah et al., 1989; Speight, 1999, 2000). Other properties may also be required for further feedstock evaluation or, more likely, for comparison between feedstocks even though they may not play any role in dictating which refinery operations are necessary. An example of such an application is the calculation of product yields for delayed coking operations by using the carbon residue and the API gravity (Table 4.6) of the feedstock.

However, it is essential that when such data are derived, the parameters employed be carefully specified. For example, the data presented in the tables were derived on the basis of straight-run residua having an API gravity less than 18°. The gas oil end point was of the order of 470–495°C (875–925°F), the gasoline end point was 205°C (400°F), and the pressure in the coke drum was standardized at 35–45 psi. Obviously there are benefits to the derivation of such specific data, but the numerical values, although representing only an approximation, may vary substantially when applied to different feedstocks (Speight, 1999, 2000).

Even with such applicability, proceeding from the raw evaluation data to full-scale production is not always the preferred step. Further evaluation of the processability of the feedstock is usually necessary through the use of a pilot-scale operation. To take the

Table 4.6 Use of Carbon Residue Test Data to Estimate
Product Yield from Delayed Coking

Wilmington

$$\text{Coke, wt\%} = 39.68 - 1.60 \times {}^\circ\text{API}$$

$$\text{Gas } (\leq C_4), \text{ wt\%} = 11.27 - 0.14 \times {}^\circ\text{API}$$

$$\text{Gasoline, wt\%} = 20.5 - 0.36 \times {}^\circ\text{API}$$

$$\text{Gas oil, wt\%} = 28.55 + 2.10 \times {}^\circ\text{API}$$

$$\text{Gasoline, vol\%} = \frac{186.5}{131.5 + {}^\circ\text{API}} \times \text{wt\% gasoline}$$

$$\text{Gas oil, vol\%} = \frac{155.5}{131.5 + {}^\circ\text{API}} \times \text{wt\% gas oil}$$

East Texas

$$\text{Coke, wt\%} = 45.76 - 1.78 \times {}^\circ\text{API}$$

$$\text{Gas } (\leq C_4), \text{ wt\%} = 11.92 - 0.16 \times {}^\circ\text{API}$$

$$\text{Gasoline, wt\%} = 20.5 - 0.36 \times {}^\circ\text{API}$$

$$\text{Gasoline, vol\%} = \frac{186.5}{131.5 + {}^\circ\text{API}} \times \text{wt\% gasoline}$$

$$\text{Gas oil, vol\%} = \frac{155.5}{131.5 + {}^\circ\text{API}} \times \text{wt\% gas oil}$$

evaluation of a feedstock one step further, it may then be possible to develop correlations between the data obtained from the actual plant operations (as well as the pilot plant data) and one or more of the physical properties determined as part of the initial feedstock evaluation.

Evaluation of petroleum from known physical properties may also be achieved by use of the refractivity intercept. Thus, if refractive indices of hydrocarbons are plotted against the respective densities, straight lines of constant slope are obtained, one for each homologous series; the intercepts of these lines with the ordinate of the plot are characteristic, and the refractivity intercept is derived from the formula

$$\text{Refractivity intercept} = n - \frac{d}{2}$$

The intercept cannot differentiate accurately among all series, which restricts the number of different types of compounds that can be recognized in a sample. The technique has been applied to nonaromatic olefin-free materials in the gasoline range by assuming additivity of the constant on a volume basis.

Following from this, an equation has been devised that is applicable to straight-run lubricating distillates if the material contains between 25% and 75% of the carbon present in naphthenic rings (denoted C_N):

$$\text{Refractivity intercept} = 1.0502 - 0.00020\% \ C_N$$

Although not specifically addressed in this chapter, the fractionation of petroleum also plays a role, along with the physical testing methods, in evaluating petroleum as a refinery feedstock and its behavior under specific refinery scenarios. For example, by careful selection of an appropriate technique it is possible to obtain a detailed overview of feedstock or product composition that can be used for process predictions. Using the adsorbent separation as an example) (Chapter 3), it becomes possible to develop one or more petroleum maps and determine how a crude oil might behave under specified process conditions.

This concept has been developed to the point where various physical parameters are used as the ordinate and abscissa. It must be recognized that such maps do not give any indication of the complex interactions that occur between, for example, such fractions as the asphaltenes and resins (Koots and Speight, 1975; Speight, 1994), but it does allow predictions of feedstock behavior. It must also be recognized that such a representation varies for different feedstocks.

In summary, evaluation of feedstock behavior from test data is not only possible but has been practiced for decades. And such evaluations will continue for decades to come. However it is essential to recognize that the derivation of an equation for predictability of behavior will not suffice (with a reasonable degree of accuracy) for all feedstocks. Many of the data are feedstock dependent because they incorporate the complex interactions between feedstock constituents. Careful testing and evaluation of the behavior of each feedstock and each blend of feedstocks is recommended. If this is not done, incompatibility or instability (Speight, 1999) can result leading to higher than predicted yields of thermal or catalytic coke.

REFERENCES

Adler, S. B., and Hall, K. R. 1988. Hydrocarbon Processing 71(11):71.

Al-Besharah, J. M., Mumford, C. J., Akashah, S. A., and Salman, O. 1989. Fuel. 68:809.

Ali, M. F., Hasan, M., Bukhari, A., and Saleem, M. 1985. Oil Gas J. 83(32):71.

ASTM. 1998. Annual Book of Standards. American Society for Testing and Materials, Philadelphia, PA.

Berkowitz, N., and Speight, J. G. 1975. Fuel 54:138.

Bland, W. F., and Davidson, R. L. 1967. Petroleum Processing Handbook. McGraw-Hill, New York. p. 12.

Butler, R. D. 1979. In: Chromatography in Petroleum Analysis. K. H. Altgelt and T. H. Gouw, eds. Marcel Dekker, New York, p. 75.

DeBruine, W., and Ellison, R. J. 1973. J. Petrol. Inst. 59:146.

Dolbear, G. E., Tang, A., and Moorehead, E. L. 1987. In: Metal Complexes in Fossil Fuels. R. H. Filby and J. F. Branthaver, eds. ACS Symp. Ser. No. 344. American Chemical Society, Washington, DC, p. 220.

Gary, J. H., and Handwerk, G. E. 1984. Petroleum Refining: Technology and Economics. Marcel Dekker, New York.

Gomez, J. V. 1989. Oil Gas J. Vol. 87, Mar. 27, p. 66.

Gomez, J. V. 1992. Oil Gas J. Vol. 90, July 13, p. 49.

Gomez, J. V. 1995. Oil Gas J. Vol. 93, Feb. 6, p. 60.

Green, L. E. 1976. Hydrocarbon Processing 55(5):205.

Green, L. E., Schmauch, L. J., and Worman, J. C. 1964. Anal. Chem. 36:1512.

Hickerson, J. F. 1975. In: Special Publication No. STP 577. American Society for Testing and Materials, Philadelphia, p. 71.

IP. 1999. Standard Methods for Analysis and Testing of Petroleum and Related Products. Institute of Petroleum, London, England.

Koots, J. A., and Speight, J. G. 1975. Fuel 54:179.

Long, R. B., and Speight, J. G. 1989. Rev. Inst. Français du Pétrole 44:205.

MacAllister, D. J., and DeRuiter, R. A. 1985. Paper SPE 14335. 60th Annual Technical Conference, Society of Petroleum Engineers, Las Vegas, Sept. 22–25.

Nelson, W. L. 1958. Petroleum Refinery Engineering. McGraw-Hill. New York.

Nelson, W. L. 1974. Oil Gas J. 72(6):72.

Noel, F. 1984. Fuel 63:931.

Roberts, I. 1989. Preprints. Div. Petrol. Chem. Am. Chem. Soc. 34(2):251.

Romanowski, L. J., and Thomas, K. P. 1985. Report No. DOE/FE/60177-2326. U.S. Department of Energy, Washington, DC.

Schwartz, H. E., Brownlee, R. G., Boduszynski, M. M., and Su, F. 1987. Anal. Chem. 59:1393.

Speight, J. G. 1990. In: Fuel Science and Technology Handbook. J. G. Speight, ed. Marcel Dekker, New York, Chapters 12–16.

Speight, J. G. 1994. In: Asphaltenes and Asphalts. Vol. I. Developments in Petroleum Science. T. F. Yen and G. V. Chilingarian, eds. Elsevier, Amsterdam, Chapter 2.

Speight, J. G. 1999. The Chemistry and Technology of Petroleum. 3rd ed. Marcel Dekker, New York.

Speight, J. G. 2000. The Desulfurization of Heavy Oils and Residua. 2nd ed. Marcel Dekker, New York.

Spragins, F. K. 1978. Development in Petroleum Science. No. 7. Bitumens, Asphalts and Tar Sands. T.F. Yen and G.V. Chilingarian, eds. Elsevier, New York, p. 92.

Stuckey, C.L. 1978. J. Chromatogr. Sci. 16:482.

Thomas, K. P., Barbour, R. V., Branthaver, J. F., and Dorrence, S. M. 1983. Fuel 62:438.

Thomas, K. P., Harnsberger, P. M., and Guffey, F. D. 1987. Report No. DOE/MC/11076-2451. U.S. Department of Energy, Washington, DC.

Vercier, P., and Mouton, M. 1979. Oil Gas J. 77(38):121.

Wallace, D., ed. 1988. A Review of Analytical Methods for Bitumens and Heavy Oils. Alberta Oil Sands Technology and Research Authority, Edmonton, AB, Canada.

Wallace, D., and Carrigy, M. A. 1988. In The Third UNITAR/UNDP International Conference on Heavy Crude and Tar Sands. R. F. Meyer, ed. Alberta Oil Sands Technology and Research Authority, Edmonton, AB, Canada, p. 95.

Wallace, D., Starr, J., Thomas, K. P., and Dorrence, S. M. 1988. Characterization of Oil Sand Resources. Alberta Oil Sands Technology and Research Authority, Edmonton, AB, Canada.

5
Reaction Stoichiometry

1. INTRODUCTION

Stoichiometry is the expression of the basic arithmetic relationship between the atoms and molecules that are produced or consumed by chemical reactions. A basic understanding of reaction stoichiometry is needed for the correct formulation of compositional changes in chemically reactive mixtures and reaction engineering problems.

Stoichiometry does not involve the reaction mechanism; therefore, stoichiometric relations can be expressed regardless of the reaction mechanisms. When the chemical reaction for the synthesis of ammonia is expressed with the stoichiometry

$$N_2 + 3H_2 \rightarrow 2NH_3$$

it does not imply any reaction mechanism for the formation of two moles of ammonia (NH_3) from one mole of nitrogen (N_2) and three moles of hydrogen (H_2). It only states the atomic or molecular relationship of the reactants and the product in ammonium synthesis.

Reaction stoichiometry is expressed by assuming that the individual reactant and product species are discrete and can be represented with chemical formulas. The heat of reaction, expressed on the basis of molar units of species consumed or produced in the reaction, can also be included in the reaction stoichiometry.

Reaction stoichiometry is the most elementary step in the analysis of chemically reactive systems. In this chapter, the basics of stoichiometry, extent of reaction and reaction rate will be reviewed. The material and nomenclature used here on stoichiometry and the theory of chemical reaction engineering draws heavily on the classical text of Aris (1969).

2. STOICHIOMETRY OF CHEMICAL REACTIONS

Regardless of the basic elementary steps and reaction mechanisms, R chemical reactions taking place among S chemical species can be represented by the relationship

$$\sum_{j=1}^{S} \alpha_{i,j} A_j = 0; \qquad i = 1, \ldots, R \tag{5.1}$$

where A_j and $\alpha_{i,j}$ are the jth chemical species (in terms of atoms or molecules) and its stoichiometric coefficient in the ith reaction. By definition, $\alpha_{i,j} > 0$ if A_j is produced and $\alpha_{i,j} < 0$ if A_j is consumed in the ith reaction ($\alpha_{i,j} = 0$ if the species does not participate in the ith reaction).

Mathematically, both sides of Eq. (5.1) can be multiplied by a given constant without changing the meaning of the stoichiometric equation. As an example, the ammonia synthesis reaction can be expressed by the reaction

$$N_2 + 3H_2 \rightarrow 2NH_3 \tag{5.2}$$

which can be expressed by the stoichiometric equation (because only one reaction is considered, the reaction index i is dropped from $\alpha_{i,j}$ and the stoichiometric relation becomes $\sum_j \alpha_j A_j = 0$):

$$-A_1 - 3A_2 + 2A_3 = 0 \tag{5.3}$$

where $A_1 = N_2$, $A_2 = H_2$, $A_3 = NH_3$, $\alpha_1 = -1$, $\alpha_2 = -3$, and $\alpha_3 = 2$. Multiplying both sides of Eq. (5.3) by $1/3$ would give the stoichiometric relation

$$-\frac{1}{3}A_1 - A_2 + \frac{2}{3}A_3 = 0 \tag{5.4}$$

which corresponds to the reaction

$$\frac{1}{3}N_2 + H_2 \rightarrow \frac{2}{3}NH_3 \tag{5.5}$$

where $\alpha_1 = -1/3$, $\alpha_2 = -1$, and $\alpha_3 = 2/3$. This example shows that there is only one independent stoichiometric relation that can be expressed for a given reaction, i.e., the stoichiometric relation expressed by Eq. (5.4) can be obtained from the stoichiometric relation expressed by Eq. (5.3).

As another example, the following chemical reactions are involved in the Claus process:

$$H_2S + 1.5O_2 \rightarrow SO_2 + H_2O \tag{5.6}$$

$$2H_2S + SO_2 \rightarrow 3S + 2H_2O \tag{5.7}$$

By defining $A_1 = H_2S$, $A_2 = O_2$, $A_3 = SO_2$, $A_4 = H_2O$, and $A_5 = S$, the stoichiometric equations corresponding to the chemical reactions of Eqs. (5.6) and (5.7) become

$$-A_1 - 1.5A_2 + A_3 + A_4 = 0 \tag{5.8}$$

and

$$-2A_1 - A_3 + 2A_4 + 3A_5 = 0 \tag{5.9}$$

On the other hand the overall reaction in the Claus process is:

$$3H_2S + 1.5O_2 \rightarrow 3S + 3H_2O \tag{5.10}$$

which corresponds to the stoichiometric equation

$$-3A_1 - 1.5A_2 + 3A_4 + 3A_5 = 0 \tag{5.11}$$

Equation (5.11) is the summation (a linear combination) of the two stoichiometrically independent equations, Eqs. (5.8) and (5.9), corresponding to chemical reactions expressed by Eqs. (5.6) and (5.7). Consequently, the stoichiometric relation expressed by Eq. 5.11 does not provide more information (stoichiometrically) than the information provided by the stoichiometric relations expressed by Eqs. (5.8) and (5.9). Similarly, the chemical reaction expressed by Eq. (5.10) does not provide more information (stoichiometrically) than the chemical reactions expressed by Eqs. (5.6) and (5.7).

3. STOICHIOMETRICALLY INDEPENDENT REACTIONS

A set of N reactions are linearly dependent if N constants $\lambda_1, \ldots, \lambda_N$ (all nonzero) can exist that satisfy the relation

$$\sum_{i=1}^{N} \lambda_i \alpha_{i,j} = 0, \qquad j = 1, \ldots, S \tag{5.12}$$

If such constants $\lambda_1, \ldots, \lambda_N$ cannot exist, then these N reactions form a set of linearly independent reactions. As a result, for a set of N independent reactions, none of the reactions can be expressed as a linear combination of the others.

As an example, for the reactions involved in the Claus process, the following relation [defined by Eq. (5.12)] exists among the stoichiometric coefficients:

$$\alpha_{1,j} + \alpha_{2,j} - \alpha_{3,j} = 0, \qquad j = 1, \ldots, 5 \tag{5.13}$$

where $\lambda_1 = \lambda_2 = 1$ and $\lambda_3 = -1$. Consequently, the three reactions expressed by Eqs. (5.6), (5.7), and (5.10) are a set of linearly dependent reactions, and any two of these reactions form a set of linearly independent reactions.

It is always practical to find a set of linearly independent reactions for material balance and energy balance calculations. For reaction kinetics, however, linearly dependent reactions should also be considered.

For a given set of R chemical reactions between S chemical species, a set of independent reactions can be found by constructing a matrix from the stoichiometric coefficients of the reactions. By using Gaussian elimination (and tracking the number of reactions), a triangular matrix with nonzero diagonal elements can be formed. The number of independent reactions is determined from the number of rows with nonzero diagonal elements, i.e., the rank of the matrix (Amundson, 1966).

4. MATERIAL BALANCE FOR A SINGLE REACTION

In chemically reactive systems, individual chemical species are measured by their mass or molar amounts. It must be remembered that in such chemically reactive systems the total mass is always conserved but the total number of moles may change. This fact has to be considered during the formulation of material balances, especially for the gaseous species.

In a reactive system, if m_j is the mass and M_j is the molecular weight for the jth species, then the number of moles N_j in the system is m_j/M_j. For a single reaction $\alpha_A A + \alpha_B B \rightarrow \alpha_C C + \alpha_D D$, the change in N_j is expressed by introducing the reaction extent X. The reaction extent, expressed in terms of moles, is the measure of progress in the reaction. By defining the reaction extent X, the mole number N_j for the jth species becomes

$$N_j = N_{j,0} + \alpha_j X \tag{5.14}$$

where $N_{j,0}$ is the initial number of moles and α_j is the stoichiometric coefficient of the jth species. By using Eq. (5.14), the reaction extent can be expressed in terms of the molar balance of any kth species involved in the reaction:

$$X = \frac{N_k - N_{k,0}}{\alpha_k} \tag{5.15}$$

and the molar balance for any jth species can be expressed in terms of the molar changes of the kth species, by inserting Eq. (5.15) into Eq. (5.14):

$$N_j = N_{j,0} + \frac{\alpha_j}{\alpha_k}\left(N_k - N_{k,0}\right) \tag{5.16}$$

If the kth species is the limiting species in the chemical reaction, it will be completely consumed ($N_k = 0$, $\alpha_k < 0$) in the reaction. In this case, the reaction extent will reach its maximum X_{max}:

$$X_{max} = \frac{-N_{k,0}}{\alpha_k} \tag{5.17}$$

which indicates that the number of moles N_j would be controlled by $N_{k,0}$ (or X_{max}) if the reaction is allowed to reach its maximum extent:

$$N_j = N_{j,0} + \alpha_j X_{max} = N_{j,0} - \frac{\alpha_j}{\alpha_k} N_{k,0} \tag{5.18}$$

If the kth species did not exist initially and is produced in the reacting system ($N_{k,0} = 0$, $\alpha_k > 0$), then Eqs. (5.15) and (5.16) become

$$X = \frac{N_k}{\alpha_k} \tag{5.19}$$

and

$$N_j = N_{j,0} + \frac{\alpha_j}{\alpha_k} N_k \tag{5.20}$$

At equilibrium, the reaction extent reaches the equilibrium extent X_e and the numbers of moles of the species reach their equilibrium values $N_{j,e}$. At equilibrium, the number of moles can be calculated in terms of the reaction extent at equilibrium, X_e:

$$N_{j,e} = N_{j,0} + \alpha_j X_e \tag{5.21}$$

Thermodynamics determine the reaction extent at equilibrium, X_e, as a function of temperature, pressure, and $N_{j,0}$. Calculation of the equilibrium mole numbers (or composition) is an important element of reaction engineering (Chapter 7, Section 4).

5. MEASURE OF CONCENTRATION

Concentration is an intensive (independent of system volume or size) property of a given system. There are several definitions of concentration commonly used in engineering practice.

If N_j moles of species A_j is present in a system of volume V, the *molar concentration c_j* is given by

$$c_j = \frac{N_j}{V} \tag{5.22}$$

The *mole fraction* y_j for species A_j is defined as

$$y_j = \frac{N_j}{\sum_j N_j} = \frac{N_j}{N} \tag{5.23}$$

Mass concentration is defined as mass density (or density). If the mass of species A_j contained in volume V is defined by m_j, then the mass concentration ρ_j is defined as

$$\rho_j = \frac{m_j}{V} \tag{5.24}$$

The *mass fraction* g_j of species A_j is defined as

$$g_j = \frac{m_j}{\sum_j m_j} = \frac{m_j}{m} = \frac{\rho_j}{\rho} \tag{5.25}$$

where $N\ (= \sum_j N_j)$, $m(= \sum_j m_j)$, and $\rho\ (= \sum_j \rho_j)$ are the total moles, total mass, and mass density of the system, respectively.

As an example, the mole fraction y_k of the kth species expressed as a function of reaction extent X is

$$y_k = \frac{N_k}{N} = \frac{N_{k,0} + \alpha_k X}{\sum\limits_{j=1}^{S} (N_{j,0} + \alpha_j X)} = \frac{N_{k,0} + \alpha_k X}{N_0 + X \sum\limits_{j=1}^{S} \alpha_j} = \frac{N_{k,0} + \alpha_k X}{N_0 + X \bar{\alpha}} \tag{5.26}$$

where, N_0 is the initial total number of moles and $\bar{\alpha} = \sum_j \alpha_j$, which is a measure of increase or decrease in the total number of moles as a result of the chemical reaction.

In chemically reactive systems the total mass m is conserved (as is the mass density ρ, if V is constant); however, the total molar concentration c may not be conserved. The drift velocity caused by the change in the total gas molar concentration is known as the Stephan flow (Frank-Kamenetskii, 1969). This principle has to be observed, especially for homogeneous gas-phase and heterogeneous gas–solid reactions.

When the gas phase is considered, concentrations are generally defined in terms of partial pressures for practical reasons. Assuming that the gas phase is an ideal gas mixture at total pressure p, the partial pressure p_j for species A_j is defined as (Dalton's law)

$$p_j = y_j p = y_j \frac{NRT}{V} = \frac{N_j}{N} \left(\frac{NRT}{V} \right) = c_j RT \tag{5.27}$$

For non-ideal gas mixtures, the compressibility factor Z_j has to be introduced into the above formulations, giving

$$p_j = c_j Z_j RT \tag{5.28}$$

6. CONCENTRATION CHANGES WITH A SINGLE REACTION

For a reacting system at constant volume V, molar balance [Eq. (5.14)] can be expressed in terms of molar concentration by dividing both sides of Eq. (5.14) by V:

$$c_j = \frac{N_j}{V} = \frac{N_{j,0}}{V} + \alpha_j \frac{X}{V} = c_{j,0} + \alpha_j \xi \tag{5.29}$$

where, $\xi\ (=X/V)$ is the intensive molar reaction extent per volume in moles per liter (mol/L). Using Eq. (5.29), the total molar concentration c becomes

$$c = \sum_j (c_{j,0} + \alpha_j \xi) = c_0 + \bar{\alpha}\xi \tag{5.30}$$

where $c_0\ (= \sum_j c_{j,0})$ is the initial molar concentration and $\bar{\alpha} = \sum_j \alpha_j$. If $\sum_j \alpha_j \neq 0$, then the mole fraction y_j of the jth species becomes

$$
\begin{aligned}
y_j &= \frac{c_j}{c} = \frac{c_{j,0} + \alpha_j \xi}{c_0 + \bar{\alpha}\xi} \\
&= \frac{c_{j,0}/c_0 + \alpha_j(\xi/c_0)}{1 + \bar{\alpha}(\xi/c_0)} = \frac{y_{j,0} + \alpha_j \xi'}{1 + \bar{\alpha}\xi'}
\end{aligned}
\tag{5.31}
$$

where $\xi'\ (=\xi/c_0)$ is the dimensionless reaction extent per unit volume (i.e., intensive dimensionless reaction extent, normalized with respect to initial mole concentration c_0) and $y_{j,0}$ is the initial mole fraction of the jth species. If $\sum_j \alpha_j = 0$, then the total molar concentration will be constant and x_j will be linear in ξ':

$$y_j = y_{j,0} + \alpha_j \xi' \tag{5.32}$$

Similarly, using Eq. (5.14), mass density and mass fraction can be expressed as functions of reaction extent [i.e., by multiplying both sides of Eq. (5.14) by M_j]:

$$m_j = M_j N_j = M_j(N_{j,0} + \alpha_j X) = m_{j,0} + \alpha_j M_j X \tag{5.33}$$

where, $m_{j,0}$ is the initial mass of the jth species in the system. Dividing both sides of Eq. (5.33) by volume V,

$$\rho_j = \rho_{j,0} + \alpha_j M_j \xi \tag{5.34}$$

where $\rho_{j,0}$ is the initial mass concentration of the jth species. Taking the summation of both sides of Eq. (5.34) over all species,

$$\rho = \sum_j \rho_j = \sum_j \rho_{j,0} + \xi\left(\sum_j \alpha_j M_j\right) = \rho_0 \tag{5.35}$$

because $\sum_j \alpha_j M_j = 0$ (i.e., total mass is conserved). Unlike the molar concentration at a constant volume, the mass concentration is conserved ($\rho = \rho_0$). Similarly, the mass fraction for the jth species g_j, as a function of reaction extent becomes

$$g_j = \frac{\rho_j}{\rho} = \frac{\rho_{j,0} + \alpha_j M_j \xi}{\rho} = \frac{\rho_{j,0}}{\rho} + \alpha_j M_j \frac{\xi}{\rho} = g_{j,0} + \alpha_j M_j \xi'' \tag{5.36}$$

where ξ'' is the normalized reaction extent per volume, in moles per unit mass:

$$\xi'' = \frac{\xi}{\rho} = \frac{X/V}{m/V} = \frac{X}{m} \tag{5.37}$$

Equation (5.36) could also be derived by dividing both sides of Eq. (5.33) by the total mass m. Because the total mass m of the system is conserved, Eq. (5.37) holds even if the volume of the system changes.

The intensive reaction extents $\xi\ (=X/V)$, $\xi'\ (=\xi/c_0)$, and $\xi''\ (=\xi/\rho)$ are associated with the reactions and are independent of the chemical species. Therefore, intensive reaction extents can be expressed in terms of selected chemical species. By eliminating the intensive reaction extents, the molar and mass concentrations can be expressed in terms of the molar and mass concentrations of the selected kth species. Using Eqs (5.29), (5.31), and (5.34), ξ can be expressed in terms of the molar concentration, mole fraction, molar weight, and mass density of the kth species as

$$\xi = \frac{c_k - c_{k,0}}{\alpha_k} = \frac{c_0(y_k - y_{k,0})}{\alpha_k - \bar{\alpha}y_k} = \frac{\rho_k - \rho_{k,0}}{\alpha_k M_k} \tag{5.38}$$

from which c_j, y_j, and ρ_j expressed by Eqs (5.29), (5.31), and (5.34) become

$$c_j = c_{j,0} + \frac{\alpha_j}{\alpha_k}(c_k - c_{k,0}) \tag{5.39}$$

$$y_j = \frac{y_{j,0}(\alpha_k - \bar{\alpha}y_k) + (y_k - y_{k,0})\alpha_j}{(\alpha_k - \bar{\alpha}y_k) + (y_k - y_{k,0})\bar{\alpha}} \tag{5.40}$$

and

$$\rho_j = \rho_{j,0} + \frac{\alpha_j M_j}{\alpha_k M_k}(\rho_k - \rho_{k,0}) \tag{5.41}$$

These relations are useful if the measurement of one of the reactive species is difficult. Using them, the concentration of an easily measured species can be used to calculate the concentration of the other reactive species. Any one of these relations can be selected for the formulation of the mass and molar concentration changes of any jth species as a function of the mass and molar concentration changes of any marked kth species.

7. CONVERSION

There is no universal definition for the conversion for a given reaction. In most cases, conversion is defined as the fraction of the initial moles of the jth species consumed in a reaction:

$$x_j = \frac{N_{j,0} - N_j}{N_{j,0}} = -\alpha_j \frac{X}{N_{j,0}} \tag{5.42}$$

In many industrial applications, conversion is also expressed as the fraction of the initial mass of the jth species consumed in a reaction:

$$x_j = \frac{m_{j,0} - m_j}{m_{j,0}} = -\alpha_j \frac{M_j X}{m_{j,0}} \tag{5.43}$$

In the petroleum industry, however, conversion in some cases is defined in terms of the volume of marketable product produced from the initial volume of the feed material. Definition of conversion on the basis of volume makes sense for the petroleum industry, because most petroleum products are fluid and their quantities are measured and valued on a volume basis.

Like concentration, conversion is an intensive (independent of system volume or size) property of a given system. Conversion measurement in some cases may be difficult. Molar balance expressed by Eq. (5.16) [as well as by Eqs. (5.39), (5.40), and (5.41)] has practical importance, if the molar or mass concentration of a specific species cannot be measured easily because of the severity of the reactor operating conditions, sampling difficulties, or unstable chemical structures or for any other reason. For example, in coal combustion, the carbon content of coal particles that are burning at temperatures above 900°C cannot be measured as easily as the CO, CO_2, and O_2 species present in the gas phase. The carbon concentration of the partially combusted coal (or char) particles can be calculated from the gas composition (i.e., CO, CO_2, and O_2) measurements.

In some applications, measurement of the inert species can be useful for conversion calculations. As an example, using the same mass balance principle, measurement of the ash content of the initial coal (or char), $g_{a,0}$, and of the partially combusted coal (or char), g_a, can be used to calculate the conversion x of coal (or char) combustion:

$$x = \frac{g_a - g_{a,0}}{g_a(1 - g_{a,0})} \tag{5.44}$$

Experimental errors made in measuring g_a and $g_{a,0}$ affect the accuracy of the calculated conversion using Eq. (5.44) in a nonlinear fashion.

8. CONCENTRATION CHANGES WITH SEVERAL REACTIONS

In the case of several reactions, concentration changes can be formulated by introducing the intensive reaction extent variables for the ith reaction:

$$\xi_i = \frac{X_i}{V}, \qquad \xi_i' = \frac{X_i}{N_0}, \qquad \xi_i'' = \frac{X_i}{m} \tag{5.45}$$

The change in the mole number N_j of the jth species in the case of several reactions becomes

$$N_j = N_{j,0} + \sum_i \alpha_{i,j} X_i \tag{5.46}$$

Similarly, multiplying both sides of Eq. (5.46) by M_j, the change in m_j becomes:

$$m_j = m_{j,0} + \sum_i \alpha_{i,j} M_j X_i \tag{5.47}$$

Following the same derivation patterns that were used for the case of a single reaction, the following relations would be obtained for molar and mass concentration variables:

$$c_j = c_{j,0} + \sum_i \alpha_{i,j} \xi_i \tag{5.48}$$

$$c = \sum_j c_j = \sum_j c_{j,0} + \sum_j \left(\sum_i \alpha_{i,j} \xi_i \right)$$

$$= c_0 + \sum_i \left(\sum_j \alpha_{i,j} \xi_i \right) = c_0 + \sum_i \bar{\alpha}_i \xi_i \tag{5.49}$$

$$y_j = \frac{c_j}{c} = \frac{y_{j,0} + \sum_i (\alpha_{i,j} \xi_i / c_0)}{1 + \sum_i (\bar{\alpha}_i \xi_i / c_0)} = \frac{y_{j,0} + \sum_i \alpha_{i,j} \xi_i'}{1 + \sum_i \bar{\alpha}_i \xi_i'} \tag{5.50}$$

$$\rho_j = M_j c_j = M_j c_{j,0} + \sum_i \alpha_{i,j} M_j \xi_i = \rho_{j,0} + \sum_i \alpha_{i,j} M_j \xi_i \tag{5.51}$$

and

$$g_j = \frac{\rho_j}{\rho} = g_{j,0} + \sum_i \frac{\alpha_{i,j} M_j \xi_i}{\rho} = g_{j,0} + \sum_i \alpha_{i,j} M_j \xi_i'' \tag{5.52}$$

9. REACTION RATE

In a closed system (i.e., no material transfer is allowed at the system boundaries, therefore $\sum_j \alpha_j A_j = 0$ holds), the extensive reaction rate R^* (mol/s) is defined as

$$R^* = \frac{dX}{dt} \tag{5.53}$$

Taking the time derivative of molar balance expressed in Eq. (5.14) and using Eq. (5.53), the change in the number of moles of the jth species can be expressed in terms of the extensive reaction rate R^*:

$$\frac{dN_j}{dt} = \alpha_j \frac{dX}{dt} = \alpha_j R^* \tag{5.54}$$

The intensive reaction rate r [mol/(s · L)] is defined as the rate of reaction per unit volume:

$$r = \frac{R^*}{V} \tag{5.55}$$

If the volume V of the reacting system is constant, then

$$r = \frac{R^*}{V} = \frac{dX/dt}{V} = \frac{d(X/V)}{dt} = \frac{d\xi}{dt} \tag{5.56}$$

and the concentration change of the jth species becomes

$$\frac{dc_j}{dt} = \frac{d(N_j/V)}{dt} = \alpha_j \frac{d(X/V)}{dt} = \alpha_j \frac{d\xi}{dt} = \alpha_j r \tag{5.57}$$

If the total volume V is not constant, the intensive reaction rate becomes

$$r = \frac{1}{V}\frac{dX}{dt} = \frac{1}{V}\frac{d(V\xi)}{dt} = \frac{d\xi}{dt} + \xi \frac{1}{V}\frac{dV}{dt} = \frac{d\xi}{dt} + \xi \frac{d\ln V}{dt} \tag{5.58}$$

Similarly, the change in c_j becomes [using Eqs. (5.14), (5.56), and (5.57)]

$$\frac{dc_j}{dt} = \frac{d(N_j/V)}{dt}$$

$$= \frac{1}{V}\frac{dN_j}{dt} - \frac{N_j}{V^2}\frac{dV}{dt} = \frac{\alpha_j}{V}\frac{dX}{dt} - \frac{N_j}{V}\frac{d\ln V}{dt} = \alpha_j r - c_j \frac{d\ln V}{dt} \tag{5.59}$$

and the rate of change of the total molar concentration c ($= \sum c_j$) becomes

$$\frac{dc}{dt} = \sum_j \frac{dc_j}{dt} = \sum_j \left(\alpha_j r - c_j \frac{d\ln V}{dt} \right) = \bar{\alpha} r - c \frac{d\ln V}{dt} \tag{5.60}$$

Using Eqs. (5.59) and (5.60), the change in molar fraction y_j becomes

$$\frac{dy_j}{dt} = \frac{d(c_j/c)}{dt} = \frac{1}{c}\frac{dc_j}{dt} - \frac{c_j}{c^2}\frac{dc}{dt} = (\alpha_j - \bar{\alpha} y_j)\frac{r}{c} \tag{5.61}$$

In chemically reactive systems the total number of moles may change, but the total mass is conserved. Becausee $\rho_j = M_j c_j$, multiplying Eq. (5.59) by M_j gives

$$\frac{d\rho_j}{dt} = M_j \alpha_j r - M_j c_j \frac{d \ln V}{dt} = M_j \alpha_j r - \rho_j \frac{d \ln V}{dt} \tag{5.62}$$

The rate of change of mass fraction g_j becomes

$$\frac{dg_j}{dt} = \frac{d(\rho_j/\rho)}{dt} = \frac{1}{\rho}\frac{d\rho_j}{dt} - \frac{\rho_j}{\rho^2}\frac{d\rho}{dt} = \frac{M_j \alpha_j r}{\rho} - g_j\left(\frac{d \ln V}{dt} + \frac{d \ln \rho}{dt}\right) \tag{5.63}$$

For an ideal gas mixture, $V\rho = \text{constant}$ (i.e., $V\rho = m$):

$$\frac{d \ln V}{dt} + \frac{d \ln \rho}{dt} = 0 \tag{5.64}$$

so Eq. (5.63) becomes

$$\frac{dg_j}{dt} = \frac{M_j \alpha_j r}{\rho} = M_j \alpha_j r'' \tag{5.65}$$

In this formulation, r'' is the intensive reaction rate per unit mass of the system. Because the total mass m of the reacting system is conserved,

$$r'' = \frac{R^*}{m} = \frac{d}{dt}\left(\frac{X}{m}\right) = \frac{d\xi''}{dt} = \frac{r}{\rho} \tag{5.66}$$

Calculation of concentration changes as described in this section requires the chemical reaction rate expressions in terms of composition as well as the temperature and pressure of the reacting system. These reaction rate expressions have to be developed from the interpretation of experimental data collected for the reaction rate measurements.

In chemical engineering practice, mass balances are generally expressed in terms of mole number N_j, mass m_j, and reaction extent X or their intensive variables c_j, ρ_j, and ξ. As an example, for the catalytic oxidation (a reversible and exothermic reaction) of SO_2 to SO_3 with air,

$$SO_2 + \frac{1}{2}O_2 \rightleftharpoons SO_3 \tag{5.67}$$

the material balances can be formulated on the basis of mole numbers N_j and reaction extent X. The molar composition of the reacting mixture (composed of SO_2, SO_3, O_2, and N_2) can be expressed as

$$N_{SO_2} = N_{SO_2,0} - X \tag{5.68}$$

$$N_{SO_3} = N_{SO_3,0} + X \tag{5.69}$$

$$N_{O_2} = N_{O_2,0} - 0.5X \tag{5.70}$$

$$N_{N_2} = N_{N_2,0} \tag{5.71}$$

The total mole number in the reacting mixture (assigning $j = 1$ for SO_2, $j = 2$ for SO_3, $j = 3$ for O_2, $j = 4$ for N_2, and $\alpha_1 = -1$, $\alpha_2 = 1$, $\alpha_3 = -0.5$, $\alpha_4 = 0$) is

$$N = \sum_{j=1}^{4} N_j = \sum_{j=1}^{4} N_{j,0} - 0.5X = N_0 - 0.5X \tag{5.72}$$

where N_0 is the initial mole number of the mixture. Assuming that the reacting mixture is an ideal mixture, then the partial pressure p_j can be expressed as a function of total pressure p and mole fraction y_j:

$$y_j = \frac{N_j}{N} = \frac{N_j}{N_0 - 0.5X} \tag{5.73}$$

$$p_j = y_j p = \frac{N_j}{N_0 - 0.5X} p \tag{5.74}$$

For constant volume operating conditions, molar concentrations c_j and intensive reaction extent ξ are more useful than mole numbers N_j and reaction extent X.

10. CONCENTRATION CHANGES IN CONTINUOUS MIXTURES

In mixtures of a very large number of species, individual species may not be distinguished. In such a system individual species A_j and their concentrations c_j are described by continuous functions $A(x)$ and $c(x)$ of an index x, such as the simulated boiling point. Similarly, the stoichiometric coefficient of the jth species, α_j, can be replaced by the continuous stoichiometric coefficient $\alpha(x)$ (Aris and Gavalas, 1966). For a single reaction, the stoichiometric relation described by Eq. (5.1) can be expressed as

$$\int_a^b \alpha(x)A(x) \, dx = 0 \tag{5.75}$$

and $\alpha(x)$ satisfies the integral

$$\int_a^b \alpha(x)M(x) \, dx = 0 \tag{5.76}$$

where $M(x)$ is the continuous molecular weight distribution function of index x. If more than one reaction can take place, these relations should hold for all individual reactions with corresponding $\alpha_j(x)$:

$$\int_a^b \alpha_j(x)A(x) \, dx = 0, \qquad j = 1, \ldots, R \tag{5.77}$$

$$\int_a^b \alpha_j(x)M(x) \, dx = 0, \qquad j = 1, \ldots, R \tag{5.78}$$

If the number of reactions is also very large, then the discrete stoichiometric coefficient $\alpha_j(x)$ can also be approximated to a continuous stoichiometric coefficient $\alpha(w, x)$, with continuous reaction index w. For this continuous approximation the stoichiometry of the reactions becomes

$$\int_a^b \alpha(w, x)A(x) \, dx = 0, \qquad c \leq w \leq d \tag{5.79}$$

and

$$\int_a^b \alpha(w, x)M(x) \, dx = 0, \qquad c \leq w \leq d \tag{5.80}$$

Similarly, the reaction extent ξ can also be expressed as a continuous function $\xi(x)$ in terms of the index x, and the concentration change becomes

$$\Delta c(x) = \int_c^d \alpha(w, x)\xi(x) \, dw \tag{5.81}$$

the time derivative of which (using $r = d\xi/dt$) is

$$\frac{dc(x, t)}{dt} = \int_c^d \alpha(w, x)r(T, P; c(x); w) \, dw \tag{5.82}$$

which is an integro-differential equation (nonlinear) for $c(x, t)$. For an isothermal continuous stirred tank reactor with reactor holding time τ and feed composition $c_f(x)$, Eq. (5.82) becomes

$$\tau\frac{dc(x, t)}{dt} = c_f(x) - c(x, t) + \tau\int_c^d \alpha(w, x)r(w) \, dw \tag{5.83}$$

where $r(w)$ is the continuous reaction rate with reaction index w. For steady-state operating conditions, Eq. (5.83) is reduced to an integral equation,

$$c(x, t) = c_f(x) + \tau\int_c^d \alpha(w, x)r(w) \, dw \tag{5.84}$$

The challenge is to find $r(w)$ for the best least-squares fit of experimental data. Significant progress has been made during the last three decades on the kinetics of complex mixtures. It is hoped that continuous representation of chemical composition, reaction stoichiometry, and reactions will find more applications in the chemical processing of petroleum products (Sapre and Krambeck, 1991).

REFERENCES

Amundson, N. R. 1966. Mathematical Methods in Chemical Engineering. Prentice-Hall, Englewood Cliffs, NJ.

Aris, R. 1969. Elementary Chemical Reactor Analysis. Prentice-Hall, Englewood Cliffs, NJ.

Aris, R., and Gavalas, G. R. 1966. On the theory of reactions in continuous mixtures. Roy. Soc. Lond., Phil. Trans. 260:351–393.

Frank-Kamenetskii, D. A. 1969. Diffusion and Heat Transfer in Chemical Kinetics. Plenum Press, New York.

Sapre, A. V., and Krambeck, F. J. 1991. Chemical Reactions in Complex Mixtures: The Mobil Workshop. Van Nostrand Reinhold, New York.

6
Chemical Kinetics

1. INTRODUCTION

The theory of chemical kinetics and the rate of chemical reactions have important applications in many branches of science and engineering. In chemical engineering applications, chemical kinetics, the rate of chemical reactions, and their dependence on temperature, pressure, concentration, and catalysts are needed for the design of chemical reactors.

Chemical kinetics is a highly complex subject. Stoichiometrically simple chemical reactions always take place through a very complex reaction path involving complex chemical species, some of which may not even be detected. It is known from experience that chemical composition, pressure, temperature, catalysts, and reactor hydrodynamics control the rate of a chemical reaction. Chemical reactions that take place in a homogeneous phase, such as the oxidation of CO to CO_2 by air oxygen, are known as homogeneous reactions. In gas–liquid reactions, reactions take place in a homogeneous liquid phase; but mass transfer takes place at the gas/liquid interface. Some chemical reactions such as CO_2 gasification of coal chars take place at the boundaries of the phases; these are known as heterogeneous reactions. Chemical reactions may also take place at the active sites of solid surfaces; examples are heterogeneous catalytic reactions. Interfacial mass transfer and adsorption (mostly chemisorption) of the reacting species on the surface of solids followed by chemical reactions are the common mechanism of these reactions.

In many cases diffusion of the chemical species through the interface and within the phases may play a significant role in the overall performance of a reaction. Rates of individual reaction paths can be influenced by the catalytic effects of homogeneous or heterogeneous substances or by the substances formed as transition species during the course of the reaction, which affect the overall reaction rate and selectivity. As a result of imperfect mixing and/or heat transfer, spatial fluctuations of composition and temperature could become controlling parameters for the overall rate of the chemical kinetics. As a result, the diffusion of mass and thermal energy (therefore hydrodynamics) becomes the controlling factor in chemical kinetics. It is safer to accept the fact that the information known today about chemical kinetics of the reactive systems may be the *apparent kinetics* than to assume that it is the *true kinetics*.

For chemical engineers, chemical kinetics are important for three reasons: (1) to estimate reactor operating conditions needed to carry out the reactions at a reasonable rate, (2) to select the reactor type and size and reactor operating mode, and (3) to predict the reactor dynamics and estimate the necessary utilities for the reactor (i.e. reactor stability, reactor control policy, and heating and cooling needs). In all of these, the cost (i.e., economics and feasibility) and environmental issues are the constraints to be satisfied; some of these are not universal and vary geographically. For these purposes, mathematical models to predict the chemical kinetics as functions of process variables would be useful. These models may not be mechanistically correct but could be practically useful by providing predictive power for the behavior of chemically reactive systems within an acceptable tolerance. Considering the complexity of the subject, in some cases even models with limited predictability are useful tools.

2. CHEMICAL KINETICS

The study of chemical kinetics concerns the evolution of chemical reactions over time or, more specifically, the interpretation of compositional changes and the measurement of chemical kinetics in chemically reacting systems. The reaction path of a chemical reaction described by a simple stoichiometric relation may be very complex, but its kinetics may be represented by relatively simple mathematical expressions.

As stated earlier, the true mechanisms of chemical reactions are complex. A reaction expressed by a simple stoichiometry may take place through a complex reaction path that involves several elemental steps. An elemental step is defined as a single irreducible step that takes place at the molecular level. Molecular adsorption, the formation and decomposition of intermediate species, and the desorption of the decomposition products are examples of elementary steps. Models or mechanisms are postulated for the path of isolated elementary steps for the course of a reaction. These models are useful tools to predict the performance of the reacting systems, not necessarily for the explanation of the true reaction path and the mechanism involved.

As an example, the stoichiometry of the hydrogenation of bromine is expressed as

$$H_2 + Br_2 \rightarrow 2HBr \tag{6.1}$$

However, it is believed that the reaction proceeds in a sequence of steps:

$$Br + H_2 \rightarrow HBr + H \tag{6.2}$$

$$H + Br_2 \rightarrow HBr + Br \tag{6.3}$$

which involves bromine atoms (Br) and hydrogen atoms (H) as intermediates.

It is commonly accepted that reactions progress through several intermediates that are formed, reach certain concentrations, and ultimately vanish. Intermediates can have lifetimes comparable to those of the initial reactants (e.g., formaldehyde, CH_2O, formed during methane oxidation) and exist in the reacting system at appreciable concentrations. Some intermediates appear in small concentrations and have short lifetimes in comparison to those of the initial reactants (e.g., hydrogen atoms and bromine atoms in the reaction of hydrogen with bromine). These intermediates play an active role in the overall reaction rate and are called active centers. Some intermediates, however, which are called transition state intermediates, cannot be isolated.

Elementary steps and intermediates are also an important part of gas–solid reactions. The stoichiometry of carbon dioxide (CO_2) gasification of carbon is expressed as

$$CO_2 + C \xrightarrow{k} 2CO \tag{6.4}$$

However, the reaction may proceed in two different sequences of steps:

$$C_s + CO_2 \xrightarrow{k_1} C(O) + CO \tag{6.5}$$

$$C(O) \xrightarrow{k_2} CO \tag{6.6}$$

$$CO + C_s \underset{k_3'}{\overset{k_3}{\rightleftharpoons}} C(CO) \tag{6.7}$$

or

$$C_s + CO_2 \underset{k_1'}{\overset{k_1}{\rightleftharpoons}} C(O) + CO \tag{6.8}$$

$$C(O) \xrightarrow{k_2} CO \tag{6.9}$$

which would predict different overall gasification rate expressions. In these suggested mechanisms C_s is the active surface concentration of carbon, C(O) and C(CO) are the concentrations of adsorbed oxygen and carbon monoxide on the carbon surface. Chemisorption of carbon dioxide on active carbon sites followed by the irreversible decomposition of adsorbed carbon dioxide to carbon monoxide is believed to be the most realistic mechanism for the carbon dioxide gasification of carbon.

3.   THERMODYNAMICS AND REACTION KINETICS

Thermodynamic free energy calculations can provide information if a chemically reacting system is in equilibrium (i.e., chemical reactions proceed in the path of negative Gibbs energy G, i.e., $\Delta G < 0$, and reach chemical equilibrium at $\Delta G = 0$). Thermodynamic free energy calculations, however, cannot provide information about the chemical reaction rate at which the chemically reactive system approaches chemical equilibrium.

Prediction of chemical reaction rates from molecular properties of the reactants and the operating conditions has been studied extensively. So far only transition state theory (or absolute rate theory) has been able to provide an acceptable semiempirical model for the kinetics of elementary steps. Even transition state theory has parameters, the pre-exponential factor and the activation energy of the rate constant, which have to be determined experimentally (Glasstone et al., 1941).

Transition state theory suggests that intermediate transition states are formed at concentrations in equilibrium with reactants, then these transient states are decomposed irreversibly into the product (Fig. 6.1). Transition state theory postulates an elementary reaction expressed with a simple stoichiometry:

$$A + B \rightarrow P \tag{6.10}$$

which goes through two consecutive elementary steps—the formation of a transition state species Z^* in equilibrium with the reactants A and B and the irreversible decomposition of the transition state species Z^* to the product P:

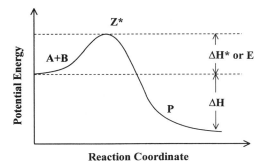

Figure 6.1 Potential energy profile suggested by the transition state theory for a simple reaction $A + B \rightarrow P$.

$$A + B \leftrightarrow Z^* \rightarrow P \tag{6.11}$$

The energy state of the transition species Z^* is much higher than the energy state of the reactants A and B. The energy difference between the transition species Z^* and the reactants A and B is the energy barrier for the reaction; it is called the activation energy and denoted by ΔH^* (or E). The energy difference between the reactants (A and B) and the product (P) is the reaction enthalpy and denoted by ΔH. If the reaction is reversible and takes place in the same reaction path, then the activation energy for the reaction $P \rightarrow A + B$ would be in the magnitude of $\Delta H^* + \Delta H$ (or $E + \Delta H$).

The concentration of the postulated transition state c_{Z^*} can be calculated from the definition of an equilibrium constant:

$$K_{Z^*} = \frac{c_{Z^*}}{c_A c_B} \tag{6.12}$$

which can be calculated using the thermodynamic relation

$$RT \ln K_{Z^*} = RT \ln \frac{c_{Z^*}}{c_A c_B} = -\Delta G^{0*} = -\Delta H^{0*} + T \, \Delta S^{0*} \tag{6.13}$$

where R is the gas constant, ΔG^{0*} is the change in the standard Gibbs free energy, and ΔH^{0*} and ΔS^{0*} are the standard enthalpy and entropy, respectively, associated with the formation of the transition state species Z^* from A and B. Using Eq. (6.13), the concentration of the transition state c_{Z^*} becomes

$$c_{Z^*} = c_A c_B \exp\left(\frac{-\Delta H^{0*}}{RT} + \frac{\Delta S^{0*}}{R}\right) = c_A c_B \exp\left(\frac{\Delta S^{0*}}{R}\right) \exp\left(\frac{-\Delta H^{0*}}{RT}\right) \tag{6.14}$$

Assuming a first-order decomposition rate for the decomposition of the transition state species Z^* to product P, with a proportionality or frequency constant v,

$$r = v c_{Z^*} = v \exp\left(\frac{\Delta S^{0*}}{R}\right) \exp\left(\frac{-\Delta H^{0*}}{RT}\right) c_A c_B = k c_A c_B \tag{6.15}$$

where k is the reaction rate constant,

$$k = v \exp\left(\frac{\Delta S^{0*}}{R}\right) \exp\left(\frac{-\Delta H^{0*}}{RT}\right) = A_p \exp\left(\frac{-\Delta H^{0*}}{RT}\right) \tag{6.16}$$

Equation (6.16) is known as the Arrhenius relation and expresses the reaction rate as a nonlinear function of temperature.

Dependence of the reaction rate constant k on temperature T in the Arrhenius form [Eq. (6.16)] is responsible for many nonlinear phenomena such as reactor stability, ignition, spontaneous combustion, and explosion. For a simple reaction, determination of the activation energy requires the measurement of reaction kinetics as a function of temperature. Activation energy can be determined from the best least-squares fit of experimentally determined k and reaction temperature T (i.e., the slope of the linear correlation of $\ln k$ versus $1/T$ would be $\Delta H^{0*}/2.303R$) to Eq. (6.16).

Chemical reactions always require high activation energies, on the order of 160–320 kJ/mol (i.e., 40–80 kcal/mol). As the reaction temperature increases, the reaction rate increases exponentially and diffusion limitations may dominate the observed overall reaction kinetics. Low activation energy for a chemical reaction, therefore, is a sign that diffusion limitations are affecting the overall rate of the reaction. Diffusion limitations for a given reaction are dependent on the size, geometry, and hydrodynamics of the reacting system. Coal combustion is an example of the effect of diffusion on a chemical reaction; an increase in combustion temperature shifts the reaction from a chemically controlled regime to mass transfer controlled in the coal particle. A further increase in temperature shifts the reaction into the boundary layer (surrounding the particle) diffusion (mass transfer) control regime. In coal combustion, that is, measurement of the activation energy of the overall combustion reaction provides information about the controlling mechanism of the combustion process.

In the reaction rate expression of Eq. (6.16), A_p is the pre-exponential factor (in s^{-1}):

$$A_p = \nu \exp\left(\frac{\Delta S^{0*}}{R}\right) \tag{6.17}$$

and ν is a universal collision frequency (it does not depend on the nature of the molecular system). Using the kinetic theory of gases, ν is calculated as

$$\nu = \frac{k_B T}{h} \tag{6.18}$$

where k_B and h are the Boltzmann constant (1.38×10^{-23} J/K) and the Planck constant (6.62×10^{-34} J·s) (Davidson, 1962). As can be seen from Eqs. (6.16) and (6.17), the pre-exponential factor A_p is a weak (i.e., linear) function of temperature.

The absolute rate theory can interpret the measured kinetics of the elementary reactions (at least in the range of limited operating conditions, in terms of reaction temperature, particle size, etc.). This theory suggests that both the energy barrier (ΔH^{0*}) and the entropy barrier (ΔS^{0*}) determine whether a collision between the reactant molecules (or between reactant molecules and solid surfaces) would result in a chemical reaction. In general, large activation energies cause slow reaction rates. If ΔH^{0*} is small, then ΔS^{0*} may have a large negative value (i.e., a decrease in the number of degrees of freedom of molecular orientations due to formation of the transition state), which may result in low reaction rates. Conversely, if ΔH^{0*} is large, then ΔS^{0*} may have a large positive value (i.e., an increase in the orientation by the formation of the transition state), which may result in high reaction rates.

The kinetic theory of gases predicts that for two hard-sphere molecules A and B with molecular diameters σ_A and σ_A and masses m_A and m_B, the rate of collision, r_{coll}, between molecules A and B is given by

$$r_{coll} = \pi\sigma^2 v_\mu c_A c_B \tag{6.19}$$

where σ is the mean molecular diameter,

$$\sigma = \frac{\sigma_A + \sigma_B}{2} \tag{6.20}$$

v_μ is the mean molecular velocity,

$$v_\mu = \left(\frac{8k_B T}{\pi\mu}\right)^{1/2} \tag{6.21}$$

and μ is the reduced mass,

$$\mu = \frac{m_A m_B}{m_A + m_B} \tag{6.22}$$

The kinetic theory of gases also predicts that for a hard-sphere molecule A, the rate of collision between molecule A and a unit surface area of a solid is given by

$$r_{coll,s} = \frac{1}{4} v_A c_A \tag{6.23}$$

where, v_A is the mean molecular velocity,

$$v_A = \left(\frac{8k_B T}{\pi m_A}\right)^{1/2} \tag{6.24}$$

The orders of magnitude of mean velocity, collision, and surface collision frequencies can thus be calculated (Table 6.1).

Using collision theory, the kinetics of an elementary reaction $A + B \rightarrow P$ and an elementary reaction between the molecules of A and a solid surface would be expressed as

$$r = \pi\sigma^2 v_\mu \exp\left(\frac{-E}{RT}\right) c_A c_B \tag{6.25}$$

and

$$r_S = \frac{1}{4} v_A \exp\left(\frac{-E}{RT}\right) c_A \tag{6.26}$$

Table 6.1 Order of Magnitude of Variables Used in Chemical Kinetics

Quantity	Functional	Order of magnitude	Units
Mean velocity	$v = (8kT/\pi m)^{1/2}$	5×10^4	cm/s
Universal frequency	kT/h	10^{13}	s^{-1}
Gas–surface collision frequency	$v/4$	10^4	cm/s
Gas–gas collision frequency	$\pi\sigma^2 v$	10^{-10}	cm^3/s

Source: Boudart, 1968.

respectively, by considering only the energy barrier E (i.e. the Boltzmann factor), which corresponds to the activation energy ΔH^{0*} of absolute rate theory.

The rate of an elementary step that takes place in the gas phase can be explained by collision theory. By introducing the probability factor f_P taking care of the fact that the molecules are not hard spheres and that each collision may not result in a chemical reaction, Eqs. (6.25) and (6.26) become

$$r = f_P \pi \sigma^2 v_\mu \exp\left(\frac{-E}{RT}\right) c_A c_B \tag{6.27}$$

$$r_S = f_P \frac{1}{4} v_A \exp\left(\frac{-E}{RT}\right) c_A \tag{6.28}$$

These two rate expressions are comparable to the rate expressions developed with absolute rate theory, suggesting that the probability factor f_P is related to the entropy of formation of the transition state species.

4. REACTION RATES AND CHEMICAL EQUILIBRIUM

The kinetics of an elementary step close to chemical equilibrium can be related to the thermodynamics of the system. A chemical reaction of an elementary step that is taking place in an ideal gas mixture can be expressed as

$$A + B \rightleftharpoons C + D \tag{6.29}$$

where, the forward and backward elementary reaction rates are

$$r_f = k_f c_A c_B \tag{6.30}$$

and

$$r_b = k_b c_C c_D \tag{6.31}$$

At equilibrium the forward and backward elementary reaction rates are equal, i.e., $r_f = r_b$. At equilibrium, a dynamic equilibrium is established between the reactants and the products:

$$k_f c_{A,e} c_{B,e} = k_b c_{C,e} c_{D,e} \tag{6.32}$$

from which the thermodynamic equilibrium constant K_e is calculated as

$$K_e = \frac{k_f}{k_b} = \frac{c_{C,e} c_{D,e}}{c_{A,e} c_{B,e}} \tag{6.33}$$

Close to chemical equilibrium, the rate of reaction approaching chemical equilibrium is given by

$$r = k_f c_A c_B - k_b c_C c_D = r_f \left[1 - \frac{1}{K_e}\left(\frac{c_C c_D}{c_A c_B}\right)\right] = r_b \left(K_e \frac{c_A c_B}{c_C c_D} - 1\right) \tag{6.34}$$

From thermodynamics the Gibbs energy of the mixture is

$$\Delta G = -RT \ln K_e - RT \ln\left(\frac{c_A c_B}{c_C c_D}\right) = -RT \ln\left(K_e \frac{c_A c_B}{c_C c_D}\right) \tag{6.35}$$

from which

$$K_e \frac{c_A c_B}{c_C c_D} = \exp\left(-\frac{\Delta G}{RT}\right) \tag{6.36}$$

and

$$\frac{1}{K_e}\left(\frac{c_C c_D}{c_A c_B}\right) = \exp\left(\frac{\Delta G}{RT}\right) \tag{6.37}$$

Using Eqs. (6.34) and (6.37), the rate expression close to equilibrium becomes

$$r = r_f\left[1 - \exp\left(\frac{\Delta G}{RT}\right)\right] = r_b\left[\exp\left(\frac{-\Delta G}{RT}\right) - 1\right] \tag{6.38}$$

Close to equilibrium, $|\Delta G| \ll RT$ and $\exp(\Delta G/RT)$ and $\exp(-\Delta G/RT)$ around $\Delta G = 0$ can be approximated as

$$\exp\left(\frac{\Delta G}{RT}\right) \approx 1 + \frac{\Delta G}{RT} \qquad \text{and} \qquad \exp\left(\frac{-\Delta G}{RT}\right) = 1 - \frac{\Delta G}{RT} \tag{6.39}$$

and reaction rates expressed by Eq. (6.38) become

$$r = r_f \frac{-\Delta G}{RT} = r_b \frac{-\Delta G}{RT} = r_e \frac{-\Delta G}{RT} \tag{6.40}$$

As can be seen from Eq. (6.40), close to equilibrium, the forward (r_f) and backward (r_b) reaction rates are equal. Also, close to equilibrium the reaction rate is linearly proportional to the driving force $-\Delta G/T$, which is called the chemical affinity. The reaction rate becomes zero when the chemical affinity vanishes (i.e., $\Delta G = 0$ at equilibrium).

This relation can be expressed more explicitly for a chemical reaction in a constant-volume ideal gas mixture,

$$\sum_j \alpha_j A_j = 0 \tag{6.41}$$

for which ($\alpha_j > 0$ for products and $\alpha_j < 0$ for reactants) the Gibbs energy is

$$\Delta G = -RT \ln\left(\frac{c_{1,e}^{\alpha_1} \cdots c_{s,e}^{\alpha_s}}{c_1^{\alpha_1} \cdots c_s^{\alpha_s}}\right) = -RT \ln\left(\frac{\Pi(c_{j,e})^{\alpha_j}}{\Pi(c_j)^{\alpha_j}}\right) \tag{6.42}$$

or

$$\Delta G = RT \ln\left[\Pi\left(\frac{c_j}{c_{j,e}}\right)^{\alpha_j}\right] = RT \sum_j \ln\left(\frac{c_j}{c_{j,e}}\right)^{\alpha_j} = RT \sum_j \ln\left(1 + \frac{c_j - c_{j,e}}{c_{j,e}}\right)^{\alpha_j} \tag{6.43}$$

Sufficiently close to equilibrium, $|(c_j - c_{j,e})/c_{j,e}| \ll 1$, and

$$\ln\left(1 + \frac{c_j - c_{j,e}}{c_{j,e}}\right) \approx \frac{c_j - c_{j,e}}{c_{j,e}} \tag{6.44}$$

and the expression for ΔG becomes

$$\Delta G = RT \sum_j \alpha_j \ln\left(1 + \frac{c_j - c_{j,e}}{c_{j,e}}\right) \approx RT \sum_j \alpha_j \frac{c_j - c_{j,e}}{c_{j,e}} \tag{6.45}$$

Using the molar balance relations,

$$c_j = c_{j,0} + \alpha_j \xi \tag{6.46}$$

$$c_{j,e} = c_{j,0} + \alpha_j \xi_e \tag{6.47}$$

from which, $c_j - c_{j,e}$ can be expressed in terms of intensive reaction extents as

$$c_j - c_{j,e} = \alpha_j(\xi - \xi_e) \tag{6.48}$$

and ΔG expressed by Eq. (6.45) becomes

$$\Delta G = RT(\xi - \xi_e) \sum_j \frac{\alpha_j^2}{c_{j,e}} \tag{6.49}$$

Using Eq. (6.49) in Eq. (6.40), the intensive rate close to equilibrium becomes

$$r = \frac{d\xi}{dt} = r_e \frac{-\Delta G}{RT} = r_e(\xi_e - \xi) \sum_j \frac{\alpha_j^2}{c_{j,e}} \tag{6.50}$$

or

$$r = \frac{d\xi}{dt} = k(\xi_e - \xi) \tag{6.51}$$

where

$$k = r_e \sum_j \frac{\alpha_j^2}{c_{j,e}} \tag{6.52}$$

In Eq. (6.52) k is the first-order reaction rate constant. This relation shows that sufficiently close to chemical equilibrium the forward and backward reaction rates are equal, $r_f = r_b = r$, and $r \, (= d\xi/dt)$ is linearly proportional (or first order) to the difference in the reaction extent from its equilibrium value $(\xi_e - \xi)$. This relation is known as the macroscopic reversibility principle (Boudart, 1968).

The macroscopic reversibility principle tells us that for a reversible reaction close to equilibrium, if the reaction rate in one direction is increased, then the reaction rate in the opposite direction is also increased. This principle has a valuable application in chemical reaction engineering. For example, if a catalyst (such as Ni, Co, Mo) is increasing the rate of hydrogenation reactions, the same catalyst should also increase the rate of dehydrogenation reactions. Therefore, if the conditions are not suitable for hydrogenation, a catalyst used to accelerate hydrogenation reactions, such as in the hydrogenation of heavy oils, may also accelerate dehydrogenation reactions, which could result in coke formation and severe deactivation of the catalyst. If the hydrogenation reactions are desired, operating conditions have to be selected such that the rate of those reactions will be dominant. These conditions can be achieved by selecting the most favorable hydrogen partial pressure and reactor operating temperature.

5. STEADY-STATE APPROXIMATION AND CATALYSIS

Chemical reactions take place through a sequence of elementary steps involving the formation and decomposition of active intermediates or active centers (free radicals, free ions, complexes at surfaces, etc.). If these active centers are reproduced by the reaction sequence it is called a closed sequence (catalysis), and if the active centers are not reproduced by the reaction sequence it is called an open sequence. Mechanisms of the chain reactions are similar to those of catalytic reactions. Both reaction mechanisms can be explained by the steady-state approximation.

The following first-order consecutive elementary reactions will be considered for the steady-state approximation (Boudart, 1968):

$$A \xrightarrow{k_1} B \xrightarrow{k_2} C \qquad (6.53)$$

where at time $t = 0$ the normalized concentrations are $c_A = 1$, $c_B = c_C = 0$. If these reactions are taking place in a batch reactor, normalized molar balances can be expressed as

$$c_A + c_B + c_C = 1 \quad \text{or} \quad \frac{dc_A}{dt} + \frac{dc_B}{dt} + \frac{dc_C}{dt} = 0 \qquad (6.54)$$

and

$$\frac{dc_A}{dt} = -k_1 c_A, \qquad \frac{dc_B}{dt} = k_1 c_A - k_2 c_B, \qquad \frac{dc_C}{dt} = k_3 c_B \qquad (6.55)$$

Solution of these coupled equations with the above initial conditions ($c_A = 1$, $c_B = c_C = 0$) gives

$$c_A = \exp(-k_1 t) \qquad (6.56)$$

$$c_B = \frac{k_1}{k_2 - k_1}[\exp(-k_1 t) - \exp(-k_2 t)] \qquad (6.57)$$

$$c_C = 1 - \frac{k_2}{k_2 - k_1}\exp(-k_1 t) + \frac{k_1}{k_2 - k_1}\exp(-k_2 t) \qquad (6.58)$$

These equations show that c_A monotonically decreases, c_B possesses a maximum value $c_{B,max}$ at t_{max} (i.e., when $dc_B/dt = 0$), and c_C monotonically increases (Fig. 6.2). The values of $c_{B,max}$ and t_{max} are

$$t_{max} = \frac{1}{k_2 - k_1}\ln\left(\frac{k_2}{k_1}\right), \qquad c_{B,max} = \left(\frac{k_1}{k_2}\right)^{k_2/(k_2 - k_1)} \qquad (6.59)$$

Also, c_C possesses an inflection point, i.e., $d^2 c_C/dt^2 = 0$ at $t = t_{max}$.

This reaction system has a special meaning when $k_2 \gg k_1$, i.e., $k_1/k_2 \rightarrow 0$. When this condition is satisfied, $t_{max} \rightarrow 0$, $c_{B\,max} \rightarrow 0$, and the solution of the normalized molar balances Eqs. (6.56), (6.57), and (6.58), are approximated to

$$c_A = \exp(-k_1 t), \qquad c_B = \frac{k_1}{k_2}c_A, \qquad c_C = 1 - \exp(-k_1 t) \qquad (6.60)$$

which are the solutions of the following equations (with the same initial conditions):

$$\frac{dc_A}{dt} = -k_1 c_A, \qquad k_1 c_A - k_2 c_B = 0, \qquad \frac{dc_C}{dt} = k_1 \exp(-k_1 t) = k_1 c_A = k_2 c_B \qquad (6.61)$$

This case is known as the steady-state approximation (i.e., $dc_B/dt = 0$). Because $k_2 \gg k_1$, the species B can be considered a highly reactive active intermediate, with a very small concentration in comparison to stable species such as A and C.

In fact, when $k_2 \gg k_1$ or $k_1/k_2 \rightarrow 0$, it can be seen from Eq. (6.57) that $k_1 c_A - k_2 c_B \rightarrow 0$, implying that

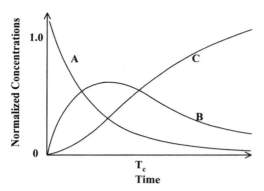

Figure 6.2 Concentration profiles as a function of time for first-order consecutive reactions A → B → C.

$$\frac{dc_B}{dt} = 0 \tag{6.62}$$

which is the expression for the steady-state approximation; that is, the time derivative of the active intermediate concentration (c_B) is equal (or approximated) to zero. To calculate the active intermediate concentration, however, Eq. (6.62) should not be integrated. The active intermediate concentration has to be calculated using the algebraic equation [Eq. (6.61)], which shows that c_B is dependent on c_A and decays in time as c_A does. In the decay of c_B, however, its time derivative dc_B/dt would always be close to zero.

Using the steady-state approximation $dc_B/dt \cong 0$, the differential concentration balance becomes

$$-\frac{dc_A}{dt} = \frac{dc_C}{dt} \tag{6.63}$$

Using the expression for c_B in Eq. (6.60), differential relations for c_A and c_C become

$$-\frac{dc_A}{dt} = \frac{dc_C}{dt} = k_2 c_B = k_1 c_A = k_1 \exp(-k_1 t) \tag{6.64}$$

which explains the relation between the rates of change of concentrations c_A and c_C.

A simple reaction A → C may consist of a sequence of reactions involving active centers (intermediates) $B_1, B_2, \ldots, B_k$. The steady-state approximation expressed by Eq. (6.62) is also applicable for this case, which means that in a sequence of reactions involving active intermediates $B_1, B_2, \ldots, B_k$ the rates of the reactions in the sequence are equal. The overall picture is that by the steady-state approximation, concentration changes during the reaction A → C can be expressed by Eq. (6.64).

The results obtained from the steady-state approximation are valid only after a relaxation time t_τ. In general t_τ is very small compared to the total reaction time; however, it must be checked. Relaxation time would be a function of the ratio k_1/k_2. For this purpose, a small deviation in c_B from its steady-state value $c_{B,s}$ is expressed by ε:

$$\varepsilon = \frac{c_B - c_{B,s}}{c_{B,s}} \quad \text{or} \quad c_B = c_{B,s}(1 + \varepsilon) \tag{6.65}$$

using which, the change in ε as a function of time can be formulated. Using Eq. (6.65) the change in c_B as a function of time can be expressed as

$$\frac{dc_B}{dt} = c_{B,s}\frac{d\varepsilon}{dt} + (1 + \varepsilon)\frac{dc_{B,s}}{dt} \tag{6.66}$$

Steady-state approximation predicts that

$$c_{B,s} = \frac{k_1}{k_2}c_A \quad \text{and} \quad \frac{dc_{B,s}}{dt} = \frac{k_1}{k_2}\frac{dc_A}{dt} = -\frac{k_1^2}{k_2}c_A \tag{6.67}$$

Using Eqs. (6.55), (6.65), and (6.67), the differential relation given by Eq. (6.66) becomes

$$\frac{d\varepsilon}{dt} + (k_2 - k_1)\varepsilon - k_1 = 0 \tag{6.68}$$

Integration of Eq. (6.68) with the initial condition $\varepsilon = -1$ (i.e., active intermediate concentration $c_B = 0$) at $t = 0$ gives

$$\varepsilon = \frac{1}{1 - K}[K - \exp(K - 1)\tau] \tag{6.69}$$

where $K = k_1/k_2$ and $\tau = k_2 t$. Since $k_2 \gg k_1$, $K \ll 1$, for sufficiently large times Eq. (6.69) can be approximated to

$$\varepsilon \approx \frac{K}{1 - K} \approx K \tag{6.70}$$

This relation tells us that at sufficiently large times the relative deviation from the steady-state concentration would be on the order of K, the ratio of the rate constants, k_1/k_2. In fact, for $K \ll 1$, at sufficiently large times, Eq. (6.69) can be approximated to

$$\varepsilon \approx K \approx -\exp(-\tau) \tag{6.71}$$

which is a decay relation; i.e., at $\tau = 1$, ε would decay to the value $\varepsilon = 1/e$. This time is called the relaxation time t_r, i.e., $t_r = 1/k_2$. The relaxation time for the reactive intermediate is equal to the inverse of its first-order decomposition rate constant or its average lifetime (Boudart, 1968).

6. CONCENTRATION CHANGE MEASUREMENTS AND REACTION RATES

Regardless of the nature of chemical reactions and their basic mechanisms, information on reaction kinetics is needed for the design of chemical reactors. Mathematical expressions are needed to express rates of reaction as functions of species concentrations, temperature, pressure, catalyst characteristics, etc. This requires experimental data collection at controlled experimental conditions and the testing of different models for the fitting of experimental data. Laboratory-scale reactors are used for experimental data collection, which provides very well controlled operating conditions at affordable cost.

In most chemically reactive systems the material being studied may have a complex chemical composition involving complex parallel and consecutive reactions. Mass and energy transfer limitations would always be coupled with reaction kinetics. The existence of catalysts would make the reacting system even more complicated to analyze. Measurement of chemical reaction rates in such complex systems is not an easy task and will continue to challenge chemical engineers in the decades to come.

From the engineering point of view, the through mechanisms of the reactions may not be needed for the design of commercial reactors. Therefore, an apparent reaction rate expression may be satisfactory enough to predict the performance of the reactor. Chemical engineers propose models (based on evidence, skills, and experience) for chemically reactive systems. These models are formulated with certain approximations. In most cases, these *certain approximations* are nothing more than the linearization of a totally nonlinear reacting system, which may result in models that predict the performance of reactors within a very narrow range of operating conditions.

Basically four types of reactors are used for the measurement of chemical kinetics, batch reactors (BRs), semi-batch reactors (SBRs), continuous flow stirred tank reactors (CSTRs), and plug flow tubular reactors (PFTRs) (Fig. 6.3). These reactors will be discussed in later chapters, but it deserves mention in this section that batch reactors and semi-batch reactors are differential reactors and tubular reactors are integral-type reactors. In batch reactors and semi-batch reactors, information obtained for compositional changes is of the differential type. In tubular reactors, the information obtained for the change of composition is of the integral type; i.e., information obtained for the change of composition consists of some kind of integral transformation of the input composition of the reactive system. Reaction rate measurements from these different types of reactors require different data manipulations as well as experimental techniques. It is recommended that numerical data always be used for the integration calculations, because the numerical error made in differential calculations is always larger.

In continuous flow stirred tank reactors, however, chemical reaction rate constants have to be determined from a set of algebraic equations (linear or nonlinear, depending on the operating conditions and the rate expressions or models). The main problem with

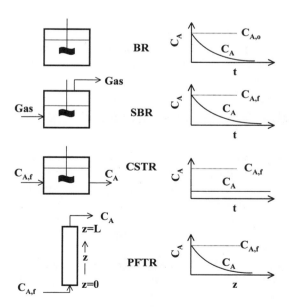

Figure 6.3 Schematics of batch (BR), semi-batch (SBR), continuous flow stirred tank (CSTR), and plug flow tubular (PFTR) reactors with concentration profiles (c_A) for a simple reaction A → P as a function of time and axial direction z ($c_{A,f}$ is the feed or initial concentration).

continuous flow stirred tank reactors, as well as with the other types of reactors, concerns the information about the mixing of the reactive species in the reactor.

In catalytic (heterogeneous) reactions and two-phase (such as gas–solid, gas–liquid) reactions, the rate of chemical reaction becomes even more difficult to measure. Measurement of operating conditions always requires sophisticated instrumentation. In these systems, mass and heat transfer are always coupled with the reaction kinetics; therefore, most of the time measured reaction kinetics is the apparent kinetics rather than the absolute kinetics.

Batch reactors or continuous flow stirred tank reactors are generally used for reaction kinetics measurements. These reactors can be operated under controlled operating conditions (temperature, pressure, velocity, hold-up, etc.), and compositions of the chemical species can be measured as a function of time. From these measurements it takes some effort to extract information about the rates of the chemical reactions.

7. REACTION SEVERITY

Time–temperature history of a chemical reaction is expressed as the reaction severity. The nonlinear Arrhenius relation between the rate constant and temperature makes the time–temperature history a significant parameter in reaction engineering. The definition of an average reaction temperature is also related to reaction severity.

Suppose that a simple reaction is taking place in a reactor of volume V in which the reactants are of almost uniform composition and spatial volumes V_z at temperatures T_z. Because of the Arrhenius relation for the dependence of kinetic rate constant on temperature, the performance of each spatial reactor element, from the reaction kinetics point of view, is different. For first-order reaction kinetics, the overall performance of the reactor can be expressed in terms of an average reaction temperature T_r:

$$VAe^{-E/RT_r}c = \sum_z V_z Ae^{-E/RT_z}c \tag{6.72}$$

This relation can be used (by canceling out c from both sides) to express the average reaction temperature T_r:

$$Ve^{-E/RT_r} = \sum_z V_z e^{-E/RT_z} \tag{6.73}$$

or

$$T_r = -\frac{E}{R\ln\left(\sum_z \frac{V_z}{V} e^{-E/RT_z}\right)} \tag{6.74}$$

Instead of average reactor temperature, an index of reaction severity (RS) can also be defined. For example, for a first-order reaction taking place in a reactor with fluctuating time–temperature history (a similar formulation can be made for a different spatial temperature in the reactor), the RS is defined as (assuming Arrhenius dependence on temperature)

$$RS = \int_0^\tau k(T(t))\,dt = \int_0^\tau Ae^{-E/RT(t)}\,dt \tag{6.75}$$

The RS described by Eq. (6.75) can be converted into a reaction severity index (RSI) by defining a standard time–temperature unit (STTU) for the reaction. The standard time–temperature unit is defined by setting temperature and activation energy in Eq. (6.75) to a set of reference values T_{ref} and E_{ref} and by setting the duration of the reaction, τ, to one unit of time:

$$\text{STTU} = \int_0^1 k(E_{ref}, T_{ref}) \, dt = \int_0^1 A e^{-E_{ref}/RT_{ref}} \, dt \tag{6.76}$$

Now that the reaction severity and the standard time–temperature unit are defined, the reaction severity index (RSI) can be defined as

$$\text{RSI} = \frac{\text{RS}}{\text{STTU}} = \frac{\int_0^\tau A e^{-E/RT(t)} \, dt}{\int_0^1 A e^{-E_{ref}/RT_{ref}} \, dt} \tag{6.77}$$

or

$$\text{RSI} = \int_0^\tau \exp\left[-\frac{1}{R}\left(\frac{E}{T(t)} - \frac{E_{ref}}{T_{ref}} \right) \right] dt \tag{6.78}$$

It is important to see that the unit of the rate constant k is $1/t$ (i.e., $1/s$ or s^{-1}) and that the reaction severity, the standard time–temperature unit, and the reaction severity index are dimensionless. The reaction severity index (RSI) can be considered a dimensionless measure of reaction severity. RSI can be used to compare the performances of two different reactors if the time–temperature histories of these reactors are known. In the evaluation of complex reactions such as hydrocracking, pyrolysis, and coking, however, interpretation of the reaction severity and the reaction severity index may be difficult. It must be remembered that the product yield distributions involving complex reactions are always *path-dependent*. Therefore, kinetic analyses of complex reactions provide a better prediction capability for the product yield distribution than similar analyses of the reaction severity or the reaction severity index measurement.

8.   PROPAGATION, BRANCHING, AND AUTOCATALYSIS

Chemical reactions always take place through a complex reaction path. In this reaction path, reactant molecules form active transient chemical species (Frank-Kamenetskii, 1969). In the propagation of a simple reaction of conversion of chemical species A to a product species P,

$$A \rightarrow P \tag{6.79}$$

the formation of an active species A^* from the reactive species A may play an important role. These active species take part in chain reactions in which the active species are regenerated or terminated. In a complex reacting system, the change in the concentration of one of these active transient species A^* can be expressed in terms of its formation (constant k_0, i.e., $k_0 = kc_A$, which decays as c_A), branching (rate constant k_b) and termination (rate constant k_t) rates:

$$\frac{dc_{A^*}}{dt} = k_0 + k_b c_{A^*} - k_t c_{A^*} \tag{6.80}$$

In a branching reaction, one of the active species can produce more than one active species, whereas in termination reactions active species react to produce chemically stable products. The overall propagation rate r_p (for a first-order kinetics) for the reaction expressed by Eq. (6.79) can be expressed as

$$r_p = k_p c_{A^*} \tag{6.81}$$

For a steady-state concentration of A^*, the propagation rate would not be affected, because an active species produces another active species and its effect on the change of active species concentration would be invisible (as discussed in Section 6.5).

It is important that Eq. (6.80) is a nonlinear equation (because of the Arrhenius form of the rate constants) and may have more than one steady-state solution satisfying the condition

$$\frac{dc_{A^*}}{dt} = 0 \tag{6.82}$$

which corresponds to the steady-state solution of Eq. (6.80):

$$c_{A_s^*} = \frac{k_0}{k_t - k_b} \tag{6.83}$$

Using the steady-state solution for c_{A^*}, the rate of propagation reaction [Eq. (6.81)] becomes

$$r_p = k_p c_{A_s^*} = k_p \frac{k_0}{k_t - k_b} \tag{6.84}$$

If $k_b > k_t$, then the time-dependent concentration of the active species A^* can be obtained from the solution of Eq. (6.80) with the initial condition $c_{A^*} = 0$ at $t = 0$:

$$c_{A^*} = \frac{k_0}{k_b - k_t}(e^{(k_b - k_t)t} - 1) \tag{6.85}$$

which is nonstationary, resulting in an exponential increase in the overall reaction rate with respect to time t.

Because Eq. (6.80) is nonlinear (Arrhenius-type rate dependence on temperature), its steady-state behavior cannot be predicted by a universal law. The stability of chemically reactive systems such as explosion, spontaneous combustion (i.e., $k_p > k_t$), and quenching (i.e., $k_p < k_t$) can be explained simply by the above-described reaction mechanisms. In fact, the stability of chemically reactive systems is much more complex than the foregoing discussion may indicate.

9. ARRHENIUS RELATION AND ITS IMPORTANCE

The nonlinear dependence of the Arrhenius relation for rate constants on temperature can explain many nonlinear and nonstationary phenomena involving chemical reactions. The Arrhenius relation shows that chemical reaction rates decay exponentially as temperature decreases but never vanish. Consequently, a reacting system may have a low initial temperature but react at a significantly high reaction rate after a sufficiently long time interval or because of a minor reaction that slightly increases the temperature of the system (i.e., spontaneous combustion and ignition).

The initial reaction rate can be assumed to be zero as a result of a sufficiently low temperature, in which case the system may be stationary but unstable. A small temperature disturbance in an initially stable system may cause chemical instability. This physical phenomenon has entered into many engineering problems such as reaction stability, spontaneous combustion, ignition, and explosion.

To describe many nonlinear phenomena that occur as a result of the Arrhenius relation, the exponent in the Arrhenius relation, E/RT, can be approximated in the neighborhood of the steady-state temperature T_s:

$$\frac{E}{RT} = \frac{E}{R(T_s + \Delta T)} = \frac{E}{RT_s}\left(\frac{T_s}{T_s + \Delta T}\right) = \frac{E}{RT_s}\left(1 - \frac{\Delta T}{T_s + \Delta T}\right) \tag{6.86}$$

because $T_s \gg \Delta T$, $T_s + \Delta T \approx T_s$, and the last term of Eq. (6.86) can be approximated to

$$\frac{\Delta T}{T_s + \Delta T} \approx \frac{\Delta T}{T_s} \tag{6.87}$$

Using this, Eq. (6.86) becomes

$$\frac{E}{RT} \approx \frac{E}{RT_s} - \frac{E\,\Delta T}{RT_s^2} \tag{6.88}$$

which is the linearization of E/RT around the steady-state temperature T_s (i.e., using the two terms in a Taylor's series):

$$\frac{E}{RT} \approx \frac{E}{RT_s} + \frac{d(E/RT)}{dT}\bigg|_{T_s}(T - T_s) = \frac{E}{RT_s} - \frac{E\,\Delta T}{RT_s^2} \tag{6.89}$$

Using the approximation expressed in Eq. (6.88), the exponential term of the Arrhenius relation becomes

$$e^{-E/RT} \approx \exp\left(-\frac{E}{RT_s} + \frac{E\,\Delta T}{RT_s^2}\right) = e^{-E/RT_s}e^{\theta} \tag{6.90}$$

where θ is the dimensionless temperature difference

$$\theta = \frac{E\,\Delta T}{RT_s^2} = \frac{E(T - T_s)}{RT_s^2} \tag{6.91}$$

Using Eq. (6.90) the Arrhenius equation is approximated in the neighborhood of the steady-state temperature T_s without losing the general nature of the nonlinearity of the Arrhenius equation. This approximation provides a relatively simple mathematical expression to study many nonlinear phenomena, whereas only numerical solutions are possible when the original form of the Arrhenius expression is used. The temperature T_s is the steady-state temperature in the neighborhood of which the behavior of the chemically reactive system is investigated. It could be the steady-state operating temperature of the reactor, the initial temperature of the spontaneous combustion, the maximum chain temperature of the flame propagation, etc.

REFERENCES

Boudart, M. 1968. Kinetics of Chemical Processes. Prentice-Hall, Englewood Cliffs, NJ.

Davidson, N. 1962. Statistical Mechanics. McGraw-Hill, New York.

Frank-Kamenetskii, D. A. 1969. Diffusion and Heat Transfer in Chemical Kinetics. Plenum Press, New York.

Glasstone, S., Laidler, K. J., and Eyring, H. 1941. The Theory of Rate Processes. McGraw-Hill, New York.

7
Thermochemistry and Chemical Equilibrium

1. INTRODUCTION

In reaction engineering problems the amount of energy associated with the chemical reaction and the equilibrium composition of the reacting system are needed at a given temperature, pressure and for an initial reactants composition (feed rates). These two basic pieces of information are related to the thermochemistry and thermodynamics of the system.

All chemical reactions are accompanied by the absorption or liberation of heat. The amount of heat exchanged between a reacting system and its environment is determined by the thermodynamic state of the reactants and the products, that is, by their enthalpies. Application of the first law of thermodynamics provides the energy balance for the reacting systems, which accounts for the enthalpies of the reactants and products and the heat of reaction. The energy balance provides the basic design parameters for the auxiliary heat transfer units.

The amount of heat exchanged between a reacting system (e.g., an exothermic reaction taking place in a batch reactor) and its surroundings affects the operating characteristics of the reactor (Fig. 7.1). In isothermal operating mode, the temperature of the reacting system is kept constant in time by a proper heat transfer mechanism between the reactor and its surroundings. In adiabatic operating mode, the reactor is thermally insulated from its surroundings and the temperature of the reactor asymptotically approaches the adiabatic operating temperature. At the adiabatic reactor operating temperature, the amount of heat generated by the chemical reaction is in balance with the amount of heat absorbed by the species of the reacting system (i.e., in a system where the heat absorbed by the reactor itself is insignificant). Most chemical reactors are designed to operate at a specified temperature (isothermal operating mode). Therefore, the amount of heat generated by the chemical reaction has to be known for the proper design of a heat exchange mechanism.

Application of the second law of thermodynamics provides information for the equilibrium composition of the reacting mixture. Information about the equilibrium composition, combined with the information about the chemical kinetics, helps in the selection of operating conditions such as the reactant composition, reactor size, and reactor temperature and pressure to achieve the desired conversion and product distribution.

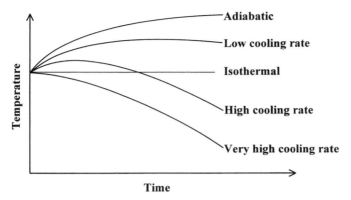

Figure 7.1 Temperature of a batch reactor as a function of time for an exothermic reaction taking place with different degrees of severity of applied cooling.

2. HEAT OF FORMATION AND HEAT OF REACTION

Chemical reactions always take place with heat release or heat intake. For a chemical reaction $\sum \alpha_j A_j = 0$ [Eq. (5.1)], the standard heat of reaction ΔH_0^0 (kJ/mol) is defined as the difference in the molar enthalpies (h_j) of the product and the reactant species at standard conditions:

$$\Delta H_{T_0}^0 = \Delta H_0^0 = \sum_{j=1}^{S} \alpha_j h_{j,T_0}^0 = \sum_{j=1}^{S} \alpha_j h_{j,0}^0 \tag{7.1}$$

where the zero subscript and superscript are used to refer to the standard temperature $(T_0 = 298.2 \, \text{K})$ and standard pressure, respectively. For a given reaction for any states of reactants and products it is the accepted convention that for endothermic reactions ΔH is positive ($\Delta H > 0$, the reacting system absorbs thermal energy) and for endothermic reactions ΔH is negative ($\Delta H < 0$, the reacting system evolves thermal energy). Therefore, the heat of reaction can be included in the stoichiometric relation. In fact, by introducing the molar enthalpies of the species into the stoichiometric relation of Eq. (5.1),

$$\sum_{j=1}^{S} \alpha_j (A_j - h_j) = \sum_{j=1}^{S} \alpha_j A_j - \sum_{j=1}^{S} \alpha_j h_j = \sum_{j=1}^{S} \alpha_j A_j - \Delta H = 0 \tag{7.2}$$

the stoichiometric relation $\sum \alpha_j A_j = 0$ [Eq. (5.1)] becomes

$$\sum_{j=1}^{S} \alpha_j A_j = \Delta H \tag{7.3}$$

It is clear from Eq. (7.2) that for a given reaction ΔH would be a function of the thermodynamic states (i.e., temperature and pressure) of the reactants and the products.

The heat of reaction can also be defined in terms of the total enthalpy H (kJ) of the reacting system. The total enthalpy of the system H is

$$H = \sum_{j=1}^{S} N_j h_j = \sum_{j=1}^{S} (N_{j,0} + \alpha_j X) h_j \tag{7.4}$$

which is a function of reaction extent. Taking the partial derivative of H with respect to X gives

$$\frac{\partial H}{\partial X} = \sum_{j=1}^{S} \alpha_j h_j + \sum_{j=1}^{S} N_j \frac{\partial h_j}{\partial X} = \sum_{j=1}^{S} \alpha_j h_j = \Delta H \tag{7.5}$$

because the second term $[\sum_j (N_j \, \partial h_j / \partial X) = 0]$ vanishes (Aris, 1966). Using Eq. (7.5), the reaction enthalpy ΔH can be calculated from the measurement of infinitesimal changes in the total enthalpy and the reaction extent.

Standard enthalpies of reactions are calculated by using the principle expressed by Eq. (7.1) and thermodynamic tables of standard heats of formation. The heat of formation $(\Delta H_f)_j$ for all species is calculated from the heats of reaction for the species formed from its elements. For the reaction stoichiometry of species A_j formed from its elements $E_1, \ldots, E_y$,

$$A_j - \sum_{e=1}^{y} \varepsilon_{j,e} E_e - (\Delta H_f)_j = 0 \tag{7.6}$$

where, $\varepsilon_{j,e}$ is the stoichiometric coefficient of the eth element in the formation of the jth species and $(\Delta H_f)_j$ is the heat of formation (like the heat of reaction) associated with the formation of species A_j. Equation (7.1) can be derived by multiplying both sides of Eq. (7.6) by α_j and summing over $j = 1, \ldots, S$ (Aris, 1966):

$$\sum_{j=1}^{S} \alpha_j A_j - \sum_{j=1}^{S} \sum_{e=1}^{y} \alpha_j \varepsilon_{j,e} E_e - \sum_{j=1}^{S} \alpha_j (\Delta H_f)_j = \sum_{j=1}^{S} \alpha_j A_j - \sum_{e=1}^{y} \sum_{j=1}^{S} \alpha_j \varepsilon_{j,e} E_e$$

$$- \sum_{j=1}^{S} \alpha_j (\Delta H_f)_j = 0 \tag{7.7}$$

The second term of Eq. (7.7) vanishes, because the reaction is balanced [Eq. (5.1)] in each elementary species and the following equality holds for all e elements:

$$\sum_{j=1}^{S} \alpha_j \varepsilon_{j,e} = 0 \tag{7.8}$$

Using Eq. (7.8), Eq. (7.7) becomes

$$\sum_{j=1}^{S} \alpha_j A_j - \sum_{j=1}^{S} \alpha_j (\Delta H_f)_j = 0 \tag{7.9}$$

Using Eq. (7.3) in Eq. (7.9) it can be seen that the heat of reaction can be calculated from the heat of formation of each species involved in a reaction:

$$\Delta H = \sum_{j=1}^{S} \alpha_j (\Delta H_f)_j \tag{7.10}$$

Thermodynamic tables are prepared by assuming that the heats of formation of all elements are zero and the standard states are 1 atm pressure and 298.2 K (subscript and superscript 0 are generally used to indicate standard temperature and standard pressure).

It must be mentioned here that direct measurement of the heat of formation of a specific species may not always be possible because of practical reasons. For example, a mixture of reactants (e.g., hydrogen, carbon, and oxygen) in a reactor may yield more than one product species. The heat of formation for such species can be calculated from the heat of combustion measurements. Because of their practical use, the standard heats of combustion are tabulated (Perry and Green, 1984). Using the linear combination of combustion reactions of chemical species, the standard heat of formation for a specific species can be calculated. This rule can be expanded for the other reactions (oxidation, hydrogenation, etc.) provided that the standard heats of reaction are known.

As an example, the standard heat of formation $(\Delta H_f)^0_{298}$ of acetaldehyde can be calculated from the measurement of the heat of combustion $(\Delta H)^0_{298}$ of acetaldehyde in the reaction

$$CH_3CHO(g) + 2.5O_2(g) \rightarrow 2CO_2(g) + 2H_2O(l) \tag{7.11}$$

which is given as -1191.73 kJ/mol. Because the standard heats of formation $(\Delta H_f)^0_{298}$ are -393.30 kJ/mol for carbon dioxide (CO_2) and -285.70 kJ/mol for water (H_2O), using Eq. (7.10), the standard heat of formation $(\Delta H_f)^0_{298}$ for acetaldehyde can be calculated as

$(\Delta H_f)^0_{298} = 1191.73 + 2(-393.30) + 2(-285.70) = -166.27$ kJ/mol $(= -39.76$ kcal/mol).

In this calculation the standard heat of formation of oxygen is taken as zero.

Because the standard heat of formation $(\Delta H_f)^0_{298}$ of acetaldehyde is known, as an another example the standard heat of reaction ΔH^0_{298} involved in the formation of acetaldehyde (CH_3CHO) by hydration of acetylene (C_2H_2), given by the reaction

$$C_2H_2(g) + H_2O(l) \rightarrow CH_3CHO(g) \tag{7.12}$$

can be calculated from the tabulated standard heats of formation $(\Delta H_f)^0_{298}$ of the reactants and the product, which are 226.61 kJ/mol for $C_2H_2(g)$; -285.70 kJ/mol for water, $H_2O(l)$; and -166.27 kJ/mol for acetaldehyde, $CH_3CHO(g)$. Using Eq. (7.10), the standard heat of the reaction expressed in Eq. (7.12) is calculated as $(\Delta H)^0_{298} = -166.27 - (226.61 - 285.70) = -107.18$ kJ/mol $(-25.63$ kcal/mol).

In general, the linear combination of chemical reactions can be used for the heat of formation (enthalpy) and heat of reaction (enthalpy) calculations. The heat of reaction of a given reaction can be calculated from the heats of reaction of the dependent reactions. As an example, the standard heats of reaction for the independent reactions:

$$C(s) + O_2(g) \rightarrow CO_2(g), \qquad \Delta H^0_0 = -393.1 \text{ kJ/mol} \tag{7.13}$$

$$C(s) + \frac{1}{2}O_2(g) \rightarrow CO(g), \qquad \Delta H^0_0 = -110.4 \text{ kJ/mol} \tag{7.14}$$

can be used to calculate the heat of reaction for the oxidation of carbon monoxide (CO) to carbon dioxide (CO_2):

$$CO(g) + \frac{1}{2}O_2(g) \rightarrow CO_2(g), \qquad \Delta H^0_0 = -282.7 \text{ kJ/mol} \tag{7.15}$$

which is the reaction stoichiometry obtained by subtracting the reaction stoichiometry expressed by Eq. (7.14) from that expressed by Eq. (7.13). The standard heat of reaction for the reaction stoichiometry expressed by Eq. (7.15) is calculated in the same way, i.e., $\Delta H_0^0 = -393.1 - (-110.4) = -282.7$ kJ/mol (-67.6 kcal/mol), using the linear combination arithmetic.

3. FIRST LAW OF THERMODYNAMICS

The first law of thermodynamics, i.e., the energy balance, will be formulated for an open reacting system (mass and energy exchange are allowed with the surroundings) (Fig. 7.2). Assuming that the system is operating at steady-state conditions, the energy balance (first law of thermodynamics) can be expressed as

$$Q - W = \sum E_{\text{exit}} - \sum E_{\text{in}} \tag{7.16}$$

where Q is the heat input into the system from its surroundings and W is the total work output from the system to its surroundings. In this notation $Q > 0$ if heat is supplied from the surroundings into the reacting system and $W > 0$ if the work is done by the system. For most chemical engineering applications W can be expressed as

$$W = W_s + P_{\text{exit}} V_{\text{exit}} - P_{\text{in}} V_{\text{in}} \tag{7.17}$$

where W_s is the shaft work (for a fuel cell reactor, electrical work should be included in W_s).

$\sum E_{\text{in}}$ and $\sum E_{\text{exit}}$ are the total (internal, potential, and mechanical) energy of the materials at the inlet and exit respectively, of the reactor,

$$\sum E_{\text{in}} = U_{\text{in}} + m\left(gz_{\text{in}} + \frac{1}{2}mv_{\text{in}}^2\right) \tag{7.18}$$

$$\sum E_{\text{exit}} = U_{\text{exit}} + m\left(gz_{\text{exit}} + \frac{1}{2}mv_{\text{exit}}^2\right) \tag{7.19}$$

Using Eqs. (7.17), (7.18), and (7.19), the energy balance expressed by Eq. (7.16) becomes

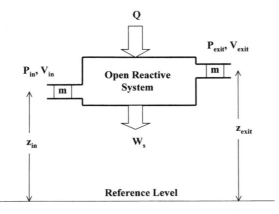

Figure 7.2 Energy balance (first law of thermodynamics) for an open reacting system.

$$Q - W_s = (U_{exit} + P_{exit}V_{exit}) - (U_{in} + P_{in}V_{in}) + mg(z_{exit} - z_{in}) + \frac{1}{2}m(v_{exit}^2 - v_{in}^2) \quad (7.20)$$

which is the most general form of the energy balance equation for an open system.

For most chemical engineering applications (i.e., for reactors), $W_s = 0$, $mg(z_{exit} - z_{in}) \ll (U_{in} + P_{in}V_{in})$ and $(U_{exit} + P_{exit}V_{exit})$, $(1/2)mg(v_{exit}^2 - v_{in}^2) \ll (U_{in} + P_{in}V_{in})$ and $(U_{exit} + P_{exit}V_{exit})$. Under these conditions the energy balance can be approximated to

$$Q = (U_{exit} + P_{exit}V_{exit}) - (U_{in} + P_{in}V_{in}) = H_{exit} - H_{in} \quad (7.21)$$

Assuming that the reactants and products are at standard (i.e., atmospheric) pressure, enthalpies of the species at the inlet and exit of the reactor can be expressed as

$$H_{in} = (H_0^0)_{React} + \int_{T_0}^{T_1} \sum (mc_p)_{React} \, dT \quad (7.22)$$

and

$$H_{exit} = (H_0^0)_{Prod} + \int_{T_0}^{T_2} \sum (mc_p)_{Prod} \, dT \quad (7.23)$$

where H_0^0 is the enthalpy of formation at standard conditions, i.e., at atmospheric pressure and 298.2 K ($T_0 = 25°C$). Using the enthalpy of the reactants and products expressed in terms of specific heats, Eq. (7.21) becomes

$$Q = (H_0^0)_{Prod} - (H_0^0)_{React} + \int_{T_0}^{T_{exit}} \sum (mc_P)_{Prod} \, dT - \int_{T_0}^{T_{in}} \sum (mc_P)_{React} \, dT \quad (7.24)$$

or

$$Q = (\Delta H_0^0)_{Reaction} + \int_{T_0}^{T_{exit}} \sum (mc_P)_{Prod} \, dT - \int_{T_0}^{T_{in}} \sum (mc_P)_{React} \, dT = (\Delta H)_{Reaction} \quad (7.25)$$

If the temperature of the reacting system is kept at the standard temperature T_0, the integral terms in Eq. (7.25) vanish:

$$Q = (H_0^0)_{Prod} - (H_0^0)_{React} = (\Delta H_0^0)_{Reaction} \quad (7.26)$$

which can be used for measurement of the standard enthalpy of reaction experimentally (i.e., heat of combustion measurements made using calorimeters).

The energy balance expressed by Eq. (7.25) is also the rate of heat generation of the chemical reaction, Q_g (i.e., $Q_g = (\Delta H)_{Reaction}$, considering the initial states of the reactants and the final states of the products). Using Eq. (7.25), for a reactor operating at a constant pressure, the differential change in Q_g would be equal to the differential change of enthalpy H in the reacting system as a result of a differential change in the extent of reaction (i.e., units are ΔH, kJ/mol; dX, mol; and dH, kJ):

$$dQ_g = -dH = (-\Delta H)\,dX \tag{7.27}$$

or, in terms of extensive reaction rate R^*,

$$\frac{dQ_g}{dt} = (-\Delta H)\frac{dX}{dt} = (-\Delta H)R^* \tag{7.28}$$

The negative sign is needed because thermal energy can be generated if the reaction is exothermic ($\Delta H < 0$). Assuming that the reactor volume V is constant during the reaction, dividing both sides of Eq. (7.28) by V gives

$$\frac{1}{V}\frac{dQ_g}{dt} = q_g = (-\Delta H)\frac{1}{V}\frac{dX}{dt} = (-\Delta H)r \tag{7.29}$$

where q_g is the heat generated (or absorbed) by the chemical reaction per unit volume of the reacting system. If more than one reaction is taking place, Eq. (7.29) becomes

$$\frac{1}{V}\frac{dQ_g}{dt} = q_g = \sum_{i=1}^{R}(-\Delta H_i)\frac{1}{V}\frac{dX_i}{dt} = \sum_{i=1}^{R}(-\Delta H_i)r_i \tag{7.30}$$

In formulating the energy balance for a chemically reactive system, the rate of chemical reaction r can be expressed in terms of the intensive reaction extent ξ [i.e., for a first-order reaction, $r = d\xi/dt$ as in Eq. (5.56)] or in terms of reactant concentration, $r = 1/\alpha_j dC_j/dt$ [as in Eq. (5.57)].

4. SECOND LAW OF THERMODYNAMICS: CHEMICAL EQUILIBRIUM

In a chemically reactive system operating at constant temperature and pressure, the composition of the system will spontaneously change in the direction of maximum entropy and minimum Gibbs free energy. The system reaches equilibrium at the maximum entropy or at the minimum Gibbs free energy.

For an adiabatic (heat exchange with the surrounding is not permitted) chemically reactive system, the first law of thermodynamics is expressed as

$$dU + dQ + dW = 0 \tag{7.31}$$

where dU is the differential change in the internal energy, dQ is the differential heat released by the chemical reaction ($dQ > 0$ for an exothermic reaction, i.e., $dQ = -\Delta H\,dX$), and dW is the differential work done by the system, and the second law of thermodynamics is expressed as

$$dG = d(U + PV - TS) \tag{7.32}$$

For a chemical reaction system operating at constant pressure and temperature, by using the first law of thermodynamics ($dU + dQ + dW = 0$), Eq. (7.32) becomes (Aris, 1969)

$$dG = d(U + PV - TS) = dU + P\,dV - T\,dS = -(dQ + T\,dS) + P\,dV - dW \tag{7.33}$$

If the work done against external pressure is the only work done by the system, then $dW = P\,dV$ and Eq. (7.33) becomes

$$dG = -(dQ + T\,dS) \tag{7.34}$$

The entropy change in the system as a result of an infinitesimal change in Q is dQ/T. The total entropy change in the system is positive (i.e., $dQ/T + dS > 0$),

$$dG \leq 0 \tag{7.35}$$

and at equilibrium

$$dG = 0 \tag{7.36}$$

Calculation of equilibrium composition in single- or multiphase systems is an important task in chemical engineering. Minimization of the Gibbs free energy [Eq. (7.36)] is commonly used for the calculation of equilibrium compositions.

For a system composed of S species, a change in the Gibbs free energy of the system is given by

$$dG = -S\,dT + V\,dP + \sum_{j=1}^{S} \mu_j\,dN_j \tag{7.37}$$

where μ_j is the molar chemical potential of A_j. Using Eq. (7.14), Eq. (7.37) becomes

$$dG = -S\,dT + V\,dP + \sum_{j=1}^{S} \mu_j \alpha_j\,dX \tag{7.38}$$

The condition for equilibrium [Eq. (7.36)] at constant temperature and pressure becomes ($dT = dP = 0$)

$$dG = \sum_{j=1}^{S} \mu_j\,dN_j = \sum_{j=1}^{S} \mu_j \alpha_j\,dX = \left(\sum_{j=1}^{S} \alpha_j \mu_j \right) dX = \Delta G\,dX = 0 \tag{7.39}$$

or

$$\left(\sum_{j=1}^{S} \alpha_j \mu_j \right) = \Delta G = 0 \tag{7.40}$$

If the system is behaving as an ideal gas mixture, chemical potential can be expressed in terms of the partial pressure p_j and the standard chemical potential μ_j^0, reducing the condition of equilibrium to

$$\sum_{j=1}^{S} \alpha_j (\mu_j^0 + RT \ln p_j) = 0 \tag{7.41}$$

or

$$RT \sum_{j=1}^{S} \alpha_j \ln p_j = -\sum_{j=1}^{S} \alpha_j \mu_j^0 = -\Delta G^0 \tag{7.42}$$

The total Gibbs free energy G^0 of the system at standard pressure is

$$G^0 = \sum_{j=1}^{S} N_j \mu_j^0 = \sum_{j=1}^{S} \mu_j^0 (N_{j,0} + \alpha_j X) \tag{7.43}$$

the derivative of which with respect to the reaction extent X gives

$$\frac{\partial G^0}{\partial X} = \sum_{j=1}^{S} \alpha_j \mu_j^0 = \Delta G^0 \tag{7.44}$$

which is the free energy change of the reaction at standard pressure [as described the total enthalpy and the reaction enthalpy by Eqs. (7.4) and (7.5)]. If the system is not ideal, then the Gibbs free energy [Eq. (7.41)] has to be expressed in terms of the partial fugacity f_j instead of the partial pressure p_j.

The equilibrium composition can be calculated from the condition of equilibrium expressed by Eq. (7.42) by rearrangement:

$$\sum_{j=1}^{S} \alpha_j \ln p_j = \sum_{j=1}^{S} \ln p_j^{\alpha_j} = \ln\left(\prod_{j=1}^{S} p_j^{\alpha_j}\right) = \ln K = -\frac{1}{RT} \sum_{j=1}^{S} \alpha_j \mu_j^0 = -\frac{\Delta G^0}{RT} \tag{7.45}$$

Using the thermodynamics relationship $\Delta G^0 = \Delta H^0 - T \Delta S^0$, the equilibrium constant K becomes

$$K = \exp\left(-\frac{\Delta G^0}{RT}\right) = \exp\left(\frac{\Delta S^0}{R}\right) \exp\left(-\frac{\Delta H^0}{RT}\right) \tag{7.46}$$

Using the thermodynamic relation

$$\frac{d}{dT}\frac{\mu_j^0}{T} = -\frac{h_j^0}{T^2} \tag{7.47}$$

the dependence of the equilibrium constant K on temperature can be expressed, using Eqs (7.45) and (7.47), as

$$\frac{d \ln K}{dT} = \frac{1}{R}\frac{d}{dT}\left(-\sum_{j=1}^{S} \alpha_j \frac{\mu_j^0}{T}\right) = \frac{1}{RT^2} \sum_{j=1}^{S} \alpha_j h_j^0 = \frac{\Delta H^0}{RT^2} \tag{7.48}$$

Integrating Eq. (7.48) from the standard temperature T_0 to temperature T gives

$$K = K(T) = K^* \exp\left(-\frac{\Delta H^0}{RT}\right) \tag{7.49}$$

Comparison of Eq. (7.46) with Eq. (7.49) shows that K^* is related to the standard entropy of the reaction:

$$K^* = \exp\left(\frac{\Delta S^0}{R}\right) \tag{7.50}$$

The equilibrium composition of a system can be calculated from the solution of a nonlinear equation expressed in Eq. (7.46) by using the thermodynamic properties of the system. For a system involving only one chemical reaction, this could be a relatively easy task.

If more than one reaction is taking place in a system at constant temperature and pressure, then the chemical equilibrium condition can be expressed [similarly to Eqs. (7.36) or (7.40)] as

$$dG = \sum_{j=1}^{S} \mu_j dN_j = \sum_{j=1}^{S} \mu_j \sum_{i=1}^{R} \alpha_{i,j} dX_i = 0 \tag{7.51}$$

where X_i is the extent of the ith reaction [as defined in Eq. (5.46)]. From Eq. (7.51), the condition for chemical equilibrium yields

$$\sum_{i=1}^{R} dX_i \sum_{j=1}^{S} \alpha_{i,j} \mu_j = 0 \tag{7.52}$$

or

$$\sum_{j=1}^{S} \alpha_{i,j} \mu_j = 0, \qquad i = 1, \ldots, R \tag{7.53}$$

which leads to R simultaneous equilibrium conditions in the form

$$K_i(T) = \exp\frac{-\sum_{j=1}^{S} \alpha_{i,j} \mu_j^0}{RT} = \exp\left(-\frac{\Delta G_i^0}{RT}\right) = \exp\left(\frac{\Delta S_i^0}{R}\right) \exp\left(-\frac{\Delta H_i^0}{RT}\right) \tag{7.54}$$

Similar to Eq. (7.49), the equilibrium constants can be expressed as

$$K_i = K_i(T) = K_i^* \exp\left(-\frac{\Delta H_i^0}{RT}\right) \tag{7.55}$$

where K_i^* is the equilibrium constant for the ith reaction at standard temperature and standard pressure:

$$K_i^* = \exp\left(\frac{\Delta S_i^0}{R}\right) \tag{7.56}$$

The equilibrium composition can be calculated from the simultaneous solution of R nonlinear equations of Eq. (7.54). This is a difficult mathematical problem to solve numerically, especially if the number of reactions is large (one of the constraints for these calculations is that all reaction extents or concentrations of chemical species must be greater than zero). The calculation of equilibrium composition by using this method is impractical if not impossible.

An alternative method for calculating equilibrium composition is to minimize the Gibbs free energy function by adjusting the moles of each species consistent with atomic constraints. This is called the method of elemental potentials (Van Zeggeren and Storey, 1970). Calculation of equilibrium composition with this method is also difficult (i.e., care must be taken to be sure that all molar concentrations calculated are non-negative quantities).

In the elemental potential method, the Gibbs free energy of the system to be minimized is (at constant temperature and pressure)

$$G = \sum_{j=1}^{S} \mu_j N_j \tag{7.57}$$

and the partial molar Gibbs function is

$$\mu_j = \mu_{j,0}(T, P) + RT \ln x_j \tag{7.58}$$

where $\mu_{j,0}(T, P)$ is the Gibbs function of the pure A_j species. Using the expression for partial molar Gibbs functions, Eq. (7.57) becomes

$$G = \sum_{j=1}^{S}[\mu_{j,0}(T, P) + RT \ln x_j]N_j = \sum_{j=1}^{S} \mu_{j,0}N_j + RTN_j \ln\left(\frac{N_j}{N_T}\right) \tag{7.59}$$

The atomic population constraints are

$$\sum_{j=1}^{S} \varepsilon_{e,j}N_j = w_e, \qquad e = 1,\ldots,y \tag{7.60}$$

where $\varepsilon_{e,j}$ is the number of eth species atoms in jth molecule, and w_e is the population of eth species atoms in the system, which is fixed. The equilibrium composition of the system at the given T and P has to minimize the Gibbs function [Eq. (7.59)] subject to the atomic constraints [Eq. (7.60)].

This minimization problem can be handled by the method of Lagrange multipliers. For convenience, G/RT can be taken as the function to be minimized (Reynolds, 1986):

$$\frac{G}{RT} = \sum_{j=1}^{S}\left(\frac{\mu_{j,0}}{RT} + \ln x_j\right)N_j \tag{7.61}$$

For an arbitrary variation in mole numbers, the change in G/RT is given by

$$d\left(\frac{G}{RT}\right) = \sum_{j=1}^{S}\left(\frac{\mu_{j,0}}{RT} + \ln x_j\right) dN_j + \sum_{j=1}^{S} N_j \frac{1}{x_j} dx_j \tag{7.62}$$

If the jth species can exist in more than one phase, then the last term in Eq. (7.62) can be rewritten considering all phases (using $x_j = N_j/N_m$ for phase m):

$$d\left(\frac{G}{RT}\right) = \sum_{j=1}^{S}\left(\frac{\mu_{j,0}}{RT} + \ln x_j\right) dN_j + \sum_{m=1}^{p} N_m\left(\sum_{j=1}^{S} dx_j\right)_m \tag{7.63}$$

where p is the total number of phases that might be present. The last term in Eq. (7.63) vanishes, however, because the mole fractions in each phase always sum to unity, and Eq. (7.63) becomes

$$d\left(\frac{G}{RT}\right) = \sum_{j=1}^{S}\left(\frac{\mu_{j,0}}{RT} + \ln x_j\right) dN_j \tag{7.64}$$

Because of the atomic constraints the dN_j are not independent, the relationship among them being obtained by differentiating the atomic constraints of Eq. (7.60):

$$\sum_{j=1}^{S} \varepsilon_{e,j} dN_j = 0, \qquad e = 1,\ldots,y \tag{7.65}$$

In this system, y of the restricted dN_j must be solved in term of $s - y$ unrestricted dN_j and substituted into Eq. (7.64) to express the variation in G/RT in terms of freely variable N_j. This process is equivalent to subtracting multiples of Eq. (7.65) from Eq. (7.64):

$$d\left(\frac{G}{RT}\right) = \sum_{j=1}^{S} \left(\frac{\mu_{j,0}}{RT} + \ln x_j\right) dN_j - \sum_{e=1}^{y} \lambda_e \sum_{j=1}^{S} \varepsilon_{e,j} \, dN_j \tag{7.66}$$

or

$$d\left(\frac{G}{RT}\right) = \sum_{j=1}^{S} \left[\frac{\mu_{j,0}}{RT} + \ln x_j - \sum_{e=1}^{y} \lambda_e \varepsilon_{e,j}\right] dN_j \tag{7.67}$$

where λ_e is the Lagrange multiplier that is required to eliminate the set of restricted dN_j from the minimization problem. To find the minimum of Eq. (7.67) [which is the condition of $d(G/RT) = 0$] the coefficients of dN_j are set to zero, which gives

$$\frac{\mu_{j,0}}{RT} + \ln x_j - \sum_{e=1}^{y} \lambda_e \varepsilon_{e,j} = 0 \tag{7.68}$$

for the restricted dN_j. With the removal of these dN_j from Eq. (7.67), the remaining dN_j can be freely varied provided that their coefficients are equal to zero. Therefore, the composition of every species become [using Eq. (7.68)]

$$x_j = \exp\left[-\frac{\mu_{j,0}}{RT} + \sum_{e=1}^{y} \lambda_e \varepsilon_{e,j}\right] \tag{7.69}$$

The molar composition of the system at equilibrium can be calculated using Eq. (7.69) as a function of its value $\mu_{j,0}/RT$, the, atomic make-up of its molecules, and a set of Lagrangian multipliers (which are determined from the atomic constraints). The Lagrangian multiplier is called the elemental potential for the eth atom (or element). This can be seen from the rearranged form of Eq. (7.68):

$$\sum_{e=1}^{y} \lambda_e \varepsilon_{e,j} = \frac{\mu_{j,0}}{RT} + \ln x_j = \frac{\mu_j}{RT} \tag{7.70}$$

in which λ_e represents the Gibbs function μ_j/RT per mole of eth species atoms. It is interesting to see that each atom of an element contributes the same amount to the Gibbs function, regardless of which molecule or phase it may be in (Reynolds, 1986).

Commercial simulation packages are available for equilibrium calculations. The commercial simulation package STANJAN was developed for equilibrium calculation in combusting systems at Stanford University (Reynolds, 1986). NASA developed a similar simulation package (Gordon and McBride, 1976). Another commercial simulation package was developed by McGill University and is called Facility for the Analysis of Chemical Thermodynamics, or F*A*C*T (Thompson et al., 1985). All of these simulation packages are valuable for chemical equilibrium calculations that can be used commercially.

Thermodynamic equilibrium calculations are always faster if the reactor operating temperature and pressure are given (or specified). Calculation of the equilibrium composition in adiabatic reactors requires iteration. In an adiabatic reactor, the equilibrium composition and the path of reaction taken to reach the equilibrium composition can be estimated (Fig. 7.3). The adiabatic reaction path can be determined from the integration of the energy balance,

$$C_p \, dT = -\Delta H \, dX \tag{7.71}$$

where C_p is the total heat capacity of the reacting system. Integration of Eq. (7.71) gives

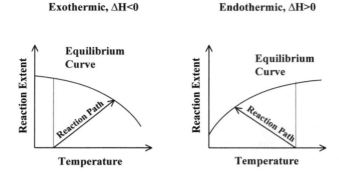

Figure 7.3 Reaction paths for exothermic ($\Delta H < 0$) and endothermic ($\Delta H > 0$) reactions at adiabatic operating conditions.

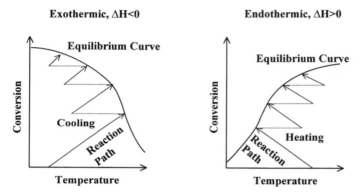

Figure 7.4 Effect of cooling and heating on the conservation of reversible exothermic ($\Delta H < 0$) and endothermic ($\Delta H > 0$) reactions.

$$\int_{T_0}^{T} dT = -\frac{\Delta H}{C_p} \int_{0}^{X_e} dX \tag{7.72}$$

which can be used as the first estimate for the adiabatic temperature and equilibrium conversion. To get a higher conversion, especially for reversible reactions, cooling is needed for exothermic reactions and heating for endothermic reactions (Fig. 7.4).

REFERENCES

Aris, R. 1969. Elementary Chemical Reactor Analysis. Prentice-Hall, Englewood Cliffs, NJ.

Gordon, S., and McBride, B. J. 1976. Computer Program for Calculation of Complex Chemical Equilibrium Compositions, Rocket Performance, Incident and Reflected Shocks, and Chapman-Jouguet Detonations. Report No. SP-273. National Aeronautics and Space Administration, Washington, DC.

Perry, R. H., and Green, D. W., eds. 1984. Perry's Chemical Engineers' Handbook. McGraw-Hill, New York, Chapter 3.

Reynolds, W. C. 1986. STANJAN, Version 3. Department of Mechanical Engineering, Stanford University, Stanford, CA.

Thompson, W. T., Pelton, A. D., and Bale, C. W. 1985. F*A*C*T Guide to Operations. École Polytechnique, Campus de l'Université de Montreal, Montreal, PQ, Canada.

Van Zeggeren, F., and Storey, S. H. 1970. The Computation of Chemical Equilibria, Cambridge Univ. Press, London, UK.

8
Kinetic Measurements and Data Analysis

1. INTRODUCTION

In almost all chemical reaction systems the true chemical kinetics and transport phenomena are coupled in a nonlinear fashion. For chemical engineers, process engineers, and process chemists, knowledge about the rate of chemical reaction and its coupling with transport phenomena is essential for the design of commercial reactors with minimum risk.

One of the main tasks for chemical engineering professionals is to design chemical reactors for commercial applications. These reactors have to accomplish process targets (i.e., conversion) while satisfying certain constraints such as construction and operating costs, control, start-up and shutdown procedures, and safety and environmental regulations. Chemical kinetics and its coupling with transport phenomena are the most needed information for the design of these reactors.

Industrial application of the coal combustion process can be presented as an example of the importance of the coupling between chemical kinetics and transport phenomena. By the beginning of the 1900s, it was observed that coal combustion kinetics is inversely proportional to the size of the coal particles. This observation led to the conclusions that coal is combusted at the external surface only and that mass transfer of oxygen (bulk to solid surface) was the controlling factor for the combustion process. As a result, grate coal combustor design was developed for combusting coal particles greater than 100 mm in diameter. By the 1940s, it was observed that coal particles combust internally also, provided that the particle size is sufficiently small. This knowledge led to the development of modern pulverized coal combustor design for combusting coal particles smaller than 0.1 mm. Since the 1950s fluidized bed combustors have been developed for combusting coal particles smaller than 10 mm (or slightly larger) in diameter. The main reason for observing different combustion rates as a function of coal particle size is that coal particle size is an important parameter for the mechanism of coal combustion. As particle size increases, coal is oxidized with carbon dioxide (CO_2), producing carbon monoxide (CO), which is combusted to carbon dioxide (CO_2) with air oxygen in the boundary layer surrounding the coal particle. As the particle size decreases, coal is directly oxidized with oxygen, producing both carbon monoxide and carbon dioxide, with the CO/CO_2 ratio increasing with increases in temperature. The oxidation of carbon with carbon dioxide is an endothermic reaction, whereas the oxidation of carbon with oxygen is exothermic, affecting the temperature of the burning coal particles and therefore the combustion kinetics. This example alone shows the importance of the reaction mechanism, i.e., coupling between the chemical reaction kinetics and transport phenomena, for commercial applications.

The first information needed for reactor design consists of the rates of chemical reactions and their couplings with transport phenomena. This information can be obtained from measurements of concentration changes of the chemical species under properly controlled operating conditions of specifically designed test reactors. In this effort, the selection of the reactor type, measurements of process variables, sampling and concentration measurements, and the design of the proper control mechanism to maintain steady-state operating conditions have to be carried out with care. Most important, the reactor must be operated in such a way that all kinds of resistance (limitations) to transport (hydrodynamics) phenomena must be eliminated or reduced to their minimum.

Practical models describing the chemical kinetics and transport phenomena have to be developed. These models are always useful tools in predicting the performance of reactors at different operating conditions such as variations in temperature, pressure, reactor size, feed composition, and feed rate. The models have to be tested using the experimentally measured data to determine the correlation between the dependent variables (state variables) and the independent variables (process variables). Collection of the experimental data and the correlation of model equations to experimental data are equally important parts of chemical kinetics studies.

2. KINETIC MODELS AND CALCULATION OF RATE CONSTANTS

In general, chemical reaction kinetics models are proposed to predict compositional changes in chemically reactive systems. For this purpose, experimental data obtained from laboratory-scale reactors have to be interpreted and reaction models have to be developed. These models are the basis for the prediction of reactor performance and for reactor design for commercial scale operations.

If the kinetics models fail to predict the performance of reactors, the consequences can be extremely costly. The main reason for the failure of these models evolves from the fact that hydrodynamic conditions in the laboratory-scale test reactors are not identical to these in commercial scale reactors. Also, most of the time the linear models are forced to interpret the phenomena of highly nonlinear chemically reactive systems.

For the modeling of chemically reacting systems, material balance will be formulated first. The material balance for a control volume V surrounded by a surface S can be expressed as

$$(\text{material})_{\text{accum}} = (\text{material})_{\text{in}} - (\text{material})_{\text{out}} + (\text{material})_{\text{gen}} \tag{8.1}$$

where $(\text{material})_{\text{accum}}$ is the accumulated material in control volume V, $(\text{material})_{\text{in}}$ and $(\text{material})_{\text{out}}$ are the amounts of material exchanged by the control volume V with its environment through the surrounding surface S, and $(\text{material})_{\text{gen}}$ is the material generated (or consumed, with a negative sign) in control volume V (Fig. 8.1). Material balance expressed by Eq. (8.1) can be formulated on a mass or molar basis, each of which may have certain advantages.

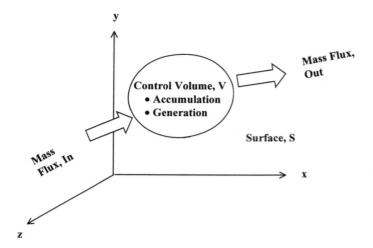

Figure 8.1 Property balance for a control volume V surrounded by surface S.

If a single chemical reaction is taking place in a closed system (no mass transfer through the system boundaries), the molar material balance expressed by Eq. (8.1) becomes

$$\frac{dN_j}{dt} = \frac{d}{dt}(N_{j,o} + \alpha_j X) = \alpha_j \frac{dX}{dt} \tag{8.2}$$

which is the derivative of the molar balance expression with respect to time. If the volume of the reactive system V is constant, Eq. (8.2) becomes

$$\frac{dc_j}{dt} = \alpha_j \frac{d(X/V)}{dt} = \alpha_j \frac{d\xi}{dt} = \alpha_j r \tag{8.3}$$

This relation indicates that for a closed constant-volume reactor system, the reaction rate can be determined from time-dependent concentration measurement of the jth species,

$$r = \frac{1}{\alpha_j V}\frac{dN_j}{dt} = \frac{1}{\alpha_j}\frac{dc_j}{dt} \tag{8.4}$$

This relation is useful for calculating the reaction rate r from concentration measurements in a closed system as a function of time. However, it does not say anything about the dependence of r on temperature, concentration, pressure, or other process variables such as catalysts, interfacial properties, and reactor hydrodynamics (i.e., in catalytic and multiphase reactions). If the reaction rate r is found to be a function of the reactor hydrodynamics, this indicates that the reaction rate is not a true reaction rate but an apparent reaction rate, i.e., chemical kinetics and transport phenomena are coupled.

The reaction rate can be expressed in terms of the concentration of the reactant species; i.e., for a first-order reaction [using $\alpha_j = -1$ in Eq. (8.4) and dropping the subscript j for simplicity],

$$r = -\frac{dc}{dt} = kc \tag{8.5}$$

where k is the reaction rate constant with Arrhenius dependence on temperature [i.e., $k = A_p \exp(-E/RT)$]. The relation expressed in Eq. (8.5) postulates that the rate of consumption of the reactant species is proportional to (i.e., first-order or linear) its concentration in the reactor. If this postulate is correct, then Eq. (8.5) should predict (or correlate) the concentration change of the reactant species as a function of time. Integration of Eq. (8.5) for initial conditions of $c = c_0$ at $t = 0$ gives

$$\ln \frac{c}{c_o} = -kt \qquad \text{or} \qquad c = c_o e^{-kt} \tag{8.6}$$

Assuming that N concentration measurements are made, i.e., $c_1, \ldots, c_N$ concentrations are measured at $t_1, \ldots, t_N$ reaction times, the unknown rate constant k can be calculated from the best least-squares fit of the kinetic model [Eq. (8.6)] to the experimental data:

$$I = \min \sum_{n=1}^{N} (c_n - c_{n,\exp})^2 \tag{8.7}$$

where, I is the risk function to be minimized, c_n is the concentration predicted using Eq. (8.6) and $c_{n,\exp}$ is the concentration measured experimentally at reaction time t_n. As a result, the unknown rate constant k is calculated from the solution of the nonlinear equation

$$\frac{\partial I}{\partial k} = \frac{\partial}{\partial k} \sum_{n=1}^{N} (c_n - c_{n,\exp})^2 = \frac{\partial}{\partial k} \sum_{n=1}^{N} (c_o e^{-kt_n} - c_{n,\exp})^2 = 0 \tag{8.8}$$

or (because c_o is constant)

$$\sum_{n=1}^{N} (c_o e^{-kt_n} - c_{n,\exp})(-t_n c_o e^{-kt_n}) = \sum_{n=1}^{N} \left(e^{-kt_n} - \frac{c_{n,\exp}}{c_o} \right)(-t_n e^{-kt_n}) = 0 \tag{8.9}$$

This problem can also be handled using the logarithmic form of the calculated concentration as a function of time. Using Eq. (8.6), the risk function I can be expressed as

$$I = \min \sum_{n=1}^{N} \left[\ln \frac{c_n}{c_o} - \ln \left(\frac{c_{n,\exp}}{c_o} \right) \right]^2 = \min \sum_{n=1}^{N} \left[-kt_n - \ln \left(\frac{c_{n,\exp}}{c_o} \right) \right]^2 \tag{8.10}$$

Similarly, the unknown rate constant k is calculated from the solution of the linear equation

$$\frac{\partial I}{\partial k} = \sum_{n=1}^{N} \left[-kt_n - \ln \left(\frac{c_{n,\exp}}{c_o} \right) \right](-t_n) = k \left(\sum_{n=1}^{N} t_n^2 \right) + \sum_{n=1}^{N} t_n \ln \left(\frac{c_{n,\exp}}{c_o} \right) = 0 \tag{8.11}$$

from which the unknown rate constant k can be calculated as

$$k = -\frac{\sum_{n=1}^{N} t_n \ln(c_{n,\exp}/c_o)}{\sum_{n=1}^{N} t_n^2} \tag{8.12}$$

which involves a much simpler computational problem. Both formulations [i.e.. Eqs. (8.9) and (8.12)] are mathematically correct; however, the logarithmic expression for the concentrations gives a simple solution for the rate constant. In both methods the error made in the measurement of the initial concentration c_o is carried into the numerical manipulation of every concentration measurement.

A second-order rate expression could also be postulated for the chemical reaction. In this case the reaction rate [Eq. (8.5)] in terms of the concentration of the reactant species would be expressed as

$$r = -\frac{dc}{dt} = kc^2 \tag{8.13}$$

where k is the reaction rate constant (the Arrhenius dependence on temperature). With a second-order rate expression, the concentration change of the reactant species is predicted by solving Eq. (8.13) for the initial conditions $t = 0$, $c = c_0$:

$$\frac{1}{c} = \frac{1}{c_0} + kt \tag{8.14}$$

Assigning a new variable z with the initial condition $z_0 = 1/c_0$,

$$z = \frac{1}{c}, \qquad z_0 = \frac{1}{c_0} \tag{8.15}$$

the nonlinear equation [Eq. (8.14)] is linearized to

$$z_n = z_0 + kt \tag{8.16}$$

The unknown rate constant k can be calculated from the best least-squares fit of the kinetic model [Eq. (8.16)] to the experimental data by minimizing the risk function (with respect to k):

$$I = \min \sum_{n=1}^{N} (z_n - z_{n,\exp})^2 \tag{8.17}$$

where $z_{n,\exp}$ ($= 1/c_{n,\exp}$) is the inverse of the nth concentration measurement. As a result, k is calculated from the solution of the linear relation

$$\frac{\partial I}{\partial k} = \frac{\partial}{\partial k} \sum_{n=1}^{N} (z_n - z_{n,\exp})^2 = \sum_{n=1}^{N} (z_0 + kt_n - z_{n,\exp})t_n = 0 \tag{8.18}$$

or

$$z_o \sum_{n=1}^{N} t_n + k \sum_{n=1}^{N} t_n^2 - \sum_{n=1}^{N} (t_n z_{n,\exp}) = 0 \tag{8.19}$$

which yields

$$k = \frac{\sum\limits_{n=1}^{N} t_n(z_{n,\exp} - z_o)}{\sum\limits_{n=1}^{N} t_n^2} \tag{8.20}$$

In calculating the second-order rate constant k, the experimental error made in the measurement of the nth concentration at elevated times (i.e., smaller c_n values) affects the numerical procedure nonlinearly [or in larger magnitudes, as can be seen from Eq. (8.15)].

A reactive system may involve S ($j = 1, \ldots, S$) chemical species and R ($i = 1, \ldots, R$) chemical reactions. In some cases, several species can be lumped into one species, based on physical properties such as boiling point, solubility, or molecular weight. Linear kinetic models are generally used to correlate the experimental data for multireaction, multicomponent systems. Several reaction mechanisms can be proposed to explain the kinetics of such systems. In any reaction model, the maximum number of proposed reaction rate constants $(k_1, \ldots, k_R)$ has to be equal to the number of independent chemical species concentrations $(c_1, \ldots, c_R)$ that are experimentally measured. Further, determinant of the Jacobian matrix formed from the mathematical expressions (solutions) of $c_1, \ldots, c_R$ has to be nonzero:

$$J = \left| \frac{\partial c_j}{\partial k_i} \right| = \begin{vmatrix} \dfrac{\partial c_1}{\partial k_1} & \dfrac{\partial c_1}{\partial k_2} & \cdots & \dfrac{\partial c_1}{\partial k_R} \\[2mm] \dfrac{\partial c_2}{\partial k_1} & \dfrac{\partial c_2}{\partial k_2} & \cdots & \dfrac{\partial c_2}{\partial k_R} \\[2mm] \cdots\cdots\cdots\cdots\cdots\cdots\cdots \\[2mm] \dfrac{\partial c_{R-1}}{\partial k_1} & \dfrac{\partial c_{R-1}}{\partial k_2} & \cdots & \dfrac{\partial c_{R-1}}{\partial k_R} \\[2mm] \dfrac{\partial c_R}{\partial k_1} & \dfrac{\partial c_R}{\partial k_2} & \cdots & \dfrac{\partial c_R}{\partial k_R} \end{vmatrix} \neq 0 \tag{8.21}$$

If the condition defined by Eq. (8.21) is satisfied, then the concentration measurements at different times can be used to determine the unknown reaction rate constants $k_1, \ldots, k_R$. For example, if the concentrations of R species, $c_1, \ldots, c_R$, are measured at different reaction times $t_1, \ldots, t_M$, then the risk function becomes

$$I = \min \sum_{m=1}^{M} \sum_{n=1}^{R} \left[c_n(t_m) - c_{n,\exp}(t_m) \right]^2 \tag{8.22}$$

from which the unknown rate constants $k_1, \ldots, k_R$ are calculated from the solution of the nonlinear equations

$$\frac{\partial I}{\partial k_r} = \frac{\partial \sum_{m=1}^{M} \sum_{n=1}^{R} \left[c_n(t_m) - c_{n,\exp}(t_m) \right]^2}{\partial k_r} = 0, \qquad r = 1, \ldots, R \tag{8.23}$$

or

$$\sum_{m=1}^{M} \sum_{n=1}^{R} \left(c_n(t_m) - c_{n,\exp}(t_m) \right) \frac{\partial c_n(t_m)}{\partial k_r} = 0, \qquad r = 1, \ldots, R \tag{8.24}$$

Solving for unknown rate constants $k_1, \ldots, k_R$ [Eq. (8.24)] with nonlinear algebraic equations could be a difficult task. Another alternative means to calculate $k_1, \ldots, k_R$ would be to find the values of rate constants that minimize the risk function I [Eq. (8.22)] (i.e., downhill simplex method). There are commercial software programs available for the solution of nonlinear equations and for searching for the minimum of a function.

The risk function I expressed by Eq. (8.7) can be used as a measure to compare the fitness of the models to correlate a set of concentration measurements. If more than one kinetic model is proposed, the model yielding the minimum I would be interpreted as the best model correlating the experimental data.

The activation energy E of a reaction rate constant k [Arrhenius relation $k = A_p \exp(-E/RT)$] [Eq. (6.16)] can be calculated from the experimentally determined reaction rate constants at different temperatures. Activation energy E and the pre-exponential constant A_p can be calculated from the best least-squares fit of the experimental data to the logarithmic form of the Arrhenius relation. In this fit, the inverse of temperature is used as the independent parameter, which linearizes the problem:

$$\ln k = \ln A_p - \frac{E}{R}\left(\frac{1}{T}\right) \tag{8.25}$$

using which, the risk function for the best least-squares fit becomes

$$I = \sum_{n=1}^{N} (\ln k - \ln k_{n,\exp})^2 = \sum_{n=1}^{N}\left[\ln A_p - \frac{E}{R}\left(\frac{1}{T_n}\right) - \ln k_{n,\exp}\right]^2 \tag{8.26}$$

The unknown kinetic parameters A_p and E can be calculated from the solution of the following linear equations:

$$\frac{\partial I}{\partial E} = \sum_{n=1}^{N} (\ln k - \ln k_{n,\exp})\frac{\partial \ln k}{\partial E}$$

$$= \sum_{n=1}^{N}\left[\ln A_p - \frac{E}{R}\left(\frac{1}{T_n}\right) - \ln k_{n,\exp}\right]\frac{1}{RT_n} = 0 \tag{8.27}$$

$$\frac{\partial I}{\partial \ln A_p} = \sum_{n=1}^{N} (\ln k - \ln k_{n,\exp})\frac{\partial \ln k}{\partial \ln A_p}$$

$$= \sum_{n=1}^{N}\left[\ln A_p - \frac{E}{R}\left(\frac{1}{T_n}\right) - \ln k_{n,\exp}\right] = 0 \tag{8.28}$$

or

$$N \ln A_p - \left(\sum_{n=1}^{N}\frac{1}{RT_n}\right)E = \sum_{n=1}^{N}\ln k_{n,\exp} \tag{8.29}$$

$$A_p = \exp\left[\frac{E}{RN}\sum_{n=1}^{N}\left(\frac{1}{T_n} + \ln k_{n,\exp}\right)\right] \tag{8.30}$$

These calculations can be performed by using commercial software programs such as linear regression software programs provided by IMSL,* RS1,† Excel, and many other software packages. In the worst case, a simple algorithm can be written for the solution for two unknowns ($\ln A_p$ and E) from two linear equations of Eqs. (8.29) and (8.30).

3. KINETIC RATE CONSTANTS IN GAS–SOLID REACTIONS

The stoichiometry of a gas–solid reaction between the gas (G) and solid (S) species producing the product gas (P) species is written as

$$G_g + \alpha_s S_s \rightarrow P_g \tag{8.31}$$

---

* IMSL, Visual Numerics, Inc. 1300 W. Sam Houston Pkwy S. Suite 150, Houston, TX 77042.
† RS1, The Olesiuk Group, P.O. Box 3453, Mission Viejo, CA 92690-3453.

where α_s is the stoichiometric coefficient associated with the solid reactant S. The rate of a gas–solid reaction expressed by the reaction stoichiometry of Eq. (8.31) can be expressed on the basis of gas or solid consumption rates, r_G and r_S:

$$r_G = \frac{\{G \text{ reacted, g}\}}{\{S, g\}\{h\}} \tag{8.32}$$

$$r_S = \frac{\{S \text{ reacted, g}\}}{\{S, g\}\{h\}} = \frac{M_S}{M_G}\alpha_s r_G \tag{8.33}$$

where M_S and M_G are the molecular weights of the solid and gas species, respectively. Reaction rate r_S defined for gas–solid reactions by Eq. (8.33) can be expressed in terms of the solid conversion x, which is defined as

$$x = \frac{m_{S,0} - m_S}{m_{S,0}} \tag{8.34}$$

where, $m_{S,0}$ is the initial mass of the solid reactant and m_S is the mass of the partially reacted solid (i.e., mass at time t). Using Eq. (8.34) in Eq. (8.33) for differential consumption of the solid species S, the reaction rate expression becomes

$$r_S = \frac{-dm_S/dt}{m_S} = \frac{m_{S,0}(dx/dt)}{m_{S,0}(1-x)} = \frac{dx/dt}{1-x} \tag{8.35}$$

where the negative sign is needed in front of dm_S/dt term to express the consumption of the solid reactant as a positive quantity.

In gas–solid reactions r_S defined by Eq. (8.35) can be determined from experimentally determined solid conversion data as a function of time. If r_G is measured as defined by Eq. (8.32) because of experimental convenience, it can be converted to r_S by using Eq. (8.33). The numerical evaluation of dx/dt needs special attention, because differentiation of numerical data is always difficult (i.e., some smoothing procedure can be adopted).

A true reaction rate r_S would be independent of reaction hydrodynamics and would be a function of temperature, concentration, pressure, and conversion. Conversion becomes a controlling factor for gas–solid reactions because the reactive surface of the solid continuously changes with the conversion. Therefore, from the reaction engineering point of view, the experimentally measured reaction rate r_S for a gas–solid reaction can be expressed in terms of temperature, pressure and composition of the gas species G, and surface characteristics of the solid species S. For this purpose, the following model can be used:

$$r_S = \frac{dx/dt}{1-x} = k'(T, p_G)f(x) \tag{8.36}$$

where $k'(T, p_G)$ is the reaction rate constant corresponding to a reference value of pore surface area for the solid reactant S and $f(x)$ is the ratio of pore surface area at solid conversion x to the reference solid surface area (i.e., initial surface area),

$$f(x) = \frac{A_s(x)}{A_{s,0}} \tag{8.37}$$

Several models are proposed for the function $f(x)$. In the volume reaction model, $f(x) = 1$, assuming that the pore surface area does not change as a function of solid conversion x. In both the shrinking-core model and the grain model, $f(x) = (1-x)^{-1/3}$ (Szekely et al., 1976). In the random pore model, $f(x)$ is expressed in the form of a complex function of x with a maximum value at a conversion level of about $0.30 < x < 0.50$ (Bhatia and Perlmutter, 1980; Gavalas, 1980).

Furthermore, the model proposed for r_S can be modified by separating the effect of true surface area, temperature, and partial pressure of gas species on $k'(T, p_G)$, i.e., by using the power law:

$$k'(T, p_G) = k(T)A_S p_G^n = k(T)A_S y_G^n p^n \tag{8.38}$$

where A_S (m^2/g) is the pore surface per unit mass of the solid, y_G is the mole fraction of species G in the gas phase, p is the total pressure, and n is the power law constant to be determined experimentally. For most gas–solid reactions the dependence of the reaction rate constant on pressure could be linear, i.e., $n = 1$. The rate expression $k(T)$ [g/(h $\cdot$ m^2 $\cdot$ atm^n] in Eq. (8.38) would be a function of temperature in the Arrhenius form,

$$k(T) = A_{p,S}e^{-E/RT} \tag{8.39}$$

which reduces r_S to the form

$$r_S = \frac{dx/dt}{1-x} = f(x)A_{S,0}A_{p,S}e^{-E/RT}y_G p \tag{8.40}$$

where $A_{p,S}$ is the pre-exponential coefficient [g/(h $\cdot$ m^2 $\cdot$ atm)], E is the activation energy (J/mol), and R is the universal gas constant [8.31 J/(K $\cdot$ mol)].

This model can interpret the concentration change measurements as long as the pressure and temperature of the system at which the reaction takes place are accurately known (measured or predicted). Temperature measurements cannot be made as simply as pressure measurements. Because of the heat of reaction, the temperature of the reacting solid may not be known with great accuracy, especially in the case of gas–solid reactions such as coal pyrolysis, coking, char combustion, and gasification. This problem is the major reason for poor interpretation of the experimental data for the gas–solid reactions.

4. MATERIAL BALANCES FOR DIFFERENT TYPES OF REACTORS

In this section material balances will be discussed for different types of reactors operating at a steady-state temperature. In these discussions homogeneous reactions will be considered, assuming that the reaction rates are totally controlled by chemical kinetics (i.e., diffusion limitations are minimized). The purpose of this review is to develop an understanding of the mathematical difficulties in the determination of rate constants using the data of different types of reactors.

4.1 Batch Reactor

The *batch reactor* is called a closed reactor; it allows energy transfer only with its surroundings (Fig. 8.2a). Material balance on a molar basis for a batch reactor has already been expressed in Eq. (8.4), which can be restated for an nth-order reaction as

$$r = \frac{1}{\alpha_j}\frac{dc_j}{dt} = kc_j^n \tag{8.41}$$

Assuming that the reaction is taking place at constant temperature operating conditions, integrating Eq. (8.41) with the initial conditions $t = 0$, $c_j = c_{j,0}$ gives

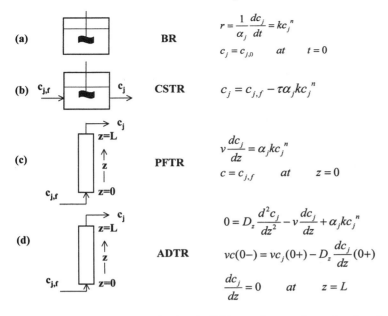

Figure 8.2 Mass balance for batch (BR), continuous flow stirred tank (CSTR), plug flow tubular (PFTR), and axially dispersed tubular (ADTR) reactors.

$$k = \frac{1}{\alpha_j t} \ln \frac{c_j}{c_{j,o}}, \qquad n = 1 \tag{8.42}$$

$$k = \frac{-1}{\alpha_j(n-1)t} \left(\frac{1}{c_j^{n-1}} - \frac{1}{c_{j,o}^{n-1}} \right), \qquad n > 1 \tag{8.43}$$

The calculation of the rate constant k from the experimentally measured concentrations as a function of time has already been discussed.

The batch reactor is a simple reactor with which to collect data for reaction kinetics studies. However, proper temperature control must be provided for highly exothermic reactions. Thermogravimetric analyzers (TGAs) are operated as batch reactors for the reactive solid sample (gas phase is not); therefore, the analysis of thermogravimetric analysis data is similar to the analysis of batch reactor data.

4.2 Continuous Flow Stirred Tank Reactor

The continuous flow stirred tank reactor (CSTR) is an open system, allowing convection mass transfer through the reactor boundaries. Ideally, mixing in the continuous flow stirred tank reactor is assumed to be perfect. The incoming fluid stream is instantly mixed with the contents of the reactor, and as a result the fluid exiting the reactor has the same composition as the fluid in the reactor. Instant mixing dilutes the concentration of the reactive species in the continuous flow stirred tank reactor, so a longer residence time is required for a given conversion (Fig. 8.2b). For a continuous flow stirred tank reactor operating at steady-state operating conditions, the material balance on a mass basis as expressed by Eq. (8.1) takes the form

$$0 = (q\rho_j)_{\text{in}} - (q\rho_j)_{\text{exit}} + VM_j\alpha_j r \tag{8.44}$$

where q is the volumetric flow rate, V is the reactor volume, α_j is the stoichiometric coefficient, r is the reaction rate, and subscripts "in" and "exit" are used to define the conditions at the reactor inlet and at the reactor exit. In Eq. (8.44) the mass density ρ_j is used for the material balance because mass density and mass are preferred in engineering practice over molar weight. To be consistent the chemical reaction term in Eq. (8.44) is multiplied by the molecular weight of the jth species, M_j.

If the total density of the fluid is constant, volumetric flow rates will also be constant (which is the case for most liquids). In this case Eq. (8.44) can be expressed in terms of mean residence time ($\tau = V/q$):

$$0 = \rho_{j,\text{in}} - \rho_{j,\text{exit}} + \tau M_j\alpha_j r \tag{8.45}$$

If the chemical reaction takes place with nth-order reaction kinetics ($r = kc_j^n$), Eq. (8.45) becomes

$$0 = \rho_{j,\text{in}} - \rho_{j,\text{exit}} + \tau M_j\alpha_j kc_j^n \tag{8.46}$$

or, dividing both sides of Eq. (8.46) by M_j (because the reactor is well mixed, $c_{j,\text{exit}} = c_j$),

$$0 = c_{j,\text{in}} - c_j + \tau\alpha_j kc_j^n \tag{8.47}$$

which is a nonlinear algebraic equation with respect to c_j (but linear with respect to τ). If a first-order reaction takes place, the mass balance becomes linear with respect to c_j. The unknown rate constants can be calculated from the best least-squares fit, as expressed by Eqs. (8.7) and (8.8).

Continuous flow stirred tank reactors are always expensive to operate and require complicated utility systems to keep constant feed and reactor operating conditions. As a result, kinetic measurements using them are always considered a secondary and complementary issue during pilot-scale tests.

4.3   Plug Flow Tubular Reactor

In a plug flow tubular reactor (PFTR) the material is assumed to flow as a plug (without axial diffusion or dispersion). The material balance on a mass basis, expressed by Eq. (8.1), applied to a differential section of a tubular reactor (Fig. 8.2c) operating at steady-state conditions can be expressed as

$$0 = (A_c v\rho_j)_z - (A_c v\rho)_{z+dz} + A_c M_j\alpha_j r \, dz \tag{8.48}$$

where A_c is the cross section of the reactor, v is the linear flow velocity (which is equal to q/A_c, the ratio of the volumetric flow rate to the cross section of the tubular reactor), and z is the distance from the reactor inlet ($A_c \, dz$ is the differential volume element). Assuming constant total mass density and therefore constant volumetric flow rate and nth-order chemical reaction kinetics, the differential mass balance becomes [approximating the second term in terms of Eq. (8.48) using Taylor series]

$$0 = -v\frac{d\rho_j}{dz} + M_j\alpha_j kc_j^n \tag{8.49}$$

or, dividing both sides of Eq. (8.49) by M_j,

$$v\frac{dc_j}{dz} = \alpha_j kc_j^n \tag{8.50}$$

Integration of Eq. (8.50) with inlet conditions of $z = 0$, $c_j = c_{j,\text{in}}$, the concentration profile of the jth species along the reactor length can be calculated as ($\xi = z/L$ is the normalized distance with respect to reactor length L)

$$\int_{c_{j,\text{in}}}^{c_j} \frac{dc}{c^n} = \frac{\alpha_j k L}{v} \int_0^1 d\xi \tag{8.51}$$

As can be seen from a comparison of Eqs. (8.41) and (8.51), the operation of batch and tubular reactors have a certain similarity (the two equations become similar when $t = z/v$). The plug flow element of the tubular reactor can be considered as a batch reactor flowing at velocity v in the tube (with a total residence time of L/v). Again, the unknown rate constants can be calculated from the best least-squares fit [as expressed by Eqs. (8.7) and (8.8)] of the solution of Eq. (8.51) to experimental data.

Like continuous flow stirred tank reactors, plug flow tubular reactors are expensive to operate and require complicated utility systems to keep constant feed and reactor operating conditions. As a result, kinetic measurement using a plug flow tubular reactor, as in the case of the continuous flow stirred tank reactor, is always considered as a secondary and a complementary issue during pilot-scale tests.

4.4 Axially Dispersed Tubular Reactor

In an axially dispersed tubular reactor (ADTR) the material is assumed to be axially dispersed as it flows through the reactor. The material balance on a mass basis, expressed by applying Eq. (8.1) to a differential section of an axially dispersed tubular reactor (Fig. 8.2d) operating at steady-state conditions, can be expressed as

$$0 = M_j D_d \frac{d^2 c_j}{dz^2} - v \frac{d\rho_j}{dz} + M_j \alpha_j k c_j^n \tag{8.52}$$

or, dividing both sides by M_j,

$$0 = D_d \frac{d^2 c_j}{dz^2} - v \frac{dc_j}{dz} + \alpha_j k c_j^n \tag{8.53}$$

where D_d is the axial dispersion coefficient (like the mass diffusion coefficient, cm²/s), v is the linear fluid velocity, α_j is the stoichiometric coefficient, z is the axial distance, and c_j is the concentration of the jth species.

Integration of Eq. (8.53) is more complex than the integration of Eq. (8.51) for the plug flow tubular reactor. The boundary conditions at the reactor inlet (discontinuous) and at the reactor exit are

$$v c_j(0-) = v c_{j,\text{in}} = v c_j(0+) - D_d \frac{dc_j}{dz}(0+) \qquad \text{at} \qquad z = 0 \tag{8.54}$$

and

$$\frac{dc_j}{dz} = 0 \qquad \text{at} \qquad z = L \tag{8.55}$$

Again, the unknown rate constant can be calculated from the best least-squares fit [as expressed by Eqs. (8.7) and (8.8)] of the solution of Eq. (8.53) with the boundary conditions of Eqs. (8.54) and (8.55) to the experimentally measured data. The main problem in these calculations is that the axial dispersion coefficient may not be known and must then be determined.

Like continuous flow stirred tank reactors and plug flow tubular reactors, axially dispersed tubular reactors are expensive to operate and require complicated utility systems to keep constant feed and reactor operating conditions. As a result, kinetic measurement using an axially dispersed tubular reactor, as in the case of continuous flow stirred tank reactors and plug flow tubular reactors, is always considered a secondary and complementary issue during pilot-scale tests.

Mass balance will be discussed in more detail in the following chapters. It must be remembered that summation of the material balance over all species must conclude that the total mass feed in and discharge out of the reactor must be balanced (because the summation of chemical reaction terms over all species has to be equal to zero). Experimental data would never satisfy such constraints, and as a result correction calculations would always be needed. In fact, correction calculations may be needed for almost all material and energy balance calculations for both laboratory and commercial plant operations.

5. EXPERIMENTAL ERRORS

In general, if a dependent variable y [$= f(x_1, \ldots, x_n)$] is a function of independent variables $x_1, \ldots, x_n$, then the errors $\Delta x_1, \ldots \Delta x_n$ made in measurements of the independent variables cause an error Δy in the dependent variable as expressed by

$$\Delta y = \frac{\partial y}{\partial x_1} \Delta x_1 + \cdots + \frac{\partial y}{\partial x_n} \Delta x_n \tag{8.56}$$

To overcome the unknown trends in the errors and their relative influences (partial derivatives), the absolute error Δy made in predicting (calculating) the dependent variable is expressed as

$$\Delta y = \left[\left(\frac{\partial y}{\partial x_1} \right)^2 (\Delta x_1)^2 + \cdots + \left(\frac{\partial y}{\partial x_n} \right)^2 (\Delta x_n)^2 \right]^{1/2} \tag{8.57}$$

In rate determination experiments, errors made in the experimental determination of concentrations (and other variables such as temperature, pressure, and time, which are the independent variables) or the dependent variables such as the unknown rate constants should be examined in the light of the above-mentioned error calculations.

The absolute error Δy made in calculating the dependent variable y predicted by Eq. (8.57) can be converted to the relative error by dividing both sides of Eq. (8.57) by y:

$$\frac{\Delta y}{y} = \Delta(\ln y) = \left[\left(\frac{\partial y}{y \partial x_1} \right)^2 (\Delta x_1)^2 + \cdots + \left(\frac{\partial y}{y \partial x_n} \right)^2 (\Delta x_n)^2 \right]^{1/2} \tag{8.58}$$

or

$$\frac{\Delta y}{y} = \Delta(\ln y) = \left[\left(\frac{\partial \ln y}{\partial x_1} \right)^2 (\Delta x_1)^2 + \cdots + \left(\frac{\partial \ln y}{\partial x_n} \right)^2 (\Delta x_n)^2 \right]^{1/2} \tag{8.59}$$

The order of the absolute errors [Eq. (8.57)] and relative errors [Eq. (8.59)] should always be estimated in all experimental studies.

6. CORRECTION CALCULATIONS

In chemical engineering applications the error analysis can be taken further by introducing correction calculations for the experimentally measured process parameters (independent variables). For the chemical kinetics rate constant determination, as an example, experimentally measured concentration variables may require correction. Corrections to be made would depend on the relative magnitudes of the experimental errors that might occur in the measurement of these variables, subject to satisfying the mass balance constraints.

As an example, for a reactor operating at steady-state operating conditions (Fig. 8.3), if two streams are coming in and one stream is going out, the experimentally measured mass flow rates of these streams have to satisfy the material balance. If the reactor is a heavy oil upgrader, the experimentally measured masses of elemental carbon, nitrogen, hydrogen, sulfur, and oxygen have to satisfy similar balance constraints. As a result, the experimentally measured mass flow rates and compositions have to be corrected to satisfy these constraints.

Correction calculations may find application not only in chemical kinetics but in other branches of engineering as well. Using correction calculations, experimental data can be checked if they make sense, and the experimental data (independent variables) can be corrected before they are used to calculate the dependent (state) variables.

In engineering practice the measured process parameters $x_1, \ldots, x_p$ (concentration, flow rate, pressure, temperature, etc.) have to satisfy certain equations for the conservation of mass, momentum, and energy. Suppose that the measured values of the independent variables $X_1, \ldots, X_p$ should satisfy a certain number of constraint equations $F_k = 0$, $k = 1, \ldots, s$ (note that x and X are used for the independent variable and its measured value, respectively),

$$F_k(x_1, \ldots, x_p) = 0, \qquad k = 1, \ldots, s \tag{8.60}$$

If the measured process variables are used, however, the F_k are found to be

$$F_k(X_1, \ldots, X_p) = z_k, \qquad k = 1, \ldots, s \tag{8.61}$$

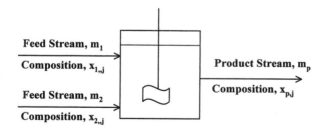

$$m_1 + m_2 = m_p$$
$$m_{1,j} + m_{2,j} = m_{p,j} \qquad \text{for} \qquad j = 1, \ldots, S$$
$$x_{k,1} + x_{k,2} + \ldots + x_{k,s} = 1 \qquad \text{for} \qquad k = 1, \ldots, p$$

Figure 8.3 Correction calculations applied to a reactor with two feeds and one effluent stream, which are all operating at steady-state conditions.

Corrections $a_1, \ldots, a_p$ have to be made on the experimentally measured process parameters $X_1, \ldots, X_p$, and the corrected independent variables are then

$$x_{i,c} = X_i + a_i, \qquad i = 1, \ldots, p \tag{8.62}$$

which would satisfy the constraints:

$$F_k(x_{1,c} \ldots x_{p,c}) = 0, \qquad k = 1, \ldots, s \tag{8.63}$$

Correction calculations suggest that the corrections $a_1, \ldots, a_p$ to be made on the process variables $x_1, \ldots, x_p$ have to be proportional to the standard deviations σ_i, $i = 1, \ldots, p$, of the experimentally measured process parameters. This policy would result in the following expression for the best corrections:

$$\sum_{i=1}^{p} \left(\frac{a_i}{\sigma_i} \right)^2 = \text{minimum} \tag{8.64}$$

This policy would allow the largest or smallest correction to be made on the process parameter measure with the largest or smallest experimental error (i.e., standard deviation), respectively.

This correction calculation problem can be solved by introducing the Lagrange multipliers. The Hamiltonian for the minimization problem is defined as

$$H = \min \left\{ \frac{1}{2} \sum_{i=1}^{p} \left(\frac{a_i}{\sigma_i} \right)^2 + \sum_{k=1}^{s} \lambda_k F_k \right\} \tag{8.65}$$

and at the minimum the gradient of the Hamiltonian would be zero:

$$\nabla H = 0 \tag{8.66}$$

or

$$\frac{\partial H}{\partial a_i} = 0, \qquad i = 1, \ldots, p \tag{8.67}$$

Using s constraint equations [Eq. (8.63)] and p equations for minimization of the Hamiltonian [Eq. (8.67)], $p + s$ unknowns ($a_1, \ldots, a_p; \lambda_1, \ldots, \lambda_s$) can be solved for.

Linearization of the constraint equations provides a solution with minimum computing effort. The constraint relations with the corrected values of the process variables can be calculated by linearizing $F_k(x_1, \ldots, x_p)$ around the experimentally measured values of the process parameters:

$$F_k(x_1, \ldots, x_p)_{X+a} = F_k(x_1, \ldots, x_p)|_X + \left. \frac{\partial F_k}{\partial x_1} \right|_X a_1 + \cdots + \left. \frac{\partial F_k}{\partial x_p} \right|_X a_p = 0 \tag{8.68}$$

where the first term on the right-hand side is z_k as expressed in Eq. (8.61), which is the discrepancy in the kth constraint. Defining $b_{k,i}$ as

$$b_{k,i} = \left. \frac{\partial F_k}{\partial x_i} \right|_X, \qquad k = 1, \ldots, s; \qquad i = 1, \ldots, p \tag{8.69}$$

the constraints [Eq. (8.68)] become

$$z_k + \sum_{i=1}^{p} b_{k,i} a_i = 0, \qquad k = 1, \ldots, s \tag{8.70}$$

Using Eq. (8.70), the Hamiltonian of the minimization problem, Eq. (8.65), becomes

$$H = \min\left\{\frac{1}{2}\sum_{i=1}^{p}\left(\frac{a_i}{\sigma_i}\right)^2 + \sum_{k=1}^{s}\lambda_k\left[z_k + \sum_{i=1}^{p}b_{k,i}a_i\right]\right\} \tag{8.71}$$

Using Eq. (8.71), the condition of minimum $\nabla H = 0$ becomes

$$\frac{\partial H}{\partial a_i} = \frac{a_i}{\sigma_i} + \sum_{k=1}^{s}\lambda_k b_{k,i} = 0, \qquad i = 1,\ldots,p \tag{8.72}$$

Equations (8.72) and (8.70) form a set of linear equations, i.e., $p + s$ equations for the solution of $p + s$ unknowns $(a_1,\ldots,a_p; \lambda_1,\ldots,\lambda_s)$, from which the solution for a_i ($i = 1,\ldots,p$) can be obtained with minimum computational effort. Once the corrections a_i are calculated, the corrected values for the measured process parameters can be recalculated using Eq. (8.62).

REFERENCES

Bhatia, S. K., and Perlmutter, D. D. 1980. AIChE J. 26(3):379–386.
Gavalas, G. R. 1980. AIChE J. 26(4):577–585.
Szekely, J., Evans, J. W., and Sohn, H. Y. 1976. Gas–Solid Reactions. Academic Press, New York.

9
Reactor Design

1. INTRODUCTION

Chemical reactors are designed to provide acceptable performance, i.e., conversion of reactant(s) to desired product(s), at minimum cost. In converting reactants to products, certain constraints have to be satisfied. These constraints are not universal and may change over time as well as by geographic and political situations. The most common constraints faced in reactor design are that the reactor must be economical to operate, reliable, controllable, safe, environmentally acceptable, easy to maintain, energy efficient, and capable of handling start-up and shutdown operations.

Reactor design is not a straightforward process; it requires skills from many branches of engineering science and economics and practical engineering experience. The design of many commercial reactors for industrially important processes has had significant inputs from experience in related fields and common sense. Significant progress has been made since the late 1800s in the science of reactor engineering and reactor design. Advances in computer technology made a big difference in reactor design by providing process simulations (mass and energy balances), thermodynamics equilibrium calculations, reactor dynamics, and control simulations at affordable costs.

The steady-state mass and energy balances are needed to determine the reactor size and steady-state reactor operating conditions and to design the utilities needed for the reactor. Unsteady-state mass and energy balances may be required for the reactor dynamics and the selection of control mechanisms. Design of the utility units and selection of the materials to handle the reactor content at the designed operating conditions (i.e., pH, temperature, and pressure) require mechanical and metallurgical engineering expertise.

In this section, mass and energy balances for batch reactors (BRs), continuous flow stirred tank reactors (CFSTRs), plug flow tubular reactors (PFTRs) and axially dispersed tubular reactors (ADTRs) will be discussed with respect to reactor design.

Mass and energy balance equations will be formulated by considering the accumulation and generation of a property (i.e., mass or energy) and its flow through the boundaries of a control volume, as expressed in the relation

$$(\text{property})_{\text{accum}} = (\text{property})_{\text{in}} - (\text{property})_{\text{out}} + (\text{property})_{\text{gen}} \qquad (9.1)$$

Applications of Eq. (9.1) to describe the performance of continuous flow stirred tank reactors, batch reactors, plug flow tubular reactors, and axially dispersed tubular reactors will be discussed. This discussion will be limited to reactor design problems; the application of Eq. (9.1) to mass, momentum, and energy transfer problems in chemically reactive systems will be discussed in Chapter 11.

2. CONTINUOUS FLOW STIRRED TANK REACTORS

Continuous flow stirred tank reactors (CSTRs) are well-stirred reactor vessels. Mixing in a continuous flow stirred tank reactor is maintained by a proper selection of impeller and baffle(s) attached to the interior of the reactor (Fig. 9.1). The feed material is continuously fed into the reactor, and the partially reacted mixture in the reactor is continuously taken out from the reactor. The product stream may be treated by some kind of separation process, and part of it may be recycled back to the reactor. Continuous flow stirred tank reactors always provide good and controlled mixing conditions; as a result they are used for liquid-phase reactions with negligible volume changes. These reactors are cost effective, easy to design and commission, and relatively simple to operate. In general, continuous flow stirred tank reactors are recommended when desired conversions are low.

The main advantage of the continuous flow stirred tank reactors is that good mixing can be achieved in the reacting fluid mixture (in comparison to most tubular reactors). Ideally, the characteristic mixing time should be negligible in comparison to the characteristic residence time; however, it may not be easy to measure either of these characteristic times. Both of them depend on the reactor geometry, fluid characteristics, feed rate, stirrer and baffle design, and mixing rate (or intensity). As a result, the duration of time in which the reactant is at intermediate concentration is negligible in comparison to the residence time (hold-up time) of the feed in the reactor.

In a continuous flow stirred tank reactor the feed material is instantly mixed with the reactor content and diluted. Low reactant concentrations decrease the rate of reaction kinetics (i.e., for a first-order reaction, $r = kc$), which results in an increase in the required reactor volume to achieve a desired conversion (i.e., in comparison to plug flow tubular

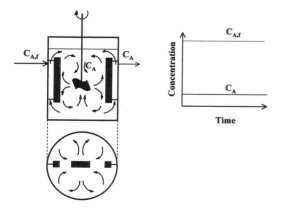

Figure 9.1 Mixing in a CSTR is maintained by the selection of proper impeller and baffle(s).

reactors, in which the reactants are reacted at higher concentrations, requiring smaller reactor volumes). The need for a larger reactor volume in continuous flow stirred tank reactor operations may raise some concern. This concern becomes more serious if a relatively high conversion level is desired.

Real continuous flow stirred tank reactors never operate at ideal operating conditions. A fraction of the feed material may not stay in the reactor long enough to react (bypassing) and another fraction of the feed material may stay in the reactor longer than expected from design considerations. The consequence of the dilution can be reduced by using two or more continuous flow stirred tank reactors in series. Changing the reactor geometry, baffle position, and mixing intensity can rectify the effect of the irregularity in the residence time distributions on reactor performance.

2.1 Mass Balance for Continuous Flow Stirred Tank Reactors

The assumption of perfect mixing in the continuous flow stirred tank reactor is the key element in formulating the material balance. For isothermal, constant-volume reactor operating conditions (i.e., constant density of the reacting mixture), if q_f and q are the volumetric flow rates (in liters per second) of the feed and effluent streams and $c_{j,f}$ and c_j are the corresponding concentrations (in moles per liter) of the jth species, then, using the principle expressed in Eq. (9.1), the material balance on a molar basis can be expressed as

$$\frac{d(Vc_j)}{dt} = q_f c_{j,f} - qc_j + V\alpha_j r(c, T, P) \tag{9.2}$$

The chemical consumption rate $r(c, T, P)$ in Eq. (9.2) is a function of the reactant concentration and the temperature and pressure of the reactor.

The volume change in the continuous flow stirred tank reactor is given by

$$\frac{dV}{dt} = q_f - q \tag{9.3}$$

If the feed and discharge volumetric flow rates are equal ($q_f = q$), then dV/dt is zero and the reactor volume becomes constant. Under these conditions, Eq. (9.2) can be expressed in terms of the mean residence time τ ($= V/q$) by dividing both sides of Eq. (9.2) by q:

$$\tau \frac{dc_j}{dt} = c_{j,f} - c_j + \tau \alpha_j r(c, T, P) \tag{9.4}$$

or, considering the relation expressed in Eqs. (5.29) and (9.4) becomes (i.e., $c_j = c_{j,f} + \alpha_j \xi$)

$$\tau \frac{d\xi}{dt} = -\xi + \tau r(\xi, T, P) \tag{9.5}$$

If the reactor is operating at steady-state operating conditions, the time derivative term of Eqs. (9.4) and (9.5) would be zero:

$$c_j = c_{j,f} + \tau \alpha_j r(c, T, P) \tag{9.6}$$

and, using the stoichiometric relation $c_j = c_{j,f} + \alpha_j \xi$ [or by setting $d\xi/dt = 0$ in Eq. (9.5)], Eq. (9.6) becomes

$$\xi = \tau r(\xi, T, P) \tag{9.7}$$

Both of the steady-state versions of the material balance expressed by Eqs. (9.6) and (9.7) are useful relations for the calculation of reactor volume, operating temperature, and number of reactors for a desired conversion.

2.2 Reactor Volume and Reactor Optimization

Suppose that a first order ($r = kc_A$) reaction

$$A \rightarrow B \tag{9.8}$$

is carried out in a continuous flow stirred tank reactor operating at steady-state and isothermal conditions. The mass balance then becomes

$$c_A = c_{A,f} - \frac{V}{q} kc_A = c_{A,f} - \tau kc_A \tag{9.9}$$

Equation (9.9) can be rearranged to express the concentration change for species A as a function of process variables V, q, and k:

$$\frac{c_{A,f} - c_A}{c_A} = \frac{x}{1-x} = \frac{kV}{q} \tag{9.10}$$

which is a useful relation for calculating the required reaction volume for a given conversion x [$= [(c_{A,f} - c_A)/c_{A,f}]$. As an example, using Eq. (9.10), to achieve 95% conversion, kV/q is calculated as

$$\frac{kV}{q} = \frac{x}{1-x} = \frac{0.95}{1-0.95} = 19 \tag{9.11}$$

For a comparison, if a plug flow tubular reactor were used, to achieve the same conversion kV/q would be equal to

$$-\int_{c_{A,f}}^{c_A} \frac{dc}{c} = \ln\left(\frac{c_{A,f}}{c_A}\right) = \frac{k}{q} A_c \int_0^L dz = \frac{k}{q} \int_0^L A_c \, dz = \frac{kV}{q} \tag{9.12}$$

from which the required kV/q for 95% conversion (i.e., $c_{A,f}/c_A$) is calculated as

$$\frac{kV}{q} = \ln\left(\frac{c_{A,f}}{c_A}\right) = \ln 20 = 3.0 \tag{9.13}$$

This example shows that the volume required to achieve 95% conversion is 6.33 ($= 19/3$) times as large in a continuous flow stirred tank reactor as that of a plug flow tubular reactor. For 50% conversion, kV/q ratios are calculated as 1 and 0.693 for continuous flow stirred tank reactors and plug flow tubular reactors, respectively, predicting a reactor volume 1.44 ($= 1/0.693$) times as large for a continuous flow stirred tank reactor as that of a plug flow tubular reactor. This discussion shows that continuous flow stirred tank reactors are not suitable reactors for achieving very high conversion levels.

To overcome the need for a large-volume continuous flow stirred tank reactor, more than one continuous flow stirred tank reactor can be used in series. If the reaction rate is known as a function of the reactant concentration (Fig. 9.2), then the reactor volume (operating at a constant temperature) can be calculated graphically. Rearranging Eq. (9.6), the reactant concentration c_A for each reactor can be calculated as

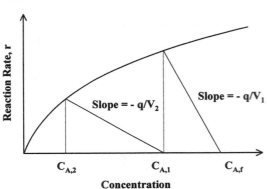

Figure 9.2 Graphical method for the optimization of total reactor volume for two or more continuous flow stirred tank reactors used in series.

$$c_A = c_{A,f} - \frac{V}{q} r(c_A) \qquad \text{or} \qquad r(c_A) = \frac{q}{V} c_{A,f} - \frac{q}{V} c_A \qquad (9.14)$$

where, $c_{A,f}$ and c_A are the concentration of the reactant species in the feed and in the exit stream, respectively, for each continuous flow stirred tank reactor in the series. For a given set of feed and exit compositions of the reactant species, a graphical method can be used (Fig. 9.2) to calculate the reactor volumes for any number of reactors. The slope of the line connecting the feed concentration and the reaction rate at the reactor concentration (the same as the reactor exit concentration) would be equal to $-(q/V)$, from which the reactor volume V (L) can be calculated for a given volumetric flow rate q (L/s).

If the reaction kinetics is first-order, i.e., $r = kc_A$, then Eq. (9.14) can be expressed as

$$c_A = \frac{c_{A,f}}{1 + (V/q)k} = \frac{c_{A,f}}{1 + \tau k} \qquad (9.15)$$

Using Eq. (9.15), if there are two isothermal reactors operating in series, then the concentration of the reactive species in the exit stream of the second reactor would be

$$c_{A,2} = \frac{c_{A,1}}{1 + \tau_2 k} = \frac{c_{A,f}}{(1 + \tau_1 k)(1 + \tau_2 k)} \qquad (9.16)$$

where τ_1 and τ_2 are the residence times in the first and second reactors. Consequently, for N isothermal reactors operating in series, the concentration of the reactive species in the exit stream of the Nth reactor would be

$$c_{A,N} = \frac{c_{A,f}}{(1 + \tau_1 k)(1 + \tau_2 k) \cdots (1 + \tau_N k)} \qquad (9.17)$$

For a constraint of a fixed total residence time τ_T,

$$\tau_T - \sum_{n=1}^{N} \tau_n = 0 \qquad (9.18)$$

a set of residence times $\tau_1, \ldots, \tau_N$ (or reactor volumes) for N continuous flow stirred tank reactors in series required to achieve maximum conversion can be calculated. Since the maximum conversion corresponds to the minimum $c_{A,N}$, this problem can be solved by the method of Lagrangian multipliers. The Hamiltonian of the minimization problem can be expressed as

$$H = \frac{c_{A,f}}{(1 + \tau_1 k)(1 + \tau_2 k) \cdots (1 + \tau_N k)} + \lambda \left(\tau_T - \sum_{n=1}^{N} \tau_n \right) \tag{9.19}$$

Conditions for the minimum would be expressed as ($\nabla H = 0$)

$$\frac{\partial H}{\partial \tau_n} = \left(\frac{1}{1 + \tau_n k} \right) \left(\frac{-k c_{A,f}}{(1 + \tau_1 k)(1 + \tau_2 k) \cdots (1 + \tau_N k)} \right) - \lambda, \qquad n = 1, \ldots, N \tag{9.20}$$

Eliminating λ from any one of these conditions, i.e., eliminating λ from the conditions for the minimum [Eq. (9.20)] for $n = 1$ and $n = 2$, gives

$$\frac{1}{1 + \tau_1 k} - \frac{1}{1 + \tau_2 k} = 0 \tag{9.21}$$

or

$$\tau_1 = \tau_2 \tag{9.22}$$

Eliminating λ from any pair of minimization conditions gives a relation identical to that expressed by Eq. (9.21). Therefore the maximum conversion (or minimum $c_{A,N}$) would be achieved when the residence times of the reactors are selected to be equal:

$$\tau_1 = \tau_2 = \cdots = \tau_N \tag{9.23}$$

which is the condition of selecting the same size reactors:

$$V_1 = V_2 = \cdots = V_N \tag{9.24}$$

As can be seen from Eq. (9.24), for a given total residence time (reactor volume) constraint, equal-volume reactors have to be selected for maximum conversion.

This method can be generalized for the selection of the best combination of temperatures and residence times for a series of N reactors carrying out complex reactions. Conversion of the feed into a certain product could be undesirable (i.e., production of gaseous hydrocarbons in the hydrocracking processes) as a strategic policy for the reactor design. As an example, if the complex reaction system involving R first-order reactions is carried out in a series on N continuous flow stirred tank reactors, conversion of the feed into undesired product ($x_T = x_1 + \cdots + x_N$) can be minimized by selecting the best combination of reactor temperatures $T_1, \ldots, T_N$ and residence times $\tau_1, \ldots, \tau_N$ for the constraint of a total residence time of τ_N. This problem can be expressed as

$$x_T = \min(x_1 + \cdots + x_N) = \min \left(\sum_{n=1}^{N} x_n(T_1, \ldots, T_N, \tau_1, \ldots, \tau_N) \right) \tag{9.25}$$

where x_n is the conversion of the feed to undesired product in the nth reactor. The total conversion x_T of the feed into undesired product would be minimized by satisfying the constraint of a fixed total residence time:

$$\tau_T - (\tau_1 + \cdots + \tau_N) = \tau_T - \sum_{n=1}^{N} \tau_n = 0 \tag{9.26}$$

Conversion in the nth reactor, expressed in Eq. (9.25), would depend on the performances of all $n - 1$ reactors. The performances of these reactors are set by the reaction rates and residence times; therefore by the temperatures and residence times in these reactors. Therefore, the Hamiltonian of the minimization problem would be

$$H = \{x_T + \lambda(\tau_T - (\tau_1, \ldots, \tau_N))\} = \left\{ \sum_{n=1}^{N} x_n(T_1, \ldots, T_n, \tau_1, \ldots, \tau_n) + \lambda\left(\tau_T - \sum_{n=1}^{N} \tau_n\right) \right\}$$

(9.27)

from which the conditions of the minimum are expressed as (i.e., $\nabla H = 0$)

$$\frac{\partial H}{\partial T_n} = \frac{\partial}{\partial T_n}\left(\sum_{k=n}^{N} x_k(T_1, \ldots, T_k, \tau_1, \ldots, \tau_k) \right) = 0, \qquad n = 1, \ldots, N \qquad (9.28)$$

$$\frac{\partial H}{\partial \tau_n} = \frac{\partial}{\partial \tau_n}\left(\sum_{k=n}^{N} x_k(T_1, \ldots, T_k, \tau_1, \ldots, \tau_k) \right) - \lambda = 0, \qquad n = 1, \ldots, N \qquad (9.29)$$

Derivatives of the Hamiltonian with respect to temperature [Eq. (9.28)] are calculated by considering the fact that the conversions are functions of the reaction rate constants $k_1, \ldots, k_R$ and the reaction rate constants are functions of temperature (Arrhenius relation). Therefore, Eq. (9.28) can be expressed as

$$\frac{\partial H}{\partial T_n} = \frac{\partial x_T}{\partial T_n} = \sum_{r=1}^{R} \frac{\partial x_T}{\partial k_r}\frac{\partial k_r}{\partial T_n} = 0, \qquad n = 1, \ldots, N \qquad (9.30)$$

where the derivative of conversion x_n (conversion in the nth reactor) cannot be a function of the temperatures of the reactors numbered $n + 1, \ldots, N$. Therefore, Eq. (9.30) can be rearranged as

$$\frac{\partial H}{\partial T_n} = \sum_{k=n}^{N}\left(\sum_{r=1}^{R} \frac{\partial x_k}{\partial k_r}\frac{\partial k_r}{\partial T_k} \right) = 0, \qquad n = 1, \ldots, N \qquad (9.31)$$

Also, the derivatives of reaction rate constants with respect to temperature are

$$\frac{\partial k_r}{\partial T_n} = A_{pr}\exp\left(-\frac{E_r}{RT_n}\right)\frac{E_r}{RT_n^2} = k_r(T_n)\frac{E_r}{RT_n^2} \qquad (9.32)$$

where $k_r(T_n)$ is the rth reaction rate constant for the nth reactor temperature, and A_{pr} and E_r are the pre-exponential constant and activation energy of the rth reaction. Using Eqs. (9.31) and (9.32), the minimization conditions expressed by Eq. (9.28) become

$$\frac{\partial H}{\partial T_n} = \sum_{k=n}^{N}\left(\sum_{r=1}^{R} \frac{\partial x_k}{\partial k_r}k_r(T_k)\frac{E_r}{RT_k^2} \right) = 0, \qquad n = 1, \ldots, N \qquad (9.33)$$

The $2N + 1$ unknowns, $T_1, \ldots, T_N$, $\tau_1, \ldots, \tau_N$, and λ can be solved from $2N + 1$ equations of Eqs. (9.26), (9.29) and (9.33).

The Hamiltonian of the minimization problem may not have an optimum in the temperature and residence time domain of interest. In this case the optimum solution would be on the boundaries of the temperature and residence time domain. If the optimum solution is found in the regime of high temperatures and short residence times, then a reactor with flash heating and a short residence time should be considered, as in the case of flash pyrolysis of heavy hydrocarbons to produce light hydrocarbons. The optimal solution can be in the low temperature and long residence time domain as in the case of mild hydrocracking (i.e., visbreaking) of heavy oil and bitumen and in the domain of high temperature and long residence time as in the case of hydrocracking of heavy oil and bitumen.

2.3 Independence of Material Balances

If more than one reaction is considered, the number of independent mass balances has to be known (Aris, 1969). As an example, mass balances for species A and B involved in first-order reversible reactions $A \rightleftharpoons B$ taking place in a continuous flow stirred tank reactor can be expressed as

$$\frac{d(Vc_A)}{dt} = c_A \frac{dV}{dt} + V \frac{dc_A}{dt} = q_f c_{A,f} - q c_A - V(kc_A - k'c_B) \tag{9.34}$$

and

$$\frac{d(Vc_B)}{dt} = c_B \frac{dV}{dt} + V \frac{dc_B}{dt} = q_f c_{B,f} - q c_B + V(kc_A - k'c_B) \tag{9.35}$$

where k and k' are the rate constants of the forward and reverse reactions.

If the volumetric flow rates of the feed and discharge streams are equal, $q_f = q$, then the volume of the reactor is also constant and Eqs. (9.34) and (9.35) can be expressed in terms of the mean residence time $\tau \; (= V/q)$,

$$\tau \frac{dc_A}{dt} = c_{A,f} - c_A - \tau(kc_A - k'c_B) \tag{9.36}$$

$$\tau \frac{dc_B}{dt} = c_{B,f} - c_B + \tau(kc_A - k'c_B) \tag{9.37}$$

If concentrations are expressed in terms of the intensive reaction extent ξ,

$$c_A = c_{A,f} - \xi, \qquad c_B = c_{B,f} + \xi \tag{9.38}$$

the mass balance equations (9.36) and (9.37) become

$$-\tau \frac{d\xi}{dt} = \xi - \tau \left[k(c_{A,f} - \xi) - k'(c_{B,f} + \xi) \right] \tag{9.39}$$

and

$$\tau \frac{d\xi}{dt} = -\xi + \tau \left[k(c_{A,f} - \xi) - k'(c_{B,f} + \xi) \right] \tag{9.40}$$

which are identical expressions. In fact, the two mass balance equations, Eqs. (9.39) and (9.40), are not independent, because only one independent reaction can exist for the two species A and B. That means that one balance equation as a function of one reaction extent ξ is sufficient to represent the mass balance for these reactions.

2.4 Energy Balance for Continuous Flow Stirred Tank Reactors

Consider a continuous flow stirred tank reactor operating at a constant volumetric flow rate of feed and discharge streams q. The unsteady-state energy balance in terms of the enthalpy of the species involved in the reaction can be expressed as (Aris, 1969)

$$\frac{d}{dt} \left(V \sum_{j=1}^{s} c_j h_j(T) \right) = q \sum_{j=1}^{s} c_{j,f} h_j(T_f) - q \sum_{j=1}^{s} c_j h_j(T) - Q^* \tag{9.41}$$

where h is the molar enthalpy (kJ/mol) and Q^* is the rate of heat exchange (kJ/s or kW) between the reactor and its surroundings ($Q^* > 0$ if the heat is taken out from the reacting system). Because the reactor contents are perfectly mixed, the temperature of the discharge stream is the same as the reactor operating temperature T, and the feed temperature is T_f. If the volumetric feed rates and the reactor volume are constant, Eq. (9.41) can be expressed in terms of the average residence time τ ($= V/q$) by dividing both sides of the equation by q:

$$\tau \frac{d}{dt}\left(\sum_{j=1}^{s} c_j h_j(T)\right) = \sum_{j=1}^{s} c_{j,f} h_j(T_f) - \sum_{j=1}^{s} c_j h_j(T) - \frac{Q^*}{q} \tag{9.42}$$

The time derivative term of Eq. (9.42) is

$$\frac{d}{dt}\left(\sum_{j=1}^{s} c_j h_j(T)\right) = \sum_{j=1}^{s} \frac{dc_j}{dt} h_j(T) + \sum_{j=1}^{s} c_j \frac{dh_j}{dT}\frac{dT}{dt} \tag{9.43}$$

The first term on the right-hand side of Eq. (9.43) can be derived by multiplying both sides of the mass balance equation for continuous flow stirred tank reactors [Eq. (9.4)] by $h_j(T)$ and summing over all (s) species:

$$\tau\left[\sum_{j=1}^{s} \frac{dc_j}{dt} h_j(T)\right] = \sum_{j=1}^{s} c_{j,f} h_j(T) - \sum_{j=1}^{s} c_j h_j(T) + \tau r \sum_{j=1}^{s} \alpha_j h_j(T) \tag{9.44}$$

The second term on the right-hand side of Eq. (9.43) can be expressed in terms of specific heats [because $dh_j(T)/dT = C_{p,j}$ is the molar specific heat of the jth species] as

$$\sum_{j=1}^{s} c_j \frac{dh_j(T)}{dT}\frac{dT}{dt} = \sum_{j=1}^{s} c_j C_{p,j} \frac{dT}{dt} = C_p \frac{dT}{dt} \tag{9.45}$$

where C_p is the enthalpy of the fluid per unit volume (because c_j is the molar concentration per unit volume). By inserting Eqs. (9.43), (9.44), and (9.45) into Eq. (9.42), and assuming R reactions are taking place the energy balance equation becomes [because $\sum_j \alpha_h h_j(T) = \Delta H$ is the reaction enthalpy]

$$\tau C_p \frac{dT}{dt} = \sum_{j=1}^{s} c_{j,f} h_f(T) - \sum_{j=1}^{s} c_j h_j(T) - \tau \sum_{i=1}^{R} (\Delta H_i) r_i - \frac{Q^*}{q} \tag{9.46}$$

If the heat capacities are assumed to be constant, then Eq. (9.46) can be expressed in terms of specific heats:

$$\tau C_p \frac{dT}{dt} = C_p(T_f - T) - \tau \sum_{i=1}^{R} (\Delta H_i) r_i - \frac{Q^*}{q} \tag{9.47}$$

which can be reduced to

$$\tau \frac{dT}{dt} = T_f - T - \tau \sum_{i=1}^{R} \left(\frac{\Delta H_i}{C_p}\right) r_i - \frac{Q^*}{qC_p} \tag{9.48}$$

or

$$\tau \frac{dT}{dt} = T_f - T + \tau \sum_{i=1}^{R} J_i r_i - Q \tag{9.49}$$

where

$$J_i = -\frac{\Delta H_i}{C_p}, \qquad Q = \frac{Q^*}{qC_p} \tag{9.50}$$

For the steady-state operating conditions, the energy balance becomes

$$T_f - T + \tau \sum_{i=1}^{R} J_i r_i - Q = 0 \tag{9.51}$$

The energy balances expressed by Eq. (9.49) and (9.51) are coupled with the mass balance expressed by Eq. (9.4) or (9.5) in a nonlinear fashion, because the chemical reaction rate is an exponential function of temperature. This nonlinear coupling makes the solution of mass and energy balances a difficult task and is responsible for many nonlinear phenomena in reaction engineering such as reactor stability, spontaneous ignition, flame stability, and explosion.

2.5 Reactor Design and Reactor Stability for Continuous Flow Stirred Tank Reactors

Mass and energy balances are useful tools for reactor design and the prediction of reactor stability. For reactor design, the steady-state forms of mass and energy balance equations [i.e., Eqs. (9.4) and (9.51)] are needed. For a simple exothermic reaction $\alpha_j A_j \rightarrow P$ taking place at constant volume, the steady-state mass and energy balance equations are (Aris, 1969)

$$c_{j,f} - c_j + \tau \alpha_j r(c_j, T, P) = 0 \tag{9.52}$$

$$T_f - T - \tau J r(c_j, T, P) - Q = 0 \tag{9.53}$$

or

$$-\xi + \tau r(\xi, T, P) = 0 \tag{9.54}$$

$$T_f - T + \tau J r(\xi, T, P) - Q = 0 \tag{9.55}$$

In reactor design, there are five variables (dependent and independent), T_f, T, ξ, τ (or V/q), and Q. Only two of these can be calculated using Eqs. (9.54) and (9.55) [or Eqs. (9.52) and (9.53)], whereas the others are given as the process operating conditions. Which variables are given and which variables are to be calculated can differ for each reactor design application. Generally, the feed composition $c_{j,f}$, feed temperature T_f, reaction extent ξ, and feed flow rate q are given, and the reaction operating temperature T and auxiliary cooling or heating Q are calculated. If reactor design calculations show that for a given set of process parameters the operation of the reactor cannot be tolerated because of practical reasons, then certain process modifications can be made (such as altering the conversion level, feed composition, feed temperature, and feed flow rate).

Because of the nonlinear coupling between the mass and heat balance equations, the steady-state operating conditions may not be stable. Small variations from the steady-state operating conditions may result in different responses depending on the nature of the reactor performance near those conditions. The nature of the steady-state operating conditions in its neighborhood of the reactor would provide information on its stability. Such an analysis may help in the design of the auxiliary cooling system as well as in the choice of control action to be taken to keep the reactor operating conditions around the desired values (i.e., steady-state values).

It is obvious that coupled time-dependent mass and energy balances are needed to study reactor stability. For a simple exothermic, constant-volume (constant-density) reaction $\alpha_j A_j \rightarrow P$, the following set of unsteady-state mass and energy balance equations have to be solved simultaneously:

$$\tau \frac{dc_j}{dt} = c_{j,f} - c_j + \tau \alpha_j r(c, T, P) \tag{9.56}$$

$$\tau \frac{dT}{dt} = T_f - T + \tau J r(c, T, P) - Q \tag{9.57}$$

or

$$\tau \frac{d\xi}{dt} = -\xi + \tau r(\xi, T, P) \tag{9.58}$$

$$\tau \frac{dT}{dt} = T_f - T + \tau J r(\xi, T, P) - Q \tag{9.59}$$

The effect of the auxiliary cooling [Q in Eq. (9.57)] on the reactor stability can be studied by formulating the heat transfer between the reactor and the cooling fluid. Circulating the cooling fluid through a heat exchanger located inside or outside of the reactor or through a jacket surrounding the reactor constitutes auxiliary cooling. If the cooling fluid is sufficiently mixed during the cooling process, then it can be assumed to be at a uniform temperature inside the cooling system. In this case the energy balance for the auxiliary cooling can be expressed as

$$Q^* = UA_h(T - T_c) = q_c C_{p,c}(T_c - T_{c,f}) \tag{9.60}$$

where U is the overall heat transfer coefficient; A_h is the heat transfer surface area of the cooling system; q_c, $C_{p,c}$, and $T_{c,f}$ are the feed rate, specific heat, and feed temperature, respectively, of the cooling fluid; T is the reactor operating temperature; and T_c is the average temperature of the cooling fluid in the cooler. Eliminating T_c from the heat balance equations expressed by Eq. (9.60), Q^* can be expressed as

$$Q^* = \frac{UA_h q_c C_{p,c}}{UA_h + q_c C_{p,c}}(T - T_{c,f}) \tag{9.61}$$

by which the energy balance equation [Eq. (9.59)] becomes

$$\tau \frac{dT}{dt} = T_f - T + \tau J r(T, \xi, P) - \kappa(T - T_{c,f}) \tag{9.62}$$

where κ is the dimensionless coefficient given by

$$\kappa = \frac{UA_h q_c C_{p,c}}{q c_p(UA_h + q_c C_{p,c})} \tag{9.63}$$

It is a difficult, if not impossible, task to find closed-form analytical solutions of the coupled mass and energy balance equations [i.e., Eqs. (9.56) and (9.62) or Eqs. (9.58) and (9.62)]. To predict the stability of the steady-state reactor operating conditions, first the steady-state solutions (i.e., time derivative terms are set to zero) can be obtained from the simultaneous solution of the nonlinear algebraic equations and then the tendency of the steady-state solution in its neighborhood can be examined. As an example, if the reactor temperature is predicted to increase or decrease naturally as a result of a small disturbance imposed on the steady-state temperature, then it can be said that the examined steady state is unstable or stable, respectively. A small disturbance in the steady-state temperature may cause a runaway or quenching in the reactor operating temperature. If proper control actions are not taken, then the reactor may perform out of control, resulting in an undesired increase (even explosion) or quenching in the reactor operating temperature.

By setting the time derivative terms to zero, the mass and energy balance equations become

$$\xi - \tau r(\xi, T, P) = 0 \tag{9.64}$$

and

$$(T - T_f) + \kappa(T - T_{c,f}) - \tau Jr(\xi, T, P) = 0 \tag{9.65}$$

The first two terms $[(T - T_f)$ and $\kappa(T - T_{c,f})]$ of Eq. (9.65) are related to the heat removal, and the third term is related to the heat generated by the exothermic reaction [this can be seen by multiplying both sides of Eq. (9.65) by qc_p]. The heat removal terms can be simplified further by introducing a hypothetical temperature T_c^* (Aris, 1969):

$$(1 + \kappa)(T - T_c^*) - \tau Jr(\xi, T, P) = 0 \tag{9.66}$$

where

$$T_c^* = \frac{T_f + \kappa T_{c,f}}{1 + \kappa} \tag{9.67}$$

Again, the first term in Eq. (9.66) is the heat removal term and the second is the heat generation term. Assigning U_r and U_g as the heat generation and heat removal terms gives

$$U_r = T - T_f + \kappa(T - T_{c,f}) = (1 + \kappa)(T - T_c^*) \tag{9.68}$$

$$U_g = \tau Jr(\xi, T, P) \tag{9.69}$$

The energy balance can be depicted graphically by plotting the relations expressed by Eqs. (9.68) and (9.69).

The heat generation U_g increases exponentially with temperature in the moderate temperature range and slows down at higher temperature regimes as a result of diffusion limitations. As a result the heat generation U_g would form an S shape (i.e., there is a temperature at which $d^2 U_g/dT^2 = 0$). The heat removal U_r can be controlled by adjusting the feed temperature T_f or cooling fluid inlet temperature $T_{c,f}$, which would be linear with reactor temperature (i.e., a straight line). Considering these facts, the effects of T_f or T_c^* and κ on U_r, and the intersections of U_r and U_g can be derived (Fig. 9.3). As can be seen from the energy balance [Eq. (9.68)], the heat removal U_g is always a linear function of temperature.

The temperatures corresponding to the intersections of U_r and U_g are the steady-state temperatures for a given set of reactor operating conditions. These steady-state operating temperatures can be examined more closely without solving the coupled mass and energy balance differential equations to gain an understanding of the nature of the steady-state temperatures.

Temperature $T_{s,1}$ (Fig. 9.3a) is a stable steady-state temperature, because any small disturbance from it will naturally bring the temperature back to $T_{s,1}$. For a reactor temperature slightly lower than $T_{s,1}$, energy generation would be greater than energy removal ($U_g > U_r$), and the reactor temperature would be pushed up to $T_{s,1}$. For a reactor tem-

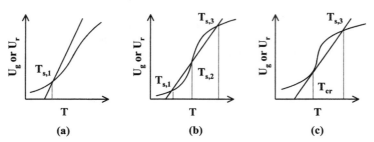

Figure 9.3 Intersections of heat generation curve U_g and heat removal (cooling) line U_r for a continuous flow stirred tank reactor.

perature slightly higher than $T_{s,1}$, energy generation would be smaller than energy removal ($U_g < U_r$), and the reactor temperature would be pulled back to $T_{s,1}$. As a result, $T_{s,1}$ is called the stable steady-state temperature, provided that the initial disturbance in $T_{s,1}$ is small enough.

Similarly, temperature $T_{s,3}$ as shown in Fig. 9.3b is also a stable steady-state temperature, because any small disturbance from it will naturally bring the temperature back to $T_{s,3}$. For a reactor temperature slightly lower than $T_{s,3}$, energy generation would be greater than energy removal ($U_g > U_r$), and the reactor temperature would be pushed up to $T_{s,3}$. For a reactor temperature slightly higher than $T_{s,3}$, energy generation would be smaller than energy removal ($U_g < U_r$), and the reactor temperature would be pulled back to $T_{s,3}$. As a result, $T_{s,3}$ is also called a stable steady-state temperature, provided that the initial disturbance in $T_{s,3}$ is small enough.

The nature of the steady-state temperature $T_{s,2}$ (Fig. 9.3b) is different than that of $T_{s,1}$ and $T_{s,3}$. For a reactor temperature slightly higher than $T_{s,2}$, energy generation would be greater than energy removal ($U_g > U_r$), and the reactor temperature would be pushed up to the next steady-state temperature $T_{s,3}$. On the other hand, for a reactor temperature slightly lower than $T_{s,2}$, energy generation would be smaller than energy removal ($U_g < U_r$), and the reactor temperature would be pulled back to the steady-state temperature $T_{s,1}$. That means that the steady-state temperature $T_{s,2}$ is an unsteady steady-state temperature; a small disturbance from $T_{s,2}$ would not rectify itself naturally and would force the reactor temperature to move to the higher or lower steady-state temperature.

As the heat removal line U_g moves toward the right-hand side (i.e., from an increase in the cooling fluid temperature), the first two steady-state temperatures $T_{s,1}$ and $T_{s,2}$ get closer to each other and approach a critical temperature T_{cr}, as shown in Fig. 9.3c. At this condition the critical temperature T_{cr} becomes unsteady, which explains many nonlinear phenomena such as spontaneous combustion and ignition.

The natural trend in the movement of the reactor operating condition toward the higher or lower steady-state operating temperature may take place violently and result in explosion and quench. Unfortunately, most industrial reactors are designed to operate at unsteady steady-state operating temperatures, thus requiring suitable action to be taken to control the reactor operating temperatures.

The effects of a change in reactor feed or in cooling fluid temperatures on reactor stability and temperature hysteresis are presented in Figs. 9.4a and 9.4b, respectively. Again, the heat generated by chemical reaction, U_g, is presented by an S-shaped curve and straight lines represent the heat removal, U_r. It can be seen that a dramatic increase

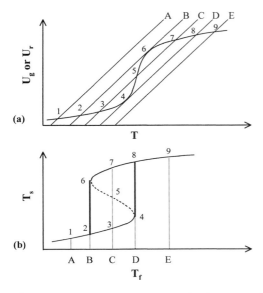

Figure 9.4 Effect of the change in the reactor feed or in the cooling fluid temperatures on reactor stability and temperature hysteresis.

and a dramatic decrease in the steady-state reactor operating temperature may take place for critically high and critically low reactor feed temperatures. For feed temperatures higher than a critical feed temperature, as in the case of line E of Fig. 9.4a, the steady-state temperature shifts to a high value T_9, corresponding to the intersection of the U_r line (line E) and the U_g curve. On the other hand, if the feed temperature is lower than a critical feed temperature, as in the case of line A of Fig. 9.4a, then the steady-state temperature shifts to a low value T_1, corresponding to the intersection of the U_r line (line A) and the U_g curve. These critical temperatures are the temperatures corresponding to two tangential lines (lines B and D, Fig. 9.4a). Reactor feed temperatures corresponding to these two critical temperatures determine the lower and upper limits of the feed temperature (Aris, 1969). Temperature hysteresis for reactor operating temperatures between the two critical temperatures is depicted in Fig. 9.4b.

In the above discussion of the stability of the steady-state reactor operating temperature, the order of the reaction rate is not specified. If the reaction kinetics are first-order (i.e., $r = -dc_A/dt = kc_A$) in the above example, the mass and energy balances [Eqs. (9.56) and (9.65)] at the steady-state operating conditions become

$$c_{A,f} - c_A - \tau k c_A = 0 \tag{9.70}$$

$$T - T_f + \kappa(T - T_{c,f}) - \tau J k c_A = 0$$

or

$$c_{A,s} = \frac{c_{A,f}}{1 + \tau k(T_s)} \tag{9.72}$$

$$U_r = T_s - T_f + \kappa(T_s - T_{c,f}) = \tau J k c_{A,s} = U_g = J c_{A,f} \frac{\tau k(T_s)}{1 + \tau k(T_s)} \tag{9.73}$$

The state parameters T_s and $c_{A,s}$ satisfying Eqs. (7.70) and (9.71) [therefore Eqs. (9.72) and (9.73)] are the steady-state values of the state parameters. For given reactor operating conditions, the two unknowns $c_{A,s}$ and T_s (the steady-state values of the state variables) can be solved from Eqs. (9.72) and (9.73) by graphical or numerical techniques (i.e., using nonlinear algebraic equations solver software). Because the solution of these equations gives the steady-state values of the concentration and temperature, these values are defined by $c_{A,s}$ and T_s.

Derivatives of U_r and U_g [Eq. (9.73), considering T_s as T] with respect to reactor operating temperature T can be evaluated by considering the Arrhenius relation for the rate constant [$k(T) = A_p \exp(-E/RT)$]:

$$\frac{dU_r}{dT} = 1 + \kappa \tag{9.74}$$

$$\frac{dU_g}{dT} = Jc_{A,f} \frac{E}{RT^2}\left(\frac{\tau k(T)}{[1 + \tau k(T)]^2}\right) = U_g \frac{E}{RT^2[1 + \tau(k(T)]} \tag{9.75}$$

As can be seen from Eqs. (9.74) and (9.75), the derivatives of U_r and U_g with respect to T are always positive. The difference is that the derivative dU_r/dT is constant, whereas the derivative dU_g/dT has a maximum, which takes place at (Aris, 1969)

$$\frac{d^2 U_g}{dT^2} = 0 \tag{9.76}$$

which corresponds to the temperature satisfying the condition [Eq. (9.73)]

$$\frac{\tau k(T)}{1 + \tau k(T)} = \frac{U_g}{Jc_{A,f}} = \frac{1}{2} - \frac{RT}{E} \tag{9.77}$$

As a result, the maximum value of dU_g/dT can be expressed by evaluating the rate constant at temperature T_m, which is the temperature satisfying the condition expressed by Eq. (9.77). If the maximum value of dU_g/dT (which is the slope of U_g versus T at T_m) is greater than dU_r/dT (which is the slope of U_r versus T and equal to $1 + \kappa$), then more than one intersection of U_g and U_r would be possible (Fig. 9.3b).

If the reacting system involves consecutive reactions, several steady-state temperatures may exist. Reactor design, on the other hand, permits the reactor to operate at one of the steady-state operating temperatures. Also, in reactor design not only the reactor operating conditions but also the reactant concentration or conversion is a concern. As a result, some kind of control action has to be taken to keep the reactor operating conditions under control to achieve the reactor design tasks. Reactor modeling would be a very useful tool that could be used to the predict reactor behavior and indicate the adaptive control action to be taken.

2.6 Stability Analysis for Continuous Flow Stirred Tank Reactors

In reactor stability analysis, the composition (or reaction extent ξ) and temperature T will be called the state (i.e., dependent) variables. The solution of the steady-state mass and energy balances would give the steady-state values of the state variables, ξ_s and T_s. Because at ξ_s and T_s the time derivatives of the state variables are zero, a set of coupled differential equations in terms of the deviations of the state variables from their steady-state values, i.e., in terms of ξ-ξ_s and T-T_s, could be studied for the stability analysis. This can be done by subtracting the steady-state form of the balance equations from their unsteady-state form. By this rearrangement, deviations of the state variables from their steady states (i.e., ξ-ξ_s and T-T_s) rather than their absolute values (i.e., ξ and T) have to be considered for the stability analysis.

For a chemical reaction A $\rightarrow$ P, the steady-state values ξ_s and T_s satisfy the steady-state mass and energy balance equations [Eqs. (9.64) and (9.66)]:

$$-\xi_s + \tau r(\xi_s, T_s, P) = 0 \tag{9.78}$$

$$-(1 + \kappa)(T_s - T_c^*) + \tau J r(\xi_s, T_s, P) = 0 \tag{9.79}$$

Using Eq. (9.68), the unsteady-state energy balance expressed by Eq. (9.62) becomes

$$\tau \frac{dT}{dt} = -(1 + \kappa)(T - T_c^*) + \tau J r(\xi, T, P) \tag{9.80}$$

By subtracting the steady-state solution of Eq. (9.78) from the unsteady-state mass balance equation [Eq. (9.58)] the following differential equation for ξ is obtained:

$$\tau \frac{d\xi}{dt} = -(\xi - \xi_s) + \tau[r(\xi, T, P) - r(\xi_s, T_s, P)]. \tag{9.81}$$

Similarly, by subtracting the steady-state solution of Eq. (9.79) from the unsteady-state energy balance equation [Eq. (9.80)] the following differential equation for T is obtained:

$$\tau \frac{dT}{dt} = -(1 + \kappa)(T - T_s) + \tau J[r(\xi, T, P) - r(\xi_s, T_s, P)] \tag{9.82}$$

Because the derivatives of state variables would not be affected by subtracting a constant (i.e., ξ_s and T_s) from the state variables,

$$\frac{d\xi}{dt} = \frac{d(\xi - \xi_s)}{dt}, \qquad \frac{dT}{dt} = \frac{d(T - T_s)}{dt} \tag{9.83}$$

by which Eqs. (9.81) and (9.82) can be expressed in terms of ξ-ξ_s and T-T_s, e.g.,

$$\tau \frac{d(\xi - \xi_s)}{dt} = -(\xi - \xi_s) + \tau[r(\xi, T, P) - r(\xi_s, T_s, P)] \tag{9.84}$$

$$\tau \frac{d(T - T_s)}{dt} = -(1 + \kappa)(T - T_s) + \tau J[r(\xi, T, P) - r(\xi_s, T_s, P)] \tag{9.85}$$

Equations (9.84) and (9.85) can be simplified by defining the new variables x and y, which are the deviations of the state variables from their steady-state values:

$$x = \xi - \xi_s \tag{9.86}$$

$$y = T - T_s \tag{9.87}$$

by which Eqs. (9.84) and (9.85) become

$$\tau \frac{dx}{dt} = -x + \tau[r(\xi_s + x, T_s + y, P) - r(\xi_s, T_s, P)] \tag{9.88}$$

$$\tau \frac{dy}{dt} = -(1 + \kappa)y + \tau J[r(\xi_s + x, T_s + y, P) - r(\xi_s, T_s P)] \tag{9.89}$$

Reactor stability can be studied by the simultaneous solution of Eqs. (9.88) and (9.89). Because of the Arrhenius type of temperature dependence of the rate constant, coupling between Eqs. (9.88) and (9.89) is nonlinear, and an analytical solution cannot be obtained. If the chemical reaction rate is not first-order, the rate expression as a function of concentration may be another source of nonlinearity that must be dealt with.

These sets of mass and energy balance equations can be linearized around the steady-state (or around the origin in x–y space) to form a set of differential equations in the form

$$\frac{dx}{dt} = a_{1,1}x + a_{1,2}y + F_1(x, y) \tag{9.90}$$

$$\frac{dy}{dt} = a_{2,1}x + a_{2,2}y + F_2(x, y) \tag{9.91}$$

which can be further reduced to the set of linear differential equations

$$\frac{dx}{dt} = a_{1,1}x + a_{1,2}y \tag{9.92}$$

$$\frac{dy}{dt} = a_{2,1}x + a_{2,2}y \tag{9.93}$$

provided that in the neighborhood of $(0, 0)$ both $F_1(x, y)$ and $F_2(x, y)$ possess the limit

$$\frac{F_1(x, y)}{d} \to 0, \qquad \frac{F_2(x, y)}{d} \qquad \text{as} \qquad d \to 0 \tag{9.94}$$

where

$$d = \sqrt{x^2 + y^2} \tag{9.95}$$

The system of coupled differential equations, Eqs. (9.90) and (9.91), satisfying the conditions of Eq. (9.94) is called an almost linear system (Boyce and DiPrima, 1965).

By linearization, mass and energy balances expressed by Eqs. (9.88) and (9.89) can also be reduced to a set of almost linear equations. Expressing the rate expression by a Taylor's series (assuming the pressure is kept constant) and neglecting the higher order terms [i.e., satisfying the conditions of Eq. (9.94)] gives

$$r(\xi_s + x, T_s + y) = r(\xi_s, T_s) + \left(\frac{\partial r}{\partial \xi}\right)_{\xi_s, T_s} x + \left(\frac{\partial r}{\partial T}\right)_{\xi_s, T_s} y \tag{9.96}$$

The partial derivatives of Eq. (9.96) can be calculated using the rate expression. For a first-order reaction $A \to P$

$$r = kc_A = A_p e^{-E/RT} c_A = A_p e^{-E/RT}(c_{A,f} - \xi) \tag{9.97}$$

where A_p is the pre-exponential coefficient and E is the activation energy.

Using Eqs. (9.96) and (9.97), the linearized form of the rate expression around the steady state (i.e., around ξ_s and T_s) becomes

$$r(\xi_s + x, T_s + y) = r(\xi_s, T_s) - A_p e^{-E/RT_s}x + (c_{A,f} - \xi_s)A_p e^{-E/RT_s}\frac{E}{RT_s^2}y \tag{9.98}$$

and using the relation $k(T_s) = A_p \exp(-E/RT_s)$,

$$r(\xi_s + x, T_s + y) - r(\xi_s, T_s) = -k(T_s)x + (c_{A,f} - \xi_s)k(T_s)\frac{E}{RT_s^2}y \tag{9.99}$$

which is linear in x and y. As a result of the linearization [i.e., using Eqs. (9.98) and (9.99)], mass and energy balances [Eqs. (9.88) and (9.89)] are reduced to the following set of coupled linear differential equations:

$$\frac{dx}{dt} = -\left[\frac{1}{\tau} + k(T_s)\right]x + (c_{A,f} - \xi_s)k(T_s)\frac{E}{RT_s^2}y \tag{9.100}$$

$$\frac{dy}{dt} = -Jk(T_s)x - \left[\frac{1+\kappa}{\tau} - J(c_{A,f} - \xi_s)k(T_s)\frac{E}{RT_s^2}\right]y \tag{9.101}$$

Equations (9.100) and (9.101) can be expressed as

$$\frac{dx}{dt} = a_{1,1}x + a_{1,2}y \tag{9.102}$$

and

$$\frac{dy}{dt} = a_{2,1}x + a_{2,2}y \tag{9.103}$$

where

$$a_{1,1} = -\left[\frac{1}{\tau} + k(T_s)\right], \qquad a_{1,2} = (c_{A,f} - \xi_s)k(T_s)\frac{E}{RT_s^2} \tag{9.104}$$

and

$$a_{2,1} = -Jk(T_s), \qquad a_{2,2} = -\left[\frac{1+\kappa}{\tau} - J(c_{A,f} - \xi_s)k(T_s)\frac{E}{RT_s^2}\right] \tag{9.105}$$

The solution of Eqs. (9.102) and (9.103) would be in the form

$$x = A_{1,1}e^{\lambda_1 t} + A_{1,2}e^{\lambda_2 t} \tag{9.106}$$

$$y = B_{1,1}e^{\lambda_1 t} + B_{1,2}e^{\lambda_2 t} \tag{9.107}$$

where λ_1 and λ_2 are the eigenvalues of the coefficient matrix of Eqs. (9.102) and (9.103):

$$\begin{pmatrix} a_{1,1} - \lambda & a_{1,2} \\ a_{2,1} & a_{2,2} - \lambda \end{pmatrix} \tag{9.108}$$

i.e., the roots of the second-order algebraic equation

$$(a_{1,1} - \lambda)(a_{2,2} - \lambda) - a_{1,2}a_{2,1} = \lambda^2 - (a_{1,1} + a_{2,2})\lambda + a_{1,1}a_{2,2} - a_{1,2}a_{2,1} = 0 \tag{9.109}$$

If the roots λ_1 and λ_2 are real, then the solution for x and y would depend on the signs and the magnitudes of λ_1 and λ_2(Fig. 9.5a). The reactor is stable if x and y (perturbations from the steady-state values) tend to approach zero as time increases ($\lambda < 0$), and it is unstable if x and y tend to increase ($\lambda > 0$) as time increases. The reactor operating conditions are oscillating if the roots are purely complex. If the roots are conjugates, then the reactor operating conditions would be oscillating with exponentially decreasing (damped) or increasing amplitudes, depending on the sign of the real part of the conjugate roots. Damped oscillation of the solution set would result in reactor stability (Fig. 9.5b).

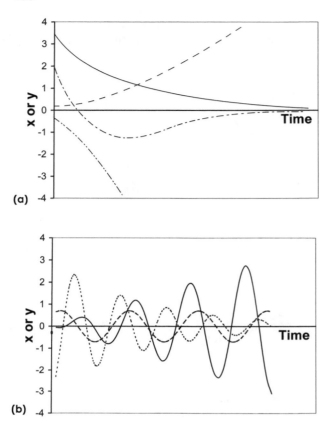

Figure 9.5 Types of solutions for the state variables (temperature and composition) for almost linear reactor mass and energy balances. (a) Real roots and (b) complex roots for the characteristic equation.

3. BATCH REACTORS

Batch reactors (BRs) are used for collecting fundamental data for the kinetics of complex reactions as well as for commercial production of specialized chemicals. Material balance for the batch reactors has already been discussed without identifying the type of reactor in formulating the reaction stoichiometry. This is obviously because the formulation of the stoichiometry was studied in a closed system (i.e., mass transfer is not allowed through the reactor boundaries), which is the batch reactor.

3.1 Mass Balance for Batch Reactors

For a simple reaction $\sum_j \alpha_j A_j = 0$, the molar balance for species A_j [the first and second terms on the right-hand side of Eq. (9.1) are zero and accumulation term becomes equal to the generation term] is

$$N_j = N_{j,0} + \alpha_j X \tag{9.110}$$

For constant volume operating conditions, Eq. (9.110) becomes

$$c_j = c_{j,0} + \alpha_j \xi \tag{9.111}$$

As defined in Chapter 5, X is the reaction extent and $\xi = X/V$ is the reaction extent per unit volume. Change in concentration can be expressed in terms of the intensive reaction rate [taking the derivative of both sides of Eq. (9.111)]:

$$\frac{dc_j}{dt} = \alpha_j \frac{d\xi}{dt} = \alpha_j r \tag{9.112}$$

If more than one reaction takes place, Eq. (9.112) should take care of the concentration changes as a result of all reactions:

$$\frac{dc_j}{dt} = \sum_{i=1}^{R} \alpha_{i,j} \frac{d\xi_i}{dt} = \sum_{i=1}^{R} \alpha_{i,j} r_i \tag{9.113}$$

where the reaction rate is a function of concentration, temperature, and pressure,

$$r_i = r_i(c_1, \ldots, c_s, T, P) \tag{9.114}$$

Change in the total mole number is important for gas-phase reactions. If the gas phase behaves as an ideal gas, the volume of the reacting system can be expressed as a function of reaction extents:

$$V = \frac{RT}{P} \sum_{j=1}^{S} N_j = \frac{RT}{P} \sum_{j=1}^{S} \left(N_{j,0} + \sum_{i=1}^{R} \alpha_{i,j} X_i \right) = \frac{RT}{P} \left(N_0 + \sum_{i=1}^{R} \bar{\alpha}_i X_i \right) \tag{9.115}$$

Expressing the initial total mole number in terms of initial conditions (i.e., P_0, T_0, and V_0), Eq. (9.115) can be restated as

$$V = \frac{RT}{P} \left(\frac{P_0 V_0}{RT_0} + \sum_{i=1}^{R} \bar{\alpha}_i X_i \right) = \frac{V_0 T P_0}{P T_0} \left(1 + \sum_{i=1}^{R} \bar{\alpha}_i \xi_i' \right) \tag{9.116}$$

where $\bar{\alpha} = \sum_j \alpha_{i,j}$ and $\xi_i' = X_i/N_0$.

3.2 Energy Balance for Batch Reactors

Similar to the expression for a continuous flow stirred tank reactor, the energy balance for a batch reactor operating at constant pressure can be expressed as

$$VC_p \frac{dT}{dt} = V \sum_{i=1}^{R} (-\Delta H_i) r_i - Q^* \tag{9.117}$$

where C_p is the specific heat per unit volume and Q^* is the heat taken out from the reactor. The energy balance expressed by Eq. (9.117) coupled with the mass balance could be used to explain many important phenomena such as spontaneous combustion and ignition, which are all related to nonlinear coupling of mass and energy balances.

If a first-order exothermic reaction $A \rightarrow P$ [i.e., $r = kc_A$ and $k = A_p \exp(-E/RT)$] takes place and the heat is removed with a cooling fluid at temperature T_c, the energy balance equation becomes

$$VC_p \frac{dT}{dt} = V(-\Delta H) r - UA_h(T - T_c) = V(-\Delta H) A_p e^{-E/RT} c_A - UA_h(T - T_c) \tag{9.118}$$

where U is the overall heat transfer coefficient and A_h is the heat transfer surface area. In Eq. (9.118) the rate of heat generation by chemical reaction, U_g, is an exponential function of temperature,

$$U_g = V(-\Delta H)r = V(-\Delta H)A_p e^{-E/RT} c_A \tag{9.119}$$

while the heat removal rate U_r is a linear function of the temperature,

$$U_r = UA_h(T - T_c) \tag{9.120}$$

Therefore, two steady-state temperatures are possible at which the heat generation and heat removal rates are equal. Because one of the steady-state temperatures, i.e., $T_{s,1}$, is stable (Figs. 9.6a and 9.6b), small perturbations imposed on $T_{s,1}$ would naturally rectify themselves and bring the temperature of the reacting system back to $T_{s,1}$. The second steady-state temperature $T_{s,2}$, however, is unsteady, because a small perturbation may pull down the reactor temperature to $T_{s,1}$ or push it up well above $T_{s,2}$. When the cooling rate curve is tangent to the heat generation rate curve, which happens at reactor operating temperature T_{cr}, two conditions would be satisfied: (1) the rate of heat generation would be equal to the rate of heat removal and (2) derivatives of the rate of heat generation and the rate of heat removal with respect to the reaction temperature would be equal. Assigning T_{cr} for the temperature where the cooling line is tangential to the heat generation curve, these two conditions can be expressed by the equations

$$U_g\big|_{T=T_{cr}} = U_r\big|_{T=T_{cr}} \tag{9.121}$$

and

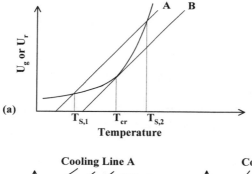

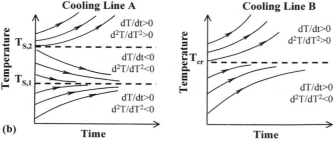

Figure 9.6 Nature of the steady-state and critical solution temperatures for the heat generation curve and heat removal (cooling) line.

$$\frac{dU_g}{dT}\bigg|_{T=T_{cr}} = \frac{dU_r}{dT}\bigg|_{T=T_{cr}} \tag{9.122}$$

or, considering Eqs. (9.119) and (9.120),

$$V(-\Delta H)A_p e^{-E/RT_{cr}} c_A = UA_h(T_{cr} - T_c) \tag{9.123}$$

$$V(-\Delta H)A_p e^{-E/RT_{cr}} c_A \frac{E}{RT_{cr}^2} = UA_h \tag{9.124}$$

Elimination of $V(-\Delta H)A_p \exp(-E/RT_{cr})$ and UA_h from Eqs. (9.123) and (9.124) would give

$$\frac{RT_{cr}^2}{E} = T_{cr} - T_c \tag{9.125}$$

from which the critical temperature T_{cr} can be calculated. Above the critical temperature T_{cr}, the reactor temperature becomes unstable and increases in an uncontrolled manner (resulting in spontaneous ignition or spontaneous combustion). Below the critical temperature T_{cr}, the reactor temperature becomes unstable and decreases in an uncontrolled manner (resulting in quenching and extinction). The lower root of the quadratic equation [Eq. (9.125)] corresponds to the ignition temperature, and the higher root corresponds to the extinction temperature (Semenov, 1935):

$$T_{cr,1} = \frac{E}{2R} - \frac{E}{2R}\left(1 - \frac{4RT_c}{E}\right)^{1/2} \qquad T_{cr,2} = \frac{E}{2R} + \frac{E}{2R}\left(1 - \frac{4RT_c}{E}\right)^{1/2} \tag{9.126}$$

Using the unsteady-state energy balance for an adiabatic batch reactor, the spontaneous ignition delay (induction time) can also be predicted. If a first-order exothermic reaction $A \rightarrow P$ [i.e., $r = kc_A$ and $k = A_p \exp(-E/RT)$] occurs in a well-mixed batch reactor, the energy balance equation becomes [by setting the heat removal term to zero in Eq. (9.118)]

$$VC_p \frac{dT}{dt} = V(-\Delta H)r = V(-\Delta H)A_p e^{-E/RT} c_A \tag{9.127}$$

which can be rearranged and integrated as in the following (T_i is the initial temperature, T_{cr} is the critical temperature for ignition, t_d is the delay time for ignition to take place):

$$\frac{C_p}{A_p c_A(-\Delta H)} \int_{T_i}^{T_{cr}} e^{E/RT} dT = \int_0^t dt \tag{9.128}$$

A closed-form solution of Eq. (9.128) cannot be obtained. To overcome this difficulty, an approximate method can be introduced by linearizing the exponential term around the initial temperature T_i:

$$\frac{E}{RT} = \frac{E}{RT_i} - \frac{E}{RT_i^2}(T - T_i) \tag{9.129}$$

and

$$e^{E/RT} = \exp\left[\frac{E}{RT_i} - \frac{E}{RT_i^2}(T - T_i)\right] = e^{E/RT_i}\exp\left[-\frac{E}{RT_i^2}(T - T_i)\right] \tag{9.130}$$

Using Eq. (9.130), the exponential integral in Eq. (9.128) becomes

$$\int_{T_i}^{T_{cr}} e^{E/RT} dT = e^{(E/RT_i)} \int_{T_i}^{T_{cr}} \exp\left[-\frac{E}{RT_i^2}(T - T_i)\right] dT = -e^{E/RT_i} \frac{RT_i^2}{E} \exp\left[-\frac{E}{RT_i^2}(T - T_i)\right]\Bigg|_{T_i}^{T_{cr}}$$

$$= \frac{e^{E/RT_i} RT_i^2}{E}\left\{-\exp\left[-\frac{E}{RT_i^2}(T_{cr} - T_i)\right] + 1\right\}$$

(9.131)

by which the integration of Eq. (9.128) becomes

$$-\exp\left[-\frac{E}{RT_i^2}(T_{cr} - T_i)\right] + 1 = -\exp\left[-\frac{E}{RT_i}\left(\frac{T_{cr}}{T_i} - 1\right)\right] + 1$$

$$= \frac{\int_0^t dt}{\dfrac{RT_i^2}{E} \dfrac{C_p e^{E/RT_i}}{A_p c_A(-\Delta H)}}$$

(9.132)

The denominator of the right-hand side of Eq. (9.132) is the characteristic time for the ignition delay:

$$t_d = \frac{RT_i^2}{E}\left(\frac{C_p e^{E/RT_i}}{A_p c_A(-\Delta H)}\right)$$

(9.133)

In fact, using t_d defined by Eq. (9.133), Eq. (9.132) can be rearranged as

$$\exp\left[-\frac{E}{RT_i}\left(\frac{T_{cr}}{T_i} - 1\right)\right] = 1 - \frac{t}{t_d}$$

(9.134)

or

$$\exp\left[\frac{E}{RT_i}\left(\frac{T_{cr}}{T_i} - 1\right)\right] = \frac{1}{1 - \frac{t}{t_d}}$$

(9.135)

which predicts that as $t \to t_d$ the critical temperature of the reacting system T_{cr} goes to infinity, which justifies the definition of t_d by Eq. (9.133).

4. PLUG FLOW TUBULAR REACTORS

In an ideal plug flow tubular reactor the reactant fluid flows in the reactor as a plug, i.e., all elements of the fluid pass through the reactor at the same velocity with no axial or radial mixing. Therefore, the flow in a plug flow tubular reactor is identical to the flow in a batch reactor moving at the fluid speed (i.e., $v = q/A_c$) in the reactor. In real operating conditions, however, axial and radial mixing take place, resulting in a shift in the performance of plug flow tubular reactors.

4.1   Mass Balance for Plug Flow Tubular Reactors

The plug flow model will be assumed for the formulation of the mass and energy balances for plug flow tubular reactors. The steady-state mass balance can be formulated by applying the principle expressed by Eq. (9.1) to a differential element of plug flow tubular reactors:

$$A_c(vc_j\big|_z - vc_j\big|_{z+dz}) + \sum_{i=1}^{s}(A_c\alpha_{i,j}r_i\big|_z)\,dz = 0 \tag{9.136}$$

Dividing both sides of Eq. (9.136) by A_c (cross-sectional area of the reactor), expressing the second term by a Taylor's series, and letting $dz \to 0$, the mass balance equation becomes

$$\frac{d(vc_j)}{dz} = \sum_{i=1}^{R}\alpha_{i,j}r_i \tag{9.137}$$

with the initial conditions

$$c_j = c_{j,0} \qquad \text{at} \qquad z = 0 \tag{9.138}$$

If the density of the reacting system is constant (in which case the volume of the system is also constant), Eq. (9.137) becomes

$$v\frac{dc_j}{dz} = \sum_{i=1}^{R}\alpha_{i,j}r_i(c, T) \tag{9.139}$$

Mass balance can also be expressed in terms of reaction extents using the relation

$$c_j = c_{j,0} + \sum_{i=1}^{R}\alpha_{i,j}\xi_i \tag{9.140}$$

where c_j and ξ_i are functions of the axial direction z. Using Eq. (9.140), Eq. (9.139) becomes

$$v\sum_{i=1}^{R}\alpha_{i,j}\frac{d\xi_i}{dz} = \sum_{i=1}^{R}\alpha_{i,j}r_i(\xi, T) \tag{9.141}$$

or

$$\sum_{i=1}^{R}\alpha_{i,j}\left(v\frac{d\xi_i}{dz} - r_i(\xi, T)\right) = 0 \tag{9.142}$$

with the initial conditions [corresponding to Eq. (9.138)]

$$\xi_i = 0 \qquad \text{at} \qquad z = 0 \tag{9.143}$$

For a reacting system with $j = 1, \ldots, S$ species, there would be S equations [Eq. (9.142)] to be solved. If the reactions considered in the reacting system are independent, the only linear combination satisfying $\sum_i \lambda_i\alpha_{ij} = 0$ for $j = 1, \ldots, S$ is the trivial set $\lambda_i = 0$ for $i = 1, \ldots, R$ (as discussed in Section 5.3). Because $\alpha_{i,j} \neq 0$, the second term in Eq. (9.142) must be zero (Aris, 1969), which results in the expression [subject to the initial conditions of Eq. (9.143)]

$$v\frac{d\xi_i}{dz} = r_i(\xi, T) \tag{9.144}$$

As a result, S equations (i.e., $j = 1, \ldots, S$) of Eq. (9.142) can be represented by R equations of Eq. (9.144).

Mass balance given by Eq. (9.144) can be expressed in terms of time. The distance traveled by a fluid element entering the reactor is

$$z = vt \quad \text{and} \quad dz = v\, dt \tag{9.145}$$

using which, the mass balance given by Eq. (9.144) can also be expressed as

$$\frac{d\xi_i}{dt} = r_i(\xi, T) \tag{9.146}$$

This relation shows that a plug flow tubular reactor is similar to a batch reactor (i.e., a fluid element entering the reactor can be considered as a fluid element in a batch reactor with a residence time of z/v).

If the density of the reacting fluid is not constant, the velocity of the fluid in the reactor will not be constant. For a simple reaction taking place in a fluid with varying density, using the principle expressed by Eq. (9.1), the steady-state mass balance equation for the j species becomes

$$\frac{d(vc_j)}{dz} = c_j \frac{dv}{dz} + v \frac{dc_j}{dz} = \alpha_j r \tag{9.147}$$

or by using the relation $c_j = c_{j,0} + \alpha_j \xi$,

$$c_j \frac{dv}{dz} + v\alpha_j \frac{d\xi}{dz} = (c_{j,0} + \alpha_j\xi)\frac{dv}{dz} + v\alpha_j \frac{d\xi}{dz} = \alpha_j r \tag{9.148}$$

Dividing both sides of Eq. (9.148) by α_j,

$$\frac{c_{j,0}}{\alpha_j}\frac{dv}{dz} + \xi\frac{dv}{dz} + v\frac{d\xi}{dz} = \frac{c_{j,0}}{\alpha_j}\frac{dv}{dz} + \frac{d(v\xi)}{dz} = r \tag{9.149}$$

or

$$\frac{d(v\xi)}{dz} = r - \frac{c_{j,0}}{\alpha_j}\frac{dv}{dz} \tag{9.150}$$

Mass balance can also be formulated on the basis of mass flux (mass flow per unit area per unit time). Because the total mass of the reacting system is conserved, Eq. (9.1) for a steady-state plug flow tubular reactor becomes

$$\frac{d(\rho v g_j)}{dz} = \alpha_j M_j r(\xi, T) \tag{9.151}$$

where g_j and M_j are the mass fraction and molecular weight of the jth species. Because the total mass flux G is fixed (G_j is the mass flux of the jth species),

$$G = \sum_{j=1}^{s} G_j = \sum_{j=1}^{s} g_j G = \rho v \tag{9.152}$$

The mass balance [Eq. (9.151)] can be expressed in terms of G and g_j as

$$\frac{dG_j}{dz} = \alpha_j M_j r(\xi, T) \tag{9.153}$$

Using the relations $G_j = g_j G$ and $g_j = j_{j,0} + \alpha_j M_j \xi''$ (where $\xi'' = \xi/\rho$),

$$G\frac{dg_j}{dz} = \alpha_j M_j r(\xi, T) \tag{9.154}$$

or, in terms of ξ'' (i.e., $dg_j/dz = \alpha_j M_j \, d\xi''/dz$),

$$G\frac{d\xi''}{dz} = r(\xi'', T) \tag{9.155}$$

with the initial condition

$$\xi'' = 0 \quad \text{at} \quad z = 0 \tag{9.156}$$

For more than one reaction taking place, the mass balance expressed by Eq. (9.154) becomes (as derived for the molar basis)

$$\sum_{i=1}^{R} \alpha_{i,j} M_j \left(G\frac{d\xi_i''}{dz} - r_i(\xi'', T) \right) = 0 \tag{9.157}$$

If the reactions considered in the reacting system are independent [as argued in the formulation of Eq. (9.144) from Eq. (9.142)], the only linear combination satisfying $\sum_i \lambda_i \alpha_{i,j}$ is the trivial set $\lambda_i = 0$ for $i = 1, \ldots, R$. Because $\alpha_{i,j} \neq 0$ and $M_j \neq 0$, the third term (the term in the large parentheses) of Eq. (9.157) must be zero:

$$G\frac{d\xi_i''}{dz} - r_i(\xi'', T) = 0 \tag{9.158}$$

or

$$G\frac{d\xi_i''}{dz} = r_i(\xi'', T) \tag{9.159}$$

by which S equations (i.e., $j = 1, \ldots, S$) of Eq. (9.157) can be represented by R equations ($i = 1, \ldots, R$) of Eq. (9.159) with the same initial conditions as those expressed in Eq. (9.156).

4.2 Energy Balance for Plug Flow Tubular Reactors

The energy balance for a plug flow tubular reactor can be formulated by applying the principle expressed by Eq. (9.1) to a differential section of the tubular reactor. Using the molar fluxes and molar enthalpies for the measurement of mass and energy, the energy balance can be expressed as

$$\sum_{j=1}^{S} A_c v c_j h_j \bigg|_z - \sum_{j=1}^{S} A_c v c_j h_j \bigg|_{z+dz} - Q^* S \, dz = 0 \tag{9.160}$$

where A_c is the cross-sectional area of the reactor, S is the perimeter of the reactor, and Q^* is the rate at which heat is extracted from the reactor per unit external area ($Q^* > 0$ when heat is removed). Expressing the second term by a Taylor series, Eq. (9.160) can be restated as

$$\frac{d\left(\sum_{j=1}^{S} v c_j h_j \right)}{dz} = -\frac{SQ^*}{A_c} \tag{9.161}$$

or

$$\sum_{j=1}^{S} h_j \frac{d(vc_j)}{dz} + \sum_{j=1}^{S} vc_j \frac{dh_j}{dT} \frac{dT}{dz} = \sum_{j=1}^{S} h_j \frac{d(vc_j)}{dz} + \sum_{j=1}^{S} vc_j C_{p,j} \frac{dT}{dz}$$

$$= -\frac{SQ^*}{A_c} \tag{9.162}$$

where $C_{p,j}$ is the specific heat of the jth species. Combining the mass balances expressed in Eqs. (9.147) and (9.162), the energy balance becomes

$$\sum_{j=1}^{S} h_j \alpha_j r + vC_p \frac{dT}{dz} = (\Delta H)r + vC_p \frac{dT}{dz} = -\frac{SQ^*}{A_c} \tag{9.163}$$

or

$$v \frac{dT}{dz} = \frac{-\Delta H}{C_p} r - \frac{SQ^*}{A_c C_p} \tag{9.164}$$

where $C_p \ (= \sum c_j C_{p,j})$ is the specific heat per unit volume of the reacting fluid. If more than one reaction is taking place, the energy balance can be expressed as

$$v \frac{dT}{dz} = \sum_{i=1}^{R} \frac{-\Delta H_i}{C_p} r_i - \frac{SQ^*}{A_c C_p} \tag{9.165}$$

Similar to the mass balance, the energy balance can also be formulated on the basis of mass flux (mass flow per unit area per unit time) and mass enthalpies. By application of the principle expressed by Eq. (9.1), the energy balance for a single reaction becomes

$$G \frac{dT}{dz} = \frac{-\Delta H}{C'_p} r - \frac{SQ^*}{A_c C'_p} \tag{9.166}$$

where G is the mass flux (per unit area per unit time) and C'_p is the specific heat of the reactant fluid per unit mass. If more than one reaction is taking place, then Eq. (9.166) becomes

$$G \frac{dT}{dz} = \sum_{i=1}^{R} \frac{-\Delta H_i}{C'_p} r_i - \frac{SQ^*}{A_c C'_p} \tag{9.167}$$

As can be seen from the mass and energy balance equations, mass and energy balances are coupled in a nonlinear fashion (i.e., r is a function of concentration and temperature).

5. AXIALLY DISPERSED TUBULAR REACTORS

In reality, unless the fluid velocity is really high, all tubular reactors operate differently than plug flow tubular reactors; that is, fluid elements in tubular reactors disperse in the axial and radial directions. Reactors operating with axial mixing of different degrees are called axially dispersed tubular reactors (ADTRs) (Levenspiel and Bischoff, 1963).

5.1 Mass Balance for Axially Dispersed Tubular Reactors

The mass balance for axially dispersed tubular reactors can be formulated using the principle expressed by Eq. (9.1) and considering convection and dispersion fluxes and chemical reactions (accumulation is zero at the steady state) for a differential section of the reactor (similar to a plug flow tubular reactor):

$$A_c\left[\left(vc_j - D_d\frac{dc_j}{dz}\right)\bigg|_z - \left(vc_j - D_d\frac{dc_j}{dz}\right)\bigg|_{z+dz}\right] + \sum_{i=1}^{R}(A_c\alpha_{i,j}r_i|_z)\,dz = 0 \tag{9.168}$$

where mass flux by axial dispersion is assumed to take place like mass flux by molecular diffusion, i.e., to be linearly proportional to the concentration gradient (the negative sign is needed because mass flux takes place in the direction of the negative gradient) with a proportionality constant D_d (dispersion coefficient). Expressing the second term in Eq. (9.168) by Taylor expansion (i.e., by dividing both sides by the cross-sectional area of the reactor, A_c) the mass balance equation becomes

$$D_d\frac{d^2c_j}{dz^2} - \frac{d[vc_j]}{dz} + \sum_{i=1}^{R}\alpha_{i,j}r_i(c, T) = 0 \tag{9.169}$$

If the reaction takes place in a constant-density system (constant v), Eq. (9.169) becomes

$$D_d\frac{d^2c_j}{dz^2} - v\frac{dc_j}{dz} + \sum_{i=1}^{R}\alpha_{i,j}r_i(c, T) = 0 \tag{9.170}$$

Using the mass balance in terms of reaction extents [i.e., by using Eq. (9.140)], Eq. (9.170) becomes

$$\sum_{i=1}^{R}\alpha_{i,j}\left(D_d\frac{d^2\xi_i}{dz^2} - v\frac{d\xi_i}{dz} + r_i(\xi, T)\right) = 0 \tag{9.171}$$

If the reactions involved are independent [using the same argument in the reduction of S equations of Eq. (9.142) to R equations of Eq. (9.144)], the expression in large parentheses in Eq. (9.171) is zero:

$$D_d\frac{d^2\xi_i}{dz^2} - v\frac{d\xi_i}{dz} + r_i(\xi, T) = 0 \tag{9.172}$$

by which S equations of Eq. (9.170) are reduced to R equations of Eq. (9.172).

Because the mass balance is expressed by a second-order differential equation, two boundary conditions are needed for the solution of mass balance equations. If the feed line is assumed as the plug flow, there will be a concentration discontinuity right at the inlet of the reactor as a result of the axial dispersion in the reactor (Fig. 9.7). As a result, the flux balance will be used as the boundary condition at the reactor inlet (which is also known as the Dankcwerts boundary condition) (Dankcwerts and Mahlmann, 1976) and zero derivatives will be used as the boundary condition at the reactor exit [i.e., the feed concentration of the jth species, $c_{j,f}$, is $c_j(0-)$]:

$$vc_j(0-) = vc_j(0+) - D_d\frac{dc_j}{dz}(0+) \qquad \text{at} \qquad z = 0 \tag{9.173}$$

$$\frac{dc_j}{dz} = 0 \qquad \text{at} \qquad z = L \tag{9.174}$$

where L is the reactor length.

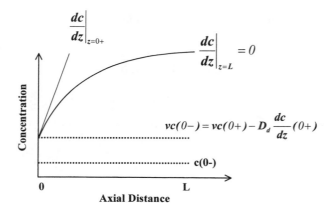

Figure 9.7 Boundary conditions (flux balance at the reactor inlet and steady conversion at the reaction exit) for an axially dispersed tubular reactor (ADTR) operating at steady-state operating conditions.

The mass balance and corresponding boundary conditions for axially dispersed tubular reactors can be expressed in terms of normalized distance Z with respect to reactor length L:

$$Z = \frac{z}{L}, \qquad dZ = \frac{dz}{L}, \qquad \text{or} \qquad \frac{d}{dz} = \frac{1}{L}\frac{d}{dZ}, \qquad \frac{d^2}{dz^2} = \frac{1}{L^2}\frac{d^2}{dZ^2} \tag{9.175}$$

by which, Eqs. (9.170) and (9.172) become

$$\frac{1}{Pe_d}\frac{d^2 c_j}{dZ^2} - \frac{dc_j}{dZ} + \frac{L}{v}\sum_{i=1}^{R}\alpha_{i,j}r_i(c,\,T) = 0 \tag{9.176}$$

and

$$\frac{1}{Pe_d}\frac{d^2 \xi_i}{dZ^2} - \frac{d\xi_i}{dZ} + \frac{L}{v}\sum_{i=1}^{R}r_i(\xi,\,T) = 0 \tag{9.177}$$

with the corresponding boundary conditions

$$c_j(0-) = c_j(0+) - \frac{1}{Pe_d}\frac{dc_j}{dZ}(0+) \qquad \text{at} \qquad Z = 0 \tag{9.178}$$

and

$$\frac{dc_j}{dZ} = 0 \qquad \text{at} \qquad Z = 1 \tag{9.179}$$

where Pe_d is the axial dispersion Peclet number, a dimensionless parameter defining the ratio of convection and dispersion mass fluxes:

$$Pe_d = \frac{vL}{D_d} \tag{9.180}$$

5.2 Energy Balance for Axially Dispersed Tubular Reactors

Similar to mass balance, an energy balance can be formulated by considering the energy balance Eq. (9.1) in a differential section of the reactor:

$$A_c\left[\sum_{j=1}^{S}\left(vc_j - D_d\frac{dc_j}{dz}\right)h_j\bigg|_z - \sum_{j=1}^{S}\left(vc_j - D_d\frac{dc_j}{dz}\right)h_j\bigg|_{z+dz}\right] - SQ^*\,dz = 0 \qquad (9.181)$$

where D_d is the axial dispersion coefficient, A_c is the cross-sectional area of the reactor, S is the perimeter of the reactor, and Q^* is the rate at which heat is extracted from the reactor per unit external area ($Q^* > 0$ when heat is removed). Expressing the second term by Taylor series, Eq. (9.81) becomes

$$\frac{d\left[\sum_{j=1}^{S}\left(vc_j - D_d\frac{dc_j}{dz}\right)h_j\right]}{dz} = -\frac{SQ^*}{A_c} \qquad (9.182)$$

or

$$\sum_{j=1}^{S}h_j\frac{d(vc_j - D_d\,dc_j/dz)}{dz} + \sum_{j=1}^{S}\left(vc_j - D_d\frac{dc_j}{dz}\right)\frac{dh_j}{dT}\frac{dT}{dz} = -\frac{SQ^*}{A_c} \qquad (9.183)$$

Combining this with the mass balance expressed in Eq. (9.169), the energy balance becomes (assuming that only one reaction is taking place and using $C_{p,j} = dh_j/dT$)

$$\sum_{j=1}^{S}h_j\alpha_j r + \sum_{j=1}^{s}vc_jC_{p,j}\frac{dT}{dz} - \sum_{j=1}^{s}D_dC_{p,j}\frac{dc_j}{dz}\frac{dT}{dz} = -\frac{SQ^*}{A_c} \qquad (9.184)$$

Equation (9.184) can be reduced to

$$(\Delta H)\,r + vC_p\frac{dT}{dz} - D_{\text{th}}C_p\frac{d^2T}{dz^2} = -\frac{SQ^*}{A_c} \qquad (9.185)$$

or

$$v\frac{dT}{dz} - D_{\text{th}}\frac{d^2T}{dz^2} = \frac{(-\Delta H)\,r}{C_p} - \frac{SQ^*}{C_pA_c} \qquad (9.186)$$

where the energy flux by axial mass dispersion in Eq. (9.184) is approximated as

$$\sum_{j=1}^{s}D_dC_{p,j}\frac{dc_j}{dz}\frac{dT}{dz} = D_{\text{th}}C_p\frac{d^2T}{dz^2} \qquad (9.187)$$

If the reactive system involves more than one reaction, the energy balance expressed by Eq. (9.186) becomes

$$v\frac{dT}{dz} - D_{\text{th}}\frac{d^2T}{dz^2} = \sum_{i=1}^{R}\frac{(-\Delta H_i)\,r_i}{C_p} - \frac{SQ^*}{C_pA_c} \qquad (9.188)$$

The energy balance for axially dispersed tubular reactors is also a second-order differential equation; therefore two boundary conditions are needed for its solution. As discussed in the formulation of the mass balance, temperature shows a discontinuity at the reactor inlet. As a result, the energy flux balance will be the inlet boundary condition, by which the set of boundary conditions can be expressed as [similar to Eqs. (9.173) and (9.174)]

$$vC_pT(0-) = vC_pT(0+) - D_{th}C_p\frac{dT}{dz}(0+) \quad \text{at} \quad z = 0 \tag{9.189}$$

$$\frac{dT}{dz} = 0 \quad \text{at} \quad z = L \tag{9.190}$$

Using the dimensionless distance as defined in Eq. (9.175), the energy balance equation (9.186) and corresponding boundary conditions can be expressed as

$$\frac{dT}{dZ} - \frac{1}{Pe_{th}}\frac{d^2T}{dZ^2} = \frac{L\sum_{i=1}^{R}(-\Delta H_i)r_i}{vC_p} - \frac{LSQ^*}{vC_pA_c} \tag{9.191}$$

$$T(0-) = T(0+) - \frac{1}{Pe_{th}}\frac{dT}{dZ}(0+) \quad \text{at} \quad Z = 0 \tag{9.192}$$

$$\frac{dT}{dZ} = 0 \quad \text{at} \quad Z = 1 \tag{9.193}$$

where Pe_{th} is the dimensionless Peclet number, which is defined as the ratio of the convection and dispersion heat fluxes:

$$Pe_{th} = \frac{vL}{D_{th}} \tag{9.194}$$

Mass and energy dispersions are dynamic properties of the reactive system. In fact, Peclet numbers corresponding to mass and energy dispersions can be expressed in terms of the Reynolds number, Re, and mass transfer and heat transfer Prandtl numbers, Pr_m and Pr_{th}:

$$Pe_d = Re\,Pr_m \quad \text{and} \quad Pe_{th} = Re\,Pr_{th} \tag{9.195}$$

where

$$Re = \frac{\rho vL}{\mu} = \frac{vL}{\gamma}, \quad Pr_m = \frac{\gamma}{D_d}, \quad Pr_{th} = \frac{\gamma}{D_{th}} \tag{9.196}$$

where γ is the kinematic viscosity ($\gamma = \mu/\rho$, whre μ is the viscosity and ρ is the density). For reactor design purposes, published experimental data as well as new measurements of Peclet numbers for the geometry, packing, and operating conditions of a given reactive system would always be needed.

REFERENCES

Aris, R. 1969. Elementary Chemical Reactor Analysis. Prentice Hall, Englewood Cliffs, NJ.

Boyce, W. E., and DiPrima, R. C. 1965. Elementary Differential Equations and Boundary Value Problems. Wiley, New York.

Dankcwerts, P. V. and Mahlmann, E. A. 1976. Chem. Eng. J. 11:19–25.

Levenspiel, O., and Bischoff, K. 1963. Adv. Chem. Eng. 4:95.

Semenov, N. N. 1935. Chemical Kinetics and Chain Reactions. Oxford Univ. Press, Oxford, England.

10

Mixing in Flow Systems and Reactors

1. INTRODUCTION

Mixing phenomena are experienced in everyday life, from food preparation to dispersion of pollutants in lakes, rivers, and the atmosphere to the blending of reactants in chemical reactors. Mixing may be caused by natural phenomena such as molecular diffusion and convection (such as in the oceans, rivers, and atmosphere) or imposed by mechanical or static agitation.

Mixing has a special meaning in chemical engineering practice. Mixing in chemical reactors brings the reacting species into contact with each other and therefore has great importance for the design of commercial reactors. Because of its industrial importance, this subject has been studied extensively and will be studied in the decades to come. There are classical works available in the literature that can be beneficial for further understanding of the phenomena (Levenspiel and Bischoff, 1963; Nauman and Buffham, 1983; Petho and Noble, 1982).

2. RESIDENCE TIME DISTRIBUTION

The concept of residence time distribution is an important element in the study of mixing in flow systems. For this purpose, a continuous closed flow system (i.e., one in which no mass flux is allowed from the system boundaries) operating at steady-state conditions (Fig. 10.1) will be considered. This flow system is connected to one inlet path and one exit path for entering the feedstream and exiting the effluent stream respectively, both operating as plug flow. In plug flow it is assumed that any fluid element may enter into the system and exit from the system only once. The volume of the flow system is assumed to be large enough that the fluid elements are slowed down in the system in comparison to the fluid elements in the feed and discharge pipes.

Any fluid element entering the system would spend some time in the system and exit through the discharge pipe. The time elapsed from the entrance of a fluid element into the system to its final departure from the system is called the residence time of the fluid element. The residence time for each of the fluid elements entering the system by the

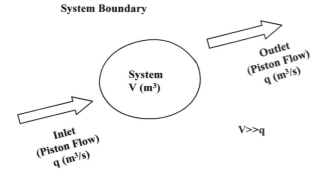

Figure 10.1 A continuous closed flow system (no mass flux is allowed from the system boundaries) connected to inlet and exit paths, both operating as piston flow.

feedstream is different and can be defined by means of a probability density function $f(t)$ of the following properties:

$$p(t, t + \Delta t) = \int_{t}^{t+\Delta t} f(\xi)\, d\xi \tag{10.1}$$

$$\int_{0}^{\infty} f(\xi)\, d\xi = 1 \tag{10.2}$$

$$F(t) = \int_{0}^{t} f(\xi)\, d\xi \tag{10.3}$$

$$W(t) = 1 - F(t) = 1 - \int_{0}^{t} f(\xi)\, d\xi = \int_{t}^{\infty} f(\xi)\, d\xi \tag{10.4}$$

where $p(t, t + \Delta t)$ is the probability that a particle has a residence time in the system between t and $t + \Delta t$, $F(t)$ is the cumulative probability density function that is the probability that a particle has a residence time in the system less than t, and $W(t)$ is the washout function, which is the probability that a particle has a residence time in the system greater than t.

These relations tell almost everything that is needed for the definition of residence time distribution. The first relation [Eq. (10.1)] tells us that the probability $p(t, t + \Delta t)$ that the residence time of a particle will be in the interval $(t, t + \Delta t)$ is equal to the integral of $f(t)$ for that time interval. Equations (10.2) and (10.3) tell us that the residence time distribution function $f(t)$ is normalized. Equation (10.4) tell us that the cumulative function $W(t)$ is the probability that a particle has a residence time greater than t, which is defined as the washout function. Based on these definitions, the limit conditions for $F(t)$ and $W(t)$ would be (see also Fig. 10.2)

$$F(0) = W(\infty) = 0 \qquad \text{and} \qquad F(\infty) = W(0) = 1 \tag{10.5}$$

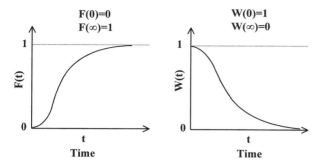

Figure 10.2 Cumulative residence time distribution $F(t)$ and washout $W(t)$ functions.

The foregoing properties of the probability density function imply that

$$f(t) = \frac{dF}{dt} = -\frac{dW}{dt} \tag{10.6}$$

All of these functions [$f(t)$, $F(t)$, and $W(t)$] are statistical information about the fluid elements (particles) that are leaving the system, i.e., exit age statistics, and do not say anything about the statistics of the fluid elements (particles) present in the system. Using the already defined statistics, the portion of the particles in the system with ages between t and $t + dt$ can be defined as the internal age distribution density function $h(t)$:

$$h(t)\, dt = \frac{W(t)\, dt}{\int_0^\infty W(\xi)\, d\xi} \tag{10.7}$$

or

$$h(t) = \frac{W(t)}{\int_0^\infty W(\xi)\, d\xi} \tag{10.8}$$

The integral in the denominator of Eq. (10.8) can be calculated by partial integration [Using Eqs. (10.5) and (10.6)]:

$$\int_0^\infty W(\xi)\, d\xi = \xi W(\xi)\Big|_0^\infty - \int_0^\infty \xi \frac{dW}{d\xi}\, d\xi = \int_0^\infty \xi f(\xi)\, d\xi = \tau \tag{10.9}$$

where τ is the expected value of the residence time. Using Eq. (10.9), the internal age distribution density function [Eq. (10.8)] becomes

$$h(t) = \frac{1}{\tau} W(t) \tag{10.10}$$

The cumulative form of the internal age distribution $h(t)$ is known as the internal age hold-up function $H(t)$:

$$H(t) = \int_t^\infty h(\xi)\, d\xi = \frac{1}{\tau} \int_t^\infty W(\xi)\, d\xi \tag{10.11}$$

which is the fraction of the particles in the system with internal ages greater than t. It can be seen that $H(t)$ is a decreasing function of t, with the limit values of $H(0) = 1$ and $H(\infty) = 0$.

In summary, for a closed-flow system (Fig. 10.1) connected to feed and exit pipes (plug flow), if the residence time distribution function is measured, all other information about the system can be derived using the foregoing relations. In this process, however, it is desired to use the experimental data for numerical integration, not for numerical differentiation (i.e., for minimization of numerical errors).

3. IDEAL FLOW SYSTEMS

Mixing in chemical reactors can be represented by two extreme cases: complete (or perfect) mixing and plug flow. Both of these flow cases are ideal, and mixing in chemical reactors hardly fits in either of these extremes.

3.1 Continuous Flow Stirred Tank Reactors with Perfect Mixing

In a continuous flow stirred tank reactor (CSTR) with perfect mixing, the fluid element entering the reactor instantly mixes with the fluid present in the reactor. As a result of perfect mixing in the continuous flow stirred tank reactor, the exit age $f(t)$ and internal age $h(t)$ distributions become identical (Nauman and Buffham, 1983):

$$h(t) = f(t) \tag{10.12}$$

Expressing both sides of Eq. (10.12) in terms of $W(t)$ [using Eqs. (10.10) and (10.6)],

$$\frac{W(t)}{\tau} = -\frac{dW(t)}{dt} \tag{10.13}$$

Integrating Eq. (10.13) with the initial condition $W(0) = 1$,

$$W(t) = e^{-t/\tau} \tag{10.14}$$

or, in terms of the residence time distribution function $f(t)$:

$$f(t) = -\frac{dW}{dt} = \frac{1}{\tau} e^{-t/\tau} \tag{10.15}$$

which is the expression for the exponential probability distribution function with parameter $1/\tau$. Since the residence time distribution function is exponential with parameter $1/\tau$, the mean (expected value) and variance of the residence time are τ and τ^2, respectively.

For a continuous flow stirred tank reactor as described above, the residence time distribution can be experimentally measured by detecting the tracers at the inlet and exit of the reactor. The simplest way of measuring the residence time distribution would be to inject a pulse of tracers (nonreactive) into the feedstream at $t = 0$ and measure the tracer concentration $c(t)$ in the reactor effluent stream as a function of time. It is important that the feed and effluent flow rates be kept constant during these measurements. The residence time probability density function $f(t)$ would be determined experimentally by using the relationship

$$f(t)\,dt = \frac{c(t)\,dt}{\int\limits_0^\infty c(\xi)\,d\xi} \tag{10.16}$$

Because the tracer is not reactive and cannot stay in the reactor forever, the integral in the denominator of Eq. (10.16) is the total amount of tracer injected into the reactor. Therefore, measurement of concentration $c(t)$ at the reactor exit as a function of time normalized by the amount of tracer injected into the reactor gives the residence time distribution function $f(t)$.

The response of a continuous flow stirred tank reactor to a tracer test can also be formulated by using the tracer mass balance (operation at constant volume and constant feed and discharge volumetric flow rates) and the principle expressed by Eq. (9.1):

$$V\frac{dc}{dt} = qc_f - qc \tag{10.17}$$

where V is the reactor volume, q is the feed and exit flow rate, c_f is the tracer concentration in the feedstream, and c is the tracer concentration in the effluent stream.

Dividing both sides of Eq. (10.17) by q,

$$\frac{V}{q}\frac{dc}{dt} = \tau\frac{dc}{dt} = c_f - c \tag{10.18}$$

where $\tau = V/q$ is the average holding time. For an impulse input $[c_f(t) = \delta(t)]$ and with no tracer initially in the reactor [at $t = 0$, $c(0) = 0$], Eq. (10.18) can be solved by the Laplace transform technique [$c_f(t) = \delta(t)$ is the Dirac delta function, and its Laplace transform is $L\{\delta(t)\} = 1$]:

$$\tau[sC(s) - c(0)] = s\tau C(s) = 1 - C(s) \tag{10.19}$$

where $C(s)$ is the Laplace transform of $c(t)$. $C(s)$ can be solved for from Eq. 10.19 as

$$C(s) = \frac{1}{1 + \tau s} = \frac{1/\tau}{s + 1/\tau} \tag{10.20}$$

or by inverse transformation of $C(s)$ from the Laplace to the time domain,

$$c(t) = f(t) = \frac{1}{\tau}e^{-t/\tau} \tag{10.21}$$

which is the mathematical form of the exponential distribution function with parameter $1/\tau$. Using Eq. (10.21) the expected mean residence time μ (or expected value of t) is calculated as

$$\mu = \int\limits_0^\infty tf(t)\,dt = \tau\int\limits_0^\infty \left(\frac{t}{\tau}\right)e^{-t/\tau}d\left(\frac{t}{\tau}\right) = \tau \tag{10.22}$$

Thus the nominal residence time $\tau = V/q$ is also the expected residence time for a perfectly mixed continuous flow stirred tank reactor.

The variance σ^2 of residence time t about the mean residence time τ is calculated as

$$\sigma_\tau^2 = \int_0^\infty (t - \tau)^2 f(t) \, dt$$

$$= \int_0^\infty (t - \tau)^2 \frac{1}{\tau} e^{-t/\tau} \, dt = \tau^2 \int_0^\infty \left(\frac{t}{\tau} - 1 \right)^2 e^{-t/\tau} \, d\left(\frac{t}{\tau} \right) = \tau^2 \qquad (10.23)$$

from which it can be seen that the standard deviation σ_τ about the mean is as large as the mean itself.

The Laplace transform method can also be used to solve Eq. (10.18) for an arbitrary tracer injection (disturbance) concentration in the feedstream. By taking the Laplace transform of Eq. (10.18), the ratio of tracer concentrations in the exit and inlet streams becomes

$$\frac{C(s)}{C_f(s)} = \psi(s) = \frac{1}{1 + \tau s} = \frac{1/\tau}{s + 1/\tau} \qquad (10.24)$$

where $C(s)$ and $C_f(s)$ are the Laplace transforms of $c(t)$ and $c_f(t)$, respectively. This ratio $C(s)/C_f(s)$ is called the *transfer function* for the perfectly mixed CSTR. The representation of the mixing in a simple flow system by Eq. (10.24) makes it simple to study mixing in complex combinations of simple flow systems (such as perfectly mixed continuous flow stirred tank reactors and piston flow tubular reactors).

3.2 Piston Flow Tubular Reactors

The opposite extreme of the perfectly mixed continuous flow stirred tank reactor is the piston flow of materials in a pipe without any dispersion in the flow (axial) direction.

For piston flow with a uniform axial velocity profile, the residence time distribution function becomes

$$f(t) = \delta(t - \tau), \qquad t = \tau, f(t) = 1; \qquad t \neq \tau, f(t) = 0 \qquad (10.25)$$

the cumulative density function of which becomes

$$F(t) = 0, \quad t < \tau; \qquad F(t) = 1, \quad t > \tau \qquad (10.26)$$

where $\delta(t - \tau)$ is the Dirac delta function. Because of the properties of the Dirac delta function,

$$\int_0^\infty c(t)\delta(t - \tau) \, dt = c(\tau) \qquad (10.27)$$

which implies that any material input to a piston flow system would be observed at the exit with a time lag of τ (i.e., $\tau = L/v$), the time required for the fluid element to travel the length L in the pipe with a uniform linear flow velocity v.

Neither perfect mixing nor plug flow can be achieved in real reactors on the laboratory, pilot, or commercial scale. Axial flow would always be accompanied by axial mixing as a result of turbulence or Taylor–Aris diffusion taking place in the flow system.

4. MEAN, MOMENTS, AND DISTRIBUTION FUNCTIONS

The residence time distribution function $f(t)$ tells us the fraction of materials $f(t)\,dt$ that spends time in the reactor in the time interval $(t, t + dt)$. Statistical values of the properties attached to a particle can be calculated by the residence time distribution or probability density function $f(t)$. As an example, if $c(t)$ is the property attached to the particles, the average (or expected value) of $c(t)$ at the reactor exit is

$$E\{c(t)\} = \int_0^\infty c(t)f(t)\,dt \tag{10.28}$$

which is the first moment of $c(t)$. The operator defined by Eq. (10.28) is called the expectation operator. Similarly, the nth moment of $c(t)$ can be calculated from the expectation operator,

$$E\{c(t)^n\} = \int_0^\infty \{c(t)\}^n f(t)\,dt \tag{10.29}$$

In the case of residence time t (or any random variable t) with a residence time distribution function $f(t)$, the nth moment of t about t_0 (μ_{n,t_0}) is calculated from

$$E\{(t - t_0)^n\} = \mu_{n,t_0} = \int_0^\infty (t - t_0)^n f(t)\,dt \tag{10.30}$$

When $t_0 = 0$, Eq. (10.30) gives the moments about the origin:

$$E\{t^n\} = \mu_n = \int_0^\infty t^n f(t)\,dt \tag{10.31}$$

If the first moment of the residence time is equal to 1, i.e., $\mu_1 = 1$, then the distribution function $f(t)$ is said to be normalized. The residence time t can be normalized by defining a new residence time t/τ [where τ is the mean, i.e., the first moment μ_1 of t, as defined by Eq. (10.31)]. The newly defined residence time t/τ is defined as the normalized residence time because it possesses a mean (the first moment μ_1) equal to 1.

Moments of the residence times can be calculated using the washout function $W(t)$ instead of the residence time probability distribution function $f(t)$. The nth moment of the residence time can be expressed in terms of $W(t)$ by using Eq. (10.6) in the integration of Eq. 10.31:

$$\mu_n = \int_0^\infty t^n f(t)\,dt = -t^n W(t)\Big|_0^\infty + \int_0^\infty n t^{n-1} W(t)\,dt = n \int_0^\infty t^{n-1} W(t)\,dt \tag{10.32}$$

For the first moment, Eq. (10.32) becomes

$$\mu_1 = \tau = \int_0^\infty W(t)\,dt \tag{10.33}$$

which is an important relation in calculating the mean residence time τ using the washout function $W(t)$.

Using Eq. (10.32), the moments for $n \geq 2$ can be expressed in terms of the hold-up function $H(t)$ [using the relations expressed by Eqs. (10.10) and (10.11)] as

$$\mu_n = n \int_0^\infty t^{n-1} W(t)\, dt$$

$$= nt^{n-1} \tau H(t)\Big|_0^\infty + n(n-1)\tau \int_0^\infty t^{n-2} H(t)\, dt \tag{10.34}$$

$$= n(n-1)\tau \int_0^\infty t^{n-2} H(t)\, dt$$

which is also an important relation in calculating the moments of residence time using the experimentally measured hold-up function $h(t)$. Equations (10.32) and (10.34) are useful in calculating all residence time statistics from the most basic measurement of the residence time distribution $f(t)$ with the minimum possible numerical error.

The relationships between the moments about the origin (μ_n) and the moments about the mean residence time τ ($\mu_{n,\tau}$) are listed without their derivations in the following:

$$\mu_{0,\tau} = \mu_0 = 1 \tag{10.35}$$

$$\mu_{1,\tau} = 0 \tag{10.36}$$

$$\mu_{2,\tau} = \mu_2 - \mu_1^2 \tag{10.37}$$

$$\mu_{3,\tau} = \mu_3 - 3\mu_1\mu_2 + 2\mu_1^2 \tag{10.38}$$

These relationships provide simplicity in numerical calculation of the nth moment about τ from the residence time distribution data. Using these relations, if the residence time distribution function $f(t)$ is measured experimentally, then the moments of residence time of any order about the origin (i.e., $t = 0$) and about the mean τ can be calculated easily. As an example, the variance (the second moment of residence time t about the mean τ) can be calculated, using Eq. (10.37), with an easy numerical procedure:

$$\sigma^2 = \mu_{2,\tau} = \int_0^\infty (t - \tau)^2 f(t)\, dt = \mu_2 - \mu_1^2 = \mu_2 - \tau^2 \tag{10.39}$$

Moments generated by the relation

$$E[\varphi(t)(t - \tau)^n] = \mu_{n,\tau}(\gamma) = \int_0^\infty \varphi(t)(t - \tau)^n f(t)\, dt \tag{10.40}$$

are called the weighted moments of $\varphi(t)$. The weighted moments for $\tau = 0$ and $\varphi(t) = \exp(-\gamma t)$,

$$E[e^{-\gamma t} t^n] = \mu_n(\gamma) = \int_0^\infty e^{-\gamma t} t^n f(t) \, dt \tag{10.41}$$

[where $\mu_n(0) = \mu_n$] are called the exponential weighted moments. For $\gamma > 0$, the tail end of the tracer (experimentally measured residence time distribution) data would have less effect on the moments, which would have special applications to interpreting the main portion of the tracer data, especially by setting γ to be equal to the mean residence time, i.e., $\gamma = \tau$. Computer codes can be developed or commercial statistical software programs can be adopted for the generation of statistical information from the experimentally measured residence time distribution measurements.

The Laplace transform $\phi(s)$ of the probability density function $f(t)$ is given by

$$\phi(s) = L\{f(t)\} = \int_0^\infty e^{-st} f(t) \, dt \tag{10.42}$$

where L is the Laplace transform operator. The Laplace transform $\phi(s)$ is also known as the characteristic function for the residence time distribution (where $s > 0$ is the Laplace variable). Moments (μ_n) of the residence time t about the origin can be calculated from the characteristic function $\phi(s)$ defined in Eq. (10.42):

$$\mu_n = E\{t^n\} = \int_0^\infty t^n f(t) \, dt = (-1)^n \lim_{s \to 0} \frac{d^n \phi(s)}{ds^n} \tag{10.43}$$

Similarly, the moments $\mu_{n,\tau}$ of the residence time t about the mean residence time τ can be calculated from the characteristic function $\phi(s + \tau)$ defined as

$$\phi(s + \tau) = L\{f(t)\} = \int_0^\infty e^{-(s+\tau)t} f(t) \, dt \tag{10.44}$$

which is the Laplace transform of $f(t)$ with the Laplace transform variable $s + \tau$. The moments $\mu_{n,\tau}$ can be calculated from Eq. (10.44):

$$\mu_{n,\tau} = E\{(t - \tau)^n\} = \int_0^\infty (t - \tau)^n f(t) \, dt = (-1)^n \lim_{s \to 0} \frac{d^n \phi(s + \tau)}{d(s + \tau)^n} \tag{10.45}$$

This property of the characteristic function is also a useful tool in the study of residence time distributions. The moments of the residence time for such a system can be calculated by using the Laplace transform of $f(t)$ [or $c(t)$]. This is a practical tool to study residence time distributions in axially dispersed (mixed) tubular reactors (Nauman and Buffham, 1983; Levenspiel and Bischoff, 1963).

5. RESIDENCE TIME DISTRIBUTION MEASUREMENT TECHNIQUES

The most practical method of measuring residence time distribution would be to inject a pulse of tracer species (or particles) into the inlet stream of the reactor (as close as possible to the reactor) and detect the tracer concentrations at the inlet and exit of the reactor. The experimental design of a reactor should not allow the tracers to enter or leave the reactor more than once. Measurement of the tracer concentration at the reactor exit for a pulse tracer input provides the residence time distribution function $f(t)$.

Measurement of the tracer concentration at the reactor exit for a positive unit step change in the input could provide the cumulative residence time distribution function $F(t)$. In many cases a unit step change (a negative unit step change is called washing) of tracer inputs may not be a practical method. Determination of $f(t)$ from experimentally measured $F(t)$ requires differentiation of numerical data, which would not be error-free. As a result, the experimental measurement of $f(t)$ is generally preferred in residence time distribution studies.

It is an important fact that the tracer must have the same properties as the incoming fluid and that the injected amount of the tracers should not disturb the hydrodynamics of the system. Most important, the detector should give a linear response to the concentration of the tracer. Radioactive isotopes are perfect tracers for many complex reacting systems. It is always advantageous to measure the tracer concentration at the reactor inlet (very close to the reactor) also. Tracer material balance is always recommended.

For a pulse input, the tracer concentration $c(t)$ measurement as a function of time at the reactor exit gives the residence time distribution function $f(t)$. For normalization [i.e., the integral of $f(t)$ for the time interval from 0 to ∞ must be equal to 1], the total amount of tracer has to be calculated by using the tracer concentration measurement either at the reactor inlet or at the reactor exit or both [i.e., $f(t)\,\Delta(t) = c(t)\,\Delta(t)/c_{in}$, where c_{in} is the total tracer injected into the system]. Formulations of residence time distribution studies should be made in such a way that the experimental data are always used for integration but not to evaluate derivatives.

6. RESIDENCE TIME DISTRIBUTION MODELS

Real flow systems such as continuous flow stirred tank reactors or piston flow reactors never behave as perfect mixers. Real flow systems can often be synthesized by connecting simple systems together.

The response of a simple flow system to a disturbance in the input stream can be predicted from the material balance [i.e., for an initially relaxed system, where $c_f(t) = 0$ for $t < 0$]

$$c(t) = \int_0^t c_f(\xi) f(t - \xi)\, d\xi \tag{10.46}$$

where c and c_f are the tracer concentrations in the feed and exit streams and $f(t)$ is the residence time distribution function for the particles in the flow system. The integral in Eq. (10.46) is the convolution integral of the functions c_f and f:

$$c(t) = \int_0^t c_f(\xi) f(t - \xi)\, d\xi = c_f(t) * f(t) \tag{10.47}$$

where the centered asterisk ($*$) is used for the convolution integral operation. The Laplace transform of Eq. (10.47) is

$$C(s) = \psi(s) C_f(s) \tag{10.48}$$

where $C(s)$, $C_f(s)$, and $\psi(s)$ are the Laplace transforms of the tracer exit concentration $c(t)$, tracer feed concentration $c_f(t)$, and the tracer residence time distribution function $f(t)$ in the flow system, respectively. As discussed in Section 3.1 of Chapter 10, $\psi(s)$ is also known as the transfer function [i.e., $\psi(s)$ in Eq. (10.24)],

$$\psi(s) = \frac{C(s)}{C_f(s)} \tag{10.49}$$

If two independent flow systems with residence time distribution functions of $f_1(t)$ and $f_2(t)$ are connected in series as depicted in Fig. 10.3, the response of the system to a disturbance in the feed system would be

$$C(s) = C_f(s)\psi_1(s)\psi_2(s) \tag{10.50}$$

where $\psi_1(s)$ and $\psi_2(s)$ are the Laplace transforms of the residence time distributions $f_1(t)$ and $f_2(t)$, respectively, of the subsystems. The response of the overall system can be expressed by the overall transfer function $\psi(s)$ in the Laplace domain:

$$C(s) = C_f(s)\psi(s) \tag{10.51}$$

where

$$\psi(s) = \psi_1(s)\psi_2(s) \tag{10.52}$$

Thus the residence time distribution function in the time domain for the overall system becomes

$$f(t) = f_1(t) * f_2(t) \tag{10.53}$$

These relations are valid for statistically independent systems. That is, the residence time distribution in one vessel is not affected by the residence time distribution in the other vessel.

Flow rates and tracer concentrations for diverging and merging flow systems (Fig. 10.4) can be obtained by using a simple material balance. Based on material balance principles, the transfer function for a composite system (Fig. 10.5) would be

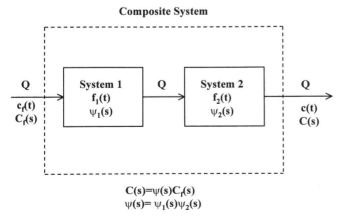

Figure 10.3 A composite flow system composed of two simple statistically independent flow systems of residence time distribution functions $f_1(t)$ and $f_2(t)$.

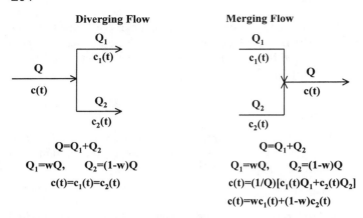

Figure 10.4 Material balances for diverging and merging flows.

$$C_1(s) = wC_f(s)\psi_1(s), \qquad C_2(s) = (1-w)C_f(s)\psi_2(s) \tag{10.54}$$

or

$$C(s) = wC_f(s)\psi_1(s) + (1-w)C_f(s)\psi_2(s) = C_f(s)[w\psi_1(s) + (1-w)\psi_2(s)] \tag{10.55}$$

or, in terms of the transfer functions,

$$\frac{C(s)}{C_f(s)} = \psi(s) = w\psi_1(s) + (1-w)\psi_2(s) \tag{10.56}$$

Using the same principles, the transfer function for a composite system with a recycle stream as in Fig. 10.6 can be expressed as

$$(1+w)C(s) = [C_f(s) + wC(s)\psi_2(s)]\psi_1(s) \tag{10.57}$$

or

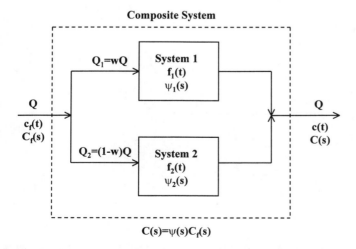

Figure 10.5 A composite flow system composed of diverging and merging flows.

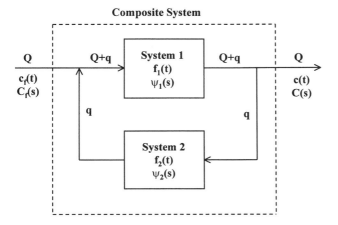

Figure 10.6   Transfer function of a composite flow system with a recycle flow.

$$\frac{C(s)}{C_f(s)} = \psi(s) = \frac{\psi_1(s)}{1 + w[1 - \psi_1(s)\psi_2(s)]} \tag{10.58}$$

where $w = q/Q$.

If the transfer functions for the subsystems are known, the overall transfer function for a composite system can be calculated in the Laplace domain. By standard mathematical manipulations, the tracer concentration at the exit of the system can be expressed as a function of time. These models can also be used to express deviations of a system from the ideal.

7.   DISPERSION MODELS FOR TUBULAR REACTORS

As discussed in Chapter 9, most industrial tubular reactors operate in flow regimes where the material in the reactor disperses in the direction of flow (axial direction). The material may also be dispersed radially. Therefore, piston flow models cannot represent the flow behavior of the material in such flow systems.

Thus, the axial dispersion in tubular flow systems (reactors) can be modeled similarly to diffusion mass flux. Assuming the validity of the model, by using the mass balance principle expressed by Eq. (9.1) the unsteady-state balance equation becomes [corresponding to the steady-state model of Eq. (9.176) for a single reaction]

$$\frac{\partial c_j}{\partial \tau} + \frac{\partial c_j}{\partial Z} = \frac{1}{\mathrm{Pe}_d} \frac{\partial^2 c_j}{\partial Z^2} + \frac{\alpha_j L}{v} r(c, T) \tag{10.59}$$

where τ and Z are the normalized time and distance, respectively, and Pe_d is the axial dispersion Peclet number

$$\tau = \frac{vt}{L} = \frac{t}{\tau_r}, \qquad Z = \frac{z}{L}, \qquad \mathrm{Pe} = \frac{vL}{D_d} \tag{10.60}$$

where D_d is the axial dispersion coefficient.

In Eq. (10.60), τ_r is the mean residence time for the piston flow (i.e., $\tau_r = L/v$, the ratio of the reactor length L to average liner flow velocity v calculated from the volumetric flow-rate and cross-sectional surface area, $v = q/A_c$). If an inert tracer is used, then the unsteady-state mass balance equation can be further simplified to

$$\frac{\partial c_j}{\partial \tau} + \frac{\partial c_j}{\partial Z} = \frac{1}{\text{Pe}_d} \frac{\partial^2 c_j}{\partial Z^2} \tag{10.61}$$

with the following boundary (Chapter 9) and initial (for a pulse input) conditions:

$$c_j(0-) = c_j(0+) - \frac{1}{\text{Pe}_d} \frac{\partial c_j}{\partial Z}(0+) \qquad \text{at} \qquad Z = 0 \tag{10.62}$$

$$\frac{dc_j}{dZ} = 0 \qquad \text{at} \qquad Z = 1 \tag{10.63}$$

$$c_j = \delta(0) \qquad \text{at} \qquad Z = 0; \qquad c_j = 0 \qquad \text{at} \qquad \tau = 0; \qquad 0 < Z < 1 \tag{10.64}$$

The axial dispersion Peclet number Pe_d can be calculated from the best least-squares fit of tracer test data to the solution of Eq. (10.61) with the boundary and initial conditions of Eqs. (10.62)–(10.64). Solution of Eq. (10.61) for any input can also be generated numerically for any initial condition expressed by Eq. (10.64). It can be seen from Eqs. (10.60) and (10.61) that the Peclet number controls the residence time distribution. As $\text{Pe}_d \to 0$ (i.e., as $D_d \to \infty$) the system becomes a perfectly mixed system, i.e., a CSTR. In this case the right-hand side of Eq. (10.61) becomes dominant in comparison to the left-hand side. As $\text{Pe}_d \to \infty$ (i.e., $D_d \to 0$) the system becomes a piston flow system. In this case the second-order derivative term in Eq. (10.61) becomes negligible, and Eq. (10.61) without the second-order derivative term represents the mass balance for a plug flow tubular reactor.

In hydrocracking reactors, the liquid- and gas-phase residence time distributions are different. Hydrogen is transferred from the gas phase into the liquid phase by interface mass transfer and is consumed in the liquid phase by catalytic hydrogenation (hydrocracking) reactions. Liquid- and gas-phase residence time distributions in these reactors can be studied by marking the liquid and gas phases with iodine-131 (^{131}I) and xenon-133 (^{133}Xe) isotopes, respectively. The liquid-phase tracer ^{131}I stays in the liquid phase because an organic species attached to ^{131}I is formed, whereas the gas-phase tracer ^{133}Xe may stay in the gas and liquid phases [^{131}I has an 8 day half-life and radiates γ rays at 364 keV (%82), whereas ^{133}Xe has a 36.4 day half-life and radiates γ rays at 203 keV (%65)]. Approximating the reaction term to zero because the time required for the tracer experiment (about 2 h) or the experimental data can be corrected for the radioactive decay, mass balances for the liquid- and gas-phase tracers can be expressed as

$$\varepsilon_l \frac{\partial c_{1,l}}{\partial t} + \varepsilon_l v_l \frac{\partial c_{1,l}}{\partial z} = \varepsilon_l D_l \frac{\partial^2 c_{1,l}}{\partial z^2} \tag{10.65}$$

$$\varepsilon_l \frac{\partial c_{2,l}}{\partial t} + \varepsilon_l v_l \frac{\partial c_{2,l}}{\partial z} = \varepsilon_l D_l \frac{\partial^2 c_{2,l}}{\partial z^2} + k_m a \left(\frac{c_{2,g}}{H} - c_{2,l} \right) \tag{10.66}$$

for the gas-phase tracer (subscript 2) in the liquid phase, and

$$(1 - \varepsilon_l) \frac{\partial c_{2,g}}{\partial t} + (1 - \varepsilon_l) v_g \frac{\partial c_{2,g}}{\partial z} = (1 - \varepsilon_l) D_g \frac{\partial^2 c_{2,g}}{\partial z^2} - k_m a \left(\frac{c_{2,g}}{H} - c_{2,l} \right) \tag{10.67}$$

for the gas-phase tracer in the gas phase. D_l and D_g are the axial dispersion coefficients for the liquid and gas phases, k_m is the interfacial mass transfer coefficient, a is the interfacial area per unit reactor volume, ε is the phase hold-up, and H is Henry's constant for the tracer gas. Boundary and initial conditions for Eqs. (10.65)–(10.67) are

$$\varepsilon_j v_j c_{i,j}(0-) = \varepsilon_j v_j c_{i,j}(0+) - \varepsilon_j D_j \frac{\partial c_{i,j}}{\partial z}(0+) \tag{10.68}$$

$$\frac{dc_{i,j}}{dz} = 0 \quad \text{at} \quad z = L \tag{10.69}$$

$$c_{i,j} = \delta_{i,j}(0) \quad \text{at} \quad Z = 0; \quad c_{i,j} = 0 \quad \text{at} \quad \tau = 0; \quad 0 < Z < L \tag{10.70}$$

where the subscripts i and j denote the tracers (1 or 2) and phases (liquid and gas), respectively.

The tracer balance equations can be expressed in terms of the dimensionless variables Z, τ_j, and Pe_j:

$$Z = \frac{z}{L}, \qquad \tau_j = \frac{v_j t}{L}, \qquad \text{Pe}_j = \frac{v_j L}{D_j} \tag{10.71}$$

Thus, the tracer balance equations become

$$\frac{\partial c_{1,l}}{\partial \tau_l} + \frac{\partial c_{1,l}}{\partial Z} = \frac{1}{\text{Pe}_l} \frac{\partial^2 c_{1,l}}{\partial Z^2} \tag{10.72}$$

$$\frac{\partial c_{2,l}}{\partial \tau_l} + \frac{\partial c_{2,l}}{\partial Z} = \frac{1}{\text{Pe}_l} \frac{\partial^2 c_{2,l}}{\partial Z^2} + \frac{k_m a L}{\varepsilon_l v_l} \left(\frac{c_{2,g}}{H} - c_{2,l} \right) \tag{10.73}$$

and

$$\frac{\partial c_{2,g}}{\partial \tau_g} + \frac{\partial c_{2,g}}{\partial Z} = \frac{1}{\text{Pe}_g} \frac{\partial^2 c_{2,g}}{\partial Z^2} - \frac{k_m a L}{(1 - \varepsilon_l) v_g} \left(\frac{c_{2,g}}{H} - c_{2,l} \right) \tag{10.74}$$

and the corresponding boundary and initial conditions become

$$c_{i,j}(0-) = c_{i,j}(0+) - \frac{1}{\text{Pe}_j} \frac{\partial c_{i,j}}{\partial Z}(0+) \tag{10.75}$$

$$\frac{dc_{i,j}}{dZ} = 0 \quad \text{at} \quad Z = 1 \tag{10.76}$$

Using tracer tests, the unknown process parameters such as Pe_l, Pe_g, $k_m a$, and ε_l (or ε_g) can be determined from the best least-squares fit of experimental data to the solution of model equations with suitable boundary conditions.

8. DISPERSION MODELS FOR AXIAL DISTRIBUTION TUBULAR REACTORS

Laplace transformation of the transient tracer balance equation [Eq. (10.61)] and corresponding boundary and initial conditions can also help the interpretation of the experimental data without the solution of the tracer balance equation in the time domain. For a single-phase flow axial distribution tubular reactor, using $C(s)$, the Laplace transform of

$c_j(t)$ in Eq. (10.61), the variance of the residence time about the mean (as discussed in Section 4 of Chapter 7) can be calculated as (Levenspiel and Bischoff, 1963)

$$\sigma^2 = \frac{2}{\text{Pe}_d^2}\left(\text{Pe}_d - 1 + e^{-\text{Pe}_d}\right) \tag{10.77}$$

For a pulse tracer input, the tracer concentration measured at the reactor exit can be used to generate the residence time distribution function $f(t)$, the first moment of the residence time (mean or μ_1), and the second moment (μ_2). As an example, if $c(t_i)$ is the tracer concentration measurement made for the time interval $t_i + \Delta t_i$, then the residence time distribution function can be derived as

$$f(t_i)\,\Delta t_i = \frac{c(t_i)\,\Delta t_i}{\sum\limits_{i=1}^{n} c(t_i)\,\Delta t_i} \tag{10.78}$$

Because $f(t)$ is determined, the first and second moments can be calculated as

$$\mu_1 = \tau = \sum_{i=1}^{n} t_i f(t_i)\,\Delta t_i \tag{10.79}$$

and

$$\mu_2 = \sum_{i=1}^{n} t_i^2 f(t_i)\,\Delta t_i \tag{10.80}$$

from which σ^2 can be calculated (i.e., $\sigma^2 = \mu_2 - \mu_1^2$) from the experimentally measured tracer response data. The calculated value of σ^2 can be used in Eq. (10.77) to calculate the axial dispersion Peclet number Pe_d

In axially dispersed tubular reactors (ADTRs) the mixing takes place independently of chemical reactions. Chemical reactions, however, depend on the concentration of the reactive species; therefore, they depend on the degree of mixing (i.e., axial dispersion). For a first-order reaction $A \rightarrow P$, which takes place in axially dispersed tubular reactors (i.e., operating at constant temperature), the material balance for the chemical species A can be derived from Eq. (10.59) as in the following:

$$\frac{\partial c_A}{\partial \tau} + \frac{\partial c_A}{\partial Z} = \frac{1}{\text{Pe}_d}\frac{\partial^2 c_A}{\partial Z^2} - \frac{Lk}{v}c_A \tag{10.81}$$

where the dimensionless group Lk/v appearing as a coefficient to c_A in the last term is called the Damkohler number Da, which is defined as the ratio of the characteristic mixing (i.e., mass transfer) time τ_m to the characteristic reaction time τ_r:

$$\text{Da} = \frac{Lk}{v} = \frac{L/v}{1/k} = \frac{\tau_m}{\tau_r} \tag{10.82}$$

As Da $\rightarrow 0$ (i.e., $k \rightarrow 0$), the effect of diffusion mass transfer on c_A would be negligible and the profile of c_A is entirely controlled by the rate of chemical reaction (chemical reaction control regime). At the other extreme, as Da $\rightarrow \infty$ (i.e., $k \rightarrow \infty$), the effect of chemical reaction on c_A would be negligible and the profile of c_A is entirely controlled by the rate of mass transfer (diffusion control regime). Mixing helps to reduce mass transfer limitations in systems involving fast chemical reactions. The mixing process is more complex than what has been discussed in this section (Danckwerts, 1958; Zwietering, 1959).

Mixing is controlled by macro- and microscale hydrodynamic processes that take place in flow systems. Mixing is always increased by an increase in energy dissipation per unit volume of the system. The physical characteristics of the system determine the size of the microscale at which mixing naturally takes place by molecular diffusion. This subject is still an attractive topic for the chemical engineering profession, and its understanding requires knowledge of the fundamentals of fluid dynamics and turbulence.

REFERENCES

Danckwerts, P. V. 1958. Chem. Eng. Sci. 8:93–102.
Levenspiel, O., and Bischoff, K. 1963. Adv. Chem. Eng. 4:95.
Nauman, E. B., and Buffham, B. A. 1983. Mixing in Continuous Flow Systems. Wiley, New York.
Petho, A., and Noble, R. D. 1982. Residence Time Distribution Theory in Chemical Engineering. Verlag Chemie, Hemsbach, Germany.
Zwietering, T. N. 1959. Chem. Eng. Sci. 11:1–15.

11

Transport Phenomena

1. INTRODUCTION

The use of a variety of reactors in refineries with different feedstocks and catalysts requires that the feedstock achieve thorough mixing of the reactants and/or efficient heat transfer and/or adequate contact with the catalyst (as examples) in order to ensure an efficient reaction. Thus, the study of transport phenomena is an important aspect of refinery and reactor engineering.

The study of transport phenomena is the branch of engineering science that deals with the transfer of mass, momentum, and energy. Because of its practical implications for many branches of engineering and our everyday life, this subject has been investigated extensively. The coupling between transport phenomena and chemical reactions (Chapter 9) is one of the most interesting subjects in chemical engineering. Considering this fact, a general discussion of transport phenomena will be presented in this chapter.

As a result of the developments made in the last four decades in transport phenomena, numerical techniques, and high-speed computers, many transport phenomena problems can be solved numerically at affordable cost. Several commercial codes that are called computational fluid dynamics (CFD) simulators are available in the market today, that could not have been even dreamed of a few decades ago. These computational fluid dynamics simulators are prepared to numerically solve complex engineering problems involving chemically reactive and nonreactive single- and multiphase flow systems. These commercial simulators are becoming standard tools in industrial and research laboratories.

The availability of commercial computational fluid dynamics simulators at affordable costs encouraged engineers and scientists to use these simulators to solve their problems. As a result, engineers and researchers are entangled with many aspects of transport phenomena with an increasing intensity. An understanding of transport phenomena is needed to use commercial computational fluid dynamics simulators efficiently as well as to judge, discuss, interpret and translate the results obtained from the simulations in terms of real-world situations. Furthermore, an in-depth knowledge of transport phenomena is needed for the overall evaluation of many engineering and scientific problems, to find practical solutions and interpret experimental observations.

There are excellent reviews of transport phenomena of a wide range of complexities (Bird et al., 1960; Brodkey, 1967; Levich 1962; Frank-Kamenetskii, 1969; Schlichting, 1955). The material presented in this section is focused on the formulation and solution of mass, momentum, and heat transfer problems. The main purpose of this chapter is to provide the basics of transport phenomena coupled with chemical reactions. As discussed in previous chapters, all chemical reactions are coupled with transport phenomena in a nonlinear fashion. In Chapter 9, it was demonstrated that many nonlinear phenomena such as stability in reacting systems, ignition, spontaneous combustion, and explosion can be explained by the coupling between transport phenomena and chemical reactions. Transport phenomena also find important applications in geology, atmospheric science, and petroleum and environmental engineering.

2. FORMULATION OF PROPERTY BALANCE EQUATIONS

Mathematical formulations of property balance equations are needed to study transport phenomena. In this section the integral form of balance equations will be used to cover the principles of the topic in a limited space. For the development of numerical methods to solve transport phenomena problems, however, differential balance equations are recommended.

The general property balance was discussed in Chapter 9 in the formulation of Eq. (9.1). The same principles will be applied in this section using a different mathematical technique. For this purpose, accumulation, flux, and generation components of the property for a control volume V surrounded by a surface S will be considered, as described in Fig. 11.1. The property balance for the control volume V can be expressed by Eq. (9.1), which is repeated here:

$$(\text{property})_{\text{accum}} = (\text{property})_{\text{in}} - (\text{property})_{\text{out}} + (\text{property})_{\text{gen}} \tag{11.1}$$

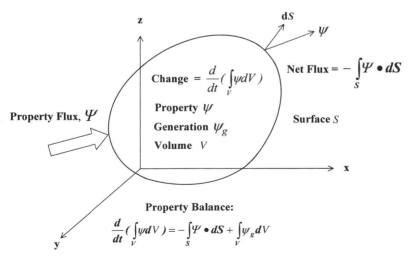

Figure 11.1 Mathematical representation of the property balance for a control volume V surrounded by surface S.

or

$$(\psi)_{\text{accum}} = (\psi)_{\text{in}} - (\psi)_{\text{out}} + (\psi)_{\text{gen}} \tag{11.2}$$

where ψ is the property (mass, momentum, or energy per unit volume), $(\psi)_{\text{in}}$ and $(\psi)_{\text{out}}$ are the amount of the property carried in and out of the control volume as a result of the property flux vector Ψ (i.e., the flux of property ψ per unit surface area per unit time) through the boundaries of the control volume, and $(\psi)_{\text{accum}}$ is the accumulation and $(\psi)_{\text{gen}}$ the generation of the property within the control volume [i.e., if ψ_{gen} is the property generation rate per unit volume, then property generation would be $(\psi)_{\text{gen}} = V\psi_{\text{gen}}$]. In this formulation the control volume can be considered as stationary or moving at velocity v with respect to a fixed coordinate system.

Using mathematical principles, the property balance given by Eq. (11.2) can be expressed as (Brodkey, 1967)

$$\frac{d}{dt}\left(\int_V \psi \, dV\right) = -\int_S \Psi \cdot dS + \int_V \psi_g \, dV \tag{11.3}$$

where the negative sign in front of the surface integral [which replaces the ψ_{in} and ψ_{out} terms of Eq. (11.2)] is needed to make the property flux positive when the property is flowing into the control volume (i.e., surface gradient and flux vectors are in opposite directions). The property ψ [Eq. (11.3)] can be a scalar (e.g., mass, energy, entropy) or vector (e.g., momentum) quantity.

The time derivative term of Eq. (11.3) can be expressed in the form

$$\frac{d}{dt}\left(\int_V \psi \, dV\right) = \int_V \frac{\partial \psi}{\partial t} \, dV + \int_S \Psi(v \cdot dS) \tag{11.4}$$

where the first term takes care of the time derivative of the property in the control volume at fixed coordinates and the second term takes care of property changes in the control volume when that volume moves with velocity v (i.e., the maximum change in property would take place when surface gradient vector dS and v vectors are in the same direction). Using the Gaussian theorem for the surface integral of a vector field Θ (Apostol, 1966),

$$\int_S (\Theta \cdot dS) = \int_V \text{div}(\Theta) \, dV = \int_V (\nabla \cdot \Theta) \, dV \tag{11.5}$$

which expresses the surface integral of a vector field in the form of a volume integral. Using the Gaussian theorem to express the surface integral in the form of a volume integral [Eqs. (11.5) and (11.4)], the property balance [Eq. (11.3)] becomes

$$\int_V \frac{\partial \psi}{\partial t} \, dV + \int_V (\nabla \cdot \psi v) \, dV = -\int_V (\nabla \cdot \Psi) \, dV + \int_V \psi_g \, dV \tag{11.6}$$

or, collecting all the terms under the volume integral operator,

$$\int_V \left(\frac{\partial \psi}{\partial t} + \nabla \cdot \psi v + \nabla \cdot \Psi - \psi_g\right) dV = 0 \tag{11.7}$$

or

$$\frac{\partial \psi}{\partial t} + \nabla \cdot \psi \boldsymbol{v} + \nabla \cdot \boldsymbol{\Psi} - \psi_g = 0 \tag{11.8}$$

which is known as the *equation of change*.

The property flux vector $\boldsymbol{\Psi}$ [the third term in Eq. (11.8)] can be expressed in terms of the gradient of the scalar property field ψ and diffusion coefficient (i.e., property flux is first order with respect to the gradient of the scalar field):

$$\boldsymbol{\Psi} = -D \, \nabla \psi \tag{11.9}$$

which reduces Eq. (11.8) to

$$\frac{\partial \psi}{\partial t} + \nabla \cdot \psi \boldsymbol{v} + \nabla \cdot (-D \, \nabla \psi) - \psi_g = 0 \tag{11.10}$$

which can be further simplified, if the diffusion coefficient D is constant, to

$$\frac{\partial \psi}{\partial t} + \nabla \cdot \psi \boldsymbol{v} - D \, \nabla^2 \psi - \psi_g = 0 \tag{11.11}$$

where $\nabla^2 (= \nabla \cdot \nabla)$ is the Laplacian operator. The mathematical expression presented as Eq. (11.11) is also called the equation of change for constant diffusivity.

This equation is the most general form of the property balance for a system exchanging mass, momentum, energy, entropy, etc. with its surroundings through the boundaries of the system. These equations are coupled and nonlinear, and their general solution may not exist for every geometry or all initial and boundary conditions. Seeking solutions for the set of mass, momentum, and energy balance equations represented by Eq. (11.11) is the subject of transport phenomena, which is one of the building blocks of reaction engineering.

3. MOMENTUM TRANSFER (FLUID DYNAMICS)

Momentum transfer can be formulated by assigning the property ψ [Eq. (11.8)] as the momentum of a fluid per unit volume $\rho \boldsymbol{v}$ (which is a vector property). Because the momentum per unit volume of the fluid is a vector, its flux can be described by the pressure tensor $\mathbf{P}$ and its generation by the external force fields acting on the mass of fluid in a unit volume ($\sum_s \rho_s \boldsymbol{F}_s$). With these substitutions, the equation of change [Eq. (11.8)] becomes (Brodkey, 1967)

$$\frac{\partial (\rho \boldsymbol{v})}{\partial t} + \nabla \cdot \rho (\boldsymbol{v}\boldsymbol{v}) + \nabla \cdot \mathbf{P} - \sum_s \rho_s \boldsymbol{F}_s = 0 \tag{11.12}$$

Using the relations

$$\frac{\partial (\rho \boldsymbol{v})}{\partial t} = \rho \frac{\partial \boldsymbol{v}}{\partial t} + \boldsymbol{v} \frac{\partial \rho}{\partial t} \tag{11.13}$$

and

$$\nabla \cdot \rho (\boldsymbol{v}\boldsymbol{v}) = (\rho \boldsymbol{v} \cdot \nabla) \boldsymbol{v} + \boldsymbol{v} (\nabla \cdot \rho \boldsymbol{v}) \tag{11.14}$$

Eq. (11.12) becomes

$$\rho \frac{\partial \boldsymbol{v}}{\partial t} + \boldsymbol{v} \left(\frac{\partial \rho}{\partial t} + \nabla \cdot \rho \boldsymbol{v} \right) + (\rho \boldsymbol{v} \cdot \nabla) \boldsymbol{v} = -\nabla \cdot \mathbf{P} + \sum_s \rho_s \boldsymbol{F}_s \tag{11.15}$$

The second term on the left-hand side of Eq. (11.15) vanishes as a result of the equation of continuity ($\partial \rho / \partial t + \nabla \cdot \rho v = 0$, which will be discussed in Section 11.5). Therefore,

$$\rho \frac{\partial v}{\partial t} + (\rho v \cdot \nabla)v = \rho \left(\frac{\partial v}{\partial t} + (v \cdot \nabla)v \right) = -\nabla \cdot \mathbf{P} + \sum_s \rho_s \mathbf{F}_s \tag{11.16}$$

or

$$\rho \frac{Dv}{Dt} = -\nabla \cdot \mathbf{P} + \sum_s \rho_s \mathbf{F}_s \tag{11.17}$$

where D/Dt is the substantial derivative operator,

$$\frac{D}{Dt} = \frac{\partial}{\partial t} + v \cdot \nabla \tag{11.18}$$

which is the expression for the time derivative operator for a reference frame moving at velocity v (i.e., the point of observation is moving with velocity v).

Two equations, (11.16) and (11.17), are known as the equation of motion. Equation (11.16) is the force balance or application of Newton's second law to a fluid element. The pressure tensor $\mathbf{P}$ of Eq. (11.16) can be expressed in terms of normal or static pressure tensor $\mathbf{p}$ and viscous stress tensor τ as

$$\mathbf{P} = \mathbf{p} + \tau \tag{11.19}$$

with which the equation of motion becomes

$$\rho \left(\frac{\partial v}{\partial t} + (v \cdot \nabla)v \right) = -\nabla \cdot \mathbf{p} - \nabla \cdot \tau + \sum_s \rho_s \mathbf{F}_s \tag{11.20}$$

For an incompressible fluid, if the viscous stress tensor τ is Newtonian ($\tau = -\mu \nabla v$), then Eq. (11.20) becomes

$$\rho \left(\frac{\partial v}{\partial t} + (v \cdot \nabla)v \right) = -\nabla \cdot \mathbf{p} + \mu \nabla^2 v + \sum_s \rho_s \mathbf{F}_s \tag{11.21}$$

which is called the Navier–Stokes equation. The equation of motion in terms of the stress tensor τ and in terms of the velocity vector v for Newtonian fluids (constant density and viscosity) in rectangular, cylindrical, and spherical coordinates is given in Tables 11.1, 11.2, and 11.3, respectively. These are the general forms of the equation of motion (i.e., momentum balance), the solution of which for specific boundary and/or initial conditions would predict the velocity profile of the fluid in a given geometry as a function of spatial position and/or time.

3.1 General Properties of the Navier–Stokes Equation and Its Solutions

The Navier–Stokes equation is the general form of the equation of motion for incompressible Newtonian fluids. It is a nonlinear partial differential equation [i.e., the second term on the left-hand side of Eq. (11.21)], and its exact solution may not exist for all flow problems. The Navier–Stokes equation is a macroscopic balance equation; therefore it describes the fluid phenomena for distances several orders of magnitude greater than the mean-free path of the molecular motion. Exact solutions for the Navier–Stokes equation can be obtained for very few flow problems.

Solutions of the Navier–Stokes equation can be classified (or summarized) as follows:

Table 11.1 Equation of Motion in Terms of τ and in Terms of Velocity Gradients for Newtonian Fluid with Constant ρ and μ in Rectangular Coordinate System of x, y, and z

In terms of τ

  x Component:

$$\rho\left(\frac{\partial v_x}{\partial t} + v_x\frac{\partial v_x}{\partial x} + v_y\frac{\partial v_x}{\partial y} + v_z\frac{\partial v_x}{\partial z}\right) = -\frac{\partial p}{\partial x} - \left(\frac{\partial \tau_{xx}}{\partial x} + \frac{\partial \tau_{yx}}{\partial y} + \frac{\partial z_{zx}}{\partial z}\right) + \rho g_x$$

  y Component:

$$\rho\left(\frac{\partial v_y}{\partial t} + v_x\frac{\partial v_y}{\partial x} + v_y\frac{\partial v_y}{\partial y} + v_z\frac{\partial v_y}{\partial z}\right) = -\frac{\partial p}{\partial y} - \left(\frac{\partial \tau_{xy}}{\partial x} + \frac{\partial \tau_{yy}}{\partial y} + \frac{\partial z_{zy}}{\partial z}\right) + \rho g_y$$

  z Component:

$$\rho\left(\frac{\partial v_z}{\partial t} + v_x\frac{\partial v_z}{\partial x} + v_y\frac{\partial v_z}{\partial y} + v_z\frac{\partial v_z}{\partial z}\right) = -\frac{\partial p}{\partial z} - \left(\frac{\partial \tau_{xz}}{\partial x} + \frac{\partial \tau_{yz}}{\partial y} + \frac{\partial z_{zz}}{\partial z}\right) + \rho g_z$$

In terms of velocity gradients for a Newtonian fluid with constant ρ and μ
  x Component:

$$\rho\left(\frac{\partial v_x}{\partial t} + v_x\frac{\partial v_x}{\partial x} + v_y\frac{\partial v_x}{\partial y} + v_z\frac{\partial v_x}{\partial z}\right) = -\frac{\partial p}{\partial x} + \mu\left(\frac{\partial^2 v_x}{\partial x^2} + \frac{\partial^2 v_x}{\partial y^2} + \frac{\partial^2 v_x}{\partial z^2}\right) + \rho g_x$$

  y Component:

$$\rho\left(\frac{\partial v_y}{\partial t} + v_x\frac{\partial v_y}{\partial x} + v_y\frac{\partial v_y}{\partial y} + v_z\frac{\partial v_y}{\partial z}\right) = -\frac{\partial p}{\partial y} + \mu\left(\frac{\partial^2 v_y}{\partial x^2} + \frac{\partial^2 v_y}{\partial y^2} + \frac{\partial^2 v_y}{\partial z^2}\right) + \rho g_y$$

  z Component:

$$\rho\left(\frac{\partial v_z}{\partial t} + v_x\frac{\partial v_z}{\partial x} + v_y\frac{\partial v_z}{\partial y} + v_z\frac{\partial v_z}{\partial z}\right) = -\frac{\partial p}{\partial z} + \mu\left(\frac{\partial^2 v_z}{\partial x^2} + \frac{\partial^2 v_z}{\partial y^2} + \frac{\partial^2 v_z}{\partial z^2}\right) + \rho g_z$$

Source: Bird et al., 1960.

1. Exact solutions
2. Approximate solutions
    a. Ideal fluid (viscous forces are neglected; i.e., $\mu \to 0$ or very high Reynolds numbers)
    b. Very slow motion (inertia forces are neglected; i.e., $v \to 0$ or very small Reynolds numbers)
    c. Boundary layer solution (viscous forces are confined to a thin layer close to stationary objects; i.e., at high velocities or at high Reynolds numbers)

The exact and approximate solutions of the Navier–Stokes equation will be briefly discussed in the following sections.

Table 11.2   Equation of Motion in Terms of τ and in Terms of Velocity Gradients for Newtonian Fluid with Constant ρ and μ in Cylindrical Coordinate System of r, θ, and z

In terms of τ

   r Component:

$$\rho\left(\frac{\partial v_r}{\partial t} + v_r\frac{\partial v_r}{\partial r} + \frac{v_\theta}{r}\frac{\partial v_r}{\partial \theta} - \frac{v_\theta^2}{r} + v_z\frac{\partial v_r}{\partial z}\right) = -\frac{\partial p}{\partial r} - \left(\frac{1}{r}\frac{\partial(r\tau_{rr})}{\partial r} + \frac{1}{r}\frac{\partial(\tau_{r\theta})}{\partial \theta} - \frac{\tau_{\theta\theta}}{r} + \frac{\partial \tau_{rz}}{\partial z}\right) + \rho g_r$$

   θ Component:

$$\rho\left(\frac{\partial v_\theta}{\partial t} + v_r\frac{\partial v_\theta}{\partial r} + \frac{v_\theta}{r}\frac{\partial v_\theta}{\partial \theta} + \frac{v_r v_\theta}{r} + v_z\frac{\partial v_\theta}{\partial z}\right) = -\frac{1}{r}\frac{\partial p}{\partial \theta} - \left(\frac{1}{r^2}\frac{\partial(r^2\tau_{r\theta})}{\partial r} + \frac{1}{r}\frac{\partial(\tau_{\theta\theta})}{\partial \theta} + \frac{\partial \tau_{\theta z}}{\partial z}\right) + \rho g_\theta$$

   z Component:

$$\rho\left(\frac{\partial v_z}{\partial t} + v_r\frac{\partial v_z}{\partial r} + \frac{v_\theta}{r}\frac{\partial v_z}{\partial \theta} + v_z\frac{\partial v_z}{\partial z}\right) = -\frac{\partial p}{\partial z} - \left(\frac{1}{r}\frac{\partial(r\tau_{rz})}{\partial r} + \frac{1}{r}\frac{\partial(\tau_{\theta z})}{\partial \theta} + \frac{\partial \tau_{zz}}{\partial z}\right) + \rho g_z$$

In terms of velocity gradients for a Newtonian fluid with constant ρ and μ
   r Component:

$$\rho\left(\frac{\partial v_r}{\partial t} + v_r\frac{\partial v_r}{\partial r} + \frac{v_\theta}{r}\frac{\partial v_r}{\partial \theta} - \frac{v_\theta^2}{r} + v_z\frac{\partial v_r}{\partial z}\right) = -\frac{\partial p}{\partial r}$$

$$+ \mu\left[\frac{\partial}{\partial r}\left(\frac{1}{r}\frac{\partial(rv_r)}{\partial r}\right) + \frac{1}{r^2}\frac{\partial^2 v_r}{\partial \theta^2} - \frac{2}{r^2}\frac{\partial v_\theta}{\partial \theta} + \frac{\partial^2 v_r}{\partial z^2}\right] + \rho g_r$$

   θ Component:

$$\rho\left(\frac{\partial v_\theta}{\partial t} + v_r\frac{\partial v_\theta}{\partial r} + \frac{v_\theta}{r}\frac{\partial v_\theta}{\partial \theta} + \frac{v_r v_\theta}{r} + v_z\frac{\partial v_\theta}{\partial z}\right) = -\frac{1}{r}\frac{\partial p}{\partial \theta}$$

$$+ \mu\left[\frac{\partial}{\partial r}\left(\frac{1}{r}\frac{\partial(rv_\theta)}{\partial r}\right) + \frac{1}{r^2}\frac{\partial^2 v_\theta}{\partial \theta^2} + \frac{2}{r^2}\frac{\partial v_r}{\partial \theta} + \frac{\partial^2 v_\theta}{\partial z^2}\right] + \rho g_\theta$$

   z Component:

$$\rho\left(\frac{\partial v_z}{\partial t} + v_r\frac{\partial v_z}{\partial r} + \frac{v_\theta}{r}\frac{\partial v_z}{\partial \theta} + v_z\frac{\partial v_z}{\partial z}\right) = -\frac{\partial p}{\partial z} + \mu\left[\frac{1}{r}\frac{\partial}{\partial r}\left(r\frac{\partial v_z}{\partial r}\right) + \frac{1}{r^2}\frac{\partial^2 v_z}{\partial \theta^2} + \frac{\partial^2 v_z}{\partial z^2}\right] + \rho g_z$$

Source: Bird et al., 1960.

Table 11.3 Equation of Motion in Terms of τ and in Terms of Velocity Gradients for Newtonian Fluid with Constant ρ and μ in Spherical Coordinate System of r, θ, and ϕ

In terms of τ

r Component:

$$\rho\left(\frac{\partial v_r}{\partial t} + v_r\frac{\partial v_r}{\partial r} + \frac{v_\theta}{r}\frac{\partial v_r}{\partial \theta} + \frac{v_\phi}{r\sin\theta}\frac{\partial v_r}{\partial \phi} - \frac{v_\theta^2 + v_\phi^2}{r}\right)$$

$$= -\frac{\partial p}{\partial r} - \left(\frac{1}{r^2}\frac{\partial(r^2\tau_{rr})}{\partial r} + \frac{1}{r\sin\theta}\frac{\partial(\tau_{r\theta}\sin\theta)}{\partial\theta} + \frac{1}{r\sin\theta}\frac{\partial\tau_{r\phi}}{\partial\phi} - \frac{\tau_{\theta\theta} + \tau_{\phi\phi}}{r}\right) + \rho g_r$$

θ Component:

$$\rho\left(\frac{\partial v_\theta}{\partial t} + v_r\frac{\partial v_\theta}{\partial r} + \frac{v_\theta}{r}\frac{\partial v_\theta}{\partial \theta} + \frac{v_\phi}{r\sin\theta}\frac{\partial v_\theta}{\partial \phi} + \frac{v_r v_\theta}{r} - \frac{v_\phi^2\cot\theta}{r}\right)$$

$$= -\frac{1}{r}\frac{\partial p}{\partial \theta} - \left(\frac{1}{r^2}\frac{\partial(r^2\tau_{r\theta})}{\partial r} + \frac{1}{r\sin\theta}\frac{\partial(\tau_{\theta\theta}\sin\theta)}{\partial\theta} + \frac{1}{r\sin\theta}\frac{\partial\tau_{\theta\phi}}{\partial\phi} + \frac{\tau_{r\theta}}{r} - \frac{\cot\theta}{r}\tau_{\phi\phi}\right) + \rho g_\theta$$

ϕ Component:

$$\rho\left(\frac{\partial v_\phi}{\partial t} + v_r\frac{\partial v_\phi}{\partial r} + \frac{v_\theta}{r}\frac{\partial v_\phi}{\partial \theta} + \frac{v_\phi}{r\sin\theta}\frac{\partial v_\phi}{\partial \phi} + \frac{v_\phi v_r}{r} + \frac{v_\theta v_\phi\cot\theta}{r}\right) = -\frac{1}{r\sin\theta}\frac{\partial p}{\partial\phi}$$

$$- \left(\frac{1}{r^2}\frac{\partial(r^2\tau_{r\phi})}{\partial r} + \frac{1}{r}\frac{\partial\tau_{\theta\phi}}{\partial\theta} + \frac{1}{r\sin\theta}\frac{\partial\tau_{\phi\phi}}{\partial\phi} + \frac{\tau_{r\phi}}{r} + \frac{2\cot\theta}{r}\tau_{\theta\phi}\right) + \rho g_\phi$$

In terms of velocity gradients for a Newtonian fluid with constant ρ and μ
r Component:

$$\rho\left(\frac{\partial v_r}{\partial t} + v_r\frac{\partial v_r}{\partial r} + \frac{v_\theta}{r}\frac{\partial v_r}{\partial \theta} + \frac{v_\phi}{r\sin\theta}\frac{\partial v_r}{\partial \phi} - \frac{v_\theta^2 + v_\phi^2}{r}\right)$$

$$= -\frac{\partial p}{\partial r} + \mu\left(\nabla^2 v_r - \frac{2}{r^2}v_r - \frac{2}{r^2}\frac{\partial v_\theta}{\partial\theta} - \frac{2}{r^2}v_\theta\cot\theta - \frac{2}{r^2\sin\theta}\frac{\partial v_\phi}{\partial\phi}\right) + \rho g_r$$

θ Component:

$$\rho\left(\frac{\partial v_\theta}{\partial t} + v_r\frac{\partial v_\theta}{\partial r} + \frac{v_\theta}{r}\frac{\partial v_\theta}{\partial \theta} + \frac{v_\phi}{r\sin\theta}\frac{\partial v_\theta}{\partial \phi} + \frac{v_r v_\theta}{r} - \frac{v_\phi^2\cot\theta}{r}\right)$$

$$= -\frac{1}{r}\frac{\partial p}{\partial \theta} + \mu\left(\nabla^2 v_\theta + \frac{2}{r^2}\frac{\partial v_r}{\partial\theta} - \frac{v_\theta}{r^2\sin^2\theta} - \frac{2\cos\theta}{r^2\sin^2\theta}\frac{\partial v_\phi}{\partial\phi}\right) + \rho g_\theta$$

(continued)

Table 11.3 (*continued*)

ϕ Component:

$$\rho\left(\frac{\partial v_\phi}{\partial t} + v_r\frac{\partial v_\phi}{\partial r} + \frac{v_\theta}{r}\frac{\partial v_\phi}{\partial \theta} + \frac{v_\phi}{r\sin\theta}\frac{\partial v_\phi}{\partial \phi} + \frac{v_\phi v_r}{r} + \frac{v_\theta v_\phi\cot\theta}{r}\right) = -\frac{1}{r\sin\theta}\frac{\partial p}{\partial \phi}$$

$$+ \mu\left(\nabla^2 v_\phi - \frac{v_\phi}{r^2\sin^2\theta} + \frac{2}{r^2\sin\theta}\frac{\partial v_r}{\partial \phi} + \frac{2\cos\theta}{r^2\sin^2\theta}\frac{\partial v_\theta}{\partial \phi}\right) + \rho g_\phi$$

where ∇^2 is the Laplacian operator in spherical coordinates, which is in the form

$$\nabla^2 = \frac{1}{r^2}\frac{\partial}{\partial r}\left(r^2\frac{\partial}{\partial r}\right) + \frac{1}{r^2\sin\theta}\frac{\partial}{\partial \theta}\left(\sin\theta\frac{\partial}{\partial \theta}\right) + \frac{1}{r^2\sin^2\theta}\left(\frac{\partial^2}{\partial \phi^2}\right)$$

Source: Bird et al., 1960

3.2 Exact Solutions (Laminar Viscous Flow)

3.2.1 Fluid Flow Along a Flat Plate

Unsteady-state flow develops when a plate is set to move at a constant velocity in an initially motionless semi-infinite incompressible fluid. A two-dimensional description of such a flow is shown in Fig. 11.2. For this flow $v_y = 0$; using the equation of continuity it can be proven that $\partial v_x/\partial x = 0$ or $v_x = v_x(y, t)$. For this flow, the fluid velocity in the x direction, v_x, can be calculated from the solution of the differential equation (which is the equation of motion for v_x):

$$\rho\frac{\partial v_x}{\partial t} = \mu\frac{\partial^2 v_x}{\partial y^2} \tag{11.22}$$

or by defining the kinematic viscosity $\gamma = \mu/\rho$,

$$\frac{\partial v_x}{\partial t} = \frac{\mu}{\rho}\frac{\partial^2 v_x}{\partial y^2} = \gamma\frac{\partial^2 v_x}{\partial y^2} \tag{11.23}$$

satisfying the initial and boundary conditions

$$t = 0, \qquad v_x = 0, \qquad \text{for all } y \tag{11.24}$$

$$t > 0, \qquad v_x = v_{x,0}, \qquad y = 0 \tag{11.25}$$

$$t > 0, \qquad v_x = 0, \qquad y = \infty \tag{11.26}$$

Defining the dimensionless variable η and the dimensionless velocity ϕ as

$$\eta = \frac{y}{2\sqrt{\gamma t}}, \qquad \phi = \frac{v_x}{v_{x,0}} \tag{11.27}$$

the partial differential equation [Eq. (11.23)] can be reduced to the second-order ordinary differential equation

$$\frac{d^2\phi}{d\eta^2} + 2\eta\frac{d\phi}{d\eta} = 0 \tag{11.28}$$

with the boundary conditions

$$\eta = 0, \qquad \phi = 1 \tag{11.29}$$

$$\eta = \infty, \qquad \phi = 0 \tag{11.30}$$

Solving Eq. (11.28) with the boundary conditions given by Eqs. (11.29) and (11.30),

$$\phi = 1 - \frac{(2/\sqrt{\pi})\int_0^\eta e^{-\xi^2}d\xi}{(2/\sqrt{\pi})\int_0^\infty e^{-\xi^2}d\xi} = 1 - \frac{2}{\sqrt{\pi}}\int_0^\eta e^{-\xi^2}d\xi \tag{11.31}$$

or, in terms of the error function (erf),

$$\phi = \frac{v_x}{v_{x,0}} = 1 - \text{erf}(\eta) = 1 - \text{erf}\frac{y}{2\sqrt{\gamma t}} = \text{erfc}\frac{y}{2\sqrt{\gamma t}} \tag{11.32}$$

The solution expressed by Eq. (11.32) can be used to define a boundary layer thickness δ as the distance y for which v_x reaches the value $0.01v_{x,0}$ This happens when $\eta = 2$, which predicts the boundary layer thickness δ as [the value of y at $\eta = 2$ in Eq. (11.27); when $\eta = 2$, $\text{erf}(\eta)$ is 0.99]:

$$\delta = 4\sqrt{\gamma t} \tag{11.33}$$

which is a very helpful relation to predict the boundary layer thickness as a function of kinematic viscosity γ [$= \mu/\rho$, as defined in Eq. (11.23)] and a characteristic time t. The solution for the velocity profile [Eq. (11.32)] is also useful for the development of laminar boundary layer theory.

3.2.2 Fluid Flow Between Two Flat Plates

Fluid flow between two flat plates (Fig. 11.3) can also be described by the exact solution of the Navier–Stokes equation. By assigning p for the dynamic pressure, i.e., the pressure difference between total pressure and the hydrostatic pressure, v_x can be found from the solution of the differential equation (which is the equation of motion for v_x, because $v_y = 0$ and the fluid is incompressible, i.e., $\nabla \cdot \boldsymbol{v} = 0$)

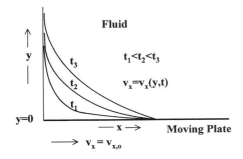

Figure 11.2 Formation of the unsteady-state flow velocity profile above a flat plate that is in motion in an initially motionless semi-infinite fluid.

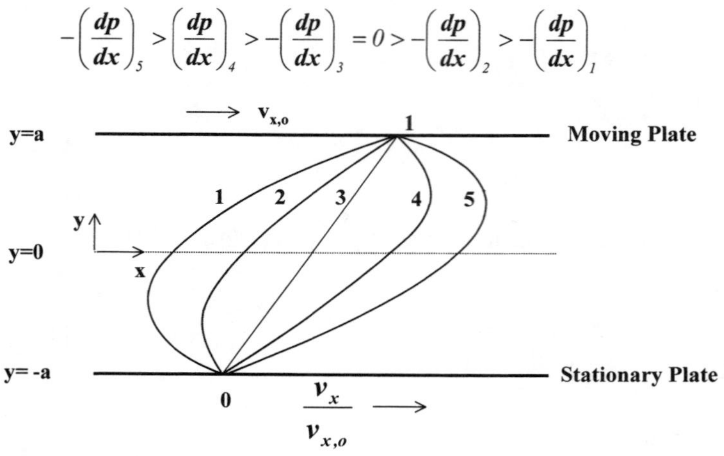

$$-\left(\frac{dp}{dx}\right)_5 > \left(\frac{dp}{dx}\right)_4 > -\left(\frac{dp}{dx}\right)_3 = 0 > -\left(\frac{dp}{dx}\right)_2 > -\left(\frac{dp}{dx}\right)_1$$

Figure 11.3 Flow between two parallel plates (Couette flow). The lower plate is stationary, and the upper plate is moving at velocity $v_{x,0}$.

$$\rho\frac{\partial v_x}{\partial t} = -\frac{dp}{dx} + \mu\frac{\partial^2 v_x}{\partial y^2} \tag{11.34}$$

by satisfying the boundary and initial conditions

$$v_x = 0, \qquad y = -a \tag{11.35}$$

$$v_x = v_{x,0}, \qquad y = a \tag{11.36}$$

$$v_x = v_0(x), \qquad t = 0 \tag{11.37}$$

If the pressure drop dp/dx is kept constant, the flow will reach the steady state, and the equation of motion [Eq. (11.34)] will be reduced to

$$\frac{d^2 v_x}{dy^2} = \frac{1}{\mu}\frac{dp}{dx} \tag{11.38}$$

The solution of Eq. (11.38) for the boundary conditions of Eqs. (11.35) and (11.36) gives

$$v_x = \frac{v_{x,0}}{2}\left(1 + \frac{y}{a}\right) - \left(\frac{a^2}{2\mu}\frac{dp}{dx}\right)\left(1 - \frac{y^2}{a^2}\right) \tag{11.39}$$

For a pressure gradient $dp/dx < 0$, the flow velocity v_x is positive everywhere, whereas for a positive pressure gradient $dp/dx > 0$, the velocity v_x can become negative, i.e., back-flow can take place. The limiting condition for reverse flow to take place is $dv_x/dy = 0$ at $y = -a$. Using Eq. (11.39), the condition for the occurrence of reverse flow can be calculated as:

$$\frac{dp}{dx} = \frac{v_{x,0}\mu}{2a^2} \tag{11.40}$$

If the pressure drop is zero ($dp/dx = 0$), the flow is known as simple Couette flow, the velocity profile of which becomes

$$v_x = \frac{v_{x,0}}{2}\left(1 + \frac{y}{a}\right) \tag{11.41}$$

which is linear in y.

3.2.3 Hagen–Poiseuille Flow (Flow in a Circular Tube)

Steady-state laminar flow of fluids in circular tubes (Fig. 11.4) is known as Hagen–Poiseuille flow. This flow is encountered in engineering and in many branches of science. The equation of motion ($v_y = 0$ and $\nabla \cdot \boldsymbol{v} = 0$ because the fluid is incompressible) and the corresponding boundary conditions (zero velocity on the tube wall and maximum velocity in the cube center) for this flow are

$$\mu\left(\frac{d^2 v_x}{dr^2} + \frac{1}{r}\frac{dv_x}{dr}\right) = \mu\frac{1}{r}\frac{d}{dr}\left(r\frac{dv_x}{dr}\right) = \frac{dp}{dx} \tag{11.42}$$

and

$$v_x = 0, \qquad r = r_0 \tag{11.43}$$

$$\frac{dv_x}{dr} = 0, \qquad r = 0 \tag{11.44}$$

the solution of which gives the velocity profile in the form

$$v_x = -\left(\frac{r_0^2}{4\mu}\frac{dp}{dx}\right)\left(1 - \frac{r^2}{r_0^2}\right) = v_{x,\max}\left(1 - \frac{r^2}{r_0^2}\right) \tag{11.45}$$

where the maximum velocity (i.e., v_x at $r = 0$), $v_{x,\max}$, is defined as

$$v_{x,\max} = -\frac{r_0^2}{4\mu}\frac{dp}{dx} \tag{11.46}$$

The average fluid velocity can be calculated from

$$v_{x,\text{ave}} = \frac{\int_0^{r_0} 2\pi r v_x(r)\, dr}{\pi r_0^2} = -\frac{r_0^2}{8\mu}\frac{dp}{dx} \tag{11.47}$$

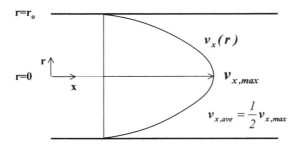

Figure 11.4 Steady-state laminar flow in circular tubes: Hagen–Poiseuille flow.

From a comparison of Eqs. (11.46) and (11.47) it can be seen that the maximum velocity is twice the average velocity,

$$v_{x,\max} = 2v_{x,\text{ave}} \tag{11.48}$$

which is one of the characteristics of laminar viscous flow. The volumetric flow rate q (volume per unit time) can be calculated by using the average flow rate and the cross-sectional area of the tube:

$$q = \pi r_0^2 v_{x,\text{ave}} = -\frac{\pi r_0^4}{8\mu}\frac{dp}{dx} \tag{11.49}$$

The relation expressed by Eq. (11.49) is known as the Hagen–Poiseuille law for laminar flow, i.e., for Reynolds numbers smaller than 2100. This relation is used for viscosity measurements from the measured volumetric flow rate q versus the pressure drop dp/dx. It is important that in these tests the laminar flow must be in the steady state (i.e., constant q or constant dp/dx) and smoothly developed (i.e., the pressure drop measurements must be made at distances at which the disturbances at the tube inlet are minimized).

3.3 Approximate Solutions

In a certain class of flow problems, some terms of the Navier–Stokes equation can be neglected. As a result, solutions of the approximate forms of the Navier–Stokes equation can be constructed. In some cases, however, even these solutions cannot be exact, so they are called approximate solutions of the approximate Navier–Stokes equations. In the following subsections, solutions of the approximate Navier–Stokes equations will be presented.

3.3.1 Ideal Flow Approximation (Potential Flow and Euler Equation)

When the viscosity of the fluid is very low, the viscous term in the Navier–Stokes equation can be neglected. This approximation is known as the potential flow approximation (i.e., zero viscosity, $\mu \to 0$). If the density of such a flow system is constant, i.e., the fluid is incompressible (constant density, $\rho = $ constant or $\nabla \cdot v = 0$), then the approximation is called the ideal flow approximation.

For the ideal flow (i.e., very low viscosity and incompressible fluid) approximation, the Navier–Stokes equation [Eq. (11.21)] takes the form

$$\rho\left(\frac{\partial v}{\partial t} + (v \cdot \nabla)v\right) = -\nabla \cdot \mathbf{p} + \sum_s \rho_s F_s \tag{11.50}$$

which is also known as the Euler equation. In real flow problems, however, the validity of zero viscosity fails in regions close to solid objects, where the viscous forces can be large even for a very small fluid viscosity. The zero viscosity assumption could adequately predict the flow behavior for high Reynolds number flows (i.e., high velocities) in regions sufficiently far from the solid objects. It is difficult to solve Eq. (11.50) for given boundary conditions, because four unknowns (three velocity components and pressure) have to be found from four equations [i.e., three equations of Eq. (11.50) and the equation of continuity $\nabla \cdot v = 0$].

The solution of the Euler equation [Eq. (11.50)] is known as the Bernoulli equation. To integrate the steady-state Euler equation, the following mathematical relation will be used for the second term in Eq. (11.50)

$$(\mathbf{v} \cdot \nabla)\mathbf{v} = \nabla \frac{1}{2}v^2 - \mathbf{v} \times (\nabla \times \mathbf{v}) \tag{11.51}$$

For a steady-state two-dimensional ideal flow problem (assuming no external forces acting on the system), using Eq. (11.51) in Eq. (11.50), the set of equations to be solved becomes

$$\nabla \cdot \mathbf{v} = 0 \tag{11.52}$$

$$\rho\left(\nabla \frac{1}{2}v^2 - \mathbf{v} \times (\nabla \times \mathbf{v})\right) + \nabla \cdot \mathbf{p} = 0 \tag{11.53}$$

Because for an ideal fluid ($\mu = 0$) the velocity vector satisfies the relation

$$\nabla \times \mathbf{v} = \mathrm{curl}(\mathbf{v}) = 0 \tag{11.54}$$

the equation of motion [Eq. (11.53)] becomes

$$\nabla\left(\rho\frac{1}{2}v^2\right) + \nabla \cdot \mathbf{p} = \nabla\left(\rho\frac{1}{2}v^2 + \mathbf{p}\right) = \nabla\left(\frac{1}{2}\rho(v_x^2 + v_y^2 + v_z^2) + p\right) = 0 \tag{11.55}$$

The integral of Eq. (11.55) is known as the Bernoulli equation:

$$\frac{1}{2}\rho(v_x^2 + v_y^2 + v_z^2) + p = C \tag{11.56}$$

where C is the integration constant (note that p can be expressed as $p = p_s + \rho g$ to take care of the system pressure p_s and the pressure resulting from gravitational forces, ρg).

For a two-dimensional ideal flow (irrotational flow, $\nabla \times \mathbf{v} = 0$), the equation of motion and the equation of continuity ($\nabla \cdot \mathbf{v} = 0$) can be expressed as

$$\nabla \times \mathbf{v} = \frac{\partial v_x}{\partial y} - \frac{\partial v_y}{\partial x} = 0 \tag{11.57}$$

$$\frac{1}{2}\rho(v_x^2 + v_y^2) + p = C \tag{11.58}$$

$$\frac{\partial v_x}{\partial x} + \frac{\partial v_y}{\partial y} = 0 \tag{11.59}$$

These three equations have to be solved to determine three unknowns, v_x, v_y, and p, which could be a difficult problem for a given geometry. For two-dimensional irrotational flow problems, it is easier to solve the velocity profiles using the stream and potential functions. When viscosity is zero, the flow becomes irrotational; i.e., in two-dimensional flow the curl of the velocity vector becomes zero [curl $\mathbf{v} = \nabla \times \mathbf{v} = 0$; Eq. (11.57)]. Because the velocity vector $\mathbf{v}$ satisfies Eq. (11.57), the velocity vector has to be the gradient of a scalar field ψ, which will be called the stream function ψ

$$\mathbf{v} = \nabla \psi \tag{11.60}$$

Because the velocity vector would also satisfy the equation of continuity ($\nabla \cdot \mathbf{v} = 0$), the stream function ψ has to satisfy the Laplacian equation,

$$\nabla \cdot \boldsymbol{v} = \nabla \cdot \nabla \psi = \nabla^2 \psi = 0 \tag{11.61}$$

where ∇^2 is the Laplacian operator. This property of the stream function ψ makes it possible to use the theory of complex functions to solve ideal flow problems. It is known that a complex function $f(z)$,

$$f(z) = \phi(x, y) + \psi(x, y)i, \qquad z = x + yi = re^{i\theta}, \qquad i = \sqrt{-1} \tag{11.62}$$

is analytical if its real and imaginary parts satisfy the Cauchy–Riemann conditions

$$\frac{\partial \phi}{\partial x} = \frac{\partial \psi}{\partial y} \qquad \text{and} \qquad \frac{\partial \phi}{\partial y} = -\frac{\partial \psi}{\partial x} \tag{11.63}$$

By taking the partial derivatives of both sides of Eq. (11.63) with respect to x and y and eliminating the cross partial derivative $\partial^2/\partial x \, \partial y$ terms, it can be shown that the real ϕ and imaginary ψ parts of the analytical function, Eq. (11.62), satisfy the Laplacian equation

$$\nabla^2 \phi = 0 \qquad \text{and} \qquad \nabla^2 \psi = 0 \tag{11.64}$$

Therefore, the real and imaginary parts of an analytical function are the stream and potential functions of some ideal flow problem (i.e., ϕ and ψ are perpendicular to each other). If the stream and potential functions are known, then the velocity vector can be calculated from the relations

$$v_x = \frac{\partial \psi}{\partial x} = \frac{\partial \phi}{\partial y} \qquad \text{and} \qquad v_y = \frac{\partial \psi}{\partial y} = -\frac{\partial \phi}{\partial x} \tag{11.65}$$

Using the linear property of the Laplacian equation, a linear combination of two potential functions ϕ_1 and ϕ_2 (i.e., $\nabla^2 \phi_1 = \nabla^2 \phi_2 = 0$),

$$\phi = a_1 \phi_1 + a_2 \phi_2 \tag{11.66}$$

is also Laplacian,

$$\nabla^2 \phi = a_1 \nabla^2 \phi_1 + a_2 \nabla^2 \phi_2 = 0 \tag{11.67}$$

Therefore, it should also represent an ideal flow problem. Using the theory of complex functions (i.e., conformal mapping), it can be shown that the analytical functions $f(z) = Az$, $f(z) = Az^2$, and $f(z) = A \ln(z)$ (where A is a numerical constant and z is a complex number) represent uniform potential flow (i.e., in the x direction), potential flow around inside of a corner or against a plate, and potential flow of sinks, sources, and circulation (irrotational), respectively.

It must be remembered that the potential flow approximation could not be used to calculate velocity profiles, boundary layer thickness, or boundary layer heat and mass transfer phenomena, etc., close to stationary solid objects in the flow field. The potential flow approximation is good only for determining the flow patterns sufficiently far from stationary solid objects.

3.3.2 Very Slow Motion (Inertial Forces Are Neglected, i.e., $v \approx 0$)

When the velocity of the fluid is very slow, the inertial force ($\rho v \cdot \nabla v$) becomes smaller in comparison to the viscous force ($\mu \nabla^2 v$). For a steady-state flow, if the inertial force is sufficiently smaller than the viscous force, then the inertial force can be neglected and the approximate form of the Navier–Stokes equation [Eq. (11.21)] becomes

$$\nabla \cdot \mathbf{p} = \mu \, \nabla^2 \boldsymbol{v} \tag{11.68}$$

This is known as the "very small motion" or "creeping motion" approximation.

It can be seen from Eq. (11.21) that if L and v_∞ are the characteristic linear dimension and linear velocity of a given flow system, then inertial and viscous force terms will be proportional to $\rho v_\infty^2/L$ and $\mu v_\infty/L^2$, respectively. The ratio of inertial to viscous forces, $\rho v_\infty L/\mu$ is the characteristic Reynolds number. Thus, the smaller the Reynolds number, the better will be the approximation of the Navier–Stokes equation by neglecting the inertial force term.

For a steady-state very slow fluid motion, the approximate form of the Navier–Stokes equation [Eq. (11.68)] and the equation of continuity ($\nabla \cdot \mathbf{v} = 0$) have to be solved simultaneously. In this approximation, the boundary conditions of the original Navier–Stokes equation have to be kept the same. As an example, for a stationary spherical object of radius r_0 in a one-dimensional fluid flowing at a very small velocity (i.e., at sufficiently low Reynolds numbers), the boundary conditions can be expressed as

$$v_x = v_y = v_z = 0 \qquad \text{at} \qquad r = r_0 \tag{11.69}$$

$$v_x = v_{x,\infty}, \qquad v_y = v_z = 0 \qquad \text{at} \qquad r = \infty \tag{11.70}$$

Solution of the approximate form of the Navier–Stokes equation can be obtained by taking the divergence of both sides of Eq. (11.68),

$$\nabla \cdot (\nabla \cdot \mathbf{p}) = \nabla^2 \mathbf{p} = \nabla \cdot (\mu \, \nabla^2 \mathbf{v}) = \mu \, \nabla^2 (\nabla \cdot \mathbf{v}) = 0 \tag{11.71}$$

or, because $\nabla \cdot \mathbf{v} = 0$ (equation of continuity for the incompressible fluid),

$$\nabla^2 \mathbf{p} = 0 \tag{11.72}$$

This proves that in a very slow motion approximation the pressure field is harmonic, i.e., it satisfies the Laplacian equation. Solution of the approximate form of the Navier–Stokes equation for very slow motion is discussed in great detail by several authors (Lamb, 1945; Happel and Brenner, 1986; Brodkey, 1967; Bird et al., 1960).

For a very slow motion (Reynolds number smaller than 0.1) of a Newtonian fluid about a spherical object of radius r_0 (Fig. 11.5), the momentum flux, pressure, and velocity distributions are given by (Bird, 1960)

$$\tau_{r,\theta} = \frac{3}{2}\left(\frac{\mu v_\infty}{r_0}\right)\left(\frac{r_0}{r}\right)^4 \sin\theta \tag{11.73}$$

$$p = p_0 - \rho g z - \frac{3}{2}\left(\frac{\mu v_\infty}{r_0}\right)\left(\frac{r_0}{r}\right)^2 \cos\theta \tag{11.74}$$

$$v_r = v_\infty\left[1 - \frac{3}{2}\left(\frac{r_0}{r}\right) + \frac{1}{2}\left(\frac{r_0}{r}\right)^3\right]\cos\theta \tag{11.75}$$

$$v_\theta = -v_\infty\left[1 - \frac{3}{4}\left(\frac{r_0}{r}\right) - \frac{1}{4}\left(\frac{r_0}{r}\right)^3\right]\sin\theta \tag{11.76}$$

where p_0 is the pressure at $z = 0$ and $\rho g z$ is the hydrostatic pressure. At the surface of the sphere, i.e., at $r = r_0$, both $v_r = 0$ and $v_\theta = 0$ are satisfied. Also, far from the surface of the sphere, $p = p_0 - \rho g z$.

Using the pressure and momentum flux distributions [Eqs. (11.73) and (11.74)], the force acting on the sphere can be calculated as

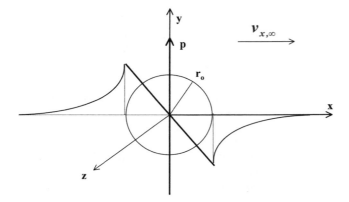

Figure 11.5 Pressure distribution for a very slow moving flow (creeping flow) over a spherical object.

$$F = \frac{4}{3}\pi r_0^3 \rho g + 2\pi \mu r_0 v_\infty + 4\pi \mu r_0 v_\infty \qquad (11.77)$$

or

$$F = \frac{4}{3}\pi r_0^3 \rho g + 6\pi \mu r_0 v_\infty \qquad (11.78)$$

where the three terms on the right-hand side of Eq. (11.77) are the buoyant, form drag (i.e., integration of normal forces), and friction drag (i.e., integration of tangential forces) forces, respectively. The second term on the right in Eq. (11.78) is the force exerted on the particles as a result of the motion of the fluid:

$$F_k = 6\pi \mu r_0 v_\infty \qquad (11.79)$$

which is known as Stokes' law. This relation is valid for flow systems with Re < 1.

Stokes' law can be used for viscosity determination from the terminal velocity of a falling sphere in a liquid. When the particle is falling at a steady-state velocity, total forces acting on the sphere would be in equilibrium:

$$\frac{4}{3}\pi r_0^3 \rho_p g = \frac{4}{3}\pi r_0^3 \rho g + 6\pi \mu r_0 v_\infty \qquad (11.80)$$

from which the viscosity of the fluid can be calculated from the measured density of the particle ρ_p, density of the fluid ρ, and falling velocity of the particle v_∞:

$$\mu = \frac{2}{9}\left(\frac{r_0^2(\rho_p - \rho)g}{v_\infty}\right) \qquad (11.81)$$

provided that the Reynolds number based on the particle diameter ($d_p = 2r_0$), i.e., $d_p \rho v_\infty/\mu$, is less than about 0.1.

Stokes' law [expressed by Eq. (11.81)] can also be used for the calculation of the sedimentation velocity of suspended particles in liquids. Using Eq. (11.81), the particle sedimentation velocity in suspensions of low solid concentrations, in which the motion of any one particle will not be disturbed by the motion of the other particles, can be calculated as

$$v_\infty = \frac{2}{9}\left(\frac{r_0^2(\rho_p - \rho)g}{\mu}\right) \tag{11.82}$$

Again, to use this relation the Reynolds number based on particle diameter should be less than 1.

3.3.3 Boundary Layer Solution

In a fluid motion, when inertial forces and viscous forces are of the same order of magnitude, then neither the ideal flow nor the very slow motion approximation is valid. In such a flow, i.e., for laminar flow at high Reynolds numbers, the boundary layer concept can be introduced as an approximation to the Navier–Stokes equation. When the Reynolds number becomes large, the effect of the viscous forces will be concentrated in a small region in which the velocity gradient will be large, i.e., close to the boundary of the stationary objects.

The simplest boundary layer that can be visualized is the two-dimensional boundary layer formed over a flat plate (Fig. 11.6). For a two-dimensional boundary layer flow, Navier–Stokes and continuity equations and their orders of magnitude based on the characteristic flow velocity and geometric dimension can be expressed as

$$\frac{\partial v_x}{\partial t} + v_x\frac{\partial v_x}{\partial x} + v_y\frac{\partial v_x}{\partial y} = -\frac{1}{\rho}\frac{dp}{dx} + \gamma\left(\frac{\partial^2 v_x}{\partial x^2} + \frac{\partial^2 v_x}{\partial y^2}\right) \tag{11.83}$$

$$\quad\;\; 1/1 \qquad\; 1\times 1/1 \qquad \delta\times 1/\delta \qquad\qquad\qquad\quad 1 \qquad 1/\delta^2$$

$$\frac{\partial v_y}{\partial t} + v_x\frac{\partial v_y}{\partial x} + v_y\frac{\partial v_y}{\partial y} = -\frac{1}{\rho}\frac{dp}{dy} + \gamma\left(\frac{\partial^2 v_y}{\partial x^2} + \frac{\partial^2 v_y}{\partial y^2}\right) \tag{11.84}$$

$$\quad\;\; \delta/1 \qquad\; 1\times \delta/1 \qquad \delta\times \delta/\delta \qquad\qquad\qquad\quad \delta \qquad 1/\delta$$

$$\frac{\partial v_x}{\partial x} + \frac{\partial v_y}{\partial y} = 0 \tag{11.85}$$

$$\quad\;\; 1/1 \qquad \delta/\delta$$

where δ is the boundary layer thickness, which is much smaller than the characteristic length L (i.e., the length along the plate, $\delta \ll L$). By analogy, v_x and v_y are assumed to be on the order of magnitude of 1 and δ, respectively. The magnitudes of the first- and second-order derivatives are also calculated accordingly. Based on the order of magnitude,

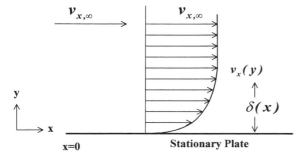

Figure 11.6 Sketch of two-dimensional boundary layer over a flat plate.

the term $\partial^2 v_x/\partial x^2$ can be neglected compared to $\partial^2 v_x/\partial y^2$ in Eq. (11.83). The kinematic viscosity γ must be on the order of δ^2 to make both sides of Eq. (11.83) the same order of magnitude. When γ is on the order of δ^2, all the terms in Eq. (11.84) will be on the order of δ or smaller and will be neglected. As a result, the derivatives of Eqs. (11.83), (11.84), and (11.85) can be approximated as

$$\frac{\partial v_x}{\partial t} + v_x \frac{\partial v_x}{\partial x} + v_y \frac{\partial v_x}{\partial y} = \gamma \frac{\partial^2 v_x}{\partial x^2} \tag{11.86}$$

$$\frac{1}{\rho}\frac{dp}{dy} = 0 \tag{11.87}$$

$$\frac{\partial v_x}{\partial x} + \frac{\partial v_y}{\partial y} = 0 \tag{11.88}$$

As can be seen from Eq. (11.86), the approximate form of the Navier–Stokes equation is also a nonlinear partial differential equation [dependent variables v_x, v_y and their derivatives $\partial v_x/\partial x$, $\partial v_y/\partial y$ appear as products on the left-hand side of Eq. (11.86)]. The boundary conditions for the approximated boundary layer equations are

$$v_x = v_y = 0 \qquad \text{at} \qquad y = 0 \qquad \text{for all } x \tag{11.89}$$

$$v_x = v_{x,\infty} \qquad \text{at} \qquad y = \infty \tag{11.90}$$

The boundary layer equations can be solved by introducing dimensionless parameter η, as it was introduced to solve the formation of the transient boundary layer problem in the flow along a flat plate [Eq. (11.27)]. In this case, however, time t will be replaced by $x/v_{x,\infty}$, which is the time required for a fluid element to reach out the distance x when it is moving at characteristic flow velocity $v_{x,\infty}$ [the numerical coefficient 2 in the denominator of Eq. (11.27) is neglected]:

$$\eta = \frac{y}{\sqrt{\gamma t}} = \frac{y}{\sqrt{\gamma x/v_{x,\infty}}} = y\sqrt{\frac{v_{x,\infty}}{\gamma x}} \tag{11.91}$$

which predicts the thickness of the boundary layer to be proportional to $(\gamma x/v_{x,\infty})^{1/2}$,

$$\delta \propto \sqrt{\frac{\gamma x}{v_{x,\infty}}} \tag{11.92}$$

or

$$\frac{\delta}{x} \propto \sqrt{\frac{\gamma}{x v_{x,\infty}}} = \frac{1}{\sqrt{\text{Re}_x}}, \qquad \text{Re}_x = \frac{x v_{x,\infty}}{\gamma} \tag{11.93}$$

Also, a stream function ψ could be assigned for the solution of the boundary layer flow problem, the derivatives of which in the x and y directions would give the components of the velocity vector (i.e., for a two-dimensional problem). Because the velocity vector would satisfy the equation of continuity [i.e., incompressible flow, Eq. 11.85)], the stream function ψ has to satisfy the relations:

$$v_x = \frac{\partial \psi}{\partial y}, \qquad v_y = -\frac{\partial \psi}{\partial x} \tag{11.94}$$

The stream function ψ can be expressed as a function of the dimensionless parameter η so that the ratio $v_x/v_{x,\infty}$ would be some function of η. To satisfy this condition, the stream function ψ has to take the form [because $d\eta/dy = (v_{x,\infty}/\gamma x)^{1/2}$]

$$\psi = \sqrt{v_{x,\infty}x\gamma}f(\eta) \tag{11.95}$$

from which, using Eq. (11.94), the velocity components can be expressed in terms of $f(\eta)$:

$$v_x = \frac{\partial\psi}{\partial y} = v_{x,\infty}f'(\eta) \tag{11.96}$$

$$v_y = -\frac{\partial\psi}{\partial x} = \frac{1}{2}\sqrt{\frac{\gamma v_{x,\infty}}{x}}[\eta f'(\eta) - f(\eta)] \tag{11.97}$$

The approximate boundary layer equation [Eq. (11.86)] can be expressed in terms of the stream function ψ:

$$\frac{\partial\psi}{\partial y}\frac{\partial^2\psi}{\partial x\,\partial y} - \frac{\partial\psi}{\partial x}\frac{\partial^2\psi}{\partial y^2} = \gamma\frac{\partial^2\psi}{\partial y^2} \tag{11.98}$$

or in terms of $f(\eta)$,

$$f(\eta)f''(\eta) + 2f'''(\eta) = 0 \tag{11.99}$$

which is a nonlinear ordinary differential equation. By considering the relations stated by Eqs. (11.96) and (11.97), the boundary conditions given by Eqs. (11.89) and (11.90) become

$$f(\eta) = f'(\eta) = 0 \qquad \text{at } \eta = 0 \tag{11.100}$$

$$f'(\eta) = 1 \qquad \text{at } \eta = \infty \tag{11.101}$$

Numerical integration of Eq. (11.99), satisfying Eqs. (11.100) and (11.101) (Fig. 11.7) shows good agreement with the experimental data (a numerical solution was first found by Blasius in 1908).

Using the numerical solution for the approximate boundary layer equation, the friction force exerted on unit area of a plate at distance x from the leading edge is calculated as

$$\tau_w = \mu\left(\frac{\partial v_x}{\partial y}\right)_{y=0} = \mu\frac{\partial(v_{x,\infty}f')}{\partial y}$$

$$= \mu v_{x,\infty}\sqrt{\frac{v_{x,\infty}}{\gamma x}}f''(0) = \rho v_{x,\infty}^2\left(\frac{v_{x,\infty}x}{\gamma}\right)^{-1/2}f''(0) \tag{11.102}$$

or

$$\tau_w = \mu v_{x,\infty}\sqrt{\frac{v_{x,\infty}}{\gamma x}}f''(0) = \rho v_{x,\infty}^2\frac{1}{\sqrt{\text{Re}_x}}f''(0) \tag{11.103}$$

where Re_x is the Reynolds number evaluated at distance x. The local drag coefficient $C_{D,x}$ is defined as

$$C_{D,x} = \frac{\tau_w}{(1/2)\rho v_{x,\infty}^2} = \frac{2f''(0)}{\sqrt{v_{x,\infty}x/\gamma}} = \frac{2f''(0)}{\sqrt{\text{Re}_x}} \tag{11.104}$$

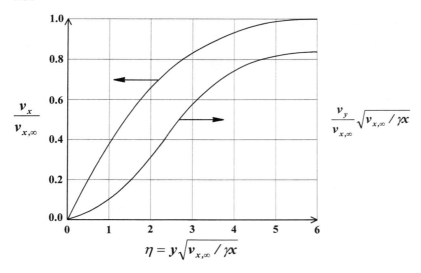

Figure 11.7 The numerical solution of the two-dimensional boundary layer approximate equations [integration of Eq. (11.99), satisfying Eqs. (11.100) and (11.101)]. (From Brodkey, 1969.)

Drag F exerted on one side of a plate of length L and width W is calculated as

$$F = W \int_0^L \mu \left(\frac{\partial v_x}{\partial y} \right)_{y=0} dx$$

$$= W \mu v_{x,\infty} f''(0) \sqrt{\frac{v_{x,\infty}}{\gamma}} \int_0^L x^{-1/2} dx = 2 W \mu v_{x,\infty} f''(0) \sqrt{\frac{v_{x,\infty} L}{\gamma}}$$

(11.105)

or

$$F = 2 W \mu v_{x,\infty} f''(0) \sqrt{\mathrm{Re}_L} = 2 W L \rho v_{x,\infty}^2 f''(0) \frac{1}{\sqrt{\mathrm{Re}_L}}$$

(11.106)

The drag coefficient C_D can also be defined using Eq. (11.106) for the total drag on one side of the plate:

$$C_D = \frac{F}{(1/2)\rho v_{x,\infty}^2 A_c}$$

$$= \frac{F}{(1/2)\rho v_{x,\infty}^2 W L} = 4 f''(0) \frac{1}{\sqrt{\mathrm{Re}_L}} = 1.33 \frac{1}{\sqrt{\mathrm{Re}_L}}$$

(11.107)

where A_c is the characteristic surface area (in this case $A_c = WL$), Re_L is the Reynolds number at distance L, and $(1/2)\rho v_{x,\infty}^2$ is, as usual, the kinetic energy per unit volume of the fluid in motion. In these derivations $f''(0) = 0.33206$, which is the second derivative of $f(\eta)$ with respect to η evaluated at $\eta = 0$ (i.e., at $y = 0$).

Using the numerical solution (Fig. 11.7) and Eq. (11.91), the boundary layer thickness δ (Fig. 11.8a) can be calculated ($\delta = y$ at η corresponding to $v_x = 0.99 v_{x,\infty}$, i.e., at $\eta = 4.99$) as

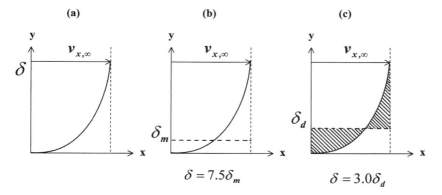

Figure 11.8 Graphical description of (a) the boundary layer thickness, δ, (b) the momentum thickness δ_m, and (c) the displacement thickness δ_d.

$$\delta = 4.99\sqrt{\frac{\gamma x}{v_{x,\infty}}} \quad \text{or} \quad \frac{\delta}{x} = 4.99\frac{1}{\sqrt{\mathrm{Re}_x}} \tag{11.108}$$

Similarly, using the numerical solution for the velocity profile, the momentum thickness δ_m (Fig. 11.8b), is calculated as (i.e., $Y \gg \delta$ and $\eta_Y \gg 4.99$)

$$\delta_m = \int_0^Y \frac{v_x}{v_{x,\infty}}\left(1 - \frac{v_x}{v_{x,\infty}}\right) dy = \sqrt{\frac{\gamma x}{v_{x,\infty}}} \int_0^{\eta_Y} f'(\eta)[1 - f'(\eta)]\, d\eta \tag{11.109}$$

or

$$\delta_m = 0.664\sqrt{\frac{\gamma x}{v_{x,\infty}}} \quad \text{or} \quad \frac{\delta_m}{x} = 0.664\frac{1}{\sqrt{\mathrm{Re}_x}} \tag{11.110}$$

Similarly, the displacement thickness δ_d (Fig. 11.8c) is calculated from the relation

$$v_{x,\infty}(Y - \delta_d) = \int_0^Y v_x\, dy \tag{11.111}$$

or, for $Y \gg \delta$ and $\eta_y \gg 4.99$,

$$\delta_d = \int_0^Y \left(1 - \frac{v_x}{v_{x,\infty}}\right) dy = \sqrt{\frac{\gamma x}{v_{x,\infty}}} \int_0^{\eta_Y} [1 - f'(\eta)]\, d\eta \tag{11.112}$$

from which δ_d is calculated as

$$\delta_d = 1.72\sqrt{\frac{\gamma x}{v_{x,\infty}}} \quad \text{or} \quad \frac{\delta_d}{x} = 1.72\frac{1}{\sqrt{\mathrm{Re}_x}} \tag{11.113}$$

As can be seen from Eqs. (11.108) and (11.113), displacement thickness δ_d is about one-third as great as the boundary layer thickness:

$$\delta \approx 3 \times \delta_d \tag{11.114}$$

To give an idea, the boundary layer approximation theory presented in this section predicts that for $v_{x,\infty} = 100 \, cm/s$, $x = 10 \, cm$, δ and δ_d are 0.63 cm and 0.21 cm for air and 0.18 cm and 0.06 cm for water, respectively, at normal temperature.

3.3.4 Von Karman's Approximate Solution for the Boundary Layer Approximation

Mass and momentum (force) balances for a differential segment dx of the boundary layer section located at a distance x from the edge of the solid object (Fig. 11.9) can be expressed by

$$\text{Mass flow} + \left(\int_0^\delta \rho v_x \, dy \right)_x - \left(\int_0^\delta \rho v_x \, dy \right)_{x+dx} = 0 \qquad (11.115)$$

where δ is the boundary layer thickness. The mass flow term in Eq. (11.115) is the flow of mass into the boundary layer from the main flow, for the boundary layer section under consideration. This is the mass of fluid that has to enter into the boundary layer (moving at velocity v_x) from the main stream (moving at velocity $v_{x,\infty}$) to keep up the form of the boundary layer. Using the Taylor's expansion of the second and third terms of Eq. (11.115), the mass flow into the differential section of the boundary layer becomes

$$\text{Mass flow} = \left(\int_0^\delta \rho v_x \, dy \right)_x + \frac{d}{dx}\left(\int_0^\delta \rho v_x \, dy \right)_x dx - \left(\int_0^\delta \rho v_x \, dy \right)_x = \frac{d}{dx}\left(\int_0^\delta \rho v_x \, dy \right)_x dx \qquad (11.116)$$

Similarly, for the same segment of the boundary layer, the momentum (force) balance can be expressed as

$$\int_0^\delta (\rho v_x^2 \, dy)\big|_x - \int_0^\delta (\rho v_x^2 \, dy)\big|_{x+dx} + v_{x,\infty}\frac{d}{dx}\left(\int_0^\delta \rho v_x \, dy \right)_x dx + \tau_w \, dx + \delta\frac{\partial p}{\partial x}\bigg|_x - \delta\frac{\partial p}{\partial x}\bigg|_{x+dx} = 0 \qquad (11.117)$$

where the terms from left to right are the momentum inflow and momentum loss by the fluid flow into the boundary layer, momentum inflow associated with the mass inflow [Eq. (11.116)] at velocity $v_{x,\infty}$, the viscous drag force, and the force exerted as a result of

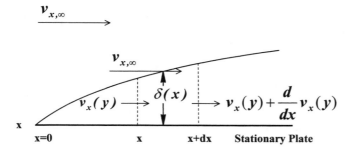

Figure 11.9 Integral momentum balance for the boundary layer developed over a plate.

pressure change in the x direction (i.e., $\partial p / \partial x$). Using the Taylor's expansion, Eq. (11.117) can be expressed as

$$\frac{d}{dx} \int_0^\delta \rho v_x(v_{x,\infty} - v_x)\, dy = -\tau_w + \delta \frac{\partial p}{\partial x} \tag{11.118}$$

which is known as the von Karman integral momentum equation. For an incompressible boundary layer flow and $\partial p / \partial x = 0$ the integral momentum (force) balance becomes

$$\tau_w = -\rho \frac{d}{dx} \int_0^\delta v_x(v_{x,\infty} - v_x)\, dy \tag{11.119}$$

Because the velocity in the boundary layer v_x approaches the velocity of the mainstream, $v_{x,\infty}$, the integral would be the same as long as the upper limit of the integral is taken as greater than the boundary layer thickness δ. Assuming that the dimensionless velocity $v_x/v_{x,\infty}$ is a function of the dimensionless distance η (i.e., $\eta = y/\delta$),

$$\frac{v_x}{v_{x,\infty}} = f(\eta) \tag{11.120}$$

Using Eq. (11.120), the integral momentum balance equation [Eq. (11.119)] becomes

$$\tau_w = -\rho v_{x,\infty}^2 \frac{d}{dx} \left[\delta \int_0^1 f(1 - f)\, d\eta \right] \tag{11.121}$$

Naturally, the boundary conditions (considering the relation $\eta = y/\delta$) for the function $f(\eta)$ are

$$\eta = 0, \quad f(\eta) = 0 \quad \text{and} \quad \eta = 1, \quad f(\eta) = 1 \tag{11.122}$$

Using the relation

$$\tau_w = -\mu \left(\frac{\partial v_x}{\partial y} \right)_w = -\mu \left(\frac{\partial v_x}{\partial f} \right)_w \left(\frac{\partial f}{\partial \eta} \right)_w \left(\frac{\partial \eta}{\partial y} \right)_w = -\mu \frac{v_{x,\infty}}{\delta} f'(0) \tag{11.123}$$

in Eq. (11.121), the closed form of the integral momentum balance becomes

$$\tau_w = -\mu \frac{v_{x,\infty}}{\delta} f'(0) = -\rho v_{x,\infty}^2 \frac{d}{dx} \left[\delta \int_0^1 f(1 - f)\, d\eta \right] \tag{11.124}$$

Integration of Eq. (11.124) with the initial condition that $\delta = 0$ at $x = 0$ gives

$$\frac{\gamma}{v_{x,\infty}} f'(0) x = \delta^2 \int_0^1 f(1 - f)\, d\eta \tag{11.125}$$

The integral on the right-hand side of Eq. (11.125) is dependent on the form of $f(\eta)$ satisfying the boundary conditions of Eq. (11.122). Boundary layer thickness δ, displacement thickness δ_d, and friction coefficient C_D calculated from the solution of Eq. (11.125)

Table 11.4 Boundary Layer Characteristics (K_1, K_2, K_3) Obtained from the Solution of Eq. (11.125) by Using Different Forms of $f(\eta)$

Solution	Blasius	η (Linear)	$1.5\eta - 0.5\eta^3$ (Prandtl)	$2\eta - 2\eta^3 + \eta^4$ (Pohlhausen)
$\int f(1-f)\,d\eta$	—	0.167	0.139	0.117
$f'(0)$	—	1.0	1.5	2.0
$\delta = K_1(\gamma x/v_{x,\infty})^{1/2}$	4.99	3.464	4.64	5.83
$C_D = K_2/\mathrm{Re}_L^{1/2}$	1.328	1.155	1.29	1.372
$\delta_d = K_3(\gamma x/v_{x,\infty})^{1/2}$	1.729	1.732	1.74	1.752

Source: Brodkey, 1967.

for different functional forms of $f(\eta)$ satisfying the boundary conditions [Eq. (11.122)] are presented in Table 11.4.

3.4 Turbulent Flow

When the flow velocity increases, the fluid flow becomes unstable in the sense that the spatial velocity becomes irregular and fluctuates in time, which is called turbulent flow (Monin and Yaglom, 1971; Hinze, 1959; Libby and Williams, 1994). In turbulent flow, similar fluctuations can also be observed in scalar characteristics of the flow field, such as fluctuations in mass, temperature, and pressure.

Turbulent flow systems are not like laminar flow systems, and their characteristics cannot be predicted by using generalized mathematical models. In most cases, turbulent characteristics have to be determined experimentally, which is a costly exercise. Turbulence measurements can be made using probes (wire, hot-wire, conductivity, or pressure probes) or optical methods (laser-Doppler, marker nephelometry, or optical). In these experimental techniques, the measured turbulent variable (i.e., velocity, concentration, pressure, or temperature) is detected in the form of an output signal $\Gamma(t)$ of an instrument (e.g., voltage output). This signal is processed to get information about the turbulence nature of the flow. A low-pass filter is always recommended to filter out the high frequency component of the signal. In filtering the signal, the cutoff frequency must be selected high enough to cover all components of the turbulence spectrum (i.e., 4 kHz could be sufficiently high; however, a power spectrum analysis should be performed to estimate the upper limit of the turbulence frequency). Marking the turbulent flow field with a marker is a common practice for the detection of the turbulent field. If particles are used for marking the flow field, these particles should be small enough to follow the smallest turbulent eddy (Becker, 1977).

The signal $\Gamma(t)$, measured as a turbulence characteristic as a function of time, can be expressed in terms the time average and fluctuation components:

$$\Gamma(t) = \overline{\Gamma} + \gamma(t) \tag{11.126}$$

The time-dependent random signal $\Gamma(t)$ has to be processed to get its time average (i.e., using digital or analog signal processing methods):

$$\overline{\Gamma} = \frac{1}{\tau}\int_0^\tau \Gamma(t)\,dt \tag{11.127}$$

The time period τ used for the averaging should be at least one order of magnitude greater than the time period for the natural frequency (oscillation) of the main flow. From Eqs. (11.126) and (11.127), the time average of the fluctuating component of the turbulence signal is zero:

$$\overline{\gamma(t)} = \frac{1}{\tau} \int_0^\tau \gamma(t) \, dt = 0 \tag{11.128}$$

However, the mean-square (or power) of the fluctuation component would not be zero:

$$\overline{\gamma^2(t)} = \frac{1}{\tau} \int_0^\tau \gamma^2(t) \, dt \neq 0 \tag{11.129}$$

The square root of the integral in Eq. (11.129) is called the root-mean-square (rms) of the fluctuation, which has the same units as the fluctuation itself:

$$\gamma' = \gamma_{\text{rms}} = \left[\overline{\gamma^2(t)} \right]^{1/2} = \left(\frac{1}{\tau} \int_0^\tau \gamma^2(t) \, dt \right)^{1/2} \tag{11.130}$$

From a comparison of Eqs. (11.129) and (11.130), the mean-square (power) of a turbulent fluctuation is the square of the root-mean-square of the fluctuation:

$$\overline{\gamma^2(t)} = \frac{1}{\tau} \int_0^\tau \gamma^2(t) \, dt = \gamma'^2 \tag{11.131}$$

The intensity of turbulence, I (dimensionless parameter), is expressed as the ratio of the root-mean-square γ' of the fluctuation to the mean of the turbulence measurement:

$$I = \frac{\left[\overline{\gamma^2(t)} \right]^{1/2}}{\overline{\Gamma}} = \frac{\gamma'}{\overline{\Gamma}} \tag{11.132}$$

The mean-square (power) of the fluctuation component of the turbulence would reflect the turbulence fluctuations of a wide range of frequencies; therefore, it can be expressed as

$$\overline{\gamma^2(t)} = \int_0^\infty E(\kappa) \, d\kappa \tag{11.133}$$

where $E(\kappa)$ is generally follows a power law relation,

$$E(\kappa) = A\kappa^{-m} \tag{11.134}$$

where κ is the wavenumber (for a turbulence fluctuation of frequency f):

$$\kappa = \frac{2\pi f}{\overline{\Gamma}} \tag{11.135}$$

The value of the exponent m is in the range $5/3 < m < 2$; the lower limit is in the case of Kolmogoroff equilibrium structure.

The relationship between the turbulence fluctuations measured at two spatial points (i.e., points A and B) is an important parameter in turbulence. This relation is expressed by the correlation coefficient C (dimensionless parameter) of the fluctuations,

$$C = \frac{\overline{\gamma_A(t)\gamma_B(t)}}{\left[\overline{\gamma_A^2(t)}\right]^{1/2}\left[\overline{\gamma_B^2(t)}\right]^{1/2}} = \frac{\overline{\gamma_A(t)\gamma_B(t)}}{\gamma_A' \gamma_B'} \tag{11.136}$$

The numerator in Eq. (11.136) is the time average of the product of two turbulent signal fluctuations (called correlation),

$$\overline{\gamma_A(t)\gamma_B(t)} = \frac{1}{\tau} \int_0^\tau \gamma_A(t)\gamma_B(t)\, dt \tag{11.137}$$

which can be generated from the property

$$\overline{\gamma_A(t)\gamma_B(t)} = \frac{1}{4}\left[\overline{[\gamma_A(t) + \gamma_B(t)]^2} - \overline{[\gamma_A(t) - \gamma_B(t)]^2}\right] \tag{11.138}$$

Processing the two signals [by standard addition, subtraction, squaring, and time-averaging procedures, Eq. (11.138)] yields the correlation between the two signal fluctuations.

Integration of the correlation coefficient [Eq. (11.136)] along a longitudinal direction is called the integral scale of turbulence λ, which is an average measure of the turbulent eddies in this direction:

$$\lambda = \int_0^\infty C(x)\, dx \tag{11.139}$$

Intermittency is the measure of the turbulent characteristic $\Gamma(t)$ existing at a given spatial location (volume) at greater than a minimum value Γ_{min} (i.e., a turbulent signal rather than a noise) at a given time. The turbulent characteristic, as an example, could be the concentration of a marked fluid, as in the case of mixing of fuel with oxygen in a turbulent flow system. Defining the probability $\beta(x)$ (which would be a function of the distance from a reference point) for the measure of the intermittency,

$$\beta(x) = \text{prob}\,(\Gamma(t) > \Gamma_{min}) \tag{11.140}$$

or the probability that the thickness of the turbulent field, δ, is greater than the distance x from a reference point,

$$\beta(x) = \text{prob}\,(\delta > x) \tag{11.141}$$

Intermittency can be determined experimentally for a given spot in the turbulent flow system by chopping (i.e., digitizing) the signal output of the measurement probe (or by chopping the light sent into the control volume, if optical methods are used) and counting the number of chopped signals satisfying Eq. (11.140). Measurement of $\beta(x)$ as a function of x could be used to identify the thickness of the turbulent flow field. As an example, the thickness of a jet, $\delta(r)$, discharged from a nozzle can be measured by setting a selected measure for the intermittency [i.e., $\beta(r) > 0.5$] as a function of distance x from the jet axis (i.e., x is the axial distance from the exit of the jet, r is the radial distance).

3.4.1 Navier–Stokes Equation for Turbulent Motion

The Navier–Stokes equation of motion should also be used for the turbulent flow because the size of the smallest eddy is much greater than that of the mean free path of the molecules of the fluid. Using the notation for time averaging (τ is a sufficiently large time scale to get a meaningful average),

$$v_x(t) = \bar{v}_x + \gamma_x(t) \tag{11.142}$$

where the time average of the velocity fluctuation, $\gamma_x(t)$, would be zero:

$$\overline{\gamma_x(t)} = \frac{1}{\tau} \int_0^\tau \gamma_x(t) \, dt = 0 \tag{11.143}$$

However, the time average of the square of the fluctuation velocity would not be zero,

$$\overline{\gamma_x^2(t)} = \gamma_x'^2 = \frac{1}{\tau} \int_0^\tau \gamma_x^2(t) \, dt \neq 0 \tag{11.144}$$

and the turbulence intensity (dimensionless parameter) I is defined as

$$I = \frac{\left[\overline{\gamma_x^2(t)}\right]^{1/2}}{\bar{v}_x} = \frac{\gamma_x'}{\bar{v}_x} \tag{11.145}$$

where γ_x' is the root-mean-square of the fluctuation component of the velocity [Eq. (11.130)]. The time average of the product of the two velocity components $v_x(t)$ and $v_y(t)$ can be expressed [because the time average of $\gamma_x(t)$ and $\gamma_y(t)$ is zero, Eq. (11.143)] as

$$\overline{v_x(t)v_y(t)} = \frac{1}{\tau} \int_0^\tau [\bar{v}_x + \gamma_x(t)][\bar{v}_y + \gamma_y(t)] \, dt = \bar{v}_x\bar{v}_y + \overline{\gamma_x(t)\gamma_y(t)} \tag{11.146}$$

where the last term is

$$\overline{\gamma_x(t)\gamma_y(t)} = \frac{1}{\tau} \int_0^\tau \gamma_x(t)\gamma_y(t) \, dt \tag{11.147}$$

The time-averaging relations

$$\overline{\bar{v}_x + \bar{v}_y} = \bar{v}_x + \bar{v}_y \tag{11.148}$$

$$\overline{\bar{v}_x v_y(t)} = \bar{v}_x\bar{v}_y \tag{11.149}$$

can be similarly derived, and

$$\frac{\overline{\partial v_x(t)}}{\partial x} = \frac{\partial \bar{v}_x}{\partial x} \tag{11.150}$$

Using the time-averaging principles, the equation of continuity and equation of motion can be expressed in terms of time-average flow characteristics. If the fluid density is one of the turbulent characteristics of the flow system,

$$\rho(t) = \bar{\rho} + \gamma_\rho(t) \tag{11.151}$$

then the equation of continuity can be expressed as

$$\frac{\partial(\overline{\rho} + \gamma_\rho(t))}{\partial t} = -\nabla \cdot \{[\overline{\rho} + \gamma_\rho(t)][\overline{\mathbf{v}} + \gamma_v(t)]\} \tag{11.152}$$

which can be expressed in terms of time-averaged flow properties [applying the time-averaging operator on both sides of Eq. (11.152)],

$$\overline{\frac{\partial(\overline{\rho} + \gamma_\rho(t))}{\partial t}} = -\overline{\nabla \cdot [\overline{\rho} + \gamma_\rho(t)][\overline{\mathbf{v}} + \gamma_v(t)]} = -(\nabla \cdot \overline{\rho\mathbf{v}}) - \overline{\nabla \cdot \gamma_\rho(t)\gamma_v(t)} \tag{11.153}$$

or

$$\frac{\partial\overline{\rho}}{\partial t} = -(\nabla \cdot \overline{\rho\mathbf{v}}) - \overline{\nabla \cdot \gamma_\rho(t)\gamma_v(t)} \tag{11.154}$$

The last term in Eq. (11.154) is called the turbulent impulse. If the flow is incompressible (i.e., ρ is constant), then this term vanishes and the equation of continuity becomes

$$\nabla \cdot \overline{\mathbf{v}} = 0 \tag{11.155}$$

which is of the same form as the equation of continuity for incompressible flow (i.e., $\nabla \cdot \mathbf{v} = 0$). Further, for an incompressible flow, because the time-dependent velocity should also satisfy the equation of continuity,

$$\nabla \cdot \mathbf{v}(t) = \nabla \cdot [\overline{\mathbf{v}} + \gamma_v(t)] = \nabla \cdot \overline{\mathbf{v}} + \nabla \cdot \gamma_v(t) = 0 \tag{11.156}$$

Using Eq. (11.155), the equation of continuity [Eq. (11.156)] yields to

$$\nabla \cdot \gamma_v(t) = 0 \tag{11.157}$$

which means that for an incompressible flow the time-dependent turbulent fluctuation velocity should also satisfy the equation of continuity for incompressible flow.

Similarly, introducing the turbulent velocity expressed in Eq. (11.142) and time-dependent pressure into the Navier–Stokes equation [Eq. (11.21)], the equation of motion for an incompressible flow (neglecting external forces) becomes

$$\rho\left(\frac{\partial\overline{\mathbf{v}} + \gamma_v(t)}{\partial t} + [\overline{\mathbf{v}} + \gamma_v(t)] \cdot \nabla[\overline{\mathbf{v}} + \gamma_v(t)]\right) = -\nabla \cdot [\overline{\mathbf{p}} + \gamma_p(t)] + \mu\nabla^2[\overline{\mathbf{v}} + \gamma_v(t)] \tag{11.158}$$

and time-averaging of both sides gives

$$\rho\frac{\partial\overline{\mathbf{v}}}{\partial t} + \rho(\overline{\mathbf{v}} \cdot \nabla)\overline{\mathbf{v}} = -\nabla\overline{\mathbf{p}} + \mu\nabla^2\overline{\mathbf{v}} - \nabla \cdot \overline{\rho\gamma_v(t)\gamma_v(t)} \tag{11.159}$$

or in the notation of the Cartesian coordinates,

$$\rho\frac{\partial\overline{\mathbf{v}}_i}{\partial t} + \rho\overline{\mathbf{v}}_j\frac{\partial\overline{\mathbf{v}}_i}{\partial x_j} = -\frac{\partial\overline{p}}{\partial x_i} + \mu\frac{\partial^2\overline{\mathbf{v}}_i}{\partial x_j^2} - \rho\frac{\partial\overline{\gamma_{v,i}(t)\gamma_{v,j}(t)}}{\partial x_j} \tag{11.160}$$

which is known as the Reynolds equation of turbulence. The last term of Eq. (11.160) is known as the Reynolds or turbulent stress τ^t (which is a tensor). In Cartesian coordinates, for a compressible flow the Reynolds stress is expressed as

$$\tau^t = \rho\overline{\gamma_{v,i}(t)\gamma_{v,j}(t)} = \rho\begin{pmatrix} \overline{\gamma_x^2(t)} & \overline{\gamma_x(t)\gamma_y(t)} & \overline{\gamma_x(t)\gamma_z(t)} \\ \overline{\gamma_y(t)\gamma_x(t)} & \overline{\gamma_y^2(t)} & \overline{\gamma_y(t)\gamma_z(t)} \\ \overline{\gamma_z(t)\gamma_x(t)} & \overline{\gamma_z(t)\gamma_y(t)} & \overline{\gamma_z^2(t)} \end{pmatrix} \tag{11.161}$$

which is a symmetrical matrix [$\gamma_x(t)$, $\gamma_y(t)$, and $\gamma_z(t)$ are the fluctuation velocities in the x, y, and z directions, respectively]. Using the Reynolds stress tensor, the Reynolds equation of motion for turbulent flow becomes (neglecting the external forces)

$$\rho \frac{\partial \bar{\mathbf{v}}}{\partial t} + \rho(\bar{\mathbf{v}} \cdot \nabla)\bar{\mathbf{v}} = -\nabla \bar{\mathbf{p}} + \mu \nabla^2 \bar{\mathbf{v}} - \nabla \cdot \tau^t \tag{11.162}$$

Reynolds stress τ^t [Eq. (11.162)] is not a property of the fluid like fluid density or viscosity; τ^t is a property of the flow itself. More important, Reynolds stress may be as large as or even larger than the viscous stress [the second term on the right hand side of Eq. (11.162)] depending on the flow conditions and the geometry of the system. To integrate the Reynolds equation, attempts were made to express the Reynolds stress as a function of flow properties; these are called phenomenological theories. Some of these theories will be mentioned in the following (Bird et al., 1960).

3.4.2 Boussinesq's Eddy Viscosity Theory

In this approach a turbulent or eddy viscosity analogy with Newton's law of viscosity was introduced. For a two-dimensional flow, the turbulent viscosity was expressed as

$$\tau_{yx}^t = -\mu^t \frac{d\bar{v}_x}{dy} \tag{11.163}$$

where the turbulent viscosity coefficient μ^t depends on the flow velocity and the spatial position.

3.4.3 Prandtl's Mixing Length Theory

Prandtl developed a concept for turbulent momentum transfer by introducing the mixing length l in analogy with the mean free path in the kinetic theory of gas. For a two-dimensional flow, the turbulent viscosity was expressed as

$$\tau_{yx}^t = -\rho l^2 \left| \frac{d\bar{v}_x}{dy} \right| \frac{d\bar{v}_x}{dy} \tag{11.164}$$

where the mixing length l is a function of position, i.e., l is linearly proportional to the distance y from a stationary solid, $l = ky$.

3.4.4 von Karman's Similarity Hypothesis

In this hypothesis, on the basis of dimensional considerations of the first and second derivatives of the mean flow velocity, von Karman suggested that for a two-dimensional flow the Reynolds stresses would be in the form

$$\tau_{yx}^t = -\rho \kappa^2 \left| \frac{(d\bar{v}_x/dy)^3}{(d^2\bar{v}_x/dy^2)^2} \right| \frac{d\bar{v}_x}{dy} \tag{11.165}$$

where κ is a universal constant ($0.36 < \kappa < 0.4$).

3.4.5 Deissler's Empirical Formula for Reynolds Stresses

Deissler proposed the following empirical formula for Reynolds stresses close to a solid object:

$$\tau_{yx}^t = -\rho n^2 \bar{v}_x y \left[1 - \exp\left(-\frac{n^2 v_x y}{\gamma} \right) \right] \frac{d\bar{v}_x}{dy} \tag{11.166}$$

where $\gamma(=\mu/\rho)$ is the kinematic viscosity of the fluid, and n is an empirically determined constant, which is suggested to be 0.124 for tube flow.

In all of these phenomenological theories, the Reynolds stress is expressed as a function of the fluid velocity and location, i.e., distance from a solid object. All of these models have to be verified by experimental data. That makes the Reynolds stress an extra unknown for the equation of motion in turbulent flow. From the engineering point of view, experimental data would be required for the turbulent flow characteristics for a given flow system unless experimental data are known for geometrically similar (i.e., comparable geometric similarity and dimensional ratios) and dynamically similar (i.e., comparable Reynolds numbers) systems.

Turbulent flow in tubular geometry has been studied extensively. Experimental measurements made in tubular geometry showed that for Reynolds numbers in the range 10^4–10^5, the time-averaged velocity profiles fit the correlation

$$\frac{\bar{v}_x(r)}{\bar{v}_{x,max}} = \left(1 - \frac{r}{r_0}\right)^{1/7} \tag{11.167}$$

where the maximum time-averaged velocity is the velocity at $r = 0$. The time-averaged velocity profile for turbulent flow is different from the velocity profile for laminar flow [i.e., Eq. (11.167) is different from Eq. (11.45)]. The velocity profile given by Eq. (11.167) predicts a different ratio of the average and maximum velocities:

$$\frac{\bar{v}_{x,ave}}{\bar{v}_{x,max}} = \frac{4}{5} \tag{11.168}$$

where

$$\bar{v}_{x,ave} = \frac{1}{\pi r_0^2} \int_0^{r_0} 2\pi r \bar{v}_x(r) \, dr \tag{11.169}$$

These relations show that in turbulent flow the major portion of the fluid flows at close to maximum velocity. Also, experimentally measured pressure drops and volumetric flow rate measurements show that the pressure drop is proportional to the 7/4 power of the volumetric flow rate q,

$$\frac{dp}{dx} \propto q^{7/4} \tag{11.170}$$

which is different from the linear relation established in laminar flow. As a result, the relation between the measured pressure drop and the volumetric flow rate could be used to predict the nature of the flow.

In turbulent flow close to solid surfaces the velocity gradient becomes large and viscous forces become dominant again. This region is called the viscous sublayer, and its flow behavior can be expressed in terms of the friction velocity v^* as

$$v^{*2} = -\gamma \frac{d\bar{v}_x}{dy}\bigg|_{y=0} = \frac{\tau_w}{\rho} \quad \text{or} \quad v^* = \sqrt{\frac{\tau_w}{\rho}} \tag{11.171}$$

and

$$v^+ = \frac{\bar{v}_x}{v^*} = \frac{1}{k}\ln y^+ + c \tag{11.172}$$

where k and c are constants (i.e., $k = 0.36$, $c = 3.8$ for $y^+ > 26$) and y^+ is the dimensionless distance from the surface of the solid object,

$$y^+ = \frac{y v^*}{\gamma} \tag{11.173}$$

For this kind of correlation, the friction factor for turbulent flow (which will be discussed in the following section) is predicted to be

$$f = 0.0791 \frac{1}{\mathrm{Re}^4} \tag{11.174}$$

3.4.6 Dimensional Analysis of the Equation of Motion

For an incompressible flow, the equation of continuity and the Navier–Stokes equation (equation of motion) can be expressed in terms of dimensionless distance, velocity, pressure, and time:

$$\boldsymbol{v}^* = \frac{\boldsymbol{v}}{V}, \qquad \boldsymbol{x}^* = \frac{\boldsymbol{x}}{L}, \qquad \mathbf{p}^* = \frac{\mathbf{p}}{\rho V^2}, \qquad t^* = \frac{t}{L/V} \tag{11.175}$$

where L is the characteristic length (i.e., tube diameter or size of an object) and V is the characteristic velocity (i.e., the average flow velocity). The equation of continuity and the Navier–Stokes equation, assuming that only the gravity force field is acting on the fluid, are

$$\nabla \cdot \boldsymbol{v} = 0, \tag{11.176}$$

$$\rho \left(\frac{\partial \boldsymbol{v}}{\partial t} + (\boldsymbol{v} \cdot \nabla) \boldsymbol{v} \right) = -\nabla \cdot \mathbf{p} + \mu \nabla^2 \boldsymbol{v} + \rho \boldsymbol{g} \tag{11.177}$$

or, in terms of dimensionless parameters [and considering the relation of the divergence operator $\nabla = (1/L)\nabla^*$ of Eq. (11.175)],

$$\nabla^* \cdot \boldsymbol{v}^* = 0 \tag{11.178}$$

$$\frac{\partial \boldsymbol{v}^*}{\partial t^*} + (\boldsymbol{v}^* \cdot \nabla^*) \boldsymbol{v}^* = -\nabla^* \cdot \mathbf{p}^* + \frac{\mu}{\rho L V} \nabla^{*2} \boldsymbol{v}^* + \frac{gL}{V^2} \left(\frac{\boldsymbol{g}}{g} \right) \tag{11.179}$$

where $\boldsymbol{g}/g$ is a unit vecter in the direction of gravity.

By defining the dimensionless groups Re (Reynolds number, ratio of the inertial to viscous forces or the ratio of characteristic times for the viscous and convection flows) and Fr (Froude number, the ratio of inertial to gravitational forces),

$$\mathrm{Re} = \frac{\rho L V}{\mu} = \frac{LV}{\gamma} = \frac{L^2/\gamma}{L/V} = \frac{\tau_{\mathrm{vis}}}{\tau_{\mathrm{con}}} \tag{11.180}$$

$$\mathrm{Fr} = \frac{V^2}{gL} \tag{11.181}$$

where τ_{vis} and τ_{con} are the characteristic time scales for viscous effects and convection (i.e., a high Reynolds number corresponds to a large ratio $\tau_{\mathrm{vis}}/\tau_{\mathrm{con}}$), Eq. (11.177) can be expressed as

$$\frac{\partial \boldsymbol{v}^*}{\partial t^*} + (\boldsymbol{v}^* \cdot \nabla^*) \boldsymbol{v}^* = -\nabla^* \cdot \mathbf{p}^* + \frac{1}{\mathrm{Re}} \nabla^{*2} \boldsymbol{v}^* + \frac{1}{\mathrm{Fr}} \left(\frac{\boldsymbol{g}}{g} \right) \tag{11.182}$$

The normalized form of the Navier–Stokes equation [Eq. (11.182)] is universal; i.e., its solution would be a function of dimensionless Reynolds and Froude numbers. Therefore, if two systems are geometrically similar and their Reynolds and Froude numbers are identical, then these two systems would have similar velocity v^* and pressure $\mathbf{p}^*$ distributions as functions of x^*, y^*, x^*, and t^*, which are called dynamically similar systems. Dimensional analysis of the differential forms of the conservation equations gives other dimensionless groups (see Table 11.5) that are the basic dimensionless groups for the similarity of flow systems (Brodky, 1967).

Using the same method as in the derivation of Eq. (11.179), dimensionless differential mass balance can be expressed as

$$\frac{\partial c^*}{\partial t^*} + (v^* \cdot \nabla^*)c^* = \frac{D}{LV}\nabla^{*2}c^* \tag{11.183}$$

or in terms of the dimensionless Peclet number, Pe($= VL/D$),

$$\frac{\partial c^*}{\partial t^*} + (v^* \cdot \nabla^*)c^* = \frac{1}{\text{Pe}}\nabla^{*2}c^* \tag{11.184}$$

where the left-hand side represents the convection and the right-hand side represents the diffusion mass transfer. The Peclet number plays the same role in convective diffusion as the Reynolds number does in fluid flow. When Peclet number (Pe) is small, convection mass transfer becomes small in comparison to the diffusion mass transfer. This can be seen in Eq. (11.184) because both sides of the equation are of the same order of magnitude and for small Pe the diffusion part has to be large and convection can be neglected.

The Peclet number can also be expressed as the product of the Reynolds and Schmidt numbers [Schmidt number (Sc$= \gamma/D$) is similar to the Prandtl number of heat transfer, which is the ratio of kinematic viscosity to thermal diffusion, γ/α]:

$$\text{Pe} = \text{Re} \times \text{Sc} = \frac{\rho VL}{\mu}\left(\frac{\gamma}{D}\right) = \frac{VL}{\gamma}\left(\frac{\gamma}{D}\right) = \left(\frac{VL}{D}\right) \tag{11.185}$$

The Reynolds number is a property of flow, whereas the Schmidt number is a property of the fluid. For gases, Sc is about 1 (mass diffusion and momentum diffusion are of the same order of magnitude). For liquids, however, Sc is on the order of 10^3 (i.e., for common liquids $\gamma \approx 10^{-2}$ cm^2/s, $D = 10^{-5}$ cm^2/s), which makes Pe sufficiently high even at rela-

Table 11.5 Dimensionless Numbers in Transport Phenomena

Number	Formula	Ratio
Reynolds	$\rho LV/\mu$	Inertia to viscous forces
Froude	$V^2/\rho L$	Inertia to gravitational forces
Peclet	VL/D	Convection to molecular diffusion
Prandtl	γ/α	Kinematic viscosity to molecular thermal diffusion
Schmidt	γ/D	Kinematic viscosity to molecular mass diffusion
Weber	$V^2L\rho/\sigma$	Inertia to surface forces
Brinkman	$\mu V^2/kT$	Heat generated by viscous dissipation to conduction

Key: L, distance; V, velocity; μ, viscosity; ρ, density; γ, kinematic viscosity; α, thermal diffusion; D, mass diffusion; σ, surface tension; k, thermal conductivity.

tively low Reynolds number (Re) flow regimes [Eq. (11.185)] that convective mass transfer becomes the dominant mass transfer mechanism.

3.4.7 Interphase Transport and Friction Factors

Pressure drops (or friction) and volumetric flow rates are the two basic flow parameters needed for most engineering applications such as flow in channels or flow around objects. These flow characteristics (in laminar or turbulent flow) can be predicted from the solution of the equation of motion coupled with the equation of continuity. Solutions of these equations in geometrically complex flow systems, however, may require numerical techniques (e.g., the use of commercially available computational fluid dynamics software packages).

As an alternative to solving flow equations, empirical correlations in terms of dimensionless numbers are developed between the pressure drop, friction force, and flow velocity (or volumetric flow rate) for geometrically similar systems. Friction factor is one of the dimensionless numbers that find many engineering applications in the prediction of flow characteristics of flow systems (Bird et al., 1960).

As an example, the force F exerted on a solid object in a fluid or on the surface of a tube, in the direction of flow, can be correlated using the relation

$$F = A\varepsilon_k f \tag{11.186}$$

where A is a characteristic surface area, ε_k is the kinetic energy per unit volume of the fluid [i.e., $(1/2)\rho v_{ave}^2$], and f is the dimensionless friction factor.

For flow in a tube of radius r_0 and length L, A is taken as the wetted surface area and Eq. (11.186) takes the form

$$F = 2\pi r_0 L \frac{1}{2}\rho v_{ave}^2 f \tag{11.187}$$

In flow experiments, however, pressure drop is measured because force measurement would be difficult. Therefore, force F [Eq. (11.186)] has to be expressed in terms of pressure drop for practical reasons, which reduces Eq. (11.187) to

$$F = \pi r_0^2 [(p_0 - p_L) + \rho g(h_0 - h_L)] = 2\pi r_0 L \frac{1}{2}\rho v_{ave}^2 f \tag{11.188}$$

where p_0 and p_L are the static fluid pressures (i.e., $p + \rho gh$ is total pressure) at the inlet and at distance L, and $\rho g(h_0 - h_L)$ is the hydrostatic pressure difference between the inlet and at distance L. Using Eq. (11.188) the friction factor (also called the Fanning friction factor) f can be expressed in terms of the tube diameter D ($D = 2r_0$) as

$$f = \frac{1}{4}\left(\frac{D}{L}\right)\left(\frac{p_0 - p_L + \rho g(h_0 - h_L)}{(1/2)\rho v_{ave}^2}\right) \tag{11.189}$$

If the heights at the fluid inlet and at distance L are equal, the hydrostatic pressure difference disappears and Eq. (11.189) becomes

$$f = \frac{1}{4}\left(\frac{D}{L}\right)\left(\frac{p_0 - p_L}{(1/2)\rho v_{ave}^2}\right) \tag{11.190}$$

The forces exerted on a solid object in the flow system can be correlated with the same methodology. As an example, the force exerted on a spherical object in a flow system can be expressed as [similar to Eqs. (11.186) and (11.187)] as

$$F = A\varepsilon_k f = \pi r_0^2 \frac{1}{2} \rho v_{\text{ave}}^2 f \tag{11.191}$$

where the characteristic surface area A is taken as the area by projecting the solid onto a plane normal to the direction of fluid flow (in the case of a sphere, $A = \pi r_0^2$). In most engineering applications the terminal velocity of the particle is measured, because the measurement of force exerted on a particle is always difficult. For example, for the steady-state fall of a spherical object in a fluid, the net force acting on the particle F is the difference between the gravitational and buoyancy forces. Using the difference in the gravitational and buoyancy forces for F, Eq. (11.191) becomes (v_{ave} will be replaced by v_∞, the terminal velocity of the particle)

$$F = \frac{4}{3} \pi r_0^3 g(\rho_s - \rho) = \pi r_0^2 \frac{1}{2} \rho v_\infty^2 f \tag{11.192}$$

where ρ_s and ρ are the densities of the solid sphere and the fluid respectively. Using Eq. (11.192), the friction factor f can be expressed in terms of the diameter of the sphere ($D = 2r_0$) as

$$f = \frac{4}{3} \left(\frac{gD}{v_\infty^2} \right) \left(\frac{\rho_s - \rho}{\rho} \right) \tag{11.193}$$

where v_∞ is the terminal velocity of the particle. This friction factor is also called the drag coefficient [i.e., $C_{D,x}$ in Eq. (11.104) and C_D in Eq. (11.107)]. In defining the friction factor by this relation, all viscous effects are combined into the terminal velocity. In tube flow, all viscous effects are combined into the pressure drop, as defined by Eqs. (11.189) and (11.190).

In tube flow, the friction factor can also be expressed in terms of viscosity and the velocity gradient on the pipe wall. The friction force exerted by the fluid on the tube wall is (for steady-state flow, constant physical properties)

$$F = 2\pi r_0 \int_0^L \left(-\mu \frac{\partial v_x}{\partial r} \right)_{r_0} dx \tag{11.194}$$

using F from Eq. (11.186), and considering that A is the area of the wetted surface, which is $2\pi r_0 L$, where, as defined before, L is the tube length, the friction factor can also be expressed as

$$\begin{aligned}
f &= \frac{F}{2\pi r_0 L(1/2)\rho v_{\text{ave}}^2} \\
&= \frac{2\pi r_0 \int_0^L (-\mu \, \partial v_x/\partial r)_{r_0} dx}{2\pi r_0 L(1/2)\rho v_{\text{ave}}^2} = \frac{2}{L} \left(\frac{\int_0^L (-\mu \, \partial v_x/\partial)_{r_0} dx}{\rho v_{\text{ave}}^2} \right)
\end{aligned} \tag{11.195}$$

The friction factor can be expressed in terms of dimensionless parameters [as discussed in the derivation of Eq. (11.182)]. By selecting D (tube diameter) and v_{ave} as the characteristic distance and velocity, Eq. (11.195) becomes (Bird et al., 1960)

$$f = 2\frac{D}{L}\left(\frac{\mu}{\rho D v_{\text{ave}}}\right)\int_0^{L/D}\left(-\frac{\partial v_x^*}{\partial r*}\right)_{r^*=1/2}dx^*$$

$$= 2\frac{D}{L}\left(\frac{1}{\text{Re}}\right)\int_0^{L/D}\left(-\frac{\partial v_x^*}{\partial r^*}\right)_{r^*=1/2}dx^* \qquad (11.196)$$

which is a valid relation for laminar and turbulent flow systems. The integral in Eq. (11.196) can be evaluated by solving the equation of continuity and equation of motion simultaneously. Assigning the dimensionless parameters $r^* = r/D$, $x^* = x/D$, $v_{\text{ave}}^* = v_x/v_{\text{ave}}$, $p^* = (p - p_0)/\rho v_x^2$, and $\text{Re} = \rho D v_{\text{ave}}/\mu$, and with the boundary conditions

$$v^* = 0 \qquad \text{at} \qquad r^* = 1/2 \qquad\qquad (11.197)$$

$$v^* = v^*(r^*) \qquad \text{at} \qquad x^* = 0 \qquad\qquad (11.198)$$

$$p^* = 0 \qquad \text{at} \qquad x^* = 0 \qquad\qquad (11.199)$$

it can be seen that the solution of Eq. (11.196) would be obtained in the form

$$f = f(L/D, \text{Re}) \qquad\qquad (11.200)$$

The expression of Eq. (11.200) tells us that the friction factor is a function of dimensionless distance L/D and Reynolds number. Furthermore, if the flow is fully developed at $x^* = 0$, then the velocity profile v^* would be independent of x^* (i.e., a function of only r^*), and the friction factor would be a function of Reynolds number only:

$$f = f(\text{Re}) \qquad\qquad (11.201)$$

In fact, the friction factor for laminar tube flow (i.e., Hagen–Poiseulle flow) is calculated as

$$f = \frac{16}{\text{Re}} \qquad\qquad (11.202)$$

where $\text{Re} = \rho D v_{\text{ave}}/\mu$. For turbulent flow, the friction factor f is calculated as

$$f = \frac{0.0791}{\text{Re}^{1/4}} \qquad\qquad (11.203)$$

for the 1/7 power law turbulent velocity profile,

$$\frac{\overline{v}_x}{v^*} = 8.56\left(\frac{(r_0 - r)v^*\rho}{\mu}\right)^{1/7} \qquad\qquad (11.204)$$

where $\text{Re} = \rho D v_{\text{ave}}/\mu$ is the Reynolds number and $v^* = (\tau_w/\rho)^{1/2}$ is the friction velocity as described by Eq. (11.171). Friction factors for laminar and turbulent tube flow are presented in Fig. 11.10 (Bird et al., 1960).

Friction factors for different flow systems, which are helpful tools for engineering design purposes, are readily available. As an example, the friction factor in a packed column is an important tool for many process and design engineering problems. Assuming uniform packing properties and the absence of channeling flow, the friction factor for packed columns can be developed by modifying the Hagen–Poiseuille flow equation [Eq. 11.47)] and the friction factor defined for laminar tube flow [Eq. (11.190)]. This is a logical approach, because the fluid flow takes place in the void space or channels of the packed column, which can be approximated to tube flow. The void

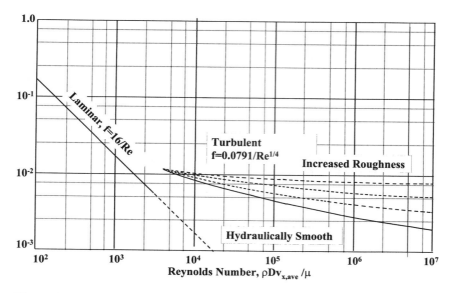

Figure 11.10 Friction factors for laminar and turbulent tube flow. (From Bird et al., 1960.)

space in the packed bed is characterized by hydraulic radius r_h, which is the ratio of the wet surface area to the wet perimeter (corresponds to $r_0/2$ in tube flow). Eventually, the hydraulic radius is related to the void fraction ε and the wetted surface area per unit volume, a, of the packed column, as derived in the equation (Bird et al., 1960)

$$r_h = \frac{A_w}{P_w} = \frac{V_w}{A_w} = \frac{V_w/V_b}{A_w/V_b} = \frac{\varepsilon}{a} \tag{11.205}$$

where A_w is the wet cross-sectional surface area available for fluid flow, P_w is the wetted perimeter in the cross section, V_w is the wet (or void) volume of the packed bed occupied by the fluid, ε is the void fraction (i.e., V_w/V_b), V_b is the packed column volume, and a is the wetted surface per unit volume of the packed column. If the particle surface area per particle volume is denoted by a_v,

$$a_v = \frac{\pi D_p^2}{(4/3)\pi D_p^3/8} = \frac{6}{D_p} \tag{11.206}$$

then

$$a = (1 - \varepsilon)a_v = (1 - \varepsilon)\frac{6}{D_p} \tag{11.207}$$

where D_p is the diameter of the spherical particles used as the packing material. Further, the actual velocity in the packed column can be expressed in terms of the superficial fluid velocity v_s (i.e., q/a_c, where q is the volumetric flow rate and a_c is the cross-sectional area of the tube),

$$v_{\text{ave}} = \frac{v_s}{\varepsilon} = \frac{q}{\varepsilon a_c} \tag{11.208}$$

because q and a_c are easy to measure. The friction factor for the packed column can be defined analogously to the friction factor defined for Hagen–Poiseuille flow [Eq. (11.189)] as

$$f = \frac{1}{4}\left(\frac{D_p}{L}\right)\left[\frac{p_0 - p_L + \rho g(h_0 - h_L)}{(1/2)\rho v_s}\right] \tag{11.209}$$

where D_p is the diameter of the packing particles and v_s is the superficial fluid velocity, [In Eq. (11.189), D is tube diameter and v_{ave} is the average fluid velocity.] Similarly, the average flow rate in the packed column can be expressed analogously to Eq. (11.47) by expressing dp/dx in terms of the hydrostatic pressure difference and replacing r_0 by $2r_h$:

$$v_{ave} = \frac{[(p_0 - p_L) + \rho g(h_0 - h_L)]r_h^2}{2\mu L} \tag{11.210}$$

Combining Eqs. (11.205)–(11.209), the superficial velocity v_s can be expressed as

$$v_s = \frac{[p_0 - p_L + \rho g(h_0 - h_L)]r_h^2 \varepsilon}{2\mu L}$$

$$= \left(\frac{p_0 - p_L + \rho g(h_0 - h_L)}{2\mu L a_v^2}\right)\left(\frac{\varepsilon^3}{(1-\varepsilon)^2}\right) \tag{11.211}$$

or, expressing a_v in terms of D_p,

$$v_s = \left(\frac{p_0 - p_L + \rho g(h_0 - h_L)}{2\mu L}\right)\left(\frac{D_p^2}{36}\right)\left(\frac{\varepsilon^3}{(1-\varepsilon)^2}\right)$$

$$= \left(\frac{p_0 - p_L + \rho g(h_0 - h_L)}{\mu L}\right)\left(\frac{D_p^2}{72}\right)\left(\frac{\varepsilon^3}{(1-\varepsilon)^2}\right) \tag{11.212}$$

In Eq. (11.212), L must be greater than the tube length because the fluid path in the axial direction in the packed bed is tortuous rather than straight. This fact can be corrected for by taking a larger value for the numerical constant (i.e., larger than 36) used in Eq. (11.212) and keeping the packed bed length L unchanged. In fact, the Blake–Kozeny equation,

$$v_s = \left(\frac{p_0 - p_L - \rho g(h_0 - h_L)}{\mu L}\right)\left(\frac{D_p^2}{150}\right)\left(\frac{\varepsilon^3}{(1-\varepsilon)^2}\right) \tag{11.213}$$

uses a value greater than 36 for the numerical constant. Eliminating the pressure term between Eqs. (11.209) and (11.213) gives

$$f = \left(\frac{(1-\varepsilon)^2}{\varepsilon^3}\right)\left(\frac{75}{D_p G_s/\mu)}\right) \tag{11.214}$$

Both Eq. (11.213) and Eq. (11.214) are suggested to correlate the superficial velocity versus pressure drop and friction factor versus superficial mass flow rate $G_s(= \rho v_s)$ for the laminar flow regime [i.e., Reynolds number based on particle diameter $D_p G_s/(1-\varepsilon)\mu < 10$]. For the large Reynolds number flow regime [i.e., $D_p G_s/(1-\varepsilon)\mu > 1000$] in packed beds, the Burke–Plummer equation is suggested, in the form

$$\frac{1}{2}\rho v_s^2 = \left(\frac{p_0 - p_L - \rho g(h_0 - h_L)}{L}\right)\left(\frac{D_p}{3.5}\right)\left(\frac{\varepsilon^3}{1 - \varepsilon}\right) \tag{11.215}$$

which correlates the friction factor by the relation

$$f = 0.875\frac{1 - \varepsilon}{\varepsilon^3} \tag{11.216}$$

The combination of the Blake–Kozeny and Burke–Plummer equations is known as the Ergun equation (Ergun, 1952):

$$\frac{150\mu v_s}{D_p^2}\left(\frac{(1 - \varepsilon)^2}{\varepsilon^3}\right) + \frac{1.75\rho v_s^2}{D_p}\left(\frac{1 - \varepsilon}{\varepsilon^3}\right) = \frac{p_0 - p_L + \rho g(h_0 - h_L)}{L} \tag{11.217}$$

or, in terms of dimensionless groups,

$$150\frac{1 - \varepsilon}{D_p G_s/\mu} + 1.75 = \left(\frac{\rho[(p_0 - p_L) + \rho g(h_0 - h_L)]}{LG_s^2}\right)\left(\frac{D_p}{L}\right)\left(\frac{1 - \varepsilon}{D_p G_s/\mu}\right) \tag{11.218}$$

Predictions of the Blake–Kozeny, Burke–Plummer, and Ergun relations are presented in Fig. 11.11.

Pressure drop measurements are the most basic data used to monitor many multiphase reactors operating at industrial scale. Pressure drop measurements in hydrotreating reactors (most of which are packed bed or trickle-bed reactors), may provide information about coke formation or plugging in the reactor.

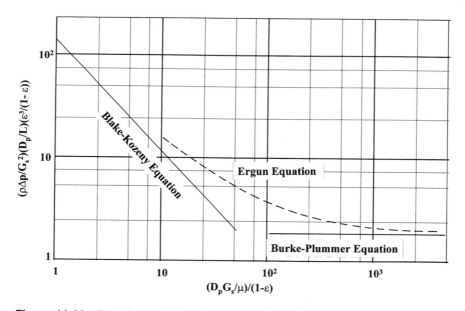

Figure 11.11 Ergun's correlation for pressure drops in packed columns in terms of dimensionless parameters. (From Ergun, 1952 and Bird et al., 1960.)

4. HEAT TRANSFER

Heat transfer is an important branch of engineering science and has been used extensively in mechanical, chemical, mineral, environmental, aerospace, and civil engineering applications. Heat transfer can take place by conduction, convection, and radiation. Radiative heat transfer becomes the dominant mechanism for heat transfer at elevated temperatures, such as in coal-fired utility boilers, because heat transfer is proportional to the fourth power of the temperature. [Classical books such as Hottel and Sarofim (1967) and Ozisik (1973) are recommended for readers who wish to study this subject in detail.]

This section will focus on the basic principles of conductive and convective heat transfer, in the hope that it will help to understand the coupling between heat transfer and chemical reactions in chemically reactive systems. The classical textbooks by Eckert and Drake (1959), Ozisik (1985), Holman (1990), and Bird et al. (1960) are referred to for further understanding of the subject.

The main task of heat transfer is to predict the amount of heat transferred (energy flux) under given boundary and initial conditions and the temperature profile in a stationary or flow system. When heat transfer is coupled with fluid flow, the equation of continuity and the equation of motion have to be simultaneously considered in addition to the equation of thermal energy (energy balance). Therefore, the energy balance equation has to be developed for the flow system. In the following sections, the energy balance will be formulated, because heat is a form of energy. In this formulation the property balance principles presented in Section 2 will be applied and solutions of the energy equation coupled with the equations of continuity and motion will be discussed.

4.1 Energy Balance (Equation of Energy)

In this section, energy flux due to only conduction and convection mechanisms will be considered, and energy generation by chemical reaction (or by any other means such as electrical or nuclear energy) will be neglected. Heat generation by chemical reactions can be added into the energy balance equation as has already been discussed in Chapter 9, Section 2.4. The main purpose of this section, however, is to formulate the energy balance and search for its solutions.

Using the property balance principles [similar to the equation of motion expressed in Eq. (11.20)], energy balance can be expressed as the sum of the internal energy ρU and kinetic energy $(1/2)\rho v^2$ per unit volume (the potential energy $\rho g h$ will be taken care of in the work term):

$$\frac{\partial}{\partial t}\rho\left(U + \frac{1}{2}v^2\right) = -\nabla \cdot \rho v\left(U + \frac{1}{2}v^2\right) - \nabla \cdot q + \rho(v \cdot g) - \nabla \cdot pv - \nabla \cdot (\tau \cdot v) \qquad (11.219)$$

where the first and second terms are the accumulation and convective transfer of the summation of internal and kinetic energies per unit volume. The remaining terms are the conductive heat transfer, the work done on the fluid by gravitational forces, the work done on the fluid by pressure forces, and the work done on the fluid by viscous forces. Introducing the equation of change for kinetic energy per unit volume $(1/2)\rho v^2$, Eq. (11.219) can be expressed as (Bird et al., 1960)

$$\frac{\partial}{\partial t}(\rho U) = -\nabla \cdot \rho U v - \nabla \cdot q - \mathbf{p}\nabla \cdot v - (\tau : \nabla v) \qquad (11.220)$$

which is also known as the equation of thermal energy. (":" is the scalar product operator for two tensors τ and ∇v). The terms in Eq. (11.220) are the accumulation and convective transfer of the internal energy per unit volume, the conductive heat transfer, and the work done on the fluid by compression and irreversible viscous dissipation. In engineering applications, it is practical to use temperature and specific heat rather than internal energy. Expressing the internal energy as a function of V and T,

$$dU = \left(\frac{\partial U}{\partial V}\right)_T dV + \left(\frac{\partial U}{\partial T}\right)_V dT = \left[-p + T\left(\frac{\partial p}{\partial T}\right)_V\right]dV + c_v\, dT \tag{11.221}$$

where c_v is the specific heat at constant volume per unit mass of the fluid. Multiplying both sides of Eq. (11.221) by ρ and combining it with Eq. (11.220) gives (in terms of thermal energy per unit volume)

$$\rho c_v \frac{\partial T}{\partial t} = -v \cdot \nabla \rho c_v T - \nabla \cdot q - T\left(\frac{\partial p}{\partial T}\right)_V (\nabla \cdot v) - (\tau : \nabla v) \tag{11.222}$$

By expressing the conductive heat flux q as a linear function of the temperature gradient ($q = -k\,\nabla T$, which is Fourier's law) and τ in terms of the velocity gradient, for a Newtonian fluid the equation of thermal energy (for constant ρ and c_v) in terms of the fluid temperature becomes [similar to the Navier–Stokes equation, Eq. (11.21)]

$$\rho c_v\left(\frac{\partial T}{\partial t} + v\nabla \cdot T\right) = k\,\nabla^2 T - T\left(\frac{\partial p}{\partial T}\right)_\rho (\nabla \cdot v) + \mu \Phi_v \tag{11.223}$$

where Φ_v is the dissipation function, expressing the production of heat per unit volume of the fluid as a result of viscous dissipation. Equation (11.223) is the most general form of the thermal energy equation. For most engineering applications the viscous dissipation term is negligible.

For an ideal gas (i.e., $pV = nRT$),

$$\left(\frac{\partial p}{\partial T}\right)_V = \frac{p}{T} \tag{11.224}$$

by which the thermal energy equation [Eq. (11.223)] becomes

$$\rho c_V\left(\frac{\partial T}{\partial t} + v \cdot \nabla T\right) = k\,\nabla^2 T - p(\nabla \cdot v) \tag{11.225}$$

For a fluid at constant pressure, using the relation

$$dU = -p\, dV + c_p\, dT \tag{11.226}$$

where c_p is the specific heat at constant pressure per unit mass of the fluid, Eq. (11.225) becomes

$$\rho c_p\left(\frac{\partial T}{\partial t} + v \cdot \nabla T\right) = k\,\nabla^2 T \tag{11.227}$$

For an incompressible fluid (and for ρ independent of T), Eq. (11.223) reduces to an equation similar to Eq. (11.227):

$$\rho c_p\left(\frac{\partial T}{\partial t} + v \cdot \nabla T\right) = k\,\nabla^2 T \tag{11.228}$$

even though they represent different physical situations.

For solids or motionless fluids (which is an ideal case), the velocity is zero and the thermal energy equation takes the form

$$\rho c_p \frac{\partial T}{\partial t} = k \, \nabla^2 T \tag{11.229}$$

In the formulation of the thermal energy balance, pure fluids are considered; therefore the heat resulting from chemical (or nuclear) reactions is not considered. The energy equation in chemically reactive systems was discussed earlier, in Chapter 9.

Energy equations in terms of energy and momentum fluxes and in terms of transport properties for rectangular, cylindrical, and spherical coordinate systems are presented in Tables 11.6 and 11.7, respectively (Bird et al., 1960).

4.2 Integration of the Thermal Energy Equation

In a given flow system, solution of the thermal energy equation provides information for the temperature profile of the system as a function of coordinates and time. Unfortunately, in most engineering applications energy transfer is coupled with mass and momentum transfer as well as with chemical reactions. Because of this coupling, it may be difficult if not impossible to find analytical solutions. As a result, the use of computational fluid dynamics software packages for the numerical integration of the energy equation is becoming a common industrial practice. Integration of the thermal energy equation for the selected geometrical and flow conditions will be summarized in the following sections.

4.3 Steady-State Heat Conduction

For the steady-state heat conduction problems in rectangular coordinates, the thermal energy equation is given by

$$\nabla^2 T = 0 \tag{11.230}$$

[i.e., with no heat generation, the steady-state form of Eq. (11.229)].

One-dimensional steady-state heat conduction in a slab of thickness L exposed to temperatures T_1 and T_2 on its opposite sides (Fig. 11.12a) can be described by

$$\frac{d^2 T}{dx^2} = 0 \tag{11.231}$$

with the boundary conditions

$$T = T_1 \quad \text{at} \quad x = 0; \qquad T = T_2 \quad \text{at} \quad x = L \tag{11.232}$$

Solution of Eq. (11.231) T with the boundary conditions of Eq. (11.232) gives

$$\frac{T - T_1}{T_1 - T_2} = \frac{x}{L} \tag{11.233}$$

which is linear in x and independent of thermal conductivity. The heat flow through the slab is calculated, using the Fourier conduction law, to be

$$q = -kA \frac{dT}{dx} = -kA \frac{T_2 - T_1}{L} = \frac{T_1 - T_2}{L/kA} \tag{11.234}$$

where L/kA (in units of Kelvin-seconds per joule or Kelvins per watt) is called the thermal resistance (i.e., identically to Ohm's law of electrical resistance). If n slabs are attached

Table 11.6 Equation of Energy in Terms of Energy and Momentum Fluxes in Rectangular, Cylindrical, and Spherical Coordinate Systems

In a rectangular coordinate system of x, y, and z:

$$\rho c_v \left(\frac{\partial T}{\partial t} + v_x \frac{\partial T}{\partial x} + v_y \frac{\partial T}{\partial y} + v_z \frac{\partial T}{\partial z} \right) =$$

$$- \left(\frac{\partial q_x}{\partial x} + \frac{\partial q_y}{\partial y} + \frac{\partial q_z}{\partial z} \right) - T \left(\frac{\partial p}{\partial T} \right)_\rho \left(\frac{\partial v_x}{\partial x} + \frac{\partial v_y}{\partial y} + \frac{\partial v_z}{\partial z} \right)$$

$$- \left\{ \tau_{xx} \frac{\partial v_x}{\partial x} + \tau_{yy} \frac{\partial v_y}{\partial y} + \tau_{zz} \frac{\partial v_z}{\partial z} \right\}$$

$$- \left\{ \tau_{xy} \left(\frac{\partial v_x}{\partial y} + \frac{\partial v_y}{\partial x} \right) + \tau_{xz} \left(\frac{\partial v_x}{\partial z} + \frac{\partial v_z}{\partial x} \right) + \tau_{yz} \left(\frac{\partial v_y}{\partial z} + \frac{\partial v_z}{\partial y} \right) \right\}$$

In a cylindrical coordinate system of r, θ, and z:

$$\rho c_v \left(\frac{\partial T}{\partial t} + v_r \frac{\partial T}{\partial r} + \frac{v_\theta}{r} \frac{\partial T}{\partial \theta} + v_z \frac{\partial T}{\partial z} \right) =$$

$$- \left(\frac{1}{r} \frac{\partial (r q_r)}{\partial r} + \frac{1}{r} \frac{\partial q_\theta}{\partial \theta} + \frac{\partial q_z}{\partial z} \right) - T \left(\frac{\partial p}{\partial T} \right)_\rho \left(\frac{1}{r} \frac{\partial (r v_r)}{\partial r} + \frac{1}{r} \frac{\partial v_\theta}{\partial \theta} + \frac{\partial v_z}{\partial z} \right)$$

$$- \left\{ \tau_{rr} \frac{\partial v_r}{\partial r} + \tau_{\theta\theta} \frac{1}{r} \left(\frac{\partial v_\theta}{\partial \theta} + v_r \right) + \tau_{zz} \frac{\partial v_z}{\partial z} \right\}$$

$$- \left\{ \tau_{r\theta} \left[r \frac{\partial}{\partial r} \left(\frac{v_\theta}{r} \right) + \frac{1}{r} \frac{\partial v_r}{\partial \theta} \right] + \tau_{rz} \left(\frac{\partial v_z}{\partial r} + \frac{\partial v_r}{\partial z} \right) + \tau_{\theta z} \left(\frac{1}{r} \frac{\partial v_z}{\partial \theta} + \frac{\partial v_\theta}{\partial z} \right) \right\}$$

In a spherical coordinate system of r, θ, ϕ:

$$\rho c_v \left(\frac{\partial T}{\partial t} + v_r \frac{\partial T}{\partial r} + \frac{v_\theta}{r} \frac{\partial T}{\partial \theta} + \frac{v_\phi}{r \sin \theta} \frac{\partial T}{\partial \phi} \right) = - \left(\frac{1}{r^2} \frac{\partial (r^2 q_r)}{\partial r} + \frac{1}{r \sin \theta} \frac{\partial (q_\theta \sin \theta)}{\partial \theta} + \frac{1}{r \sin \theta} \frac{\partial q_\phi}{\partial \phi} \right)$$

$$- T \left(\frac{\partial p}{\partial T} \right)_\rho \left(\frac{1}{r^2} \frac{\partial (r^2 v_r)}{\partial r} + \frac{1}{r \sin \theta} \frac{\partial (v_\theta \sin \theta)}{\partial \theta} + \frac{1}{r \sin \theta} \frac{\partial v_\phi}{\partial \phi} \right)$$

$$- \left\{ \tau_{rr} \frac{\partial v_r}{\partial r} + \tau_{\theta\theta} \left(\frac{1}{r} \frac{\partial v_\theta}{\partial \theta} + \frac{v_r}{r} \right) + \tau_{\phi\phi} \left(\frac{1}{r \sin \theta} \frac{\partial v_\phi}{\partial \phi} + \frac{v_r}{r} + \frac{v_\theta \cot \theta}{r} \right) \right\}$$

$$- \left\{ \tau_{r\theta} \left(\frac{\partial v_\theta}{\partial r} + \frac{1}{r} \frac{\partial v_r}{\partial \theta} - \frac{v_\theta}{r} \right) + \tau_{r\phi} \left(\frac{\partial v_\phi}{\partial r} + \frac{1}{r \sin \theta} \frac{\partial v_r}{\partial \phi} - \frac{v_\phi}{r} \right) \right.$$

$$\left. + \tau_{\theta\phi} \left(\frac{1}{r} \frac{\partial v_\phi}{\partial \theta} + \frac{1}{r \sin \theta} \frac{\partial v_\theta}{\partial \phi} - \frac{\cot \theta}{r} v_\phi \right) \right\}$$

where c_v is the specific heat capacity at constant volume of the fluid per unit mass, and the terms in braces { } are the heat generated by viscous dissipation, which may be for systems with small velocity gradients.

Source: Bird et al., 1960.

Table 11.7 Equation of Energy in Terms of Transport Properties in Rectangular, Cylindrical, and Spherical Coordinate Systems

In a rectangular coordinate system of x, y, and z:

$$\rho c_v \left(\frac{\partial T}{\partial t} + v_x \frac{\partial T}{\partial x} + v_y \frac{\partial T}{\partial y} + v_z \frac{\partial T}{\partial z} \right) =$$

$$k \left(\frac{\partial^2 T}{\partial x^2} + \frac{\partial^2 T}{\partial y^2} + \frac{\partial^2 T}{\partial z^2} \right) + 2\mu \left\{ \left(\frac{\partial v_x}{\partial x} \right)^2 + \left(\frac{\partial v_y}{\partial y} \right)^2 + \left(\frac{\partial v_z}{\partial z} \right)^2 \right\}$$

$$+ \mu \left\{ \left(\frac{\partial v_x}{\partial y} + \frac{\partial v_y}{\partial x} \right)^2 + \left(\frac{\partial v_x}{\partial z} + \frac{\partial v_z}{\partial x} \right)^2 + \left(\frac{\partial v_y}{\partial z} + \frac{\partial v_z}{\partial y} \right)^2 \right\}$$

In a cylindrical coordinate system of r, θ, and z:

$$\rho c_v \left(\frac{\partial T}{\partial t} + v_r \frac{\partial T}{\partial r} + \frac{v_\theta}{r} \frac{\partial T}{\partial \theta} + v_z \frac{\partial T}{\partial z} \right) = k \left[\frac{1}{r} \frac{\partial}{\partial r} \left(r \frac{\partial T}{\partial r} \right) + \frac{1}{r^2} \frac{\partial^2 T}{\partial \theta^2} + \frac{\partial^2 T}{\partial z^2} \right]$$

$$+ 2\mu \left\{ \left(\frac{\partial v_r}{\partial r} \right)^2 + \left[\frac{1}{r} \left(\frac{\partial v_\theta}{\partial \theta} + v_r \right) \right]^2 + \left(\frac{\partial v_z}{\partial z} \right)^2 \right\}$$

$$+ \mu \left\{ \left(\frac{\partial v_\theta}{\partial z} + \frac{1}{r} \frac{\partial v_\theta}{\partial \theta} \right)^2 + \left(\frac{\partial v_z}{\partial r} + \frac{\partial v_r}{\partial z} \right)^2 + \left[\frac{1}{r} \frac{\partial v_r}{\partial \theta} + r \frac{\partial}{\partial r} \left(\frac{v_\theta}{r} \right) \right]^2 \right\}$$

In a spherical coordinate system of r, θ, and ϕ:

$$\rho c_v \left(\frac{\partial T}{\partial t} + v_r \frac{\partial T}{\partial r} + \frac{v_\theta}{r} \frac{\partial T}{\partial \theta} + \frac{v_\phi}{r \sin \theta} \frac{\partial T}{\partial \phi} \right) =$$

$$k \left[\frac{1}{r^2} \frac{\partial}{\partial r} \left(r^2 \frac{\partial T}{\partial r} \right) + \frac{1}{r^2 \sin \theta} \frac{\partial}{\partial \theta} \left(\sin \theta \frac{\partial T}{\partial \theta} \right) + \frac{1}{r^2 \sin^2 \theta} \frac{\partial^2 T}{\partial \phi^2} \right]$$

$$+ 2\mu \left\{ \left(\frac{\partial v_r}{\partial r} \right)^2 + \left(\frac{1}{r} \frac{\partial v_\theta}{\partial \theta} + \frac{v_r}{r} \right)^2 + \left(\frac{1}{r \sin \theta} \frac{\partial v_\phi}{\partial \phi} + \frac{v_r}{r} + \frac{v_\theta \cot \theta}{r} \right)^2 \right\}$$

$$+ \mu \left\{ \left[r \frac{\partial}{\partial r} \left(\frac{v_\theta}{r} \right) + \frac{1}{r} \frac{\partial v_r}{\partial \theta} \right]^2 + \left[\frac{1}{r \sin \theta} \frac{\partial v_r}{\partial \phi} + r \frac{\partial}{\partial r} \left(\frac{v_\phi}{r} \right) \right]^2 \right.$$

$$\left. + \left[\frac{\sin \theta}{r} \frac{\partial}{\partial \theta} \left(\frac{v_\theta}{\sin \theta} \right) + \frac{1}{r \sin \theta} \frac{\partial v_\theta}{\partial \phi} \right]^2 \right\}$$

where c_v is the specific heat capacity at constant volume of the fluid per unit mass, and the terms in braces { } are the heat generated by viscous dissipation, which may be for the systems with small velocity gradients.

Source: Bird et al., 1960.

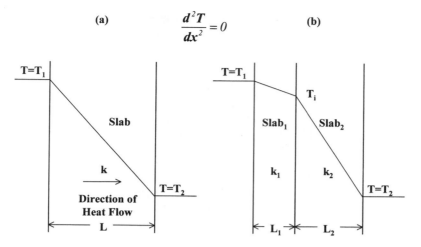

(a) $\dfrac{d^2T}{dx^2} = 0$ **(b)**

Figure 11.12 One-dimensional steady-state heat conduction problem (a) in a slab and (b) in two slabs.

together and exposed to temperatures T_1 and T_2 on opposite sides, using the Fourier conduction law the steady-state heat transfer becomes (Fig. 11.12b)

$$q = \frac{T_1 - T_2}{\sum\limits_{j=1}^{n} L_j/k_j A} \tag{11.235}$$

where L_j is the thickness and k_j the thermal conductivity of the jth slab.

For steady-state radial heat conduction in a tube exposed to T_i and T_o at the inner $(r = r_i)$ and outer $(r = r_o)$ surfaces, the thermal energy equation [Eq. (11.230)] becomes (heat conduction in the radial direction only in cylindrical geometry, Fig. 11.13a).

$$\frac{d^2T}{dr^2} + \frac{1}{r}\frac{dT}{dr} = 0 \tag{11.236}$$

with the boundary conditions

$$T = T_i \quad \text{at} \quad r = r_i; \qquad T = T_o \quad \text{at} \quad r = r_o \tag{11.237}$$

Solution of Eq. (11.236) with the boundary conditions of Eq. (11.237) gives

$$\frac{T - T_o}{T_i - T_o} = \frac{\ln(r/r_o)}{\ln(r_i/r_o)} \tag{11.238}$$

Using the Fourier conduction law, heat transferred across the tube wall (at steady-state conditions) for a tube of length L is

$$q = -kA(r)\frac{dT}{dr} = -2\pi rkL\frac{dT}{dr} \tag{11.239}$$

the integration of which for the same boundary conditions of Eq. (11.237) gives

$$q = \frac{T_i - T_o}{(1/2\pi kL)\ln(r_o/r_i)} \tag{11.240}$$

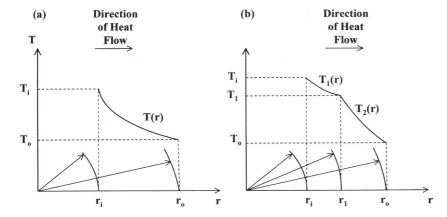

Figure 11.13 Steady-state heat conduction in a radial direction (a) in a tube and (b) in two coaxial tubes.

where $(1/2\pi k L)\ln(r_o/r_i)$ is called the thermal resistance (again, in units of $K \cdot s/J$, identical to Ohm's law of electrical resistance). If there are n layers of tubes, exposed to temperatures T_i and T_o on the inside and outside surfaces, respectively, steady-state heat transfer in the radial direction is given by (Fig. 11.13b)

$$q = \frac{T_i - T_o}{\sum\limits_{j=1}^{n} (1/2\pi k_j L)\ln(r_o/r_i)_j} \tag{11.241}$$

where L is the length of the tube and k_j and $(r_o/r_i)_j$ are the thermal conductivity and the ratio of the outer and inner diameters, respectively, of the jth tube layer.

For steady-state heat conduction for a thin rod attached to a heated wall (Fig. 11.14), the thermal energy equation takes the form

$$kA_c\frac{d^2T}{dx^2} = kA_c\frac{d^2(T - T_f)}{dx^2} = hP(T - T_f) \tag{11.242}$$

where A_c is the cross-sectional area, P is the perimeter, k is the thermal conductivity [W/(m · K)] of the rod, and h is the convection heat transfer coefficient [W/(m^2 · K)] for the heat loss from the rod to the fluid at temperature T_f. A realistic set of boundary conditions would be

$$T = T_w \quad \text{at} \quad x = 0; \qquad \frac{dT}{dx} = 0 \quad \text{at} \quad x = L \tag{11.243}$$

where T_w is the temperature of the heated wall. The second boundary condition of Eq. (11.243) requires that the rod temperature asymptotically approach a steady temperature profile as $x \to L$. Introducing the dimensionless temperature ψ, distance $\xi = x/L$, and heat transfer coefficient K,

$$\Psi = \frac{T - T_f}{T_w - T_f}, \qquad \xi = \frac{x}{L}, \qquad K = \left(\frac{hPL^2}{kA_c}\right)^{1/2} \tag{11.244}$$

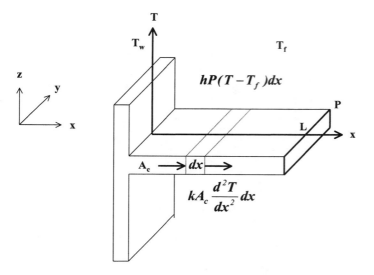

Figure 11.14 Steady-state heat conduction for a thin rod attached to a heated wall.

and the thermal energy equation [Eq. (11.242)] becomes

$$\frac{d^2\Psi}{d\xi^2} - K^2\Psi = 0 \tag{11.245}$$

with the boundary conditions

$$\psi = 1 \quad \text{at} \quad \xi = 0; \qquad \frac{d\psi}{d\xi} = 0 \quad \text{at} \quad \xi = 1 \tag{11.246}$$

Solution of Eq. (11.245) with the boundary conditions of Eq. (11.246) gives

$$\psi = \cosh(K\xi) - \tanh(K)\sinh(K\xi) \tag{11.247}$$

or

$$\psi = \frac{\cosh[K(1 - \xi)]}{\cosh(K)} \tag{11.248}$$

Based on the temperature profile [Eq. (11.247) or (11.248)], the efficiency of the fin η, the ratio of actual heat transferred to the hypothetical value of the heat transferred if the rod surface temperature was at T_w, is given by

$$\eta = \frac{\text{heat lost}}{\text{potential for heat lost}} = \frac{\int_0^L hP(T - T_f)\, dx}{hPL(T_w - T_f)} = \int_0^1 \psi\, d\xi \tag{11.249}$$

which gives

$$\eta = \int_0^1 \frac{\cosh[K(1 - \xi)]}{\cosh(K)}\, d\xi = \frac{1}{\cosh(K)}\left(-\frac{1}{K}\sinh(K(1 - \xi)) \Big|_0^1 \right) = \frac{\tanh(K)}{K} \tag{11.250}$$

This relation is identical to the efficiency for the chemical reaction coupled with diffusion (*K* replaces the dimensionless Thiele number). In practical applications, a large number of shorter rods are preferred to increase the total heat loss (i.e., for cooling). If the rods are too close, forced convection may also be needed for effective cooling.

4.4 Unsteady-State Heat Conduction

Solutions of the thermal energy equation [Eq. (11.229)] for unsteady-state (transient state) heat conduction problems have been discussed by Carslaw and Jaeger (1959). Unsteady-state heat conduction is used particularly in heating or cooling of objects. Considering its specific importance, solution of the thermal energy equation for such problems will be discussed briefly.

An example of unsteady-state heat transfer is the case of a slab of thickness *L* initially at a uniform temperature *T* insulated from one side and exposed to cooling (or heating) by a fluid at temperature T_f at the other side (Fig. 11.15). To predict the temperature profile of the slab as a function of time and distance, the unsteady-state heat conduction equation [which can be derived from Eq. (11.229)]

$$\frac{\partial \phi}{\partial t} = \alpha \frac{\partial^2 \phi}{\partial x^2} \tag{11.251}$$

has to be solved for the boundary and initial conditions

$$\frac{\partial \phi}{\partial x} = 0 \quad \text{at} \quad x = 0 \tag{11.252}$$

$$\left. \frac{\partial \phi}{\partial x} \right|_{x=L} = -\frac{h}{k} \phi \Big|_{x=L} \quad \text{at} \quad x = L \tag{11.253}$$

$$\phi = \phi_0 \quad \text{at} \quad t = 0 \tag{11.254}$$

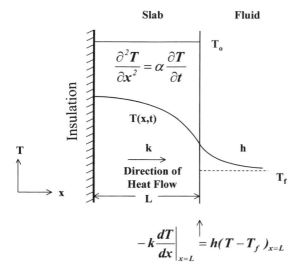

Figure 11.15 Unsteady-state heat transfer for a slab of thickness *L* initially at uniform temperature T_0 insulated from one side and exposed to cooling (or heating) by a fluid at temperature T_f at the other side.

where $\phi(= T - T_f)$ is the temperature difference, $\phi_0(= T_0 - T_f)$ is the initial temperature difference, h is the convective heat transfer coefficient on the surface exposed to the fluid, and $\alpha(= k/\rho c_p)$ and c_p are the thermal diffusion coefficient and the specific heat capacity (per unit mass) of the slab, respectively.

Using the method of separation of variables, the solution of Eq. (11.251) is given by

$$\phi = Ce^{-\lambda^2 \alpha t} \cos(\lambda x) \tag{11.255}$$

where C (integration constant) and λ will be obtained from the boundary conditions. Using the boundary condition of Eq. (11.253),

$$\left(\frac{\partial \phi}{\partial x}\right)_{x=L} = -Ce^{-\lambda^2 \alpha t} \lambda \sin(\lambda L) = -\frac{h}{k}\phi\Big|_{x=L} = -\frac{h}{k}Ce^{-\lambda^2 \alpha t} \cos(\lambda L) \tag{11.256}$$

or

$$\cot(\lambda L) = \frac{\lambda k}{h} \tag{11.257}$$

which gives more than one solution for λ. These solutions are called the eigenvalues of the linear boundary value problem (Fig. 11.16). As a result, the solution for the time dependent temperature profile [Eq. (11.255)] could be expressed as a linear combination of the solutions corresponding to each λ (calculated from Eq. (11.257)],

$$\phi = \sum_{n=1}^{\infty} C_n e^{-\lambda_n^2 \alpha t} \cos(\lambda_n x) \tag{11.258}$$

where C_n is the integration constant to be calculated corresponding to λ_n. Using the initial condition expressed by Eq. (11.254),

$$\phi_0 = \sum_{n=1}^{\infty} C_n \cos(\lambda_n x) \tag{11.259}$$

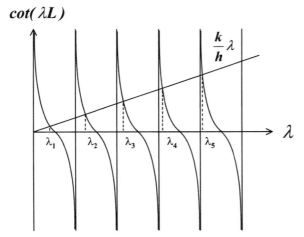

Figure 11.16   Calculated eigenvalues from the solution of Eq. (11.257) corresponding to the heat transfer problem depicted in Fig. 11.15.

Using the orthogonal property of the functions $\cos(\lambda_n x)$,

$$\int_0^L \cos(\lambda_n x)\,\cos(\lambda_m x)\,dx = 0 \qquad n \neq m \tag{11.260}$$

A_n can be calculated by multiplying both sides of Eq. (11.259) and integrating from 0 to L:

$$\int_0^L \phi_0 \cos(\lambda_n x)\,dx = \int_0^L C_n \cos^2(\lambda_n x)\,dx \tag{11.261}$$

or

$$C_n = \frac{2\phi_0 \sin(\lambda_n L)}{\lambda_n L + \sin(\lambda_n L)\cos(\lambda_n L)} \tag{11.262}$$

Using C_n from Eq. (11.262) in Eq. (11.258), the solution for the time-dependent temperature profile becomes

$$\phi(x, t) = \sum_{n=1}^{\infty} e^{-\lambda_n^2 \alpha t} \frac{2\phi_0 \sin(\lambda_n L)}{\lambda_n L + \sin(\lambda_n L)\cos(\lambda_n L)} \cos(\lambda_n x) \tag{11.263}$$

The heat loss dq for the slab in time interval dt, which is also time-dependent, is calculated using the Fourier conduction law:

$$\begin{aligned} dq(t) &= -kA \frac{d\phi(x, t)}{dx}\bigg|_{x=L} dt \\ &= \frac{2kA\phi_0}{L} \sum_{n=1}^{\infty} e^{-\lambda_n^2 \alpha t} \frac{\lambda_n L \sin^2(\lambda_n L)}{\lambda_n L + \sin(\lambda_n L)\cos(\lambda_n L)} dt \end{aligned} \tag{11.264}$$

where A is the surface area exposed to cooling and $\alpha t / L^2$ is the Fourier (dimensionless) number, which is the normalized time with respect to the characteristic time for thermal diffusion. Integrating Eq. (11.264) from $t = 0$ to $t = \tau$ gives the total heat lost up to time τ:

$$q(\tau) = \frac{2kA\phi_0}{L} \int_0^\tau \sum_{n=1}^{\infty} e^{-\lambda_n^2 \alpha t} \frac{\lambda_n L \sin^2(\lambda_n L)}{\lambda_n L + \sin(\lambda_n L)\cos(\lambda_n L)} dt \tag{11.265}$$

or

$$q(\tau) = \frac{2kAL\phi_0}{\alpha} \sum_{n=1}^{\infty} -e^{-\lambda_n^2 \alpha t} \frac{\lambda_n L \sin^2(\lambda_n L)}{(\lambda_n L)^2 [\lambda_n L + \sin(\lambda_n L)\cos(\lambda_n L)]}\bigg|_0^\tau \tag{11.266}$$

Using Eq. (11.266), the heat lost per unit surface area is given by

$$\frac{q(\tau)}{A} = \frac{2kL\phi_0}{\alpha} \sum_{n=1}^{\infty} \frac{\sin^2(\lambda_n L)(1 - e^{-\lambda_n^2 \alpha \tau})}{(\lambda_n L)^2 + \lambda_n L \sin(\lambda_n L)\cos(\lambda_n L)} \tag{11.267}$$

It is important that the solution of the time-dependent temperature profile is determined by the solution of Eq. (11.257), which can be rearranged as

$$(\lambda_n L)\tan(\lambda_n L) = \frac{hL}{k} \tag{11.268}$$

where the dimensionless group hL/k, which is the ratio of convective to conductive heat transfer rates, is called the Biot number. As deduced from Eq. (11.268), $\lambda_n L$ is some function of the Biot number:

$$\lambda_n L = \varphi\left(\frac{hL}{k}\right) \tag{11.269}$$

as a result, the solutions expressed by Eqs. (11.263), (11.266), and (11.267) can be expressed in terms of normalized temperature ϕ/ϕ_0, Biot number hL/k, Fourier number $\alpha t/L^2$, and normalized distance x/L (in cylindrical and spherical coordinate systems the normalized radius r/r_0). These solutions are presented in the form of graphs in standard textbooks and engineering handbooks, plotting the dimensionless temperature as a function of Fourier number (i.e., as a function of time), for selected x/L (or r/r_0) and Biot numbers, for linear, cylindrical, and spherical geometries (Perry and Green, 1984).

4.5 Convective Heat Transfer in Laminar Flow

In the laminar flow regime, the mechanism of convective heat transfer is similar to the mechanism of momentum transfer. When a solid body at temperature T_w is submerged in a flow at temperature T_f, a temperature field builds up in the surrounding medium that is called the thermal boundary layer. The thickness of the thermal boundary layer, δ_t, and of the hydrodynamic boundary layer, δ, defined in momentum transfer (velocity boundary layer or hydrodynamic boundary layer) are different. In the thermal boundary layer, heat conduction is equally as important as heat convection; outside the thermal boundary layer, heat conduction becomes too small (or negligible) in comparison to heat convection. The temperature profile in the thermal boundary layer is needed for calculating the forced convective heat transfer coefficient in laminar flow.

Energy transfer in the thermal boundary layer can be formulated using methods similar to those used to formulate fluid motion in the boundary layer. In the first method, the orders of magnitude of the terms in the general form of the thermal energy equation [Eq. (11.228)] are compared for its simplification. In the second method, an approximate heat flow equation in the thermal boundary layer is developed and solved (i.e., similar to von Karman's approximation made for laminar flow). All of these discussions are limited to the case $\delta_t < \delta$.

4.5.1 Integration of the Thermal Energy Equation for Forced Convection in Laminar Flow

Consider a steady-state thermal boundary layer developed on a flat plate stationed in an incompressible laminar flow (Fig. 11.17). It will be assumed that thermal and hydrodynamic boundary layers start developing at the same reference point (i.e., at the tip of the plate). Within the thermal boundary layer, the thermal energy equation [Eq. (11.228)] and the order of magnitude of each term can be expressed (provided that the thermal boundary layer thickness δ_t is smaller than the hydrodynamic boundary layer thickness δ) as

$$\rho c_p\left(v_x \frac{\partial T}{\partial x} + v_y \frac{\partial T}{\partial y}\right) = k\left(\frac{\partial^2 T}{\partial x^2} + \frac{\partial^2 T}{\partial y^2}\right) + \mu \Phi_v \tag{11.270}$$

$$\begin{array}{cccc} 1 \times 1/1 & \delta \times 1/\delta & \delta^2 & 1 & 1/\delta^2 \end{array}$$

where the viscous dissipation function Φ_v of Eq. (11.223) is simplified to (Schlichting, 1955)

$$\delta_t(x) < \delta(x)$$

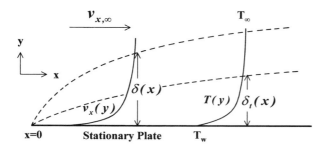

Figure 11.17 Steady-state thermal boundary layer developed on a flat plate stationed in an incompressible laminar flow.

$$\Phi_v = \left(\frac{\partial v_x}{\partial y}\right)^2 \tag{11.271}$$

which is negligibly small for laminar viscous flow. By comparing the orders of magnitude of the terms [see Eq. (11.270)], the thermal energy equation can be reduced to

$$v_x \frac{\partial T}{\partial x} + v_y \frac{\partial T}{\partial y} = \alpha \frac{\partial^2 T}{\partial y^2} \tag{11.272}$$

where $\alpha(= k/\rho c_p)$ is the thermal conductivity of the fluid. The boundary conditions for Eq. (11.272) are

$$T = T_w \quad \text{at} \quad y = 0; \qquad T = T_\infty \quad \text{and} \quad \frac{\partial T}{\partial y} = 0 \quad \text{at} \quad y = \delta_t \tag{11.273}$$

where T_w is the surface temperature of the object (i.e., thermal equilibrium is established between the solid surface and the fluid) and T_∞ is the temperature of the fluid outside the thermal boundary layer (the boundary condition would not be changed by taking the upper limit for y as δ_t or ∞). As can be seen from an examination of Eq. (11.270), the thermal conductivity k of the fluid has to be two orders of magnitude smaller than the velocity or temperature to make both sides of the thermal energy equation comparable.

In Eq. (11.272), v_x and v_y are known from the simultaneous solution of the equation of continuity,

$$\frac{\partial v_x}{\partial x} + \frac{\partial v_y}{\partial y} = 0 \tag{11.274}$$

and the equation of motion,

$$v_x \frac{\partial v_x}{\partial x} + v_y \frac{\partial v_x}{\partial y} = \gamma \frac{\partial^2 v_x}{\partial y^2} \tag{11.275}$$

Assuming that v_x and v_y are known, the thermal energy equation becomes linear in temperature. Solution of the thermal energy equation would give better results as more realistic flow velocities are introduced into the equation. The form of Eq. (11.272) (without the viscous dissipation term) is similar to the equation of motion (without the pressure term) for a laminar boundary layer. Furthermore, when the kinematic viscosity $\gamma(=\mu/\rho)$ and thermal diffusivity $\alpha(=k/\rho c_p)$ are of the same order of magnitude, the equation of motion and the thermal energy equation become identical.

The solution of the thermal energy equation will be discussed for the case where hydrodynamic and thermal boundary layer thicknesses are of the same order of magnitude. This means that the heat transfer Prandtl number $(\text{Pr} = \mu c_p/k = \rho\mu c_p/k\rho = \gamma/\alpha,$ which is also defined as the Schmidt number for heat transfer) is of the order of magnitude of 1 (or larger), which is the case for most liquids.

At low flow velocities and at $y = 0$, the steady-state thermal energy equation [Eq. (11.272)] becomes

$$\frac{\partial^2 T}{\partial y^2} = 0 \quad \text{at} \quad y = 0 \tag{11.276}$$

Combining this condition with the first three conditions [Eq. (11.273) makes four conditions] allows one to suggest a temperature profile in the form of a third-order polynomial (i.e., one with four coefficients):

$$T = a_0 + a_1 y + a_2 y^2 + a_3 y^3 \tag{11.277}$$

For practical reasons, Eq. (11.272) will be expressed in terms of dimensionless temperature ψ:

$$\Psi = \frac{T - T_\infty}{T_w - T_\infty} \tag{11.278}$$

using which, Eqs. (11.272), (11.273), and (11.276) can be expressed as

$$v_x \frac{\partial\Psi}{\partial x} + v_y \frac{\partial\Psi}{\partial y} = \alpha \frac{\partial^2\Psi}{\partial y^2} \tag{11.279}$$

$$\Psi = 1 \quad \text{at} \quad y = 0; \qquad \Psi = 0 \quad \text{and} \quad \frac{\partial\Psi}{\partial y} = 0 \quad \text{at} \quad y = \delta_t \tag{11.280}$$

$$\frac{\partial^2\Psi}{\partial y} = 0 \quad \text{at} \quad y = 0 \tag{11.281}$$

which gives the dimensionless temperature profile in the form

$$\Psi = \frac{T - T_\infty}{T_w - T_\infty} = 1 - \frac{3}{2}\left(\frac{y}{\delta_t}\right) + \frac{1}{2}\left(\frac{y}{\delta_t}\right)^3 \tag{11.282}$$

Using Eq. (11.282), the rate of convective heat transfer is calculated as

$$q = -k\left(\frac{\partial T}{\partial y}\right)_w = h(T_w - T_\infty) \tag{11.283}$$

or

$$h = \frac{3}{2}\left(\frac{k}{\delta_t}\right) = \frac{3}{2}\left(\frac{k}{\varsigma\delta}\right) \tag{11.284}$$

where $\zeta(= \delta_t/\delta)$ is the ratio of the thickness of the thermal boundary layer δ_t to the thickness of the hydrodynamic boundary layer δ. As can be seen from Eq. (11.284), the convective heat transfer coefficient is inversely proportional to the thickness of the thermal boundary layer. Further, the calculated value of ζ is given as (Eckert and Drake, 1959)

$$\zeta = \frac{\delta_t}{\delta} = \frac{1}{1.026(\text{Pr})^{1/3}} \tag{11.285}$$

Using the classical predictions for the hydrodynamic boundary layer thickness [i.e., $\delta = 4.64(\gamma x/v_\infty)^{1/2}$], the convective heat transfer coefficient is calculated as

$$h = 0.332k \, \text{Pr}^{1/3} \left(\frac{v_\infty}{\gamma x}\right)^{1/2} \tag{11.286}$$

or

$$\frac{hx}{k} = \text{Nu} = 0.332 \, \text{Pr}^{1/3} \left(\frac{v_\infty x}{\gamma}\right)^{1/2} = 0.332 \, \text{Pr}^{1/3} \, \text{Re}_x^{1/2} \tag{11.287}$$

where Nu is the Nusselt number and Re_x is the Reynolds number for the distance x from the edge of the thermal boundary layer.

For many viscous fluids, the Prandtl number is on the order of 1000. For these fluids the thickness of the thermal boundary layer δ_t is [by Eq. (11.285)] one-tenth of the thickness of the hydrodynamic boundary layer. For gases Pr is about 1; then ζ would be of the same order of magnitude. Therefore, for all of these materials the forced convective heat transfer calculations in laminar flow regimes should give acceptable results. For molten metals the Prandtl number is very small; therefore, the above results are not valid for molten metals. When Pr is very small (i.e., k is very large), the heat transfer mechanism is dominated by thermal conduction, and the convective heat transfer mechanism does not play a major role (i.e., the assumption that k is on the order of δ^2 is not valid in molten metals).

Solutions of the energy equation coupled with continuity and the equation of motion are also known for flows in different geometries. These solutions (exact or approximate) are presented as correlations of Nu versus Re_x Pr (x/L), or Re_D Pr (x/D) for the flow in cylindrical geometry, where D is the diameter of the tube (i.e., $D = 2r_0$). These solutions are used to correlate laminar flow convective heat transfer data and are published in classical textbooks as well as engineering handbooks (Perry and Green, 1984).

4.5.2 Integration of Thermal Energy Equation for Forced Convection in Turbulent Flow

Similar to momentum transfer, in high Reynolds number flows the transfer of heat is dominated by turbulent fluctuation velocities (similar to Reynolds stresses). Heat transfer by turbulent mixing could be much greater than by both conductive and convective heat transfer. The mechanism of turbulent mixing is not understood; however, turbulence measurements and phenomenological theories developed for momentum transfer can be useful for the modeling of turbulent heat transfer.

A fluid flow over a flat plate each at different temperatures (Fig. 11.18) will be considered again. As the Reynolds number of the flow increases, a thin viscous zone is formed in the neighborhood of the solid, in which heat is transferred mainly by conduction. In the main turbulent core, heat is transferred by turbulent eddy mass fluctuations. A buffer zone is established between the two extreme zones, in which combinations of the two mechanisms transfer heat. The rate of heat transfer with turbulent eddies is faster than

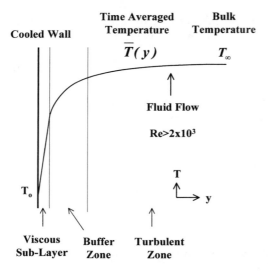

Figure 11.18 Heat transfer mechanism close to a flat plate (wall) placed in a high Reynolds number flow regime, the viscous sublayer, and buffer and turbulent transfer zones.

the rate of heat transfer in the buffer and viscous sublayer zones, and as a result a large temperature drop would take place through the thin viscous sublayer.

In the turbulent core, as a result of eddy activities, the temperature at a given set of coordinates fluctuates over time. As discussed earlier, in turbulent flow time-dependent temperature can be expressed as (Bird et al., 1960)

$$T(t) = \overline{T} + \gamma_T(t) \tag{11.288}$$

where $\overline{T}$ is the time-averaged temperature and $\gamma_T(t)$ is the temperature fluctuation, which may be as high as 10% of the time-averaged temperature. The time average of the fluctuation temperature is zero, as defined earlier; however, the time average of the product of fluctuation velocity and fluctuation temperature would not be zero:

$$\overline{\gamma_T(t)} = 0, \qquad \overline{\gamma_{v_x}(t)\gamma_T(t)} \neq 0, \qquad \overline{\gamma_{v_y}(t)\gamma_T(t)} \neq 0, \qquad \overline{\gamma_{v_z}(t)\gamma_T(t)} \neq 0 \tag{11.289}$$

The thermal energy equation [Eq. (11.223)] expressed in terms of turbulent velocity $\mathbf{v}(t) = \bar{v} + \gamma_v(t)$ and tempeature $T(t) = \overline{T} + \gamma_T(t)$ and then averaged over time (as is done for the equation of motion) yields

$$\rho c_p \left(\frac{\partial(\overline{v_x}\overline{T})}{\partial x} + \frac{\partial(\overline{v_y}\overline{T})}{\partial y} + \frac{\partial(\overline{v_z}\overline{T})}{\partial z} \right) = k \left(\frac{\partial^2 \overline{T}}{\partial x^2} + \frac{\partial^2 \overline{T}}{\partial y^2} + \frac{\partial^2 \overline{T}}{\partial z^2} \right)$$

$$- \rho c_P \left(\frac{\partial \overline{\gamma_{v_x}(t)\gamma_T(t)}}{\partial x} + \frac{\partial \overline{\gamma_{v_y}(t)\gamma_T(t)}}{\partial y} + \frac{\partial \overline{\gamma_{v_z}(t)\gamma_T(t)}}{\partial z} \right)$$

$$+ \mu\Phi_v + \mu\Phi_v^{(t)} \tag{11.290}$$

where the second term on the right-hand side is the divergence of the turbulent energy flux $\mathbf{q}^{(t)}$, i.e.,

$$q_x^{(t)} = \rho c_p \overline{\gamma_{v_x}(t) \gamma_T(t)}, \qquad q_y^{(t)} = \rho c_p \overline{\gamma_{v_y}(t) \gamma_T(t)}, \qquad q_z^{(t)} = \rho c_p \overline{\gamma_{v_z}(t) \gamma_T(t)} \tag{11.291}$$

Thermal energy generated as a result of viscous dissipation in laminar flow [Eq. (11.223)] is small in comparison to the turbulent dissipation, which is

$$\Phi_v' = \mu \sum_{i=1}^{3} \sum_{j=1}^{3} \left(\overline{\frac{\partial \gamma_{vi}(t)}{\partial x_i} \frac{\partial \gamma_{vi}(t)}{\partial x_i}} + \overline{\frac{\partial \gamma_{vi}(t)}{\partial x_j} \frac{\partial \gamma_{vj}(t)}{\partial x_i}} \right) \tag{11.292}$$

For most engineering applications the turbulent viscous dissipation is also negligible and can be dropped.

Turbulent energy fluxes [Eqs. (11.290) and (11.291)] are not the property of the fluid itself, they are the property of the flow system and could strongly depend on the geometry of the flow system. Semiempirical relations are suggested to correlate turbulent heat flux in terms of time-averaged characteristics of the flow.

The eddy thermal conductivity is suggested to correlate the turbulent heat flux as a function of time-averaged temperature as

$$q_y^{(t)} = -k^{(t)} \frac{d\overline{T}}{dy} \tag{11.293}$$

which is in analogy with the Fourier law for heat conduction. The linearity coefficient $k^{(t)}$ is called the turbulent eddy conductivity. The turbulent eddy kinematic viscosity,

$$\gamma^{(t)} = \frac{\mu^{(t)}}{\rho} \tag{11.294}$$

and the turbulent eddy thermal diffusivity,

$$\alpha^{(t)} = \frac{k^{(t)}}{\rho c_p} \tag{11.295}$$

have the same dimensions and meanings. Experimental data show that the ratio of turbulent eddy kinematic viscosity to eddy thermal diffusion $\gamma^{(t)}/\alpha^{(t)}$ is in the range of $0.5 < \gamma^{(t)}/\alpha^{(t)} < 1$.

Prandtl's mixing length theory should also be applied to turbulent eddy thermal conductivity, because the mechanisms of momentum and heat transfer are the same. Using Prandtl's mixing length theory, which assumes that $\gamma^{(t)}/\alpha^{(t)} = 1$, the turbulent heat flux is correlated by

$$q^{(t)} = -\rho c_p l^2 \left| \frac{d\overline{v}_x}{dy} \right| \frac{d\overline{T}}{dy} \tag{11.296}$$

where l is the Prandtl mixing length, which is dependent on the distance y from the solid surface, i.e., $l = k_1 y$.

Von Karman's similarity hypothesis, which assumes that $\gamma^{(t)}/\alpha^{(t)} = 1$, correlates the turbulent heat flux by

$$q^{(t)} = -\rho c_p k_2^2 \left| \frac{(d\overline{v}_x/dy)^3}{(d^2\overline{v}_x/dy^2)^2} \right| \frac{d\overline{T}}{dy} \tag{11.297}$$

where k_2 is a constant.

Deissler suggested that close to solid walls, turbulent heat flux is correlated by

$$q^{(t)} = -\rho c_p n^2 \bar{v}_x y \left[1 - \exp\left(-n^2 \frac{\bar{v}_x y}{\gamma} \right) \right] \frac{d\bar{T}}{dy} \qquad (11.298)$$

where n is an empirical constant (0.124 for tube flow) and y is the distance from the wall.

For most engineering applications, it is much easier to use empirical relations to predict turbulent heat flux. These empirical correlations are developed based on dimensional analysis of the equation of motion and the equation of thermal energy. Dimensional analysis suggests that heat flux would be expressed in the form of one dimensionless group as a function of other dimensionless groups. As an example, the Nusselt number (Nu) can be correlated as a function of Reynolds number (Re), Prandtl number (Pr), and dimensionless distance L/D (L is the tube length, D is the tube diameter) or x/L (x is the distance from a reference point, L is the length of an object):

$$\text{Nu} = f(\text{Re}, \text{Pr}, L/D) \qquad (11.299)$$

For highly turbulent, incompressible tube flow, all turbulent heat flux correlations for different values of L/D converge to a single correlation in the form

$$\text{Nu} = 0.026 \, \text{Re}^{0.8} \text{Pr}^{1/3} \left(\frac{\mu_b}{\mu_0} \right)^{0.14} \qquad (11.300)$$

where $(\mu_b/\mu_0)^{0.14}$ is another dimensionless correction factor for the viscosity of the fluid at an average bulk fluid temperature and at an average wall temperature, and Re and Pr are evaluated using the viscosity at an average bulk fluid temperature. Similar correlations are developed for different flow geometries and are published in engineering handbooks (Perry and Green, 1984).

Convective heat transfer can also take place as a result of fluid motion due to the density differences upon heating or cooling and is then called natural convective heat transfer. Analysis of natural convective heat transfer, similar to forced convective heat transfer, showed that the Nusselt number is a function of Grashof (Gr) and Prandtl (Pr) numbers (k is the thermal conductivity of the fluid),

$$\text{Nu}_x = \frac{hx}{k} = f(\text{Gr}, \text{Pr}) \qquad (11.301)$$

where Gr is

$$\text{Gr}_x = \frac{x^3 g \beta (T_w - T_\infty)}{\gamma^2} \qquad (11.302)$$

x is the characteristic distance, g is the gravitational acceleration, β [$= (1/V)(\partial V/\partial T)_p$] is the thermal expansion coefficient, T_w is the surface temperature of the object, T_∞ is the ambient fluid temperature, and γ is the kinematic viscosity of the fluid.

The role of the Grashof number in natural convective heat transfer and the role of the Reynolds number in forced convective heat transfer are similar. For example, the natural convective heat transfer coefficient from a vertical plate to a fluid is correlated by (Eckert and Drake, 1959)

$$\text{Nu}_x = 0.508 \text{Pr}^{1/2}(0.952 + \text{Pr})^{-1/4} \text{Gr}_x^{1/4} \qquad (11.303)$$

or for air (Pr = 0.7) as the fluid by

$$\text{Nu}_x = 0.378 \, \text{Gr}_x^{1/4} \qquad (11.304)$$

The natural convective heat transfer coefficient from a single sphere to a fluid, for $Gr^{1/4}Pr^{1/3} < 200$, is correlated by (Bird et al., 1960)

$$Nu = \frac{hD}{k} = 2.0 + 0.60 \ Gr^{1/4} \ Pr^{1/3} \tag{11.305}$$

and for from a long horizontal cylinder to fluid, for $GrPr > 10^4$, by

$$Nu = \frac{hD}{k} = 0.525(Gr \ Pr)^{1/4} \tag{11.306}$$

For a very large range of GrPr values, $GrPr > 10^9$, heat transfer takes place by the turbulent heat transfer mechanism.

The heat transfer mechanism in boiling and condensation is related to formation of the second phase (gas phase in bubbling and liquid phase in condensation). The heat transfer mechanism may be associated with phase change (moving boundaries, such as solidification). These heat transfer mechanisms can be studied by applying the principles presented in this section.

The basic theory of heat transfer has been summarized here, but many important heat transfer mechanisms have not been discussed at all. An understanding of the basics of transport phenomena is essential to an understanding of the heat transfer mechanism in any system. From a practical engineering point of view, the expected heat transfer coefficients for different fluids and different heat transfer processes are necessary for consideration (Table 11.8) (Bird et al., 1960).

5. MASS TRANSFER

An understanding of mass transfer coupled with momentum and heat transfer is essential to understand many industrially important processes and to solve process problems in chemical, energy, mineral, and environmental technologies. Heat and mass transfer coupled with chemical reactions can explain many industrially important phenomena such as combustion, ignition, explosion, catalytic reactions, and purification.

Table 11.8 Expected Heat Transfer Coefficients for Different Fluids and Different Heat Transfer Systems

Heat transfer mechanism	h $[kJ/(m^2 \cdot K)]$	h $[Btu/(ft^2 \cdot h \cdot {}^{\circ}F)]$
Free convection		
Gases	10^{-3}–10^{-2}	1–4
Liquids	0.1–1.0	20–120
Boiling water	1–10	10^2–10^3
Forced convection		
Gases	10^{-2}–10^{-1}	2–20
Viscous fluid	0.1–1.0	10–10^2
Water	1–10	10^2–10^3
Condensing vapors	1–10^2	10^2–10^4

Source: Bird et al., 1960.

In this section, mass balance (also known as the equation of continuity) will be formulated and its solution for specific applications will be discussed. In formulating mass (molar) balance, the principles of property balance [Eq. (11.8)] will also be used, by assigning the scalar property ψ as the mass (or mole) density and the vector $\mathbf{\Psi}$ as the mass (or molar) flux. In this effort, no chemical reaction is assumed to take place; therefore, the mass (molar) generation term ψ_g is set to zero. If needed, the chemical reaction term ψ_g will be introduced into the system, as demonstrated in Chapter 9. Property ψ and its flux $\mathbf{\Psi}$ will be defined on a mass basis as well as on a molar basis, because each derivation may have specific advantages in engineering applications. Several classical books can be referred to for further discussion of mass transfer with and without chemical reaction (Astarita, 1967; Bird et al., 1960; Danckwerts, 1970; Frank-Kamenestkii, 1969; Levich; 1962; Treybal, 1980).

For the derivation of the mass balance equation similar to momentum and thermal energy balances, the definition of mass flux has to be understood. Mass flux $\mathbf{n}_j$ and molar flux $\mathbf{N}_j$ with respect to a fixed reference frame can be defined based on mass average and molar average velocities [the nomenclature of Bird et al. (1960) will be used as much as possible]:

$$\mathbf{n}_j = v_j \rho_j \tag{11.307}$$

$$\mathbf{N}_j = v_j c_j \tag{11.308}$$

where ρ_j is the mass density and v_j is the velocity of the jth species of the system. Mass average velocity v and molar average velocity v^* are defined as

$$\mathbf{v} = \frac{\sum_j v_j \rho_j}{\sum_j \rho_j} = \frac{\sum_j v_j \rho_j}{\rho} \tag{11.309}$$

$$\mathbf{v}^* = \frac{\sum_j v_j c_j}{\sum_j c_j} = \frac{\sum_j v_j c_j}{c} \tag{11.310}$$

Mass and molar diffusion fluxes are defined based on a reference frame moving with mass average velocity or molar average velocity,

$$\mathbf{j}_j = \rho_j (v_j - v) \tag{11.311}$$

$$\mathbf{J}_j = c_j (v_j - v) \tag{11.312}$$

and

$$\mathbf{j}_j^* = \rho_j (v_j - v^*) \tag{11.313}$$

$$\mathbf{J}_j^* = c_j (v_j - v^*) \tag{11.314}$$

Diffusion fluxes expressed on the basis of molar average velocity v^* [Eqs. (11.313) and (11.314)] can be expressed in terms of the gradients of mass density and molar density ($\rho_j = g_j \rho$ and $c_j = x_j c$) as

$$\mathbf{j}_j^* = \rho_j (\mathbf{v}_j - \mathbf{v}^*) = -D_j \nabla \rho_j = -\rho D_j \nabla g_j \tag{11.315}$$

$$\mathbf{J}_j^* = c_j (\mathbf{v}_j - \mathbf{v}^*) = -D_j \nabla c_j = -c D_j \nabla x_j \tag{11.316}$$

provided that mass and molar fluxes are linearly proportional to the gradients of mass and molar concentrations. The relations given by Eqs. (11.315) and (11.316) are known as

Fick's first law of diffusion. Temperature and pressure gradients and external forces can also cause diffusion fluxes; however, in most engineering applications these fluxes are negligible in comparison to the diffusion flux as a result of the molar concentration gradient.

The total molar flux $(\sum_j J_j^*)$ is zero for all multicomponent systems [Eq. (11.314)]:

$$\sum_j J_j^* = \sum_j c_j(v_j - v^*) = \sum_j c_j v_j - v^* \sum_j c_j = cv^* - cv^* = 0 \tag{11.317}$$

For a binary system of species A and B, Eq. (11.317) shows that $J_A^* = -J_B^*$. Furthermore, the relation between the J_A^* and N_A can be derived [using Eqs. (11.308) and (11.310)] as

$$J_A^* = c_A(v_A - v^*) = c_A v_A - \frac{c_A}{c}(N_A + N_B) = N_A - x_A(N_A + N_B) \tag{11.318}$$

which shows that the molar flux of species A referenced to a frame moving with molar average velocity v^* is the difference between the molar flux N_A (referenced to a fixed frame) and the molar flux caused by the molar bulk flow (also referenced to a fixed frame). Similarly, the relation between j_A and n_A can be developed [using Eqs. (11.307) and (11.309)] as

$$j_A = \rho_A(v_A - v) = \rho_A v_A - \frac{\rho_A}{\rho}(\rho_A v_A + \rho_B v_B) = n_A - g_A(n_A + n_B) \tag{11.319}$$

which shows that the mass flux of species A referenced to a frame moving with mass average velocity v is the difference between the molar flux n_A (referenced to a fixed frame) and the mass flux caused by the mass bulk flow (also referenced to a fixed frame).

5.1 Equation of Changes in Multicomponent Systems

In a chemically reactive multicomponent system the property ψ, the property flux vector $\boldsymbol{\psi}$ and property generation ψ_g [Eq. (11.8)] for the jth species are represented by the mass density (ρ_j), mass flux vector n_j (with respect to a fixed reference frame), and mass generation rate r_j per unit volume. For a binary system $(j = A$ and $B)$, using Eq. (11.8), the mass balance equation becomes (Bird et al., 1960)

$$\frac{\partial \rho_j}{\partial t} + \nabla \cdot n_j = r_j \tag{11.320}$$

Taking the summation of both sides of Eq. (11.320) over j [since $n_A + n_B = \rho v$, using Eqs. (11.307) and (11.309) and $r_A + r_B = 0$, i.e., total mass is conserved] gives

$$\frac{\partial \rho}{\partial t} + \nabla \cdot (\rho v) = \frac{\partial \rho}{\partial t} + \rho(\nabla \cdot v) + v \cdot (\nabla \rho) = 0 \tag{11.321}$$

which is known as the equation of continuity. Equations of continuity in terms of mass density ρ_A of species A and total fluid mass density ρ in rectangular, cylindrical, and spherical coordinate systems are presented in Table 11.9.

For constant density fluids (i.e., incompressible fluids), Eq. (11.321) becomes

$$\nabla \cdot v = 0 \tag{11.322}$$

For a binary system, using Eq. (11.8), a similar formulation can be made for the molar balances, using molar concentrations, molar fluxes, and molar generation $(j = A, B)$:

Table 11.9 Equation of Continuity in Terms of Mass Density ρ_A of Species A and Total Fluid Mass Density ρ in Rectangular, Cylindrical, and Spherical Coordinate Systems

In terms of species A mass density ρ_A

In a rectangular coordinate system of x, y, and z:

$$\frac{\partial \rho_A}{\partial t} + \frac{\partial(\rho_A v_x)}{\partial x} + \frac{\partial(\rho_A v_y)}{\partial y} + \frac{\partial(\rho_A v_z)}{\partial z} = r_A$$

In a cylindrical coordinate system of r, θ, and z:

$$\frac{\partial \rho_A}{\partial t} + \frac{1}{r}\frac{\partial(\rho r_A v_r)}{\partial r} + \frac{1}{r}\frac{\partial(\rho_A v_\theta)}{\partial \theta} + \frac{\partial(\rho_A v_z)}{\partial z} = r_A$$

In a spherical coordinate system of r, θ, and ϕ:

$$\frac{\partial \rho_A}{\partial t} + \frac{1}{r^2}\frac{\partial}{\partial r}(\rho_A r^2 v_r) + \frac{1}{r \sin\theta}\frac{\partial(\rho_A v_\theta \sin\theta)}{\partial \theta} + \frac{1}{r \sin\theta}\frac{\partial(\rho_A v_\phi)}{\partial \phi} = r_A$$

In terms of total fluid mass density ρ
In a rectangular coordinate system of x, y, and z:

$$\frac{\partial \rho}{\partial t} + \frac{\partial(\rho v_x)}{\partial x} + \frac{\partial(\rho v_y)}{\partial y} + \frac{\partial(\rho v_z)}{\partial z} = 0$$

In a cylindrical coordinate system of r, θ, and z:

$$\frac{\partial \rho}{\partial t} + \frac{1}{r}\frac{\partial(\rho r v_r)}{\partial r} + \frac{1}{r}\frac{\partial(\rho v_\theta)}{\partial \theta} + \frac{\partial(\rho v_z)}{\partial z} = 0$$

In a spherical coordinate system of r, θ, and ϕ:

$$\frac{\partial \rho}{\partial t} + \frac{1}{r^2}\frac{\partial}{\partial r}(\rho r^2 v_r) + \frac{1}{r \sin\theta}\frac{\partial(\rho v_\theta \sin\theta)}{\partial \theta} + \frac{1}{r \sin\theta}\frac{\partial(\rho v_\phi)}{\partial \phi} = 0$$

where r_A is the mass generation rate of species A per unit volume.

Source: Bird et al., 1960.

$$\frac{\partial c_j}{\partial t} + \nabla \cdot N_j = R_j \tag{11.323}$$

Taking the summation of both sides of Eq. (11.323) over j [because $N_A + N_B = cv^*$, and using Eqs. (11.308) and (11.310)],

$$\frac{\partial c}{\partial t} + \nabla \cdot (cv^*) = R_A + R_B \tag{11.324}$$

which is the equation of continuity on a molar basis. Unlike total mass, total moles may not be conserved, therefore, $R_A + R_B$ may not be equal to zero. For a fluid of constant molar density c (total molar concentration is constant, $\partial c/\partial t = 0$), Eq. (11.324) becomes

$$\nabla \cdot \mathbf{v}^* = \frac{1}{c}(R_A + R_B) \tag{11.325}$$

In a reaction system, if the total moles produced as a result of the chemical reaction is not equal to zero, there may be a volume change, therefore a convection current in the reacting system. This convection current is added to the diffusion flux, affecting the diffusion rate as well as the heat transfer rate. These phenomena are known as Stephan flow.

Mass $\mathbf{n}_j$ and molar $\mathbf{N}_j$ fluxes expressed in mass and molar balance equations [Eqs. (11.313) and (11.316) can be expressed in terms of mass and molar concentration gradients and diffusion coefficients. For a binary system composed of A and B species, Eqs. (11.320) and (11.323) become

$$\frac{\partial \rho_A}{\partial t} + \nabla \cdot (\rho_A \mathbf{v}) = \nabla \cdot (\rho D_{AB} \nabla g_A) + r_A \tag{11.326}$$

$$\frac{\partial c_A}{\partial t} + \nabla \cdot (c_A \mathbf{v}^*) = \nabla \cdot (c D_{AB} \nabla x_A) + R_A \tag{11.327}$$

Equations of continuity in terms of molar fluxes and concentration gradients for constant ρ and D_{AB} in rectangular, cylindrical, and spherical coordinate systems are listed in Tables 11.10 and 11.11 (Bird et al., 1960).

Mass and molar balance equations [Eqs. (11.326) and (11.327)] can be derived from the general property balance relation, Eq. (11.8). If the property balance [in the form of Eq. (11.8)] is used, then $\mathbf{v}$ will be the mass average ($\mathbf{v}$) or molar average ($\mathbf{v}^*$) velocity. Further, ψ will be the mass ($\mathbf{j}_A^*$) or molar ($\mathbf{J}_A^*$) diffusion flux vector (with respect to a reference frame moving at velocity $\mathbf{v}$ or $\mathbf{v}^*$) and ψ_g will be the rate of generation of the chemical species A or B per unit volume on a mass (r_A) or molar (R_A) basis. If the coordinate system is kept fixed, then ψ will be the mass flux ($\mathbf{n}_A$) or molar flux ($\mathbf{N}_A$).

For a binary system of constant mass density ρ (i.e., $\nabla \cdot \mathbf{v} = 0$) and diffusivity D, Eq. (11.326) becomes

Table 11.10 Equation of Continuity of Species A in Terms of Molar Fluxes in Rectangular, Cylindrical, and Spherical Coordinate Systems

In a rectangular coordinate system of x, y, and z:

$$\frac{\partial c_A}{\partial t} + \left(\frac{\partial N_{Ax}}{\partial x} + \frac{\partial N_{Ay}}{\partial y} + \frac{\partial N_{Az}}{\partial z} \right) = R_A$$

In a cylindrical coordinate system of r, θ, and z:

$$\frac{\partial c_A}{\partial t} + \frac{1}{r}\frac{\partial}{\partial r}(r N_{Ar}) + \frac{1}{r}\frac{\partial N_{A\theta}}{\partial \theta} + \frac{\partial N_{Az}}{\partial z} = R_A$$

In a spherical coordinate system of r, θ, and ϕ:

$$\frac{\partial c_A}{\partial t} + \frac{1}{r^2}\frac{\partial}{\partial r}(r^2 N_{Ar}) + \frac{1}{r \sin \theta}\frac{\partial (N_{A\theta} \sin \theta)}{\partial \theta} + \frac{1}{r \sin \theta}\frac{\partial N_{A\phi}}{\partial \phi} = R_A$$

Source: Bird et al., 1960.

Table 11.11 Equation of Continuity of Species A in Terms of Concentration Gradients for Constant ρ and D_{AB} in Rectangular, Cylindrical, and Spherical Coordinate Systems

In a rectangular coordinate system of x, y, and z:

$$\frac{\partial c_A}{\partial t} + v_x \frac{\partial c_A}{\partial x} + v_y \frac{\partial c_A}{\partial y} + v_z \frac{\partial c_A}{\partial z} = D_{AB} \left(\frac{\partial^2 c_A}{\partial x^2} + \frac{\partial^2 c_A}{\partial y^2} + \frac{\partial^2 c_A}{\partial z^2} \right) + R_A$$

In a cylindrical coordinate system of r, θ, and z:

$$\frac{\partial c_A}{\partial t} + v_r \frac{\partial c_A}{\partial r} + \frac{v_\theta}{r} \frac{\partial c_A}{\partial \theta} + v_z \frac{\partial c_A}{\partial z} = D_{AB} \left[\frac{1}{r} \frac{\partial}{\partial r} \left(r \frac{\partial c_A}{\partial r} \right) + \frac{1}{r^2} \frac{\partial^2 c_A}{\partial \theta^2} + \frac{\partial^2 c_A}{\partial z^2} \right] + R_A$$

In a spherical coordinate system of r, θ, and ϕ:

$$\frac{\partial c_A}{\partial t} + v_r \frac{\partial c_A}{\partial r} + \frac{v_\theta}{r} \frac{\partial c_A}{\partial \theta} + \frac{v_\phi}{r \sin \theta} \frac{\partial c_A}{\partial \phi} =$$

$$D_{AB} \left[\frac{1}{r^2} \frac{\partial}{\partial r} \left(r^2 \frac{\partial c_A}{\partial r} \right) + \frac{1}{r^2 \sin \theta} \frac{\partial}{\partial \theta} \left(\sin \theta \frac{\partial c_A}{\partial \theta} \right) + \frac{1}{r^2 \sin^2 \theta} \frac{\partial^2 c_A}{\partial \phi^2} \right] + R_A$$

Source: Bird et al., 1960.

$$\frac{\partial \rho_A}{\partial t} + \boldsymbol{v} \cdot \nabla \rho_A = D_{AB} \nabla^2 \rho_A + r_A \tag{11.328}$$

or, dividing both sides by the molecular weight M_A of species A,

$$\frac{\partial c_A}{\partial t} + \boldsymbol{v} \cdot \nabla c_A = D_{AB} \nabla^2 c_A + R_A \tag{11.329}$$

which is used for dilute solutions at constant temperature and pressure.

Similarly, for a system of constant molar concentration and diffusivity, Eq. (11.327) becomes

$$\frac{\partial c_A}{\partial t} + c_A \nabla \cdot \boldsymbol{v}^* + \boldsymbol{v}^* \cdot \nabla c_A = D_{AB} \nabla^2 c_A + R_A \tag{11.330}$$

or, using Eq. (11.325) for a fluid of constant molar density c,

$$\frac{\partial c_A}{\partial t} + \boldsymbol{v}^* \cdot \nabla c_A = D_{AB} \nabla^2 c_A + R_A - x_A(R_A + R_B) \tag{11.331}$$

which is generally used for low density gases at constant temperature and pressure.

If no chemical reaction takes place and average mass velocity $\boldsymbol{v}$ is zero, then Eq. (11.329) becomes

$$\frac{\partial c_A}{\partial t} = D_{AB} \nabla^2 c_A \tag{11.332}$$

which is known as Fick's second law of diffusion and is applicable for diffusion in stationary media (such as solids and quiescent liquids) and for equimolar counterdiffusion [i.e., $v^* = 0$ in Eq. (11.331)] in gases.

5.2   Similarity in Mass, Momentum, and Energy Transfer

There is a similarity between the mass, momentum, and heat transfer (except for heat transfer by radiation). For example, one-dimensional diffusion transport of mass, momentum, and thermal energy has already been expressed (Fick's, Newton's, and Fourier's laws) in terms of the products of gradient of mass concentration, momentum, and temperature and the proportionality constants,

$$j_A = -D_{AB} \frac{d\rho_A}{dy} \tag{11.333}$$

$$\tau_{yx} = -\nu \frac{d(\rho v_x)}{dy} \tag{11.334}$$

$$q_y = -\alpha \frac{d(\rho c_p T)}{dy} \tag{11.335}$$

where the proportionality constants are the mass diffusion coefficient, kinematic viscosity, and thermal diffusion coefficient, and all are in units of cm^2/s. These equations state that the diffusion flux of mass [$\text{g}/(\text{cm}^2 \cdot \text{s})$], momentum ($\text{dyn}/\text{cm}^2$ or N/m^2, which is Pa), and thermal energy [$\text{kcal}/(\text{m}^2 \cdot \text{s})$ or $\text{kJ}/(\text{m}^2 \cdot \text{s})$, which is kW/m^2] take place in proportion to the first power (linear) of the gradients of mass concentration, momentum, and thermal energy, respectively. It can be seen that mass and thermal energy balances form three sets of equations (mass and thermal energy are scalar, but their fluxes are vector quantities), whereas momentum balance forms nine equations (momentum is a vector, and its transfer is a tensor quantity).

The forms of thermal energy and mass balance equations are similar. Assuming that no chemical reaction is taking place, unsteady-state diffusion and steady-state diffusion in quiescent liquids and solids and convective flow thermal energy balances are

$$\frac{\partial T}{\partial t} = \alpha \, \nabla^2 T \tag{11.336}$$

$$\nabla^2 T = 0 \tag{11.337}$$

$$\frac{\partial T}{\partial t} + v \cdot \nabla T = \alpha \, \nabla^2 T \tag{11.338}$$

and the corresponding equations for the mass balance for the jth chemical species are

$$\frac{\partial c_j}{\partial t} = D\nabla^2 c_j \tag{11.339}$$

$$\nabla^2 c_j = 0 \tag{11.340}$$

$$\frac{\partial c_j}{\partial t} + v \cdot \nabla c_j = D\nabla^2 c_j \tag{11.341}$$

If a solution is known for a given geometry and given boundary (and initial) conditions for a specific type of problem, then this solution can be generalized. That is, if a thermal energy equation is solved, then the solution would be valid for the corresponding mass balance equation. Steady- and unsteady-state diffusion problems, therefore, will not be discussed, because the solutions are discussed for the case of thermal energy balance in diffusive heat transfer. Unsteady-state diffusive and convective mass transfer find applications in process and environmental engineering (e.g., dispersion of pollutants in rivers, lakes, and oceans and in the atmosphere).

5.3   Forced Convective Mass Transfer in Laminar Viscous Flow

Diffusive mass transfer in a laminar viscous flow system, called forced convective mass transfer, finds many applications in process engineering. Forced convective mass transfer will be discussed for the case of well-defined laminar viscous flow, such as a falling liquid film on a solid wall (Fig. 11.19) (Bird, 1960). It is assumed that the surface of the falling liquid is exposed to gaseous species A and that chemical equilibrium is established between the liquid and gas at their interface. Species A penetrates by diffusion into the liquid much smaller distances than the velocity boundary layer thickness δ of the falling liquid film (for that reason δ can be considered infinite). The velocity profile of the falling liquid film (an incompressible laminar viscous flow) is given by

$$v_z(x) = v_{max}\left[1 - \left(\frac{x}{\delta}\right)^2\right] \tag{11.342}$$

Because the penetration by diffusion is much smaller than δ, Eq. (11.342) can be approximated to a constant velocity flow [i.e., $v_z(x) = v_{max}$]. Because no chemical reaction is taking place in the liquid phase, the mass balance equation for species A in terms of molar concentrations can be given [using the steady-state form of Eq. (11.329) and taking $\delta = \infty$ in Eq. (11.342)] by

$$v_{max}\frac{\partial c_A}{\partial z} = D_{AB}\frac{\partial^2 c_A}{\partial x^2} \tag{11.343}$$

with the boundary conditions

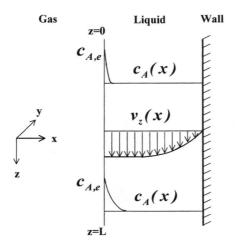

Figure 11.19   Forced convection mass transfer in a laminar viscous falling liquid film on a solid wall.

$$c_A = 0 \qquad \text{at} \quad z = 0 \tag{11.344}$$

$$c_A = c_{A,e} \qquad \text{at} \quad x = 0 \tag{11.345}$$

$$\frac{\partial c_A}{\partial x} = 0 \qquad \text{at} \quad x = \infty \tag{11.346}$$

where $c_{A,e}$ is the equilibrium liquid concentration of species A at the gas/liquid interface. The solution of Eq. (11.343) with these boundary conditions is

$$\frac{c_A}{c_{A,e}} = 1 - \frac{2}{\sqrt{\pi}} \int_0^\eta e^{-\xi^2} d\xi = 1 - \text{erf}(\eta) = \text{erfc}(\eta) \tag{11.347}$$

where η is the dimensionless distance (normal to the gas/liquid interface)

$$\eta = \frac{x}{(4D_{AB}z/v_{\max})^{1/2}} \tag{11.348}$$

Using the solution expressed by Eq. (11.347), the local mass flux in the x direction can be calculated as

$$N_{A,x}\big|_{x=0} = -D_{AB} \frac{\partial c_A}{\partial x}\bigg|_{x=0} = c_{A,e} \left(\frac{D_{AB}v_{\max}}{\pi z}\right)^{1/2} \tag{11.349}$$

and the total mass flux from gas to liquid is calculated (in terms of moles) as

$$m_A = W \int_0^L N_{A,x}\big|_{x=0} \, dz = W c_{A,e} \left(\frac{D_{AB}v_{\max}}{\pi}\right)^{1/2} \int_0^L z^{-1/2} \, dz = W L c_{A,e} \left(\frac{4D_{AB}v_{\max}}{\pi L}\right)^{1/2} \tag{11.350}$$

or

$$m_A = W L c_{A,e} \left(\frac{4D_{AB}}{\pi \tau_d}\right)^{1/2} \tag{11.351}$$

where τ_d is the total exposure time for diffusion

$$\tau_d = L/v_{\max} \tag{11.352}$$

As can be seen from Eq. (11.351), the mass transfer rate is proportional to the square root of the diffusivity and inversely proportional to the square root of the exposure time τ_d for diffusion.

Results obtained from the forced convective mass transfer in laminar viscous flow can also be applied to mass transfer during the rise of a small bubble in a quiescent liquid (Fig. 11.20). Because of the difference between the densities of the liquid and gas phases, the gas in the bubble is in continuous circulating motion and the liquid in the neighborhood of the bubble behaves like a falling liquid film, forming a laminar viscous boundary layer around the stationary bubble. Molar flux for the mass transfer of species A from bubble to liquid could be expressed by replacing L by the diameter D of the bubble and $v_{\max}$ by the terminal bubble rise velocity v_{rise} (Higbie, 1935):

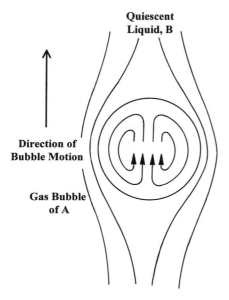

Quiescent Liquid, B

Direction of Bubble Motion

Gas Bubble of A

Figure 11.20 Mass transfer from a gas bubble rising in a quiescent liquid.

$$N_A = c_{A,e}\left(\frac{4D_{AB}v_{rise}}{\pi D}\right)^{1/2} = c_{A,e}\left(\frac{4D_{AB}}{\pi \tau_d}\right)^{1/2} \tag{11.353}$$

An attempt was made to improve the predictive power of Higbie's model by assuming that surface elements randomly renew themselves during the rise of the bubble. Assuming a surface element renewal life distribution function $\psi(\tau_d)$ such that

$$\int_0^\infty \Psi(\tau_d)\, d\tau_d = 1 \tag{11.354}$$

the molar flux for the mass transfer of species A expressed by Eq. (11.353) becomes

$$N_A = \int_0^\infty c_{A,e}\left(\frac{4D_{AB}}{\pi \tau_d}\right)^{1/2} \Psi(\tau_d)\, d\tau_d \tag{11.355}$$

Danckwerts (1970) proposed that the rate of disappearance of a surface element is given by

$$-\frac{d\Psi(\tau_d)}{d\tau_d} = s\Psi \tag{11.356}$$

where s has physical meaning as the rate of surface renewal, i.e., $1/s$ has the meaning of the average lifetime of the surface in the renewal process. Solution of Eq. (11.356) satisfying Eq. (11.354) gives

$$\Psi = se^{-s\tau_d} \tag{11.357}$$

Using Eq. (11.357) for ψ, Eq. (11.355) becomes

$$N_A = \int_0^\infty c_{A,e} \left(\frac{4D_{AB}}{\pi \tau_d}\right)^{1/2} s e^{-s\tau_d} \, d\tau_d = c_{A,e}\sqrt{4D_{AB}s} = c_{A,e}\left(\frac{4D_{AB}}{1/s}\right)^{1/2} \qquad (11.358)$$

which does not provide any information different from that of Eq. (11.353), because $1/s$ in Eq. (11.358) has the physical meaning of surface renewal time.

5.4 Forced Convective Mass Transfer in Turbulent Flow

The mechanism of mass transfer in turbulent flow is similar to the mechanisms of momentum and thermal energy transfer. As a result, mass transfer in turbulent flow can be expressed by using the same mathematical formulation as was used for momentum and thermal energy transfer.

In a turbulent flow system, the molar concentration c_A of a chemical species A can be expressed as (Bird et al., 1960)

$$c_A(t) = \bar{c}_A + \gamma_c(t) \qquad (11.359)$$

where $c_A(t)$ is the time-dependent molar concentration of species A, $\bar{c}_A$ is the time-averaged molar concentration, and $\gamma_c(t)$ is the time-dependent turbulent molar concentration fluctuation. For an incompressible convective flow system with nth-order chemical reaction (i.e., A → Product), the mass balance expression on a molar basis is

$$\frac{\partial c_A}{\partial t} + \mathbf{v} \cdot \nabla c_A = D \, \nabla^2 c_A - k c_A^n \qquad (11.360)$$

where k is the kinetic rate constant (depending on temperature with the Arrhenius relation). In the high Reynolds number flow regime, the flow becomes turbulent and the concentration of species A would be the time-dependent turbulent concentration [Eq. (11.359)]. By expressing the velocity and concentration in terms of their time-average and fluctuation components, Eq. (11.360) becomes (for an incompressible fluid)

$$\frac{\partial(\bar{c}_A + \gamma_c)}{\partial t} + \left[\bar{v} + \gamma_v(t)\right] \cdot \nabla[\bar{c}_A + \gamma_c(t)] = D \, \nabla^2[\bar{c}_A + \gamma_c(t)] - k[\bar{c}_A + \gamma_c(t)]^n \qquad (11.361)$$

The time-average form of Eq. (11.361) is

$$\frac{\partial \bar{c}_A}{\partial t} = -\bar{v} \cdot \nabla \bar{c}_A - \nabla\overline{[\gamma_v(t) \cdot \gamma_c(t)]} + D \, \nabla^2 \bar{c}_A - k\overline{[\bar{c}_A + \gamma_c(t)]^n} \qquad (11.362)$$

The second term on the right-hand side, the time average of the product of fluctuating components of velocity and concentration, is called the turbulent mass flux (on a molar basis) $\mathbf{J}_A^{(t)}$,

$$\mathbf{J}_A^{(t)} = \overline{\gamma_v(t) \cdot \gamma_c(t)} \qquad (11.363)$$

which is similar to turbulent momentum flux [Eq. (11.162)] and turbulent thermal energy flux [Eq. (11.290)].

The form of the time average of the reaction term with turbulent concentration would be dependent on the order of reaction. The last term in Eq. (11.362) would be

$$k\overline{[\bar{c}_A + \gamma_c(t)]^n} = k\bar{c}_A \qquad \text{for} \quad n = 1 \qquad (11.364)$$

and

$$k\overline{[\bar{c}_A + \gamma_c(t)]^n} = k\bar{c}_A^2 + k\overline{\gamma_c^2(t)} \qquad \text{for} \quad n = 2 \tag{11.365}$$

Turbulent mass flux is not a property of the fluid (like diffusion flux), it is a property of the flow. That is, it is a function of Reynolds number and flow geometry. For integration of the mass balance equation for turbulent flow, turbulent mass flux has to be expressed in terms of $\bar{c}_A$ or its gradient. Using the similarity between the turbulent transport of mass, momentum, and thermal energy, semiempirical relations are developed to express turbulent mass transfer as a function of mean concentration measurements (Bird et al., 1960).

By analogy with Fick's first law of diffusion, the turbulent mass flux is correlated by

$$J_A^{(t)} = -D_A^{(t)} \frac{d\bar{c}_A}{dy} \tag{11.366}$$

where $D_A^{(t)}$ is the turbulent diffusion coefficient. Also by analogy, $D_A^{(t)}/\alpha^{(t)}$ and $D_A^{(t)}/\gamma^{(t)}$ should be on the order of 1.

The Prandtl mixing length model, by analogy with turbulent momentum transfer, suggests that turbulent mass flux can be correlated by

$$J_A^{(t)} = -l^2 \left| \frac{d\bar{v}_x}{dy} \right| \frac{d\bar{c}_A}{dy} \tag{11.367}$$

where $l(= ky$, where y is the distance from the solid wall) is the mixing length.

Von Karman's similarity hypothesis, by analogy with turbulent momentum transfer, suggests that turbulent mass flux can be correlated by

$$J_A^{(t)} = -k^2 \left| \frac{(d\bar{v}_x/dx)^3}{(d^2v_x/dy^2)^2} \right| \frac{d\bar{c}_A}{dy} \tag{11.368}$$

where $k(= cy$, where c is a constant and y is the distance from the solid wall) is a constant.

Deissler, by analogy with turbulent momentum transfer, suggests that turbulent mass flux close to a wall can be correlated by

$$J_A^{(t)} = -n^2\bar{v}_x y \left[1 - \exp\left(\frac{-n^2\bar{v}_x y}{\gamma} \right) \right] \frac{d\bar{c}_A}{dy} \tag{11.369}$$

where n is a constant to be determined experimentally.

In engineering applications, however, the mass transfer rate is correlated by using dimensionless numbers. In these correlations, the mass transfer coefficient k_c (not mass flux) is used; it is defined (e.g., for mass transfer from a soluble solid to a flowing fluid) as

$$N_A = k_c(c_{A,e} - c_A) \tag{11.370}$$

where N_A is the diffusion flux [mol/(s · cm^2)] of species A, k_c is the mass transfer coefficient (cm/s), $c_{A,e}$ is the concentration of species A at the solid/fluid interface (i.e., equilibrium concentration), and c_A is the concentration of species A in the ambient fluid. For gaseous systems, the mass transfer coefficient k_G [mol/(cm^2 · atm · s)] is defined as [similar to Eq. (11.370)]

$$N_A = k_G(p_{A,0} - p_A) \tag{11.371}$$

where $p_{A,0}$ is the partial pressure of the diffusing substance (i.e., species A) at the interface (i.e., equilibrium partial pressure) and p_A is the partial pressure of the diffusing substance in the ambient fluid. The following relation exists between k_c and k_G:

$$k_G = \frac{k_c}{RT} \tag{11.372}$$

where R is the universal gas constant [82.05 atm $\cdot$ cm^3/(mol $\cdot$ K) or 8.31 J/(mol $\cdot$ K)].

The dimensionless number representing the mass transfer rate is the Sherwood number, Sh (corresponding to the Nusselt number in heat transfer), which is expressed as

$$Sh = \frac{k_c L}{D_A} \tag{11.373}$$

where k_c is the mass transfer coefficient of Eq. (11.370), L is a characteristic length (size of a plate, diameter of a tube, etc.), and D_A is the diffusion coefficient. [The Sherwood number can be considered as $(L^2/D_A)/(L/k_c)$, i.e., the ratio of the characteristic time for diffusive mass transfer to the characteristic time for convective mass transfer.] By analogy to heat transfer [i.e., Nu is correlated as a function of Reynolds and Prandtl numbers and the dimensionless distance L/D, Eq. (11.299)], the Sherwood number is correlated as a function of Reynolds and Schmidt (Sc $= \mu/\rho D_A = \gamma/D_A$) numbers and dimensionless distance L/D (L is the tube length and D is the tube diameter) or x/L (x is the distance from a reference point and L is the length of an object):

$$Sh = f(Re, Sc, L/D) \tag{11.374}$$

For highly turbulent, incompressible tube flow, similar to heat transfer, the Sherwood number is

$$Sh = 0.026 Re^{0.8} Sc^{1/3} \tag{11.375}$$

Because there is a similarity between heat and mass transfer, they must be correlated by the same relation (Chilton and Colburn, 1934). For similar geometry and boundary conditions, heat (j_H) and mass (j_D) transfer Chilton and Colburn factors are defined as

$$j_H = \frac{Nu}{Re\ Pr^{1/3}} \quad \text{and} \quad j_D = \frac{Sh}{Re\ Sc^{1/3}} \tag{11.376}$$

For most engineering problems of similar geometry, it can be expected that the Chilton and Colburn factors j_H and j_D are identical,

$$j_H = j_D \tag{11.377}$$

which is a very useful relation for engineering applications.

6. COMPUTATIONAL FLUID DYNAMICS

Most engineering problems are involved with simultaneous mass, momentum, and energy transfer coupled with chemical reactions, in some cases in more than one phase. The momentum balance equation expressing momentum transfer is nonlinear because of the convection term, which is in the form of the products of velocity and velocity gradients. For chemically reactive flow systems, the property generating terms Ψ_g in the mass and thermal energy balances are also coupled in a nonlinear fashion. This nonlinearity evolves from the dependence of the reaction rate [$\Psi_g = r(T, P, c)$] and heat generation $\Psi_g =$

$r(T, P, c)(-\Delta H)]$ terms on temperature in the Arrhenius form, i.e., an exponential function of temperature $[r = A \exp(-E/RT)c^n$, another source of nonlinearity if $n \neq 1]$. The coupling and nonlinearity make the chemically reactive systems interesting but at the same time very difficult to study. These problems are faced in reactor design, process dynamics, process control, spontaneous combustion, explosion and quenching. A closed-form analytical solution of many of these problems is not possible even if the couplings in the mass, momentum, and energy balances are well defined and well formulated. Linearization around the steady-state or perturbation methods may provide some information about chemically reactive flow systems.

Another alternative to the study of chemically reactive flow systems is to numerically solve mass, momentum, and energy balance equations (and equations of state) simultaneously. Chemical reactions can be introduced into the mass balance as well as thermal energy balance equations by specifying the rate constants as a function of temperature and concentrations (i.e., reaction orders). All physical characteristics of the flow system can be introduced as inputs.

In any computational fluid dynamics procedure simulating a chemically reactive flow problem in one phase, fluid properties such as density, viscosity, thermal conductivity, and diffusivity and thermodynamic properties such as pressure, temperature, enthalpy, internal energy, specific entropy, specific heat at constant pressure, and specific heat at constant volume are needed. In addition, the equation of state for compressible fluid is described by using one of the classical equations of state (e.g., the van der Waals equation), which requires thermodynamic data such as critical temperature and critical pressure of the species. Mass, momentum, and energy balances are defined by considering the turbulent motion as well.

To simulate a chemically reactive system composed of S chemical species (a one-phase problem), $S + 5$ unknowns (S compositions, temperature, pressure, and three components of flow velocity) have to be solved for for each node of discretization. To solve for $S + 5$ unknowns, $S + 5$ equations (S equations for material balances, the equation of state, the thermal energy equation, and three equations of motion) are needed. It is always better to reduce the dimensions of the problem; a three-dimensional problem should be reduced to a two-dimensional problem by carefully selecting a symmetry plane, which may result in saving a significant amount of computational time.

REFERENCES

Apostol, T. 1966. Advanced Calculus. Academic Press, New York.

Astarita, G. 1967. Mass Transfer with Chemical Reaction. Elsevier, New York.

Becker, H. A. 1977. Mixing concentration fluctuations and marker nephelometry. In: Studies in Convection. B. Launder, ed. Academic Press, New York.

Bird, R. B., Stewart, W. E., and Lightfoot, E. N. 1960. Transport Phenomena. Wiley, New York.

Brodkey, R. 1967. The Phenomena of Fluid Motion. Addison-Wesley, Menlo Park, CA.

Carslaw, H. S., and Jaeger, J. C. 1959. Conduction of Heat in Solids. Oxford Univ. Press, Oxford, England.

Chilton, T. H., and Colburn, A. P. 1934. Ind. Eng. Chem. 26, 1183.

Danckwerts, P. V. 1970. Gas Liquid Reactions. McGraw-Hill, New York.

Eckert, E. R. G., and Drake, R. M. 1959. Heat and Mass Transfer. McGraw-Hill, New York.

Ergun, S. 1952. Fluid Flow Through Packed Columns. Chem. Eng. Progr. 48:89–94.

Frank-Kamenestkii, D. A. 1969. Diffusion and Heat Transfer in Chemical Kinetics. Plenum Press, New York.

Happel, J., and Brenner, H. 1986. Low Reynolds Number Hydrodynamics. Martinus Nijhoff, Boston, MA.

Higbie, R. 1935. Trans. AIChE. 31:365–389.

Hinze, J. O. 1959. Turbulence. McGraw-Hill, New York.

Holman, J. P. 1990. Heat Transfer. 7th ed. McGraw-Hill, New York.

Hottel, H. C., and Sarofim, A. F. 1967. Radiative Transfer. McGraw-Hill, New York.

Lamb, H. 1945. Hydrodynamics. Dover Publications, New York.

Levich, V. G. 1962. Physicochemical Hydrodynamics. Prentice-Hall, Englewood Cliffs, NJ.

Libby, P. A., and Williams, F. A. 1994. Turbulent Reactive Flows. Academic Press, San Diego, CA.

Monin, A. S., and Yaglom, A. M. 1971. Statistical Fluid Mechanics. MIT Press, Cambridge, MA.

Ozisik, M. N. 1973. Radiative Transfer and Interactions with Conduction and Convection. Wiley, New York.

Ozisik, M. N. 1985. Heat Transfer: A Basic Approach. McGraw-Hill, New York.

Perry, R. H., and Green, D. W. 1984. Perry's Chemical Engineers' Handbook. McGraw-Hill, New York.

Schlichting, H. 1955. Boundary Layer Theory. McGraw-Hill, New York.

Treybal, R. E. 1980. Mass Transfer Operations. McGraw-Hill, New York.

12
Heterogeneous Gas–Solid Reactions and Heterogeneous Catalysis

1. INTRODUCTION

Many industrially important reactions involve fluid species reacting with the surface of solids; these are called heterogeneous reactions. The solids may also act only as catalysts (i.e. active sites of the solid surface) for reactants that are originally present in the fluid phase; in this case the reaction is called heterogeneous catalysis.

In heterogeneous reactions the solid (i.e., its surface) may act as one of the reactants as in the case of coal combustion or coal gasification. In these cases the reactions between the fluid-phase and solid-phase species are coupled with mass and heat transfer within the solid as well as the gas film surrounding the solid particles. As a result, particle size, pore structure, and reactor hydrodynamics may result in profound differences in the observed or apparent rate of reaction.

Solid–fluid interactions are also important for heterogeneous catalysis. These interactions can also be important in purely physical processes such as crystallization and in physical-chemical processes such as those in fluid coking reactors used by the petroleum industry to produce light hydrocarbons from heavy fractions such as vacuum bottoms or bitumen produced by the thermal coking process. In crystallization reactors solid–fluid interactions are important for the mass and heat transfer mechanisms in the liquid phase attached to the crystal surface, which affects the crystal growth rate as well as morphological stability of the crystal surface (therefore the purification efficiency). In the fluid coking process (Chapter 13), these interactions control thickness of the heavy crude film on the coke particles and therefore the heat and mass transfer rates between the coke and the fluid phase, which affects the product yield of the coking process. In all of these processes differences in particle–fluid interactions may cause significant differences in the performance of the process.

The definition of reaction rates in gas–solid reactions (Chapter 8, Section 3) and elementary steps involved in coal gasification (Chapter 6, Section 2) have already been discussed. All of these discussions were devoted to noncatalytic gas–solid reactions and for the case when the solid becomes one of the reactants. In these discussions, it was also assumed that the diffusion mass transfer in the solid pore structure or in the fluid boundary layer surrounding the particle was not the rate-limiting step.

2. REACTION REGIMES IN HETEROGENEOUS GAS–SOLID REACTIONS

In principle, a gas–solid reaction is chemically controlled at low temperatures, and as temperature increases the diffusion limitations affect the apparent rate of the gas–solid reaction. As temperature increases, the rate of the gas–solid reaction also increases in an exponential fashion, resulting in shortages of the gas reactant in the solid (Fig. 12.1). In the low temperature range, the chemical reaction is slow (i.e., rate-controlling or rate-limiting step) in comparison to diffusion of mass, and as a result no concentration gradient exists either in the boundary layer or throughout the solid particle. This reaction regime, in which the solid is consumed internally, is called Regime I (Walker et al., 1959). As temperature increases, the diffusion mass transfer limitations start to contribute to the rate-controlling mechanism. Concentration gradients start to exist both in the boundary layer and within the pores of the solid. This is called Regime II and is the regime in which the diffusion of the reactant becomes the rate-controlling (or rate-limiting) step. In this reaction regime the solid is consumed externally as well as internally. As temperature increases, the consumption of the reactant by chemical reaction in the solid increases exponentially. At sufficiently high temperatures, the chemical reaction may be so fast that boundary layer diffusion becomes the rate-controlling step. This regime, in which the solid is mainly consumed externally, is called Regime III.

Apparent activation energy measurements, such as ln k versus $1/T$ plots, are indications of the reaction regimes (Fig. 12.2). In Regime I, because the process is controlled by chemical reaction, the apparent activation energy would be high, in the range of 240–320 kJ/mol (60–80 kcal/mol) or even higher.

In Regime II, pore diffusion limitation becomes visible as a result of the coupling of chemical reaction with pore diffusion. In this regime the diffusion and chemical reaction

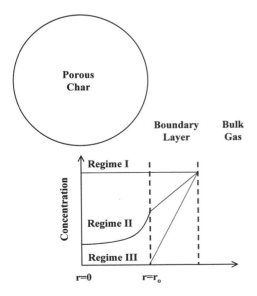

Figure 12.1 Reactant profiles in the boundary layer and in the porous solid in gas–solid reactions in regimes I, II, and III.

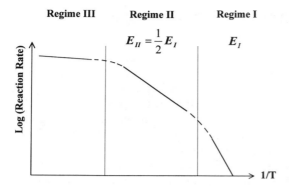

Figure 12.2 Apparent activation energies (i.e., $\ln k$ versus $1/T$ plot) for gas–solid reactions in regimes I, II, and III. (From Walker et al., 1959.)

rates are comparable. To examine the effect of the coupling between the chemical reaction and pore diffusion on the overall apparent chemical reaction rate, a reaction between a gas and a semi-infinite solid (Fig. 12.3) will be considered. Because solid is continuously consumed by the reaction, no steady-state operating condition can exist. However, to simplify the problem, solid composition will be assumed to be constant. The steady-state mass balance equation for the reacting gas species can be expressed for a first-order reaction as

$$D_e \frac{d^2 c}{dx^2} = k_s \sigma c \tag{12.1}$$

with the boundary conditions

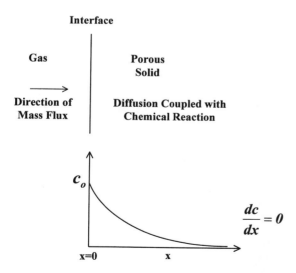

Figure 12.3 Coupling of diffusion and chemical reaction for a gas reacting with a semi-infinite solid.

$$c = c_0 \quad \text{at} \quad x = 0, \qquad \frac{dc}{dx} = 0 \quad \text{at} \quad x = \infty \tag{12.2}$$

where k_s is the surface reaction rate constant (cm/s), D_e is the effective diffusivity (cm^2/s), c_0 is the molar concentration (mol/cm^3) of the reactant at the outer surface of the solid (i.e., at $x = 0$), c is the molar concentration of the reactant species in the pores of the solid, σ is the specific pore surface area of the solid (in square centimeters of pore surface area per cubic centimeter of solid, cm^{-1}), and x (cm) is the distance from the outer surface of the solid. The solution of Eq. (12.1) with the boundary conditions of Eq. (12.2) is in the form

$$c = c_0 \exp\left[-\left(\frac{k_s\sigma}{D_e}\right)^{1/2} x\right] \tag{12.3}$$

which predicts the molar flux of the reactant per unit of outer surface area of the slab as

$$N = -D_e\frac{dc}{dx}\bigg|_{x=0} \tag{12.4}$$

or

$$N = -D_e c_0 - \left(\frac{k_s\sigma}{D_e}\right)^{1/2} \exp\left[-\left(\frac{k_s\sigma}{D_e}\right)^{1/2}\right]\bigg|_{x=0} = c_0\sqrt{D_e k_s \sigma} \tag{12.5}$$

Because the molar flux of the reactant would be consumed by chemical reaction, N is also related to the apparent reaction rate (dc/dt for the reactant gas or dm/dt for the consumed solid) of the gas–solid reaction. From Eq. (12.5) it can be seen that the apparent activation energy E_a of the overall process (i.e., dm/dt) for Regime II is

$$E_a = \frac{1}{2}(E_{\text{dif}} + E_{\text{react}}) \approx \frac{1}{2} E_{\text{react}} \tag{12.6}$$

which would be equal to almost one-half of the apparent activation energy of Regime I, because the activation energy of the diffusion is much smaller than the activation energy of the chemical reaction, $E_{\text{diff}} \ll E_{\text{react}}$. Equation (12.6) can be derived from Eq. (12.5) because the activation energy terms for diffusion and chemical reaction are in the exponential form.

As temperature increases further, the consumption of the reactant in the pores of the solid would be increased further and further, making the mass transfer of the reactant in the boundary layer surrounding the solid particle, which is the reaction regime defined as Regime III, the rate-limiting step. Also, in this regime the chemical reaction would take place practically entirely on the outer layer of the solid particle.

It is clear that the volume reaction model for reactions in Regime I can represent the gas–solid reactions. Gas–solid reactions in Regime II can be represented by the diffusion–chemical reaction model, which will be discussed next. Gas–solid reactions of Regime III can be modeled by the shrinking-core model (Chapter 8, Section 3).

The diffusion–reaction model of Regime II will be discussed for a spherical particle undergoing a first-order reaction with a gas species (Fig. 12.4). As in the case of the gas–solid reaction in a semi-infinite solid, the solid is continuously consumed by the reaction and a true steady-state operating condition may not exist. A steady-state condition will be assumed, however, to simplify the problem without losing the principles of coupling between diffusion and chemical reaction. For a spherical geometry, the steady-state diffu-

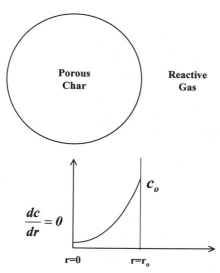

Figure 12.4 Diffusion–reaction model (Regime II) for a spherical particle reacting with a gaseous species in a first-order reaction.

sion–reaction model for a reactive gaseous species can be expressed [for the gas–solid reaction stoichiometry of $G_{(g)} + \alpha_s S \rightarrow P_{(g)}$] as

$$D_e \frac{d^2 c}{dr^2} + \frac{2}{r} \frac{dc}{dr} - k_s \sigma c = 0 \tag{12.7}$$

with the boundary conditions

$$c = c_0 \quad \text{at} \quad r = r_0, \qquad \frac{dc}{dr} = 0 \quad \text{at} \quad r = 0 \tag{12.8}$$

where r_0 is the radius of the spherical solid particle. Introducing the normalized variables

$$C = \frac{c}{c_0}, \qquad \xi = \frac{r}{r_0}, \qquad \phi = r_0 (k_s \sigma / D_e)^{1/2} \tag{12.9}$$

where ϕ is the Thiele module [which is the ratio of the characteristic time for diffusion τ_d to that for reaction τ_r, i.e., $\phi = \tau_d / \tau_r$, which is the same as the Damkohler number (Chapter 10, Section 8)], the mass balance equations (12.7) and (12.8) become

$$\frac{d^2 C}{d\xi^2} + \frac{2}{\xi} \frac{dC}{dr} - \phi^2 C = 0 \tag{12.10}$$

with the boundary conditions

$$C = 1 \quad \text{at} \quad \xi = 1; \qquad \frac{dC}{d\xi} = 0 \quad \text{at} \quad \xi = 0 \tag{12.11}$$

Solution of Eq. (12.10) using the boundary conditions given by Eq. (12.11) (i.e., by defining a new dependent variable $W = C/\xi$, the differential equation can be solved for W and the solution can be expressed in C) is

$$C = \frac{\sinh(\xi\phi)}{\xi\sinh(\phi)} \tag{12.12}$$

The effectiveness factor η can be defined as the ratio of the actual mass of gas reactant species consumed (which is the molar flux at $r = r_0$) to the maximum possible mass of gas reactant species that could have been consumed if there were no diffusion limitation present. Because the reactant species concentration profile in the solid is known, the effectiveness factor becomes

$$\eta = \frac{4\pi r_0^2[-D_e(dc/dr)|_{r=r_0}]}{(4/3)\pi r_0^3 k_s \sigma c_0} \tag{12.13}$$

or

$$\eta = \frac{3}{\phi}\left(\frac{1}{\tanh(\phi)} - \frac{1}{\phi}\right) = \frac{3}{\phi\tanh(\phi)}\left(1 - \frac{\tanh(\phi)}{\phi}\right) \tag{12.14}$$

As deduced [Eq. (12.14)], as $\phi \to 0$ the effectiveness factor $\eta \to 1$. As $\phi \to \infty$, the effectiveness factor η becomes inversely proportional to ϕ; i.e., as $\phi \to \infty$, $\eta \approx 3/\phi$ [because $\tanh(\phi) = 0$ at $\phi = 0$ and $\tanh(\phi) \to 1$ as $\phi \to \infty$]. For example, it is safe to assume that the effectiveness factor η is 0.6 when the Thiele module is about 5 and that η is 0.3 when the Thiele module is about 10 (i.e., $\eta = 3/\phi$) (Fig. 12.5). In summary, for a gas–solid reaction taking place in Regime II (the diffusion–chemical reaction regime), the effectiveness factor can be improved by reducing the Thiele modules, which can be achieved by reducing the particle size.

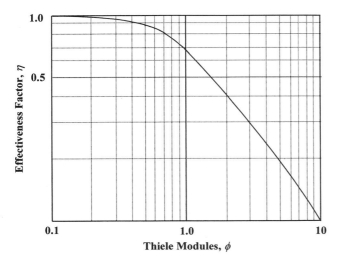

Figure 12.5 Effectiveness factor as a function of Thiele number for a spherical solid reacting with a gas.

3.   COMBUSTION AND GASIFICATION

In any gas–solid reaction, coupling of chemical reaction with the diffusion of the reactant both in the boundary layer and within the solid pore structure affects the apparent reaction rate. Coupling between the chemical reaction and mass diffusivity, which finds applications in coal combustion and gasification, was discussed in detail for char reactivity by Walker et al. (1959). Pulverized coal combustion is probably the most industrially important application of heterogeneous gas–solid reactions and serves as an example for the effect of particle size and reactor hydrodynamics on reaction rate.

In fact, coal combustion and gasification are often considered to be representative of gas–solid reactions and will be used here as being illustrative (not typical) of the reactions that might occur between gaseous feedstocks and solid catalysts during refinery processes. Remembering, of course, that examples are only as good as the imagination will allow! Thus, the following discussion should be taken as being illustrative of gas–solid reactions but not typical of reactions that occur in many refinery reactors.

In coal combustion (also in gasification), when coal is exposed to a hot oxidative environment, coal devolatilization reactions (thermal pyrolysis) take place as a result of rapid heating, producing volatiles (hydrocarbons) and char from the feed coal. Therefore, when coal is introduced into the hot oxidative environment, simultaneous reactions of coal devolatilization and combustion of volatiles and char take place. Coal particle size, reaction temperature, coal type (i.e., amount of volatiles and char pore characteristics), coal firing method, and boiler geometry are controlling factors for the rate of combustion and gasification processes, flame stability, and extinction (Essenhigh, 1981).

The first coal combustion model, proposed by Nusselt (1924) and also known as the square-law model, states that the rate of coal (or char) combustion is controlled by the diffusion of oxygen to the particle surface through a stagnant boundary layer. Later, in the 1930s, it was observed that chemical reaction also controls the rate of coal combustion (Parker and Hottel, 1936; Hottel and Stewart, 1939). In the 1960s, more systematic studies on the combustion of pulverized coal particles proved that even at 300 μm particle diameter, the combustion rate is strongly controlled by the diffusion of oxygen through the boundary layer surrounding the particle. There are experimental evidences that this limit can be taken even as small as 100 μm. As particle diameter gets smaller than 100 μm, the contribution of chemical reaction to control char combustion becomes visible. As a result, char particle combustion takes place by the diffusion–chemical reaction mechanism unless the temperature of the particle gets too high, i.e., approaching 1725°C (3140°F; 2000 K), the temperature range in which chemical reaction becomes too fast and diffusion becomes the rate-limiting step.

A significant amount of work was devoted to coal combustion research by the British Coal Utilization Research Association (BCURA) and by the Commonwealth Scientific and Industrial Research Organization (CSIRO) in Australia. In fact, the laminar drop tube reactor was used by BCURA (Field et al., 1967). At BCURA, char samples for combustion tests were produced in an inert gas atmosphere containing 1% by volume oxygen (to burn the volatiles and prevent the pores from plugging). In these combustion tests, particle size was in the range of 28–105 μm and burning particle temperature was in the range 925–1725°C (1700–3140°F; 1200–2000 K). Partially combusted char samples were analyzed to measure the char combustion reaction extent (burn-off), and the measured combustion rates were expressed in terms of the surface reaction rates. The laminar drop tube reactor used by BCURA is similar to thermogravimetric analyzers (TGAs) which can be very informative in combustion research.

The laminar drop tube reactor was also used by CSIRO (Mulcahy and Smith, 1969; Ayling and Smith, 1972). Char samples were produced in the flame. In these combustion tests, particle size was in the range of 4–90 μm and burning particle temperature was in the range of 725–2025°C (1340–3680°F; 1000–2300 K). Gas composition measurements were used to calculate the combustion reaction extent (burn-off), and the measured combustion rates were expressed in terms of the surface reaction rates [i.e., r_s in kg/(m$^2 \cdot$ s)]. In laminar drop tube reactors the char temperature during the combustion process is not known, but it can be predicted by applying the thermal balance.

Further, in CSIRO a two-color pyrometer was also used to measure the burning char temperature. In this technique, radiation emitted from a burning char particle is optically divided into two paths and each path is filtered at wavelengths λ_1 and λ_2; the intensity of radiation is then measured using two photomultiplier tubes. The intensity of the radiation I_λ as a function of λ and source temperature T is given by Planck's distribution law:

$$I_\lambda = C_s \left(\frac{2\pi c^2 h}{\lambda^5} \right) \left(\frac{1}{e^{ch/k\lambda T} - 1} \right) \tag{12.15}$$

where C_s is a constant specific to the experimental system, h is Planck's constant (6.62×10^{-27} erg $\cdot$ s), k is Boltzmann's constant [1.38×10^{-16} erg/(mol $\cdot$ K)], and c is the speed of light (2.9979×10^{10} cm/s). If two radiation intensities are measured at wavelengths λ_1 and λ_2, from the ratio of the radiation intensities [by which C_s in Eq. (12.15) is eliminated],

$$\frac{I_{\lambda_1}}{I_{\lambda_2}} = \left(\frac{\lambda_2}{\lambda_1} \right)^5 \left(\frac{e^{ch/k\lambda_2 T} - 1}{e^{ch/k\lambda_1 T} - 1} \right) \tag{12.16}$$

from the solution of which (nonlinear algebraic equation) the source temperature T can be solved (predicted). Some laboratories use a three-color pyrometer, which provides three independent radiation intensity measurements I_{λ_1}, I_{λ_2}, and I_{λ_3} and three independent estimates for the source temperature.

Measurement of the burning char particle temperature can provide information on the rate of char combustion. Applying the steady-state energy balance to the burning char particle,

$$\rho_s(-\Delta H) = h(T - T_g) + \varepsilon\sigma(T^4 - T_w^4) \tag{12.17}$$

where ρ_s is the particle burning rate per unit external surface area [mol char/(cm$^2 \cdot$ s)], T_g and T_w are the gas and wall temperatures, h is the transfer coefficient between particle and surrounding gas, ΔH [cal/(mol $\cdot$ C)] is the enthalpy of the combustion reaction, ε is the emissivity of the surface of the burning particle, and σ is the Stefan–Boltzmann constant [5.666×10^{-12} W/(K$^4 \cdot$ cm^2) or 1.355×10^{-12} cal/(K$^4 \cdot$ cm$^2 \cdot$ s)].

All of these studies showed that for coal particles larger than 100 μm, combustion is controlled mainly by the mass transfer in the boundary layer surrounding the particle for particle temperatures in the range of 925–2725°C (1700–4940°F; 1200–3000 K). As particle diameter gets smaller than 100 μm, combustion takes place within the pores and the chemical reaction partially controls the overall combustion rate. Based on independent measurements of burning char temperature and char combustion rate, it was concluded that carbon monoxide is the main combustion product, with the carbon monoxide/carbon dioxide ratio increasing as the particle burning temperature increases.

Later, optical pyrometers were modified for simultaneous measurements of burning char particle size, flow velocity, and temperature in an entrained flow reactor. These studies confirmed that the boundary layer combustion of carbon monoxide with oxygen, producing carbon dioxide, also contributes to the equilibrium temperature of the burning char particle.

In coal combustion, char particle size also affects the mechanism of char combustion. As char particle size gets larger than 100 μm, char combustion takes place primarily by the oxidation of char with carbon dioxide, especially at elevated temperatures (> 1425°C; > 2600°F; > 1700 K), probably because the rate of the carbon–carbon dioxide reaction is slower than that of the carbon–oxygen reaction, i.e., by the chemical reaction

$$C + CO_2 \rightarrow 2CO, \qquad \Delta H^0 = 17,237\,\text{kJ/mol (41.220 kcal/mol)} \tag{12.18}$$

followed by the boundary layer combustion of carbon monoxide to carbon dioxide by oxygen,

$$CO + \tfrac{1}{2}O_2 \rightarrow CO_2, \qquad \Delta H^0 = -28,284\,\text{kJ/mol } (-67.636\,\text{kcal/mol}) \tag{12.19}$$

On the other hand, as particle size decreases, the combustion mechanism shifts toward the direct oxidation of char to carbon monoxide by the reaction

$$C + \tfrac{1}{2}O_2 \rightarrow CO, \qquad \Delta H^0 = -110,466\,\text{kJ/mol } (-26.416\,\text{kcal/mol}) \tag{12.20}$$

or to carbon dioxide by the reaction

$$C + O_2 \rightarrow CO_2, \qquad \Delta H^0 = -393,307\,\text{kJ/mol } (-94.052\,\text{kcal/mol}) \tag{12.21}$$

As a result, when char particle size increases, the burning temperature decreases. However, as particle size decreases, the burning temperature of the char first increases and passes through a maximum and cools down again because the creation of excessive surface area promotes cooling. The maximum of the char burning temperature may occur at about 10 μm or even smaller particle size.

The reaction order for the combustion of char (i.e., its dependence on the partial pressure of oxygen), which provides information on the domination of adsorption and desorption mechanisms, is not certain. Experimental evidence indicates that the reaction order changes from 1/2 to 1 with increases in temperature (even becoming zero order at very low temperature). Activation energies for char combustion kinetics are reported in the range of 80–240 kJ/mol, which can be considered an indication of chemical reaction control in char combustion. Char reactivity largely depends on char surface area and the accessibility of the surface area to oxygen. High reactivity chars (pore size greater than 1300 nm) burn as much internally as externally, compared to less reactive chars (much smaller macropores), which burn externally (shrinking core mode). Char particles burning internally have higher flame emissivity than char particles burning externally only and therefore maintain a higher rate of heat transfer to unburned char.

Char reactivity is important because it influences ignition, extinction, flame stability, flame temperature, heat flux, and fuel burn-out in utility boilers. These are critical issues for the sustainability of the flame in tangentially fired utility boilers, where the jet injection velocity of the fuel with primary air is in the range of 20–30 m/s. Char reactivity, however, is dependent on the coal type (coal rank). Therefore, coal rank and coal particle size are the basic parameters in coal combustion, along with the combustor geometry and operating conditions.

Selection of the combustor (boiler) type and its operating conditions is related to coal type (rank) and coal size. In the grate combustor, which generally operates in counterflow mode, coal is utilized as large lumps, about 100 mm in size. The heating rate of the coal is about $1°C/s$ ($1.8°F/s$), the flame propagation speed is about 10^{-5}–10^{-3} m/s, and the characteristic reaction time (i.e., residence time) is about 10^2 s for the volatiles and about 10^3 s for the char.

The fluid bed combustor generally operates in the diffusion-controlled regime (particles are well mixed, gas is plug flow). Coal is crushed to smaller than about 20 mm size. The heating rate of the coal particles is about 1000–10,000°C/s (1800–18,000°F/s). It is difficult to define a flame propagation speed in a fluid bed combustor. The characteristic reaction time (i.e., residence time) is about 10 s for the volatiles and about 300 s for the char. The operating temperature of the fluid bed combustor is below 1000°C (1830°F), which provides certain advantages, especially for the utilization of unconventional solid fuels.

In the pulverized coal combustor, coal is crushed smaller than 0.1 mm ($100\,\mu$m), and for a typical tangentially fired utility boiler 80% of the feed is smaller than 0.074 mm (200 mesh; $74\,\mu$m). The heating rate of the coal particles is in the range of 1000–1,000,000°C/s (1800–1,800,000°F/s). The flame propagation speed in the pulverized coal combustor is in the range of 0.1–10^3 m/s (10–20 m/s in a typical utility boiler). The characteristic reaction time (i.e., residence time) is about 0.1 s for the volatiles and about 1 s for the char. The operating temperature of the pulverized coal combustor is much higher than the operating temperature of the fluid bed combustor.

For coals with low reactivity (such as low volatile bituminous coals), down-shut boilers are used. These boilers provide a longer residence time (about 3–4 s) for the low reactivity coal char to combust (i.e., above 90% conversion or burn-off). Pulverized coal combustion in diesel engines and turbines is achieved by crushing the coal below 0.01 mm ($10\,\mu$m) particle size and using excessive washing to liberate coal mineral matter from the combustibles.

There are similarities between coal combustion and coal gasification reactions (Laurendeau, 1978). Coal rank, coal particle size, and char characteristics are the most important parameters in both coal combustion and coal gasification, controlling the reaction rate (along with the ignition, flame temperature, flame stabilization, and heat flux). Therefore, coal particle size and char characteristics affect the residence time required for the combustion and gasification reactions to take place at acceptable conversion (burn-out) levels. Many coal gasification processes are commercially available for the gasification of a wide range of feedstocks (including petroleum pitch). Selection of the type of coal gasification process (i.e., dry or wet, air-fired or oxygen-fired) is very much dependent on the intended use of the product gas. Most commercial coal gasifiers use pressurized fluid or entrained flow reactors operating at or above 1000°C (1830°F; 1275 K) and above 435 psi (3 Mpa). (An exception is the Lurgi gasifier.)

4. HETEROGENEOUS CATALYSTS

By definition, if the catalyst is present in a phase different from that of the reaction medium, then it is called a heterogeneous catalyst. Most industrially important processes such as ammonia synthesis, petroleum refining, and petrochemical production are achieved by the use of heterogeneous catalysts; the catalysts used are solids, whereas the reactants are either gases or liquids (Satterfield, 1991; Matar et al., 1989).

A catalyst is a substance that accelerates the rate of a chemical reaction so that the reacting system approaches chemical equilibrium at a faster rate. Catalysts do not appear in the stoichiometric equation of the reaction and do not affect the chemical equilibrium. The first observation of catalytic activity of certain substances was recorded by Berzelius in 1835. He recognized the *catalytic force* of certain species in *decomposition of bodies*. Since then, catalysts have been developed and commercially used for several processes in the chemical, petrochemical, energy, and environmental industries.

Catalysts can be homogeneous or heterogeneous and affect the rate of selected reactions (catalysts do not affect chemical equilibrium). Homogeneous catalysts are chemical intermediates or transition state species that are formed homogeneously during the course of a reaction. These homogeneously formed transition state species carry out a reaction along a pathway by which certain desirable products can be obtained from the reactants.

Heterogeneous catalysts are substances that are present in one or more phases distinct from the phases containing the reactants and the products. Because of the phase differences, interactions between the reactants or products and the catalyst play an important role in the performance of heterogeneous catalysts.

Generally, chemical reactions involving heterogeneous catalysts take place at selected sites on the surface of the catalyst. Therefore, a higher rate of reaction can be obtained with a higher surface area of the catalyst, provided that the surface area of the catalyst is accessible to the reactant and products can easily be released from the catalyst surface. Mechanical strength, surface area, pore size, and catalyst size are the factors (in addition to other characteristics such as activity and selectivity) to be considered in the manufacture of heterogeneous catalysts.

In any chemical reaction involving a heterogeneous catalyst the following steps have to be considered:

1. Transfer of the reactants from the bulk phase to catalyst particles
2. Diffusion of reactants into the pore structure of the catalyst coupled with their chemisorption on the surface of the catalyst
3. Chemical reactions taking place between the chemisorbed species and reactants
4. Diffusion of the products in the pores of the catalyst coupled with their desorption
5. Transfer of the products from the catalyst surface into the bulk phase

The slowest step among these consecutive (in some cases simultaneous) steps controls the overall rate of the process and is called the *rate-limiting step*. In reactions involving heterogeneous reactions, the rate-limiting step is affected by the properties of the catalyst (size, pore structure, active sites, etc.) as well as the operating conditions such as temperature, pressure, chemical composition, and the hydrodynamics of the bulk phase.

5. INTERACTION OF REACTANTS WITH HETEROGENEOUS CATALYSTS

In heterogeneous catalysis, the interaction between the reactants and the catalyst is always chemisorption. Chemisorption is related to the electronic properties of the catalyst. Reactants attach to the catalyst surface as a result of chemical interactions such as electron transfer from or to the surface. The tendency of the reactant and the catalyst to exchange electrons or form chemical bonds plays an important role in chemisorption (Satterfield, 1991).

The reactions involved in the attachment of the reactants to the catalyst surface, which are called chemisorption, obey the general principles of chemical reactions. As a result, the interaction of the reactants and the catalyst surface always has an activation energy barrier. There is no clear understanding of what actually happens between the reactant and the catalyst surface; however, theoretical chemistry and advanced analytical instruments provide information to enable speculation in this matter.

Models can be developed based on the physical and chemical processes taking place between the reactants and the catalyst. One speculation is that, most likely, chemically unstable *chemical intermediates* or *transitory complexes* are formed as a result of chemical interaction between the reactant molecules and the catalyst surface (i.e., chemisorption). These intermediates then decompose to form the products, transforming the catalyst surface back to its original state.

Heterogeneous catalysts may go through chemical and structural changes during the course of the reactions. For example, a metal oxide catalyst may be reduced to its metal form. During the reaction, the metal/metal oxide ratio and the surface roughness of the catalyst may change. The important thing about the catalysts is that the stoichiometry of these changes is not visible in the overall chemical stoichiometry of the catalyzed chemical reaction.

The energy of formation (ΔH_{ad}) of the intermediates plays an important role in the performance of heterogeneous catalysts. If the energy of formation of these intermediates is low, the affinity between the catalyst surface and the reactants is also low; therefore, the rate of formation of the intermediates will be unavoidably low. In this case the overall process is controlled by the rate of formation of the intermediates. On the other hand, if the energy of formation of these intermediates is high, the affinity between the catalyst surface and the reactants is also high. This high affinity will result in a high rate of formation of chemically stable intermediates. In this case a slow rate of decomposition (desorption) of the intermediates becomes the rate-controlling (limiting) step for the overall process. It is logical, therefore, to expect a maximum rate for the overall process when the bonds between the intermediate and the catalyst surface are neither too weak nor too strong.

More than one type of intermediate may form as a result of chemical interaction between the reactants and the catalyst surface. The strength of the bonds between the intermediates and the catalyst surface determines the rate of chemical reaction and therefore determines the selectivity of the catalyst. It can be visualized easily that the selectivity of a heterogeneous catalyst is not a static (constant) property. Catalyst selectivity is a dynamic property; it may change with operating temperature, pressure, and reactant concentration.

Any knowledge of the affinity between the intermediates and the catalyst surface is a help in the design of commercial heterogeneous catalysts. Most commercial heterogeneous catalysts operate in the temperature range of 50–500°C (125–930°F). To predict the behavior of a heterogeneous catalyst, the relation between the surface properties and catalyst selectivity has to be correlated. For commercial operations the desired characteristics for the catalyst to be considered are long-time operability, mechanical strength, activity, selectivity, replacement frequency, and regeneration and replacement costs.

From a practical point of view, if the presence of a small amount of a substance can make a big difference in the conversion of reactants to a certain desired product or products, then this substance can be called a catalyst. Some industrially important processes that are based on heterogeneous catalysts are (Satterfield, 1991)

1. Production of sulfuric acid by the oxidation of sulfur dioxide (SO_2) to sulfur trioxide (SO_3) using a platinum catalyst (1875)
2. Production of sulfuric acid by the lead chambers method, by the oxidation of sulfur dioxide (SO_2) to sulfur trioxide (SO_3) using vanadium pentoxide (V_2O_5) as a catalyst
3. Partial oxidation of methanol to formaldehyde using silver oxide (Ag_2O) or iron molybdate [$Fe_2(MoO_4)_3$] as a catalyst (1890)
4. Oxidation of ammonia to nitrogen oxides using platinum catalyst for nitric acid production (Ostwald, in 1903)
5. Ammonia synthesis from hydrogen (H_2) and nitrogen (N_2) using an iron catalyst (Haber and Bosch, in 1908)
6. Methanol synthesis by the hydrogenation of carbon monoxide (CO) using a cupric oxide–zinc oxide (CuO-ZnO) catalyst (1923)
7. Hydrocarbon synthesis by the hydrogenation of carbon monoxide (CO) using iron or cobalt catalysts (Fischer–Tropsch method, 1930)

6. MECHANISM OF HETEROGENEOUS CATALYSIS

The actual mechanism of heterogeneously catalyzed chemical reactions may not be known in full detail. The basic concept of heterogeneous catalysis is that a small amount of the additive results in a large change in the product composition without being appreciably consumed in the process. Based on this definition, many polymerization catalysts (also called initiators) such as organic peroxides will not be considered catalysts. Consumption of these initiators is nonstoichiometric, but they are completely consumed during the polymerization and cannot be regenerated as true catalysts. Similarly, the reaction between hydrogen and oxygen is influenced by radiation and bombardment with high energy particles, but these processes are not considered catalysts either, for similar reasons.

In some cases, free radicals are formed on the surface of a heterogeneous catalyst as a result of reactant–catalyst interaction. These free radicals may desorb from the surface and enter into chemical reactions in the homogeneous fluid phase. These types of reactions are more common in high temperature (400–800°C; 750–1470°F) gas-phase reactions such as steam reforming, synthesis gas production, and partial oxidation reactions (e.g., the oxidative coupling of methane). Because of high temperature operating conditions, thermal and catalytic reactions may proceed simultaneously, resulting in unwanted deposition of carbon on the active sites of the heterogeneous catalysts.

Some additives act as a negative catalyst (reaction inhibitor), i.e., they slow down certain chemical reactions. Negative catalysts can be seen especially in free radical reactions. For example, in hydrocarbon combustion the negative catalyst reduces the number of free radicals in the chain of combustion reactions, resulting in a slowdown in the rate of the overall combustion process. Lead alkyls such as tetraethyllead are used in gasoline to suppress uncontrolled pressure increases (knock) in internal combustion engines. Tetraethyllead reacts with the free radicals formed in the combusting mixture, reduces their reactivity, and causes a delay in the formation of the combustion wave.

6.1 Active Sites

Macroscopic properties of all heterogeneous catalysts are nonuniform in the sense that physical and chemical properties of the surface change with the location on the surface (lattice defects, corners, edges, etc.). Variations of the heat of adsorption and the activation energy of adsorption as a function of temperature and percent surface covered are the signs of nonuniformity on the catalyst surfaces. Furthermore, observation of more than one maximum for chemisorption isobars indicates that more than one type of chemisorption reaction may take place between the reactant and the catalyst.

Some reactions taking place on metal surfaces are independent of the surface characteristics. These reactions are called structure-insensitive, and their rates are proportional to the number of atoms per unit area of the catalysts (e.g., on metal surfaces there are about 10^{15} atoms per square centimeter). Some reactions, however, are dependent on the nature of the surface structure; these are called structure-sensitive reactions. Intermediates formed between the reactants and the catalyst may move to the neighboring active sites by surface diffusion and be transformed into different intermediates that may produce different products. These types of chemical reactions are structure-sensitive.

It is difficult to measure the true number of active sites of a heterogeneous catalyst. Physical and chemical adsorption techniques may provide information on the total number of active sites, but some of these active sites may not be effective for the desired reaction. The number of active sites that promotes the desired reaction (also known as the turnover number) is a function of temperature, pressure, and the composition of the reacting fluid.

6.2 Selectivity and Catalytic Activity

Catalytic activity refers to the role of a catalyst in increasing the rate of a chemical reaction carrying the reacting system toward chemical equilibrium. The performance of industrial reactors with heterogeneous catalysts is measured by *space–time yield* (STY), the quantity of product formed per unit time per unit volume of the reactor. As an example, the molar balance for a fixed-bed reactor operating at steady-state conditions is given by

$$N_p = \alpha_p V r \tag{12.22}$$

where N_p is the molar flow rate (mol/h) of the product in the exit stream (assuming the feedstream does not contain the product species), r is the intensive reaction rate [mol/(m$^3 \cdot$ h)], V is the reactor volume (m^3), and α_p is the stoichiometric coefficient for the product. Using Eq. (12.22), STY can be expressed as

$$\text{STY} = \frac{N_p}{V} = \alpha_p r \tag{12.23}$$

which is commonly used in industrial practice. High STY is desired for the economics of commercial operations.

Catalyst activity can be measured by using one of the model compounds as a feed and comparing it to other catalysts [i.e., using Eq. (12.23)] provided that the rates are measured at standard operating conditions (temperature, pressure, feed, and hydrodynamic conditions). The temperature required to achieve a given conversion is also considered a measure of catalytic activity.

Selectivity of a heterogeneous catalyst is a measure of the effectiveness of the catalyst in accelerating a certain reaction to form one or more of the desired products. Formation of specific intermediates would result in an increase in the extent of the formation of certain products. It is always recommended that the product composition at chemical equilibrium be predicted by using the thermodynamic principle of minimization of Gibbs free energy (see Section 4 of Chapter 7). Thermodynamic equilibrium calculations would give sufficient information about the maximum yield that could be obtained with the presence of heterogeneous catalysts. Selectivity is a property of the catalyst and usually changes with operating conditions, i.e., temperature, pressure, reactant composition, and reaction extent (or conversion).

Selectivity and functionality are two characteristics of a heterogeneous catalyst that are related to and controlled by equilibrium conditions (Satterfield, 1991). For example, ethanol would dominantly be subject to dehydrogenation on a copper catalyst,

$$C_2H_5 - OH \rightarrow CH_3 - CHO + H_2 \tag{12.24}$$

and to dehydration reactions on alumina catalyst,

$$C_2H_5 - OH \rightarrow C_2H_4 + H_2O \tag{12.25}$$

$$2C_2H_5 - OH \rightarrow C_2H_5 - O - C_2H_5 + H_2O \tag{12.26}$$

because of the affinity of copper to hydrogen and that of alumina to water molecules. Selective adsorption of hydrogen on copper makes it a mild hydrogenation/dehydrogenation catalyst (as discussed in microscopic reversibility, Chapter 6, Section 4).

On the other hand, adsorption of water on alumina makes it a dehydration catalyst. Functionality, as demonstrated in this example, is characteristic of selective adsorption, i.e., hydrogen on copper and water on alumina affect the selectivity of the catalyst. Based on these schematics, hydrogenation/dehydrogenation selectivity of a catalyst can be estimated by measuring its hydrogen chemisorption characteristics. Isomerization of hydrocarbons takes place on acidic sites of catalyst surfaces; therefore, measurement of acidic sites of a catalyst (by adsorption of ammonia or organic amine) can give a good estimate of its selectivity for isomerization reactions.

In some cases two types of functionalities may be needed. This can be achieved by blending two types of catalysts of different selectivities or implanting two types of functionalities on the same catalyst. For example, isomerization of n-paraffins to isoparaffins takes place by consecutive reactions of dehydrogenation of n-paraffins to n-olefins, of n-olefins to isoolefins, and of isoolefins to isoparaffins. In this mechanism, a platinum catalyst promotes the dehydrogenation and hydrogenation reactions and acidic sites implanted on the platinum catalyst promote the isomerization reaction.

An increase in temperature causes an increase in catalytic activity (like most chemical reactions); however, it causes a decrease in catalyst life. An increase in temperature may also increase the rate of undesired reactions, resulting in a decrease in the selectivity of the catalyst. For example, in the consecutive reactions

$$A \rightarrow B \rightarrow C \tag{12.27}$$

if B is the desired product, then maximum conversion of reactant A into B would be obtained at some intermediate temperature, with maximum selectivity at the lowest conversion. On the other hand, if C is the desired product, an increase in temperature would be beneficial (reactor optimization is also discussed elsewhere, Chapter 9, Section 2.2).

As demonstrated in the above example, selectivity is related closely to the percent of the intermediates formed during the chemisorption of the reactants on the catalyst surface, from which the desired product can be obtained. The yield is the measure of product quality, which is expressed as the quantity of product formed from the reactant. In most industrial practices, the yield is expressed on a weight basis. Yield can be greater than 100%, as it is in partial oxidation processes. In the fuel industry, products are sold on a volume basis. In this case yield may also exceed 100%, because most of the fuel products are lighter in density than the crude oil.

6.3 Thermodynamics of Adsorption

For a simple reversible reaction between the reactants and products, R $\rightleftharpoons$ P, the standard Gibbs free energy of formation is expressed as

$$\Delta G^0 = -RT \ln K = -RT \ln\left(\frac{c_P}{c_R}\right) \tag{12.28}$$

where K is the equilibrium constant, R is the universal gas constant, T is the temperature, and c_P and c_R are the concentrations of the product and reactant. The presence of a solid catalyst cannot change ΔG^0 and therefore cannot change the ratio c_P/c_R.

The selectivity of a catalyst is related to its ability to promote the formation of certain intermediates by chemisorption, which results in the formation of different products. In this process, only the selected reactions are carried out to their equilibrium, while the undesired reactions are allowed to progress at their normal slow rates.

As an example, the hydrogenation of carbon monoxide on a copper–zinc oxide–alumina ($Cu/ZnO/Al_2O_3$) catalyst produces methanol (CH_3OH), even though the formation of paraffins, olefins, and higher alcohols is more favorable thermodynamically under the process conditions. On the other hand, hydrogenation of carbon on iron or cobalt catalysts (Fischer–Tropsch synthesis) produces paraffins and olefins even at lower pressures and temperatures than those used for the formation of methanol. Selectivity of the catalyst is responsible for the formation of different products (i.e., alcohol versus paraffins).

Because the chemisorbed intermediates or reaction pathways control the products in the dynamic state of the chemical reactions (i.e., far from the equilibrium state), considering an alternative reaction pathway is also a possibility. In this alternative, a different reaction pathway can be determined and an appropriate catalyst can be searched for. As an example, the isomerization of an olefin such as 1-butene on an acid catalyst will be considered. A number of transformation pathways are possible. Among these, cis–trans isomerization on a mild acid catalyst followed by double-bond migration can be selected.

Certain reactions can also be promoted to their thermodynamically allowable limits by removing a product from the reacting system either physically or chemically. For example, in dehydrogenation reactions, the concentration of hydrogen in the product can be lowered by adding oxygen to convert it to water to promote the overall reaction.

The microscopic reversibility principle (Chapter 6, Section 4) can be applied to a single reaction step of a heterogeneous reaction. If a catalyst accelerates the forward reaction it should also accelerate the reverse reaction. This can be concluded from the fact that catalysts cannot change the equilibrium constant K, which is the ratio k_f/k_r where k_f and k_r are the rate constants of the forward and reverse reactions, respectively (i.e,. at equilibrium $k_f c_R = k_r c_P$, then $K = c_P/c_R$ or $K = k_f/k_r$). This principle is useful for screening catalysts for certain activity. In some cases, however, side reactions may cause a significant reduction in catalyst activity. For example, a nickel catalyst is highly active in the hydrogenation of organic compounds. It may not be effective for dehydrogenation reactions because the coke formed as a result of a side reaction causes a significant reduction in catalytic activity.

6.4 Reaction Pathways in Heterogeneous Catalysis

A heterogeneously catalyzed reaction between two reactants may take place at a considerably faster rate than its homogeneous equivalent. The difference in the rates is attributed to the fact that a heterogeneously catalyzed reaction passes through several elementary reaction steps with smaller activation energies, resulting in an increase in the overall reaction rate.

Heterogeneously catalyzed reactions involve three elementary steps:

1. Adsorption
2. Formation and decomposition of the activated complexes (which is basically the heterogeneous reaction)
3. Desorption of the products

The rates of these elementary steps are controlled by the number of active sites, the other reactant concentrations on the catalyst surface, and the corresponding activation energy (Fig. 12.6). As a result, a homogeneous reaction progressing at a slow reaction rate (because of the large activation energy) may take place at a much faster rate when a heterogeneous catalyst is used because of the much smaller activation energy of each step). Because heterogeneously catalyzed reactions involve these elementary steps, the apparent reaction rates may not be correlated by the Arrhenius relation over a wide range of temperature (i.e., $\ln k$ versus $1/T$). The rate of the heteroge-

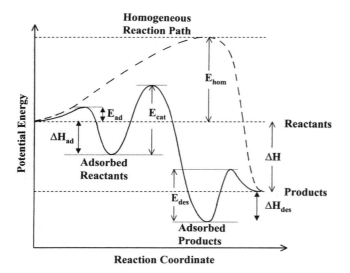

Figure 12.6 Reaction paths and associated activated energies suggested for homogeneous reactions and a heterogeneous catalyst.

neously catalyzed reaction is proportional to the accessible active surface area; as a result, it is generally desirable for a heterogeneous catalyst to have a surface area of $100 \, m^2/g$ or more.

In the case of some endothermic reactions, high temperature operating conditions are required because of the thermodynamic restrictions. When the reactor is operated at an excessively high temperature, no significant increase in the rate of selected reactions can be achieved by using any catalyst. These reactions are thermal reactions rather than catalytic reactions. For example, in the dehydrogenation of ethane (CH_3CH_3) to ethylene ($CH_2 = CH_2$), a high temperature (approximately 725°C; 1335°F) is required for 50% conversion at atmospheric pressure. Because of the high temperature operating conditions, this reaction is carried out industrially as a homogeneous thermal reaction. Another example is the dehydrogenation of methane to carbon black, which is carried out industrially above 85 0°C (1560°F) as a homogeneous thermal reaction.

On the other hand, the dehydrogenation of higher paraffins such as that of butane to butene gives a considerable conversion at relatively lower temperature operating conditions. A substantial increase in the rate of this reaction is obtained by using chromia–alumina (Cr_2O_3-Al_2O_3) catalyst. Similarly, it is preferable for hydrocracking reactors to be operated at the lowest possible temperature, because at higher operating temperatures thermal reactions would dominate the catalytic reactions, which would result in reducing the selectivity and in excessive coke formation (Satterfield, 1991).

7. ADSORPTION

When a fluid is exposed to a solid, interactions take place between the fluid and the solid. These interactions are functions of the solid and fluid characteristics, temperature, pressure, and the hydrodynamics of the fluid–solid system.

Attachment of fluid molecules on the solid surface as a result of molecular collisions is defined as adsorption. The strength of the adsorption is determined by the nature of the attachment of the fluid molecules on the surface. The strength of adsorption causes two distinct types of adsorption, which are called physical adsorption and chemisorption (chemical adsorption).

Physical adsorption takes place as a result of weak van der Waals and dipole–dipole interactions, resulting in bringing fluid molecules together in a quasi-liquid layer on the solid surface. Physical adsorption is a reversible process and is similar to condensation of vapor molecules onto a liquid of the same composition. Many chemical reactions take place at high temperatures and pressures above the boiling points of the reactants. Therefore, physical adsorption cannot play a role in heterogeneous catalytic reactions. However, physical adsorption is important because it provides information about the surface area, pore size, and pore size distribution of the catalyst.

Chemisorption is basically a chemical reaction between the fluid reactant and the catalyst surface, which results in the bonding of the reactant onto the catalyst surface. An understanding of chemisorption would help us to interpret the mechanism of heterogeneous catalysis of chemical reactions. Chemisorption may also be useful for characterization of the catalyst surface. Chemisorption can be characterized by the surface electric potential or work function, surface electrical conductivity, and collective paramagnetism (for paramagnetic absorbers) methods.

The correlation between the catalytic activity and chemisorption of one or more of the reactants on the catalyst surface is an indication that most heterogeneously catalyzed chemical reactions and chemisorption are interrelated. Furthermore, forces involved with physical adsorption are not strong enough to overcome the forces of chemical bonding. Unlike physical adsorption, chemisorption is an irreversible process.

The main difference between physical and chemical adsorption is the heat of adsorption. In physical adsorption, the heat of adsorption for the formation of a monolayer of adsorbed molecules on the solid surface exceeds the heat of liquefaction, probably because of the interaction between the catalyst surface and the adsorbed molecules. For relatively small molecules such as carbon monoxide, nitrogen, and methane (CH_4), the heat of physical adsorption (ΔH_{ad}) is on the order of $-10\,kJ/mol$ (exothermic, $\Delta H_{ad} < 0$). The heat of physical adsorption may be slightly higher if the solid behaves as a molecular sieve (e.g., zeolites or certain carbon-based adsorbents) in which the adsorbed molecules are surrounded by the solid in almost all directions. On the other hand, the heats of chemical adsorption (chemisorption) are comparable to the heats of chemical reaction, in the range of -80 to $-200\,kJ/mol$ (-20 to $-50\,kcal/mol$) and may be as high as $-600\,kJ/mol$ ($-145\,kcal/mol$).

Physical adsorption is always exothermic ($\Delta H_{ad} < 0$), but chemical adsorption may be exothermic or endothermic ($\Delta H_{ad} > 0$). From thermodynamics principles, a spontaneous process such as adsorption can take place if the Gibbs free energy is negative (Satterfield, 1991):

$$\Delta G_{ad} = \Delta H_{ad} - T\,\Delta S_{ad} < 0 \tag{12.29}$$

In most chemisorption processes, $\Delta S_{ad} < 0$ ($-T\,\Delta S_{ad} > 0$), because the molecules attached to the solid surface would lose their freedom of orientation in comparison to the molecules in the fluid state. Then, in these systems the absolute value of ΔH_{ad} must be greater than $T\,\Delta S_{ad}$ so that $\Delta G_{ad} < 0$ (i.e., $\Delta H_{ad} - T\,\Delta S_{ad} < 0$) and the adsorption process proceeds.

In some cases, the chemisorption process takes place with the dissociation of the adsorbed molecules. This results in an increase in the two-dimensional freedom of the molecules, which could make $\Delta S_{ad} > 0$ and $\Delta H_{ad} > 0$ (endothermic), still satisfying the condition $\Delta G_{ad} < 0$ [Eq. (12.29)]. In this case the dissociation energy of the absorbed molecules must be greater than the energy of formation of the bonds between the absorbent molecules and the adsorbate. Adsorption of hydrogen molecules on iron contaminated with sulfide is an example of endothermic adsorption, but endothermic adsorption is rare.

Both physical adsorption and chemisorption are dynamic processes; therefore they both take place at certain rates. Physical adsorption, like condensation, requires no activation energy, so it can take place as fast as molecules strike a surface. A slow uptake observed in a fine-pore adsorbent (e.g., zeolite and some carbons) is caused by the diffusion of absorbent molecules into the fine pores (the rate of adsorption is controlled by the diffusion, not by the adsorption). Chemisorption exhibits activation energy, just like chemical reactions. As a result, chemisorption takes place at an appreciable rate only above a certain temperature range. If the surface is highly active, however, chemisorption may occur rapidly even at low temperatures.

Temperature plays an important role in all adsorption processes. Physical adsorption may take place under conditions below the critical temperature. In physical adsorption, the amount of gas adsorbed decreases as temperature increases. The amount of molecules physically adsorbed is measured by relative pressure, p/p_0, where p is the partial pressure of the vapor and p_0 is the vapor pressure of the pure adsorbent at the same temperature.

When $p/p_0 < 0.01$, the amount of physical adsorption is negligible. When p/p_0 is about 0.1, the amount adsorbed corresponds to monolayer adsorption, and as p/p_0 is increased, multilayer adsorption occurs until essentially a bulk liquid is reached at $p/p_0 = 1$.

Because chemisorption is similar to a chemical reaction, the chemisorption process is always slower than the physical adsorption process, especially at low temperatures. Chemisorption takes place in a monolayer, on only a small fraction of the surface, provided that the surface is accessible and capable of forming a chemical bond with the adsorbent (unlike physical adsorption). In many cases, it is difficult to distinguish chemisorption and chemical reaction between the adsorbent and the solid (e.g., adsorption of CO_2 on char). Chemisorption can take place at temperatures greatly above the boiling temperature or above the critical temperature of the adsorbent. The effect of temperature on chemisorption is complex and cannot be generalized. At low temperatures, the rate of chemisorption is so slow that saturation cannot be achieved. Frequently, chemisorption reaches a saturation state over a certain temperature, and in that state the amount of chemisorbed material becomes constant. Also, more than one kind of chemisorption may take place simultaneously at different rates over a wide range of temperature.

Physical adsorption is a reversible process, and equilibrium is established if pore diffusion limitations do not exist. Chemisorption, however, may or may not be reversible. Chemical change observed in the adsorbate is generally a sign of chemisorption. Oxygen chemisorbed on char can be recovered in the form of carbon monoxide or carbon dioxide; hydrogen chemisorbed on oxides can be recovered as water, and hydrogen chemisorbed on char can be recovered as methane upon heating. Hydrogen–deuterium exchange is a common test to detect chemisorption of hydrogen. If hydrogen–deuterium is detected in the desorbed gas after hydrogen and deuterium (D_2) adsorption, chemisorption must have occurred.

Physical adsorption is not surface-specific and will occur on most surfaces provided that p/p_0 is sufficiently large, whereas chemisorption is surface-specific and will occur only if a chemical bond can be formed between the adsorbate and the adsorbent. Chemisorption isotherms show more complex behavior as a result of the heterogeneous structure of the surface, and some of the chemisorption may even be reversible. Two or more types of chemisorption at different rates may take place at different temperatures, promoting different reactions (i.e., affecting the selectivity). At constant pressure, more than one chemisorption maximum can be obtained as temperature increases.

Uncontrolled chemisorption of water vapor on coal (i.e., coal dried below its capacity moisture level) may release sufficient heat to cause excessive local heating in the coal pile. Uncontrolled oxidation may progress at these locally heated spots in a nonlinear fashion, which is called spontaneous combustion of the coal.

Spontaneous coal combustion is a serious industrial problem (especially for coals of lower ranks, high volatile bituminous and low volatile subbituminous coals) that can be controlled by limiting the coal moisture content above the capacity moisture level in the coal-drying process. Another option would be to plug the coal pores with an additive (or partial pyrolysis of the coal) to prevent the penetration of humidity into the coal pore structure.

In principle, an increase of fluid pressure and decrease in temperature increase the extent of adsorption on solids. An *adsorption isotherm* is the relationship at constant temperature between the partial pressure of the adsorbate and the amount adsorbed at equilibrium, which varies from zero at $p/p_0 = 0$ to infinity at $p/p_0 = 1$. At low pressures, when the adsorbent is exposed to the solid, the adsorbed molecules on the solid surface migrate into the smallest pores. This is explained by Kelvin's relation for the vapor pressure of liquids in a capillary tube:

$$\ln\frac{p_c}{p} = -\frac{2\sigma V_M \cos\theta}{r_c RT} \tag{12.30}$$

where p and p_c are the vapor pressure of the liquid (adsorbed phase) over flat and curved surfaces, respectively, V_M is the molecular volume of the liquid, r_c is the radius of curvature, σ is the surface tension of the liquid, θ is the contact angle, and R is the universal gas constant. As the ratio p/p_0 increases, pores with larger diameters are filled with the adsorbed molecules.

Heat of adsorption is a characteristic property of the type of adsorption, the extent of adsorption, and the degree of heterogeneity of the surface. For a reversible adsorption isotherm, the differential heat of adsorption can be calculated from the Clausius–Clapeyron equation, as a function of the adsorbed gas volume v (Glasstone and Lewis, 1963),

$$\left(\frac{\partial p}{\partial T}\right)_v = \frac{L}{T(V_g - V_{ad})} \tag{12.31}$$

$$\left(\frac{\partial \ln p}{\partial T}\right)_v = \frac{L}{RT^2} \tag{12.32}$$

where V_g and V_{ad} are the molar gas and liquid volumes of the absorbent ($V_g \gg V_{ad}$) and L is the isosteric heat of adsorption. A calorimetric method can also be used to measure the integral value of the heat of adsorption, which is the average value over the surface coverage studied.

Physical adsorption provides information about the pore surface area and pore size distribution; therefore, it finds important applications in heterogeneous catalysis and gas–solid reactions. In this section some of the recognized adsorption isotherms will be presented. From the modeling point of view, the derivation of adsorption isotherms is exactly the same for physical and chemical adsorption, provided the adsorption process is reversible (i.e., no chemical change in the adsorbate) and the equilibrium condition is reached.

The *Langmuir adsorption isotherm* assumes that monolayer adsorption takes place on the surface of the solid and that the energy of adsorption is independent of surface coverage. The Langmuir adsorption isotherm also assumes that adsorption is linearly proportional to the fraction of the free surface and the partial pressure of the adsorbed molecules in the gas phase. This assumption implies that the solid surface has uniform activity for adsorption to take place and that adsorbed molecules do not exhibit forces of attraction or repulsion. For a solid adsorbing pure molecules of A, the rates of adsorption r_{ab} and desorption r_{des} are given by

$$r_{ad} = \left(\frac{dn_A}{dt}\right)_{ad} = k_{ad}(1 - \theta_A)p_A \tag{12.33}$$

$$r_{des} = \left(\frac{dn_A}{dt}\right)_{des} = k_{des}\theta_A \tag{12.34}$$

where n_A is the number of molecules of A adsorbed on the solid, θ_A is the fraction of the surface being occupied by the adsorbed molecules, and p_A is the partial pressure of the adsorbed molecules in the gas phase. At equilibrium, the rate of adsorption and the rate of desorption are equal:

$$k_{ad}(1 - \theta_A)p_A = k_{des}\theta_A \tag{12.35}$$

from which the Langmuir adsorption isotherm can be expressed in terms of θ_A as

$$\theta_A = \frac{k_{ad}p_A}{k_{des} + k_{ad}p_A} = \frac{K_A p_A}{1 + K_A p_A} \tag{12.36}$$

where $K_A = k_{ad}/k_{des}$ is the adsorption equilibrium constant. K_A is like a thermodynamic property, and its dependence on temperature can be expressed as

$$K_A = K_{A,0}\exp(-\Delta H_{ad}/RT) \tag{12.37}$$

where $K_{A,0}$ is a preexponential coefficient and ΔH_{ad} is the heat of adsorption, which is similar to the isosteric heat of adsorption L of the Clausius–Clapeyron equation [(Eq. (12.31)]. A large value for K_A implies strong adsorption or a larger fraction of surface coverage. For exothermic adsorption, $\Delta H_{ad} < 0$, K_A decreases exponentially with an increase in temperature. The Langmuir adsorption isotherm can be approximated to a linear relation between the percent surface covered θ_A and p_A ($\theta_A = K_A p_A$) provided that p_A or K_A is low enough to satisfy $K_A p_A \ll 1$.

The Langmuir adsorption isotherm fails to correlate many adsorption isotherms. As the surface coverage increases, the heat of adsorption decreases and the repulsive forces between the adsorbed molecules increase, which are not considered in the Langmuir model. In the use of the Langmuir adsorption isotherm to correlate catalytic reactions, the variation in the heat of adsorption as a function of the surface covered does not introduce a serious error. Molecules adsorbed onto the surface with a large heat of adsorption (a strong molecule–solid bonding) and with small heat of adsorption (a weak molecule–solid bonding) would not participate in the catalytic reactions. The molecules bonded to the surface with a moderate heat of adsorption would play the most important role in catalytic reactions.

The *Freundlich adsorption isotherm* is an empirical model expressed as

$$\theta_A = cp_A^{1/n} \tag{12.38}$$

where c and n ($n > 1$) are the proportionality and power constants. Both of these constants decrease with increases in temperature.

The *Temkin adsorption isotherm* assumes a linear decrease in adsorption energy with the surface occupied by the molecules:

$$q_{ad} = q_{ad,0}(1 - \alpha\theta) \tag{12.39}$$

With this energy distribution, the Langmuir adsorption isotherm can be expressed as

$$\theta = \frac{RT}{q_0\alpha}\ln A_0 p \tag{12.40}$$

where q is the differential heat of adsorption (q_0 is its value at zero surface coverage), $A_0 = a_0\exp(-q_0/RT)$, and a_0 and α are constants. Like the Freundlich isotherm, the Temkin isotherm has two adjustable constants, a_0 and α.

Measurement of the amount of adsorbent adsorbed on a solid sample can be used to predict the surface area and surface characteristics of the solid. The *BET* (Brunauer–Emmett–Teller) *method* (Brunauer et al., 1938) almost became the standard of many catalyst laboratories. In this approach, in the first layer of adsorption the heat of adsorption is assumed to be independent of the surface occupied by the adsorbent molecules. The heat of adsorption beyond the first layer is assumed to be the same as the heat of condensation. The BET isotherm gives the following relation for the amount (or volume) of molecules adsorbed as a function of pressure:

$$\frac{p}{V(p_0 - p)} = \frac{1}{V_m C} + \frac{C - 1}{CV_m}\left(\frac{p}{p_0}\right) \qquad (12.41)$$

where V is the volume of the gas adsorbed (at pressure p), V_m is the volume of the gas adsorbed in a monolayer, p and p_0 are the pressure and saturation pressure of the adsorbed gas, and C is a constant given by

$$C = \exp\left(\frac{q_{ad} - q_{con}}{RT}\right) \qquad (12.42)$$

where q_{ad} is the heat of adsorption, q_{con} is the heat of condensation, and R is the universal gas constant. It is clear from Eq. (12.41) that a plot of $p/[V(p_0 - p)]$ versus p/p_0 should give a straight line (a linear relation), from the slope [i.e., $(C - 1)/CV_m$] and intercept (i.e., $1/CV_m$) of which V_m and C can be calculated. Measuring C at different temperatures would provide information [using Eq. (12.42)] about the heat of adsorption q_{ad}.

The BET adsorption model correlates many adsorption data for the p/p_0 ratio in the range 0.05–0.3, which is suitable for surface area measurements provided that the adsorbed molecules are sufficiently larger than the pore size of the solid.

8. CHEMISORPTION AND REACTION RATE IN HETEROGENEOUS CATALYSTS

Chemisorption, unlike physical adsorption, is much like a chemical reaction between the fluid molecules and the surface of the solid. A better understanding of chemisorption may result in better design of heterogeneous catalysts. Chemisorption would provide information about the rate of chemical adsorption (and desorption), surface heterogeneity, and the strength of bonds between the absorbent and adsorbate. The information obtained about the pore surface area (e.g., using the BET method) would be useful for the modeling of chemical kinetics dealing with heterogeneous catalysis.

8.1 Langmuir–Hinshelwood Model

In the *Langmuir–Hinshelwood model*, catalytic reactions are assumed to occur between the adsorbed (chemisorbed) species on the catalyst surface. Reactant and product molecules that have an affinity with the catalyst surface compete with each other for chemisorption. Adsorption (chemisorption) isotherms are assumed to obey the Langmuir adsorption isotherm, and the adsorption process is in equilibrium (Satterfield, 1991).

On the basis of these assumptions, reaction rate expressions can be developed that consider adsorption affinities of the reactant and the product molecules with the catalyst surface and the molecular states of the adsorbed molecules. In heterogeneous catalysis, the basic mechanisms of chemisorption, chemical reaction of the chemisorbed molecules (or the products of the chemisorption) and desorption of the products, affect the overall or apparent rate of the chemical reaction. More important, these steps control the selectivity of the catalytic process.

When the decomposition products are not adsorbed, the molecules of A (the reactant) are chemisorbed on the catalyst surface and react on the catalyst sites, producing the product P,

$$A \rightarrow P \tag{12.43}$$

where the product P is not adsorbed on the catalyst surface. Using the relation for the Langmuir isotherm $[\theta_A = K_A p_A/(1 + K_A p_A)]$ and assuming that the reaction rate is proportional to the quantity of adsorbed A molecules (i.e., the fraction of surface sites θ_A occupied by A molecules), the overall rate of the chemical reaction can be expressed as

$$r_s = k_s \theta_A = \frac{k_s K_A p_A}{1 + K_A p_A} \tag{12.44}$$

where k_s is the rate constant with Arrhenius-type temperature dependence. The units of r_s depend on the units of the rate constant k_s; if k_s is defined in terms of moles per square meter per second $[\mathrm{mol}/(\mathrm{m}^2 \cdot \mathrm{s})]$ then r_s is also. If r_s and the specific surface area of the catalyst a_s (in m^2/g) are known, then the reaction rate r_m per unit mass of catalyst, in terms of moles per gram of catalyst per second $[\mathrm{mol}/(\mathrm{g} \text{ catalyst} \cdot \mathrm{s})]$ can be calculated as

$$r_m = r_s a_s \tag{12.45}$$

For $K_A p_A \ll 1$, i.e., for species A weakly adsorbed on the solid, Eq. (12.44) is approximated to

$$r_s = k_s K_A p_A \qquad \text{for} \qquad K_A p_A \ll 1 \tag{12.46}$$

where the reaction rate expression becomes linear (first order) with respect to p_A. For $K_A p_A \gg 1$, i.e., for species A strongly adsorbed on the solid, Eq. (12.44) is approximated to

$$r_s = k_s p_A^0 = k_s \neq f(p_A) \qquad \text{for} \qquad K_A p_A \gg 1 \tag{12.47}$$

where the rate expression becomes zero order (independent) with respect to p_A.

When the decomposition products are adsorbed, the molecules of A (the reactant) are chemisorbed on the catalyst surface and decompose to products B and C on the catalyst sites:

$$A \rightarrow B + C \tag{12.48}$$

where species A, B, and C are all adsorbed on the surface of the catalysts. Because all three species have a tendency to be adsorbed by the catalyst surface, the Langmuir adsorption isotherm for molecules of A can be expressed as

$$k_{A,ad}[1 - (\theta_A + \theta_B + \theta_C)]p_A = k_{A,des}\theta_A \tag{12.49}$$

or

$$\theta_A = K_A p_A[1 - (\theta_A + \theta_B + \theta_C)] \tag{12.50}$$

where $K_A = k_{A,ad}/k_{A,des}$. Similarly, the following relations can be developed:

$$\theta_B = K_B p_B[1 - (\theta_A + \theta_B + \theta_C)] \tag{12.51}$$

and

$$\theta_C = K_C p_C[1 - (\theta_A + \theta_B + \theta_C)] \tag{12.52}$$

Adding Eqs. (12.50), (12.51), and (12.52) gives

$$\theta_A + \theta_B + \theta_C = [1 - (\theta_A + \theta_B + \theta_C)](K_A p_A + K_B p_B + K_C p_C) \tag{12.53}$$

and subtracting both sides of Eq. (12.53) from unity gives

$$[1 - (\theta_A + \theta_B + \theta_C)] = 1 - [1 - (\theta_A + \theta_B + \theta_C)](K_A p_A + K_B p_B + K_C p_C) \tag{12.54}$$

which can be solved for $1 - (\theta_A + \theta_B + \theta_C)$ as

$$1 - (\theta_A + \theta_B + \theta_C) = \frac{1}{1 + K_A p_A + K_B p_B + K_C p_C} \tag{12.55}$$

by which θ_A of Eq. (12.50) becomes

$$\theta_A = \frac{K_A p_A}{1 + K_A p_A + K_B p_B + K_C p_C} \tag{12.56}$$

If it is postulated that an unoccupied site is also needed for the adsorption of products B and C produced from reactant A, the rate expression could be expressed [using Eqs. (12.55) and (12.56)] in the form

$$r_s = k_s \theta_A [1 - (\theta_A + \theta_B + \theta_C)] = \frac{k_s K_A p_A}{(1 + K_A p_A + K_B p_B + K_C p_C)^2} \tag{12.57}$$

If it is speculated that the reaction rate at the surface of the catalyst is first-order with respect to θ_A, then the rate of decomposition of A becomes [using Eq. (12.56)]

$$r_s = k_s \theta_A = \frac{k_s K_A p_A}{1 + K_A p_A + K_B p_A + K_C p_C} \tag{12.58}$$

Bimolecular reactions, such as the reaction between reactant species A and B,

$$A + B \rightarrow C \tag{12.59}$$

may take place on the catalyst surface in bimolecular fashion. If the reactant and the product molecules are adsorbed on the catalyst surface, the rate of reaction can be expressed as

$$r_s = k_s \theta_A \theta_B = \frac{k_s K_A K_B p_A p_B}{(1 + K_A p_A + K_B p_B + K_C p_C)^2} \tag{12.60}$$

the derivation of which would be similar to that of the previous reaction rate expressions.

Chemisorption of A_2 molecules may take place by *dissociation* of A_2 molecules into two A molecules, and the *desorption* of two A molecules may take place by association of two A molecules:

$$A_2 \rightleftharpoons 2A \rightarrow P \tag{12.61}$$

This reaction would require two unoccupied sites adjacent to one another. Up to fairly high fractional conversions, the number of pairs of adjacent sites is proportional to the square of the number of single sites. Then the rate of adsorption may be given by

$$r_{ad} = k_{ad} p_A (1 - \theta_A)^2 \tag{12.62}$$

Assuming that desorption involves the interaction of two neighboring adsorbed molecules,

$$r_{des} = k_{des} \theta_A^2 \tag{12.63}$$

At equilibrium, the rate of adsorption r_{ad} and the rate of desorption r_{des} are equal, from which θ_A can be solved as

$$\theta_A = \frac{(K_A p_A)^{1/2}}{1 + (K_A p_A)^{1/2}} \tag{12.64}$$

where $K_A (= (K_{ad}/K_{des})$ is the adsorption equilibrium constant for molecule A_2, as defined earlier.

If the rate of formation of the products is first-order with respect to dissociated molecules of A, the rate expression becomes

$$r_s = k_s \theta_A = \frac{k_s (K_A p_A)^{1/2}}{1 + (K_A p_A)^{1/2}} \tag{12.65}$$

If the rate of formation of the products is second-order with respect to dissociated molecules of A, the rate expression becomes

$$r_s = k_s \theta_A^2 = \frac{k_s^2 K_A p_A}{\left[1 + (K_A p_A)^{1/2}\right]^2} \tag{12.66}$$

For a catalytic reaction taking place between A_2 and B as a result of the reaction between two A molecules (formed by dissociation of A_2) and B,

$$A_2 + B \rightleftharpoons 2A + B \rightarrow P \tag{12.67}$$

and assuming that only A and B are adsorbed on the surface, the reaction rate [Eq. (12.66)] becomes

$$r_s = k_s \theta_A^2 \theta_B = \frac{k_s K_A K_B p_A p_B}{\left[1 + (K_A p_A)^{1/2} + K_B p_B\right]^3} \tag{12.68}$$

the derivation of which is similar to the previous derivations of rate expressions.

Chemisorption of a hydrogen molecule (H_2) on most metals is the most common example of dissociative adsorption. If the adsorption of hydrogen on a metal catalyst is not strong in comparison to that of the other reactant, then the rate of the hydrogenation reaction may show first-order dependence on the hydrogen partial pressure.

For a chemical reaction *between the independently chemisorbed reactants* A and B,

$$A + B \rightarrow P \tag{12.69}$$

if the molecules of A and B are selectively adsorbed on different sites on the catalyst surface, the Langmuir isotherms for each molecule (or each type of active site) are

$$\theta_A = \frac{K_A p_A}{1 + K_A p_A} \tag{12.70}$$

and

$$\theta_B = \frac{K_B p_B}{1 + K_B p_B} \tag{12.71}$$

If the rate of chemical reaction on the catalyst surface is first-order with respect to chemisorbed molecules of A and B, the rate expression becomes

$$r_s = k_s \theta_A \theta_B = \frac{k_s K_A K_B p_A p_B}{(1 + K_A p_A)(1 + K_B p_B)} \tag{12.72}$$

which is a typical rate expression for dual-function catalysts.

It can be seen from the foregoing discussion of rate expressions that the rate of reaction dealing a the heterogeneous catalyst would be a different function of the reactant concentration (i.e., partial pressure) for different adsorption and reaction mechanisms.

8.2 Apparent Activation Energies

For a simple reaction and low surface area coverage (i.e., $K_A p_A \ll 1$), the apparent activation energy of the reaction would be independent of temperature. The rate expression for this specific example can be approximated to (Satterfield, 1991)

$$r_s = k_s K_A p_A = k_{app} p_A \tag{12.73}$$

where the apparent rate constant k_{app} would be equal to $k_s K_A$. Considering the temperature dependence of the adsorption coefficient,

$$K_A = K_{A,0} e^{q_{ad}/RT} \tag{12.74}$$

where q_{ad} is the heat of chemisorption, which can be taken to be independent of temperature. The surface reaction rate is dependent on temperature also in an exponential fashion (Arrhenius relation):

$$k_s = k_{s,0} e^{-E_s/RT} \tag{12.75}$$

where E_s is the activation energy of the surface reaction, which can also be taken to be independent of temperature. Introducing Eqs. (12.74) and (12.75) into Eq. (12.73), the overall rate expression becomes

$$r_s = A_{app} e^{-(E_s - q_{ad})/RT} p_A = A_{app} e^{-E_{App}/RT} p_A \tag{12.76}$$

where

$$E_{App} = E_s - q_{ad} \tag{12.77}$$

is the apparent (or experimentally determined) activation energy, E_s is the true activation energy of the chemical reaction taking place on the surface of the catalyst among the chemisorbed molecules, and q_{ad} is the heat of chemisorption. Because E_s and q_{ad} are normally positive, the apparent activation energy E_{App} would be lower than the true activation energy E_s.

It is clear that for a chemical reaction

$$A \rightarrow B + C \tag{12.78}$$

with A and B weakly adsorbed while C is strongly adsorbed, the rate expression can be expressed as

$$r_s = \frac{k_s K_A p_A}{(K_C p_C)^2} \tag{12.79}$$

and the apparent activation energy would be

$$E_{App} = E_s - q_A + 2q_C \tag{12.80}$$

where q_A and q_C are the heat of chemisorption for molecules A and C, respectively. The apparent activation energy can also be lower than E_s, provided that $q_A > 2q_C$.

It is the role of heterogeneous catalysts to lower the apparent activation energy for the overall reaction, which increases the selectivity. It is hoped that the advances in inorganic and coordination chemistry will help in the development of new catalysts for commercial applications in the decades to come.

REFERENCES

Ayling, A. B., and Smith, I. W. 1972. Combust. Flame 18:173–184.

Brunauer, S., Emmett, P. H., and Teller, E. 1938. J. Am. Chem. Soc. 60:309.

Essenhigh, R. H. 1981. Fundamentals of coal combustion. In: Chemistry of Coal Utilization. 2nd ed. M. A. Elliott, ed. Wiley, New York.

Field, M. A., Gill, D. W., Morgan, B. B., and Hawksley, P. G. W. 1967. Combustion of Pulverized Coal. British Coal Utilization Research Association, Leatherhead, Surrey, England.

Glasstone, S., and Lewis, D. 1963. Elements of Physical Chemistry. Macmillan, London, England.

Hottel, H. C., and Stewart, I. C. 1939. Ind. Eng. Chem. 32:719.

Laurendeau, N. M. 1978. Prog. Energy Combust. Sci. 4:221–270.

Matar, S., Mirbach, M. J., and Tayim, H. A. 1989. Catalysis in Petrochemical Processes. Kluwer, Dordrecht, The Netherlands.

Mulcahy, M. F. R., and Smith, I. W. 1969. Rev. Pure Appl. Chem. 19:81–108.

Nusselt, W. 1924. Ver. Dent. Ing. 68:124–128.

Parker, A. S., and Hottel, H. C. 1936. Ind. Eng. Chem. 28:1334–1341.

Satterfield, C. N. 1991. Heterogeneous Catalysis in Industrial Practice. Krieger, Malabar, FL.

Walker, P. L., Rusinko, R. F., and Austin, L. G. 1959. Adv. Catal. 11:133–221.

13

Pretreatment and Distillation

1. INTRODUCTION

Petroleum in the unrefined state is of limited value and of limited use. Refining is required to obtain products that are attractive to the marketplace. Thus petroleum refining is a series of steps by which the crude oil is converted into salable products with the desired qualities and in the amounts dictated by the market.

Modern petroleum refineries are much more complex operations (Fig. 13.1) than refineries of the early 1900s and even of the years immediately following World War II (Nelson, 1958; Gruse and Stevens, 1960; Bland and Davidson, 1967; Hobson and Pohl, 1973; Aalund, 1975; Jones, 1995; Speight, 1999). Early refineries were predominantly distillation units, perhaps with ancillary units to remove objectionable odors from the various product streams. The refinery of the 1930s was somewhat more complex but was essentially a distillation unit (Speight, 1999). At this time cracking and coking units were starting to appear in the scheme of refinery operations. These units were not what we imagine today as a cracking and coking unit but were the forerunners of today's units. Also at this time, asphalt was becoming a recognized petroleum product. Finally, current refineries (Fig. 13.1) are a result of major evolutionary trends and are highly complex operations. Most of the evolutionary adjustments to refineries have occurred during the decades since the commencement of World War II. In the petroleum industry, as in many other industries, *supply* and *demand* are key factors in efficient and economic operation. Innovation is also a key.

A refinery is an integrated group of manufacturing plants (Fig. 13.1) that vary in number with the variety of products. Refinery processes must be selected and products manufactured to give a balanced operation; that is, crude oil must be converted into products according to the rate of sale of each. For example, the manufacture of products from the lower boiling portion of petroleum automatically produces a certain amount of higher boiling components. If the latter cannot be sold as, say, heavy fuel oil, they accumulate until refinery storage facilities are full. To prevent the occurrence of such a situation, the refinery must be flexible and able to change operations as needed. This usually means more processes to accommodate the ever-changing demands of the market (Nelson, 1958; Hobson and Pohl, 1973; Speight, 1999). This could be reflected by the inclusion of a

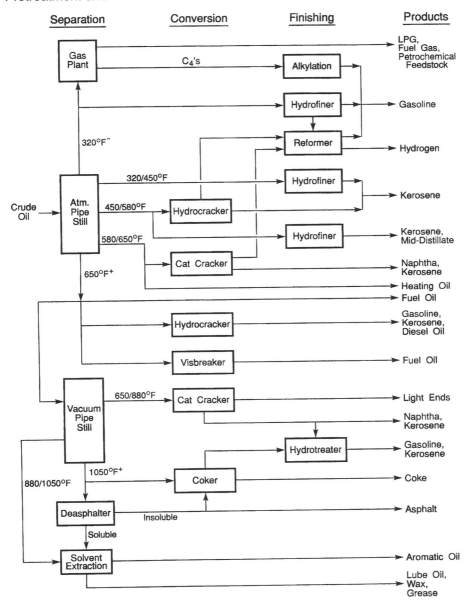

Figure 13.1 General schematic of a refinery.

cracking process to change an excess of heavy fuel oil into more gasoline, with coke as the residual product, or the inclusion of a vacuum distillation process to separate the heavy oil into lubricating oil stocks and asphalt.

Thus, to accommodate the sudden changes in market demand, a refinery must include the following (Kobe and McKetta, 1958):

1. All necessary nonprocessing facilities
2. Adequate tank capacity for storing crude oil, intermediate, and finished products

3. A dependable source of electrical power
4. Material-handling equipment
5. Workshops and supplies for maintaining a continuous 24 hours a day, 7 days a week operation
6. Waste disposal and water-treating equipment
7. Product-blending facilities

Petroleum refining as we know it is a very recent science, and many innovations evolved during the twentieth century. It is the purpose of this chapter to illustrate the initial processes that are applied to a feedstock in a refinery. The first processes are focused on the cleanup of the feedstock, particularly the removal of the troublesome brine constituents. This is followed by distillation to remove the volatile constituents with the concurrent production of a residuum that can be used as a cracking (coking) feedstock or as a precursor to asphalt. In the case of tar sand bitumen, the distillation step is unnecessary because of the low amount of feedstock that would be volatile under distillation conditions. Current methods of bitumen processing (Speight, 1999, 2000) involve direct use of the bitumen as feedstock for delayed or fluid coking (Chapter 14).

Other methods of feedstock treatment that involve the concept of volatility are also included here even though some of them (such as stripping, rerunning, and the like) might also be used for product purification.

2. PRETREATMENT

Even though distillation is, to all appearances, the first step in crude oil refining, it should be recognized that crude oil that is contaminated by salt water either from the well or during transportation to the refinery must be treated to remove the emulsion. If salt water is not removed, the materials of construction of the heater tubes and column intervals are exposed to chloride ion attack and the corrosive action of hydrogen chloride, which may be formed at the temperature of the column feed (*Hydrocarbon Processing*, 1996).

Most crude oils contain traces of salt from the salt water produced with crude oil at some time (Burris, 1992). Desalting operations are necessary to remove salt from the brines that are present with the crude oil after recovery. The salt can decompose in the heater to form hydrochloric acid, which can corrode the fractionator overhead equipment. To remove the salt, water is injected into the partially preheated crude and the stream is thoroughly mixed so that the water extracts practically all of the salt from the oil. The mixture of oil and water is separated in a desalter, which is a large vessel in which the water settles out of the oil, a process that may be accelerated by the addition of chemicals or by electrical devices. The salt-laden water is automatically drained from the bottom of the desalter. Failure to reduce the amount of brine to acceptable levels can result in the production of unacceptable levels of hydrogen chloride during refining. The hydrogen chloride will cause the equipment to corrode and weaken it even to the point of causing fires and explosions.

Generally, removal of this unwanted water has been fairly straightforward, involving wash tanks or a heater. Removal of this water along with reduction of the salt concentration presents a completely different set of problems. These can be overcome with relative ease if the operator is willing to spend many thousands of dollars each year for water, fuel, power, and chemical additives. On the other hand, if some time is spent in the initial design stages to determine the best ways to remove water to achieve lower bottom sediment and water (BS&W) remnants, mixing, and injection, then a system can be designed that will operate at a greatly reduced annual cost. In most cases, yearly savings on operational expenses will pay for the complete installation in 1–3 years.

The practice of desalting crude oil is an old process and can be carried out at the wellhead or (depending on the level of salt in the crude oil) at the refinery (Burris, 1992). Indeed, refineries have been successfully desalting crude oil to less than 5 lb/1000 bbl for many years, and mechanical and electrostatic desalting have been improved greatly. However, very little attention has been given to the use of dilution water, probably due to the general availability of both fresh water and wastewater in and near refineries. Three general approaches have been taken to the desalting of crude petroleum (Fig. 13.2). Numerous variations of each type have been devised, but the selection of a particular process depends on the type of salt dispersion and the properties of the crude oil.

The salt or brine suspensions may be removed from crude oil by heating to 90–150°C (200–300°F) under a pressure of 50–250 psi, which is sufficient to prevent vapor loss, and then allowing the material to settle in a large vessel. Alternatively, coalescence is aided by passage through a tower packed with sand, gravel, or the like.

The common removal technique is to dilute the original brine with fresher water so that the salt content of the water that remains after separation treatment is acceptable, perhaps 10 lb or less per thousand barrels of crude oil. In areas where freshwater supplies are limited, the economics of this process can be critical. However, crude oil desalting techniques in the field have improved with the introduction of the electrostatic coalescing

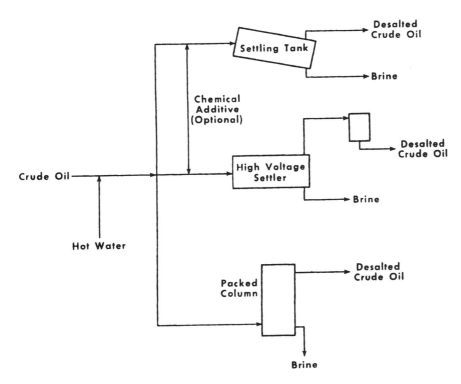

Figure 13.2   General methods for desalting crude oil.

process (Fig. 13.2). Even when adequate supplies of fresh water are available for desalting operations, preparation of the water for dilution purposes may still be expensive.

Requirements for dilution water ratios based on water salinity calculations (Table 13.1; Fig. 13.3) can be calculated as a material balance, and by combining the arithmetic mean of material balance and water injection and dispersion for contact efficiency, very low dilution water use rates can be achieved. This can be highly significant in an area where production rates of 100,000 bbl/day are common and the supply of fresh water is limited.

Desalter units will generally produce a dehydrated stream containing like amounts of bottom sediment and water (BS&W) from each stage. Therefore, bottom sediment and water can be considered as *pass-through volume*, and the dilution water added is the amount of water to be recycled. The recycle pump, however, is generally oversized to compensate for difficult emulsion conditions and upsets in the system. *Dilution water* calculations for a two-stage system using recycle are slightly more complicated than for a single-stage process (Table 13.2; Fig. 13.4) or a two-stage process without recycle (Table 13.3; Fig. 13.5).

The common approach to desalting crude oil involves use of the two-stage desalting system (Fig. 13.5) in which dilution water is injected between stages after the stream water content has been reduced to a very low level by the first stage. Further reduction is achieved by adding the second-stage recycle pump. The *second-stage water* is much lower in sodium chloride (NaCl) than the produced stream inlet water due to addition of dilution water. By recycling this water to the first stage, both salt reduction and dehydration are achieved in the first stage. The water volume to be recycled is assumed to be the same as the dilution water injection volume.

A very low bottom sediment and water content at the first-stage desalter exit requires a high percentage of dilution water to properly contact the dispersed produced water droplets and achieve the desired reduction of salt concentration. This percentage of dilution varies with the strength of the water–oil emulsion and oil viscosity. Empirical data show that the range is from 4.0% to as high as 10%. Obviously, this indicates that the mixing efficiency of 80% is not valid when low water contents are present. Additional field data show that the use rates for low dilution water can be maintained and still meet the required mixing efficiencies. The problem encountered with very low bottom sediment and water content concerns the size of the produced water droplets and their dispersement in

Table 13.1   Pounds of Salt per 1000 bbl of Crude Oil

Salt content of water (ppm)	Volume percent water content in oil (bbl water)				
	1.00 (10)	0.50 (5)	0.20 (2)	0.10 (1)	0.05 (0.5)
10,000	35.00	17.50	7.00	3.50	1.75
20,000	70.00	35.00	14.00	7.00	3.50
30,000	105.00	52.50	21.00	10.00	5.25
40,000	140.00	70.00	28.00	14.00	7.00
50,000	175.00	87.50	35.00	17.50	8.75
100,000	350.00	175.00	70.00	35.00	17.50
150,000	525.00	262.50	105.00	52.50	26.25
200,000	700.00	300.00	140.00	70.00	35.00

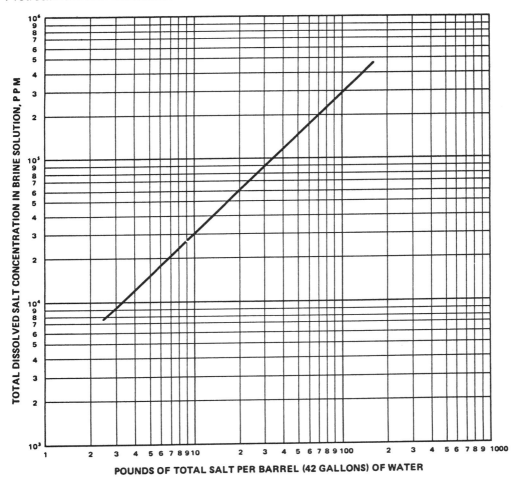

Figure 13.3 Pounds of salt per barrel (42 U.S. gallons) of water.

the crude oil. When 99.9% of the produced water has been removed, the remaining 0.1% consists of thousands of very small droplets more or less evenly distributed throughout the oil. To contact them would require either the dispersal of a large amount of dilution water in the oil or better droplet dispersion of a somewhat smaller amount. Whatever the required amount, it can be attained without exceeding the dilution water rates shown in the earlier example.

Water contained in each desalter unit is an excellent source of volume ratio makeup through the use of a recycle option (Fig. 13.6). The rate at which water is recycled to the first stage must be the same as the dilution water injection rate to maintain the water level in the second-stage unit. The volume of water recycled to the second-stage inlet can be any amount, because it immediately rejoins the controlled water volume in the lower portion of the desalter unit (*internal recycle*). An additional pump may be used (Fig. 13.6) for recycling first-stage water to the first-stage inlet. The first-stage internal recycle is not required for each installation and depends on the amount of produced water present in

Table 13.2 Calculation for Required Amount of Dilution Water[a]

$$Z = BK_3$$

$$K_3 = \frac{AK_1 + YK_2E}{A + YE}$$

where

$A = \dfrac{1000X_1}{1 - X_1} = $ water in inlet stream, bbl

$B = \dfrac{1000x_2}{1 - X_2} = $ water in clean oil, bbl

$C = A + Y = $ water to desalter inlet, bbl

$Y = $ injection water (varies with each problem), bbl

$V = A + Y - B = $ water to disposal, bbl

$E = $ mixing efficiency of Y with A (as a function); assume 80%

$K_1 = $ salt per barrel of water in produced oil stream, 1b

$K_2 = $ salt per barrel of dilution water, lb

$K_3 = \dfrac{AK_1 + YK_2E}{A + YE} = $ salt per barrel of water to desalter inlet, 1b

$X_1 = $ fraction of water in produced oil stream

$X_2 = $ fraction of water in clean oil outlet

$Z = $ salt in outlet clean oil per 1000 bbl of net oil, lb

[a] See Fig. 13.4.

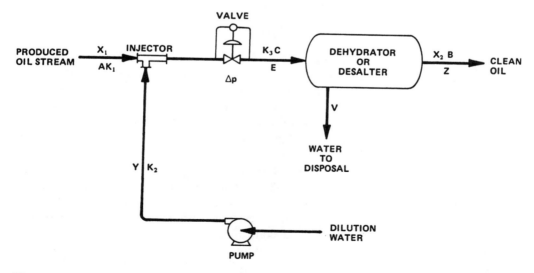

Figure 13.4 A single-stage desalting system.

Table 13.3 Calculation for Required Amount of Dilution Water[a]

$$VK_5 = \frac{BAK_1(R - C) + SYK_3(R - E_2 C)}{SR + BE_1(C - R)}$$

$$V = B + Y - C$$

$$Z = BK_2 + YK_3 - VK_5$$

where

$A = \dfrac{1000X_1}{1 - X_1}$ = water in inlet stream to facility, bbl

$B = \dfrac{1000X_2}{1 - X_2}$ = water in effluent of first stage, bbl

$C = \dfrac{1000X_3}{1 - X_3}$ = water in effluent of second-stage desalter, bbl

Y = injection water of lower salinity than inlet water (A), bbl

V = recycle water to injection in first-stage inlet line, bbl

E_1 = mixing efficiency of V with A (as a fraction), assumed 80%

E_2 = mixing efficiency of Y with B (as a fraction), assumed 80%

X_1 = fraction of water in inlet stream to facility

X_2 = fraction of water in first-stage outlet oil

X_3 = fraction of water in second-stage outlet oil

K_1 = salt per barrel of water to facility, lb

$K_2 = \dfrac{AK_1 + VE_1 K_5}{A + VE_1}$ = salt per barrel of water to first-stage desalter, lb

K_3 = salt per barrel of dilution water, lb

$K_4 = \dfrac{BK_2 = YE_2 K_3}{B + E_2 Y}$ = salt per barrel of water to second-stage desalter, lb

K_5 = salt per barrel of water to recycle injection into inlet line to first stage, lb

$S = A + E_1 V$ = water to first-stage desalter, bbl

$R = B + E_2 Y$ = water to second-stage desalter, bbl

Z = salt in outlet per 1000 bbl of net oil, lb

Note: All water volumes are per 1000 bbl net oil, all salt contents are pounds of total dissolved salts per barrel of water, and Y varies with each individual problem.
[a] See Fig. 13.5.

the inlet stream. In terms of water requirements, internal recycle can be ignored because it does not increase the salt or water volume in the stream.

Emulsions can also be broken by the addition of treating agents such as soaps, fatty acids, sulfonates, and long-chain alcohols. When a chemical is used for emulsion breaking during desalting, it may be added at one or more of three points in the system. First, it may be added to the crude oil before it is mixed with fresh water. Second, it may be added to

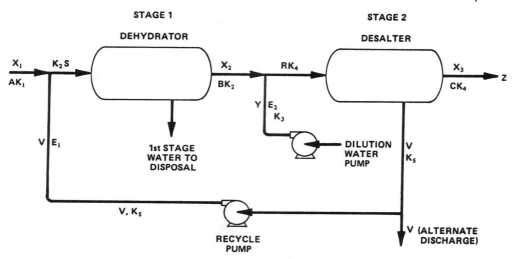

Figure 13.5 A two-stage desalting system.

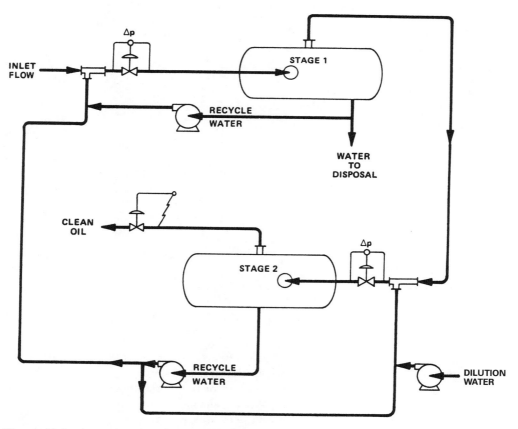

Figure 13.6 A two-stage desalting system with recycle.

the fresh water before it is mixed with the crude oil. Third, it may be added to the mixture of crude oil and water. A high potential field across the settling vessel also aids coalescence and breaks emulsions, in which case dissolved salts and impurities are removed with the water.

If the oil entering the desalter is not hot enough, it may be too viscous to permit proper mixing and complete separation of the water and the oil, and some of the water may be carried into the fractionator. If, on the other hand, the oil is too hot, some vaporization may occur, and the resulting turbulence can result in improper separation of the oil and water. The desalter temperature is therefore quite critical, and normally a bypass is provided around at least one of the exchangers so that the temperature can be controlled. The optimum temperature depends upon the desalter pressure and the quantity of light material in the crude but is normally about 120°C (250°F), being lower for low pressures and light crude oils (100°C; 212°F). The average water injection rate is 5% of the charge.

Regular laboratory analyses will monitor the desalter performance, and the desalted crude should normally not contain more than 1 kg of salt per 1000 bbl of feed.

Good desalter control is indicated by the chloride content of the overhead receiver water, which should be on the order of 10–30 ppm chlorides. If the desalter operation appears to be satisfactory but the chloride content in the overhead receiver water is greater than 30 ppm, then caustic should be injected at the rate of 1–3 lb per 1000 bbl of charge to reduce the chloride content to the range of 10–30 ppm. Salting out will occur below 10 ppm, and severe corrosion above 30 ppm.

Another controlling factor with respect to the overhead receiver water is pH, which should be between 5.5 and 6.5. The injection of ammonia into the tower top section can be used as a control for this.

In addition to electrical methods, desalting can also be achieved by using the concept of a packed column (Fig. 13.2) that facilitate the separation of the crude oil and brine through the agency of an adsorbent.

Finally, *flashing* the crude oil feed can frequently reduce corrosion in the principal distillation column. In the flashing operation, desalted crude is heat exchanged against other heat sources that are available to recover maximum heat before the crude is charged to the heater, which ultimately supplies all the heat required for operation of the crude unit.

Adjusting the heater transfer temperature to offset the flow of fuel to the burners controls the heat input. The heater transfer temperature is merely a convenient control, and the actual temperature, which has no great significance, will vary from 320°C (610°F) to as high as 430°C (805°F), depending on the type of crude oil and the pressure at the bottom of the fractionating tower.

3. DISTILLATION

Distillation has remained a major refinery process. Of the 81,500,000 bbl/day that was taken into refineries worldwide in 1998, 26,700,000 bbl/day (32.8%) goes through vacuum (and by inference atmospheric) distillation units. The United States, which has the largest total refinery capacity in the world, 16,540,990 bbl/day crude oil (154 refineries), can accommodate 7,376,895 bbl/day (44.6%) through vacuum distillation units (Radler, 1999). The continued importance of distillation is evidenced by the fact that older units

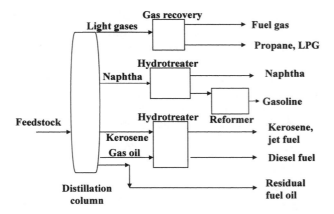

Figure 13.7 A hydroskimming refinery (cf. Fig. 13.8).

continue to be modernized and new units are under construction (Hood et al., 1999; *Hydrocarbon Processing*, 1999).

In the modern sense, distillation was the first method by which petroleum was refined (Speight, 1999). As petroleum refining evolved, distillation became a formidable means by which various products were separated. Further evolution saw the development of topping or skimming or hydroskimming refineries (Fig. 13.7) and conversion refineries (Fig. 13.8),

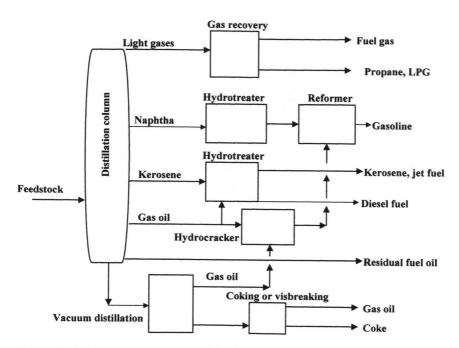

Figure 13.8 A conversion refinery (cf. Fig. 13.7).

named for the manner in which petroleum was treated in each case. Many of these configurations still exist in the world of modern refining.

The simplest refinery configuration, called a *topping refinery*, is designed to prepare feedstocks for petrochemical manufacture or for the production of industrial fuels in remote oil production areas. It consists of tankage, a distillation unit, recovery facilities for gases and light hydrocarbons, and the necessary utility systems (steam, power, and water treatment plants).

Topping refineries produce large quantities of unfinished oils and are highly dependent on local markets, but the addition of hydrotreating and reforming units to this basic configuration results in a more flexible hydroskimming refinery (Fig. 13.7), which can also produce desulfurized distillate fuels and high octane gasoline. Still, these refineries may produce up to half of their output as residual fuel oil, and they face increasing economic hardship as the demand for high sulfur fuel oils declines. Indeed, few older hydroskimming refineries survived the precipitous reduction in worldwide demand for petroleum products that followed the sharp rise in crude oil prices in 1973 and 1979. Those that were not retired from service found it economical to invest in more sophisticated processing facilities in order to increase their yield of gasoline, jet fuel, and diesel oils and to curtail the production of residual fuels.

The most versatile refinery configuration today is known as the conversion refinery (Fig. 13.8). A conversion refinery incorporates all the basic building blocks found in both the topping and hydroskimming refineries, but it also features gas oil conversion plants such as catalytic cracking and hydrocracking units, olefin conversion plants such as alkylation or polymerization units, and, frequently, coking units for sharply reducing or eliminating the production of residual fuels. Modern conversion refineries may produce two-thirds of their output as unleaded gasoline, with the balance distributed between high quality jet fuel, liquefied petroleum gas (LPG), low sulfur diesel fuel, and a small quantity of petroleum coke. Many such refineries also incorporate solvent extraction processes for manufacturing lubricants and petrochemical units with which to recover high purity propylene, benzene, toluene, and xylenes for further processing into polymers.

A multitude of separations are accomplished by distillation, but its most important and primary function in the refinery is the use of the distillation tower and the temperature gradients therein (Fig. 13.9) for the separation of crude oil into fractions that consist of varying amounts of different components (Table 13.4; Fig. 13.10) (Gruse and Stevens, 1960; Speight, 1999, 2000). Thus it is possible to obtain products ranging from gaseous materials taken off the top of the distillation column to a nonvolatile atmospheric residuum (*bottoms, reduced crude*) with correspondingly lower boiling materials (gas, gasoline, naphtha, kerosene, and gas oil) taken off at intermediate points, with each crude oil providing different amounts of the various fractions (Fig. 13.11) (Nelson, 1958; Charbonnier et al., 1969; Coleman et al., 1978; Gary and Handwerk, 1984; Gruse and Stevens, 1960; Hobson and Pohl, 1973; Bland and Davidson, 1967; Diwekar, 1995; Jones, 1995; Speight, 1999, 2000).

The reduced crude may then be processed by vacuum or steam distillation to separate the high-boiling lubricating oil fractions without the danger of decomposition, which occurs at high temperatures (> 350°C, 660°F) (Speight, 1999, 2000). Indeed, atmospheric distillation may be terminated with a lower boiling fraction (*boiling cut*) if it is thought that vacuum or steam distillation will yield a better quality product or if the process appears to be economically more favorable.

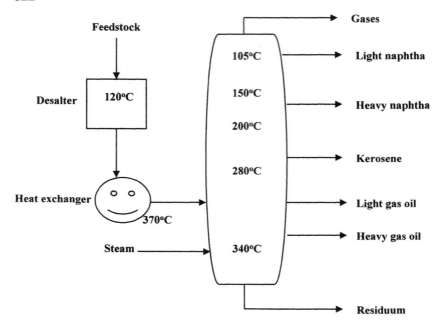

Figure 13.9 Temperature variation in an atmospheric distillation tower.

A very approximate estimation of the potential for thermal decomposition of various feedstocks can be made using the Watson characterization factor (Speight, 1999), K_w,

$$K_w = \frac{T_b^{1/3}}{\text{sp gr}}$$

Table 13.4 Boiling Fractions of Petroleum

Fraction	Boiling range		Uses
	0°C	°F	
Fuel gas	−160 to −40	−260 to −40	Refinery fuel
Propane	−40	−40	Liquefied petroleum gas (LPG)
Butane(s)	−12 to −1	11–30	Increases volatility of gasoline, advantageous in cold climates
Light naphtha	−1 to 150	30–300	Gasoline components, may be (with heavy naphtha) reformer feedstock
Heavy naphtha	150–205	300–400	Reformer feedstock; with light gas oil, jet fuels
Gasoline	−1 to 180	30–355	Motor fuel
Kerosene	205–260	400–500	Fuel oil
Stove oil	205–290	400–550	Fuel oil
Light gas oil	260–315	500–600	Furnace and diesel fuel components
Heavy gas oil	315–425	600–800	Feedstock for catalytic cracker
Lubricating oil	> 400	> 750	Lubrication
Vacuum gas oil	425–600	800–1100	Feedstock for catalytic cracker
Residuum	> 600	> 1100	Heavy fuel oil, asphalts

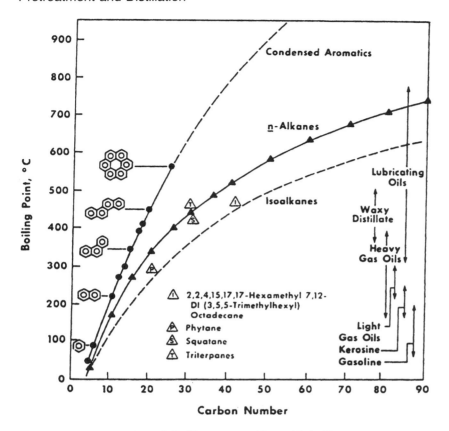

Figure 13.10 Variation of distillate composition with boiling range.

where T_b is the mean average boiling point in degrees Rankine (490 + °F) and "sp gr" is the specific gravity. The characterization factor ranges from about 10 for paraffinic crude oil to about 15 for highly aromatic crude oil. On the assumption that the components of paraffinic crude oil are more thermally labile that the components of aromatic crude oil, it might be supposed that there is a viable relationship between the characterization factor and temperature. The relationship is broad and may not be sufficiently accurate to help the refiner.

It should be noted at this point that not all crude oils yield the same distillation products. In fact, the nature of the crude oil dictates the processes that may be required for refining. Petroleum can be classified according to the nature of the distillation residue, which in turn depends on the relative content of hydrocarbon types: paraffins, naphthenes, and aromatics. For example, a *paraffin-base crude oil* produces distillation cuts with a higher proportions of paraffins than asphalt base crude. The converse is also true; that is, an *asphalt-base crude oil* produces materials with higher proportions of cyclic compounds. A paraffin-base crude oil yields wax distillates rather than the lubricating distillates produced by the naphthenic-base crude oils. The residuum from paraffin-base petroleum is referred to as *cylinder stock* rather than *asphaltic bottoms*, which is the name often given to the residuum from distillation of *naphthenic crude oil*. It is emphasized that in these cases

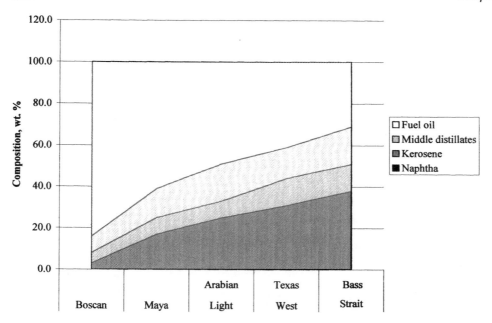

Figure 13.11 Distillation fractionation of different feedstocks.

the terms are not a matter of the use of archaic terminology but a reflection of the nature of the product and the petroleum from which it is derived.

3.1 Atmospheric Distillation

The present-day petroleum distillation unit is, in fact, a collection of distillation units that enable a fairly efficient degree of fractionation to be achieved. In contrast to the early units, which consisted of separate stills, a tower is used in the modern-day refinery. In fact, of all the units in a refinery, the distillation unit (Fig. 13.12) is required to have the greatest flexibility in terms of variable quality of feedstock and range of product yields.

Thus, crude oil can be separated into gasoline, kerosene, diesel oil, gas oil, and other products by distillation at atmospheric pressure. Distillation is an operation in which vapors rising through fractionating decks in a tower are intimately contacted with liquid descending across the decks so that higher boiling components are condensed and concentrate at the bottom of the tower while the lighter ones are concentrated at the top or pass overhead.

The desalted crude oil feedstock is generally pumped to the unit directly from a storage tank, and it is important that charge tanks be drained completely free of water before charging to the unit. The crude feedstock is heat exchanged against whatever other heat sources are available to recover maximum heat before crude is charged to the heater, which ultimately supplies all the heat required for operation of the distillation unit.

If water is entrained in the charge, it will vaporize in the exchangers and in the heater and cause a high pressure drop through that equipment. If a slug of water is charged to the unit, the quantity of steam generated by its vaporization is so much greater than the

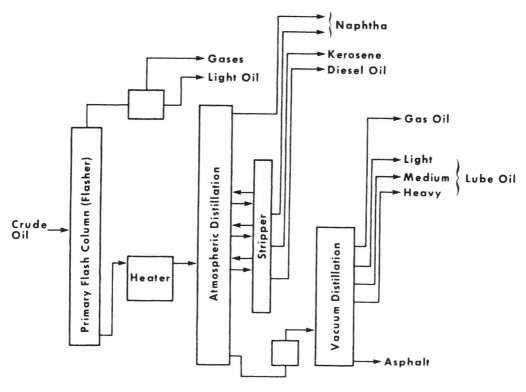

Figure 13.12 A refinery distillation section.

quantity of vapor obtained from the same volume of oil that the decks in the fractionating column can be damaged. Water expands in volume 1600-fold upon vaporization at 100°C at atmospheric pressure.

The feed to a fractional distillation tower is heated by flow through piping arranged within a large furnace. The heating unit is known as a *pipestill heater* or *pipestill furnace*, and the heating unit and the fractional distillation tower make up the essential parts of a distillation unit or pipe still (Fig. 13.12). The pipestill furnace heats the feed to a predetermined temperature, usually a temperature at which a calculated portion of the feed changes into vapor. The vapor is held under pressure in the pipestill furnace until it discharges as a foaming stream into the fractional distillation tower. Here the vapors pass up the tower to be fractionated into gas oil, kerosene, and naphtha while the non-volatile or liquid portion of the feed descends to the bottom of the tower to be pumped away as a bottom product.

Heat exchangers are used to preheat the crude oil feedstock before it enters the distillation unit. To reduce the cost of operating a crude unit, as much heat as possible is recovered from the hot streams by heat exchanging them with the cold crude charge. The number of heat exchangers within the crude unit and cross exchange of heat with other units will vary with unit design. A record should be kept of heat exchanger outlet temperatures so that fouling can be detected and possibly corrected before the capacity of the unit is affected.

Heat exchangers are bundles of tubes arranged within a shell so that a stream passes through the tubes in the opposite direction of a stream passing through the shell. Thus cold crude oil passes through a series of heat exchangers where hot products from the distillation tower are cooled before it enters the furnace, and saving of heat in this manner may be a major factor in the economical operation of refineries.

Crude entering the flash zone of the fractionating column flashes into the vapor that rises up the column and the liquid residue that drops downward. This flash is a very rough separation; the vapors contain appreciable quantities of heavy ends that must be rejected downward into reduced crude, whereas the liquid contains lighter products that must be stripped out. In the distillation of crude petroleum, light naphtha and gases are removed as vapor from the top of the tower, heavy naphtha, kerosene, and gas oil are removed as sidestream products, and reduced crude is taken from the bottom of the tower.

The theoretical maximum permissible temperature of the feedstock in the vaporizing furnace is the factor limiting the range of products in a single-stage (atmospheric) column. Thermal decomposition or *cracking* of the constituents begins as the temperature of the oil approaches 350°C (660°F), and the rate increases markedly above this temperature (Speight, 1999). This thermal decomposition is generally regarded as being undesirable because the coke-like material produced tends to be deposited on the tubes with consequent formation of hot spots and eventual failure of the affected tubes. In the processing of lubricating oil stocks an equally important consideration in the avoidance of these high temperatures is the deleterious effect on the lubricating properties. However, there are occasions when cracking distillation might be regarded as beneficial and the still temperature will be adjusted accordingly. In such a case, the products will be named accordingly using the adjective *cracked*, e.g., *cracked residuum*.

Having the heater transfer temperature reset, flow of fuel to the burners controls the heat input. The heater transfer temperature is merely a convenient control, and the actual temperature, which has no great significance, will vary from 320°C (610°F) to as high as 430°C (805°F), depending on the type of crude and the pressure at the bottom of the fractionating tower. It is noteworthy that if the quantity of gasoline and kerosene in a crude is reduced, the transfer temperature required for the same operation will be increased even though the *lift* is less. However, at such temperatures, the residence time of the crude oil and its fractions exerts considerable influence on the potential for racking reactions to occur. This is particularly important in determining the properties of the atmospheric residuum.

The tower is divided into a number of horizontal sections by metal trays or plates, and each is the equivalent of a still. The more trays, the more redistillation and hence the better the fractionation or separation of the mixture fed into the tower. A tower for fractionating crude petroleum may be 13 ft in diameter and 85 ft high according to the general formula

$$c = 220d^2r$$

where c is the capacity in barrels per day, d is the diameter in feet, and r is the amount of residuum expressed as a fraction of the feedstock (Fig. 13.13) (Nelson, 1943).

A tower for stripping unwanted volatile material from gas oil may be only 3 or 4 ft in diameter and 10 ft high with fewer than 20 trays (Table 13.5). Towers used in the distillation of liquefied gases are only a few feet in diameter but may be up to 200 ft in height. A tower used in the fractionation of crude petroleum may have 16 to 28 trays, but one used in the fractionation (superfractionation) of liquefied gases may have 30–100 trays. The feed to a typical tower enters the vaporizing or flash zone, an area without trays. The

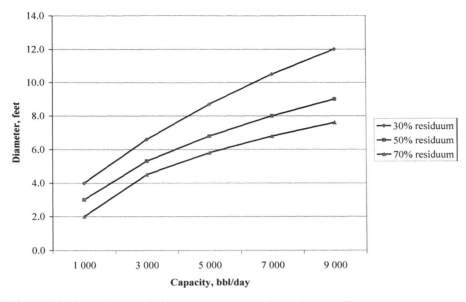

Figure 13.13 Relationship between tower capacity and tower diameter.

majority of the trays are usually located above this area. The feed to a bubble tower, however, may be at any point from top to bottom, with trays above and below the entry point depending on the kind of feedstock and the characteristics desired in the products.

Liquid collects on each tray to a depth of, say, several inches, with the depth controlled by a dam or weir. As the liquid level rises, excess liquid spills over the weir into a channel (downspout), which carries the liquid to the tray below. The temperature of the trays is progressively cooler from bottom to top. The bottom tray is heated by the incoming heated feedstock, although in some instances a steam coil (reboiler) is used to supply additional heat. As the hot vapors pass upward in the tower, they condense onto the trays until refluxing (simultaneous boiling of a liquid and condensation of the vapor) occurs on the trays. Vapors continue to pass upward through the tower, whereas the liquid on any particular tray spills onto the tray below, and so on until the heat at a particular point is too intense for the material to remain liquid. It then becomes vapor and joins the other vapors passing upward through the tower. The whole tower thus simulates a collection of several (or many) stills, with the composition of the liquid at any one point or on any one tray remaining fairly consistent. This allows part of the refluxing liquid to be tapped off at various points as sidestream products.

Table 13.5 Number of Trays Required According to Tower Function

Degree of rectification	No. of trays	Ratio of vapor to feed
1. Stripping still	10–20	Vapor = 20% of feed
2. Primary fractionator	20–40	Vapor = 35–40% of feed
3. Secondary fractionator	40–50	Feed = 50% of vapor
4. Splitter	50–70	Feed = 25% of vapor
5. Superfractionator	70–100	Feed = 10% of vapor

The efficient operation of the distillation, or fractionating, tower requires the rising vapors to mix with the liquid on each tray. This is usually achieved by installing a short chimney on each hole in the plate and a cap with a serrated edge (*bubble cap*, hence *bubble-cap tower*) over each chimney (Fig. 13.14). The cap forces the vapors to go below the surface of the liquid and bubble up through it. Because the vapors may pass up the tower at substantial velocities, the caps are held in place by bolted steel bars.

Perforated trays are also used in fractionating towers. Such a tray is similar to the bubble-cap tray but has smaller holes [$\sim \frac{1}{4}$ in. (6 mm) versus 2 in. (50 mm)]. The liquid spills back to the tray below through weirs and is prevented from returning through the holes to the tray below by the velocity of the rising vapors. Needless to say, a minimum vapor velocity is required to prevent the return of the liquid through the perforations.

As a result, flashed vapors rise up the fractionating column through the trays and countercurrent to the internal reflux flowing down the column. The lightest product, which is generally gasoline, passes overhead and is condensed in the overhead receiver. If the crude oil contains any noncondensable gas it will leave the receiver as a gas and can be recovered by other equipment, which should be operated to obtain the minimum flash zone pressure. The temperature at the top of the fractionator is a good measure of the endpoint of the gasoline and is controlled by returning some of the condensed gasoline (as

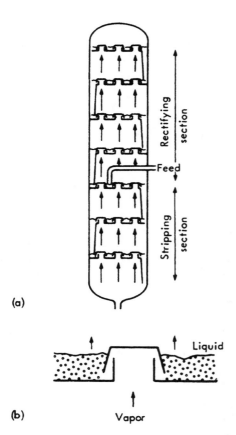

Figure 13.14   Cross section of (a) a distillation tower and (b) a bubble cap.

reflux) to the top of the column. Increasing the reflux rate lowers the top temperature and results in the net overhead product having a lower endpoint. The loss in net overhead product must be removed on the next lower draw tray. This decreases the initial boiling point of material from this tray. Increasing the heater transfer temperature increases the heat input and demands more reflux to maintain the same top temperature.

External reflux that is returned to the top of the fractionator passes downward against the rising vapors. Lighter components of the reflux are revaporized and return to the top of the column, whereas the heavier components in the rising vapors are condensed and return down the column. Thus, there is an internal reflux stream flowing from the top of the fractionator all the way back to the flash zone and becoming progressively heavier as it descends.

The products heavier than the net overhead are obtained by withdrawing portions of the internal reflux stream. The endpoint of a sidestream fraction (*side cut*) will depend on the quantity withdrawn. If the sidestream fraction withdrawal rate is increased, the extra product is material that was formerly flowing down the fractionator as internal reflux. Because the internal reflux below the *draw-off* is reduced, heavier vapors can now rise to that point, which results in a heavier product. The manner in which sidestream fractions are kept on endpoint specifications changes the draw-off rate. The temperature of the draw-off decks is an indication of the endpoint of the product drawn at that point, and the draw-off rate can be controlled to hold a constant deck temperature and therefore a specification product.

The degree of fractionation is generally judged by measuring the number of degrees Celsius between the 95% point of the lighter product and the 5% point of the heavier product. The initial boiling point (IBP) and the final boiling point (FBP) can be used, but the initial boiling point varies with the intensity or efficiency of the stripping operation. Fractionation can be improved by increasing the reflux in the fractionator, which is done by raising the transfer temperature. There may be occasions when the internal reflux necessary to achieve satisfactory fractionation between the heavier products is so great that if it were supplied from the top of the fractionator the upper decks would flood. An *intermediate circulating reflux* (ICR) solves this problem. Some internal reflux is withdrawn, pumped through a cooler or exchanger, and returned colder a few decks higher in the column. This cold oil return condenses extra vapors to liquid and increases the internal reflux below that point. Improvement in the fractionation between the light and heavy gas oil can be achieved by increasing the heater transfer temperature, which would cause the top reflux to increase, then restoring the top reflux to its former rate by increasing the circulating reflux rate. Even though the heater transfer temperature is increased, the extra heat is recovered by exchange with crude oil feedstock, and as a result the heater duty will increase only slightly.

Sometimes a fractionator will be *pulled dry* insofar as the rate at which a product is being withdrawn is greater than the quantity of internal reflux in the fractionator. All the internal reflux then flows to the stripper, the decks below the draw-off run dry, and therefore no fractionation takes place, while at the same time there is insufficient material to maintain the level in the stripper, and the product pump will tend to lose suction. It is necessary then to either lower the product withdrawal rate or increase the internal reflux in the tower by raising the transfer temperature or by reducing the rate at which the next lightest product is being withdrawn.

Pipestill furnaces vary greatly in size, shape, and interior arrangement and accommodate 25,000 bbl or more of crude petroleum per day. The walls and ceiling are insulated with firebrick, and gas or oil burners are inserted through one or more walls. The interior of the furnace is partially divided into two sections: a smaller convection section where the oil first enters the furnace and a larger section into which the burners discharge and where the oil reaches its highest temperature.

It is common practice to use furnaces to heat the feedstock only when distillation temperatures above 205°C (400°F) are required. Lower temperatures (such as that used in the redistillation of naphtha and similar low-boiling products) are provided by heat exchangers and/or steam reboilers.

Steam reboilers may take the form of a steam coil in the bottom of the fractional distillation tower or in a separate vessel. In the latter case, the bottom product from the tower enters the reboiler, where part of it is vaporized by heat from the steam coil. The hot vapor is directed back to the bottom of the tower and provides part of the heat needed to operate the tower. The nonvolatile product leaves the reboiler and passes through a heat exchanger, where its heat is transferred to the feed to the tower. Steam may also be injected into a fractional distillation tower not only to provide heat but also to induce boiling at lower temperatures. Reboilers generally increase the efficiency of fractionation, but a satisfactory degree of separation can usually be achieved more conveniently by the use of a stripping section.

The *stripping section* is the part of the tower below the point at which the feed is introduced and in which the more volatile components are stripped from the descending liquid. Above the feed point (the rectifying section), the concentration of the less volatile component in the vapor is reduced. The stripping section is necessary because the flashed residue in the bottom of the fractionator and the sidestream products have been in contact with lighter boiling vapors. These vapors must be removed to meet flash point specifications and to drive the light ends into lighter and (usually) more valuable products.

Steam, usually superheated steam, is used to strip these light ends. Generally, sufficient steam is used to meet a flash point specification, and although a further increase in the quantity of steam may raise the initial boiling point of the product slightly, the only way to substantially increase the initial boiling point of a specific product is to increase the yield of the next lighter product. Provided, of course, the fractionator has enough internal reflux to accomplish an efficient separation of the feedstock constituents.

All of the stripping steam that is condensed in the overhead receiver must be drained off because refluxing water will upset the balance of activities in the fractionator. If the endpoint of the overhead product is very low, water may not pass overhead and will accumulate on the upper decks and cause the tower to flood, thereby reducing the tower's efficiency and perhaps even shutting it down. If the latter occurs, and if distillation is the first process (other than desalting) to which a crude oil is subjected in a refinery, the economic consequences for the refinery operation can be substantial.

In simple refineries, cut points can be changed slightly to vary yields and balance products, but the more common practice is to produce relatively narrow fractions and then process (or blend) them to meet product demand. Because all these primary fractions are equilibrium mixtures, they all contain some proportion of the lighter constituents characteristic of a lower boiling fraction and so are stripped of these constituents, or stabilized, before further processing or storage. Thus gasoline is stabilized to a controlled butanes–pentanes content, and the overhead may be passed to superfractionators, towers with a large number of plates that can produce nearly pure C_1–C_4 hydrocarbons (methane to butanes, CH_4 to C_4H_{10}), the successive columns termed de-ethanizers, depropanizers, and debutanizers.

Kerosene and gas oil fractions are obtained as sidestreams from the atmospheric tower (*primary tower*) and are treated in stripping columns (i.e., vessels of a few bubble trays) into which steam is injected, and the volatile overhead from the stripper is returned to the primary tower. Steam is usually introduced by the stripping section of the primary column to lower the temperature at which fractionation of the heavier ends of the crude can occur.

The specifications for most petroleum products make it extremely difficult to obtain marketable material by distillation only. In fact, the purpose of atmospheric distillation is considered the provision of fractions that serve as feedstocks for intermediate refining operations and for blending. Generally it is carried out at atmospheric pressure, although light crude oils may be *topped* at an elevated pressure and the residue then distilled at atmospheric pressure.

The *topping* operation differs from normal distillation procedures insofar as the majority of the heat is directed to the feedstream rather than to reboiling the material in the base of the tower. In addition, products of volatility intermediate between that of the overhead fractions and bottoms (residua) are withdrawn as sidestreams. Furthermore, steam is injected into the base of the column and the sidestream strippers to adjust and control the initial boiling range (or point) of the fractions. Topped crude oil must always be stripped with steam to elevate the flash point or to recover the final portions of gas oil. The composition of the topped crude oil is a function of the temperature of the vaporizer (or flasher).

All products are cooled before being sent to storage. Low boiling products should be below 60°C (140°F) to reduce vapor losses in storage, but heavier products need not be as cold. If a product is being charged to another unit, there may be an advantage in sending it out hot. A product must never leave a unit at a temperature over 100°C (212°F) if there is any possibility of it entering a tank with water bottoms. The hot oil could readily boil the water and cause the roof to detach from the tank, perhaps violently!

3.2   Vacuum Distillation

The temperature at which the residue starts to decompose or crack limits the boiling range of the highest boiling fraction cut obtainable at atmospheric pressure. If the stock is required for the manufacture of lubricating oils, further fractionation without cracking may be desirable, and this may be achieved by distillation under vacuum.

Vacuum distillation as applied to the petroleum refining industry is truly a technique of the twentieth century and has seen wide use in petroleum refining. Vacuum distillation evolved because of the need to separate the less volatile products, such as lubricating oils, from the petroleum without subjecting these high boiling products to cracking conditions. The boiling point of the heaviest cut obtainable at atmospheric pressure is limited by the temperature (approximately 350°C; ~ 660°F) at which the residue starts to decompose or *crack*, unless *cracking distillation* is preferred. When the feedstock is required for the manufacture of lubricating oils, further fractionation without cracking is desirable, and this can be achieved by distillation under vacuum conditions.

To maximize the production of gas oil and lighter components from the residuum of an atmospheric distillation unit (reduced crude), the residuum can be further distilled in a vacuum distillation unit (Fig. 13.12). Residuum distillation is conducted at a low pressure in order to avoid thermal decomposition or cracking at high temperatures. A stock that boils at 400°C (750°F) at 0.1 psi (50 mm Hg) would not boil until about 500°C (930°F) at atmospheric pressure, and petroleum constituents start to thermally decompose (crack) at about 350°C (660°F) (Speight, 1999, 2000 and references cited therein). In the vacuum unit, almost no attempt is made to fractionate the products. It is only desired to vaporize the gas oil, remove the entrained residuum, and condense the liquid product as efficiently as possible. Vacuum distillation units that produce lubricating oil fractions are completely different in both design and operation.

In the vacuum tower, the reduced crude is charged through a heater into the vacuum column in the same manner as whole crude is charged to an atmospheric distillation unit. However, whereas the flash zone of an atmospheric column may be at 14–18.5 psi (760–957 mm Hg), the pressure in a vacuum column is very much lower, generally less than 0.8 psi (< 40 mm Hg) in the flash zone to less 0.2 psi (< 10 mm Hg) at the top of the vacuum tower. The vacuum heater transfer temperature is generally used for control, even though the pressure drop along the transfer line makes the temperature at that point somewhat meaningless. The flash zone temperature has much greater significance.

The heater transfer and flash zone temperatures are generally varied to meet the vacuum bottoms specification, which is probably either a gravity (or viscosity) specification for fuel oil or a *penetration specification* for asphalt. The penetration of an asphalt is the depth, in hundredths of a centimeter, to which a needle carrying a 100 g weight sinks into a sample at 25°C (77°F) in 5 s (ASTM D-5); the smaller the penetration, the heavier the residuum or asphalt. If the flash zone temperature is too high, the crude can start to crack and produce gases that overload the ejectors and break the vacuum. It is then necessary to lower the temperature, and if a heavier residuum product is still required an attempt should be made to obtain a better vacuum.

Slight cracking may occur without seriously affecting the vacuum, and the occurrence of cracking can be established by a positive result from the Oliensis spot test. This is a convenient laboratory test that indicates the presence of cracked components as sediment by the separation of the sediment when a 20% solution of asphalt in naphtha is dropped on a filter paper. However, some crude oils yield a residuum or asphalt that exhibit a positive Oliensis test regardless of process conditions. If a negative Oliensis is required, operation at the highest vacuum and lowest temperature should be attempted. Because the degree of cracking depends on both temperature and the time (residence time in the hot zone) for which the oil is exposed to that temperature, the level of the residuum in the bottom of the tower should be held at a minimum, and its temperature should be reduced by recirculating some of the residuum from the outlet of the residuum–crude oil heat exchanger to the bottom of the column. Quite often, when the level of the residuum rises, the column vacuum falls because of cracking due to long residence time.

The flash zone temperature will vary widely and is dependent on the source of the crude oil, residuum specifications, the quantity of product taken overhead, and the flash zone pressure, and temperatures from below 315°C (600°F) to more than 425°C (800°F) have been used in commercial operations. Some vacuum distillation units are provided with facilities to strip the residuum with steam, which tends to lower the temperature necessary to meet an asphalt specification, but an excessive quantity of steam will overload the jets.

The distillation of high-boiling lubricating oil stocks may require pressures as low as 0.29–0.58 psi (15–30 mm Hg), but operating conditions are more usually 0.97–1.93 psi (50–100 mm Hg). Volumes of vapor at these pressures are large, and pressure drops must be small to maintain control, so vacuum columns are necessarily of large diameter. Differences in the vapor pressure of different fractions are relatively larger than for lower boiling fractions, and relatively few plates are required. Under these conditions, a heavy gas oil may be obtained as an overhead product at temperatures of about 150°C (300°F). Lubricating oil fractions may be obtained as sidestreams at temperatures of 250–350°C (480–660°F), the feedstock and residue temperatures being kept below 350°C (660°F), above which the rate of thermal decomposition (cracking) increases (Speight, 1999, 2000). The partial pressure of the hydrocarbons is effectively reduced even further by the injection of steam. The steam added to the column, principally for the stripping of bitumen in the base of the column, is superheated in the convection section of the heater.

When trays similar to those used in the atmospheric column are used in vacuum distillation, the column diameter may be extremely large, up to 45 ft (14 m). To maintain low pressure drops across the trays, the liquid seal must be minimal. The low hold-up and the relatively high viscosity of the liquid limit the tray efficiency, which tends to be much lower than in the atmospheric column. The vacuum is maintained in the column by removing the noncondensable gas that enters the column by way of the feed to the column or by leakage of air.

Prior to 1960, most of the trays in a vacuum tower were conventionally designed to provide as low a pressure drop as possible. Many of these standard trays have been replaced by grid packing that provides very low pressure drops as well as high tray efficiency. Formerly, flash zone temperature reduction was enhanced by steam stripping of the residuum, but with the new grid packing the use of steam to enhance flash temperature has been eliminated and most modern units are dry vacuum units.

The fractions obtained by vacuum distillation of reduced crude depend on whether the run is designed to produce lubricating or vacuum gas oils. In the former case, the fractions include

1. *Heavy gas oil*, an overhead product that is used as catalytic cracking stock or, after suitable treatment, a light lubricating oil
2. *Lubricating oil* (usually three fractions: light, intermediate, and heavy), obtained as a sidestream product
3. *Residuum*, the nonvolatile product that may be used directly as asphalt or to asphalt

The residuum can also be used as a feedstock for a coking operation or blended with gas oils to produce a heavy fuel oil.

If the reduced crude is not required as a source of lubricating oils, then the lubricating and heavy gas oil fractions are combined or, more likely, removed from the residuum as one fraction and used as a catalytic cracking feedstock.

The continued use of atmospheric and vacuum distillation was a major part of refinery operations in the twentieth century and no doubt will continue to be employed, at least for a few decades, as the primary refining operation.

The vacuum residuum (vacuum bottoms) must be handled more carefully than most other refinery products, because the pumps that handle hot heavy material have a tendency to lose suction. Recycling cooled residuum to the column bottom, thereby reducing the tendency of vapor to form in the suction line, can minimize this potential. It is also important that the residuum pump be sealed in such a manner as to prevent the entry of air. In addition, because most vacuum residua are solid at ambient temperature, all vacuum residua handling equipment must either be kept active or flushed out with gas oil when it is shut down. Steam tracing alone may be inadequate to keep the residuum fluid, but where it is done a high pressure steam should be used.

The vacuum residuum from a vacuum tower is sometimes cooled in open box units, because shell-and-tube units are not efficient in this service. It is often desirable to send residuum to storage at high temperature to facilitate blending. If it is desired to increase the temperature of the residuum, it is better to do so by lowering the level of water in the open box, not by lowering the water temperature. If the water in the box is too cold, the residuum can solidify on the inside wall of the tube and insulate the hot residuum in the central core from the cooling water. Lowering the water temperature can thus actually result in a hotter product. When the residuum is sent to storage at a temperature over 100°C (212°F), care should be taken to ensure that the tank is absolutely free of water. Residuum coolers should always be flushed out with gas oil immediately once the residuum flow stops, because melting the contents of a cooler is a slow process.

The vapor rising above the flash zone will entrain some residuum that cannot be tolerated in cracking unit charge. The vapor is generally washed with gas oil product sprayed into the *slop wax* section. The mixture of gas oil and entrained residuum is known as slop wax, and it is often circulated over the decks to improve contact, although the circulation rate may not be critical. The final stage of entrainment removal is obtained by passing the rising vapors through a metallic mesh demister blanket through which the fresh gas oil is sprayed.

Most of the gas oil spray is revaporized by the hot rising vapors and returned up the column. Some slop wax must be yielded in order to reject the captured entrainment. The amount of spray to the demister blanket is generally varied so that the yield of slop wax necessary to maintain the level in the slop wax pan is about 5% of the charge. If the carbon residue or the metals content of the heavy vacuum gas oil is high, a greater percentage of slop wax must be withdrawn or circulated. Variation in the color of the gas oil product is a valuable indication of the effectiveness of entrainment control.

Slop wax can be recirculated through the heater to the flash zone and reflashed. If, however, a crude contains volatile metal compounds, these will be recycled with the slop wax and can finally rise into the gas oil. Where volatile metals are a problem, it is necessary either to yield slop wax as a product or to make lighter asphalt, which will contain the metal compounds returned with the slop wax.

The scrubbed vapor rising above the demister blanket is the product, and no further fractionation is required. It is only desired to condense this vapor as efficiently as possible. This could be done in a shell-and-tube condenser, but such a condenser is inefficient at low pressures, and the high pressure drop through it would raise the flash zone pressure. The most efficient method is to contact the hot vapors with liquid product that has been cooled by pumping through heat exchangers.

Finally, confusion often arises because of the different scales used to measure the vacuum. Positive pressures are commonly measured as kilograms per square centimeter gauge, which is kilograms per square centimeter above atmospheric pressure, which is 1.035 kg/cm^2 or 14.7 psi. Another measure is millimeters of mercury (mm Hg); atmospheric pressure (sea level) is 760 mm Hg absolute, whereas a perfect vacuum is 0 mm Hg absolute.

Variations in both the atmospheric and vacuum distillation protocols are claimed to improve process efficiency and economics. For example, the D2000 process (Fig. 13.15) (*Hydrocarbon Processing*, 1998, p. 66) uses progressive distillation to minimize the total energy consumption required for separation. This process is normally applied for new topping units or newly integrated topping/vacuum units. Incorporation of a vacuum

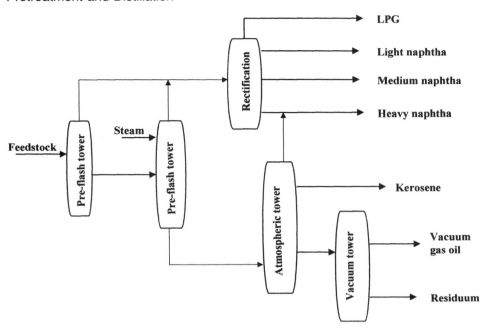

Figure 13.15 The D2000 distillation process.

flasher into the distillation circuit (Fig. 13.16) is claimed to produce a greater yield of distillate materials as well as the usual vacuum residuum (*Hydrocarbon Processing*, 1998, p. 66). Finally, in the integration of a crude distillation unit, the use of a hydrodesulfur-

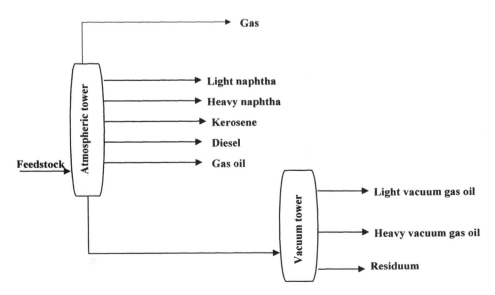

Figure 13.16 Distillation with the inclusion of a vacuum flasher.

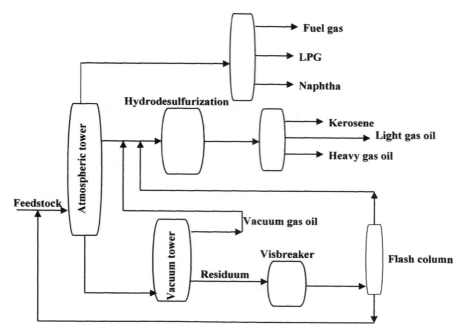

Figure 13.17 Distillation followed by high vacuum distillation, hydrodesulfurization, and vis-breaking.

ization unit, a high vacuum unit, and a visbreaker (Fig. 13.17) (Chapter 14) also improves efficiency (*Hydrocarbon Processing*, 1998, p. 67).

4. OTHER PROCESSES

Atmospheric distillation and vacuum distillation provide the primary fractions from crude oil to use as feedstocks for other refinery processes for conversion into products. Many of these subsequent processes involve fractional distillation, and some are so specialized and used with such frequency that they are identified here by name.

4.1 Stripping

Stripping is a fractional distillation operation carried out on each sidestream product immediately after it leaves the main distillation tower. Because perfect separation is not accomplished in the main tower, unwanted components are mixed with those of the side-stream product. The purpose of stripping is to remove the more volatile components and thus reduce the flash point of the sidestream product. Thus a sidestream product enters at the top tray of a stripper, and as it spills down the four to six trays, steam injected into the bottom of the stripper removes the volatile components. The steam and volatile components leave the top of the stripper to return to the main tower. The stripped sidestream product leaves at the bottom, is cooled in a heat exchanger, and goes to storage. Because strippers are short, they are arranged one above another in a single tower, each stripper, however, operates as a separate unit.

4.2 Rerunning

Rerunning is a general term covering the redistillation of any material and indicating, usually, that a large part of the material is distilled overhead. Stripping, in contrast, removes only a relatively small amount of material as an overhead product. A rerun tower may be associated with a crude distillation unit that produces wide boiling range naphtha as an overhead product. By separating the wide-cut fraction into light and heavy naphtha, the rerun tower acts in effect as an extension of the crude distillation tower.

The product from the chemical treatment of various fractions may be rerun to remove the treating chemical or its reaction products. If the volume of material being processed is small, a *shell still* may be used instead of a continuous fractional distillation unit. The same applies to gas oils and other fractions from which the *front end* or *tail* must be removed for special purposes.

4.3 Stabilization and Light End Removal

The gaseous and more volatile liquid hydrocarbons produced in a refinery are collectively known as *light hydrocarbons* or *light ends* (Table 13.6). Light ends are produced in relatively small quantities from crude petroleum and in large quantities when gasoline is manufactured by cracking and reforming. When a naphtha or gasoline component at the time of its manufacture is passed through a condenser, most of the light ends do not condense and are withdrawn and handled as a gas. A considerable part of the light

Table 13.6 Hydrocarbon Constituents of the Light Ends Fraction

| Hydrocarbon | Carbon atoms | Mol. wt. | Boiling range | | Uses |
			0°C	°F	
Methane	1	16	−182	−296	Fuel gas
Ethane	2	30	−89	−128	Fuel gas
Ethylene	2	28	−104	−155	Fuel gas, petrochemicals
Propane	3	44	−42	−44	Fuel gas, LPG
Propylene	3	42	−48	−54	Fuel gas, petrochemicals, polymer gasoline
Isobutane	4	58	−12	11	Alkylate, motor gasoline
n-Butane	4	58	−1	31	Automotive gasoline
Isobutylene	4	56	−7	20	Synthetic rubber and chemicals, polymer gasoline, alkylate, motor gasoline
Butylene-1[a]	4	56	−6	21	Synthetic rubber and chemicals,
Butylene-2[a]	4	56	1	34	alkylate, polymer gasoline, motor gasoline
Isopentane	5	72	28	82	Automotive and aviation gasolines
n-Pentane	5	72	36	97	Automotive and aviation gasolines
Pentylenes	5	70	30	86	Automotive and aviation gasolines
Isohexane	6	86	61	141	Automotive and aviation gasolines
n-Hexane	6	86	69	156	Automotive and aviation gasolines

[a] Numbers refer to the positions of the double bond. For example, butylene-1 (or butene-1 or but-1-ene) is $CH_3CH_2CH{=}CH_2$, and butylene-2 (or butene-2 or but-2-ene) is $CH_3CH{=}CHCH_3$.

ends, however, can remain dissolved in the condensate, thus forming a liquid with a high vapor pressure.

Liquids with high vapor pressures may be stored in refrigerated tanks or in tanks capable of withstanding the pressures developed by the gases dissolved in the liquid. The more usual procedure, however, is to separate the light ends from the liquid by a distillation process generally known as *stabilization*. Enough of the light ends is removed to make a stabilized liquid, that is, a liquid with a low enough vapor pressure to permit its storage in ordinary tanks without loss of vapor. The simplest stabilization process is stripping. Light naphtha from a crude tower, for example, may be pumped into the top of a tall, small-diameter fractional distillation tower operated under a pressure of 50–80 psi. Heat is introduced at the bottom of the tower by a steam reboiler. As the naphtha cascades down the tower, the light ends separate and pass up the tower to leave as an overhead product. Because reflux is not used, considerable amounts of liquid hydrocarbons pass overhead with the light ends.

Stabilization is usually a more precise operation than that just described. An example of more precise stabilization can be seen in the handling of the mixture of hydrocarbons produced by cracking. The overhead from the atmospheric distillation tower that fractionates the cracked mixture consists of light ends and cracked gasoline with light ends dissolved in it. If the latter is pumped to the usual type of tank storage, the dissolved gases cause the gasoline to boil, with consequent loss of the gases and some of the liquid components. To prevent this, the gasoline and the gases dissolved in it are pumped to a stabilizer maintained under a pressure of approximately 100 psi and operated with reflux. This fractionating tower makes a cut between the highest boiling gaseous component (butane) and the lowest boiling liquid component (pentane). The bottom product is thus a liquid free of all gaseous components, including butane; hence the fractionating tower is known as a *debutanizer* (Fig. 13.18). The debutanizer bottoms (gasoline constituents) can be safely stored, whereas the overhead from the debutanizer contains the butane, propane, ethane, and methane fractions. The butane fraction, which consists of all the hydrocarbons containing four carbon atoms, is particularly needed to give easy starting characteristics to gasoline. It must be separated from the other gases and blended with gasoline in amounts that vary with the season: more in the winter and less in the summer. Separation of the butane fraction is effected by another distillation in a fractional distillation tower called a *depropanizer* because its purpose is to separate propane and the lighter gases from the butane fraction.

The depropanizer is very similar to the debutanizer, except that it is smaller in diameter because of the smaller volume being distilled and is taller because of the larger number of trays required to make a sharp cut between the butane and propane fractions. Because the normally gaseous propane must exist as a liquid in the tower, a pressure of 200 psi is maintained. The bottom product, known as the butane fraction, stabilizer bottoms, or refinery casinghead, is a high vapor pressure material that must be stored in refrigerated tanks or pressure tanks. The depropanizer overhead, consisting of propane and lighter gases, is used as a petrochemical feedstock or as a refinery fuel gas, depending on its composition.

A depentanizer is a fractional distillation tower that removes the pentane fraction from a debutanized (butane-free) fraction. Depentanizers are similar to debutanizers and were introduced recently to segregate the pentane fractions from cracked gasoline and reformate. The addition of the pentane fraction to a premium gasoline makes the gasoline extraordinarily responsive to the demands of an engine accelerator.

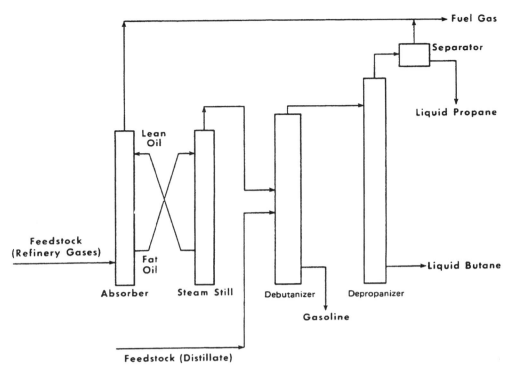

Figure 13.18 A light ends plant.

The gases produced as overhead products from crude distillation, stabilization, and depropanization units may be delivered to a gas absorption plant for the recovery of small amounts of butane and higher boiling hydrocarbons. The gas absorption plant consists essentially of two towers. One tower is the absorber where the butane and higher boiling hydrocarbons are removed from the lighter gases. The gas mixture enters at the bottom of the tower and rises to the top. As it does this, it contacts the lean oil, which absorbs the butane and higher boiling hydrocarbons but not the lower boiling hydrocarbons. The latter leave the top of the absorber as dry gas. The lean oil that has become enriched with butane and higher boiling hydrocarbons is now termed *fat oil*. This is pumped from the bottom of the absorber into the second tower, where fractional distillation separates the butane and higher boiling hydrocarbons as an overhead fraction and the oil, once again lean oil, as the bottom product.

The condensed butane and higher boiling hydrocarbons are included with the refinery *casinghead bottoms* or *stabilizer bottoms*. The dry gas is frequently used as fuel gas for refinery furnaces. It contains propane and propylene, however, which may be required for liquefied petroleum gas for the manufacture of polymer gasoline or petrochemicals. Separation of the propane fraction (propane and propylene) from the lighter gases is accomplished by further distillation in a fractional distillation tower similar to those previously described and particularly designed to handle liquefied gases. Further separation of hydrocarbon gases is required for petrochemical manufacture.

4.4 Superfractionation

The term *superfractionation* is sometimes applied to the process carried out in a highly efficient fractionating tower used to separate ordinary petroleum products. For example, to increase the yield of furnace fuel oil, heavy naphtha may be redistilled in a tower that is capable of making a better separation of the naphtha and the fuel oil components. The latter, obtained as a bottom product, is diverted for use as furnace fuel oil.

Fractional distillation as normally carried out in a refinery does not completely separate one petroleum fraction from another. One product overlaps another, depending on the efficiency of the fractionation, which in turn depends on the number of trays in the tower, the amount of reflux used, and the rate of distillation. Kerosene, for example, normally contains a small percentage of hydrocarbons that (according to their boiling points) belong in the naphtha fraction and a small percentage that should be in the gas oil fraction. Complete separation is not required for the ordinary uses of these materials, but certain materials, such as solvents for particular purposes (hexane, heptane, and aromatics), are required as essentially pure compounds. Because they occur in mixtures of hydrocarbons they must be separated by distillation, with no overlap of one hydrocarbon with another. This requires highly efficient fractional distillation towers specially designed for the purpose and referred to as *superfractionators*. Several towers with 50–100 trays operated with high reflux ratios may be required to separate a single compound with the necessary purity.

4.5 Azeotropic Distillation and Extractive Distillation

Azeotropic distillation is the use of a third component to separate two close-boiling components by forming an azeotropic mixture between one of the original components and the third component to increase the difference in the boiling points and facilitate separation by distillation.

Sometimes the separation of a desired compound calls for azeotropic distillation. All compounds have definite boiling temperatures, but a mixture of chemically dissimilar compounds sometimes causes one or both of the components to boil at a temperature other than that expected. The separation of these components of similar volatility may become economical if an *entrainer* can be found that effectively changes their relative volatility. It is also desirable that the entrainer be reasonably cheap, stable, nontoxic, and readily recoverable from the components. In practice it is probably this last criterion that severely limits the application of extractive and azeotropic distillation. The majority of successful processes, in fact, are those in which the entrainer and one of the components separate into two liquid phases on cooling if direct recovery by distillation is not feasible.

A further restriction in the selection of an azeotropic entrainer is that the boiling point of the entrainer be 10–40°C (18–72°F) below that of the components. Thus, although the entrainer is more volatile than the components and distills off in the overhead product, it is present in a sufficiently high concentration in the rectification section of the column.

Extractive distillation is the use of a third component to separate two close-boiling components by extracting one of the original components in the mixture and retaining it in the liquid phase to facilitate separation by distillation.

Using acetone–water as an extractive solvent for butanes and butenes, butane is removed as overhead from the extractive distillation column with acetone–water charged at a point close to the top of the column. The bottom product of butenes and the extractive solvent are fed to a second column, where the butenes are removed as overhead. The acetone–water solvent from the base of this column is recycled to the first column.

Extractive distillation (Fig. 13.19) can also be used for the continuous recovery of individual aromatics, such as benzene, toluene, or xylene(s), from the appropriate petroleum fractions. *Prefractionation* concentrates a single aromatic cut into a close-boiling cut, after which the aromatic concentrate is extractively distilled with a solvent (usually phenol) to recover the benzene or toluene. Mixed cresylic acids (cresols and methylphenols) are used as the solvent for xylene recovery.

In general, none of the fractions or combinations of fractions separated from crude petroleum is suitable for immediate use as a petroleum product. Each must be separately refined by treatments and processes that vary with the impurities in the fraction and the properties required in the finished product. The simplest treatment is to wash a fraction with a lye solution to remove sulfur compounds. The most complex is a series of treatments—solvent treating, dewaxing, clay treating or hydrorefining, and blending—required to produce lubricating oils. On rare occasions no treatment of any kind is required. Some crude oils yield a light gas oil fraction that is suitable as furnace fuel oil or as a diesel fuel.

Two processes illustrate the similarities and differences between azeotropic distillation and extractive distillation. Both have been used for the separation of C_4 hydrocarbons (Figs. 13.20 and 13.21). Thus, butadiene and butene may be separated by the use of liquid ammonia, which forms an azeotrope with butene. The ammonia–butene azeotrope overhead from the azeotropic distillation is condensed, cooled, and allowed to separate into a butene layer and a heavier ammonia layer. The butene layer is fed to a second column, where the ammonia is removed as a butene–ammonia azeotrope, and the remaining butene is recovered as bottom product. The ammonia layer is returned to the lower section of the first azeotropic distillation column. Butadiene is recovered as bottom product from this column.

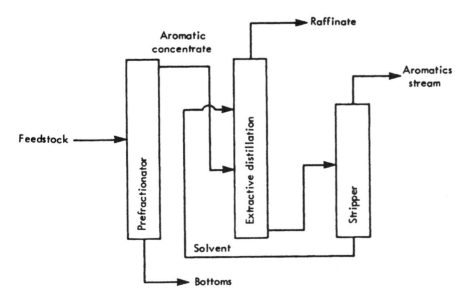

Figure 13.19 Extractive distillation for aromatics recovery.

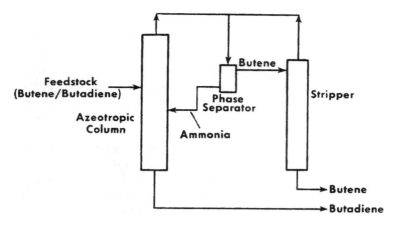

Figure 13.20 Separation of butene and butadiene by azeotropic distillation.

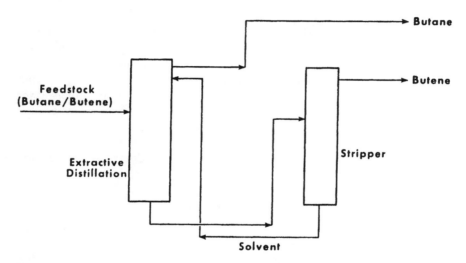

Figure 13.21 Separation of butane and butene by extractive distillation.

REFERENCES

Aalund, L. R. 1975. Oil Gas J. 77(35):339.

Bland, W. F., and Davidson, R. L. 1967. Petroleum Processing Handbook. McGraw-Hill, New York.

Burris, D. R. 1992. In Petroleum Processing Handbook. J. J. McKetta, ed. Marcel Dekker, New York, p. 666 et seq.

Charbonnier, R. P., Draper, R. G., Harper, W. H., and Yates, Y. 1969. Analyses and Characteristics of Oil Samples from Alberta. Information Circular No. IC 232. Department of Energy Mines and Resources, Mines Branch, Ottawa, ON, Canada.

Coleman, H. J., Shelton, E. M., Nicholls, D. T., and Thompson, C. J. 1978. Analysis of 800 Crude Oils from United States Oilfields. Report No. BETC/RI-78/14. Technical Information Center, Department of Energy, Washington, DC.

Diwekar, U. M. 1995. Batch Distillation: Simulation, Optimal Design, and Control. Taylor & Francis, Philadelphia, PA.

Gary, J. H., and Handwerk, G. L. 1984. Petroleum Refining: Technology and Economics. 2nd ed. Marcel Dekker, New York.

Gruse, W. A., and Stevens, D. R. 1960. Chemical Technology of Petroleum. McGraw-Hill, New York.

Hobson, G. D., and Pohl, W. 1973. Modern Petroleum Technology. Applied Science Publishers, Barking, Essex, England.

Hood, A. R., Stander, A., Lilburne, G., and Soydaner, S. 1999. Hydrocarbon Process. 78(11):63.

Hydrocarbon Processing. 1996. 75(11).

Hydrocarbon Processing. 1998. 77(11):53 et seq.

Hydrocarbon Processing. 1999. 78(10): HPI Construction Boxscore.

Jones, D. S. J. 1995. Elements of Petroleum Processing. Wiley, Chichester, England.

Kobe, K. A., and McKetta, J. J. 1958. Advances in Petroleum Chemistry and Refining. Interscience, New York.

Nelson, W. L. 1943. Oil Gas J. 41(16):72.

Nelson, W. L. 1958. Petroleum Refinery Engineering. McGraw-Hill, New York, p. 226 et seq.

Radler, M. 1999. Oil Gas J. 97(51):45 et seq.

Speight, J. G. 1999. The Chemistry and Technology of Petroleum. 3rd ed. Marcel Dekker, New York.

Speight, J. G. 2000. The Desulfurization of Heavy Oils and Residua. 2nd ed. Marcel Dekker, New York.

14
Thermal Cracking

1. INTRODUCTION

Distillation (Chapter 13) has remained a major refinery process and a process to which just about every crude oil that enters the refinery is subjected. However, not all crude oils yield the same distillation products. In fact, the nature of the crude oil dictates the processes that may be required for refining. Balancing product yield with demand is a necessary part of refinery operations.

The balancing of product yield and market demand without manufacturing large quantities of fractions of low commercial value has long required processes for the conversion of hydrocarbons of one molecular weight range and/or structure into those of some other molecular weight range and/or structure. Basic processes for this are still the so-called cracking processes in which relatively high boiling constituents are cracked, that is, thermally decomposed into lower molecular weight, smaller, lower boiling molecules, although reforming, alkylation, polymerization, and hydrogen refining processes have wide applications in making premium quality products (Corbett, 1990; Trash, 1990).

After World War I, the demand for automotive (and other) fuels began to outstrip the market requirements for kerosene, and refiners, needing to stay abreast of the market pull, were pressed to develop new technologies to increase gasoline yields. Because there were finite amounts of straight-run distillate fuels in crude oil, refiners felt an urgency to develop processes to produce additional amounts of these fuels. The ability to convert coal and oil shale to liquid through the agency of cracking had been known for centuries, and the ability to produce various spirits from petroleum though thermal methods had been known since at least the inception of Greek fire many centuries earlier.

The knowledge that higher molecular weight (higher boiling) materials could be decomposed to lower molecular weight (lower boiling) products was used to increase the production of kerosene; the process was called *cracking distillation* (Kobe and McKetta, 1958). Thus a batch of crude oil was heated until most of the kerosene was distilled from it and the overhead material became dark in color. At this point the still fires were lowered, the rate of distillation was decreased, and the heavy oils were held in the hot zone for a time during which some of the large hydrocarbons were decomposed and rearranged into lower molecular weight products. After a suitable time, the still fires were increased and distillation continued in the normal way. The overhead product, however, was light oil suitable for kerosene instead of the heavy oil that would otherwise have been produced. Thus, it was not surprising that such technologies were adapted for the fledgling petroleum industry.

The earliest process, called thermal cracking, consisted of heating heavier oils (for which there was a low market requirement) in pressurized reactors and thereby cracking, or splitting, their large molecules into the smaller ones that form the lighter, more valuable fractions such as gasoline, kerosene, and light industrial fuels.

Gasoline manufactured by thermal cracking processes performed better in automobile engines than gasoline derived from straight distillation of crude petroleum. The development of more powerful aircraft engines in the late 1930s gave rise to a need to increase the combustion characteristics of gasoline to improve engine performance. Thus during World War II and the late 1940s, improved refining processes involving the use of catalysts led to further improvements in the quality of transportation fuels and further increased their supply. These improved processes, including catalytic cracking of residua and other heavy feedstocks (Chapter 19), alkylation (Chapter 20), polymerization (Chapter 20), and iso-merization (Chapter 20), enabled the petroleum industry to meet the demands of high performance combat aircraft and, after World War II, to supply increasing quantities of transportation fuels.

The 1950s and 1960s brought a large-scale demand for jet fuel and high quality lubricating oils. The continuing increase in demand for petroleum products also heightened the need to process a wider variety of crude oils into high quality products. Catalytic reforming of naphtha (Chapter 20) replaced the earlier thermal reforming process and became the leading process for upgrading fuel qualities to meet the needs of higher compression engines. Hydrocracking, a catalytic cracking process conducted in the presence of hydrogen (Chapter 17), was developed to be a versatile manufacturing process for increasing the yields of either gasoline or jet fuels.

In the early stages of thermal cracking development, processes were generally classified as either liquid-phase, high pressure (350–1500 psi; 2413–10,342 kPa), low temperature (400–510°C; 750–950°F) or vapor-phase, low pressure (< 200 psi; < 1379 kPa), high temperature (540–650°C, 1000–1200°F). In reality, the processes were mixed phase, with no process being entirely liquid- or vapor-phase, but the classification (like many classifications of crude oil and related areas) was still used as a matter of convenience (Speight, 1999).

The processes described in the preceding paragraphs were classified as liquid-phase processes. These liquid-phase processes had the following advantages over vapor-phase processes:

1. Large yields of gasoline of moderate octane number
2. Low gas yields
3. Ability to use a wide variety of charge stocks
4. Long cycle time due to low coke formation
5. Flexibility and ease of control

The vapor-phase processes had the advantages of operation at lower pressures and the production of a higher octane gasoline due to the increased production of olefins and light aromatics. However, there were many disadvantages that curtailed the development of vapor-phase processes. High temperatures were required that the steel alloys available at the time could not tolerate. The operations had to be very closely controlled. There were high gas yields and resulting losses, because the gases were normally not recovered. There was a high production of olefinic compounds that created a gasoline with poor stability (increased tendency to form undesirable gum) (Mushrush and Speight, 1995; Speight, 1999) and required subsequent treating of the gasoline to stabilize it against gum formation. The vapor-phase processes were not considered suitable for the production of large quantities of gasoline, but they did find application in petrochemical manufacture owing to the high concentration of olefins produced.

It is generally recognized that the most important part of any refinery, after the distillation units, is the gasoline (and liquid fuel) manufacturing facility; other facilities are added to manufacture additional products as indicated by technical feasibility and economic gain. More equipment is used in the manufacture of gasoline, the equipment is more elaborate, and the processes are more complex than for any other product. Among the processes that have been used for liquid fuel production are thermal cracking, catalytic cracking, thermal reforming, catalytic reforming, polymerization, alkylation, coking, and distillation of fractions directly from crude petroleum (Fig. 14.1). Each of these processes can be carried out in a number of ways that differ in details of operation or essential equipment or both (Bland and Davidson, 1967).

Thermal processes are essentially processes that decompose, rearrange, or combine hydrocarbon molecules by the application of heat. The major variables involved are feedstock type, time, temperature, and pressure and as such are usually considered in promoting cracking (thermal decomposition) of the heavier molecules to lighter products and in minimizing coke formation. Thus, one of the earliest processes used in the petroleum industry is *thermal cracking*, the noncatalytic conversion of higher boiling petroleum stocks into lower boiling products.

The thermal decomposition (cracking) of high molecular weight hydrocarbons to lower molecular weight, normally more valuable, hydrocarbons has long been practiced in the petroleum refining industry. Although catalytic cracking has generally replaced thermal cracking, noncatalytic cracking processes using high temperature to achieve the decomposition are still in operation. In several cases, thermal cracking processes to produce specific desired products or to dispose of specific undesirable charge streams are being operated or installed. The purpose of this chapter is to provide basic information to assist the practicing engineer/petroleum refiner in

1. Determining if a particular thermal cracking process would be suitable for a specific application and could fit into the overall operation
2. Developing a basic design for a thermal cracking process
3. Operating an existing or proposed process

Conventional *thermal cracking* is the thermal decomposition, under pressure, of high molecular weight constituents (higher molecular weight and higher boiling than gasoline constituents) to form lower molecular weight (and lower boiling) species. Thus, the thermal cracking process is designed to produce gasoline from higher boiling charge stocks, and any unconverted or mildly cracked charge components (compounds that have been partially decomposed but are still higher boiling than gasoline) are usually recycled to extinction to maximize gasoline production. A moderate quantity of light hydrocarbon gases is also formed. As thermal cracking proceeds, reactive unsaturated molecules are formed that continue to react and can ultimately create higher molecular weight species that are relatively hydrogen deficient and readily form coke. These species cannot be

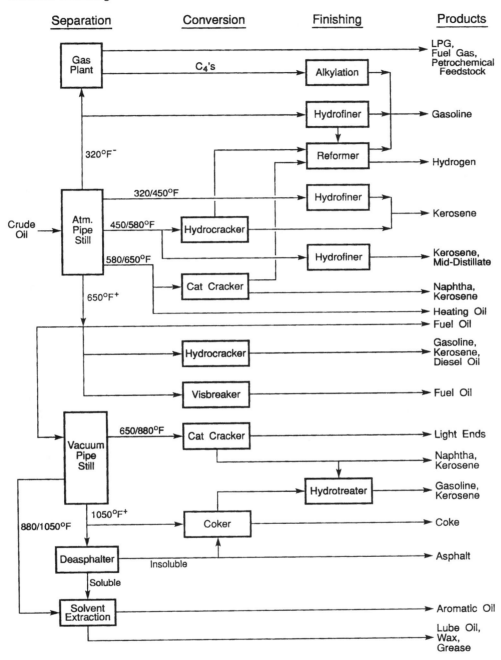

Figure 14.1 Schematic representation of a refinery, showing placement of the various conversion units.

recycled without excessive coke formation and are therefore removed from the system as cycle fuel oil.

When petroleum fractions are heated to temperatures over 350°C (660°F), thermal decomposition is caused to proceed at a significant rate (Speight, 1999, 2000). Thermal decomposition does not require the addition of catalyst, so this approach is the oldest technology available for residue conversion. The severity of thermal processing determines the type of conversion and the product characteristics. Thermal treatment of residues ranges from mild treatment for reduction of viscosity to *ultrapyrolysis* (high temperature cracking at very short residence time) for complete conversion to olefins and light ends. The higher the temperature, the shorter the time required to achieve a given conversion, but in many cases there is a change in the chemistry of the reaction. The *severity* of the process conditions refers to the combination of reaction time and temperature to achieve a given conversion.

Thermal reactions, however, can give rise to a variety of reactions, so selectivity for a given product changes with temperature and pressure. The mild and high severity processes are frequently used for processing of residua, whereas conditions similar to *ultrapyrolysis* (high temperature and very short residence time) are only used commercially for cracking ethane, propane, butane, and light distillate feeds to produce ethylene and higher olefins.

Sufficiently high temperatures convert oils entirely to gases and coke. Cracking conditions are controlled to produce as much as possible of the desired product, which is usually gasoline but can be cracked gases for petrochemicals or a lower viscosity oil for use as a fuel oil. The feedstock, or cracking stock, can be almost any fraction obtained from crude petroleum, but the greatest amount of cracking is carried out on gas oils, the portion of crude petroleum that boils between the fuel oils (kerosene and/or stove oil) and the residuum. Residua are also cracked, but the processes are somewhat different from those used for gas oils.

Thus, thermal conversion processes are designed to increase the yield of lower boiling products obtainable from petroleum either directly (by means of the production of gasoline components from higher boiling feedstocks) or indirectly (by production of olefins and the like, which are precursors of the gasoline components). These processes may also be characterized by the physical state (liquid and/or vapor phase) in which the decomposition occurs. The state depends on the nature of the feedstock as well as on conditions of pressure and temperature (Nelson, 1976; Trimm, 1984; Vermillion and Gearhart, 1983; Thomas et al., 1989).

From the chemical viewpoint the products of cracking are very different from those obtained directly from crude petroleum. When a 12-carbon atom hydrocarbon typical of straight-run gas oil is cracked or broken into two parts, one may be a six-carbon paraffin hydrocarbon and the other a six-carbon olefin hydrocarbon:

$$CH_3(CH_2)_{10}CH_3 \rightarrow CH_3(CH_2)_4CH_3 + CH_2{=}CH(CH_2)_3CH_3$$

The paraffin may be the same as is found in straight-run (distilled) gasoline, but the olefin is new. Furthermore, the paraffin has an octane number approaching zero, but the olefin has an octane number approaching 100 (Chapter 20). Hence naphtha formed by cracking (*cracked gasoline*) has a higher octane number than straight-run gasoline. In addition to a large variety of olefins, cracking produces high octane aromatic and branched-chain hydrocarbons in higher proportions than are found in straight-run gasoline. Diolefins are produced, but in relatively small amounts; they are undesirable in gasoline because they readily combine to form gum.

The hydrocarbons with the least thermal stability are the paraffins, and the olefins produced by the cracking of paraffins are also reactive. Cycloparaffins (naphthenes) are less easily cracked, their stability depending mainly on any side chains present, but ring splitting may occur, and dehydrogenation can lead to the formation of unsaturated naphthenes and aromatics. Aromatics are the most stable (*refractory*) hydrocarbons, their stability depending on the length and stability of side chains. Very severe thermal cracking of high molecular weight constituents can result in the production of excessive amounts of coke.

The higher boiling oils produced by cracking are light and heavy gas oils as well as a residual oil, which in the case of thermal cracking is usually (erroneously) called tar and in the case of catalytic cracking is called *cracked fractionator bottoms*. The residual oil may be used as heavy fuel oil, and gas oils from catalytic cracking are suitable as domestic and industrial fuel oils or as diesel fuels if blended with straight-run gas oils. Gas oils from thermal cracking must be mixed with straight-run (distilled) gas oils before they become suitable for use as domestic fuel oils and diesel fuels.

The gas oils produced by cracking are an important source of gasoline, and in a once-through cracking operation all of the cracked material is separated into products and may be used as such. The cracked gas oils are more resistant to cracking (more refractory) than straight-run gas oils but can still be cracked to produce more gasoline. This is done in a recycling operation in which the cracked gas oil is combined with fresh feed for another trip through the cracking unit. The operation may be repeated until the cracked gas oil is almost completely decomposed (*cracking to extinction*) by recycling (*recycling to extinction*) the higher boiling product, but it is more usual to withdraw part of the cracked gas oil from the system according to the need for fuel oils. The extent to which recycling is carried out affects the amount or yield of cracked gasoline resulting from the process.

The gases formed by cracking are particularly important because of their chemical properties and their quantity. Only relatively small amounts of paraffinic gases are obtained from crude oil, and these are chemically inactive. Cracking produces both paraffinic gases (e.g., propane, C_3H_8) and olefinic gases (e.g., propene, C_3H_6); the latter are used in the refinery as the feed for polymerization plants where high octane polymer gasoline is made. In some refineries the gases are used to make alkylate, a high octane component for aviation gasoline and automotive gasoline. In particular, the cracked gases are the starting points for many petrochemicals (Speight, 1999).

The importance of solvents in coking has been recognized for many years (Speight, 1999, and references cited therein), but their effects have often been ascribed to hydrogen donor reactions rather than phase behavior. The separation of the phases depends on the solvent characteristics of the liquid. Addition of aromatic solvents will suppress phase separation (Chapters 3 and 19), whereas addition of paraffins will enhance separation. Microscopic examination of coke particles often shows evidence for the presence of a *mesophase*, spherical domains that exhibit the anisotropic optical characteristics of liquid crystal. This phenomenon is consistent with the formation of a second liquid phase; the mesophase liquid is denser than the rest of the hydrocarbon, has a higher surface tension, and likely wets metal surfaces better than the rest of the liquid phase. The mesophase characteristic of coke diminishes as the liquid phase becomes more compatible with the aromatic material (Speight, 1990, 2000).

Thermal cracking of higher boiling materials to produce motor gasoline is now becoming an obsolete process, because the antiknock requirement of modern automobile engines has outstripped the ability of the thermal cracking process to supply an economical source of high quality fuel. New units are rarely installed, but a few refineries still operate thermal cracking units built in previous years (Bland and Davidson, 1967).

In summary, the cracking of petroleum constituents can be visualized as a series of simple thermal conversions. The reactions involve the formation of transient highly reactive species that may react further in several ways to produce the observed product slate (Germain, 1969; Speight, 1999, 2000). Thus, even though chemistry and physics can be used to explain feedstock reactivity, the main objective of feedstock evaluation (Chapter 4) is to allow a degree of predictability of feedstock behavior in thermal processes (Speight, 1999, 2000). And in such instances chemical principles must be combined with engineering principles to understand feedstock processability and the predictability of feedstock behavior.

In the simplest sense, process planning can be built on an understanding of the following three parameter groups:

1. *Feedstock parameters* such as carbon residue (potential coke formation), sulfur content (hydrogen needs for desulfurization), metallic constituents (catalyst rejuvenation), nitrogen content (catalyst rejuvenation), naphthenic or paraffinic character through use of a characterization factor or similar indicator (potential for cracking in different ways to give different products), and, to a lesser extent, asphaltene content (coke formation), because this last parameter is related to several of the previous parameters

2. *Process parameters* such as time–temperature–pressure relationships (distillate and coke yields), feedstock recycle ratio (distillate and coke yields plus overall conversion), and coke formation (lack of liquid production when liquids are the preferred products)

3. *Equipment parameters* such as batch operation, semicontinuous operation, or continuous operation (residence time and contact with the catalyst, if any), coke removal, and unit capacity which also dictates residence time

It is not the purpose of this text to present the details of these three categories, but they should be borne in mind when considering and deciding upon the potential utility of any process presented in this and subsequent chapters.

2. THERMAL CRACKING

The majority of regular thermal cracking processes use temperatures of 455–540°C (850–1005°F) and pressures of 100–1000 psi.

The Dubbs process, in which a reduced crude is charged, employs the concept of recycling in which the gas oil is combined with fresh feedstock for further cracking. In a typical application of conventional thermal cracking (Fig. 14.2), the feedstock (reduced crude, i.e., residuum or flashed crude oil) is preheated by direct exchange with the cracked products in the fractionating columns. Cracked gasoline and middle distillate fractions are removed from the upper section of the column. Light and heavy distillate fractions are removed from the lower section and are pumped to separate heaters. Higher temperatures are used to crack the more refractory light distillate fraction. The streams from the heaters are combined and sent to a soaking chamber, where additional time is provided to complete the cracking reactions. The cracked products are then separated in a low pressure

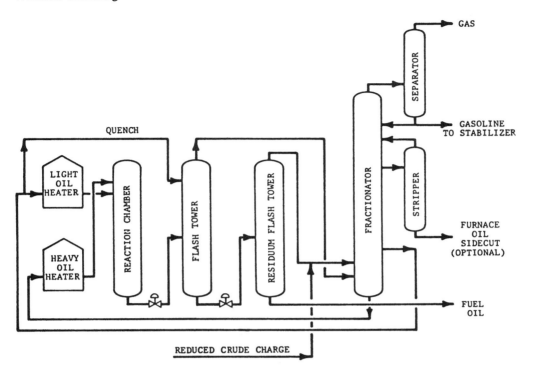

Figure 14.2 A thermal cracking unit.

flash chamber, where a heavy fuel oil is removed as bottoms. The remaining cracked products are sent to the fractionating columns.

Low pressures (< 100 psi) and temperatures in excess of 500°C (930°F) tend to produce lower molecular weight hydrocarbons than those produced at higher pressures (400–1000 psi) and at temperatures below 500°C (930°F). The reaction time is also important; light feeds (gas oils) and recycle oils require longer reaction times than the readily cracked heavy residues. Recycle of the light oil (middle distillate or fuel oil) fraction also affects the product slate of the thermal cracker (Table 14.1). Mild cracking conditions (defined here as a low conversion per cycle) favor a high yield of gasoline components with low gas and coke production, but the gasoline quality is not high, whereas more severe conditions give increased gas and coke production and reduced gasoline yield (but of higher quality). With limited conversion per cycle, the heavier residues must be recycled. However, the recycled oils become increasingly refractory upon repeated cracking, and if they are not required as a fuel oil stock they may be subjected to a coking operation to increase gasoline yield or refined by means of a hydrogen process.

3. VISBREAKING

Although new thermal cracking units are under development (Chapter 19), processes that can be regarded as having evolved from the original concept of thermal cracking are visbreaking and the various coking processes (Table 14.2). At its peak in 1930, gas oil

Table 14.1 Example of the Thermal Cracking of a Reduced
(Flasher Residuum) Crude Oil

Feedstock	
API gravity	25.0
IBP	227.0°C; 440.0°F
Cracking parameters	
Temperature	500.0°; 930.0°F
Soaker pressure	225.0 psig; 1550.0 kPa
Product yields, vol%	
Gasoline[a]	
Naphtha	57.5
Heating oil (light oil)	0.0
Residuum	37.5
API gravity	7.4
Heating oil[b]	
Gas	1.0
Naphtha	42.0
Heating oil (light oil)	23.0
Residuum	34.0
API gravity	8.0

[a] Light oil recycle.
[b] No recycle of light oil.

Table 14.2 Summary of the Various Thermal
Cracking Processes

Visbreaking
 Mild heating (880–920°F) at 50–200 psig
 Reduce viscosity of fuel oil
 Low conversion (10%) to 430°F
 Heated coil or drum
Delayed coking
 Moderate heating (900–960°F) at 90 psig
 Soak drums (845–900°F)
 Coked until drum solid
 Coke (removed hydraulically) 20–40% on feed
 Yield at 430°F, 30%
Fluid coking
 Severe heating (900–1050°F) at 10 psig
 Oil contacts refractory coke
 Bed fluidized with steam, even heating
 Higher yields of light ends ($< C_5$)
 Less coke yield

thermal cracking represented about 55% of the total crude oil capacity, or over 2,000,000 bbl/day. Thereafter, a rapid decline of gas oil thermal cracking capacity occurred due to the construction and operation of catalytic cracking units. Many of the original thermal cracking units were converted to other operations, such as visbreaking, or were completely dismantled. Visbreaking capacity has been varied over the decades but is currently in excess of 2,500,000 bbl/day. *Thermal operations* for worldwide refineries are on the order of 3,775,000 bbl/day from a total refinery capacity of 81,500,000 per day (Radler, 1999); visbreaking is believed to be the major operative thermal process. For refineries in the United States, the total capacity is approximately 16,500,000 bbl per day, of which thermal operations account for 56,300 bbl/day (Radler, 1999), and it is assumed that the major *thermal operation* is visbreaking.

Visbreaking, an abbreviated term for *viscosity breaking* or *viscosity lowering*, is a relatively mild, liquid-phase thermal cracking process used to convert heavy, high viscosity petroleum stocks to lower viscosity fractions suitable for use in heavy fuel oil. This ultimately results in less production of fuel oil, because less cutter stock (low viscosity diluent) is required for blending to meet fuel oil viscosity specifications. The cutter stock no longer required in fuel oil can then be used in more valuable products. A secondary benefit from the visbreaking operation is the production of gas oil and gasoline streams that usually have higher product values than the visbreaker charge. Visbreaking produces a small quantity of light hydrocarbon gases and a larger amount of gasoline.

Visbreaking, unlike conventional thermal cracking. normally does not employ a recycle stream. Conditions are too mild to crack a gas oil recycle stream, and the unconverted residual stream, if recycled, would cause excessive heater coking. The boiling range of the product residual stream is extended by visbreaking so that light and heavy gas oils can be fractionated from the product residual stream, if desired. In some applications the heavy gas oil stream is recycled and cracked to extinction in a separate higher temperature heater.

Visbreaking (viscosity reduction, viscosity breaking) is a relatively mild thermal cracking operation used to reduce the viscosity of residua through low conversion and the production of products that are lower boiling than the original feedstock (Table 14.3) (Ballard et al., 1992; Dominici and Sieli, 1997). Residua are sometimes blended with lighter heating oils to produce fuel oils of acceptable viscosity. By reducing the viscosity of the nonvolatile fraction, visbreaking reduces the amount of the more valuable light heating oil that is required for blending to meet the fuel oil specifications. The process is also used to reduce the pour point of a waxy residue.

The visbreaking process (*Hydrocarbon Processing*, 1996; 1998, p. 111) uses the approach of mild thermal cracking as a relatively low cost and low severity approach to improving the viscosity characteristics of the residue without attempting significant conversion to distillates. Short residence times are required to avoid polymerization and coking reactions, although additives can help to suppress coke deposits on the tubes of the furnace.

Two visbreaking processes are commercially available (Fig. 14.3): the coil or furnace type and the soaker type.

The *coil visbreaking process* (Fig. 14.4) (*Hydrocarbon Processing*, 1998, p. 111) achieves conversion by high temperature cracking within a dedicated soaking coil in the furnace. With conversion achieved primarily as a result of temperature and residence time, coil visbreaking is described as a high temperature, short residence time route. The main advantage of the coil-type design is the two-zone fired heater that provides better control

Table 14.3 Examples of Product Yields and Properties for Visbreaking Various Feedstocks

	Feedstock									
	Louisiana vacuum residuum	Arabian Light atmospheric residuum	Arabian Light atmospheric residuum	Arabian Light atmospheric residuum	Arabian Light vacuum residuum	Arabian Light vacuum residuum	Arabian Light vacuum residuum	Kuwait atmospheric residue	Iranian Light vacuum residue	Athabasca bitumen
Feedstock										
Gravity, API	11.9	15.9	16.9	16.9	7.1	6.9	6.9	14.4	8.2	8.6
Carbon residue[a]	10.6	8.5			20.3			9.4	22.0	13.5
Sulfur, wt%	0.6	3.0	3.0	3.0	4.0	4.0	4.0	4.1	3.5	4.8
Product yields,[b] vol%										
Naphtha ($<425°F$; $<220°C$)	6.2	8.0	7.8	7.8	6.0	8.1	8.1	4.4	3.5	4.8
Light gas oil ($425–645°F$; $220–340°C$)	6.3	15.0	11.9		16.0	10.5	8.1	16.9	13.1	21.0
Heavy gas oil ($645–1000°F$; $340–545°C$)				70.8		20.8				35.0
Residuum	88.4	79.7	20.9	76.0	60.5	91.8	76.6	79.9	34.0	
Gravity, API	11.4	14.7	1.3		3.5	0.8	7.2	11.0	5.5	
Carbon residue[a]	15.0	3.2								
Sulfur, wt%	0.6	3.5	5.0		4.7	4.6	4.0	4.4	3.8	

[a] Conradson.
[b] A blank product yield line indicates that the yield of the lower boiling product has been included in the yield of the higher boiling product.

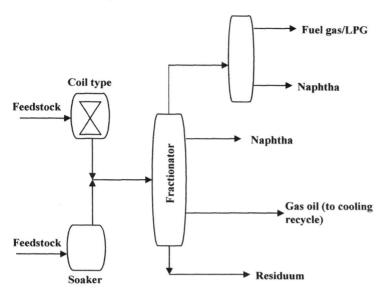

Figure 14.3 General representation of a visbreaking unit.

of the material being heated, and with the coil-type design decoking of the heater tubes is accomplished more easily by the use of steam–air decoking.

The alternative *soaker visbreaking process* (Fig. 14.5) (*Hydrocarbon Processing*, 1998, p. 112) achieves some conversion within the heater but the majority of the conversion occurs in a reaction vessel or soaker that holds the two-phase effluent at an elevated

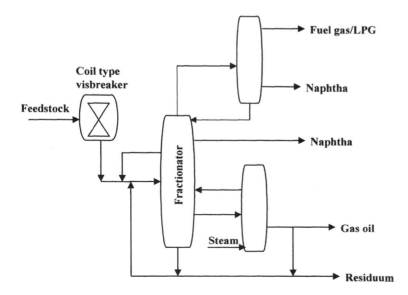

Figure 14.4 A coil-type visbreaker.

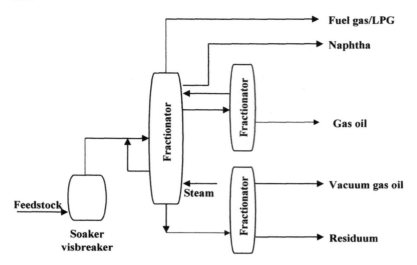

Figure 14.5   A soaker-type visbreaker.

temperature for a predetermined length of time. Soaker visbreaking is described as a low temperature, high residence time route. Product quality and yields from the coil and soaker drum design are essentially the same at a specified severity, being independent of visbreaker configuration. By providing the residence time required to achieve the desired reaction, the soaker drum design allows the heater to operate at a lower outlet temperature (thereby saving fuel), but there are disadvantages. The main disadvantage is the decoking operation of the heater and soaker drum, and although decoking of the soaker drum design is not required as frequently as with the coil-type design, the soaker design requires more equipment for coke removal and handling. The customary practice of removing coke from a drum is to cut it out with high pressure water, thereby producing a significant amount of coke-laden water that needs to be handled, filtered, and then recycled for reuse.

Other variations of visbreaking technology include the Tervahl T and Tervahl H processes. The Tervahl T alternative (Fig. 14.6) includes only the thermal section to produce a synthetic crude oil with better transportability by having reduced viscosity and greater stability. The Tervahl H alternative adds hydrogen that also increases the extent of the desulfurization and decreases the carbon residua. The Aquaconversion process (Fig. 14.7) (*Hydrocarbon Processing*, 1998, p. 112) is a new hydrovisbreaking technology that uses catalyst-activated transfer of hydrogen from water added to the feedstock. Reactions that lead to coke formation are suppressed, and there is no separation of asphaltene-type material (Marzin et al., 1998).

Visbreaking conditions range from 455 to 520°C (850–950°F) at a short residence time and from 50 to 300 psi at the heating coil outlet. It is the short residence time that renders visbreaking a mild thermal reaction. This is in contrast to, for example, the delayed coking process, where residence times are much longer and the thermal reactions are allowed to proceed to completion. The visbreaking process uses a quench operation to terminate the thermal reactions. Liquid-phase cracking takes place under these low severity conditions to produce some naphtha as well as material in the kerosene and gas oil boiling range. The gas oil may be used as additional feed for catalytic cracking units or as heating oil.

Sell your books at
World of Books!
Go to sell.worldofbooks.com
and get an instant price quote.
We even pay the shipping - see
what your old books are worth
today!

000795159323

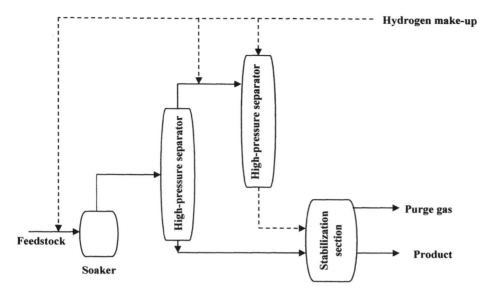

Figure 14.6   The Tervahl T and Tervahl H process configurations.

Atmospheric and vacuum residua are the usual feedstocks to a visbreaker, although tar sand bitumen has long been considered a likely feedstock. The atmospheric and vacuum residua will typically achieve a conversion to gas, gasoline, and gas oil on the order of 10–50% by weight, depending on the severity and feedstock characteristics. The conversion of the residua to distillate (low-boiling products) is commonly used as a measure of the severity of the visbreaking operation, and the conversion is determined as the amount of $345°C^+$ ($650°F^+$) material present in the atmospheric residuum or the $482°C^+$

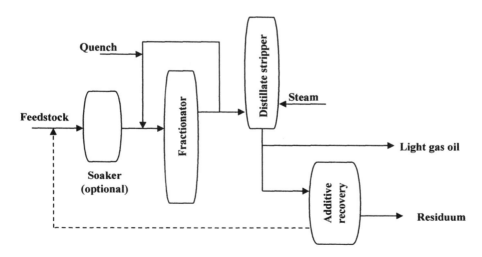

Figure 14.7   The aquaconversion process.

($900°F^+$) material present in the vacuum residuum that is converted (*visbroken*) into lower boiling components.

The extent of feedstock conversion is limited by a number of feedstock characteristics, such as asphaltene content (Fig. 14.8), which varies with the type of feedstock and hence the type of residuum and, more particularly, carbon residue (Fig. 14.9). In very general terms, paraffinic feedstocks will have a low heptane-asphaltene content (0–8% by weight), whereas naphthenic feedstock will have a much higher heptane-asphaltene content (10–20% by weight), with the mixed crude oils having intermediate values. Of course, when the heptane-asphaltenes are concentrated in the residua (through distillation) the proportions of the asphaltenes will be much higher. Thus, feedstocks with a high heptane-asphaltene content will result in an overall lower conversion than feedstocks a lower heptane-asphaltene content while maintaining production of a stable fuel oil from the visbreaker bottoms. Minimizing the sodium content to almost a negligible amount and minimizing the Conradson carbon weight percent will result in longer cycle run lengths.

In addition, variations in feedstock quality will impact the level of conversion obtained at a specific severity. For example, for a given feedstock, as the severity is increased, the viscosity of the $205°C^+$ ($400°F^+$) visbroken residue (often referred to as *visbroken tar* or *visbreaker tar*) initially decreases, and then at higher severity levels it increases dramatically, indicating the formation of coke precursors and their initial phase separation as sediment. The point at which this viscosity reversal occurs differs from feedstock to feedstock but can be estimated from the amount of low molecular weight hydrocarbon gases ($\leq C_3$) (Dominici and Sieli, 1997).

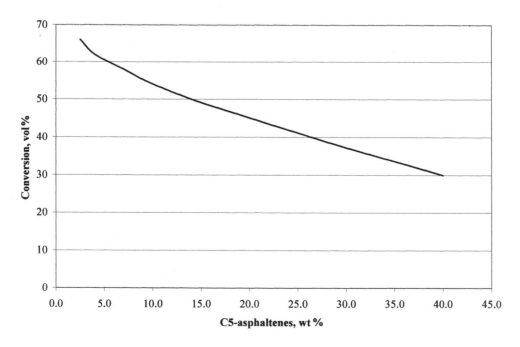

Figure 14.8 Relationship of visbreaker conversion to asphaltene content.

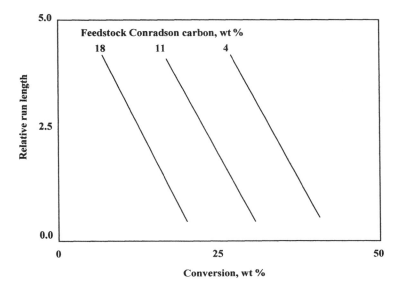

Figure 14.9 Relationship of visbreaker conversion time to feedstock carbon residue.

Thus, a crude oil residuum is passed through a furnace where it is heated to a temperature of 480°C (895°F) under an outlet pressure of about 100 psi. The heating coils in the furnace are arranged to provide a soaking section of low heat density, where the charge remains until the visbreaking reactions are completed. The cracked products are then passed into a flash distillation chamber. It is advisable to maintain the flash zone temperature as low as possible to minimize the potential for coking. Under fixed flashing conditions, increasing the yield of the residuum will reduce this temperature.

The overhead material from this chamber is then fractionated to produce a low quality gasoline as an overhead product and light gas oil as bottoms. The liquid products from the flash chamber are cooled with a gas oil flux and then sent to a vacuum fractionator. This yields a heavy gas oil distillate and a residuum of reduced viscosity. A quench oil may also be used to terminate the reactions and will also influence the temperature of the flash zone.

A 5–10% conversion of atmospheric residua to naphtha is usually sufficient to afford at least an approximately five-fold reduction in viscosity. Reduction in viscosity is also accompanied by a reduction in the pour point. However, the reduction in viscosity of distillation residua tends to reach a limiting value with conversion, although the total product viscosity can continue to decrease. The minimum viscosity of the unconverted residue can lie outside the range of allowable conversion if sediment begins to form. When shipment of the visbreaker product by pipeline is the process objective, addition of a diluent such as gas condensate can be used to achieve a further reduction in viscosity.

Conversion of residua in visbreaking follows first-order reaction kinetics (Speight, 2000, and references cited therein). The minimum viscosity of the unconverted residue can lie outside the range of allowable conversion if sediment begins to form. When pipelining the visbreaker product is the process objective, addition of a diluent such as gas condensate can be used to achieve a further reduction in viscosity.

The severity of the visbreaking operation is generally limited by the stability of the visbroken fuel oil produced. If overcracking occurs, the resulting fuel oil may form excessive deposits in storage or when used as a fuel in a furnace. One method of measuring the thermal stability of the fuel oil is the U.S. Navy special fuel oil thermal stability test (ASTM D-1661). The visbreaking correlations presented are based on operating to levels where the fuel oil quality will be limited by this test. This severity level is well within the operating limits that would be imposed by excessive coke formation in properly designed visbreaking furnaces.

The main limitation of the visbreaking process, and for that matter all thermal processes, is that the products can be unstable. Thermal cracking at low pressure gives olefins, particularly in the naphtha fraction. These olefins give a very unstable product that tends to undergo secondary reactions to form gum and intractable residua. Product stability of the visbreaker residue is a main concern in selecting the severity of the visbreaker operating conditions. Severity, or the degree of conversion, can cause phase separation of the fuel oil even after cutter stock blending. Increasing visbreaking severity and percent conversion will initially lead to a reduction in the visbroken fuel oil viscosity. However, visbroken fuel oil stability will decrease as the level of severity—and hence conversion—is increased beyond a certain point, dependent on feedstock characteristics.

The instability of the visbroken fuel oil is related to the asphaltene constituents in the residuum. Asphaltenes are heavy nonvolatile compounds that can be classified according to their solubility in various solvents. The asphaltene constituents can be thermally altered during visbreaking operations. In addition, during visbreaking some of the high molecular weight constituents, including some of the asphaltene constituents, are converted to lower boiling and medium boiling paraffinic components, some of which are removed from the residuum. The asphaltenes, being unchanged, are thus concentrated in the product residuum (which may contain new paraffinic material), and if the extent of the visbreaking reaction is too high, the asphaltene constituents or altered asphaltene constituents will tend to precipitate in the product fuel oil, creating an unstable fuel oil.

A common method of measuring the amount of asphaltenes in petroleum is to add a low-boiling liquid hydrocarbon such as n-pentane (ASTM D-893, ASTM D-2006, ASTM D-2007) or n-heptane (ASTM D-3279, ASTM D-4124, IP 143), which causes the asphaltenes to separate as a solid (Speight, 1999, 2000). Because the amount of asphaltenes in the visbreaking unit charge residuum may limit the severity of the visbreaking operations, the normal pentane insolubles content of the charge residuum is used as the correlating parameter in various visbreaking correlations. However, these correlations can be visbreaker- and feedstock-dependent, and their application from one unit to another or from one feedstock to another can be misleading.

Sulfur in the visbroken residuum can also be an issue, because the sulfur content of the visbreaker residuum is often higher (approximately 0.5 wt% greater) than that in the feedstock. Therefore it can be difficult to meet the commercial sulfur specifications of the refinery product residual fuel oil, and blending with low sulfur cutter stocks may be required.

Visbreaking, like thermal cracking, is a *first-order reaction*. However, due to the visbreaking severity limits imposed by fuel oil instability, operating conditions do not approach the level where secondary reactions—leading to coke formation—occur to any significant extent. The first-order reaction rate equation altered to fit the visbreaking reaction is

$$K = \frac{1}{t} \ln \frac{100}{X_1}$$

where K is the first-order reaction velocity constant, s^{-1}; t is the time at thermal conversion conditions, s; $X_1 = 900°F^+$ visbroken residuum yield, vol%. A simplified graphical representation of the yields of the various products with conversion can be constructed (Fig. 14.10) with the understanding that different feedstocks will require different graphical presentations.

The visbreaking reaction is first-order, and velocity constant data as a function of visbreaking furnace outlet temperature can be presented graphically (Fig. 14.11). The thermal conversion reactions are generally assumed to start at 425°C (800°F), although some visbreaking occurs below this temperature. Therefore, the residence time in the 425–450°C (800–865°F) reaction zone should be 613 s.

The central piece of equipment in any thermal process is, with no exception the heater. The heater must be adequate to efficiently supply the heat required to accomplish the desired degree of thermal conversion. A continually increasing temperature gradient designed to give most of the temperature increase in the front part of the heater tubes with only a slow rate of increase near the outlet is preferred. Precision control of time and temperature is usually not critical in the processes covered in this chapter. Usually, all that is required is to design to some target temperature range and then adjust actual operations to achieve the desired cracking. In the higher temperature processes (e.g., ethylene manufacture), temperature control does become of prime importance due to equilibrium considerations.

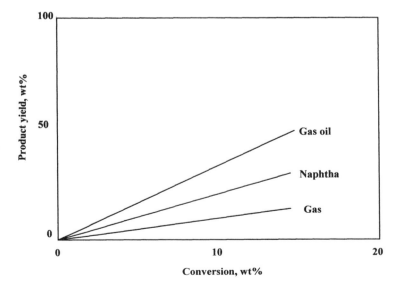

Figure 14.10 Trends for visbreaker product yields with feedstock conversion.

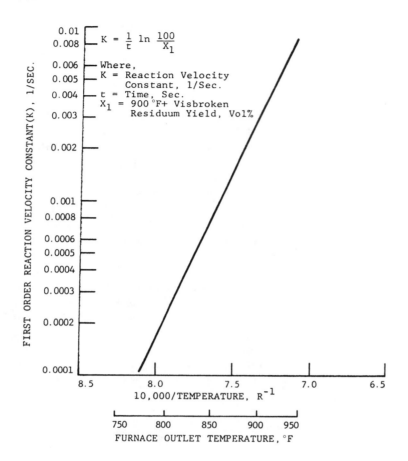

Figure 14.11   First-order reaction velocity constants for visbreaking.

Equipment design to minimize coke formation is important. The excessive production of coke adversely affects the thermal cracking process because it

1.   Reduces heat transfer rates
2.   Increases pressure drops
3.   Creates overheating
4.   Reduces run time
5.   Requires the expense of removing the coke from the equipment

The metallurgy of the equipment, specifically the heater tubes and pumps, in the high temperature, corrosive environments must be adequate to prevent the destruction and consequent expensive replacement of equipment. In the early days of thermal cracking, the metallurgy of the heater tubes was not of sufficient quality to permit extended periods of high temperatures. Modern improvements in the quality of steel have extended the durability of thermal cracking equipment.

The advances in heater design have reached a point where very efficient furnace and heating tube arrangements can be built that give the refiner the desired thermal cracking operation. The practice of the refiner is generally to set the specifications the heater is expected to meet for the specific application and have a heater manufacturer prepare a suitable design. Proper *tube size* selection depends upon minimizing pressure drop while obtaining good turbulence for proper heat transfer, which is also dependent upon the charge rate (Table 14.4), which in turn ultimately affects the residence time and therefore the extent of the conversion.

The charge stock *liquid velocity* should be sufficient to provide enough turbulence to ensure a good rate of heat transfer and to minimize coking. A minimum linear cold (15.6°C; 60°F) velocity of 5 ft/s for a 100% liquid charge rate should be sufficient. The maximum velocity would be limited to about 10 ft/s due to excessive pressure drop. The velocities at the higher cracking temperatures would, of course, be greater due to the partial vaporization of the charge.

Most of the *heat supplied* to the charge stock is radiant heat. The convection section of the heater is used primarily to preheat the charge prior to the main heating in the radiant section. The *heat transfer rate* in the convection section will range from 3000 to 10,000 Btu per square foot of tube outside area per hour with an average rate of 5000 Btu/(ft$^2 \cdot$ h). The heating rates in the radiant section will range from 8000 to 20,000 Btu/(ft$^2 \cdot$ h) depending upon the charge stock, with heavier oil generally requiring the lower heating rate.

The heating tube *outlet temperature* will depend upon the charge stock being processed and the degree of thermal conversion required. The outlet temperature will vary from a minimum of 425°C (800°F) for visbreaking to a maximum of 595°C (1100°F) for thermal reforming. The combustion chamber temperature will range from 650 to 870°C (from 1200°F to 1600°F) at a point about 1 ft below the radiant tubes. Flue gas temperatures are usually high (425–595°C; 800–1100°F), particularly since the heavy charge stock is usually entering the heater at a high temperature from a fractionating tower. An exception to the charge entering at a high temperature would be when gasoline is being changed to a thermal reformer. However, because thermal reforming requires high temperatures, flue gas temperatures will also be high.

Because it is desirable to maintain different temperature increase rates throughout the charge heating, i.e., rapid increase at the beginning of the heating coil and a lower rate

Table 14.4 Relationship of Feedstock Flow Rate to Tube Diameter

Total charge rate (fresh feed + recycle) (bbl/day)	Internal diameter of tube (in.)
3000	2–3
6000	3–4
12,000	4–4$\frac{1}{2}$[a]

[a] Parallel tubes of smaller diameters would be preferable to one large tube. In this case, two 3 in. diameter parallel tubes may be preferred.

near the outlet, zone temperature control within the furnace is desired. A three-zone furnace is preferred, with the first zone giving the greatest rate of temperature increase and the last zone the least.

Coke formation limits the operation of the heater, and techniques should be employed to minimize coke formation in the heater tubes. Coking occurs on the walls of the tubes, particularly where turbulence is low and temperature is high. Maintaining sufficient turbulence assists in limiting coke formation. Baffles within the tubes are sometimes used, but water injection into the charge stream is the preferred method. Water is usually injected at the inlet but also may be injected at additional points along the heater tubes. The water, in addition to providing turbulence in the heater tubes as it is vaporized to steam, also provides a means to control temperature. The optimum initial point of water injection into the heater tubes is at the point of incipient cracking where coke would start to form. An advantage to this injection point is the elimination of the additional pressure drop that would have been created by the presence of water between the heater inlet and the point of incipient cracking.

The preferred method for decoking the heater tubes is to burn off the coke using a steam–air mixture. The heater tubes, therefore, should be capable of withstanding temperatures up to 760°C (1400°F) at low pressures for limited time periods. The heater tubes along with the tube supports should be designed to handle the thermal expansion extremes that would be encountered. Mechanical means such as drills can also be used to remove coke, but most modern heaters use the steam–air combustion technique. Parallel heaters may be employed so that one can be decoked while cracking is permitted to proceed in the other(s).

The metallurgy of thermal cracking units is variable, but alloy steel tubes of 7–9% chromium are usually satisfactory to resist sulfur corrosion in thermal cracking heaters. If the hydrogen sulfide content of the cracked products exceeds 0.1 mole% in the cracking zone, a higher alloy steel may be required. Stabilized stainless steel, such as Type 321 or 347, would be suitable in this case. Other alloys, such as the Inconel or Incoloy alloys, could also be used. Seamless tubes with welded return bends are now normally used in heaters. Flanged return bends were used in earlier thermal cracking units to facilitate cleaning. However, the use of steam and air to burn out the coke essentially eliminates the need for flanged fittings, which in turn reduces the possibility of dangerous leaks.

A useful tool to aid in the design and operation of thermal cracking units is the soaking volume factor (SVF). This factor combines time, temperature, and pressure of thermal cracking operations into a single numerical value. The SVF is defined as the *equivalent* coil volume in cubic feet per daily barrel of charge (fresh plus recycle) if the cracking reaction had occurred at 800°F and 750 psig (5171 kPa).

$$\text{SVF}_{750\text{psi}/800\text{F}} = \frac{1}{F} R K_\text{p} \, dV$$

where $\text{SVF}_{750\text{psi}/800\text{F}}$ is the SVF at base reaction conditions of 750 psi gauge pressure and 800°F, in cubic feet of coil volume per total charge throughput in barrels per day; F is the charge (fresh plus recycle) throughput rate, in barrels per day; R is the ratio of the reaction velocity constant at temperature Y to the reaction velocity constant at 800°F, K_T/K_{800}; K_p is the pressure correction factor for pressures other than 750 psi gauge; and dV is the incremental coil volume, in cubic feet.

The ratio of reaction velocity constants that can be obtained from graphical formats (Fig. 14.12) should not be obtained from such plots of reaction velocity constants because a correction for the effect of temperature on the volume of the reacting material must be incorporated into the data (Fig. 14.13). The pressure correction factor, K_p, can be obtained graphically (Fig. 14.14).

When an additional soaking drum is used, the SVF for the soaking drum should be added to the coil SVF. The SVF for the drum can be determined from

$$\mathrm{SVF}_D = \frac{DV}{(F K_{TD} K_p)}$$

where SVF_D is the SVF of the drum; DV is the volume of the drum, ft^3; F is the charge (fresh plus recycle) throughput rate, bbl/day; K_{TD} is the reaction velocity constant for the mean drum temperature; and K_p is the pressure correction factor for the mean drum pressure.

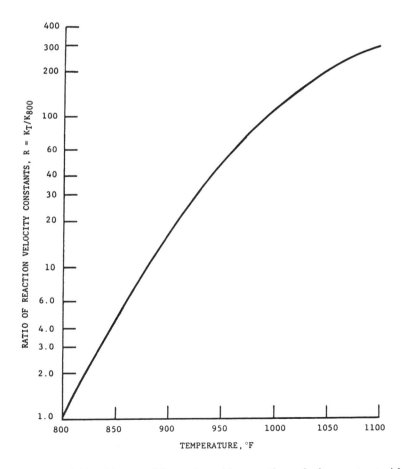

Figure 14.12 Change of thermal cracking reaction velocity constant with temperature in excess of 425°C (800°F).

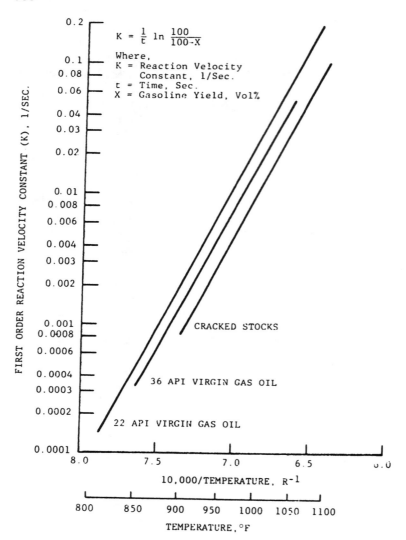

Figure 14.13 First-order reaction velocity constants for thermal cracking.

The SVF will range from 0.03 for visbreaking of heavy residual stocks to about 1.2 for light gas oil cracking. The SVF is a numerical expression of cracking rate and thus can be correlated with product yield and quality. SVF can also be translated into cracking coils and still volumes of known dimensions under design conditions of temperature and pressure.

A cracking unit seldom operates very long at design conditions. Charge stock quality changes, desired product yields and qualities change, or additional capacity is required. These changes require an SVF that is different from the design SVF. The SVF can be varied by

1. Varying pressure at constant temperature and feed rate

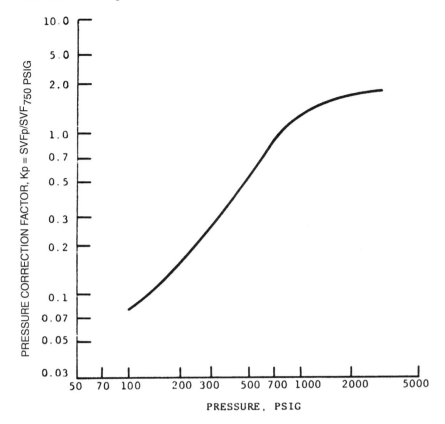

Figure 14.14 Pressure correction for the soaking volume factor (SVF) with pressure in excess of 750 psig (5170 kPa).

2. Varying temperature at constant feed rate, the pressure gradient varying with the effect upon cracking rate and fluid density in the cracking coil
3. Varying soaking volume at constant temperature and pressure by varying heater feed rate and/or varying the number of tubes in the section above 425°C (800°F)

With the advent of more efficient heaters with higher firing rates, the use of external *soaking drums* to provide longer reaction time became of less importance in thermal cracking operations. In modern units the coil in the heater is usually sufficient to provide the necessary temperature–time relationships. A possible exception would be the case where it is desirable to crack a considerable amount of heavy residual stock. The temperature required probably could not be successfully obtained in a heater coil without excessive coking. A reaction chamber (soaking drum) is employed where the hotter, cleaner light gas oil is used to supply heat to the heavier dirty oil stream in a soaking drum. A low temperature light gas oil stream is also frequently used to wet the walls of the soaking drum to minimize coking. Parallel soakers could be used to allow one to be decoked while the other is used for the cracking operations.

The *pumps* used in thermal cracking operations must be capable of operation for extended periods handling a corrosive liquid at high temperature [above 230°C (450°F) and up to 345°C (650°F)]. In addition, because coke particles are formed in thermal cracking, the pumps must be able to withstand the potential erosion of the metal parts by the coke particles. In the early days of thermal cracking, reciprocating pumps were commonly used. In later units centrifugal pumps were used. A preferred centrifugal pump would be of the coke-crushing type or may have open impellers with case wear plates substituted for the front rings. The metal should be 12% chromium steel alloy or a higher alloy if serious corrosion is possible.

Heat exchangers should be constructed to facilitate cleaning, because the high temperatures and coke particles can create extensive fouling of the exchangers.

The downstream processing equipment (*flash drums, separators, fractionating towers*) would be of standard design, and no special design specifications are required other than minimizing potential coke buildup. This can be accomplished by designing the equipment so there would be no significant hold-up or dormant spots in the process equipment where coke could accumulate.

In thermal cracking operations there is a considerable amount of *excess heat* that cannot be economically utilized within the cracking unit itself. When a thermal cracking unit is being considered, it is desirable to construct the unit in conjunction with some other unit, such as a crude still, that could use the excess heat to preheat the crude oil charge. Alternatively, the excess heat could be used in steam generation facilities.

4. COKING

Coking is a thermal cracking type of operation used to convert low grade feedstocks such as straight-run and cracked residua to coke, gas, and distillates. Two types of petroleum coking processes are presently operating: (1) delayed coking, which uses multiple coking chambers to permit continuous feed processing wherein one drum is making coke and one drum is being decoked; and (2) fluid coking, which is a fully continuous process where product coke can be withdrawn as a fluidized solid.

Installed capacities for coking (delayed coking, fluid coking, flexicoking) on a worldwide basis are 3,745,020 bbl/day of a total worldwide refinery capacity 81,549,796 bbl/day. On the other hand, the coking capacity of refineries in the United States is 2,022,490 bbl/day of a total refinery capacity of 16,540,990 bbl per day (Radler, 1999).

Crude oil residua obtained from the vacuum distillation tower as a nonvolatile (bottoms) fraction are the usual charge stocks to coking. Atmospheric tower bottoms (long residua) can be charged to coking units but it is generally not attractive to thermally degrade the gas oil fraction contained in the longer residua. Other charge stocks to coking are deasphalter bottoms and tar sand bitumen, and cracked residua (thermal tars). The products are gases, naphtha, fuel oil, gas oil, and coke. The gas oil may be the major product of a coking operation and serves primarily as a feedstock for catalytic cracking units. The coke obtained is usually used as fuel, but processing for specialty uses such as electrode manufacture and the production of chemicals, and metallurgical coke is also possible and increases the value of the coke. For these uses, the coke may require treatment to remove sulfur and metal impurities.

Furthermore, the increasing attention paid to reducing atmospheric pollution has also served to direct some attention to coking, because the process not only concentrates such pollutants as feedstock sulfur in the coke but also usually yields products that can be conveniently subjected to desulfurization processes.

Coking processes generally utilize longer reaction times than thermal cracking processes. To accomplish this, drums or chambers (reaction vessels) are employed, but it is necessary to use two or more such vessels so that coke removal can be accomplished in vessels that are not on-stream to avoid interrupting the semicontinuous nature of the process.

Coking processes have the virtue of eliminating the residue fraction of the feed but at the cost of forming a solid carbonaceous product. The yield of coke in a given coking process tends to be proportional to the carbon residue content of the feed (measured as the Conradson carbon residue; see Chapter 4). The data in Table 14.5 illustrate how the yield of coke from delayed and fluid coking varies with the Conradson carbon residue of the feed.

The formation of large quantities of coke is a severe drawback unless the coke can be put to use. Calcined petroleum coke can be used for making anodes for aluminum manufacture and a variety of carbon or graphite products such as brushes for electrical equipment. These applications, however, require a coke that is low in mineral matter and sulfur.

If the feedstock produces a high sulfur, high ash, high vanadium coke, one option for use of the coke is combustion to produce process steam (and large quantities of sulfur dioxide unless the coke is first gasified or the combustion gases are scrubbed). Another option is stockpiling.

For some feedstocks, particularly those from heavy oil, the combination of poor coke properties for anode use, limits on sulfur dioxide emissions, and loss of liquid product volume have tended to relegate coking processes to a strictly secondary role in any new upgrading facilities.

4.1 Delayed Coking

Delayed coking is the oldest, most widely used cooking process and has changed very little over its 60 year history. It is a semicontinuous (semi-batch) process in which the heated charge is transferred to large coking (or soaking) drums that provide the long residence

Table 14.5 Relationship of Coke Yield to API Gravity

Carbon residue (wt%)	°API	Coke yield, delayed coker	Weight % fluid coker
1		0	
5	26	8.5	3
10	16	18	11.5
15	10	27.5	17
20	6	35.5	23
25	3.5	42	29
30	2		34.5
40	−2.5		46

Source: Nelson, 1976.

time needed to allow the cracking reactions to proceed to completion (McKinney, 1992; *Hydrocarbon Processing*, 1996; Feintuch and Negin, 1997).

The delayed coking process (Fig. 14.15) is widely used for treating residua and is particularly attractive when the green coke produced can be sold for the manufacture of anode or graphitic carbon or when there is no market for fuel oils. The process uses long reaction times in the liquid phase to convert the residue fraction of the feed to gases, distillates, and coke. The condensation reactions that give rise to the highly aromatic coke product also tend to retain sulfur, nitrogen, and metals so that the coke is enriched in these elements relative to the feed.

In the process (Fig. 14.15), the feedstock is charged to the fractionator and subsequently charged with an amount of recycle material (usually about 10%, but as much as 25%, of the total feedstock) from the coker fractionator through a preheater and then to one of a pair of coke drums; the heater outlet temperature varies from 480 to 515°C (895–960°F) to produce the various products (Table 14.6). The cracked products leave the drum as overheads to the fractionator, and coke deposits form on the inner surface of the drum. The majority of the sulfur originally in the feedstock remains in the coke (Table 14.7). A pair of coke drums are used so that while one drum is on stream the other is being cleaned, allowing continuous processing; the drum operation cycle is typically 48 h. (Table 14.8). The temperature in the coke drum ranges from 415 to 450°C (780–840°F) at pressures from 15 to 90 psi (103–620 kPa).

Delayed coking units fractionate the coke drum overhead products into fuel gas (low molecular weight gases up to and including ethane), propane and

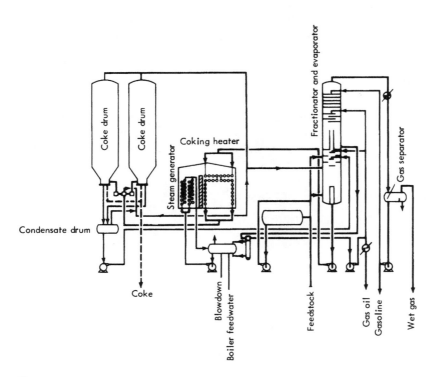

Figure 14.15 The delayed coking process.

Table 14.6 Examples of Product Yields and Product Properties for Delayed Coking of Various Feedstocks[a]

Feedstock	Louisiana residuum	Kuwait residuum	Kuwait residuum[c]	West Texas residuum	West Texas residuum[c]	Oklahoma residuum	Oklahoma residuum	California residuum	California residuum	Midcontinent residuum	Middle East residuum	Middle East residuum	Venezuela residuum
API gravity	12.3	6.7	16.1	8.9	15.2	13.0	16.8	12.0	12.0	12.3	7.4	8.2	2.6
Carbon residue[b]	13.0	19.8	9.1	17.8	9.3	14.1	8.0	9.4	9.6	11.3	20.0	15.6	23.3
Sulfur, wt%	0.7	5.2	0.7	3.0	0.6	1.2		1.6	1.6	0.4	4.2	3.4	4.4
Product yields, vol%													
Naphtha (95–925°F; 35–220°C)	22.8	26.7	22.0	28.9	20.1		10.7	22.5	15.7	16.0	12.6	17.4	10.0
Light gas oil (425–645°F; 220–340°C)	18.4	28.0	41.9	16.5	31.7	20.4	36.5	36.5					
Heavy gas oil (645–1000°F; 340–540°C)	37.6	18.4	19.1	26.4	27.5	57.2	16.7	16.7	72.3	56.5	50.8	48.5	50.3
Coke	23.7	30.2	18.5	28.4	20.7	23.6	20.8	19.1	21.6	21.0	28.7	24.9	31.0
Sulfur, wt%	1.3	7.5	1.7	4.5	1.6								

Feedstock	North Africa residuum	North Africa residuum	SE Asia residuum	Arkansas residuum	Tia Juana residuum	Alaska NS residuum[d]	Alaska NS residuum	Arabian Light residuum	Arabian Light residuum	Mexican residuum	Santa Maria	Athabasca bitumen
API gravity	15.2	12.8	17.1	15.3	8.5	7.4	7.4	16.9	6.9	4.0	7.2	7.3
Carbon residue[b]	16.7	5.2	11.1	11.5	22.0	18.1	18.1			22.0	14.8	17.9
Sulfur, wt%	0.7	0.6	0.5	2.8	2.9	2.0	2.0	3.0	4.0	5.3	6.7	5.3
Product yields, vol%												
Naphtha (95–925°F; 35–220°C)	19.9	18.5	20.4	13.5	25.6	15.0	12.5	14.2	19.1	21.4	22.4	20.3
Light gas oil (425–645°F; 220–340°C)					26.4						16.2	
Heavy gas oil (645–1000°F; 340–540°C)	46.0	65.3	54.5	63.0	13.8	44.9	51.2	70.6	48.4	33.0	36.8	58.8
Coke	26.4	10.0	17.7	22.6	33.0	30.2	27.2	15.4	32.8	35.1	19.8	21.0
Sulfur, wt%						2.6	2.6	4.8	5.6			8.0

[a] A blank product line indicates that the yield of the lower boiling product has been included in the yield of the higher boiling product. [b] Conradson. [c] Hydrodesulfurized. [d] 35 psig compared to ~14–18 psig for the other delayed cokers.

Table 14.7   Relationship of Feedstock Sulfur to Coke Sulfur

Feedstock	API gravity	Sulfur in feed (wt%)	Sulfur in coke (wt%)	%S in coke/ %S in feedstock
Elk Basin, WY, residuum	2.5	3.5	6.5	1.83
Hawkins, TX, residuum	4.5	4.5	7.0	1.55
Kuwait, residuum	6.0	5.37	10.8	2.01
Athabasca (Canada), bitumen	7.3	5.3	5.65	1.06
West Texas, residuum	—	3.5	3.06	0.875
Boscan (Venezuela), crude oil	10.0	5.0	5.0	1.0
East Texas, residuum	10.5	1.26	2.57	2.04
Texas Panhandle, residuum	18.9	0.6	0.6	1.00

propylene ($CH_3CH_2CH_3$, $CH_3CH=CH_2$), butane and butene ($CH_3CH_2CH_2CH_3$, $CH_3CH_2CH=CH_2$), naphtha, light gas oil, and heavy gas oil. Yields and product quality vary widely due to the broad range of feedstock types charged to delayed coking units (Table 14.6).

Coker naphthas have boiling ranges up to 220°C (430°F), are olefinic, and must be upgraded by hydrogen processing for removal of olefins and sulfur. They are then used conventionally for reforming to gasoline or as chemical feedstocks. Middle distillates, boiling in the range of 220–360°C (430–680°F), are also hydrogen treated for improved storage stability, sulfur removal, and nitrogen reduction. They can then be used for either diesel or burner fuels or thermally processed to lower boiling naphtha. The gas oil boiling up to about 510°C (950°F) endpoint can be charged to a fluid catalytic cracking unit immediately or after hydrogen upgrading when low sulfur is a requirement.

As noted (Table 14.8), the coke drums are on a 48 h cycle. The coke drum is usually on-stream for about 24 h before becoming filled with porous coke, after which time the coke is removed by the following procedure:

1.   The coke deposit is cooled with water.

Table 14.8   Typical Time Cycle of Coke Drums in a Delayed Coker

Operation	Time (h)
Coking	24
Decoking	24
Switch drums	0.5
Steam, cool	6.0
Drain, unhead decoke	7.0
Reheat, warm-up	9.0
Spare time, contingency	1.5

2. One of the heads of the coking drum is removed to permit the drilling of a hole through the center of the deposit.
3. A hydraulic cutting device, which uses multiple high pressure water jets, is inserted into the hole and the wet coke is removed from the drum.

Normally, 24 h is required to complete the cleaning operation and to prepare the coke drum for subsequent use on-stream (Table 14.8).

A well-designed delayed coker will have an operating efficiency of better than 95%, although delayed coking units are generally scheduled for shutdown for cleaning and repairs on a 12–18 month schedule, depending on what is most economical for the refinery.

Obviously, the feedstock heater and the coke drums are the most critical parts of the delayed coking process. The function of the heater or furnace is to preheat the charge quickly to the required temperature to avoid preliminary decomposition. Because coking is endothermic, the furnace outlet temperature must be about 55°C (100°F) higher than the coke drum temperature to provide the necessary process heat. The heater run length is a function of coke laydown in heater tubes, and careful design is necessary to avoid premature shutdown with cycle lengths preferably at least one year. When the charge stock is derived from crude distillation, double desalting is desirable, because salt deposits will shorten heater cycles.

The heater for a delayed coking unit does not require as broad an operating range as a thermal cracking or visbreaking heater, where both contact time and temperature can be varied to achieve the desired level of conversion. The coker heater must reach a fixed outlet temperature for the required coke drum temperatures. Thus the coker heater requires a short residence time, high radiant heat flux, and good control of heat distribution.

The function of the coke drum is to provide the residence time required for the coking reactions to proceed to completion and to accumulate the coke. In sizing coke drums, a superficial vapor velocity in the range of 0.3–0.5 ft/s is used; coke drums with heights of 97 ft (30 m) have been constructed, which approach a practical limit for hydraulic coke cutting. Drum diameters up to 26 ft (8 m) have been commonly used, and larger drums are feasible for efficient processing. Various types of level detectors are used to permit drum filling to within 7–8 ft (2–2.5 m) of the upper tangent line of the drum monitor coke height in the drum during on-stream service.

Hydraulic cutters are used to remove coke from the drum, and the first step is to bore a vertical pilot hole through the coke, after which cutting heads with horizontally directed nozzles undercut the coke and drop it out of the bottom of the drum. Hydraulic pressures in the range of 3000–3600 psi (20.7–24.8 MPa) are used in the 26 ft (8 m) diameter coking drums.

In regard to the process parameters and product yields, an increase in the coking temperature (1) decreases coke production, (2) increases liquid yield, and (3) increases gas oil endpoint. On the other hand, increasing pressure and/or recycle ratio (1) increases gas yield, (2) increases coke yield, (3) decreases liquid yield, and (4) decreases the gas oil endpoint. As an example, increasing the pressure from the currently designed 15 psig (103 kPa) to 35 psig (241 kPa) (Table 14.6) causes the higher boiling products to remain in the hot zone longer, causing further decomposition and an increase in the yield of the naphtha fraction, a decrease in the yield of the middle distillate–gas oil fraction, and an increase in the yield of coke.

4.2 Fluid Coking and Flexicoking

Throughout the history of the refining industry, with only short-term exceptions, there has been a considerable economic driving force for upgrading residua. This has led to the development of processes to reduce residua yields such as thermal cracking, visbreaking, delayed coking, vacuum distillation, and deasphalting.

As a brief history, in the late 1940s and early 1950s there was a strong incentive to develop a continuous process to convert heavy vacuum residua into lighter, more valuable products. During this period, fluid coking using the principle of fluidized solids was developed and contact coking, using the principle of a moving solids bed, was also developed, and the first commercial fluid coker went on-stream in late 1954. During the late 1960s, environmental considerations indicated that, in many areas, it would no longer be possible to use high sulfur coke as a boiler fuel. This and other environmental considerations resulted in the development of flexicoking to convert the coke product from a fluid coker into clean fuel. The first commercial flexicoker went on-stream in 1976.

Fluid coking (Fig. 14.16) is a continuous process that uses the fluidized solids technique to convert residua, including vacuum residua and cracked residua, to more valuable products (Table 14.9) (Roundtree, 1997). This coking process allows improvement in the yield of distillates by reducing the residence time of the cracked vapors and also allows simplified handling of the coke product. Heat for the process is supplied by partial combustion of the coke, with the remaining coke being drawn as product. The new coke is deposited in a thin fresh layer ($\sim 0.005\,\mathrm{mm}$; $\sim 5\,\mu\mathrm{m}$) on the outside surface of the circulating coke particle, giving an onionskin effect.

The equipment for the fluid coking process is similar to that used in fluid catalytic cracking (Chapter 15) and follows comparable design concepts except that the fluidized

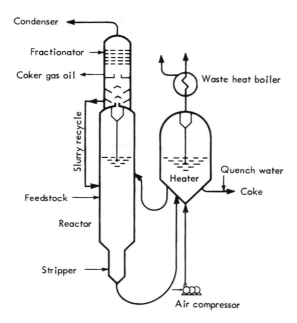

Figure 14.16 The fluid coking process.

Table 14.9 Examples of Product Yields and Product Properties for Fluid Coking Various Feedstocks

	LA Basin vacuum residuum	LA Basin visbreaker residuum	LA Basin deasphalter bottoms	Texas vacuum residuum	Kuwait vacuum residuum	Louisiana atmospheric residuum	Louisiana vacuum residuum	Hawkins vacuum residuum	Middle East vacuum residuum	Tia Juana vacuum residuum
Feedstock										
API gravity	6.7	−5.0	−1.0	17.3	5.6	17.8	11.6	4.2	5.1	8.5
Carbon residue[a]	17.0	41.0	33.0	11.0	21.8	5.0	13.0	24.5	21.4	22.0
Sulfur, wt%	2.1	2.1	2.3	0.7	5.5	0.5	0.6	4.3	3.4	2.9
Product yields,[b] vol%										
Naphtha (95–425°F; 35–220°C)	17.0	14.0	18.0	21.0	21.0	17.0	21.0	19.5	15.4	20.7
Gas oil (425–1000°F; 220–540°C)	62.0	32.0	45.0	69.0	48.0	74.0	61.0	53.0	55.1	48.3
Coke, wt%	21.0	48.0	36.0	12.0	28.0	8.0	17.0	27.5	26.4	20.0

	Tia Juana vacuum residuum	Arabian Light vacuum residuum	Arabian Heavy vacuum residuum	Arabian Heavy vacuum residuum	Arabian Heavy vacuum residuum	Iranian Heavy vacuum residuum	Bachaquero vacuum residuum	Bachaquero vacuum residuum	Zaca vacuum residuum	West Texas bitumen
Feedstock										
API gravity	7.9	6.5	3.2	4.4	3.3	5.1	2.6	2.6	4.7	−0.2
Carbon residue[a]	23.3	19.2	28.5	24.4	27.8	21.4	21.4	26.5	19.0	34.0
Sulfur, wt%	3.0	4.3	5.6	5.3	6.0	3.4	3.7	3.7	7.8	4.6
Product yields,[b] vol%										
Naphtha (95–425°F; 35–220°C)	15.1	15.5	14.4	15.0	17.0	15.4	14.7	14.8	20.5	13.4
Gas oil (425–1000°F; 220–540°C)	52.2	58.0	37.5	47.7	45.0	55.1	48.3	47.7	61.0	36.9
Coke, wt%	29.2	23.4	35.2	30.4	34.0	26.4	32.9	33.2	17.5	43.9

[a] Conradson.
[b] A blank product yield line indicates that the yield of the lower boiling product has been included in the yield of the higher boiling product.

coke solids replace catalyst. Small particles of coke made in the process circulate in a fluidized state between the vessels and are the heat transfer medium; thus the process requires no high temperature preheat furnace.

Fluid coking uses two vessels—a reactor and a burner. Coke particles are circulated between these to transfer heat (generated by burning a portion of the coke) to the reactor (Fig. 14.16) (Blaser, 1992). The reactor holds a bed of fluidized coke particles, and steam is introduced at the bottom of the reactor to fluidize the bed. The feed coming from the bottom of a vacuum tower at, for example, 260–370°C (500–700°F), is injected directly into the reactor. The temperature in the coking vessel ranges from 480°Cto 565°C (900–1050°F), with short residence times on the order of 15–30 s, and the pressure is substantially atmospheric, so the incoming feed is partly vaporized and partly deposited on the fluidized coke particles. The material on the particle surface then cracks and vaporizes, leaving a residue that dries to form coke. The vapor products pass through cyclones that remove most of the entrained coke.

Vapor products leave the bed and pass through cyclones, which are necessary for removal of the entrained coke. The cyclones discharge the vapor into the bottom of a scrubber, any coke dust remaining after passage through the cyclones is scrubbed out with a pump around stream, and the products are cooled to condense the heavy tar. The resulting slurry is recycled to the reactor. The scrubber overhead vapors are sent to a fractionator, where they are separated into wet gas, naphtha, and various gas oil fractions. The wet gas is compressed and further fractionated into the desired components.

In the reactor the coke particles flow down through the vessel into the stripping zone. The stripped coke then flows down a standpipe and through a slide valve that controls the reactor bed level. A riser carries the cold coke to the burner. Air is introduced to the burner to burn part of the coke to provide reactor heat. The hot coke from the burner flows down a standpipe through a slide valve that controls coke flow and thus the reactor bed temperature. A riser carries the hot coke to the top of the reactor bed. Combustion products from the burner bed pass through two stages of cyclones to recover coke fines and return them to the burner bed.

Coke is withdrawn from the burner to keep the solids inventory constant. To aid in keeping the coke from becoming too coarse, large particles are selectively removed as product in a quench elutriator drum, and coke fines are returned to the burner. The product coke is quenched with water in the quench elutriator drum, and pneumatically transported to storage. A simple jet attrition system in the reactor provides additional seed coke to maintain a constant particle size within the system.

Due to the higher thermal cracking severity used in fluid coking compared to delayed coking, the products are somewhat more olefinic than the products from delayed coking. In general, products from both coking processes are handled for upgrading in a comparable manner.

Coke, being a product of the process, must be withdrawn from the system to keep the solids inventory from increasing. The net coke produced is removed from the burner bed through a quench elutriator drum, where water is added for cooling and cooled coke is withdrawn and sent to storage. During the course of the coking reaction the particles tend to grow in size. The size of the coke particles remaining in the system is controlled by a grinding system within the reactor.

The coke product from the fluidized process is a laminated sphere with an average particle size of 0.17–0.22 mm (170–220 μm), readily handled by fluid transport techniques.

It is much harder and denser than delayed coke and in general is not as desirable for manufacturing formed products.

The yields of products are determined by the feed properties, the temperature of the fluid bed, and the residence time in the bed. The lower limit on operating temperature is set by the behavior of the fluidized coke particles. If the conversion to coke and light ends is too slow, the coke particles agglomerate in the reactor, a condition known as *bogging*. The use of a fluidized bed reduces the residence time of the vapor-phase products in comparison to delayed coking, which in turn reduces cracking reactions. The yield of coke is thereby reduced, and the yield of gas oil and olefins is increased. An increase of 5°C (9°F) in the operating temperature of the fluidized bed reactor typically increases gas yield by 1wt% and naphtha by about 1wt%.

The disadvantage of burning the coke to generate process heat is that sulfur from the coke is liberated as sulfur dioxide (SO_2). The gas stream from the coke burner also contains carbon monoxide (CO), carbon dioxide (CO_2), and nitrogen (N_2). An alternative approach is to use a coke gasifier to convert the carbonaceous solids to a mixture of CO, CO_2, and hydrogen (H_2).

Delayed coking and fluid coking have been the processes of choice for conversion of Athabasca bitumen to liquid products for more than three decades (Spragins, 1978; Speight, 1990, 1999, 2000). Both processes are termed *primary conversion* processes for the tar sand plants in Ft. McMurray, Alberta, Canada. The unstable liquid product streams are hydrotreated before recombining to form synthetic crude oil.

Flexicoking (Fig. 14.17) is a direct descendant of fluid coking (Fig. 14.16) and uses the same configuration as the fluid coker but includes a gasification section in which excess coke can be gasified to produce refinery fuel gas (Fig. 14.17) (Roundtree, 1997). The flexicoking process was designed and modified during the late 1960s and the

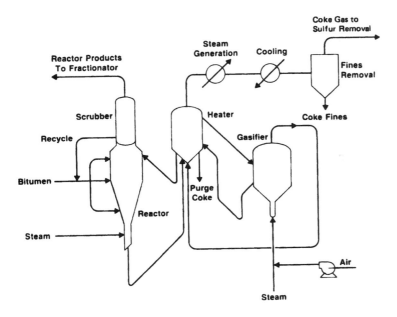

Figure 14.17 The flexicoking process.

1970s as a means by which excess coke could be reduced in view of the gradual incursion of the heavier feedstocks into refinery operations. Such feedstocks are notorious for producing high yields of coke (> 15% by weight) in thermal and catalytic operations.

In the process, excess coke is converted to a low heating value gas in a fluidized bed gasifier with steam and air. The air is supplied to the gasifier to maintain temperatures of 830–1000°C (1525–1830°F) but is insufficient to burn all of the coke. Under these reducing conditions, the sulfur in the coke is converted to hydrogen sulfide, which can be scrubbed from the gas prior to combustion. A typical gas product, after removal of hydrogen sulfide, contains carbon monoxide (CO, 18%), carbon dioxide (CO_2, 10%), hydrogen (H_2, 15%), nitrogen (N_2, 51%), water (H_2O, 5%), and methane (CH_4, 1%). The heater is located between the reactor and the gasifier, and it serves to transfer heat between the two vessels.

Yields of liquid products from flexicoking are the same as from fluid coking, because the coking reactor is unaltered. The main drawback of gasification is the requirement for a large additional reactor, especially if high conversion of the coke is required. Units are designed to gasify 60–97% of the coke from the reactor. Even with the gasifier, the product coke will contain more sulfur than the feed, which limits the attractiveness of even the most advanced coking processes.

The flexicoking process produces a clean fuel gas with a heating value of about 90 Btu/ ft^3 (800 kcal/m^3; 3351 kJ/m^3) or higher. The coke gasification can be controlled to burn about 95% of the coke to maximize production of coke gas or at a reduced level to produce both gas and a coke that has been desulfurized by about 65%. This flexibility permits adjustment for coke market conditions over a considerable range of feedstock properties. Fluid coke is currently being used in power plant boilers.

Fluid coking and flexicoking are versatile processes that are applicable to a wide range of heavy feedstocks and provide a variety of products (Fig. 14.18). The feedstock should have a carbon residue of more than 5%, and there is no upper limit on the carbon residue. Suitable feedstocks include vacuum residua, asphalt, tar sand bitumen, and visbreaker residuum.

The liquid products from the coker can, after cleanup via commercially available gas oil hydrodesulfurization technology (Chapter 16), provide large quantities of low sulfur fuel (< 0.2 wt% sulfur). The incentive for fluid coking or flexicoking increases relative to alternative types of processing, such as direct hydroprocessing, as feedstock quality (Conradson carbon, metals, sulfur, nitrogen, etc.) decreases. Changes in yields and product quality result from a change from a low cut point, high reactor temperature operation to a high cut point operation with a lower reactor temperature (Table 14.10).

Fluid coke is used in electrodes for aluminum manufacture, in silicon carbide manufacture, in ore sintering operations, and as fuel. The coke from a feedstock containing a large amount of contaminants may not be suitable for these uses, from the standpoint of either product contamination or environmental considerations. The flexicoking process overcomes this problem by converting part of the gross coke to a gas that can be burned in process furnaces and boilers. The coke fines from a flexicoker contain most of the metals in the feedstock and may be suitable for metals recovery.

The fluid coking processes can be used to produce a high yield of low sulfur fuel oil as well as to completely eliminate residual fuel and asphalt from the refinery product slate (Table 14.11). The different distributions are obtained by varying the fluid coker/flexicoker

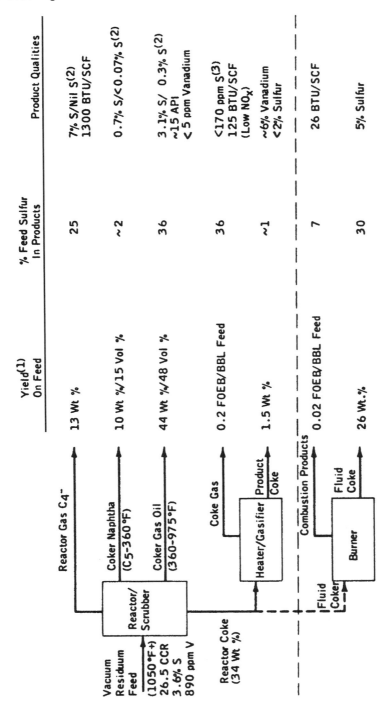

Figure 14.18 Product slates and product qualities for fluid coking and flexicoking.

Table 14.10   Flexibility of Operations in Fluid Cokers Allows Changes to Product Quality

Feed characteristics

Conradson carbon, wt%	15.5
Gravity, °API	6.4
LV below 1000°F, %	8.0
Sulfur, wt%	2.6
Nitrogen, wt%	1.0
Nickel, ppm	283
Vanadium, ppm	126

	Low reactor temperature, high cut point maximum gas oil		High reactor temperature, low cut point low metals gas oil	
Yields	Wt%	LV%	Wt%	LV%
Hydrogen sulfide	0.5		0.7	
Hydrogen	0.1		0.2	
C_1–C_3	8.0		9.6	
C_4	1.6	2.8	2.0	3.5
C_5 to 215°F (ASTM)	4.2	6.2	5.1	7.5
215–400°F	8.6	11.4	10.4	13.9
400°F to endpoint	58.4	62.5	51.8	56.3
Gross coke	18.5		20.2	
Net coke	10.0		10.6	
Product qualities				
C_5 to 215 naphtha				
Gravity, °API	71.6		71.6	
Sulfur, wt%	0.41		0.43	
Nitrogen, wt%	0.009		0.009	
215–400°F				
Gravity	52.3		52.2	
Sulfur, wt%	0.66		0.69	
Nitrogen, wt%	0.036		0.025	
Gas oil				
Gravity, °API	16.9		18.4	
Sulfur, wt%	2.36		2.28	
Nitrogen, wt%	0.73		0.63	
Nickel, ppm	2.7		0.43	
Vanadium, ppm	0.5		0.05	
Conradson carbon, wt%	2.3		1.0	
Aniline point, °F	140		110	
Coke				
Sulfur, wt%	3.4		3.4	
Nickel, ppm	1520		1400	
Vanadium, ppm	680		620	

Table 14.11   Effect of Flexicoking on Product Yields

	Crude composition	Flexicoking gas oil hydrodesulfurization	Flexicoking gas oil hydrodesulfurization and cat. cracking
Yields, LV% crude			
Gas, FOEB	3	7	12
Naphtha	16	23	47
Middle distillate	26	31	35
Gas oil (LSFO)	25	34	
Residuum	30		
Total	100	95	94

operating conditions and changing the downstream processing of the coker reactor products. In fact, there are many process variations that can be used to adapt the process to particular refining situations. Once-through or partial recycle coking can be used where there is a small market for heavy fuel oil or where a quantity of high sulfur material can be blended into the fuel oil pool.

In reference to the process parameters, the *reactor temperature* is normally set at 510–540°C (950–1000°F). Low temperature favors high liquid yields and reduces the unsaturation of the gas but increases the reactor holdup requirements. The burner temperature is normally 55–110°C (100–200°F) above the reactor temperature. Regulating the amount of coke sent to the reactor from the burner controls the reactor temperature. Burner temperature is controlled by the air rate to the burner.

Low pressure provides a maximum gas oil recycle cut point, minimizes steam requirements, and reduces air blower horsepower. Reactor pressure normally adjusts to the gas compressor suction pressure but is made higher by the pressure drop through the piping, the condenser, the fractionation tower, and the reactor cyclone. The unit pressure balance required for coke circulation and is normally controlled at a fixed differential pressure relative to the reactor.

The *reactor coke level* is controlled by the cold coke slide valve on the transfer line from the reactor to the burner, and the *burner coke level* is controlled by the coke withdrawal rate.

In the flexicoking process, the heater temperature is controlled by the rate of coke circulation between the heater and the gasifier. Adjusting the air rate to the gasifier controls the unit inventory of coke, and the gasifier temperature is controlled by steam injection to the gasifier.

In all coking processes, product yields are a function of feed properties, the severity of the operation, and the recycle cut point (Table 14.12). Severity is a function of time and temperature, because low severity and a high gas oil cut point favor a high liquid yield whereas high severity and a low gas oil cut point increase coke and gas yields (Tables 14.5 and 14.7–14.9; Fig. 14.19). Data indicate that the gross coke yield is directly related to feedstock Conradson carbon residue. Coke quality (Table 14.13) and gas quality (Table 14.14) are also important.

Table 14.12 Examples of Product Yields and Product Properties for Flexicoking Various Feedstocks

Vacuum residuum properties	Arabian Light	Iranian Heavy	Arabian Heavy	Bachaquero	West Texas sour asphalt
Gravity, °API	6.5	5.1	4.4	2.6	−0.2
Conradson carbon, wt%	19.2	21.4	24.4	26.5	34.0
Sulfur, wt%	4.29	3.43	5.34	3.66	4.6
Nitrogen, wt%	0.34	0.77	0.41	0.81	0.65
V + Ni, ppm	90	525	225	1040	137
Flexicoking yields on vacuum residuum					
C_3 gas, wt%	9.8	9.9	10.7	10.6	11.3
C_4 saturates, wt%	0.6	0.6	0.6	0.7	0.7
C_4 unsaturates, wt%	1.3	1.3	1.4	1.4	1.5
C_5–360°F naphtha					
wt%	11.2	11.0	10.6	10.4	9.2
LV%	15.5	15.4	15.0	14.8	13.4
360–975°F gas oil,					
wt%	53.7	50.8	46.3	43.7	33.4
LV%	58.0	55.1	50.4	47.7	36.9
Gross coke, wt%	23.4	26.4	30.4	33.2	43.9
Purge coke, wt%	1.1	1.2	1.4	1.5	2.0
Coke gas, FOE vol%	13.1	15.5	18.3	21.3	30.0

In most cases, high liquid yield and minimum coke and gas yields are required and, in theory, two cracking rates should be considered. The first is the rate at which the liquid cracks and vaporizes after initially laying down on the coke particles. The second is that vapor-phase cracking determines the distribution of the products between gas, naphtha, and gas oil. The vapor residence time can be determined from the reactor volume and the volume flow of hydrocarbon vapor and steam and can be divided into time in the fluid bed and time in the dispersed phase. The former is a function of the coke hold-up or weight space velocity which is normally expressed as reciprocal hours. For maximum liquid yield, the secondary cracking time should be kept at a minimum, and thus it is normally desirable to design the unit for the maximum operable weight space velocity.

The maximum rate at which feed can be injected into a fluid coker is limited by *bogging*. The conditions required to avoid bogging are

1. The feedstock must be uniformly distributed over the entire surface of the heat transfer medium.
2. The layer of feed material on the particles should not be too great; the thickness of the sticky plastic layer depends on the specific flow rate of feedstock, its coking factor, and the recirculation rate of the heat transfer medium.
3. The bed temperature and the initial temperature of the heat transfer medium should be sufficiently high that the first stage of the process is completed in a short time.

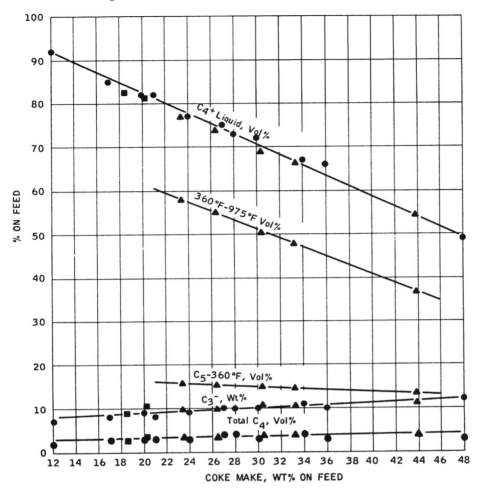

Figure 14.19 Fluid coking yields and coke make.

Table 14.13 Representative Properties of Fluid Coke and Flexicoke

	Flexicoke	Fluid coke
Bulk density, lb/ft^3	50	60
Particle density, lb/ft^3	85	95
Surface area, m^2/g	70	< 12
Average particle size, μm	120	170–240
Sulfur, wt%	2.0	6.0

Table 14.14   Representative Gas Composition from
Flexicoking Operations

	After particulate removal	After sulfur removal
H$_2$S, vppm	7100	< 10
COS, vppm	150	< 15
NH$_3$, vppm[a]	< 3	< 3
HCN, vppm	< 3	Nil
Solids, lb/Mscf	0.0042	Nil
Sulfur, wt% FOE basis	9.7	< 0.04

[a] Below detectable limits of 3 vppm.

4. The heat transfer medium should not consist of particles that are too fine. The
    heat reserve of the granules should be sufficient to cover the entire energy
    requirements in connection with heating the feedstock, supplying the energy
    for the endothermic cracking reaction, and evaporating the decomposition pro-
    ducts.

Thus, if the feed injection rate exceeds the vaporization rate for an extended period of
time, the thickness of the tacky oil film on the particles will increase until the particles
rapidly agglomerate, causing the bed to lose fluidity. When fluidization is lost, the heat
transfer rate is greatly reduced, further aggravating the condition. Coke circulation cannot
be maintained due to the loss of reactor fluidization.

For comparative purposes, although the processes are similar, there are some notable
differences between the operation of a fluid coker and a fluid catalytic cracking unit
(FCCU), and some of these differences tend to make the fluid coker easier to operate.
The fluid coker heat balance is very easy to maintain, because there is always an excess of
carbon to burn, whereas a fluid catalytic cracking unit has a sensitive interaction between
heat balance and intensity balance and therefore between carbon burned and carbon
produced, which complicates control, especially during operating changes, start-up, and
shutdown.

In addition, recovery from upsets caused by loss of utilities such as steam and air is
normally easier and faster with a fluid coker than with a fluid catalytic cracking unit. The
fluid coker normally operates well at low feed rates, and turndown to low rate is normally
limited by the ability of the tower to maintain fractionation of the products. The fluid
coker proper can operate at any feed rate that will provide enough coke to maintain a heat
balance.

However, the fluid coker has some inherent features that can create problems if proper
precautions are not followed. The heavy residuum can set up if the lines are not properly
heat traced and insulated. A low reactor temperature results in reactor bogging. If the
particle size of the circulating coke is not properly controlled, it can increase to the point
that coke circulation problems are encountered. The feed nozzles must be maintained and
occasionally cleaned to prevent poor feed distribution followed by excessive agglomerate
formation Control of the reactor bed level is critical, because an excessively high bed level

will flood the reactor cyclone and allow coke be carried to the scrubber, where it will plug the heavy oil circuits.

5. OTHER PROCESSES

The *decarbonizing* thermal process is designed to minimize coke and gasoline yields but at the same time to produce maximum yields of gas oil. The process is essentially the same as the delayed coking process but employs lower temperatures and pressures. For example, pressures range from 10 to 25 psi, heater outlet temperatures may reach 485°C (905°F), and coke drum temperatures may be on the order of 415°C (780°F).

Decarbonizing in this sense of the term should not be confused with *propane decarbonizing*, which is essentially a solvent deasphalting process.

Low pressure coking is a process designed for a once-through low pressure operation. The process is similar to delayed coking except that recycling is not usually practiced and the coke chamber operating conditions are 435°C (815°F), 25 psi. Excessive coking is inhibited by the addition of water to the feedstock in order to quench and restrict the reactions of the reactive intermediates.

High temperature coking is a semicontinuous thermal conversion process designed for high melting asphaltic residua that yields coke and gas oil as the primary products. The coke may be treated to remove sulfur to produce a low sulfur coke ($\leq 5\%$), even though the feedstock contained as much as 5wt% sulfur.

Thus, the feedstock is transported to the pitch accumulator, then to the heater [370°C, (700°F), 30 psi], and finally to the coke oven, where temperatures may be as high as 980–1095°C (1800–2000°F). Volatile materials are fractionated, and after the cycle is complete, the coke is collected for sulfur removal and quenching before storage.

Mixed-phase cracking (also called *liquid-phase cracking*) is a continuous thermal decomposition process for the conversion of heavy feedstocks to products boiling in the gasoline range. The process generally employs rapid heating of the feedstock (kerosene, gas oil, reduced crude, or even whole crude), after which it is passed to a reaction chamber and then to a separator, where the vapors are cooled. Overhead products from the flash chamber are fractionated to gasoline components and recycle stock, and flash chamber bottoms are withdrawn as a heavy fuel oil. Coke formation, which may be considerable at the process temperatures (400–480°C, 750–900°F), is minimized by use of pressures in excess of 350 psi.

Vapor-phase cracking is a high temperature (545–595°C; 1000–1100°F), low pressure (< 50 psi) thermal conversion process that favors dehydrogenation of feedstock (gaseous hydrocarbons to gas oils) components to olefins and aromatics. Coke is often deposited in heater tubes, causing shutdowns. Relatively large reactors are required for these units.

Selective cracking is a thermal conversion process that utilizes different conditions depending on the nature of the feedstock. For example, a heavy oil may be cracked at 494–515°C (920–960°F) and 300–500 psi; a lighter gas oil may be cracked at 510–530°C (950–990°F) and 500–700 psi (Fig. 14.20).

Each feedstock has its own particular characteristics that dictate the optimum conditions of temperature and pressure for maximum yields of the products. These factors are used in selectively combining cracking units in which the more refractory feedstocks are

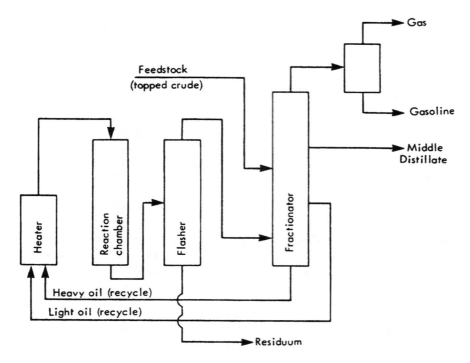

Figure 14.20 Selective thermal cracking.

cracked for longer periods of time or at higher temperatures than the less stable feed-stocks.

The process eliminates the accumulation of stable low-boiling material in the recycle stock and also minimizes coke formation from high temperature cracking of the higher boiling material. The end result is the production of fairly high yields of gasoline, middle distillates, and olefin gases.

The thermal cracking of naphtha involves the upgrading of low octane fractions of catalytic naphtha to higher quality material. The process is designed, in fact, to upgrade the heavier portions of naphtha, which contain virgin feedstock, and to remove naphthenes as well as paraffins. Some heavy aromatics are produced by condensation reactions, and substantial quantities of olefins occur in the product streams.

REFERENCES

Ballard, W. P., Cottington, G. I., and Cooper, T. A. 1992. In: Petroleum Processing Handbook. J. J. McKetta, ed. Marcel Dekker, New York, p. 309.

Bland, W. F., and Davidson, R. L. 1967. Petroleum Processing Handbook. McGraw-Hill, New York.

Blaser, D. E. 1992. In: Petroleum Processing Handbook. J. J. McKetta, ed. Marcel Dekker, New York, p. 255.

Corbett, R. A. 1990. Oil Gas J. 88(13):49.

Dominici, V. E., and Sieli, G. M. 1997. In: Handbook of Petroleum Refining Processes. 2nd ed. R. A. Meyers, ed. McGraw-Hill, New York, Chapter 12.3.

Feintuch, H. M., and Negin, K. M. 1997. In: Handbook of Petroleum Refining Processes. 2nd ed. R. A. Meyers, ed. McGraw-Hill, New York, Chapter 12.2.

Germain, J. E. 1969. Catalytic Conversion of Hydrocarbons. Academic Press, New York.

Hydrocarbon Processing. 1996. 74(11):51 et seq.

Hydrocarbon Processing. 1998. 77(11):53 et seq.

Kobe, K. A., and McKetta, J. J. 1958. Advances in Petroleum Chemistry and Refining. Interscience, New York.

Marzin, R., Pereira, P., McGrath, M. J., Feintuch, H. M., and Thompson, G. 1998. Oil Gas J. 97(44):79.

McKinney, J. D. 1992. In: Petroleum Processing Handbook. J. J. McKetta, ed. Marcel Dekker, New York, p. 245.

Mushrush, G. W., and Speight, J. G. 1995. Petroleum Products: Instability and Incompatibility. Taylor & Francis, Philadelphia, PA.

Nelson, W. L. 1976. Oil Gas J. 74(21):60.

Radler, M. 1999. Oil Gas J. 97(51):45 et seq.

Roundtree, E. M. 1997. In: Handbook of Petroleum Refining Processes. 2nd ed. R. A. Meyers, ed. McGraw-Hill, New York, Chapter 12.1.

Speight, J. G. 1990. In: Fuel Science and Technology Handbook. J. G. Speight, ed. Marcel Dekker, New York, Chapters 12–16.

Speight. J. G. 1999. The Chemistry and Technology of Petroleum. 3rd ed. Marcel Dekker, New York.

Speight. J. G. 2000. The Desulfurization of Heavy Oils and Residua. 2nd ed. Marcel Dekker, New York.

Spragins, F. K. 1978. In: Bitumens, Asphalts, and Tar Sands. G. V. Chilingarian and T. F. Yen, eds. Elsevier, Amsterdam, p. 92.

Thomas, M., Fixari, B., Le Perchec, P., Princic, Y., and Lena, L. 1989. Fuel 68:318.

Trash, L. A. 1990. Oil Gas J. 88(13):77.

Trimm, D. L. 1984. Chem. Eng. Proc. 18:137.

Vermillion, W. L., and Gearhart, W. 1983. Hydrocarbon Process. 62(9):89.

15
Catalytic Cracking

1. INTRODUCTION

Catalytic cracking is a conversion process (Fig. 15.1) that can be applied to a variety of feedstocks ranging from gas oil to heavy oil. It is one of several practical applications used in a refinery that employ a catalyst to improve process efficiency (Table 15.1). The original incentive to develop cracking processes arose from the need to increase gasoline supplies, and, because cracking could virtually double the volume of gasoline from a barrel of crude oil its use was wholly justified.

The growth of catalytic cracking capacity worldwide from 1945 until 1978, except for the (then) Soviet Union, Eastern Europe, or China, for which published data were not available is shown in Table 15.2, Luckenbach et al. (1992). Even excluding these countries and regions, current data show that worldwide catalytic cracking capacity has grown to approximately 12,000,000 bbl/day (Radler, 1999).

In total, as of December 1998, on a worldwide basis, catalytic cracking accounts for approximately 13,750,000 bbl/day out of the total refining capacity of approximately 81,600,000 bbl/day; i.e., 16.9% of the refining capacity is devoted to catalytic cracking. In the United States, approximately 5,500,000 bbl/day of feedstock is passed through catalytic crackers out of a total refining capacity of approximately 16,500,000 bbl/day; i.e., one-third (or double the worldwide proportion) of the feedstocks in U.S. refineries passes through catalytic cracking units (Radler, 1999).

In the 1930s, thermal cracking units produced about half the total gasoline manufactured, the octane number of which was about 70 compared to 60 for straight-run gasoline. This gasoline were usually blended with light ends and sometimes with polymer gasoline and reformatted to form a gasoline base stock with an octane number of about 65. The addition of *tetraethyllead* (*ethyl fluid*) increased the octane number to about 70 for the regular grade gasoline and 80 for premium grade gasoline. The thermal reforming and polymerization processes that were developed during the 1930s could be expected to further increase the octane number of gasoline to some extent, but something new was needed to break the octane barrier that threatened to stop the development of more powerful automobile engines. In 1936 a new cracking process opened the way to higher octane gasoline; this process was catalytic cracking. Since that time the use of catalysts in

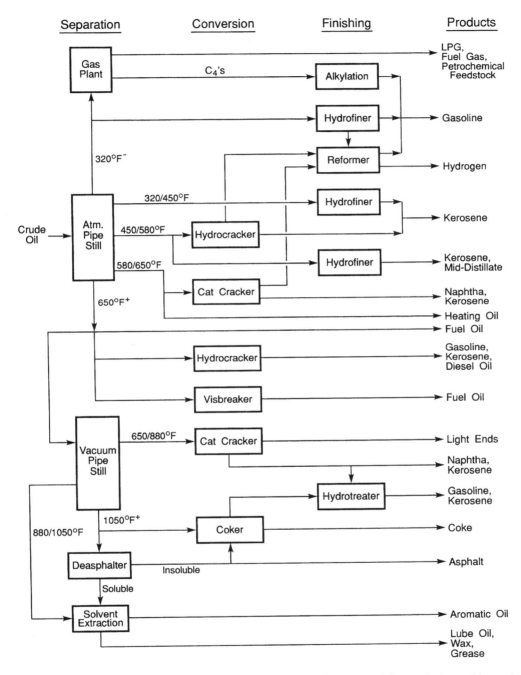

Figure 15.1 Generalized refinery layout showing relative placement of the catalytic cracking units.

Table 15.1 Refinery Processes That Employ Catalyst to Enhance Reactivity

Process	Materials charged	Products recovered	Temperature of reaction	Type of reaction
Cracking	Gas oil, fuel oil, heavy feedstocks	Gasoline, gas, and fuel oil	875–975°F (470–525°C)	Dissociation or splitting of molecules
Hydrogenation	Gasoline to heavy feedstocks	Low-boiling products	400–850°F (205–455°C)	Mild hydrogenation; cracking; removal of sulfur, nitrogen, oxygen, and metallic compounds
Reforming	Gasolines, naphthas	High octane gasolines, aromatics	850–1000°F (455–535°C)	Dehydrogenation, dehydroisomerization, isomerization, hydrocracking, dehydrocyclization
Isomerization	Butane, C_4H_{10}	Isobutane, C_4H_{10}		Rearrangement
Alkylation	Butylene and isobutane, C_4H_8 and C_4H_{10}	Alkylate, C_8H_{18}	32–50°F (0–10°C)	Combination
Polymerization	Butylene, C_4H_8	Octene, C_8H_{16}	300–350°F (150–175°C)	Combination

Table 15.2 Growth of Catalytic Cracking Capacity

Year	Capacity (bbl/day)	Comments
1945	1,000,000	Excludes Soviet Union, Eastern Europe, and China
1950	1,700,000	Excludes Soviet Union, Eastern Europe, and China
1960	5,200,000	Excludes Soviet Union, Eastern Europe, and China
1970	8,300,000	Excludes Soviet Union, Eastern Europe, and China
1978	9,900,000	Excludes Soviet Union, Eastern Europe, and China
1993	11,000,000	Excludes former Soviet Union, Eastern Europe, and China
	12,400,000	Includes Russia, Eastern Europe, and China
1998	12,000,000	Excludes former Soviet Union, Eastern Europe, and China
	13,800,000	Includes Russia, Eastern Europe, and China

the petroleum industry has spread to other processes (Table 15.1) (see also Bradley et al., 1989).

The next 50 years saw substantial advances in the development of catalytic processes (Luckenbach et al., 1992). This involved not only rapid advances in the chemistry and physics of the catalysts themselves but also major engineering advances in reactor design, for example the evolution of the design of the catalyst beds from *fixed beds* to *moving beds* to *fluidized beds*. Catalyst chemistry and physics and bed design allowed major improvements in process efficiency and product yields (Sadeghbeigi, 1995).

Catalytic cracking is basically the same as thermal cracking but uses a catalyst that is not (in theory) consumed in the process (Table 15.3). The catalyst directs the course of the cracking reactions to produce more of the desired products that can be used for the

Table 15.3 Summary of Catalytic Cracking Processing Conditions

Conditions
  Solid acid catalyst (silica-alumina, zeolite, others)
  900–1000°F (solid/vapor contact)
  10–20 psig
Feeds
  Virgin naphthas to atmospheric residua
  Pretreated to remove salts (metals)
  Pretreated to remove asphalts
Products
  Lower molecular weight components
  C_3–C_4 gases > C_2 gases
  Isoparaffins
  Coke (fuel)
Variations
  Fixed bed
  Moving bed
  Fluidized bed

production of better quality gasoline and other liquid fuels (Table 15.4) (Avidan et al., 1990).

Catalytic cracking has a number of advantages over thermal cracking. The gasoline produced by catalytic cracking has a higher octane number and consists largely of isoparaffins and aromatics. The isoparaffins and aromatic hydrocarbons have high octane numbers and greater chemical stability than monoolefins and diolefins. The olefins and diolefins are present in much greater quantities in thermally cracked gasoline. Furthermore, olefins (e.g., $RCH=CH_2$, where $R=H$ or an alkyl group) and smaller quantities of methane (CH_4) and ethane (CH_3CH_3) are produced by catalytic cracking and are suitable for petrochemical use (Speight, 1999). Sulfur compounds are changed in such a way that the sulfur content of gasoline produced by catalytic cracking is lower than that of gasoline produced by thermal cracking. Catalytic cracking produces less residuum and more of the useful gas oil constituents than does thermal cracking. Finally, the process has considerable flexibility, permitting the manufacture of both motor gasoline and aviation gasoline and a variation in the gas oil production to meet changes in the fuel oil market (Speight, 1986).

The usual commercial catalytic cracking process involves contacting a feedstock (usually a gas oil fraction) with a catalyst under suitable conditions of temperature, pressure, and residence time. By this means, a substantial part (> 50%) of the feedstock is converted into gasoline and lower boiling products, usually in a single-pass operation

Table 15.4 Comparison of Thermal Cracking and Catalytic Cracking

Hydrocarbon	Catalytic cracking	Thermal cracking
n-Paraffins	Extensive breakdown to C_2 and larger fragments. Product largely in C_2–C_6 range and contains many branched aliphatics. Few normal α-olefins above C_4.	Extensive breakdown to C_2 fragments, with much C_1 and C_2. prominent amounts of C_4–C_{n-1} normal α-olefins. Aliphatics largely unbranched.
Isoparaffins	Cracking rate relative to n-paraffins increased considerably by presence of tertiary carbon atoms.	Cracking rate increased to a relatively small degree by presence of tertiary carbon atoms.
Naphthenes	Crack at about same rate as paraffins with similar numbers of tertiary carbon atoms. Aromatics produced, with much hydrogen transfer to unsaturates.	Crack at lower rate than normal paraffins. Aromatics produced with little hydrogen transfer to unsaturates.
Unsubstituted aromatics	Little reaction; some condensation to biaryls.	Little reaction; some condensation to biaryls.
Alkyl aromatics (substituents C_3 or larger)	Entire alkyl group cracked next to ring and removed as olefin. Crack at much higher rate than paraffins.	Alkyl group cracked to leave one or two carbon atoms attached to ring. Crack at lower rate than paraffins.
n-Olefins	Product similar to that from n-paraffins but more olefinic.	Product similar to that from n-paraffins but more olefinic
All olefins	Hydrogen transfer is an important reaction, especially with tertiary olefins. Crack at much higher rate than corresponding paraffins.	Hydrogen transfer is a minor reaction, with little preference for tertiary olefins. Crack at about same rate as corresponding paraffins.

(Bland and Davidson, 1967). However, during the cracking reaction, carbonaceous material is deposited on the catalyst, which markedly reduces its activity, and removal of the deposit is necessary. The carbonaceous deposit arises from the presence of high molecular weight polar species (Speight, 1999, 2000) in the feedstock. The deposit is usually removed from the catalyst by burning in the presence of air until catalyst activity is reestablished.

The reactions that occur during catalytic cracking are complex (Germain, 1969), but there is a measure of predictability now that catalyst activity is better understood. The major catalytic cracking reaction exhibited by paraffins is carbon–carbon bond scission to form a lighter paraffin and olefin. Bond rupture occurs at certain definite locations on the paraffin molecule, rather than randomly as in thermal cracking. For example, paraffins tend to crack toward the center of the molecule, the long chains cracking in several places simultaneously. Normal paraffins usually crack at γ carbon–carbon bonds or still nearer the center of the molecule. On the other hand, isoparaffins tend to rupture between carbon atoms that are, respectively, β and γ to a tertiary carbon. In either case, catalytic cracking tends to yield products containing three or four carbon atoms rather than the one- or two-carbon atom molecules produced in thermal cracking.

As in thermal cracking (Chapter 14), high molecular weight constituents usually crack more readily than small molecules unless there has been some recycle and the constituents of the recycle stream have become more refractory and are less liable to decompose. Paraffins having more than six carbon atoms may also undergo rearrangement of their carbon skeletons before cracking, and a minor amount of dehydrocyclization also occurs, yielding aromatics and hydrogen.

Olefins are the most reactive class of hydrocarbons in catalytic cracking and tend to crack from 1000 to 10,000 times faster than in thermal processes. Severe cracking conditions destroy olefins almost completely, except for those in the low-boiling gasoline and gaseous hydrocarbon range, and, as in the catalytic cracking of paraffins, isoolefins crack more readily than *n*-olefins. The olefins tend to undergo rapid isomerization and yield mixtures with an equilibrium distribution of double-bond positions. In addition, the chain-branching isomerization of olefins is fairly rapid and often reaches equilibrium. These branched-chain olefins can then undergo hydrogen transfer reactions with naphthenes and other hydrocarbons. Other olefin reactions include polymerization and condensation to yield aromatic molecules, which in turn may be the precursors of coke formation.

In catalytic cracking, the cycloparaffin (naphthene) species crack more readily than paraffins but not as readily as olefins. Naphthene cracking occurs by both ring and chain rupture and yields olefins and paraffins, but formation of methane and the C_2 hydrocarbons (ethane, CH_3CH_3; ethylene, $CH_2{=}CH_2$; and acetylene, $CH{\equiv}CH$) is relatively minor.

Aromatic hydrocarbons exhibit wide variations in their susceptibility to catalytic cracking. The benzene ring is relatively inert, and condensed-ring compounds, such as naphthalene, anthracene, and phenanthrene, crack very slowly. When these aromatics crack, a substantial part of their *conversion* is reflected in the amount of coke deposited on the catalyst. Alkylbenzenes with attached groups of C_2 or larger primarily form benzene and the corresponding olefins, and heat sensitivity increases as the size of the alkyl group increases.

The several processes currently employed commercially in catalytic cracking differ mainly in the method of catalyst handling, although there is an overlap with regard to catalyst type and the nature of the products. The catalyst, which may be an activated natural or synthetic material, is employed in bead, pellet, or microspherical form and can be used in *fixed-bed, moving-bed,* or *fluid-bed* configurations.

The *fixed-bed process* was the first of these processes to be used commercially and uses a static bed of catalyst in several reactors, which allows a continuous flow of feedstock to be maintained. Thus the cycle of operations consists of (1) flow of feedstock through the catalyst bed, (2) discontinuance of feedstock flow and removal of coke from the catalyst by burning, and (3) insertion of the reactor on-stream.

The *moving-bed process* uses a reaction vessel in which cracking takes place and a kiln in which the spent catalyst is regenerated. Catalyst movement between the vessels is provided by various means.

The *fluid-bed process* differs from the fixed-bed and moving-bed processes in that the powdered catalyst is circulated essentially as a fluid with the feedstock (Sadeghbeigi, 1995). The several fluid catalytic cracking processes in use differ primarily in mechanical design (see, e.g., Hemler, 1997; Hunt, 1997; Johnson and Niccum, 1997; Ladwig, 1997). A side-by-side reactor–regenerator configuration or a configuration where the reactor is either above or below the regenerator are the main mechanical variations. From a flow standpoint, all fluid catalytic cracking processes contact the feedstock and any recycle streams with the finely divided catalyst in the reactor.

Feedstocks may range from naphtha fractions (included in normal heavier feedstocks for upgrading) to an atmospheric residuum (*reduced crude*). Feed preparation (to remove metallic constituents and high molecular weight nonvolatile materials) is usually carried out through the application of any one of several other processes: coking, propane deasphalting, furfural extraction, vacuum distillation, visbreaking, thermal cracking, and hydrodesulfurization (Speight, 2000).

The major process variables are temperature, pressure, catalyst/oil ratio (ratio of the weight of catalyst entering the reactor per hour to the weight of oil charged per hour), and space velocity (weight or volume of the oil charged per hour per unit weight or unit volume of catalyst in the reaction zone). Wide flexibility in product distribution and quality is possible through control of these variables along with the extent of internal recycling. Conversion can be increased by applying a higher temperature or higher pressure. Alternatively, lower space velocity and higher catalyst/oil ratio will also contribute to an increase in conversion.

When cracking is conducted in a single stage, the more reactive hydrocarbons may be cracked, with a high conversion to gas and coke, in the reaction time necessary for reasonable conversion of the more refractory hydrocarbons. However, in a two-stage process, gas and gasoline from a short reaction time, high temperature cracking operation are separated before the main cracking reactions take place in a second-stage reactor. For the short time of the first stage, a flow line or vertical riser may act as the reactor, and some conversion is effected with minimal coke formation. Cracked gases are separated and fractionated; the catalyst and residue, together with recycle oil from a second-stage fractionator, pass to the main reactor for further cracking. The products of this second-stage reaction are gas, gasoline and gas oil streams, and recycle oil.

2. FIXED-BED PROCESSES

Although fixed-bed catalytic cracking units have been phased out of existence, they represented an outstanding chemical engineering commercial development by incorporating a fully automatic instrumentation system that provided a short-time reactor–

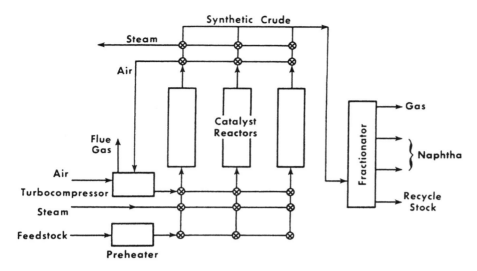

Figure 15.2 The Houdry fixed-bed catalytic cracking process.

purge–regeneration cycle, a novel molten salt heat transfer system, and a flue gas expander for recovering power to drive the regeneration air compressor. Historically, the Houdry fixed-bed process (Fig. 15.2), which went on-stream in June 1936, was the first of the modern catalytic cracking processes. Only the McAfee batch process, which employed a metal halide catalyst but has long since lost any commercial significance, preceded it.

In a *fixed-bed process*, the catalyst in the form of small lumps or pellets was made up in layers or beds in several (four or more) catalyst-containing drums (Fig. 15.2). Feedstock that had been vaporized at about 450°C (840°F) and less than 7–15 psi pressure was passed through one of the converters, where the cracking reactions took place. After a short time, deposition of coke on the catalyst rendered the catalyst ineffective, and using a synchronized valve system the feedstream was passed into a converter while the catalyst in the first converter was regenerated by carefully burning the coke deposits with air. After approximately 10 min, the catalyst was ready to go on-stream again.

The requirement of complete vaporization necessarily limited feeds to those with a low boiling range, and higher boiling feedstock constituents were retained in a separator before the feed was passed into the bottom of the upflow fixed-bed reactors. The catalyst consisted of a pelletized natural silica alumina catalyst and was held in reactors or *cases* about 11 ft (3.4 m) in diameter and 38 ft (11.6 m) in length for a 15,000 bbl/day unit. Cracked products were passed through the preheat exchanger, condensed, and fractionated in a conventional manner. The reactors operated at about 30 psi (207 kPa) gauge and 480°C (900°F).

The heat of reaction is provided by circulating a molten salt through vertical tubes distributed through the reactor beds. The reaction cycle of an individual reactor was about 10 min, after which the feed was automatically switched to a new reactor that had been regenerated. The reactor was purged with steam for about 5 min and then isolated by an automatic cycle timer. Regeneration air was introduced under close control, and carbon was burned off at a rate at which the recirculating salt stream could control the bed temperature. This stream comprised a mixture of potassium nitrate (KNO_3) and sodium nitrate ($NaNO_3$), which melts at 140°C (284°F), and was cooled in the reactors through which feed was being processed. The regeneration cycle lasted about 10 min. The regenerated bed was then purged of oxygen and automatically cut back into cracking service. There were three to six reactors in a unit. Gasoline yields diminished over the life of the catalysts (18 months) from 52% by volume to 42% by volume, based on fresh feedstock.

Equilibrium was never reached in this cyclic process. The gas oil conversion, i.e., the amount of feed converted to lighter components, was high at the start of a reaction cycle and progressively diminished as the carbon deposit accumulated on the catalyst until regeneration was required. Multiple parallel reactors were used to approach a steady-state process. However, the resulting process flows were still far from steady state. The reaction bed temperature varied widely during reaction and regeneration periods, and the temperature differential within the bed during each cycle was considerable.

Fixed-bed catalytic cracking units have generally been replaced by moving-bed or fluid-bed processes.

3. MOVING-BED PROCESSES

The fixed-bed process had obvious capacity and mechanical limitations that needed improvement and thus was replaced by a moving-bed process in which the hot salt systems were eliminated. The catalyst was lifted to the top of the reactor system and flowed by gravity down through the process vessels. The plants were generally limited in size to units processing up to about 30,000 bbl/day. These units have been essentially replaced by larger fluid solids units.

In the moving-bed processes, the catalyst is in a pelletized form [beads of approximately 0.125 in. (3 mm) in diameter] that flows by gravity from the top of the unit through a seal zone to the reactor, which operates at about 10 psi (69 kPa) gauge and 455–495°C (850–925°F). The catalyst then flows down through another seal and countercurrent through a stripping zone to a regenerator or kiln that operates at a pressure that is close to atmospheric.

In early moving-bed units, built around 1943, bucket elevators were used to lift the catalyst to the top of the structure. In units built about 1949, a pneumatic lift was used. This pneumatic lift permitted higher catalyst circulation rates, which in turn permitted the injection of all liquid feedstocks and the use of feedstocks that had a higher boiling range. A primary airstream was used to convey the catalyst (Fig. 15.3). A secondary airstream was injected through an annulus into which the catalyst could flow. Varying the secondary air rate varied the circulation rate.

The lift pipe is tapered to a larger diameter at the top to minimize erosion and catalyst attrition at the top. This taper is also designed so that total collapse of circulation will not occur instantaneously when a specific concentration or velocity of solids, below which particles tend to drop out of the flowing gas stream, is experienced. The taper can be designed so that this potential separation of solids is preceded by a pressure instability that can alert the operators to take corrective action.

The *Airlift Thermofor catalytic cracking (Socony Airlift TCC) process* (Fig. 15.4) is a moving-bed, reactor-over-generator continuous process for conversion of heavy gas oils into lighter high quality gasoline and middle distillate fuel oils. Feed preparation may

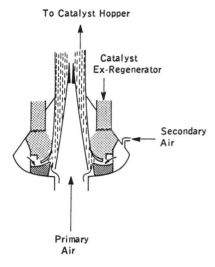

Figure 15.3 Catalyst pick-up system.

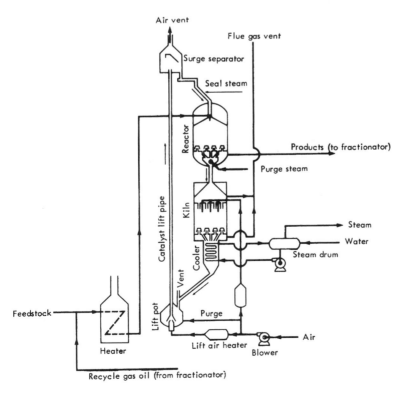

Figure 15.4 The Airlift Thermofor catalytic cracking process.

consist of flashing in a tar separator to obtain vapor feed, and the tar separator bottoms may be sent to a vacuum tower from which the liquid feed is produced.

The gas–oil vapor–liquid mixture flows downward through the reactor concurrently with the regenerated synthetic bead catalyst. The catalyst is purged by steam at the base of the reactor and gravitates into the kiln, or regeneration is accomplished by the use of air injected into the kiln. Approximately 70% of the carbon on the catalyst is burned in the upper kiln burning zone and the remainder in the bottom burning zone. Regenerated, cooled catalyst enters the lift pot, where low pressure air transports it to the surge hopper above the reactor for reuse.

The *Houdriflow catalytic cracking process* (Fig. 15.5) is a continuous moving-bed process that employs an integrated single vessel for the reactor and regenerator kiln. The charge stock, sweet or sour, can be any fraction of the crude boiling between naphtha

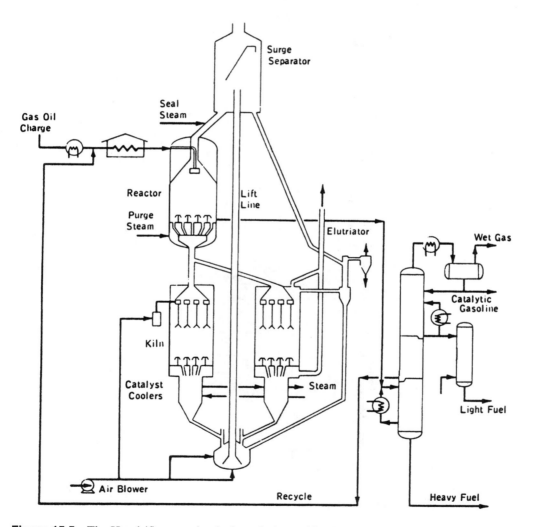

Figure 15.5 The Houdriflow moving-bed catalytic cracking process.

and atmospheric residua. The catalyst is transported from the bottom of the unit to the top in a gas lift that operates with compressed flue gas and steam.

The reactor feed and catalyst pass concurrently through the reactor zone to a disengager section in which vapors are separated and directed to a conventional fractionation system. The spent catalyst, which has been steam purged of residual oil, flows to the kiln for regeneration, after which steam and flue gas are used to transport the catalyst to the reactor.

The *Houdresid catalytic cracking process* (Fig. 15.6) is a process that uses a variation of the continuously moving catalyst bed designed to obtain high yields of high octane gasoline and light distillate from reduced crude charge. Residuum cuts ranging from crude tower bottoms to vacuum bottoms, including residua high in sulfur or nitrogen, can be employed as the feedstock, and the catalyst is synthetic or natural. Although the equipment employed is similar in many respects to that used in Houdriflow units, novel process features modify or eliminate the adverse effects and catalyst and product selectivity that usually result when heavy metals—iron, nickel, copper, and vanadium—are present in the fuel. The Houdresid catalytic reactor and catalyst-regenerating kiln (Fig. 15.6) are contained in a single vessel. Fresh feed plus recycled gas oil is charged to the top of the unit in a partially vaporized state and mixed with steam.

The *Suspensoid catalytic cracking process* (Fig. 15.7) was developed from the thermal cracking process carried out in tube-and-tank units. Small amounts of powdered catalyst or a mixture with the feedstock are pumped through a cracking coil furnace. Cracking temperatures are 550–610°C (1025–1130°F), with pressures of 200–500 psi. After leaving the furnace, the cracked material enters a tar separator, where the catalyst and tar are left behind. The cracked vapors enter a bubble tower, where they are separated into two parts, gas oil and pressure distillate. The latter is separated into gasoline and gases. The spent catalyst is filtered from the tar, which is used as a heavy industrial fuel oil. The process is actually a compromise between catalytic and thermal cracking. The main effect of the catalyst is to allow a higher cracking temperature and to assist mechanically in keeping

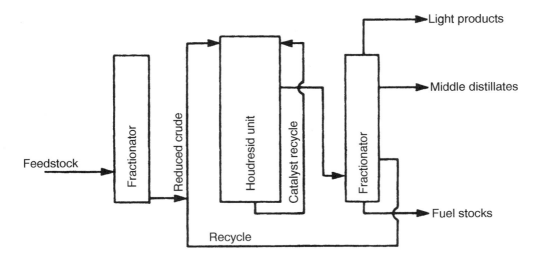

Figure 15.6 The Houdresid catalytic cracking process.

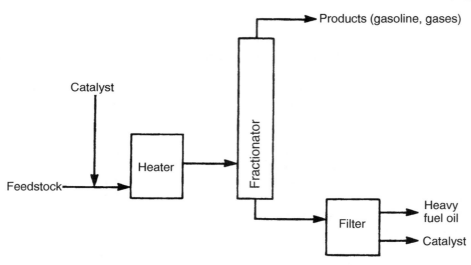

Figure 15.7   The Suspensoid catalytic cracking process.

coke from accumulating on the walls of the tubes. The catalyst is usually spent clay obtained from the contact filtration of lubricating oils (2–10 lb per barrel of feed).

4.   FLUID-BED PROCESSES

The application of fluidized solids techniques to catalytic cracking resulted in a major process breakthrough. It was possible to transfer all of the regeneration heat to the reaction zone. Much larger units could be built, and higher boiling feedstocks could be processed. In fact, the improvement in catalysts and unit configurations permitted the catalytic cracking of higher boiling (poorer quality) feedstocks such as residua. Presently, there are a number of processes that allow catalytic cracking of heavy oils and residua (*Hydrocarbon Processing*, 1998, p. 78).

The first fluid catalytic cracking units were Model I upflow units in which the catalyst flowed up through the reaction and regeneration zones in a riser type of flow regime (Fig. 15.8). Originally, the Model I unit was designed to feed a reduced crude to a vaporizer furnace where all of the gas oil was vaporized and then fed as vapor to the reactor. The nonvolatile residuum (bottoms) bypassed the cracking section. The original Model I upflow design (1941) was superseded by the Model II downflow design (1944) (Fig. 15.9) followed by the Model III (1947) balanced pressure design and later (1952) the Model IV low elevation design (Fig. 15.10).

Of the catalytic cracking process concepts, the fluid catalytic cracking process (Bartholic and Haseltine, 1981) is the most widely used and is characterized by the use of a finely powdered catalyst that is moved through the processing unit (Fig. 15.11). The catalyst particles are of such a size that when *aerated* with air or hydrocarbon vapor, the catalyst behaves like a liquid and can be moved through pipes. Thus, vaporized feedstock and fluidized catalyst flow together into a reaction chamber, where the catalyst, still dispersed in the hydrocarbon vapors, forms beds and the cracking reactions take place. The cracked vapors pass through cyclones located in the top of the reaction chamber, and

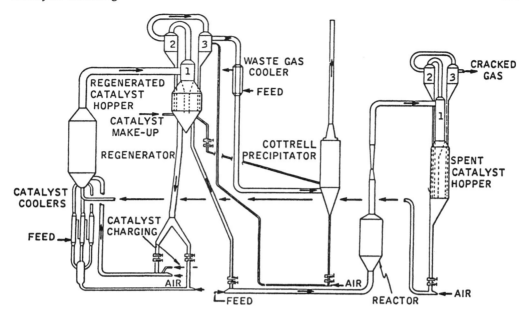

Figure 15.8   The Model I catalytic cracking unit.

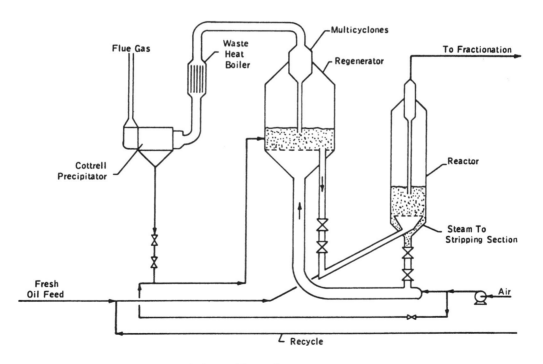

Figure 15.9   The Model II catalytic cracking unit.

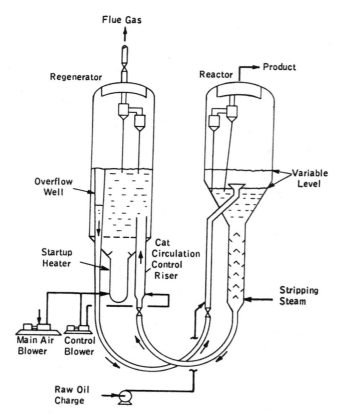

Figure 15.10   The Model IV catalytic cracking unit.

the catalyst powder is thrown out of the vapors by centrifugal force. The cracked vapors then enter the bubble towers, where they are fractionated into light and heavy cracked gas oils, cracked gasoline, and cracked gases.

Because the catalyst in the reactor becomes contaminated with coke, it is continuously withdrawn from the bottom of the reactor and lifted by means of a stream of air into a regenerator, where the coke is removed by controlled burning. The regenerated catalyst then flows to the fresh feed line, where the heat in the catalyst is sufficient to vaporize the fresh feed before it reaches the reactor, where the temperature is about 5 10°C (950°F).

The *Model IV fluid-bed catalytic cracking unit* (Fig. 15.10) employs a process in which the catalyst is transferred between the reactor and regenerator by means of U-bends and the catalyst flow rate can be varied in relation to the amount of air injected into the spent-catalyst U-bend. Regeneration air, other than that used to control circulation, enters the regenerator through a grid, and the reactor and regenerator are mounted side by side.

The *Orthoflow fluid-bed catalytic cracking process* (Fig. 15.12) (*Hydrocarbon Processing*, 1998, p. 80) uses the unitary vessel design, which provides a straight-line flow of catalyst and thereby minimizes the erosion encountered in pipe bends.

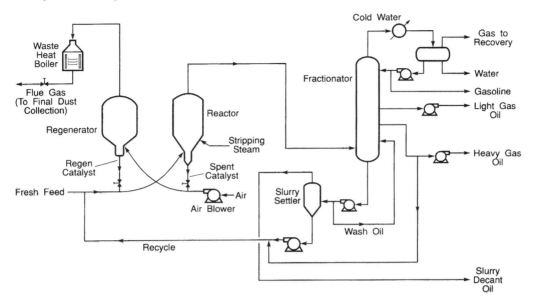

Figure 15.11   The fluid-bed catalytic cracking process.

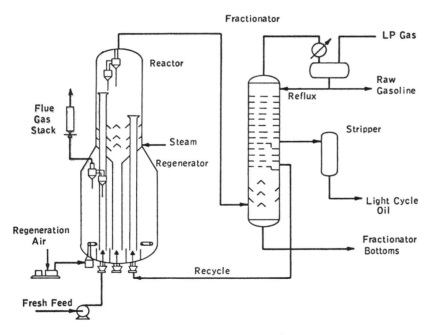

Figure 15.12   The Orthoflow catalytic cracking unit.

Commercial Orthoflow designs are of three types: Models A and C, with the regenerator beneath the reactor, and Model B, with the regenerator above the reactor. In all cases, the catalyst-stripping section is located between the reactor and the regenerator. All designs employ the heat-balanced principle, incorporating fresh feed–recycle feed cracking.

The *Universal Oil Products (UOP) fluid-bed catalytic cracking* process (Fig. 15.13) is adaptable to the needs of both large and small refineries. The major distinguishing features of the process are (1) elimination of the air riser with its attendant large expansion joints, (2) elimination of considerable structural steel supports, and (3) reduction in regenerator and in air-line size through the use of 15–18 psi pressure operation.

The *Shell two-stage fluid-bed catalytic cracking process* (Fig. 15.14) was devised to permit greater flexibility in shifting product distribution to satisfy market demand. Thus, feedstock is first contacted with cracking catalyst in a riser reactor—that is, a pipe in which fluidized catalyst and vaporized oil flow concurrently upward—and the total contact time in this first stage is of the order of seconds. High temperatures (470–565°C; 875–1050°F) are employed to reduce undesirable coke deposits on the catalyst without destruction of gasoline by secondary cracking. Other operating conditions in the first stage are a pressure of 16 psi (110 kPa) and a catalyst/oil ratio of 3:1 to 50:1, and volume conversion ranges between 20% and 70% have been recorded.

All or part of the unconverted or partially converted gas oil product from the first stage is then cracked further in the second-stage fluid-bed reactor. Operating conditions are 480–540°C (900–1000°F) and 16 psi with a catalyst/oil ratio of 2:1 to 12:1.

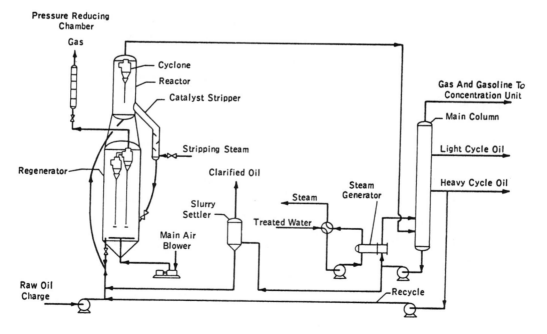

Figure 15.13   A stacked catalytic cracking unit.

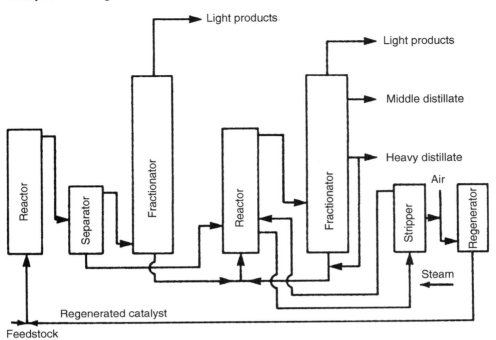

Figure 15.14 A two-stage fluid catalytic cracking process.

Conversion in the second stage varies between 15% and 70%, with an overall conversion range of 50–80%.

The residuum cracking unit of M. W. Kellogg Co. (Phillips Petroleum Co.) (Fig. 15.15) offers conversions up to 85 by weight of atmospheric residua or equivalent feedstocks (Table 15.5). The unit is similar to the Orthoflow C unit of Fig. 15.12, but there are some differences that enhance performance on residua. The catalyst flows from the regenerator through a plug valve that controls the flow to hold the reactor temperature. Steam is injected upstream of the feed point to accelerate the catalyst and disperse it so as to avoid high rates of coke formation at the feed point. The feedstock, atomized with steam, is then injected into this stream through a multiple nozzle arrangement. The flow rates are adjusted to control the contact time in the riser, because the effects of metals poisoning on yields are claimed to be largely a function of the time that the catalyst and oil are in contact. Passing the mix through a rough cut cyclone stops the reaction.

The *Gulf residuum process* consists of cracking a residuum that has been previously hydrotreated to low sulfur and metals levels. In this case, high conversions are obtained but coke yield and hydrogen yield are kept at conventional levels by keeping the metals level on catalyst low.

Other processes include the deep catalytic cracking (DCC) process (Fig. 15.16), which is designed for selective conversion of gas oil and paraffinic residual feedstocks to C_2–C_5 olefins, aromatic-rich naphtha, and other distillates (*Hydrocarbon Processing*, 1998, p. 69). There is also the flexicracking process (Fig. 15.17), which is designed for conversion of gas oils, residua, and deasphalted oils to distillates (Ladwig, 1997; *Hydrocarbon Processing*, 1998 p. 80).

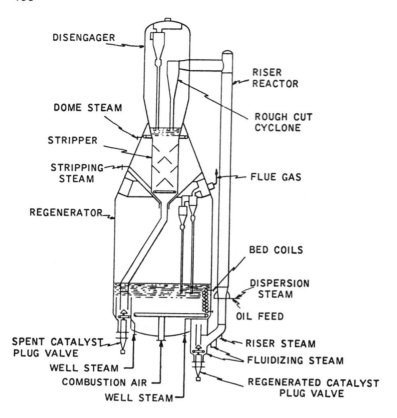

Figure 15.15   A residuum catalytic cracking unit.

5.   PROCESS VARIABLES

5.1   Feedstock Quality

Generally, the ability of any single unit to accommodate wide variations in feedstock is an issue related to the flexibility of the process. Initially, catalytic cracking units were designed to process gas oil feedstocks, but many units have been modified successfully, and new units designed, to handle more complex feedstocks and feedstock blends containing residua (*Hydrocarbon Processing*, 1998).

Vacuum gas oil [ibp: 315–345°C (600–650°F); fbp: 510–565°C (950–1050°F)], as produced by vacuum flashing or vacuum distillation, is the usual feedstock, with the final boiling point being limited by the carbon forming constituents (measured by the Conradson carbon residue) or metals content, because both of these properties have adverse effects on cracking characteristics. The vacuum residua (565°C$^+$; 1050°F$^+$) are occasionally included in catalytic cracker feed when the units (residuum catalytic crackers) are capable of handling such materials (Fig. 15.18) (*Hydrocarbon Processing*, 1998, p. 108). In such cases, if the residua are relatively low in terms of carbon-forming constituents and metals (as, for example, a residuum from a waxy crude), the effects of these properties are relatively small. Many units also recycle slurry oil (455°C$^+$; 850°F$^+$) and a heavy cycle oil

Table 15.5 Representative Product Slate for Residuum
Catalytic Cracking (81.7 vol% conversion at 400°F)

	Vol%		Wt%
Product yield			
C$_5$/400	59.7		48.2
Light cycle oil	13.3		13.3
Decant oil	5.0		5.7
Coke		13.3	
Product quality[a]			
Gasoline, °API		−56	
Sulfur, wt%		0.25	
Octane			
RON		90	
MON		79	
Light cycle oil, °API			
Sulfur, wt%		2.2	
Decant oil, °API			
Sulfur, wt%		5.5	

[a] RON = research octane number; MON = motor octane number.

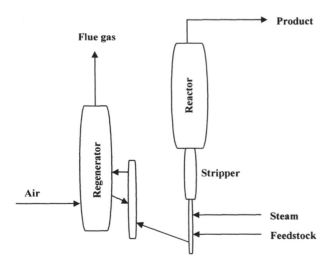

Figure 15.16 The deep catalytic cracking process.

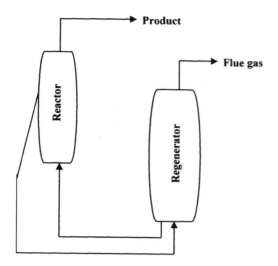

Figure 15.17   The flexicracking process.

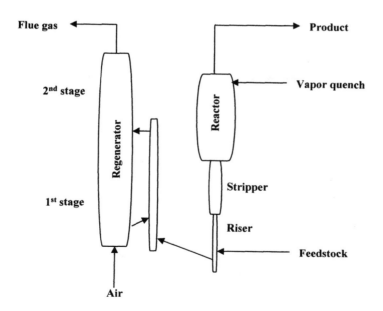

Figure 15.18   A residuum catalytic cracking unit.

stream. Gas oils from thermal cracking or coking operations (Chapter 14), hydrotreated gas oils, and deasphalted oils are often included in catalytic cracker feedstocks.

The general feedstock quality effects can be indicated by a characterization factor [characterization factor = $(MABP)^{13}$/specific gravity at $60°F/60°F$, where MABP is the mean average boiling point expressed in degrees Rankin ($°R =° F + 460$)]. However, a single parameter such as this can reflect only general trends (Table 15.6), and even then the accuracy and meaningful nature of the data may be very questionable. Generally speaking, coke yield increases as the characterization factor (K) decreases or as the feed becomes less paraffinic, the API gravity decreases, and the conversion increases (Fig. 15.19) (Maples, 2000). With straight-run gas oils, gasoline yield increases as the characterization factor (paraffinicity) decreases, but just the opposite effect is obtained with cracked stocks or cycle oils.

Molecular weight, average boiling point, and feed boiling range are important feed-stock characteristics in determining catalytic cracking yields and product quality. In general, for straight-run fractions, thermal sensitivity (degree of thermal decomposition or cracking) increases as molecular weight increases; coke and gasoline production (at constant processing conditions) also increase with the heavier feedstocks but, not to become too enthusiastic about the word *increase*, coke yield also increases (Fig. 15.20).

However, the characterization factor and feed boiling range are generally insufficient to characterize a feedstock for any purpose other than approximate comparisons. A more detailed description of the feedstock is needed to reflect and predict the variations in feedstock composition and cracking behavior (Speight, 1999, 2000).

Irrespective of the source of the high-boiling feedstocks, a number of issues typically arise when these materials are processed in a fluid catalytic cracking unit, although the magnitude of the problem can vary substantially.

Table 15.6 General Indications of Feedstock Cracking

Characterization factor, K	Naphtha yield (vol%)	Coke yield (wt%)	Feedstock type	Relative reactivity (relative crackability)
11.0[a]	35.0	13.5	Aromatic	Refractory; estimated coke yield
11.2[b]	49.6	12.5	Aromatic	Refractory
11.2[a]	37.0	11.5	Aromatic	Refractory
11.4[b]	47.0	9.1	Aromatic-naphthenic	Intermediate
11.4[a]	39.0	9.0	Aromatic-naphthenic	Intermediate
11.6[b]	45.0	7.1	Naphthenic	Intermediate
11.6[a]	40.0	7.2	Naphthenic	Intermediate
11.8[b]	43.0	5.3	Naphthenic-paraffinic	High
11.8[a]	41.0	6.0	Naphthenic-paraffinic	High
12.0[b]	41.5	4.0	Naphthenic-paraffinic	High
12.0[a]	41.5	5.3	Naphthenic-paraffinic	High
12.2[b]	40.0	3.0	Paraffinic	High

[a] Cycle oil/cracked feedstocks; 60% conversion.
[b] Straight-run/uncracked feedstocks; 60% conversion.

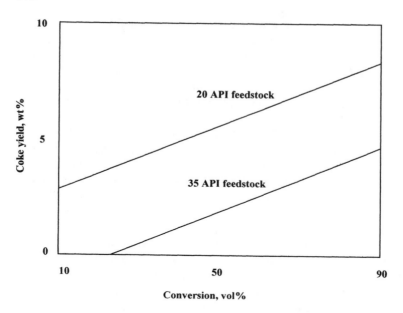

Figure 15.19 Relationship of coke yields to conversion and feedstock API gravity.

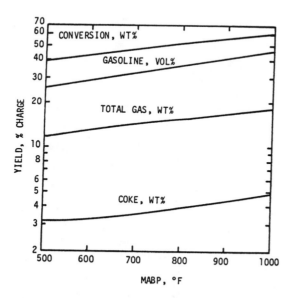

Figure 15.20 Relationship of product slate to feedstock boiling point.

Heavy feedstocks have high levels of contaminants, such as carbon-forming constituents that yield high levels of (Conradson) carbon residue, and the overall coke production (as carbon on the catalyst) is high. Burning this coke requires additional regeneration air, a constraint that might limit the capacity of the unit.

Metals in the heavy feedstocks also deposit (almost quantitatively) on the catalyst, causing two significant effects. First, the deposited metals can accelerate certain metal-catalyzed dehydrogenation reactions, thereby contributing to light gas (hydrogen) production and to the formation of additional coke. A second, and more damaging, effect is the situation in which the deposition of the metals causes a decline in catalyst activity because of the limited access to the active catalytic sites. This latter effect is normally controlled by catalyst makeup practices (adding and withdrawing catalyst).

The amounts of sulfur and nitrogen in the products, waste streams, and flue gas generally increase when high-boiling feedstocks are processed, because these feed components typically have higher sulfur and nitrogen contents than gas oil. However, in the case of nitrogen, the issue is one not only of higher nitrogen levels in the products but also (because of the feedstock nitrogen is basic in character) of catalyst poisoning that reduces the useful activity of the catalyst.

Heat balance control may be the most immediate and most troublesome aspect of processing high-boiling feedstocks. As the contaminant carbon increases, the first response is usually to increase the regenerator temperature. Adjustments in operating parameters can be made to assist in this control, but eventually a point will be reached for heavier feedstocks when the regenerator temperature is too high for good catalytic performance. At this point, some external heat must be removed from the regenerator, necessitating a mechanical modification such as a catalyst cooler.

For the last two decades, demetallized oil (produced by the extraction of a vacuum tower bottoms stream using a light paraffinic solvent) has been included as a component of the feedstock in fluid catalytic cracking units. Modern solvent extraction processes, such as the Demex process (Salazar, 1986; Thompson, 1997), provide a higher demetallized oil yield than is possible in the propane-deasphalting process that has been used to prepare fluid catalytic cracker feedstock and demetallized feedstocks for other processes. Consequently, the demetallized oil is more heavily contaminated. In general, demetallized oils are still good cracking stocks, but most feedstocks can be further improved by hydrotreating to reduce contaminant levels and to increase their hydrogen content, thereby becoming more presentable and processable (Shorey et al., 1999).

In many cases, atmospheric residua have been added as a blended component to feedstocks for existing fluid catalytic cracking units as a means of converting the highest boiling constituents of crude oil. In fact, in some cases the atmospheric residuum has ranged from a relatively low proportion of the total feed all the way to a situation in which it represents the entire feed to the unit. To improve the handling of these high-boiling feedstocks, several units have been revamped to upgrade them from their original gas oil designs, whereas other units have taken a stepwise approach to residuum processing whereby modifications to the operating conditions and processing techniques are made as more experience is gained in the processing of residua.

5.2 Feedstock Preheating

In a heat-balanced commercial operation, increasing the temperature of the feed to a cracking reactor reduces the heat that must be supplied by combustion of the coked catalyst in the regenerator. Feedstock preheating is usually supplied by heat exchange with hot product streams, a fired preheater, or both. When feed rate, recycle rate, and reactor temperature are held constant as feed preheat is increased, the following changes in operation result:

1. The catalyst/oil ratio (catalyst circulation rate) is decreased to hold the reactor temperature constant.
2. Conversion and all conversion-related yields. including coke, decline owing to the decrease in catalyst/oil ratio and severity.
3. The regenerator temperature will usually increase. Although the total heat released in the regenerator and the amount of air required by the regenerator are reduced by the lower coke yield, the lower catalyst circulation usually overrides this effect and results in an increase in regenerator temperature.
4. As a result of the lower catalyst circulation rate, residence time in the stripper and overall stripper efficiency are increased, liquid recovery is increased, and a corresponding decrease in coke usually results.

Advantage is usually taken of these feed preheat effects, including the reduced air requirement, by increasing the total feed rate until coke production again requires all of the available air (Fig. 15.21).

5.3 Feedstock Pressure

Catalytic cracker pressures are generally set slightly above atmospheric by balancing the yield and quality debits of high pressure plus increased regeneration air compression costs against improved cracking and regeneration kinetics, the lower cost of smaller vessels, plus, in some cases, power recovery from the regenerator stack gases. Representative yield and product effects (Table 15.7) show, at the same conversion level, that coke and gasoline

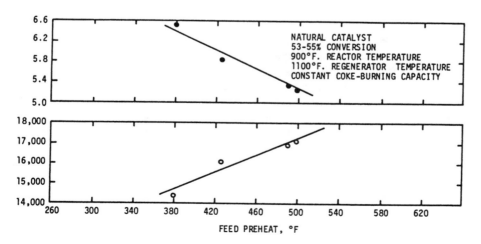

Figure 15.21   The effect of feedstock preheating on coke yield.

yields are increased marginally and light gas yields are reduced at the higher pressure. The sulfur content of the gasoline fraction is reduced.

Pressure levels in commercial units are generally in the range of 15–35 psi (103–241 kPa) gauge. Lowering the partial pressure of the reacting gases with steam will improve yields somewhat, but the major beneficial effect of feed injection steam is that it atomizes the feed to small droplets that will vaporize and react quickly. If feed is not atomized, it will soak into the catalyst and possibly crack to a higher coke make.

Both pressure and partial pressure of the feedstock, or steam/feedstock ratio, are generally established in the design of a commercial unit and thus are usually not available as independent variables over any significant range. However, in some units, injector steam is varied over a narrow range to balance carbon production with regeneration carbon burn off.

5.4 Feedstock Conversion

All of the independent variables in catalytic cracking have a significant effect on conversion that is truly a dependent variable but can be shown as a function of API gravity as well as a variety of other functions (Maples, 2000). The detailed effects of changing conversion depend on how the conversion is changed, i.e., by temperature, space velocity,

Table 15.7 Yield Variations with Pressure

Total pressure, psi gauge	12	50
Conversion, vol%	65–67	65–66
Yields		
H_2, wt%	0.2	0.1
C_1, wt%	1.9	1.2
C_2^{2-}, wt%	1.4	0.9
C_2, wt%	1.2	1.5
C_3^{2-}, wt%	5.3	3.9
C_3, wt%	3.9	4.4
iC_4^{2-}, vol%	11.8	10.7
nC_4^{2-}, vol%	2.0	1.2
iC_4, vol%	4.4	3.5
nC_4, vol%	2.4	3.2
iC_5, vol%	6.3	5.6
C_5^{2-}, vol%	3.1	3.0
nC_5, vol%	1.4	1.6
C_6, 284°F, 90% pt, vol%	16.7	19.6
Heavy naphtha, vol%	8.2	7.2
C_6, 410°F, EP gasoline, vol%	24.9	26.8
Cycle oil, vol%	32.9	33.3
Carbon, wt%	6.1	6.8
Quality, C_6, 410°F, gasoline		
°API	39	44
Sulfur, wt%	0.14	0.09
RON clear	98	98
ASTM clear	86	84

or catalyst/oil ratio, and catalyst activity. Increasing conversion increases yields of gasoline and all light products up to a conversion level of 60–80% by volume in most cases. At this high conversion level, secondary reactions become sufficient to cause a decrease in the yields of olefins and gasoline. However, the point at which this occurs is subject to the feedstock, operating conditions, catalyst activity, and other parameters.

5.5 Reactor Temperature

The principal effects of increasing reactor temperature at constant conversion are to decrease gasoline yield and coke yield and increase yields of methane (CH_4), ethane (C_2H_6), propane ($CH_3CH_2CH_3$), and total butanes (C_4H_{10}); yields of the pentanes (C_5H_{12}) and higher molecular weight paraffin decrease, whereas olefin yields are increased. The effect of reactor temperature on a commercial unit is, of course, considerably more complicated, because variables other than temperature must be changed to maintain heat balance. For example, to increase the temperature of a reactor at a constant fresh feed rate, interrelated changes of recycle rate, space velocity, and feedstock preheat are required to maintain heat balance on the unit by increasing circulation rate and coke yield. The combined effects of higher reactor temperature and higher conversion result in the following additional changes:

1. Yields of butanes and propane increased.
2. Gasoline yield increases.
3. Yield of light catalytic cycle oil decreases.

Thus the effects of an increase in reactor temperature on an operating unit reflect not only the effects of temperature per se but also the effects of several concomitant changes such as increased conversion and increased catalyst/oil ratio.

5.6 Recycle Rate

With most feedstocks and catalyst, gasoline yield increases with increasing conversion up to a point, passes through a maximum, and then decreases. This phenomenon (*overcracking*) is due to the increased thermal stability (*refractory character*) of the unconverted feed as conversion increases and the destruction of gasoline through secondary reactions, primarily cracking of olefins. The onset of secondary reactions and the subsequent leveling off or decrease in gasoline yield can be avoided by recycling a portion of the reactor product, usually a fractionator product with a boiling point on the order of 345–455°C (650–850°F).

Other tests showed the following effects of increasing the recycle rate when space velocity was simultaneously adjusted to maintain conversion constant:

1. Gasoline yield increased significantly.
2. Coke yield decreased appreciably.
3. The yield of dry gas components, propylene, and propane decreased.
4. Butane yields decreased, and butylene yields increased slightly.
5. Yields of light catalytic cycle oil and clarified oil increased, but that of heavy catalytic cycle oil decreased.

With the introduction of high activity zeolite catalysts, it was found that in once-through cracking operations with no recycle the maximum in gasoline yield corresponded higher conversions. In effect, the higher activity catalysts were allowing higher conversions to be obtained at severity levels that significantly reduced the extent of secondary reactions (*overcracking*). Thus, on many units employing zeolite catalysts, recycle has been eliminated or reduced to less than 15% of the fresh feed rate.

5.7 Space Velocity

The role of space velocity as an independent variable arises from its relation to *catalyst contact time* or *catalyst residence time*. Thus,

$$\Theta = 60/(WHSV \times C/O)$$

where Θ = catalyst residence time (min), WHSV = weight hourly space velocity on a total weight basis, and C/O = catalyst/oil weight ratio.

The catalyst/oil ratio is a dependent variable, so catalyst time becomes directly related to the weight hourly space velocity. When catalyst contact time is low, secondary reactions are minimized; thus gasoline yield is improved and light gas and coke yields are decreased. In dense bed units the hold-up of catalyst in the reactor can be controlled within limits, usually by a slide valve in the spent catalyst standpipe, and the feed rate can also be varied within limits. Thus, there is usually some freedom to increase space velocity and reduce catalyst residence time. In a riser-type reactor, hold-up and feed rate are not independent, and space velocity is not a meaningful term. Nevertheless, with both dense-bed and riser-type reactors, contact times are usually minimized to improve selectivity. An important step in this direction was the introduction of high activity zeolite catalysts. These catalysts require short contact times for optimum performance and have generally moved cracking operations in the direction of minimum hold-up in dense-bed reactors or replacement of dense-bed reactors with short contact time riser reactors.

Strict comparisons of short contact time riser cracking versus the longer contact time dense-bed mode of operations are generally not available due to differences in catalyst activity, carbon content of the regenerated catalyst, or factors other than contact time but inherent in the two modes of operation. However, in general, improvements in catalyst activity have resulted in the need for less catalyst in the reaction zone.

Many units are designed with only riser cracking; i.e., no dense-bed catalyst reactor cracking occurs and all cracking is done in the catalyst/oil transfer lines leading into the reactor cyclone vessel. However, in some of these instances the reactor temperature must be increased to the 550–565°C (1020–1050°F) range in order to increase the intensity of cracking conditions to achieve the desired conversion level. This is because not enough catalyst can be held in the riser zone, because the length of the riser is determined by the configuration and elevation of the major vessels in the unit. Alternatively, superactive catalysts can be used to achieve the desired conversion in the riser.

Significant selectivity disadvantages have not been shown if a dispersed catalyst phase or even a very small dense bed is provided downstream of the transfer line riser cracking zone. In this case the cracking reaction can be run at a lower temperature, say 510°C (950°F), which will reduce light gas make and increase gasoline yield compared to the higher temperature (550–565°C; 1020–1050°F) operation.

5.8 Catalyst Activity

Catalyst activity as an independent variable is governed by the ability of the unit to control the carbon content of the spent catalyst and the quantity and quality of fresh catalyst that can be continuously added to the unit. The carbon content of the regenerated catalyst is generally maintained at the lowest practical level to obtain the selectivity benefits of low carbon on the catalyst. Thus catalyst addition is, in effect, the principal determinant of catalyst activity.

The deliberate withdrawal of catalyst over and above the inherent loss rate through regenerator stack losses and decant oil or clarified oil and a corresponding increase in fresh catalyst addition rate is generally not practiced as a means of increasing the activity level of the circulating catalyst. If a higher activity is needed, the addition of a higher activity fresh catalyst to the minimum makeup rate to maintain inventory is usually the more economical route.

The general effects of increasing activity are to permit a reduction in severity and thus reduce the extent of secondary cracking reactions. Higher activity typically results in more gasoline and less coke. In other cases, higher activity catalysts are employed to increase the feed rate at essentially constant conversion and constant coke production so that the coke burning or regenerator air compression capacities are fully utilized.

5.9 Catalyst/Oil Ratio

The catalyst/oil ratio, a dependent variable, is established by the unit heat balance and coke make, which in turn are influenced by almost every independent variable. Because the catalyst/oil ratio changes are accompanied by one or more shifts in other variables, the effects of the catalyst/oil ratio are generally associated with other effects. A basic relation, however, in all catalyst/oil ratio shifts is the effect on conversion and carbon yield. At constant space velocity and temperature, increasing the catalyst/oil ratio increases conversion (Fig. 15.22). In addition to increasing conversion, higher catalyst/oil ratios generally increase coke yield at constant conversion. This increase in coke is related to the hydrocarbons entrapped in the pores of the catalyst and carried through the stripper to the regenerator. Thus, this portion of the catalyst/oil ratio effect is highly variable and depends not only on the change in the ratio but also on the catalyst porosity and stripper conditions (Fig. 15.23).

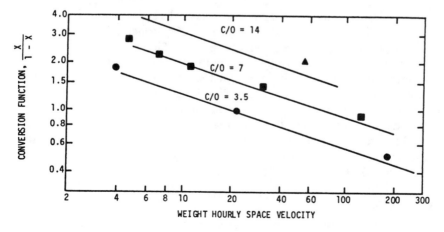

Figure 15.22 Effect of catalyst/oil ratio (C/O) and space velocity on feedstock conversion.

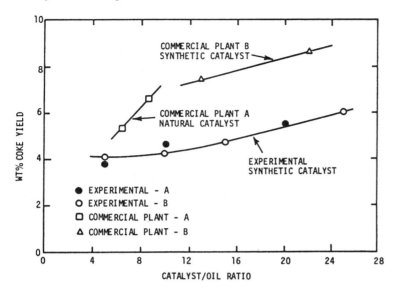

Figure 15.23 Effect of catalyst/oil ratio on coke yields at constant conversion.

The following changes, in addition to increased coke yield, accompany an increase in the catalyst/oil ratio in the range of 5–20 at constant conversion, reactor temperature, and catalyst activity:

1. Decreased hydrogen yield
2. Decreased methane to butane(s) yields
3. Little effect on the gasoline yield or octane number

5.10 Regenerator Temperature

Catalyst circulation, coke yield, and feedstock preheating are the principal determinants of regenerator temperature, which is generally allowed to respond as a dependent variable within limits. Mechanical or structural specifications in the regenerator section generally limit the regenerator temperature to a maximum value specific to each unit. However, in some cases the maximum temperature may be set by catalyst stability. In either case, if the regenerator temperature is too high it can be reduced by decreasing the feedstock preheat; catalyst circulation is then increased to hold a constant reactor temperature, and the increased catalyst circulation will carry more heat from the regenerator and lower the regenerator temperature. The sequence of events is actually more complicated, because the shift in the catalyst/oil ratio and, to a lesser extent, the shift in carbon content of the regenerated catalyst will change the coke make and the heat release in the regenerator.

5.11 Regenerator Air Rate

The amount of air required for regeneration depends primarily on coke production. Regenerators have been operated with only a slight excess of air leaving the dense phase. With less air, the carbon content of the spent catalyst increases and a reduction in coke yield is required, or air burning of carbon monoxide (CO) to carbon dioxide (CO_2)

will occur above the dense bed. This *afterburning* must be controlled as the catalyst will be subjected to extremely high temperature in the absence of the heat sink.

In terms of catalyst regeneration, more stable catalysts are available, and regenerator temperatures can be increased by 38–65°C (100–150°F) up to the 720–745°C (1325–1370°F) range without significant thermal damage to the catalyst. At these higher temperatures the oxidation of carbon monoxide to carbon dioxide is greatly accelerated, and the regenerator can be designed to absorb the heat of combustion in the catalyst under controlled conditions. The carbon burning rate is improved at the higher temperatures, and there are usually selectivity benefits associated with the lower carbon on the regenerated catalyst; this high temperature technique usually results in levels of 0.05 wt% or less of carbon on the regenerated catalyst. In addition, the use of catalysts containing promoters for the oxidation of carbon monoxide to carbon dioxide produce a major effect of high temperature regeneration, i.e., low carbon monoxide content regenerator stack gases and the resultant regenerator conditions may result in the regenerated catalyst having a lower carbon content.

6. CATALYSTS

6.1 Catalyst Types

The cracking of crude oil fractions occurs over many types of catalytic materials, and cracking catalysts can differ markedly both in their activity to promote the cracking reaction and in the quality of the products obtained from cracking the feedstocks (Gates et al., 1979; Wojciechowski and Corma, 1986; Stiles and Koch, 1995; Cybulski and Moulijn, 1998; Occelli and O'Connor, 1998). Activity can be related directly to the total number of active (acid) sites per unit weight of catalyst and also to the acidic strength of these sites. Differences in activity and acidity regulate the extent to which various secondary reactions occur and thus differences in product quality. The acidic sites are considered to be Lewis- or Brønsted-type acid sites, but there is much controversy as to which type of site predominates.

The first cracking catalysts were acid-leached *montmorillonite clays*. The acid leach was done to remove various metal impurities, principally iron, copper, and nickel, that could have adverse effects on the cracking performance of the catalyst. The catalysts first used in fixed-bed and moving-bed reactor systems were in the form of shaped pellets. Later, with the development of the fluid catalytic cracking process, clay catalysts were made in the form of a ground, sized powder. Clay catalysts are relatively inexpensive and have been used extensively for many years.

The desire to have catalysts that were uniform in composition and catalytic performance led to the development of *synthetic catalysts*. The first synthetic cracking catalyst consisted of 87% silica (SiO_2) and 13% alumina (Al_2O_3) and was used in pellet form in fixed-bed units in 1940. Catalysts of this composition were ground and sized for use in fluid catalytic cracking units. In 1944 catalysts in the form of beads about 2.5–5.0 mm in diameter were introduced that comprised about 90% silica and 10% alumina and were extremely durable. One version of these catalysts contained a minor amount of chromia (Cr_2O_3) to act as an oxidation promoter.

Neither silica (SiO_2) nor alumina (Al_2O_3) alone is effective in promoting catalytic cracking reactions. In fact, they (and also activated carbon) promote the thermal decomposition of hydrocarbons. Mixtures of anhydrous silica and alumina ($SiO_2 \cdot Al_2O_3$) or anhydrous silica with hydrated alumina ($2SiO_2 \cdot 2Al_2O_3 \cdot 6H_2O$) are also essentially ineffective. A catalyst having appreciable cracking activity is obtained only by preparation from hydrous oxides followed by partial dehydration (*calcining*). The small amount of water remaining is necessary for proper functioning.

Commercial synthetic catalysts are amorphous and contain more silica than is called for by the preceding formulas; they are generally composed of 10–15% alumina (Al_2O_3) and 85–90% silica (SiO_2). The natural materials montmorillonite, a nonswelling bentonite, and halloysite are hydrosilicates of aluminum with a well-defined crystal structure and approximate composition $Al_2O_3 \cdot 4Si_2O \cdot xH_2O$. Some of the newer catalysts contain up to 25% alumina and are reputed to have a longer active life.

The catalysts are porous and highly adsorptive, and their performance is affected markedly by the method of preparation. Two catalysts that are chemically identical but have pores of different sizes and distributions may differ in their activity, selectivity, temperature coefficient of reaction rate, and response to poisons. The intrinsic chemistry and catalytic action of a surface may be independent of pore size, but small pores appear to produce different effects because of the manner and time in which hydrocarbon vapors are transported into and out of the interstices.

In addition to synthetic catalysts comprising silica-alumina, other combinations of *mixed oxides* were found to be catalytically active and were developed during the 1940s. These systems included silica (SiO_2), magnesia (MgO), silica-zirconia (SiO_2-ZrO), silica-alumina-magnesia, silica-alumina-zirconia, and alumina-boria (Al_2O_3-B_2O_3). Of these, only silica-magnesia was used in commercial units, but operating difficulties developed with the regeneration of the catalyst, which at the time demanded a switch to another catalyst. Further improvements in silica-magnesia catalysts have since been made. High yields of desirable products are obtained with hydrated aluminum silicates. These may be either activated (acid-treated natural clays of the bentonite type) or synthesized silica-alumina or silica-magnesia preparations. Both the natural and the synthetic catalysts can be used as pellets or beads and also in the form of powder; in either case, replacement is necessary because of attrition and gradual loss of efficiency (DeCroocq, 1984; Le Page et al., 1987; Thakur, 1985).

During the period 1940–1962, the commercial cracking catalysts used most widely were the aforementioned acid-leached clays and silica-alumina. The latter was made in two versions; *low alumina* (about 13% Al_2O_3) and *high alumina* (about 25% $A1_2O_3$). High alumina content catalysts showed a higher equilibrium activity level and greater surface area.

During 1958–1960, *semisynthetic catalysts* of silica-alumina catalyst were used in which approximately 25–35% kaolin was dispersed throughout the silica-alumina gel. These catalysts could be offered at a lower price and therefore were disposable, but they were marked by a lower catalytic activity and greater stack losses because of increased attrition rates. One virtue of the semisynthetic catalysts was that a lesser amount of adsorbed, unconverted, high molecular weight products on the catalyst was carried over to the stripper zone and regenerator. This resulted in a higher yield of more valuable products and also smoother operation of the regenerator because local hot spots were minimized.

Catalysts must be stable to physical impact loading and thermal shocks and must withstand the action of carbon dioxide, air, nitrogen compounds, and steam. They should also be resistant to sulfur compounds; the synthetic catalysts and certain selected clays appear to be better in this regard than average untreated natural catalysts.

Commercially used cracking catalysts are *insulator catalysts* that possess strong protonic (acidic) properties. They function as catalysts by altering the cracking process mechanisms through an alternative mechanism involving *chemisorption* by *proton donation* and *desorption*, resulting in cracked oil and, theoretically, restored catalyst. Thus it is not surprising that all cracking catalysts are poisoned by proton-accepting vanadium.

The catalyst/oil volume ratios range from 5:1 to 30:1 for the different processes. Most processes are operated to a ratio of 10:1, but for moving-bed processes the catalyst/oil volume ratios may be substantially lower than 10:1.

Crystalline *zeolite catalysts* with molecular sieve properties were introduced as selective adsorbents in the 1955–1959 period. Within a relatively short time, all of the cracking catalyst manufacturers were offering their versions of zeolite catalysts to refiners. The intrinsically higher activity of the crystalline zeolites vis-à-vis conventional amorphous silica-alumina catalysts coupled with the much higher yields of gasoline and decreased yields of coke and light ends served to revitalize research and development in the mature refinery process of catalytic cracking.

A number of zeolite catalysts, such as synthetic faujasite (X and Y types), offretite, mordenite, and erionite, have been mentioned as having catalytic cracking properties. Of these, the faujasites have been most widely used commercially. Although faujasite is synthesized in the sodium form, base exchange removes the sodium with other metal ions, which, for cracking catalysts, include magnesium, calcium, rare earths (mixed or individual), and ammonium. In particular, mixed rare earths alone or in combination with ammonium ions have been the most commonly used forms of faujasite in cracking catalyst formulations. Empirically, X-type faujasite has a stoichiometric formula of $Na_2O \cdot Al_2O_3 \cdot 2.5SiO_2$, and Y-type faujasite $Na_2O \cdot Al_2O_3 \cdot 4.8SiO_2$. Slight variations in the silica/alumina (SiO_2/Al_2O_3) ratio exist for each of the types. Rare earth exchanged Y-type faujasite retains much of its crystallinity after steaming at 825°C (1520°F) for 12 h; rare earth form X-faujasite, although thermally stable in dry air, will lose its crystallinity at these temperatures in the presence of steam.

6.2 Catalyst Manufacture

Each manufacturer has developed proprietary procedures for making silica-alumina catalyst, but the general procedure consists of (1) gelling dilute sodium silicate solution ($Na_2O \cdot 3.25SiO_2 \cdot xH_2O$) by the addition of an acid (H_2SO_4, CO_2) or an acid salt such as aluminum sulfate; (2) aging the hydrogel under controlled conditions; (3) adding the prescribed amount of alumina as aluminum sulfate and/or sodium aluminate; (4) adjusting the pH of the mixture; and (5) filtering the composite mixture. After filtering, the filter cake can be either (1) washed free of extraneous soluble salts by a succession of slurrying and filtration steps and spray dried or (2) spray dried and then washed free of extraneous soluble salts before flash drying the finished catalyst.

There are a number of critical areas in the preparative processes that affect the physical and catalytic properties of the finished catalyst. Principal among them are the concentration and temperature of the initial sodium silicate solution, the amount of acid added to effect gelation, the length of time the gel is aged, the method and conditions of adding the aluminum salt to the gel, and its incorporation therein. Under a given set of conditions the product catalyst is quite reproducible in both physical properties and catalytic performance.

During the period 1940–1962, a number of improvements were made in silica-alumina catalyst manufacture. These included the use of continuous production lines versus batch-type operation, the introduction of spray drying to eliminate grinding and sizing of the catalyst and reduce catalyst losses as fines, improvement of catalyst stability by controlling pore volume, and improvement in wash procedures to remove extraneous salts from high alumina content catalysts to improve equilibrium catalyst performance.

Zeolite cracking catalysts are made by dispersing or embedding the crystals in a matrix. The matrix is generally amorphous silica-alumina gel and may also contain finely divided clay. The zeolite content of the composite catalyst is generally in the range of 5–16% by weight. If clay (e.g., kaolin) is used in the matrix, it is present in an amount of 25–45% by weight, the remainder being the silica-alumina hydrogel *glue* that binds the composite together. The zeolite may be pre-exchanged to the desired metal form and calcined to lock the exchangeable metal ions into position before compositing with the other ingredients. In an alternative scheme, sodium-form zeolite is composited with the other components, washed, and then treated with a dilute salt solution of the desired metal ions before the final drying step.

As stated above, the matrix generally consists of silica-alumina, but several catalysts have been commercialized that contain (1) silica-magnesia and kaolin or (2) synthetic montmorillonite-mica and/or kaolin as the matrix for faujasite.

6.3 Catalyst Selectivity

In the catalytic cracking process, the most abundant products are those having three, four, or five carbon atoms. On a weight basis the four-carbon atom fraction is the largest. The differences between the catalysts of the mixed oxide type lie in the relative action toward promoting the individual reaction types included in the overall cracking operation. For example, a silica-magnesia catalyst under a given set of cracking conditions will give a higher conversion to cracked products than a silica-alumina catalyst. However, the products from a silica-magnesia (SiO_2-MgO) catalyst have a higher average molecular weight, hence a lower volatility and lesser amounts of highly branched/acyclic isomers but more olefins among the gasoline boiling range products (C_4, 220°C; 430°F) than the products from a silica-alumina catalyst. With these changes in composition, the gasoline from cracking with a silica magnesia catalyst is of lower octane number.

These differences between catalysts can also be described as differences in the intensity of the action at the individual active catalytic centers. That is, a catalyst such as silica-alumina would result in a more intense of reaction than silica-magnesia, as observed from the nature and yields of the individual cracked products and the motor gasoline octane number. Titration of these two catalysts shows silica-alumina to have a lower acid titer than silica-magnesia but the acid strength of the silica-alumina sites is higher.

Although each of the individual component parts in these catalysts is essentially nonacidic, when mixed together properly they give rise to a titratable acidity as described above. Many of the secondary reactions occurring in the cracking process can also be promoted with strong mineral acids, such as concentrated sulfuric and phosphoric acids, aluminum halides, hydrogen fluoride, and hydrogen fluoride–boron trifluoride (BF_3) mixtures. This parallelism lends support to the concept of the active catalytic site as being acidic. Zeolites have a much higher active site density (titer) than the amorphous mixed oxides, which may account in large part for their extremely high cracking propensity. In addition, these materials strongly promote complex hydrogen transfer reactions among the primary products so that the recovered cracked products have a much lower olefin content and higher paraffin content than are obtained with the amorphous mixed oxide catalysts. This hydrogen transfer propensity of zeolites to saturate primary cracked product olefins to paraffins minimizes the reaction of polymerizing the olefins to form a coke deposit, thus accounting in part for the much lower coke yields with zeolite catalysts than with amorphous catalysts.

Activity of the catalyst varies with faujasite content, as does selectivity of the catalyst to coke and naphtha. As the faujasite content drops below 5% by weight, the catalyst starts to show some of the cracking properties of the matrix, whereas for zeolite contents of 10% by weight or higher very little change in selectivity patterns is noted. The various ion-exchanged forms of the faujasite can result in slightly different cracking properties; e.g., the use of high cerium content mixed rare earths improves carbon burning rates in the regenerator, the use of H-form faujasite improves selectivity to propane–pentane fractions, and the use of a minor amount of copper-form faujasite increases light olefin yield and naphtha octanes.

6.4 Catalyst Deactivation

A cracking catalyst should maintain its cracking activity with little change in product selectivity as it ages in a unit. A number of factors contribute to degradation of the catalyst: (1) the combination of high temperatures, steam partial pressure, and time; (2) impurities present in the fresh catalyst; and (3) impurities picked up by the catalyst from the feed while in use. Under normal operating conditions, the catalyst experiences temperatures of 480–515°C (900–960°F) in the reactor and steam stripper zones and temperatures of 620–720°C (1150–1325°F) and higher in the regenerator accompanied by a substantial partial pressure of steam. With mixed oxide amorphous gel catalysts, the plastic nature of the gel is such that the surface area and pore volume decrease rather sharply in the first few days of use and then at a slow inexorable rate thereafter. This plastic flow also results in a reduction in the number and strength of the active catalytic sites.

Zeolite catalysts comprising both amorphous gel and crystalline zeolite degrade from instability of the gel, as stated above, and also from loss of crystallinity. The latter also results from the combined effects of time, temperature, and steam partial pressure. When crystallinity is lost, the amorphous residue is relatively low in activity, approximating that of the amorphous gel matrix. The rate of degradation of the amorphous gel component may not be the same as that of the zeolite crystals; e.g., the gel may degrade rapidly and, through thermoplastic flow, effectively coat the crystals and interfere with the diffusion of hydrocarbons to the catalytic sites in the zeolite. Manufacturers of zeolite catalysts try to combine high stability in the matrix with highly stable zeolite crystals.

Residual impurities in freshly manufactured catalysts are principally sodium and sulfate. These result from the use of sodium silicate and aluminum sulfate in making the silica-alumina gel matrix and subsequent washing of the composite catalyst with ammonium sulfate to remove sodium. Generally, the sodium content of the amorphous gel is < 0.1 wt% (as Na_2O), and the sulfate content < 0.5 wt%.

With zeolite catalysts the residual sodium may be primarily associated with the zeolite, so sodium levels may range from about 0.2% to 0.80% for the composite catalyst. Sulfate levels in zeolite catalysts are still less than 0.5%. An excessive amount of sodium reacts with the silica in the matrix under regenerator operating conditions and serves as a flux to increase the rate of loss of surface area and pore volume. Sodium faujasite is not as hydrothermally stable as other metal-exchanged (e.g., mixed rare earth) forms of faujasite. It is most desirable to reduce the sodium content of the faujasite component to less than 5.0% by weight (as Na_2O) with rare earths or with mixtures of rare earths and ammonium ions.

Finally, catalysts can degrade as a result of impurities picked up from the feed being processed. These impurities are sodium, nickel, vanadium, iron, and copper. Sodium laid down on the catalyst not only acts to neutralize active acid sites, reducing catalyst activity, but also acts as a flux to accelerate matrix degradation. Freshly deposited metals are effective poisons to cracking catalysts because of the loss of active surface area by metal deposition (Fig. 15.24) (Otterstedt et al., 1986). Zeolite catalysts are less responsive to metal contaminants than amorphous gel catalysts. Hence equilibrium catalysts can tolerate low levels of these metals as long as they have enough time to become buried. A sudden deposition of fresh metals can have an adverse effect on unit performance. Metals levels on equilibrium catalysts reflect the metals content of the feeds being processed; typical ranges are 200–1200 ppm V, 150–500 ppm Ni, and 5–45 ppm Cu. Sodium levels are in the range of 0.25–0.8% by weight (as Na_2O).

6.5 Catalyst Stripping

Catalyst leaving the reaction zone is fluidized with reactor product vapors that must be removed and recovered with the reactor product. In order to accomplish this, the catalyst is passed into a stripping zone where most of the hydrocarbon is displaced with steam.

Stripping is generally done in a countercurrent contact zone where shed baffles or contactors are provided to ensure equal vapor flow up through the stripper and efficient contact. Stripping can be accomplished in a dilute catalyst phase. Generally, a dense phase is used, but with lighter feeds or higher reactor temperature and high conversion operations, a significant portion of the contacting can be done in a dilute phase.

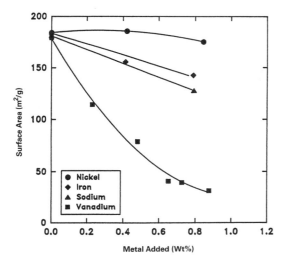

Figure 15.24 Effect of metals deposition on available surface area of the catalyst.

The amount of hydrocarbon carried to the regenerator is dependent upon the amount of stripping steam used per pound of catalyst and the pressure and temperature at which the stripper operates.

Probes at the stripper outlet have been used to measure the composition of the hydrocarbon vapors leaving the stripper. When expressed as percent of coke burned in the regenerator, the strippable hydrocarbon is only 2–5 wt%. Very poor stripping is shown when the hydrogen content of the regenerator coke is 10% by weight or higher. Good stripping is shown by hydrogen levels of 6–9 wt%.

The proper level of stripping is found in many operating units by reducing the stripping steam until there is a noticeable effect or a rise in regenerator temperature. Steam is then marginally increased above this rate. In some units, stripping steam is used as a control variable to control the carbon burning rate or differential temperature between the regenerator bed and cyclone inlets.

In summary, the catalytic cracking unit is an extremely dynamic unit, primarily because there are three major process flow streams (the catalyst, hydrocarbon, and regeneration air), which interact with each other. Problems can arise in the equipment and flowing streams that are sometimes difficult to diagnose because of the complex effects they can create (Table 15.8).

6.6 Catalyst Treatment

The latest technique developed by the refining industry to increase gasoline yield and quality is to treat the catalysts from the cracking units to remove metal poisons that accumulate on them. Nickel, vanadium, iron, and copper compounds contained in catalytic cracking feedstocks are deposited on the catalyst during the cracking operation, thereby adversely affecting both catalyst activity and selectivity. Increased catalyst metal contents affect catalytic cracking yields by increasing coke formation, decreasing gasoline, butane, and butylene production; and increasing hydrogen production.

The recent commercial development and adoption of cracking catalyst-treating processes definitely improve the overall economics of the catalytic cracking process.

6.6.1 Demet

A cracking catalyst is subjected to two pretreatment steps (Fig. 15.25). The first step effects vanadium removal and the second, nickel removal, to prepare the metals on the catalyst for chemical conversion to compounds (chemical treatment step) that can be readily removed through water washing (catalyst wash step). The treatment steps include the use of a sulfur compound followed by chlorination with an anhydrous chlorinating agent (e.g., chlorine gas) and washing with an aqueous solution of a chelating agent (e.g., citric acid). The catalyst is then dried and further treated before being returned to the cracking unit.

6.6.2 Met-X

The Met-X process consists of cooling, mixing, and ion-exchange separation, filtration, and resin regeneration. Moist catalyst from the filter is dispersed in oil and returned to the cracking reactor in a slurry (Fig. 15.26). On a continuous basis, the catalyst from a cracking unit is cooled and then transported to a stirred reactor and mixed with an ion-exchange resin (introduced as slurry). The catalyst–resin slurry then flows to an elutriator for separation. The catalyst slurry is taken overhead to a filter, and the wet filter cake is

(text continues on p. 437)

Table 15.8 General Commentary on the Fluid Catalytic Cracking Process

Problem	Symptoms	Causes	Data required	Corrective action
	Fresh Feed Quality Problems			
(a) Meals in feed	(1) High H_2 make (2) General increase in light ends (3) Excessive coke production (4) Higher reactor velocities (5) Overloaded gas compressor	(1) Pitch entrainment at atmosphere and vacuum feed preparation unit (A&V) (2) Feed type change (3) Abnormal A&V operation	(1) Feed nickel equivalent (2) Feed CCR (3) H_2 make SCF/bbl fresh feed (4) Actual vs. predicted yields (5) A&V operating conditions (6) Fresh feed color	(1) Lower metals in feed. (2) Feed segregation. (3) Catalyst replacement program
(b) Feed contamination with heavy hydrocarbons	(1) Excessive coke make (2) Unexplained increase in air requirements at same conversion (3) Poor weight balance	(1) Leak in exchanger train (2) Partly open valves (3) Fractionator bottoms entrainment in heart cut recycle	(1) Feed CCR (2) Feed RI and °API (3) Recycle stream inspections	(1) Isolate leaking exchangers. (2) Minimize possibility of leaks by pressure balance.
(c) Feed contamination with light hydrocarbons	(1) Unsteady preheat header pressures and flows (2) Poor weight balance (3) Shift in yields distribution	(1) Leak in exchanger train (2) Partly open valves	(1) Feed °API (2) Feed front end distillation	(1) As above.
(d) Water in feed	(1) Vibration in preheat system (2) Unsteady flows and temperatures (3) Severe upset if large amounts of water in feed	(1) Water in feed tankage (2) Leaks from steam-out connections (3) Trapped water from idle equipment	(1) Water in feed	(1) Check for steam leaks on tank heaters. (2) Isolate and swing to high suction on catalyst feed tank. (3) Ensure that idle equipment is well drained and hot before being brought on-stream

(*continued*)

Table 15.8 (*continued*)

Problem	Symptoms	Causes	Data Required	Corrective Action
		Catalyst Problems		
(a) Catalyst contamination	(1) See metals in feed	(1) Low catalyst replacement rate (2) Metals in feed	(1) Metals on catalyst (2) H_2 production (3) Product yield distribution	(1) Consider lowering metals in feed; A&V operation. (2) Catalyst replacement program. (3) Feed segregation.
(b) Sodium on catalyst	(1) See sintering	(1) Salt in feed (2) Treated boiler feedwater used in regenerator sprays	(1) Historical Na content of catalyst	(1) Minimize sources of Na input to system.
(c) Sintering of catalyst	(1) Apparent decrease in catalyst activity (2) Increase in carbon on regenerated catalyst (3) Change in product yield distribution	(1) High Na and V on catalyst (2) Excessive regenerator temperatures (3) Low bed stability (4) Excessive or prolonged use of torch oil (5) Localized high temperatures	(1) Sintering index (2) EASC total and burnable carbon (3) Metals on catalyst (4) Yield distribution (5) Predicted and measured activity (6) Regenerator operating conditions	(1) Catalyst replacement. (2) Minimize Na input with seawater or salt in feed, or with regenerator sprays. (3) Consider lower V content in feed. (4) Use high stability catalysts. (5) Review regenerator operations.
(d) Coarse catalyst	(1) Poor circulation (2) Poor regeneration (3) Change in yield distribution (4) Poorer stripping	(1) Loss of fines	(1) Roller analysis (2) H/C ratio (3) Yield distribution (4) Operating conditions in general	(1) Minimize catalyst losses by lowering regenerator velocity. (2) Change to fiber catalyst. (3) Consider use of attriter.

	Cause	Indication	Check	Remedy
(e) Attrition	(1) High velocity stream into dense phase (2) Fragile catalyst	(1) Higher catalyst losses with increasing fines content	(1) Check unit for high velocity streams exceeding 200 ft/s into catalyst (2) Missing ROs, blast steam, partly open valves, etc. (3) Check fresh catalyst properties (4) Study catalyst loss pattern	(1) Eliminate or reduce high velocity fluid injection. (2) Follow up fresh catalyst supplies.

Fresh Feed Quality Problems

	Cause	Indication	Check	Remedy
(a) Reactor cyclone failure	(1) Erosion and/or corrosion (2) Pressure surges	(1) High catalyst in fractionator bottoms (2) Frequent fractionator bottoms pump plugging (3) Loss of catalyst fines (4) Catalyst losses become progressively higher	(1) Catalyst content in fractionator bottoms (2) Cyclone pressure drop	(1) Minimize reactor velocity. (2) Review new methods in cyclone design and repair. (3) Review operating procedures and history of past occurrences.
(b) Eroded or plugged grid holes	(1) Lumps of coke or refractory in catalyst (2) Failure of grid hole inserts (3) Hole velocity too high (4) Feed injector velocity too low	(1) Change in grid ΔP (2) Decline in reactor efficiency (3) Change in yield distribution (4) Low overflow well level if grid eroded, high if plugged	(1) Grid pressure drop (2) Product yield pattern (3) Declining catalyst activity as determined by unit tracking despite adequate catalyst replacement rate	(1) Review methods in grid design and repair. (2) Change reactor–regenerator differential pressure controller (DPRC) setting and/or control air to maintain circulation and normal overflow well level. (3) Check feed injector operation.

(continued)

Table 15.8 (continued)

Problem	Symptoms	Causes	Data Required	Corrective Action
(c) Unsteady bed temperature and reactor pressure				See Water in feed
(d) Coking in overhead line	(1) Rise in ΔP between reactor outlet and fractionator inlet	(1) Condensation of reactor products in overhead line	(1) Pressure survey	(1) Ensure that overhead line insulation is in good condition.
(e) Poor catalyst stripping	(1) Unexplained increase in coke (2) Higher H/C ratio	(1) Insufficient stripping steam (2) Poor catalyst/steam contacting (3) Catalyst properties (4) Low reactor temp. (5) Inaccurate steam flow controller (FRC)	(1) H/C ratios (2) Stripper tests	(1) Check stripping steam rate. (2) Consider higher reactor temperature. (3) Consider lower circulation rate. (4) Review stripper design.

Regenerator Problems

Problem	Symptoms	Causes	Data Required	Corrective Action
(a) Cyclone failure	(1) Increase in catalyst losses (2) Catalyst losses become progressively higher	(1) Excessive temperatures and allowable stresses exceeded (2) Erosion	(1) Pressure differential between dilute phase and plenum (2) Measure of catalyst losses (3) Regenerator velocity	(1) Minimize catalyst feed. (2) Minimize regenerator velocity. (3) Review new methods in cyclone design and repair. (4) Review operating procedures and history.
(b) Plenum chamber failure	(1) Increase in catalyst losses above normal level	(1) Excessive temperatures and stress cracking (2) Impingement of plenum sprays	(1) As above (2) Check conditions of spray nozzles	(1) As above.

(c)	Failure of internal seals	(1) Uneven bed and cyclone inlet temperatures (2) Uneven O_2 distribution in dilute phase (3) Salt and pepper catalyst (4) Surging of catalyst bed (5) Drop in grid Δp (6) Increase in catalyst losses (7) Afterburning	(1) Erosion (2) Pressure bump (3) Stresses too high (4) Abnormal conditions with auxiliary burner on start-up	(1) Grid Δp (2) Catalyst appearance (3) O_2 analysis of gas at dilute phase sprays (4) Historical operating data	(1) Avoid pressure surges on start-up. (2) Maximize air to grid (minimize control air). (3) Low regenerator pressure to increase velocity through grid. (4) Review operating history and seal design.
(d)	Hole in overflow well	(1) Unstable overthrow well level and catalyst circulation (2) Unsteady overflow well density (3) High regenerator hold-up (4) Uneven temperatures (5) Uneven O_2 distribution	(1) Abnormal stresses (2) Erosion	(1) Overflow well level and densities (2) Historical regenerator operator conditions, particularly hold-up and temperatures	(1) Alter standpipe aeration. (2) Adjust DPRC and circulation. (3) Maximize control air (minimize air to grid). (4) Review design.
(e)	Grid hole plugging and or erosion	(1) Uneven bed and cyclone temperatures (2) Uneven O_2 distribution (3) Increased catalyst losses (4) Surging of catalyst bed (5) Increased sintering	(1) Lumps of catalyst or refractory in catalyst bed (2) Low bed stability	(1) Catalyst sintering index (2) Bed stability (3) O_2 analyses of dilute phase	(1) Maintain grid pressure drop at 30% bed pressure drop. (2) Maximize air through grid. (3) Review grid design. (4) Check condition of catalyst hopper to ensure that debris does not enter regenerator.

(continued)

Table 15.8 (*continued*)

Problem		Symptoms		Causes		Data Required		Corrective Action
(f) Stuck or failed trickle valves	(1)	High catalyst losses but no increase with time	(1)	Binding of hinge rings by	(1)	Careful examination of trickle valves on shutdown	(1)	Revise or replace as required to meet specifications.
				(a) Clearances not large enough				
	(2)	Loss of catalyst fines; increase in coarseness index		(b) Oxidation scale or pieces of lining				
				(c) Nonuniform wear of moving parts				
			(2)	Installation angles not correct				
			(3)	Improper material				
(g) Plugged diplegs	(1)	As above	(1)	Spalled refractory forming partial and eventually final plug	(1)	Careful examination of diplegs or use of probes, lights, balls, etc. to ensure diplegs are free before start-up	(1)	Lower bed level to allow catalyst to flow out of dipleg.
			(2)	Air-out periods with a lot of water/steam in vessel				
(h) Excessive input of steam or air	(1)	Higher level of catalyst losses	(1)	Missing restriction orifices	(1)	Equipment survey to make sure all restriction orifices (ROs) are in place, etc.	(1)	Make careful detailed check of unit for and correct equipment conditions listed under causes.
	(2)	High cyclone pressure drop	(2)	Large restriction orifices				
	(3)	Indication of attrition (higher catalyst fines content despite higher losses)	(3)	Partially open valves on steam, water, or air lines				
			(4)	Malfunctioning steam traps				
			(5)	Metering errors				

	Problem	Symptoms	Possible causes	Detection	Remedial action
(i)	Bed stability too low	(1) Increase in temperature below the grid	(1) Insufficient air through grid to support catalyst bed (2) Eroded grid holes	(1) Grid ΔP (2) Historical records of temperature below grid	(1) Maximize air through grid (2) If problem expected to persist, redesign grid for lower air rates.
(j)	Surging of catalyst bed	(1) Erratic or cycling instrument records on hold-up, density, and overflow well (2) Unsteady circulation and heat balance (3) Catalyst sintering (4) Uneven catalyst regeneration	(1) Seal failures (2) Grid hole erosion (3) Hole in overflow well (4) Poor bed stability (5) Poor DPRC control	(1) Pressure survey (2) Operating condition survey (3) Bed stability calculation	(1) See action required for seal failures, grid hole erosion, poor bed stability, poor DPRC control.
(m)	Unsteady pressure differential control	(1) Fluctuating regenerator pressure (2) Unsteady circulation and overflow well level (3) Unsteady rector temperature (4) Catalyst shifts between reactor and regenerator	(1) Poor stack slide valve performance (a) Sticky slide valves (b) Poor slide valve instrument performance (2) Unsteady circulation	(1) Unit performance records	(1) Check out slide valves and associated instrumentation. (2) Check adequacy of U-bend or standpipe aeration.
(n)	Regenerator hot spots	(1) High temperatures on vessel shells or U-bends	(1) Damaged refractory	(1) Change in paint color (2) Glow at night (3) Surface temperature measurements	(1) Cool the spot with steam or water, avoiding abrupt changes in temperature due to interpretation of cooling.

(continued)

Table 15.8 (*continued*)

Problem	Symptoms	Causes	Data Required	Corrective Action
(o) Rough catalyst circulation	(1) Unsteady regenerator and reactor temperatures and hold-ups (2) Unsteady control air (3) Fluctuating overflow well level and U-bend densities (4) U-bend vibration	(1) Improper aeration (2) Coarse catalyst (3) Fluctuating DPRC (4) Water in aeration medium (5) Poor performance of control air system	(1) U-bend aeration pattern and pressure survey (2) Standpipe and U-bend densities (3) Circulation rate (4) Catalyst roller analysis	(1) Change aeration. (2) Put control air on manual to determine if it is the cause. (3) Check DPRC system (4) Make sure aeration medium is water-free.
(k) High carbon on catalyst. Carbon buildup	(1) Dark catalyst (2) Dilute phase temperature decreases relative to dense bed temperature (3) Low excess O_2 (4) Apparent loss in catalyst activity	(1) Excessive coke from (a) Increase in operating intensity (b) Poorer feedstock (c) Poor catalyst stripping (d) Heavier recycle (e) Leakage of fraction bottoms into feed (2) Low excess O_2 (3) Poor air distribution (4) False O_2 recorder readings (5) Feed and recycle meter errors	(1) Carbon on catalyst analysis (2) Feed quality (3) Recycle boiling range and CCR (4) Check O_2 levels in regenerator (5) Check O_2 recorder for accuracy (6) Check stripper performance	(1) Lower intensity of operation. (2) Increase air to regenerator. (3) Inject lighter feed. (4) Improve normal feed and/or recycle if possible. (5) Improve catalyst stripping. (6) Check meter accuracy.

Problem	Indications	Possible Causes	Checks/Tests	Remedy
(l) Afterburning	(1) Excessive dilute phase and cyclone temperatures (2) Trend to lighter catalyst (3) Increase in dilute phase temperature relative to dense bed (4) High excess O_2 (5) Higher CO_2/CO ratios	(1) Insufficient coke production: (a) Decrease of intensity (b) Swing to better feed (c) Lighter recycle (2) Too much excess air (3) False O_2 recorder readings (4) Feed and recycle meter errors (5) Cyclone steam meter errors	(1) Same as in (k) above except (6)	(1) Decrease air. (2) Marginal use of torch oil. (3) Increase operating intensity. (4) Check cyclone stream rate.
(p) Low catalyst circulation rate	(1) Inability to lower regenerator temperature (2) Excessive feed preheat requirements	(1) Partial blockage of U-bends (2) Too much stripping steam (3) Improper aeration (4) Control air too low (5) DPRC setting inadequate	(1) As in (o) above	(1) Change DPRC setting. (2) Check control air system and increase control air rate. (3) Ensure adequate aeration. (4) Check unit design.

Fractionator Problems

Problem	Indications	Possible Causes	Checks/Tests	Remedy
(a) Poor split between LCGO and recycle	(1) High overlaps between heating oil and recycle stream distillation (2) Shift in yield pattern (3) Increase in recycle volume at constant operating conditions	(1) Inadequate steam to recycle stripper (2) Improper tray loading in tower (3) Stripper malfunction	(1) Heating oil and recycle distillations (2) Fractionator pressure survey (3) Pumparound heat removal (4) Detailed fractionator and stripper analysis (5) Consider equipment X-rays	(1) Ensure adequate stripping steam rates. (2) Adjust pumparound heat duties. (3) Consider equipment changes.

(*continued*)

Table 15.8 (*continued*)

Problem	Symptoms	Causes	Data Required	Corrective Action
(b) Fractionator bottoms too light	(1) High bottoms API (2) Excessive bottoms yield (3) Dark color of recycle stream	(1) Too much heat removal in bottoms pumparound (2) Tower bottom liquid too cold (3) Poor liquid–vapor contact in shed section (4) Lower operating intensity (5) Heart cut recycle rate too low (6) Leaks into bottoms system	(1) Fractionator bottoms inspections (2) Tower operating conditions in the lower half (3) Check on cracking intensity (4) Check recycle flow rate; meter accuracy	(1) Lower heat removal from tower bottoms consistent with coking considerations. (2) Maximize heart cut recycle. (3) Consider equipment changes.
(c) Coking	(1) Frequent plugging of exchangers and pumps with coke (2) Poor heat transfer in exchangers (3) High insolubles in fractionator bottoms	(1) Excessive temperatures in tower bottom (2) Low bottoms API (3) High liquid residence time in tower bottoms	(1) History of bottoms API and bottoms sediment and water content (2) Tower bottoms operating conditions (3) Calculations of bottoms liquid residence time (4) History of equipment fouling	(1) Review recommended limitations of tower bottoms operating conditions and adjust operation accordingly. (2) Consider changing pump screen size.

	Problem	Symptoms	Probable cause	Checks	Remedy
(d)	Salt fouling of fractionator top	(1) High pressure drop in top trays (2) Flooding in the top section (3) Poor split between overhead and first sidestream (4) Salt in TPA pumps (5) Chlorides in overhead water	(1) Salts in fresh feed (2) TPA return too cold	(1) Pressure and temperature survey of tower top (2) TPA operating conditions (3) Chlorides in fresh feed (4) Chlorides in distillate drum water (5) Split between overhead product and first sidestream	(1) Cautiously inject water to TPA. (2) Adjust tower temperatures. (3) Adjust TPA return temperature. (4) Consider installing screens in TPA pump suctions.

Miscellaneous Problems

	Problem	Symptoms	Probable cause	Checks	Remedy
(a)	Air blower low turbine efficiency	(1) Excessive steam usage (2) Unusual exhaust steam conditions (3) Decrease in rpm at maximum steam	(1) Turbine fouling (2) Turbine blade wear (3) Quality of steam supply	(1) Steam supply and exhaust conditions (2) Calculation of turbine efficiency (3) Blower operation conditions	(1) Improve steam quality. (2) Consider turbine wash or shutdown for cleaning.
(b)	Surplus heat in unit	(1) High regeneration temperature (2) Low control air blower rate (see also low circulation rate)	(1) Too much feed preheat (2) Reactor temperature set too low	(1) Unit temperatures and heat balances (2) Circulation rate (3) Coke make	(1) Back off on feed injection temperature. (2) Increase circulation rate. (3) Increase reactor temperature and lower hold-up. (4) Adjust DPRC. (5) Use dilute phase sprays.

(continued)

Table 15.8 (*continued*)

Problem	Symptoms	Causes	Data Required	Corrective Action
(c) Insufficient heat in unit	(1) Low regenerator temperature (2) High control air if on automatic setting	(1) Insufficient feed preheat (2) Reactor temperature set too high (3) Not enough carbon produced: (a) Catalyst activity low (b) Lighter recycle (c) Lighter feed (d) Lower reactor hold-up	(1) Unit temperatures and heat balances (2) Operating intensity level (3) Circulation rate (4) Coke production (5) Feed and recycle inspections	(1) Increase preheat if possible. (2) Increase fresh catalyst additions. (3) Take out sprays if any. (4) Use torch oil.

Source: McKetta, 1992.

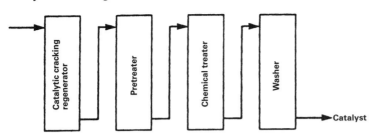

Figure 15.25 The Demet process.

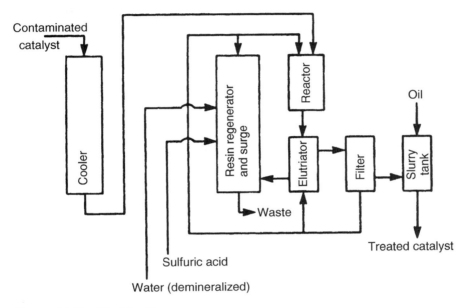

Figure 15.26 The Met-X process.

slurried with oil and pumped into the catalytic cracked feed system. The resin leaves the bottom of the elutriator and is regenerated before returning to the reactor.

REFERENCES

Avidan, A. A., Edwards, M., and Owen, H. 1990. Oil Gas J. 88(2):33.

Bartholic, D. B., and Haseltine, R. P. 1981. Oil Gas J. 79(45):242.

Bland, W. F., and Davidson. R. L. 1967. Petroleum Processing Handbook. McGraw-Hill, New York.

Bradley, S. A., Gattuso, M. J., and Bertolacini, R. J. 1989. Characterization and Catalyst Development. ACS Symp. Ser. No. 411. Am. Chem. Soc., Washington, DC.

Cybulski, A., and Moulijn, J. A., eds. 1998. Structured Catalysts and Reactors. Marcel Dekker, New York.

DeCroocq, D. 1984. Catalytic Cracking of Heavy Petroleum Hydrocarbons. Editions Technip, Paris.

Gates, B. C., Katzer, J. R., and Schuit, G. C. A. 1979. Chemistry of Catalytic Processes. McGraw-Hill, New York.

Germain, G. E. 1969. Catalytic Conversion of Hydrocarbons. Academic Press, New York.

Hemler, C. L. 1997. In: Handbook of Petroleum Refining Processes. R. A. Meyers, ed. McGraw-Hill, New York, Chapter 3.3.

Hunt, D. A. 1997. In: Handbook of Petroleum Refining Processes. R. A. Meyers, ed. McGraw-Hill, New York, Chapter 3.5.

Hydrocarbon Processing. 1998. 77(11):53 et seq.

Johnson, T. E., and Niccum, P. K. 1997. In: Handbook of Petroleum Refining Processes. R. A. Meyers, ed. McGraw-Hill, New York, Chapter 3.2.

Ladwig, P. K. 1997. In: Handbook of Petroleum Refining Processes. 2nd ed. R. A. Meyers, ed. McGraw-Hill, New York, Chapter 3.1.

Le Page, J. F., Cosyns, J., Courty, P., Freund, E., Franck, J. P., Jacquin, Y., Juguin, B., Marcilly, C., Martino, G., Miguel, J., Montarnal, R., Sugier, A., and von Landeghem, H. 1987. Applied Heterogeneous Catalysis. Editions Technip, Paris.

Luckenbach, E. C., Worley, A. C., Reichle, A. D., and Gladrow, E. M. 1992. In: Petroleum Processing Handbook. J. J. McKetta, ed. Marcel Dekker, New York, p. 349.

Maples, R. E. 2000. Petroleum Refinery Process Economics. 2nd ed. PennWell Corporation, Tulsa, OK.

McKetta, J. J. 1992. Petroleum Processing Handbook. Marcel Dekker, New York.

Occelli, M. L., and O'Connor, P. 1998. Fluid Cracking Catalysts. Marcel Dekker, New York.

Otterstedt, J. E., Gevert, S. B., Jaras, S. G., and Menon, P. G. 1986. Appl. Catal. 22:159.

Radler, M. 1999. Oil Gas J. 97(51):41 et seq.

Sadeghbeigi, R. 1995. Fluid Catalytic Cracking: Design, Operation, and Troubleshooting of FCC Facilities. Gulf Publ. Co., Houston, TX.

Salazar, J. R. 1986. In: Handbook of Petroleum Refining Processes. R. A. Meyers, ed. McGraw-Hill, New York, Chapter 8.5.

Shorey, S. W., Lomas, D. A., and Keesom, W. H. 1999. Hydrocarbon Process. 78(11):43.

Speight, J. G. 1986. Annu. Rev. Energy 11:253.

Speight, J. G. 1999. The Chemistry and Technology of Petroleum. 3rd ed. Marcel Dekker, New York.

Speight, J. G. 2000. The Desulfurization of Heavy Oils and Residua. 2nd ed. Marcel Dekker, New York.

Stiles, A. B. and Koch, T. A. 1995. Catalyst Manufacture. Marcel Dekker, New York.

Thakur, D. S. 1985. Appl. Catal. 15:197.

Thompson, G. J. 1997. In: Handbook of Petroleum Refining Processes. 2nd ed. R. A. Meyers, ed. McGraw-Hill, New York, Chapter 8.3.

Wojciechowski, B. W., and Corma, A. 1986. Catalytic Cracking: Catalysts, Chemistry, and Kinetics. Marcel Dekker, New York.

16
Hydrotreating

1. INTRODUCTION

Hydroprocessing (which covers the process terms *hydrotreating* and *hydrocracking*) is a refining technology in which a feedstock is thermally treated with hydrogen under pressure (Bridge, 1997). It can affect a refinery's product slate by strategic placement within the overall refinery processes (Fig. 16.1).

The use of hydrogen in thermal processes is perhaps the single most significant advance in refining technology during the twentieth century (Bridge et al., 1981; Scherzer and Gruia, 1996; Dolbear, 1998). As an example of the popularity or necessity of hydrotreating, data (Radler, 1998) show that of a total worldwide daily refinery capacity of approximately 81,500,000 bbl of oil, approximately 8,500,000 bbl/day is dedicated to catalytic hydrorefining and approximately 28,100,000 bbl/day is dedicated to catalytic hydrotreating. On a national (United States) basis, the data are approximately 1,780,000 bbl/day hydrorefining and approximately 1,900,000 bbl/day catalytic hydrotreating out of a daily total refining capacity of approximately 16,500,000 bbl/day.

The hydrotreating process makes use of the principle that the presence of hydrogen during *mild* thermal treatment reaction of a petroleum feedstock removes the heteroatoms and metals. On the other hand, the presence of hydrogen during cracking (*hydrocracking*) terminates many of the coke-forming reactions and enhances the yields of the lower boiling components, such as gasoline, kerosene, and jet fuel. However, hydrocracking (Chapter 17) should not be regarded as a competitor for catalytic cracking (Chapter 15). Catalytic cracking units normally use gas oil distillates as feedstocks, whereas hydrocracking feedstock usually consists of refractive gas oils derived from cracking and coking operations. Hydrocracking is a supplement to, rather than a replacement for, catalytic cracking.

In hydrotreating, the feedstock is reacted with hydrogen at elevated temperatures, in the range of 300–450°C (570–840°F), and elevated pressures, in the range of 120–2200 psi (0.7–15 MPa) under the presence of a hydrogenation catalyst, typically cobalt-molybdenum (Co-Mo) or nickel-molybdenum (Ni-Mo) on γ-alumina (γ-Al_2O_3). The catalysts are produced in the oxide forms and are sulfided before their use in the process. During hydrotreating the heteroatoms are removed in the form of hydrogen sulfide (H_2S), ammo-

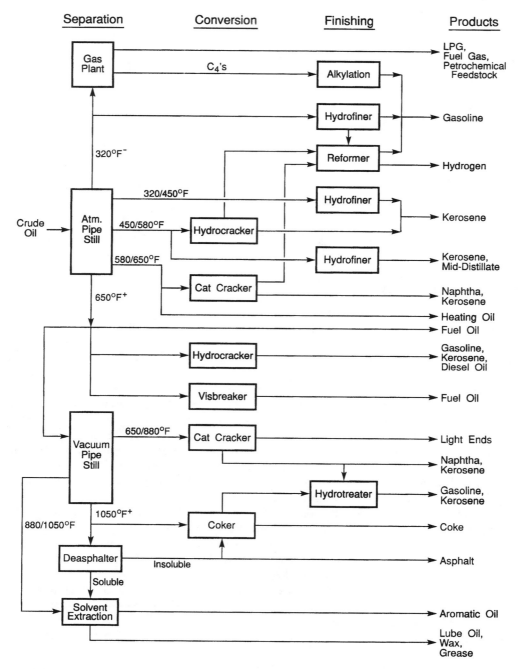

Figure 16.1 Refinery schematic showing the relative placement of hydrotreating units.

nia (NH₃), and water (H₂O), and metal species such as vanadium (V) and nickel (Ni) are simultaneously removed by hydrodemetallization reactions. In hydrotreating, hydrogen is also added to the feedstock, which in the case of the heavy feedstocks reduces the tendency for coke formation during subsequent thermal or catalytic cracking processes. Fouling of hydrotreating catalysts by metal deposits as well as by coke deposition becomes unavoidable, and may cause expensive plant shutdowns for replacement of the catalyst. Catalyst reactivation and replacement of poisoned catalyst by the fresh catalyst are important elements of reactor design.

The distinguishing feature of the hydroprocesses is that although the composition of the feedstock is relatively unknown and a variety of reactions may occur simultaneously, the final product may actually meet all the required specifications for its particular use. Thus, the purposes of refinery hydroprocesses are

1. To improve existing petroleum products
2. To enable petroleum products to meet market specifications
3. To develop new products or even new uses for existing products
4. To convert inferior or low grade materials into valuable products
5. To transform near-solid residua into liquid fuels

There is a rough correlation between the quality of petroleum products and their hydrogen content (Dolbear, 1998). It so happens that desirable aviation gasoline, kerosene, diesel fuel, and lubricating oil are made up of hydrocarbons that contain high proportions of hydrogen (Fig. 16.2). In addition, it is usually possible to convert olefins and higher molecular weight constituents to paraffins and monocyclic hydrocarbons by hydrogen addition processes. These facts have for many years encouraged attempts to employ hydrogenation for refining operations. Despite considerable technical success, such processes were not economically possible until low-priced hydrogen became available

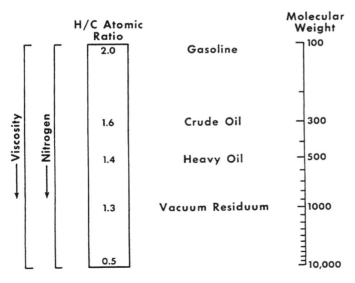

Figure 16.2 Atomic hydrogen/carbon ratios of various feedstocks.

as a result of the rise of reforming, which converts naphthenes to aromatics with the release of hydrogen.

As already noted, hydroprocesses for the conversion of petroleum fractions and petroleum products may be classified as *nondestructive* and *destructive*. Nondestructive hydrogenation is characterized by the removal of heteroatom constituents as the hydrogenated analogs:

$$R-S-R^1 + H_2 \rightarrow RH + R^1H + H_2S$$

$$2R-N(R^2)-R^1 + 3H_2 \rightarrow 2RH + 2R^1H + 2R^2H + 2NH_3$$

$$R-O-R^1 + H_2 \rightarrow RH + R^1H + H_2O$$

as well as for the saturation of olefin-type bonds in products from thermal processes:

$$R-CH=CH-R^1 + H_2 \rightarrow R-CH_2CH_2-R^1$$

Aromatics may also be saturated to produce cycloaliphatics (naphthenes) by this treatment:

$$\underset{\text{benzene}}{C_6H_6} + 3H_2 \rightarrow \underset{\text{cyclohexane}}{C_6H_{12}}$$

Any metals (usually nickel and vanadium) present in the feedstock are usually removed during hydrogen processing at the cost of hydrogen consumption (Scott and Bridge, 1971; Aalund, 1975) and not, of course, as their hydrogen analogs but by deposition on the catalyst through changes in the chemical properties of the metal-containing constituents by the high temperatures and the presence of hydrogen. These two process parameters progressively affect the ability of the organic structures to retain the metals within the organic matrix, and deposition ensues.

Thus, hydrotreating of distillates may be defined simply as the removal of sulfur, nitrogen, and oxygen compounds as well as olefinic compounds by selective hydrogenation. The hydrotreating catalysts are usually cobalt plus molybdenum sulfides (CoS-MoS) or nickel plus molybdenum sulfides (NiS-MoS) impregnated on an alumina (Al_2O_3) base. The hydrotreating operating conditions, 1000–2000 psi (6895–13,790 kPa) hydrogen and about 370°C (700°F), are such that appreciable hydrogenation of aromatics does not occur. The desulfurization reactions are invariably accompanied by small amounts of hydrogenation and hydrocracking, the extent of which depends on the nature of the feedstock and the severity of desulfurization.

In summary, hydrotreating (nondestructive hydrogenation) is generally used for the purpose of improving product quality without appreciably altering the boiling range. Mild processing conditions are employed so that only the more unstable materials are attacked.

Destructive hydrogenation (*hydrogenolysis* or *hydrocracking*) is characterized by the cleavage of carbon–carbon linkages accompanied by hydrogen saturation of the fragments to produce lower boiling products.

$$R-CH_2CH_2-R^1 + H_2 \rightarrow R-CH_3 + CH_3-R^1$$

Such treatment requires thermal processing regimes that are similar to those used in catalytic cracking (Chapter 15) and the use of high hydrogen pressures to minimize the reactions that lead to coke formation (Stanislaus and Cooper, 1994).

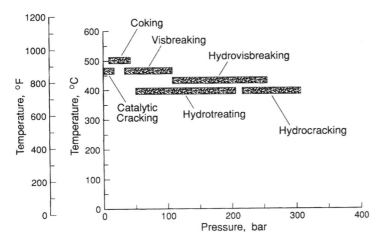

Figure 16.3 Temperature and pressure ranges for various processes.

The major differences between hydrotreating and hydrocracking are the time for which the feedstock remains at reaction temperature and the extent of the decomposition of the non-heteroatomic constituents and products. The upper limits of hydrotreating conditions may overlap with the lower limits of hydrocracking conditions (Fig. 16.3), with a lower overall conversion for hydrotreating (Fig. 16.4). And where the reaction conditions overlap, feedstocks to be hydrotreated will generally be exposed to the reactor temperature for shorter periods; hence the reason hydrotreating conditions may be referred to as *mild*. All is relative.

Unsaturated compounds, such as olefins, are not indigenous to petroleum, are produced during cracking processes, and need to be removed from product streams because of the tendency of unsaturated compounds and heteroatomic polar compounds to form gum and sediment (Speight, 1999). On the other hand, aromatic compounds are indigenous to petroleum and some may be formed during cracking reactions. The most likely explanation is that the aromatic compounds present in product streams are related to the aromatic

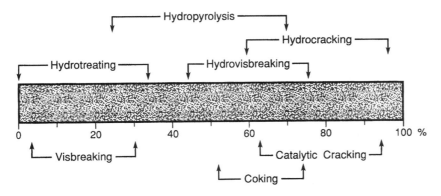

Figure 16.4 Feedstock conversion in various processes.

compounds originally present in petroleum but have shorter alkyl side chains. Thus, in addition to olefins, many product streams will contain a range of aromatic compounds that have to be removed to enable them to meet product specifications.

Of the aromatic constituents, the polycyclic aromatics are first partially hydrogenated before cracking of the aromatic nucleus takes place. The sulfur and nitrogen atoms are converted to hydrogen sulfide and ammonia, but a more important role of the hydrogenation is probably to hydrogenate the coke precursors rapidly and prevent their conversion to coke.

One of the problems in the processing of high sulfur and high nitrogen feeds is that large quantities of hydrogen sulfide and ammonia are produced. Substantial removal of both compounds from the recycle gas can be achieved by the injection of water in which, under the high pressure conditions employed, both hydrogen sulfide and ammonia are very soluble compared with hydrogen and hydrocarbon gases. The solution is processed in a separate unit for the recovery of anhydrous ammonia and hydrogen sulfide.

Hydrotreating is carried out by charging the feed to the reactor together with a portion of all the hydrogen produced in the catalytic reformer. Suitable catalysts are tungsten-nickel sulfide, cobalt-molybdenum-alumina, nickel oxide-silica-alumina, and platinum-alumina. Most processes employ cobalt-molybdenum catalysts, which generally contain about 10% molybdenum oxide and less than 1% cobalt oxide supported on alumina. The temperatures employed are in the range of 300–345°C (570–850°F), and the hydrogen pressures are about 500–1000 psi.

The reaction generally takes place in the vapor phase but, depending on the application, may be a mixed-phase reaction. The reaction products are cooled in a heat exchanger and led to a high pressure separator where hydrogen gas is separated for recycling. Liquid products from the high pressure separator flow to a low pressure separator (stabilizer), where dissolved light gases are removed. The product may then be fed to a reforming or cracking unit if desired.

Generally, it is more economical to hydrotreat high sulfur feedstocks before catalytic cracking than to hydrotreat the products from catalytic cracking. The advantages are that

1. The products require less finishing.
2. Sulfur is removed from the catalytic cracking feedstock, and corrosion is reduced in the cracking unit.
3. Carbon formation during cracking is reduced, and higher conversions result.
4. The catalytic cracking quality of the gas oil fraction is improved.

One of the chief problems with the processing of residua is the deposition of metals, in particular vanadium, on the catalyst. It is not possible to remove vanadium from the catalyst, which must therefore be replaced when deactivated, and the time taken for catalyst replacement can significantly reduce the unit time efficiency. Fixed-bed catalysts tend to plug because of solids in the feed or carbon deposits during processing of residual feeds. As mentioned previously, the highly exothermic reaction at high conversion is responsible for difficult reactor design problems in heat removal and temperature control.

The physical and chemical composition of a feedstock plays a large part in determining not only the nature of the products that arise from refining operations but also the precise manner by which a particular feedstock should be processed (Speight, 1986). Furthermore, it is apparent that the conversion of heavy oils and residua requires new lines of thought to develop suitable processing scenarios (Celestinos et al., 1975). Indeed, the use of thermal (*carbon rejection*) processes and of hydrothermal (*hydrogen addition*) processes, which were inherent in the refineries designed to process lighter feedstocks, have been a particular cause for concern. This has brought about the evolution of processing schemes that accommodate the heavier feedstocks (Chapter 19) (Corbett, 1989; Benton et al., 1986; Boening et al., 1987; Wilson, 1985; Khan, 1998). As a point of reference, an example of the former option is the delayed coking process in which the feedstock is converted to overhead with the concurrent deposition of coke, for example that used by Suncor, Inc., at their oil sands plant (Chapter 14).

The hydrogen addition concept is illustrated, in part, by the hydrotreating processes in which hydrogen is used in an attempt to remove the heteroatoms and metals as well as to *stabilize* the reactive fragments produced by the low degree of hydrocracking, thereby decreasing their potential for recombination to heavier products and ultimately to coke. The choice of processing schemes for a given hydrotreating application depends upon the nature of the feedstock as well as the product requirements (Aalund, 1981; Nasution, 1986; Suchanek and Moore, 1986; Murphy and Treese, 1979). The process can be simply illustrated as a single-stage or two-stage operation (Fig. 16.5).

The petroleum industry often employs two-stage processes in which the feedstock undergoes both hydrotreating and hydrocracking. In the first, or pretreating, stage the main purpose is to convert nitrogen compounds in the feed to hydrocarbons and ammonia by hydrogenation and mild hydrocracking. Typical conditions are 340–390°C (650–740°F), 150–2500 psi (1–17 MPa), and a catalyst contact time of 0.5–1.5 h. Up to 1.5 wt% hydrogen is absorbed, partly by conversion of the nitrogen compounds but chiefly by aromatic compounds that are hydrogenated. Thus, the single-stage process can be used to facilitate hydrotreating, and the two-stage process can then be used for hydrocracking, primarily to produce distillates from high boiling feedstocks. Both processes use an extinction–recycling technique to maximize the yields of the desired product. Significant conversion of heavy feedstocks can be accomplished by including the second stage and

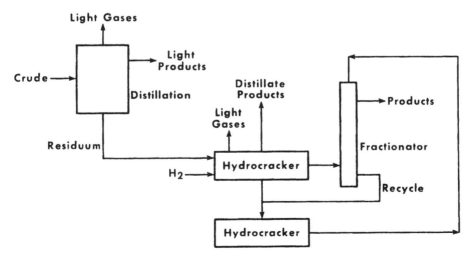

Figure 16.5 A single-stage and two-stage hydroprocessing configuration.

Table 16.1   Effect of Recycle on Product Distribution

	Once-through	Recycle
Conversion, vol%	90	95
Conversion per pass, vol%	90	70
LHSV, based on fresh feed	Base	Base
Hydrogen partial pressure	Base	−20%
Catalyst concentration, wt%	Base	+60%
Hydrogen consumption, wt%	Base	−13%
C_1–C_4 yield, wt%	Base	−10%
Liquid yield, wt%	Base	+3–6%
vol%	Base	+3/6%

hydrocracking at high severity (Howell et al., 1985). For some applications, the products boiling up to 340°C (650°F) can be blended to give the desired final product.

Product distribution and quality vary considerably depending on the nature of the feedstock constituents as well as on the process. Different process configurations will produce variations in the product slate from any one particular feedstock, and the *feedstock recycle* option adds another dimension to variations in product slate (Table 16.1) (Muñoz et al., 1986).

In modern refineries, hydrotreating is one of several process options that can be applied to the production of liquid fuels from the heavier feedstocks (Speight, 2000). A most important aspect of the modern refinery operation is the desired product slate, which dictates the matching of a process with any particular feedstock to overcome differences in feedstock composition. Hydrogen consumption is also a parameter that varies with feedstock composition (Tables 16.2 and 16.3), thereby indicating the need for a thorough understanding of the feedstock constituents if the process is to be employed to maximum efficiency.

Table 16.2   Hydrogen Consumption During Hydrotreating of Various Feedstocks

	°API	Sulfur (wt%)	Carbon residue (Conradson) (wt%)	Nitrogen (wt%)	Hydrogen (scf/bbl)
Venezuela, atmospheric	15.3–17.2	2.1–2.2	9.9–10.4	—	425–730
Venezuela, vacuum	4.5–7.5	2.9–3.2	20.5–21.4	—	825–950
Boscan (whole crude)	10.4	5.6	—	0.52	1100
Tia Juana, vacuum	7.8	2.5	21.4	0.52	490–770
Bachaquero, vacuum	5.8	3.7	23.1	0.56	1080–1260
West Texas, atmospheric	17.7–17.9	2.2–2.5	8.4	—	520–670
West Texas, vacuum	10.0–13.8	2.3–3.2	12.2–14.8	—	675–1200
Khafji, atmospheric	15.1–15.7	4.0–4.1	11.0–12.2	—	725–800
Khafji, vacuum	5.0	5.4	21.0	—	1000–1100
Arabian light, vacuum	8.5	3.8	—	—	435–1180
Kuwait, atmospheric	15.7–17.2	3.7–4.0	8.6–9.5	0.20–0.23	470–815
Kuwait, vacuum	5.5–8.0	5.1–5.5	16.0	—	290–1200

Table 16.3 Additional Hydrogen Consumption Caused by Nitrogen and Metals During Hydrodesulfurization

Metals	
V + Ni (ppm)	Corrections (%) to hydrogen consumption
0–100	−2
200	1
300	2.5
400	4
500	6.5
600	9
700	12
800	16
900	21
1000	28
1100	38
1200	50

Nitrogen compounds	Additional hydrogen required	
	(mol H_2/compound)	scf/bbl feed
Saturated amines	1	83
Pyrrolidine	2	167
Nitriles, pyrroline, alkyl cyanides	3	250
Pyrrole, nitroparaffins	4	334
Analine, pyridine	5	417
Indole	7	584

Source: Nelson, 1977.

A convenient means of understanding the influence of feedstock on the hydrotreating process is through a study of the hydrogen content (H/C atomic ratio) and molecular weight (carbon number) of the various feedstocks or products (Fig. 16.6). These data show the carbon number and/or the relative amount of hydrogen that must be added to generate the desired heteroatom removal. In addition, it is also possible to use data for hydrogen usage in residuum processing (Fig. 16.7), where the relative amount of hydrogen consumed in the process can be shown to be dependent upon the sulfur content of the feedstock (Bridge et al., 1975, 1981).

The commercial processes for treating or finishing petroleum fractions with hydrogen all operate in essentially the same manner with similar parameters. The feedstock is heated and passed with hydrogen gas through a tower or reactor filled with catalyst pellets. The reactor is maintained at a temperature of 260–425°C (500–800°F) at pressures from 100 to 1000 psi, depending on the particular process, the nature of the feedstock, and the degree of hydrogenation required. After leaving the reactor, excess hydrogen is separated from the treated product and recycled through the reactor after removal of hydrogen sulfide. The liquid product is passed into a stripping tower, where steam removes dissolved hydro-

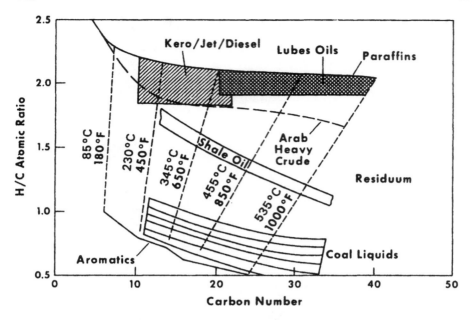

Figure 16.6 Representation of process chemistry and hydrogen requirements.

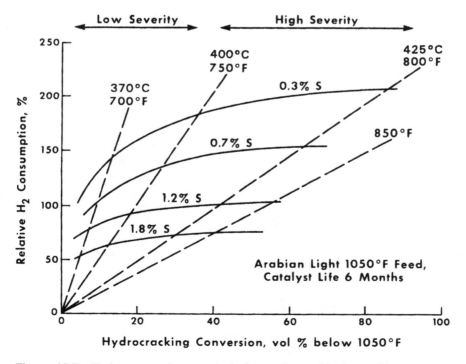

Figure 16.7 Hydrogen requirements in hydrotreating and hydrocracking processes.

gen and hydrogen sulfide, and after cooling the product is run to finished product storage or, in the case of feedstock preparation, pumped to the next processing unit.

It is most important to reduce the nitrogen content of the product oil to less than 0.001wt% (10 ppm). This stage is usually carried out with a bifunctional catalyst containing hydrogenation promoters such as nickel and tungsten or molybdenum sulfides on an acid support such as silica-alumina. The metal sulfides hydrogenate aromatics and nitrogen compounds and prevent deposition of carbonaceous deposits; the acid support accelerates nitrogen removal as ammonia by breaking carbon–nitrogen bonds. The catalyst is generally used as 1/8 × 1/8 in. (0.32 × 0.32 cm) or 1/16 × 1/8 in. (0.16 × 0.32 cm) pellets, formed by extrusion.

Most of the hydrocracking is accomplished in the second stage, which resembles the first but uses a different catalyst. Ammonia and some gasoline are usually removed from the first-stage product, and then the remaining oil, which is low in nitrogen compounds, is passed over the second-stage catalyst. Again, typical conditions are 300–370°C (600–700°F), 1500–2500 psi (10–17 MPa) hydrogen pressure, and 0.5–1.5 h contact time; 1–1.5% by weight hydrogen may be absorbed. Conversion to gasoline or jet fuel is seldom complete in one contact with the catalyst, so the lighter oils are removed by distillation of the products and the heavier, higher boiling product is combined with fresh feed and recycled over the catalyst until it is completely converted.

2. HYDRODESULFURIZATION

At this point, and in the context of this chapter, it is pertinent that there be a discussion of the techniques for desulfurization and concurrent demetallization of various feedstocks.

By way of introduction, sulfur content and metals content vary with crude oil type (Fig. 16.8) and may or may not present a problem to the refiner, depending on the amount of metals present and the downstream processing required. Vanadium (V) and nickel (Ni), the primary metals found in petroleum, can range from less than 1 part per million (ppm) by weight in some crude oils to as high as 1100 ppm vanadium and 85 ppm nickel for Boscan (Venezuela) crude oil. Vanadium is usually present in higher concentrations than nickel for Middle East and Venezuelan crude oils. However, for many U.S. crude oils, particularly those from California, the nickel content may be higher than the vanadium content. Sodium and iron are also found in quantities up to 100 and 60 ppm, respectively, though usually much lower, along with other metals.

A number of methods are available for segregating metals from crude oil. For example, deasphalting removes metals insofar as they appear in the separated asphalt (Speight, 2000), and coking processes cause the metals to concentrate in the coke (Speight, 2000); there is a similar segregation of sulfur but it is not as dramatic as the metal segregation (Speight, 2000). Hydrodemetallization and hydrodesulfurization processes are the most effective and will be discussed here.

Hydrotreating processes, in particular the hydrodesulfurization of petroleum residua, are catalytic processes. Hydrocarbon feedstock and hydrogen are passed through a catalyst bed at elevated temperatures and pressures. Some of the sulfur atoms attached to hydrocarbon molecules react with hydrogen on the surface of the catalyst to form hydrogen sulfide (H_2S), and thermodynamic equilibrium calculations show that these reactions can be driven to almost 100% completion (Speight, 2000).

Hydrodesulfurization and demetallization occur simultaneously on the active sites within the catalyst pore structure. Sulfur and nitrogen occurring in residua are converted

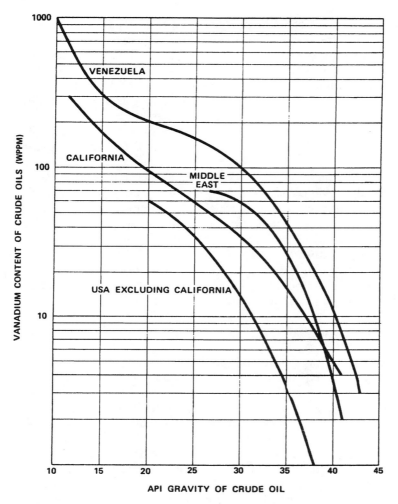

Figure 16.8 General representation of metal content of various feedstocks.

to hydrogen sulfide and ammonia in the catalytic reactor, and these gases are scrubbed out of the reactor effluent gas stream. The metals in the feedstock are deposited on the catalyst in the form of metal sulfides, and cracking of the feedstock to distillate produces a lay-down of carbonaceous material on the catalyst; both events poison the catalyst and activity or selectivity suffers. The deposition of carbonaceous material is a fast reaction that soon equilibrates to a particular carbon level and is controlled by hydrogen partial pressure within the reactors. On the other hand, metal deposition is a slow reaction that is directly proportional to the amount of feedstock passed over the catalyst.

The life of a catalyst used to hydrotreat petroleum residua is dependent on the rate of carbon deposition and the rate at which organometallic compounds decompose and form metal sulfides on the surface. Several different metal complexes exist in the asphaltene fraction of the residuum, and an explicit reaction mechanism of decomposition that would be a perfect fit for all of the compounds is not possible. However, in general terms, the reaction can be described as hydrogen (A) dissolved in the feedstock contacting an orga-nometallic compound (B) at the surface of the hydrotreating catalyst and producing a metal sulfide (C) and a hydrocarbon (D):

$$A + B \rightarrow C + D$$

Different rates of reaction may occur with various types and concentrations of metallic compounds. For example, a medium metal content feedstock will generally have a lower rate of demetallization than a high metal content feedstock. And, although individual organometallic compounds decompose according to both first- and second-order rate expressions, for reactor design a second-order rate expression is applicable to the decomposition of residuum as a whole.

2.1 Process Configuration

All hydrodesulfurization processes react a hydrocarbon feedstock with hydrogen to produce hydrogen sulfide and a desulfurized hydrocarbon product (Fig. 16.5). The feedstock is preheated and mixed with hot recycle gas containing hydrogen, and the mixture is passed over the catalyst in the reactor section at temperatures of 290–445°C (550–850°F) and pressures of 150–3000 psig (1–20.7 MPa gauge). The reactor effluent is then cooked by heat exchange, and desulfurized liquid hydrocarbon product and recycle gas are separated at essentially the same pressure as that used in the reactor. The recycle gas is then scrubbed and/or purged of the hydrogen sulfide and light hydrocarbon gases, mixed with fresh hydrogen makeup, and preheated prior to being mixed with hot hydrocarbon feedstock.

The recycle gas scheme is used in the hydrodesulfurization process to minimize physical losses of expensive hydrogen. Hydrodesulfurization reactions require a high hydrogen partial pressure in the gas phase to maintain high desulfurization reaction rates and to suppress carbon laydown (catalyst deactivation). The high hydrogen partial pressure is maintained by supplying hydrogen to the reactors at several times the chemical hydrogen consumption rate. The majority of the unreacted hydrogen is cooled to remove hydrocarbons, recovered in the separator, and recycled for further utilization. Hydrogen is physically lost in the process by dissolution in the desulfurized liquid hydrocarbon product and from losses during the scrubbing or purging of hydrogen sulfide and light hydrocarbon gases from the recycle gas.

2.2 Downflow Fixed-Bed Reactor

The reactor design commonly used in hydrodesulfurization of distillates is the fixed-bed reactor design in which the feedstock enters at the top of the reactor and the product leaves at the bottom of the reactor (Fig. 16.9). The catalyst remains in a stationary position (fixed bed), with hydrogen and petroleum feedstock passing in a downflow direction through the catalyst bed. The hydrodesulfurization reaction is exothermic, and the temperature rises from the inlet to the outlet of each catalyst bed. With high hydrogen consumption and subsequent large temperature rise, the reaction mixture can be quenched with cold recycled gas at intermediate points in the reactor system. This is achieved by dividing the catalyst charge into a series of catalyst beds and the quenching effluent from each catalyst bed to the inlet temperature of the next catalyst bed.

The extent of desulfurization is controlled by raising the inlet temperature to each catalyst bed to maintain constant catalyst activity over the course of the process. Fixed-

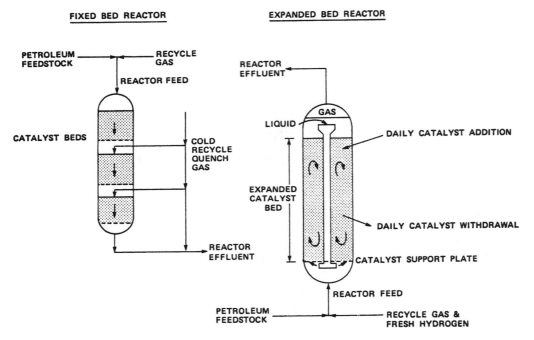

Figure 16.9   Reactor design for hydrodesulfurization processes.

bed reactors are mathematically modeled as plug-flow reactors with very little back-mixing in the catalyst beds. The first catalyst bed is poisoned with vanadium and nickel at the inlet to the bed and may be a cheaper catalyst (*guard bed*). As the catalyst is poisoned in the front of the bed, the temperature exotherm moves down the bed and the activity of the entire catalyst charge declines, thus requiring the reactor temperature to be raised over the course of the process sequence. After catalyst regeneration, the reactors are opened and inspected, and the high metal content catalyst layer at the inlet to the first bed may be discarded and replaced with fresh catalyst. The catalyst loses activity after a series of regenerations, and consequently after a series of regenerations it is necessary to replace the complete catalyst charge. In the case of very high metal content feedstocks (such as residua), it is often necessary to replace the entire catalyst charge rather than regenerate it. This is due to the fact that the metal contaminants cannot be removed by economical means during rapid regeneration and the metals have been reported to interfere with the combustion of carbon and sulfur, catalyzing the conversion of sulfur dioxide (SO_2) to sulfate (SO_4^{2-}), which has a permanent poisoning effect on the catalyst.

Fixed-bed hydrodesulfurization units are generally used for distillate hydrodesulfurization and can also be used for residuum hydrodesulfurization but require special precautions in processing. The residuum must undergo two-stage electrostatic desalting so that salt deposits do not plug the inlet to the first catalyst bed, and the residuum must be low in vanadium and nickel content to avoid plugging the beds with metal deposits. Hence the need for a guard bed in residuum hydrodesulfurization reactors.

During the operation of a fixed-bed reactor, contaminants entering with fresh feed are filtered out and fill the voids between catalyst particles in the bed. The buildup of contaminants in the bed can result in the channeling of reactants through the bed, reducing the hydrodesulfurization efficiency. As the flow pattern becomes distorted or restricted, the pressure drop throughout the catalyst bed increases. If the pressure drop becomes high enough, physical damage to the reactor internals can result. When high pressure drops are observed throughout any portion of the reactor, the unit is shut down and the catalyst bed is skimmed and refilled.

With fixed-bed reactors, a balance must be reached between reaction rate and pressure drop across the catalyst bed. As catalyst particle size is decreased, the desulfurization reaction rate increases but so does the pressure drop across the catalyst bed. Expanded-bed reactors do not have this limitation, and small (1/32 in. 0.8 mm) extrudate catalysts or fine catalysts can be used without increasing the pressure drop.

2.3 Upflow Expanded-Bed Reactor

Expanded-bed reactors are applicable to distillates, but are commercially used for very heavy, high metals content, and/or dirty feedstocks that have extraneous fine solids material. They operate in such a way that the catalyst is in an expanded state so that the extraneous solids pass through the catalyst bed without plugging. They are isothermal, which conveniently handles the high temperature exotherms associated with high hydrogen consumption. Because the catalyst is in an expanded state of motion, it is possible to treat the catalyst as a fluid and to withdraw and add catalyst during operation.

Expanded beds of catalyst (Fig. 16.10) are referred to as particulate fluidized beds, because the feedstock and hydrogen flow upward through an expanded bed of catalyst, with each catalyst particle in independent motion. Thus, the catalyst migrates through the entire reactor bed. Expanded-bed reactors are mathematically modeled as back-mix reactors, with the entire catalyst bed at one uniform temperature. Spent catalyst can be withdrawn and replaced with fresh catalyst on a daily basis. Daily catalyst addition and withdrawal eliminates the need for costly shutdowns to change out catalyst and also results in constant equilibrium catalyst activity and product quality. The catalyst is withdrawn daily with a vanadium, nickel, and carbon content that is representative on a

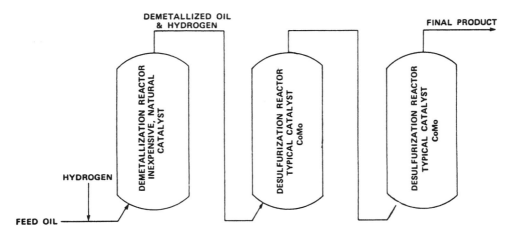

Figure 16.10 Representation of guard bed placement for hydrodemetallization and hydrodesulfurization.

macroscale of what is found throughout the entire reactor. On a microscale, individual catalyst particles have ages from that of fresh catalyst to as old as the initial catalyst charge to the unit, but the catalyst particles of each age group are so well dispersed in the reactor that the reactor contents appear to be uniform.

In the unit (Fig. 16.9), the feedstock and hydrogen recycle gas enter the bottom of the reactor, pass up through the expanded catalyst bed, and leave from the top of the reactor. Commercial expanded-bed reactors normally operate with 1/32 in. (0.8 mm) extrudate catalysts that provide a higher rate of desulfurization than the larger catalyst particles used in fixed-bed reactors. With extrudate catalysts of this size, the upward liquid velocity based on fresh feedstock is not sufficient to keep the catalyst particles in an expanded state. Therefore, for each part of the fresh feed, several parts of product oil are taken from the top of the reactor, recycled internally through a large vertical pipe to the bottom of the reactor, and pumped back up through the expanded catalyst bed. The amount of catalyst bed expansion is controlled by the recycle of product oil back up through the catalyst bed.

The expansion and turbulence of gas and oil passing upward through the expanded catalyst bed are sufficient to cause almost completely random motion in the bed (particulate fluidized). This effect produces the isothermal operation. It also causes almost complete back-mixing. Consequently, in order to effect nearly complete sulfur removal (over 75%), it is necessary to operate with two or more reactors in series. The ability to operate at a single temperature throughout the reactor or reactors and to operate at a selected optimum temperature rather than an increasing temperature from the start to the end of the run results in more effective use of the reactor and catalyst contents. When all these factors are put together—use of a smaller catalyst particle size; isothermal, fixed temperature throughout the run; back-mixing; daily catalyst addition; and constant product quality—the reactor size required for an expanded bed is often smaller than that required for a fixed bed to achieve the same product goals. This is generally true when the feeds have high initial boiling points and/or the hydrogen consumption is very high.

2.4 Demetallization Reactor (Guard Bed Reactor)

Feedstocks that have relatively high metals contents (> 300 ppm) substantially increase catalyst consumption because the metals poison the catalyst, thereby requiring frequent catalyst replacement. The usual desulfurization catalysts are relatively expensive for these consumption rates, but there are catalysts that are relatively inexpensive and can be used in the first reactor to remove a large percentage of the metals. Subsequent reactors downstream of the first reactor would use normal hydrodesulfurization catalysts. Because the catalyst materials are proprietary, it is not possible to identify them here. However, it is understood that such catalysts contain little or no metal promoters, i.e., nickel, cobalt, molybdenum. Metals removal on the order of 90% has been observed with these materials.

Thus, one method of controlling demetallization is to employ separate smaller *guard reactors* just ahead of the fixed-bed hydrodesulfurization reactor section. The preheated feed and hydrogen pass through the guard reactors, which are filled with an appropriate catalyst for demetallization that is often the same as the catalyst used in the hydrodesulfurization section. The advantage of this system is that it enables replacement of the most contaminated catalyst (guard bed), where pressure drop is highest, without having to replace the entire inventory or shut down the unit. The feedstock is alternated between guard reactors while catalyst in the idle guard reactor is being replaced.

When the expanded-bed design is used, the first reactor could employ a low cost catalyst (5% of the cost of Co/Mo catalyst) to remove the metals, and subsequent reactors can use the more selective hydrodesulfurization catalyst (Fig. 16.10). The demetallization catalyst can be added continuously without taking the reactor out of service, and the spent demetallization catalyst can be loaded to more than 30% vanadium, which makes it a valuable source of vanadium.

2.5 Catalysts

Hydrodesulfurization catalysts consist of metals impregnated on a porous alumina support. Almost all of the surface area is found in the pores of the alumina (200–300 m^2/g), and the metals are dispersed in a thin layer over the entire alumina surface within the pores. This type of catalyst displays a huge catalytic surface for a small weight of catalyst.

Cobalt (Co), molybdenum (Mo), and nickel (Ni) are the most commonly used metals for desulfurization catalysts. The catalysts are manufactured with the metals in an oxide state. In the active form they are in the sulfide state, which is obtained by sulfiding the catalyst either prior to use or with the feed during use. Any catalyst that exhibits hydrogenation activity will catalyze hydrodesulfurization to some extent. However, the group VIB metals (chromium, molybdenum, and tungsten) are particularly active for desulfurization, especially when promoted with metals from the iron group (iron, cobalt, nickel).

Cobalt-molybdenum catalysts are by far the most popular choice for desulfurization, particularly for straight-run petroleum fractions. Nickel-molybdenum catalysts are often chosen instead of cobalt-molybdenum catalysts when higher activity is required for saturation of polynuclear aromatic compounds or removal of nitrogen or when more refractory sulfur compounds such as those in cracked feedstocks must be desulfurized. In some applications, nickel-cobalt-molybdenum (Ni-Co-Mo) catalysts appear to offer a useful balance of hydrotreating activity. Nickel-tungsten (Ni-W) is usually chosen only when very high activity for aromatics saturation is required along with activity for sulfur and nitrogen removal. Catalysts are available in several different compositions (Table 16.4).

Cobalt-molybdenum (Co-Mo) and nickel-molybdenum (Ni-Mo) catalysts resist poisoning and are the most universally applied catalysts for hydrodesulfurization of everything from naphthas to residua. In addition, they promote both demetallization and desulfurization. The vanadium deposition rate at a given desulfurization level is a function of the pore structure of the alumina support and the types of metals on the support (Fig. 16.11), whereas a catalyst support that has small pores preferentially removes sulfur with a low degree of demetallization (Fig. 16.12).

Table 16.4 Representative Metal Content of Catalyst

	Co-Mo	Ni-Mo	Ni-Co-Mo	Ni-W
Metal (wt%)				
Cobalt	2.5		1.5	
Nickel		2.5	2.3	4.0
Molybdenum	10.0	10.0	11.0	
Tungsten				16.0

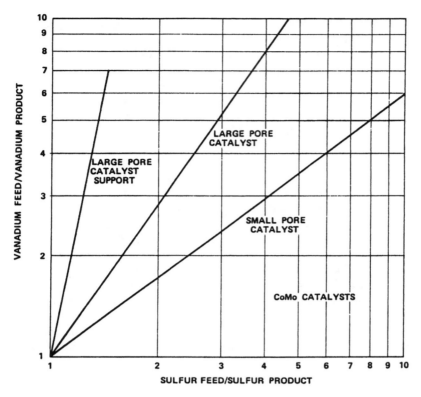

Figure 16.11 General relationship between vanadium and sulfur removal for different catalysts.

3. DISTILLATE HYDRODESULFURIZATION

3.1 Processes

Hydrotreating (catalytic hydrodesulfurization) of naphtha is widely applied to prepare charge for catalytic reforming and isomerization processes. It is accomplished by passing a feedstock together with hydrogen over a fixed catalyst bed at elevated temperature and pressure (Fig. 16.13). Although the main purpose is to remove sulfur, denitrogenation, deoxygenation, and olefin saturation reactions occur simultaneously with desulfurization. These reactions are also beneficial, because the noble metal catalysts used in reforming and isomerization can be poisoned by olefins, oxygen, and nitrogen compounds as well as by sulfur in the feedstocks. Sulfur and nitrogen limitations can in some cases be 0.5 ppm or less. Failure to adequately remove these contaminants can lead to poor yields, low catalyst activity, and brief catalyst life.

In terms of specific processes, *hydrofining* can be used to upgrade low quality, high sulfur naphtha, thus increasing the supply of catalytic reformer feedstock, solvent naphtha, and other naphtha-type materials. The sulfur content of kerosene can be reduced, improving the color, odor, and wick-char characteristics of the kerosene. The tendency of kerosene to form smoke is not affected, because aromatics, which cause smoke, are not affected by the mild hydrofining conditions. Cracked gas oil with a high

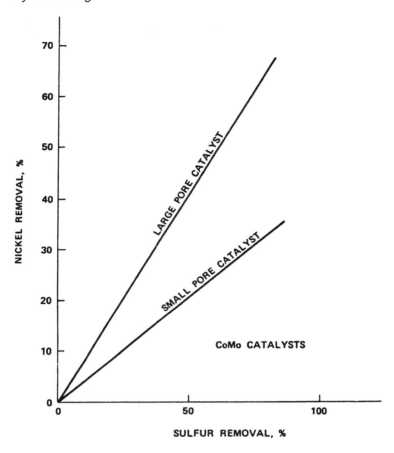

Figure 16.12 General relationship between nickel and sulfur removal for different catalysts.

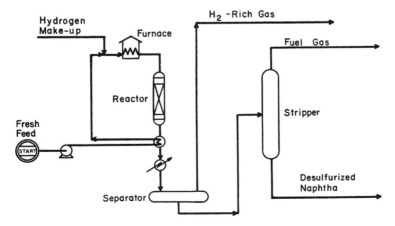

Figure 16.13 Representation of a naphtha hydrotreater.

sulfur content can be converted to excellent fuel oil and diesel fuel by reducing its sulfur content and eliminating the components that form gum and carbon residues.

This process can be applied to lubricating oil, naphtha, and gas oil. The feedstock is heated in a furnace and passed with hydrogen through a reactor containing a suitable metal oxide catalyst such as cobalt and molybdenum oxides or alumina (Fig. 16.14). Hydrogen is obtained from catalytic reforming units. Reactor operating conditions range from 205°C to 425°C (400–800°F) and from 50 to 800 psi, depending on the kind of feedstock and the degree of treatment required. Higher boiling feedstocks, high sulfur content, and maximum sulfur removal require higher temperatures and pressures.

After passing through the reactor, the treated oil is cooled and separated from the excess hydrogen, which is recycled through the reactor. The treated oil is pumped to a stripper tower, where hydrogen sulfide formed by the hydrogenation reaction is removed by steam, vacuum, or flue gas, and the finished product leaves the bottom of the stripper tower. The catalyst is not usually regenerated; it is replaced after about a year's use.

The *autofining* process differs from other hydrorefining processes in that an external source of hydrogen is not required. Sufficient hydrogen to convert sulfur to hydrogen sulfide is obtained by dehydrogenation of naphthenes in the feedstock. The processing equipment is similar to that used in hydrofining (Fig. 16.15). The catalyst is cobalt and molybdenum oxides on alumina, and operating conditions are usually 340–425°C (650–800°F) at pressures of 100–200 psi. Hydrogen formed by dehydrogenation of naphthenes in the reactor is separated from the treated oil and is then recycled through the reactor. The catalyst is regenerated with steam and air at 200–1000 h intervals, depending on whether light or heavy feedstocks have been processed. The process is used for the same purpose as hydrofining but is limited to fractions with endpoints no higher than 370°C (700°F).

Early desulfurization processes, such as treatment with caustic, amine, and clay, were not as successful with middle distillates as they were with naphthas and lighter feedstocks. More sophisticated extraction processes utilizing hydrogen fluoride or sulfur dioxide contacting were somewhat more applicable to the removal of sulfur from diesel, gas oils, and cycle oils. Such processes did produce satisfactory distillate fuels, although substantial reductions in yield were inherent.

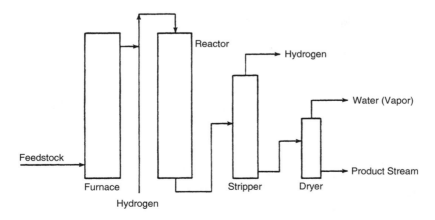

Figure 16.14 The hydrofining process.

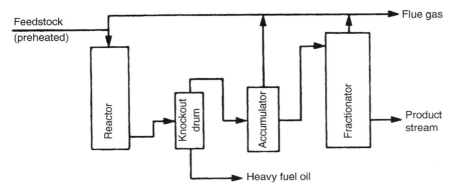

Figure 16.15 The autofining process.

The availability of cheaper hydrogen from catalytic reforming led to the development of many middle distillate hydrotreating (hydrodesulfurization) processes in the 1950s. These processes result in superior treatment of distillates to improve color, odor, corrosion properties, thermal stability, and burning characteristics in addition to accomplishing essentially complete removal of sulfur.

In a typical middle distillate hydrodesulfurization process (Fig. 16.16), the feedstock is mixed with fresh and recycled hydrogen and heated under pressure to the proper reactor temperature. The feedstock–hydrogen mixture is charged to the reactor, passing downflow through the catalyst. In the reactor, fresh feed is hydrotreated, and a limited amount of hydrogenation, isomerization, and cracking occur to produce a small amount of C_1–C_5 paraffins. In addition, sulfur compounds are converted to hydrogen sulfide and nitrogen compounds are converted to ammonia. Olefins are also saturated. These reactions are

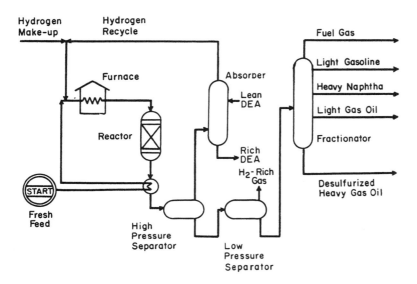

Figure 16.16 Representation of a gas oil hydrotreater.

exothermic, and, in the cases of vacuum gas oils or unsaturated feedstocks, reactor temperature is regulated by the use of cold recycle gas quench.

Reactor effluent is cooled and enters the high pressure separator, where the oil is separated from the hydrogen sulfide and hydrogen-rich gas. Hydrogen sulfide is scrubbed from the gas (optional for light distillate units), and the hydrogen-rich gas is recycled. The liquid is passed through a low pressure separator and stripper to remove the remaining light ends and dissolved hydrogen sulfide. Fractionation of the liquid product is sometimes employed, especially on higher boiling feedstocks, and cracked stocks would show upgrading and product slate similar to that from gas oil (Table 16.5) but would entail much greater hydrogen consumption.

The *unifining* process is a regenerative, fixed-bed, catalytic process to desulfurize and hydrogenate refinery distillates of any boiling range. Contaminating metals, nitrogen compounds, and oxygen compounds are eliminated, along with sulfur. The catalyst is a cobalt-molybdenum-alumina type that can be regenerated in situ with steam and air.

Ultrafining is a regenerative, fixed-bed, catalytic process to desulfurize and hydrogenate refinery stocks from naphtha up to and including lubricating oil. The catalyst is cobalt-molybdenum on alumina and can be regenerated in situ using an air–steam mixture. Regeneration requires 10–20 h and can be repeated 50–100 times for a given batch of catalyst; catalyst life is 2–5 years depending on the feedstock.

The *Isomax* process is a two-stage, fixed-bed catalyst system (Fig. 16.17) that operates, for example, under hydrogen pressures of 500–1500 psi in a temperature range of 205–3 70°C (400–700°F) with middle distillate feedstocks. Exact conditions depend on the feedstock and product requirements, and hydrogen consumption is on the order of 1000–1600 ft^3/bbl of feed processed. Each stage has a separate hydrogen recycling system. Conversion

Table 16.5 Yield and Product Properties for Desulfurization of a Kuwait Crude Oil Vacuum Gas Oil[a]

Yield, % of charge (average for cycle)			
H$_2$S, wt%		2.64	
NH$_3$, wt%		0.03	
C$_1$–C$_4$, wt%		0.70	
C$_5$–375°F, vol%		1.08	
375–650°F, vol%		12.05	
650°F + gas oil, vol%		87.74	

Property	Feed	C$_5$–375°F	375–650°F	650°F$^+$
Gravity, °API	24.1	51.0	35.0	29.0
Sulfur, wt%	2.60	< 0.01	0.02	0.13
Nitrogen, wt%	0.07	—	0.005	0.05
Aniline point, °F	174.4	—	—	186.3
Distillation, °F				
10%	682	—	358	659
50%	810	—	464	791
90%	961	—	600	954

[a] Hydrogen consumption, 380 scf/bbl.

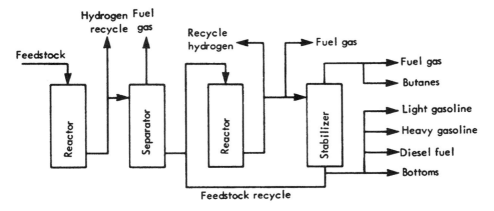

Figure 16.17 The Isomax process.

can be balanced to provide products for variable requirements, and recycling can be taken to extinction if necessary. Fractionation can also be handled in a number of different ways to yield desired products.

3.2 Process Parameters

The operating conditions in distillate hydrodesulfurization are dependent upon the stock to be charged as well as the desired degree of desulfurization or quality improvement. Kerosene and light gas oils are generally processed at mild severity and high throughput, whereas light catalytic cycle oils and thermal distillates require slightly more severe conditions. Higher boiling distillates, such as vacuum gas oils and lube oil extracts, require the most severe conditions (Table 16.6).

The principal variables affecting the required severity in distillate desulfurization are

1. Hydrogen partial pressure
2. Space velocity
3. Reaction temperature
4. Feedstock properties

3.2.1 Hydrogen Partial Pressure

The important effect of *hydrogen partial pressure* is the minimization of coking reactions. If the hydrogen pressure is too low for the required duty at any position within the

Table 16.6 Representative Process Parameters for Hydrotreating Distillates

	Kerosene and light gas oils	Heavy gas oils and lube oil extracts
Total pressure, psi gauge	100–1000	500–1500
Reactor temperature, °F	450–800	650–800
Space velocity, V/h/V	2–10	1–3

reaction system, premature aging of the remaining portion of catalyst will be encountered. In addition, the effect of hydrogen pressure on desulfurization varies with feed boiling range. For a given feed there exists a threshold level above which hydrogen pressure is beneficial to the desired desulfurization reaction. Below this level, desulfurization drops off rapidly as hydrogen pressure is reduced.

3.2.2 Space Velocity

As the *space velocity* is increased, desulfurization is decreased, but increasing the hydrogen partial pressure and/or the reactor temperature can offset the detrimental effect of increasing space velocity.

3.2.3 Reaction Temperature

A higher *reaction temperature* increases the rate of desulfurization at constant feed rate, and the start-of-run temperature is set by the design desulfurization level, space velocity, and hydrogen partial pressure. The capability to increase temperature as the catalyst deactivates is built into most process or unit designs. Temperatures of 415°C (780°F) and above result in excessive coking reactions and higher than normal catalyst aging rates. Therefore, units are designed to avoid the use of such temperatures for any significant part of the cycle life.

3.2.4 Catalyst Life

Catalyst life depends on the charge stock properties and the degree of desulfurization desired. The only permanent poisons to the catalyst are metals in the feedstock that deposit on the catalyst, usually quantitatively, causing permanent deactivation as they accumulate. However, this is usually of little concern except when deasphalted oils are used as feedstocks, because most distillate feedstocks contain low amounts of metals. Nitrogen compounds are a temporary poison to the catalyst but there is essentially no effect on catalyst aging except that caused by a higher temperature requirement to achieve the desired desulfurization. Hydrogen sulfide can be a temporary poison in the reactor gas, and recycle gas scrubbing is employed to counteract this condition.

Provided that pressure drop buildup is avoided, cycles of 1 year or more and ultimate catalyst life of 3 years or more can be expected. The catalyst, employed can be regenerated by normal steam–air or recycle combustion gas–air procedures. The catalyst is restored to nearly fresh activity by regeneration during the early part of its ultimate life. However, permanent deactivation of the catalyst occurs slowly during use and repeated regeneration, so replacement becomes necessary.

3.2.5 Feedstock Effects

The character of the *feedstock properties*, especially the feed boiling range, has a definite effect on the ultimate design of the desulfurization unit and process flow. In agreement, there is a definite relationship between the percent by weight sulfur in the feedstock and the hydrogen requirements (Fig. 16.18) (Maples, 2000).

In addition, the reaction rate constant in the kinetic relationships decreases rapidly with increasing average boiling point in the kerosene and light gas oil range but much more slowly in the heavy gas oil range. This is attributed to the difficulty in removing sulfur from ring structures present in the entire heavy gas oil boiling range.

The hydrodesulfurization of light (low-boiling) distillate (naphtha) is one of the more common catalytic hydrodesulfurization processes bcause it is usually used as a pretreat-

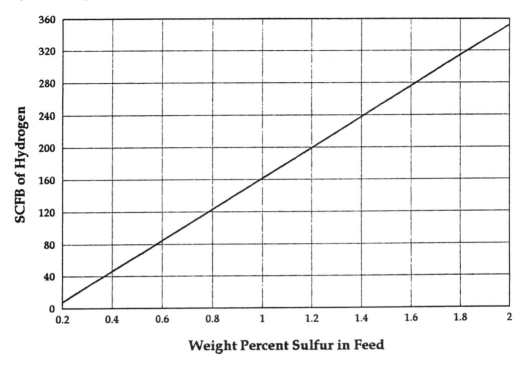

Figure 16.18   General relationship of feedstock sulfur to hydrogen requirements.

ment of such feedstocks prior to catalytic reforming. Hydrodesulfurization of such feed-stocks is required because sulfur compounds poison the precious metal catalysts used in reforming, and desulfurization can be achieved under relatively mild conditions and is nearly quantitative (Table 16.7). If the feedstock arises from a cracking operation, hydro-desulfurization will be accompanied by some degree of saturation, resulting in increased hydrogen consumption.

The hydrodesulfurization of low boiling (naphtha) feedstocks is usually a gas-phase reaction and may employ the catalyst in fixed beds, and (with all of the reactants in the

Table 16.7   Hydrodesulfurization of Various Naphtha Fractions

Feedstock	Boiling range		Sulfur (wt%)	Desulfurization (%)
	°C	°F		
Visbreaker naphtha	65–220	150–430	1.00	90
Visbreaker-coker naphtha	65–220	150–430	1.03	85
Straight-run naphtha	85–170	185–340	0.04	99
Catalytic naphtha (light)	95–175	200–350	0.18	89
Catalytic naphtha (heavy)	120–225	250–440	0.24	71
Thermal naphtha (heavy)	150–230	300–450	0.28	57

[a] Process conditions: Co-Mo on alumina, 260–370°C (500–700°F), 200–500 psi (1380–3440 kPa) hydrogen.

gaseous phase) only minimal diffusion problems are encountered within the catalyst pore system. It is, however, important that the feedstock be completely volatilized before it enters the reactor, because there may be the possibility of pressure variations (leading to less satisfactory results) if some of the feedstock enters the reactor in the liquid phase and is vaporized within the reactor.

In applications of this type, the sulfur content of the feedstock may vary from 100 ppm to 1%, and the necessary degree of desulfurization to be effected by the treatment may vary from as little as 50% to more than 99%. If the sulfur content of the feedstock is particularly low, it will be necessary to presulfide the catalyst. For example, if the feedstock has only 100–200 ppm sulfur, several days may be required to sulfide the catalyst as an integral part of the desulfurization process even with complete reaction of all of the feedstock sulfur to, say, cobalt and molybdenum (catalyst) sulfides. In such a case, presulfiding can be conveniently achieved by adding sulfur compounds to the feedstock or by adding hydrogen sulfide to the hydrogen.

Generally, hydrodesulfurization of naphtha feedstocks to produce catalytic reforming feedstocks is carried to the point where the desulfurized feedstock contains less than 20 ppm sulfur. The net hydrogen produced by the reforming operation may actually be sufficient to provide the hydrogen consumed in the desulfurization process.

The hydrodesulfurization of middle distillates is also an efficient process (Table 16.8), and applications include predominantly the desulfurization of kerosene, diesel fuel, jet fuel, and heating oils that boil over the general range 250–400°C (480–750°F). However, with this type of feedstock, hydrogenation of the higher boiling catalytic cracking feedstocks has become increasingly important where hydrodesulfurization is accomplished alongside the saturation of condensed-ring aromatic compounds as an aid to subsequent processing.

Under the relatively mild processing conditions used for the hydrodesulfurization of these particular feedstocks, it is difficult to achieve complete vaporization of the feed. Process conditions may dictate that only part of the feedstock be in the vapor phase and that sufficient liquid phase be maintained in the catalyst bed to carry the larger molecular constituents of the feedstock through the bed. If the amount of liquid phase is insufficient for this purpose, molecular stagnation (leading to carbon deposition on the catalyst) will occur.

Hydrodesulfurization of middle distillates causes a more marked change in the specific gravity of the feedstock, and the amount of low-boiling material is much more significant than with the naphtha-type feedstock. In addition, the somewhat more severe reaction conditions (leading to a designated degree of hydrocracking) also lead to an overall increase in hydrogen consumption when middle distillates are employed as feedstocks in place of the naphtha.

High-boiling distillates, such as the atmospheric and vacuum gas oils, are not usually produced as refinery products but merely serve as feedstocks to other processes for conversion to lower boiling materials. For example, gas oils can be desulfurized to remove more than 80% of the sulfur originally in the gas oil, with some conversion of the gas oil to lower boiling materials (Table 16.9). The treated gas oil (which has less carbon residue as well as lower sulfur and nitrogen contents than the untreated material) can then be converted to lower boiling products in, say, a catalytic cracker, where improvements in catalyst life and volumetric yield may be noted.

Table 16.8 Hydrodesulfurization of Middle Distillates

	Straight-run middle distillate	Cracked middle distillate
Feedstock		
Specific gravity	0.844	0.901
°API	36.2	25.5
ASTM distillation, °C		
IBP	238	227
10 vol%	265	242
50 vol%	288	256
90 vol%	332	277
EP	368	293
Sulfur, wt%	1.20	2.34
Aniline point, °C	76.0	32.4
Pour point, °C	−7	−28
Conradson carbon (10% residuum)	0.024	0.34
Viscosity, cSt (38°C)	4.14	2.38
Process conditions		
Catalyst	Cobalt molybdate on alumina	Cobalt molybdate on alumina type
Temperature, °C	315–430	315–430
Pressure, psi	250–1000	250–1000
Product		
Specific gravity	0.832	0.876
°API	38.6	30.0
ASTM distillation, °C		
IBP	222	203
10 vol%	259	227
50 vol%	287	250
90 vol%	330	275
EP	336	293
Sulfur, wt%	0.09	0.35
Aniline point, °C	79.5	36.0
Pour point, °C	−6	−28
Conradson carbon (10% residuum)	0.022	0.05
Viscosity, cSt (38°C)	3.77	2.22

The conditions used for the hydrodesulfurization of a gas oil may be somewhat more severe than those employed for the hydrodesulfurization of middle distillates with, of course, the feedstock in the liquid phase.

In summary, the hydrodesulfurization of the low-, middle-, and high-boiling distillates can be achieved quite conveniently using a variety of processes. One major advantage of this type of feedstock is that the catalyst does not become poisoned by metal contaminants in the feedstock, because only negligible amounts of these contaminants will be present. Thus, the catalyst may be regenerated several times, and on-stream times between catalyst regeneration (while varying with the process conditions and application) may be of the order of 3–4 years (Table 16.10).

Table 16.9 Hydrodesulfurization of Gas Oil

Feedstock		
Source	Kuwait	Khafji
Boiling range, °C	370–595	370–595
Specific gravity	0.935	0.929
°API	19.8	20.2
Sulfur, wt%	3.25	3.05
Process conditions		
Temperature, °C		370–430
Pressure, psi		100–500
Hydrogen consumption, scf/bbl	420	400
Product		
Naphtha, C_5–205°C		
Yield, vol%	1.7	1.7
Specific gravity	0.802	0.802
°API	45.0	45.0
Sulfur, wt%	0.02	0.02
High boilers (> 205°C)		
Yield, vol%	99.6	99.5
Specific gravity	0.897	0.893
°API	26.3	27.0
Sulfur, wt%	0.50	0.48

4. RESIDUUM HYDRODESULFURIZATION

4.1 Processes

Advances made in hydrotreating processes have made the utilization of heavier feedstock almost a common practice for many refineries. Upgrading processes can be used for the upgrading of atmospheric (650°F⁺, 345°C⁺) and vacuum (1050°F⁺, 565°C⁺) residua, heavy oil, and tar sand bitumen. However, there is no hydroprocessing process that can be universally applicable to the upgrading of all heavy feedstocks. As a result, several hydroprocessing processes have been developed for different commercial applications, and many other processes are in their development stages.

In refining heavy feedstocks, hydrodesulfurization (HDS) and hydrodemetallization (HDM) processes are used to reduce or eliminate poisoning of sophisticated and expensive catalysts that are used in the downstream refining steps [e.g., fluid catalytic cracking (FCC), reforming, and hydrotreating]. The hydrodemetallization process is a pretreatment process for the heavy feedstock by which metals and part of the heteroatomic contaminants are removed along with conversion of the residue to a lighter fractions. The hydrodemetallization process uses low cost catalysts in either a fixed-bed or moving-bed reactor operating at moderate temperatures and pressures (580–2900 psi; 4–20 MPa) and at relatively high liquid hourly space velocity (LHSV).

Processes for the direct desulfurization of residua (Fig. 16.19) have a flow similar to that of distillate hydrodesulfurization but with distinguishing features such as the catalyst compositions and shapes employed. Examples of such processes are the RDS/VRDS process for hydrotreating atmospheric residua and vacuum residua (Fig. 16.20) and the

Table 16.10 Process Parameters for Hydrodesulfurization of Various Feedstocks

| Feedstock | Boiling range | | | Process condition | | | | | | | Catalyst life | |
| | | | Temperature | | Hydrogen pressure | | Hydrogen rate (scf/bbl) | LHSV | | | | |
	°C	°F	°C	°F	psi	kg/cm^{-2}					Months	bbl/lb
Naphtha	70–170	160–340	300–370	570–700	100–450	7–31.5	250–1500	5–8			36–48	500–1200
Kerosene	160–240	320–465	330–370	625–700	150–500	10.5–35	500–1500	4–6			36–48	300–600
Gas oil	240–350	465–660	340–400	645–750	150–700	10.5–49	1000–2000	2–6			36–48	200–400
Vacuum gas oil	350–650	660–1200	360–400	680–750	450–800	31.5–56	1000–4000	1–3			36–48	50–350
Residua	> 650	> 1200	370–450	700–840	750–2250	52.5–157.5	1500–10,000	0.5–2			12–24	2–50

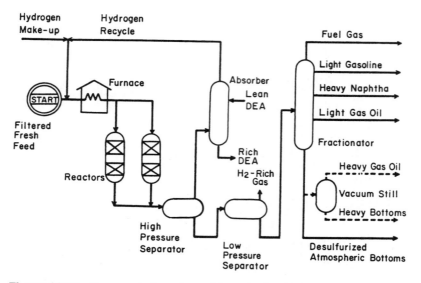

Figure 16.19 Representation of a residuum hydrotreater.

residfining process (Fig. 16.21) (a derivative of the hydrofining process) (Brossard, 1997; *Hydrocarbon Processing*, 1998, pp. 90–96; Speight, 2000).

The RDS and VRDS processes are designed to remove sulfur, nitrogen, asphaltene, and metal contaminants from residua and are also capable of accepting whole crude oils or topped crude oils as feedstocks (Brossard, 1997; *Hydrocarbon Processing*, 1998). The major product of the processes is a low sulfur fuel oil, and the amounts of gasoline and middle distillates are maintained at a minimum to conserve hydrogen. The basic elements of each process are similar (Fig. 16.20), consisting of once-through treatment of the feed-

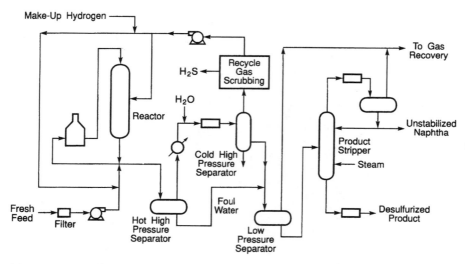

Figure 16.20 Representation of the RDS and VRDS processes.

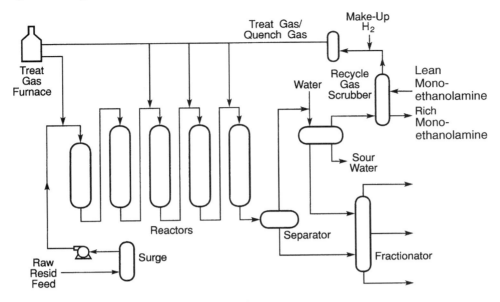

Figure 16.21 Representation of the residfining process.

stock coming into contact with hydrogen and the catalyst in a downflow reactor that is designed to maintain activity and selectivity in the presence of deposited metals. Moderate temperatures and pressures are employed to reduce the incidence of hydrocracking and hence minimize production of low-boiling distillates (Table 16.11). The combination of a desulfurization step and a vacuum residuum desulfurizer (VRDS) is often seen as an

Table 16.11 Process Data for the RDS and VRDS Processes

	Arabian light		Arabian heavy	
	Atmospheric	Vacuum	Atmospheric	Vacuum
Feed properties				
Gravity, °API	17.7	6.5	16.8	6.1
Sulfur, wt%	3.2	4.1	3.9	5.1
Distillation, vol%				
Boiling below 650°F	5	—	16	—
Boiling below 1000°F	67	1	60	10
H_2 consumption (chemical), scf/bbl	560	750	780	960
Product yields				
$\leq C_4$, wt%	3.5	—	4.7	—
C_5–350°F, vol%	1.4	5.7	1.4	6.4
350–650°F, vol%	11.1		25.0	
650–1000°F, vol%	64.4	18.1	43.4	26.6
1000°F+, vol%	24.3	78.4	31.4	69.9
Sulfur content, wt%				
650°F+	0.46	1.0	0.55	1.0
1000°F+	0.97	1.2	0.88	1.25

attractive alternative to the atmospheric residuum desulfurizer (RDS) because the combination route uses less hydrogen for a similar investment cost.

Both the RDS and VRDS processes can be coupled with other processes (such as delayed coking, fluid catalytic cracking, and solvent deasphalting) to achieve the most optimum refining performance.

The residfining process is a catalytic fixed-bed process for the desulfurization and demetallization of residua (Table 16.12) (*Hydrocarbon Processing*, 1998, p. 94). The process can also be used to pretreat residua to suitably low contaminant levels prior to catalytic cracking. In the process (Fig. 16.21), liquid feed to the unit is filtered, pumped to pressure, preheated, and combined with treat gas prior to entering the reactors. A small guard reactor would typically be employed to prevent plugging or fouling of the main reactors. Provisions are employed to periodically remove the guard while keeping the main reactors on-line. The temperature rise associated with the exothermic reactions is controlled by using either a gas or liquid quench. A train of separators is employed to separate the gas and liquid products. The recycle gas is scrubbed to remove ammonia and H_2S. It is then combined with fresh makeup hydrogen before being reheated and combined with fresh feed. The liquid product is sent to a fractionator, where the product is fractionated.

The different catalysts allow other minor differences in operating conditions and peripheral equipment. Primary differences include the use of higher purity hydrogen makeup gas (usually 95% or greater), inclusion of filtration equipment in most cases, and facilities to upgrade the offgases to maintain higher concentration of hydrogen in the recycle gas. Most of the processes utilize downflow operation over fixed-bed catalyst systems, but exceptions to this are the H-Oil and LC-Fining processes (which are predominantly conversion processes), which employ upflow designs and ebullating catalyst systems with continuous catalyst removal capability, and the Shell Process (a conversion process), which may involve the use of a *bunker flow* reactor (Fig. 16.22) ahead of the main reactors to allow periodic changeover of catalyst.

The primary objective in most of the residue desulfurization processes is to remove sulfur with minimum consumption of hydrogen. Substantial percentages of nitrogen, oxygen, and metals are also removed from the feedstock. However, complete elimination of other reactions is not feasible, and hydrocracking, thermal cracking, and aromatic saturation reactions also occur to some extent. Certain processes, e.g., H-Oil (Fig. 16.23), which

Table 16.12   Residfining Data for Two Atmospheric Residua

	Gach Saran ($650°F^+$)	Arabian heavy ($650°F^+$)
Feedstock		
°API	15.0	12.3
Sulfur, wt%	2.50	4.19
Products		
°API	19.6	20.7
Sulfur, wt%	0.3	0.3
$C_5/400°F$, vol%	3.4	6.0
$400°F^+$, vol%	98.1	96.4
Chemical H_2 consumption, scf/bbl		

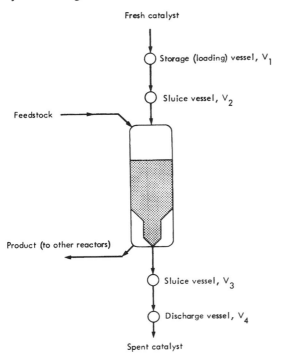

Figure 16.22 Representation of the bunker reactor for a residuum hydrotreater.

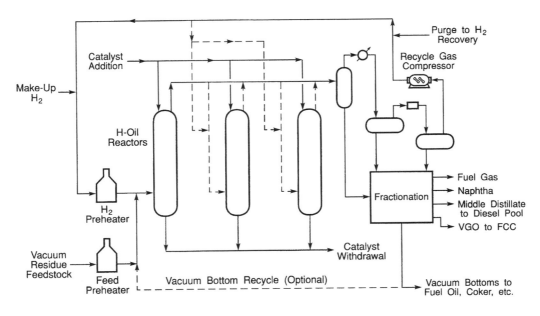

Figure 16.23 The H-Oil process.

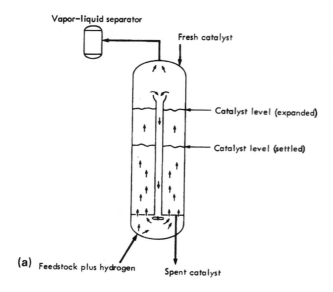

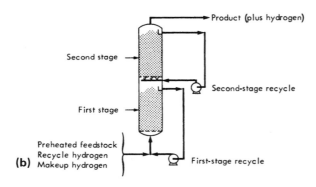

Figure 16.24   (a) Single-stage and (b) two-stage reactors for H-Oil process.

uses a single-stage or two-stage reactor (Fig. 16.24), and LC-Fining (Fig. 16.25), which uses an expanded-bed reactor, can be designed to accomplish greater amounts of hydrocracking to yield larger quantities of lighter distillates at the expense of desulfurization.

Removal of nitrogen is much more difficult than removal of sulfur. For example, nitrogen removal may be only about 25–30% when sulfur removal is at a 75–80% level. Metals are removed from the feedstock in substantial quantities and are mainly deposited on the catalyst surface, existing as metal sulfides at processing conditions. As these deposits accumulate, the catalyst pores eventually become blocked and inaccessible; thus catalyst activity is lost.

Desulfurization of residua is considerably more difficult than desulfurization of distillates (including vacuum gas oil), because many more contaminants are present and very large, complex molecules are involved. The most difficult portion of feed in residue desulfurization is the asphaltene fraction, which forms coke readily, and it is essential that these large molecules be prevented from condensing with each other to form coke, which deac-

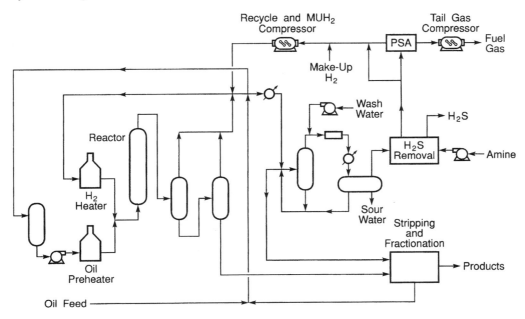

Figure 16.25 The LC-Fining process.

tivates the catalyst. This is accomplished by selecting proper catalysts, using adequate hydrogen partial pressure, and ensuring intimate contact between the hydrogen-rich gases and oil molecules in the process design.

4.2 Process Parameters

4.2.1 Catalyst Types

The general *catalyst types* used in residuum desulfurization are combinations of metal oxides on alumina (Al_2O_3) or silica-stabilized alumina (SiO_2-Al_2O_3) supports. Molybdenum always seems to be one of the metals, with cobalt and/or nickel being used in combination with the molybdenum in many cases. The supports are usually tailored to the process objectives, because different supports can be made to accomplish particular goals of a specific unit. For example, smaller pored catalysts will tend to remove less metals than larger pored catalysts and be more active for desulfurization reactions. However, the metal-holding capacity of the small-pored catalyst will also be less than that of the large-pored catalyst, which results in a sacrifice of catalyst life. The better catalyst selection would depend on the combination of these two characteristics that will allow the best life-activity relationship for a given application. In some cases, critical combinations of large and small pore sizes are used to arrive at the best catalyst for a given feedstock and operating conditions. The objectives of the process unit are also important, because metals removal would be more critical in an application to produce residual feedstock for cata-lytic cracking than in an application to produce residual fuel.

Catalyst size and shape are also important factors in residue desulfurization processes. Smaller size contributes to improved desulfurization and demetallization, but pressure drop considerations become more important.

4.2.2 Metals Accumulation

Many of the conventional design criteria in distillate desulfurization (hydrogen partial pressure, degree of desulfurization, gas circulation rates) must be considered in residuum desulfurization, and an additional important criterion is the effect of *metals accumulation* on the catalyst. The effective life of a particular catalyst will vary depending on its pore structure and total pore volume. It is also dependent upon the particular feedstock being processed and the operating conditions employed. As mentioned previously, many of the competitive processes use different catalyst characteristics that have been tailored to achieve the objectives of most concern to the individual licensor. Thus some processes remove more metals from the feedstock and others reject metals to a greater extent.

In general terms, catalysts that show better selectivity for metals removal will also hold more total metals before they become inoperable for the required desulfurization duty. This holding capacity for metals has been defined as a *saturation level* that increases with decreasing size for a given catalyst. However, the selectivity for demetallization over desulfurization reactions also increases with decreasing catalyst particle size. The combination of these effects results in an optimum particle size to maximize the cycle life.

4.2.3 Catalyst Activity

A gradual loss of *catalyst activity* occurs during normal operation of the residue process. Therefore a gradual increase in catalyst temperature is required throughout the cycle to maintain the desired sulfur content in the product. This loss in activity is caused by deposition of coke and metals (from nickel and vanadium in the feedstock) on the catalyst surface and in the catalyst pores. Ultimate *catalyst life* is directly related to the total metals tolerance of the catalyst, which is a function of particle size and shape and pore size and volume. Deposited metals cause permanent deactivation of the catalyst and preclude the restoration of catalyst activity by normal regeneration procedures. Spent catalysts are either discarded or returned to reclaimers for recovery of the various metals of value. The amount of coke deposited on the catalyst depends primarily on hydrogen partial pressure but is also influenced by the asphaltene content of the feedstock. Higher hydrogen pressure decreases coking, whereas higher asphaltene content increases coking.

In addition to hydrogen partial pressure having an effect on *catalyst aging*, adequate amounts of hydrogen must be maintained within the system to prevent undesirable side reactions. Circulation rates and purity requirements are set to avoid a shortage of hydrogen anywhere in the reaction system.

4.2.4 Temperature and Space Velocity

Temperature and *space velocity* are very important variables that influence the operation of the process. In a given system, a reduction in space velocity without an appropriate reduction in temperature will result in feedstock overtreating. This will lead to irreversible premature aging of the catalyst by virtue of increased coking and incremental metals deposition.

Because catalyst temperature is increased over the length of an operating cycle, both yields of lighter materials and properties of the remaining feedstock are affected. The magnitude of the variations will depend on the catalyst selected for the operation and the operating conditions employed (Table 16.13). The product properties affected to the greatest extent are viscosity and pour point, and, together with changes in distillate yields, their values indicate that cracking reactions are increasing as the run progresses (due to temperature increases, which lead to a greater degree of thermal cracking).

Table 16.13 Yields and Properties for Desulfurization of Kuwaiti Crude Oil Atmospheric Residuum

Yield (% of HDS charge)	Feed	1% Sulfur 650°F+ fuel[a]	0.3% Sulfur 375°F+ fuel[a]	0.1% Sulfur 375°F+ fuel[a]
H₂S, wt%		3.07–3.14 (3.10)	3.73–3.74 (3.73)	3.93–3.94 (3.93)
NH₃, wt%		0.08–0.07 (0.08)	0.13–0.12 (0.12)	0.17–0.17 (0.17)
C₁–C₄, wt%		0.27–1.10 (0.62)	0.33–1.67 (0.89)	0.40–2.07 (1.14)
C₅–375°F naphtha, vol%		1.4–4.1 (2.6)	2.0–6.4 (3.8)	2.5–7.6 (4.6)
375–650°F distillate, vol%		6.6–11.2 (8.5)	8.9–17.4 (12.5)	9.1–20.8 (14.0)
650°F+ residue, vol%		92.7–86.5 (90.1)	90.4–78.9 (85.6)	89.9–74.6 (83.5)
Chemical hydrogen consumption, scf/bbl		(497)	(650)	(725)
Residue product properties				
Gravity, °API	16.6	20.5–22.1 (21.2)	22.0–23.9 (22.8)	22.5–24.4 (23.3)
Sulfur, wt%	3.8	1.0	0.32–0.35 (0.33)	0.10–0.11 (0.11)
Carbon residue, wt%	9.0	5.6–6.1 (5.8)	3.6–4.1 (3.8)	3.0–3.6 (3.2)
Nitrogen, wt%	0.22	0.17–0.19 (0.18)	0.13–0.15 (0.14)	0.09–0.11 (0.10)
Pour point, °F	+60	+65–+55 (+60)	+70–+20 (+35)	+60–0 (+15)
Nickel, ppm	15	4.5–4.8 (4.6)	1.3–1.6 (1.5)	0.4–0.5 (0.5)
Vanadium, ppm	45	7.9–8.6 (8.2)	2.4–2.8 (2.6)	1.0–1.3 (1.2)
Viscosity, SUS at 210°F	250	122–88 (108)	104–70 (90)	94–64 (81)

[a] Values in parentheses indicate average values for cycle.

4.2.5 Feedstock Effects

The problems encountered in hydrotreating heavy feedstocks can be related directly to the amount of higher boiling constituents (Speight, 2000). Processing these feedstocks is not just a matter of applying know-how derived from the refining of conventional crude oils, which often uses hydrogen-to-carbon (H/C) atomic ratios as the main criterion for determining process options, but also requires knowledge of several other properties (Table 16.14). The materials are complex not only in terms of the carbon number and boiling point ranges but also because a large part of this *envelope* (Fig. 16.26) falls into a range in which very little is known about model compounds. It is also established that the majority of the higher molecular weight materials produce coke (with some liquids), whereas the majority of the lower molecular weight constituents produce liquids (with some coke). Hydrotreating is concerned with the latter.

Thus, there is the potential for the application of more efficient conversion processes to heavy feedstock refining. Hydroprocessing, i.e., hydrotreating (in the present context), is probably the most versatile petroleum refining process because of its applicability to a wide range of feedstocks. In fact, hydrotreating can be applied to the removal of heteroatoms from heavier feedstocks as a first step in preparing the feedstock for other process options.

In practice, the reactions that are used to chemically define the processes i.e., hydrodesulfurization and hydrodenitrogenation, and the hydrocracking reactions (Chapter XX) can occur (or be encouraged to occur) in any particular process. Thus, hydrodesulfurization is in all likelihood accompanied by a degree of hydrocracking as determined by the refiner, thereby producing not only products that are low in sulfur but also low-boiling products. Thus, the choice of processing schemes for a given hydroprocess depends upon the nature of the feedstock as well as the product requirements (Nasution, 1986; Suchanek and Moore, 1986). The process can be simply illustrated as a single-stage or two-stage operation (Fig. 16.5).

The single-stage process can be used to produce gasoline but is more often used to produce middle distillate from heavy vacuum gas oils and may be used to remove the heteroatoms from residua with a specified degree of hydrocracking. The two-stage process was developed primarily to produce high yields of gasoline from straight-run gas oils, and the first stage may be a purification step to remove sulfur-containing and nitrogen-containing organic materials. Both processes use an extinction–recycle technique to maximize the yields of the desired product. Significant conversion of heavy feedstocks can be accomplished by hydrocracking at high severity (Howell et al., 1985). For some applications, the products boiling up to 340°C (650°F) can be blended to give the desired final product.

In reality, no single bottom-of-the-barrel processing scheme is always the best choice. Refiners must consider the potential of proven processes, evaluate the promise of newer ones, and choose one according to the situation. The best selection will always depend on the kind of crude oil, the market for products, and financial and environmental considerations. Although there are no simple solutions, the available established processes and the growing number of new ones under development offer some reasonable choices. The issue then becomes how to most effectively handle the asphaltene fraction of the feedstock at the most reasonable cost. Solutions to this processing issue can be separated into two broad categories: (1) conversion of asphaltenes into another, salable product and (2) concentration of the asphaltenes into a marketable, or usable, product such as asphalt.

Table 16.14 Properties of Residua

Feedstock	Gravity (°API)	Sulfur (wt%)	Nitrogen (wt%)	Nickel (ppm)	Vanadium (ppm)	Asphaltenes (heptane) (wt%)	Carbon residue (Conradson) (wt%)
Arabian light, > 650°F	17.7	3.0	0.2	10.0	26.0	1.8	7.5
Arabian light, > 1050°F	8.5	4.4	0.5	24.0	66.0	4.3	14.2
Arabian heavy, > 650°F	11.9	4.4	0.3	27.0	103.0	8.0	14.0
Arabian heavy, > 1050°F	7.3	5.1	0.3	40.0	174.0	10.0	19.0
Alaska, North Slope, > 650°F	15.2	1.6	0.4	18.0	30.0	2.0	8.5
Alaska, North Slope, > 1050°F	8.2	2.2	0.6	47.0	82.0	4.0	18.0
Lloydminster (Canada), > 650°F	10.3	4.1	0.3	65.0	141.0	14.0	12.1
Lloydminster (Canada), > 1050°F	8.5	4.4	0.6	115.0	252.0	18.0	21.4
Kuwait, > 650°F	13.9	4.4	0.3	14.0	50.0	2.4	12.2
Kuwait, > 1050°F	5.5	5.5	0.4	32.0	102.0	7.1	23.1
Tia Juana, > 650°F	17.3	1.8	0.3	25.0	185.0		9.3
Tia Juana, > 1050°F	7.1	2.6	0.6	64.0	450.0		21.6
Taching, > 650°F	27.3	0.2	0.2	5.0	1.0	4.4	3.8
Taching, > 1050°F	21.5	0.3	0.4	9.0	2.0	7.6	7.9
Maya, > 650°F	10.5	4.4	0.5	70.0	370.0	16.0	15.0

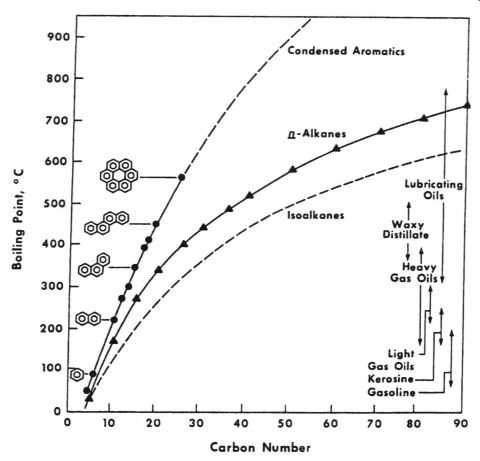

Figure 16.26 Relationship of carbon number and boiling range for various organic compounds.

The hydrodesulfurization process variables (Speight, 2000) usually require some modification to accommodate the various feedstocks that are submitted for this particular aspect of refinery processing (Table 16.10). The main point of this section is to outline the hydrotreating process with particular reference to the heavier oils and residua. However, some reference to the lighter feedstocks is warranted. This will serve as a base point to indicate the necessary requirements for heavy oil and residuum hydrodesulfurization.

One particular aspect of the hydrotreating process that needs careful monitoring, with respect to feedstock type, is the exothermic nature of the reaction. The heat of the reaction is proportional to the hydrogen consumption and with the more saturated lower boiling feedstocks where hydrocracking may be virtually eliminated, the overall heat production during the reaction may be small, leading to a more controllable temperature profile. However, with heavy feedstocks where hydrogen consumption is appreciable (either by virtue of the hydrocracking that is necessary to produce a usable product or by virtue of the extensive hydrodesulfurization that must occur), it may be desirable to provide internal cooling of the reactor. This can be accomplished by introducing cold recycle gas to the catalyst bed to compensate for excessive heat. One other generalization may apply to the lower boiling feedstocks in the hydrodesulfurization process. The process may actually have very little effect on the properties of the feedstock (assuming that hydrocracking reactions are negligible), removal of sulfur will cause some drop in specific gravity, which could give rise to volume recoveries approaching (or even above) 100%. Furthermore, with the assumption that cracking reactions are minimal, there may be a slight lowering of the boiling range due to sulfur removal from the feedstock constituents. However, the production of lighter fractions is usually small and may amount to only about 1–5% by weight of the products boiling below the initial boiling point of the feedstock.

One consideration for heavy feedstocks is that it may be more economical to hydrotreat and desulfurize high sulfur feedstocks before catalytic cracking than to hydrotreat the products after. This approach (Speight and Moschopedis, 1979; Decroocq, 1984; Speight, 2000) has the potential for several advantages; for example,

1. The products require less finishing.
2. Sulfur is removed from the catalytic cracking feedstock, and corrosion is reduced in the cracking unit.
3. Coke formation is reduced.
4. Feedstock conversions are higher.
5. There is a potential for better quality products.

The downside is that many of the heavier feedstocks act as hydrogen sinks in terms of their ability to interact with the expensive hydrogen. A balance of the economic advantages and disadvantages must be struck on an individual feedstock basis.

In terms of the feedstock composition, it must be recognized that when catalytic processes are employed for heavy feedstocks, complex molecules (such as those that may be found in the original asphaltene fraction or those formed during the process) are not sufficiently mobile. They are also too strongly adsorbed by the catalyst to be saturated by the hydrogenation component and therefore continue to react and eventually degrade to coke. These deposits deactivate the catalyst sites and eventually interfere with the hydroprocess by causing a decrease in the relative rate of hydrodesulfurization (Fig. 16.27) (Skripek et al., 1975).

Heavy feedstocks, such as residua, require more severe hydrodesulfurization conditions to produce low sulfur liquid product streams that can then, as is often now desired, be employed as feedstocks for other refining operations. Hydrodesulfurization of the heavier feedstocks is normally accompanied by a high degree of hydrocracking, and thus the process conditions required to achieve 70–90% desulfurization will also effect substantial conversion of the feedstock to lower boiling products. In addition, the extent of hydrodesulfurization of heavy feedstocks is dependent upon the temperature, and the reaction rate increases with increase in temperature.

In contrast to the lighter feedstocks that may be subjected to the hydrodesulfurization operation, the process catalysts are usually susceptible to poisoning by nitrogen (and oxygen) compounds and metallic salts (in addition to the various sulfur compounds), which tend to be concentrated in residua or exist as an integral part of the heavy oil matrix. Thus, any processing sequence devised to hydrodesulfurize a residuum must be

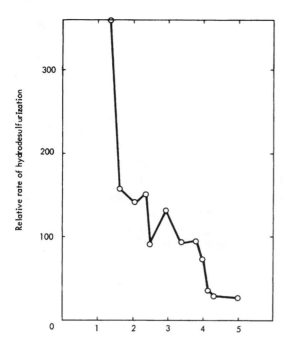

Figure 16.27 Relationship of hydrodesulfurization rate to feedstock sulfur.

capable of accommodating the constituents that adversely affect the ability of the catalyst to function in the most efficient manner possible.

The conditions employed for the hydrodesulfurization of the heavier feedstocks may be similar to those applied to the hydrodesulfurization of gas oil fractions but often with higher pressures. However, carbon deposition on, and metal contamination of, the catalyst is much greater when residua and heavy oils are employed as feedstocks, and unless a low level of desulfurization is acceptable, frequent catalyst regeneration is necessary.

A wide choice of commercial processes are available for the catalytic hydro-desulfurization of residua. The suitability of any particular process depends not only on the nature of the feedstock but also on the degree of desulfurization that is required. There is also a dependence on the relative amounts of the lower boiling products that are to be produced as feedstocks for further refining and generation of liquid fuels.

One aspect of feedstock properties has not yet been discussed fully, and that is feed-stock composition. This particular aspect of the nature of the feedstock is, in fact, related to the previous section, where the influence of various feedstock types on the hydro-desulfurization process was noted, but it is especially relevant when residua and heavy oils from various sources are to be desulfurized.

One of the major drawbacks to defining the influence of the feedstock on the process is that research with respect to feedstocks has been fragmented. In every case, a conventional catalyst has been used, and the results obtained are valid only for the particular operating conditions, reactor system, and catalyst used in that case.

More rigorous correlation is required and there is a need to determine the optimum temperature for each type of sulfur compound. To obtain a useful model, the intrinsic kinetics of the reaction for a given catalyst should also be known. In addition, other factors that influence the desulfurization process, such as (1) catalyst inhibition or deactivation by hydrogen sulfide, (2) effect of nitrogen compounds, and (3) the effects of various solvents, should also be included in order to obtain a comprehensive model that is independent of the feedstock. The efficacy of other catalytic systems on various feedstocks also needs to be evaluated.

Residua and other heavy feedstocks contain impurities other than sulfur, nitrogen, and oxygen, and the most troublesome of these impurities are the organometallic compounds of nickel and vanadium. The metals content of a residuum can vary from several parts per million to more than 1000 ppm (Table 16.15), and there does seem to be a more than a chance relationship between the metals content of a feedstock and its physical properties (Reynolds, 1998; Speight, 1999, 2000). In the hydrodesulfurization of the heavier feedstocks, the metals (nickel plus vanadium) are an important factor, because large amounts (over 150 ppm) will cause rapid deterioration of the catalyst. The free metals or their sulfides deposit on the surface of the catalyst and within its pores, thereby poisoning the catalyst by making the more active catalyst sites inaccessible to the feedstock and the hydrogen. This results in frequent replacement of an expensive process commodity unless there are adequate means by which the catalyst can be regenerated.

The problem of metal deposition on the hydrodesulfurization catalysts has generally been addressed using one of three methods:

1. Suppressing deposition of the metals on the catalyst
2. Development of a catalyst that will accept the metals and can tolerate high levels of metals without marked reduction in its hydrodesulfurization capabilities
3. Removal of the metal contaminants before the hydrodesulfurization step

The first two of these methods involve a careful and deliberate choice of the process catalyst as well as the operating conditions. However, these methods may be viable only for feedstocks with less than 150 ppm total metals, because the decrease in catalyst activity is directly proportional to the metals content of the feedstock. There are, however, catalysts that can tolerate substantial proportions of metals within their porous structure before the desulfurizing capability drops to an unsatisfactory level. Unfortunately, data on such catalysts are extremely limited because of their proprietary nature, and details are not always available, but tolerance levels for metals that are equivalent to 15–65% by weight of the catalyst have been quoted.

The third method may be especially applicable to feedstocks with a high metals content and requires a separate demetallization step just prior to the hydrodesulfurization reactor by use of a guard bed reactor. Such a step might involve passage of the feedstock through a demetallization chamber that contains a catalyst with a high selectivity for metals but whose activity for sulfur removal is low. Nevertheless, demetallization applied as a separate process can be used to generate low metal feedstocks and will allow a more active and more stable desulfurization system so that a high degree of desulfurization can be achieved on high metal feedstocks with an acceptable duration of operation (Table 16.16) (Fig. 16.10).

Table 16.15 Sulfur Content and Metal Content of Various Feedstocks

Feedstock	°API	V + Ni (ppm)	% S
Alaska, Simpson	~ 4	100–212	2.4–3.1
Alaska, Hurl State	~ 6	72–164	1.0–1.9
Arabian light, atmospheric	16.1–17.3	34–42	2.8–3.0
Arabian light, vacuum	6.1–8.2	101–112	4.0–4.2
Arabian light, vacuum	11.2	71	3.71
Arabian medium, atmospheric	~ 15.0	51–140	4.0–4.3
Arabian heavy, atmospheric	11.0–17.2	102–140	4.2–4.6
Safaniya	11.0–12.7	102–130	4.3
Khursaniya	15.1	41	4.0
Khafji, atmospheric	14.7–15.7	95–125	4.0–4.3
Khafji, vacuum	5.0	252	5.4
Wafra, Eocene	9.7	125	5.6
Wafra, Ratawi	14.2	100	4.7–5.2
Agha Jari	16.6	92–110	2.3–2.5
Agha Jari, vacuum	7.5	274	3.8
Iranian light, vacuum	9.5	150–316	3.1–3.2
Iranian heavy, atmospheric	15.0–19.8	179–221	2.5–2.8
Alaska, Kuparup	~ 11	64–153	1.5–2.4
Alaska, Put River	~ 10	34–83	1.5–2.1
Alaska, Sag River	~ 8	52–129	1.6–2.2
California, Coastal	4.3	300	2.3
California, atmospheric	13.7	134	1.73
California, deasphalted	13.8	24	0.95
Delta Louisiana	11.3	25	0.93
Mid-continent, vacuum	14.5	25	1.2
Wyoming, vacuum	6.3	292	4.56
Salamanca, Mexico	5.4–10.7	182–250	3.2–3.8
Gach Saran	15.8–17.0	210–220	2.4–2.6
Gach Saran, vacuum	9.2–10.0	329–359	3.1–3.3
Darius, vacuum	12.8	61	4.7
Sassan, atmospheric	18.3	33	3.3
Kuwait, atmospheric	10.2–16.7	60–65	3.8–4.1
Kuwait, vacuum	5.5–8.3	110–199	5.1–5.8
Zubair, Iraq	18.1	43	3.3
Khurais	14.6	31	3.3
Murban	24.0–25.0	2–10	1.5–1.6
Qatar marine	25.2	35	2.77
Middle East, vacuum	8.3–17.6	40–110	3.0–5.8
Pilon (Tar belt)	9.7	608	3.92

Feedstock	°API	V + Ni (ppm)	% S
Monogus (Tar belt)	12.0	254	2.5
Morichal (Tar belt)	9.6–12.4	233–468	2.1–4.1
Melones II (Tar belt)	9.9	424	3.3
Tucupita	16.0	129	1.03
Bachaquero (Lake)	5.8–14.6	415–970	2.2–3.5
Boscan (Lake)	9.5–10.4	1185–1750	5.2–5.9
Lagunillas (Lake)	~ 16.5	271–400	2.2–2.6
Taparita (Lake)	16.9	490	2.4
Tia Juana (lake)	6.5–7.8	380–705	2.4–3.5
Tia Juana (Lake)	~ 8.0	327–534	2.6–4.0
Urdaneta	11.7	480	2.68
Venezuela, atmospheric	11.8–17.2	200–460	2.1–2.8
Venezuela, vacuum	4.5–7.5	690–760	2.9–3.2
West Texas, atmospheric	15.4–17.9	39–40	2.5–3.7
West Texas, vacuum	7.3–10.5	65–88	3.2–4.6

Table 16.16 Process Data for Desulfurization and Demetallization of a Residuum

Feedstock data	Type of operation	
	Desulfurization	Demetallization/desulfurization
Gravity, °API	11.8	12.6
Sulfur, wt%	2.8	2.8
Vanadium, ppm	375	398
Nickel, ppm	55	57
975°F$^+$, vol%	55	55

Product yield	Desulfurization			Demetallization/ desulfurization		
	Yield (vol%)	Sulfur (wt%)	Vanadium (ppm)	Yield (vol%)	Sulfur (wt%)	Vanadium (ppm)
C$_4$ endpoint	103.5	0.64	170	103.3	0.64	88
Chemical H$_2$ consumption, scf/bbl	—	720	—	—	680	—
Desulfurization, %[a]	—	77	—	—	77	—

[a] For two Co-Mo reactors in series.

REFERENCES

Aalund, L. R. 1975. Oil Gas J. 77(35):339.

Aalund, L. R. 1981. Oil Gas J. 79(13):70; 79(37):69.

Boening, L. G., McDaniel, N. K., Petersen, R. D., and van Dreisen, R. P. 1987. Hydrocarbon Process. 66(9):59.

Bridge, A. G. 1997. In: Handbook of Petroleum Refining Processes. R. A. Meyers, ed. McGraw-Hill, New York, Chapter 14.1.

Bridge, A. G., Reed, E. M., and Scott, J. W. 1975. Paper presented at the API Midyear Meeting, May.

Bridge, A. G., Gould, G. D., and Berkman, J. F. 1981. Oil Gas J. 79(3):85.

Brossard, D. N. 1997. In: Petroleum Refining Processes. 2nd ed. R. A. Meyers, ed. McGraw-Hill, New York, Chapter 8.1.

Celestinos, J. A., Zermeno, R. G., Van Dreisen, R. P., and Wysocki, E. D. 1975. Oil Gas J. 73(48):127.

Corbett, R. A. 1989. Oil Gas J. 87(26):33.

DeCroocq, D. 1984. Catalytic Cracking of Heavy Petroleum Hydrocarbons. Editions Technip, Paris.

Dolbear, G. E. 1998. In: Petroleum Chemistry and Refining. J. G. Speight, ed. Taylor & Francis, Washington, DC, Chapter 7.

Howell, R. L., Hung, C., Gibson, K. R., and Chen, H. C. 1985. Oil Gas J. 83(30):121.

Hydrocarbon Processing. 1998. 57(11): p. 53 et seq.

Khan, M. R. 1998. In: Petroleum Chemistry and Refining. J. G. Speight, ed. Taylor & Francis, Washington, DC, Chapter 6.

Maples, R. E. 2000. Petroleum Refinery Process Economics. 2nd ed. PennWell Corp., Tulsa, OK.

McConaghy, J. R. 1987. U.S. Patent 4,698,147.

Murphy, J. R., and Treese, S. A. 1979. Oil Gas J. 77(26):135.

Nasution, A. S. 1986. Preprints. Div. Petrol. Chem. Am. Chem. Soc. 31(3):722.

Nelson, W. L. 1977. Oil Gas J., Feb. 28, p. 127.

Radler, M. 1999. Oil Gas J. 97(51):445 et seq.

Reynolds, J. G. 1998. In: Petroleum Chemistry and Refining. J. G. Speight, ed. Taylor & Francis, Washington, DC, Chapter 3.

Reynolds, J. G., and Beret, S. 1989. Fuel Sci. Technol. Int. 7:165.

Scherzer, J., and Gruia, A. J. 1996. Hydrocracking Science and Technology. Marcel Dekker, New York.

Scott, J. W., and Bridge, A. G. 1971. In: Origin and Refining of Petroleum. H. G. McGrath and M. E. Charles, eds. Adv. Chem. Ser. 103. Am. Chem. Soc., Washington, DC, p. 113.

Skripek, M., Alley, S. K., Helfrey, P. F., and Leuth, P. F. 1975. Proceedings 79th National Meeting. American Institute of Chemical Engineers.

Speight, J. G. 1986. Annu. Rev. Energy 11:253.

Speight, J. G. 1999. The Chemistry and Technology of Petroleum. 3rd ed. Marcel Dekker, New York.

Speight, J. G. 2000. The Desulfurization of Heavy Oils and Residua. 2nd ed. Marcel Dekker, New York.

Speight, J. G., and Moschopedis, S. E. 1979. Fuel Process. Technol. 2:295.

Stanislaus, A., and Cooper, B. H. 1994. Catal. Rev.—Sci. Eng. 36(1):75.

Suchanek, A. J., and Moore, A. S. 1986. Oil Gas J. 84(31):36.

Wilson, J. 1985. Energy Proc. 5:61.

17

Hydrocracking

1. INTRODUCTION

Hydrocracking is a refining technology that, like hydrotreating (Chapter 16), falls under the general umbrella of *hydroprocessing*. The outcome is the conversion of a variety of feedstocks to a range of products (Table 17.1), and units to accomplish this goal can be found at various points in a refinery (Fig. 17.1). Throughout this chapter reference will be cited and cross references made to other chapters, as dictated by need for comparisons and the context of the issue being discussed.

Hydrocracking is a more recently developed process than thermal cracking, visbreaking, and coking. In fact, the use of hydrogen in thermal processes was perhaps the single most significant advance in refining technology during the twentieth century (Bridge et al., 1981; Scherzer and Gruia, 1996; Dolbear, 1998). As noted in Chapter 16, hydrotreating, which includes catalytic hydrorefining (approximately 8,500,000 bbl/day), and is dedicated to catalytic hydrotreating (approximately 28,100,000 bbl/day), is a substantial portion of the total worldwide daily refinery capacity (approximately 81,500,000 bbl of oil). On a national basis, U.S. data are approximately 1,780,000 bbl/day hydrorefining and approximately 1,900,000 bbl/day catalytic hydrotreating out of a daily total refining capacity of approximately 16,500,000 bbl/day (Radler, 1999). Add to these the respective worldwide and national totals of catalytic hydrocracking (approximately 4,020,000 bbl/day worldwide and approximately 1,423,000 bbl/day nationally), and the importance of hydrocracking is placed into correct perspective as an extremely important aspect of refinery technology.

The ability of refiners to cope with the renewed trend toward distillate production from heavier feedstocks with low atomic hydrogen/carbon ratios (Fig. 17.2) has created a renewed interest in hydrocracking. Without the required conversion units, heavier crude oils produce lower yields of naphtha and middle distillate. To maintain current gasoline and middle distillate production levels, additional conversion capacity is required because of the differences between the amount of distillates produced from light crude oil and the amount of distillate products produced from heavier crude oil (Fig. 17.3).

The concept of hydrocracking allows the refiner to produce products having a lower molecular weight with a higher hydrogen content and a lower yield of coke. In summary,

Table 17.1 Refinery Processes That Employ Hydrocracking

Feedstock	Process characteristics						Products
	Hydro-cracking	Aromatics removal	Sulfur removal	Nitrogen removal	Metals removal	Olefins removal	
Naphtha			✓	✓		✓	Reformer feedstock Liquefield petroleum gas (LPG)
Gas oil	✓						Naphtha
Atmospheric	✓	✓	✓	✓			Catalytic cracker feedstock Diesel fuel Kerosene
Vacuum	✓	✓	✓		✓		Jet fuel Naphtha LPG Lubricating oil
Residuum	✓	✓					Diesel fuel (others)

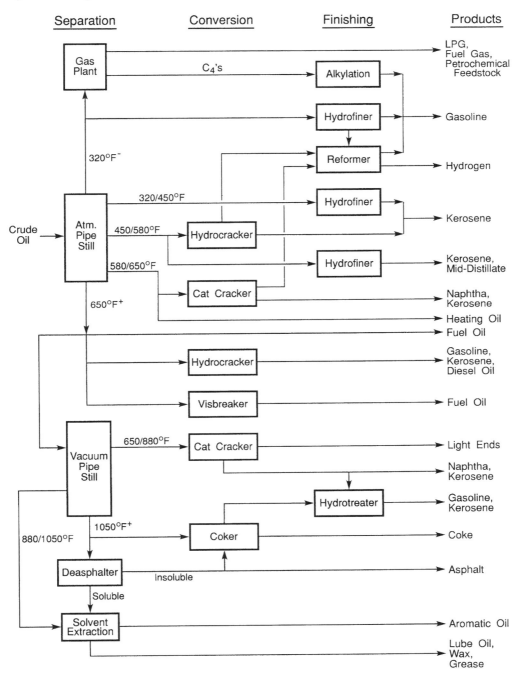

Figure 17.1 Refinery schematic showing relative positions of hydrocracking units.

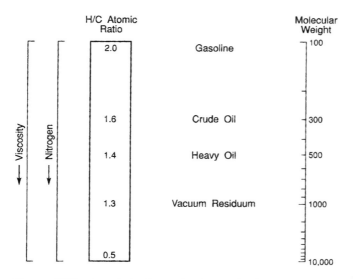

Figure 17.2 Atomic hydrogen/carbon ratios for various feedstocks.

hydrocracking facilities add flexibility to refinery processing and to the product slate. Hydrocracking is more severe than hydrotreating (Chapter 16), there being the intent, in hydrocracking processes, to convert the feedstock to lower boiling products rather than to treat the feedstock only to remove heteroatoms and metals. Process parameters (Figs. 17.4 and 17.5) emphasize the relatively severe nature of the hydrocracking process.

The older hydrogenolysis type of hydrocracking practiced in Europe during and after World War II used tungsten sulfide (WS_2) or molybdenum sulfide (MoS) as catalysts. These processes required high reaction temperatures and operating pressures, sometimes in excess of about 3000 psi (20,684 kPa), for continuous operation. The modern hydrocracking

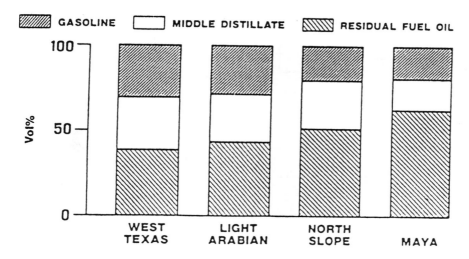

Figure 17.3 Hydrocracking product yields from various feedstocks.

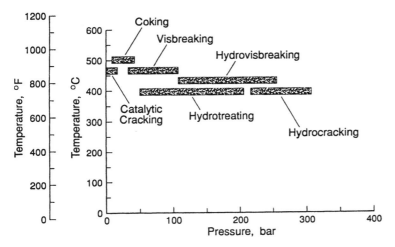

Figure 17.4 Temperature and pressure parameters for various processes.

processes were initially developed for converting refractory feedstocks to gasoline and jet fuel; process and catalyst improvements and modifications have made it possible to yield products ranging from gases and naphtha to furnace oils and catalytic cracking feedstocks.

Hydrocracking is an extremely versatile process that can be utilized in many different ways, and one of its advantages is its ability to break down the high-boiling aromatic stocks produced by catalytic cracking or coking. To take full advantage of hydrocracking, the process must be integrated in the refinery with other process units (Fig. 17.1). In gasoline production, for example, the hydrocracker product must be further processed in a catalytic reformer because it has a high naphthene content and relatively low octane number. The high naphthene content makes the hydrocracker gasoline an excellent feed for catalytic reforming, and good yields of high octane number gasoline can be obtained.

If high molecular weight petroleum fractions are *pyrolyzed*, that is, if no hydrogenation occurs, progressive cracking and condensation reactions generally lead to the final products. These products are usually

1. Gaseous and low-boiling liquid compounds of high hydrogen content

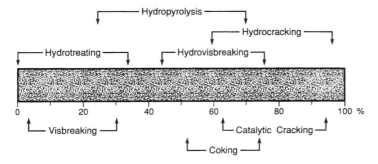

Figure 17.5 Feedstock conversion in various processes.

2. Liquid material of intermediate molecule weight with a hydrogen/carbon atomic ratio differing more or less from that of the original feedstock, depending on the method of operation

3. Material of high molecular weight, such as coke, possessing a lower hydrogen/ carbon atomic ratio than the starting material

Highly aromatic or refractory recycle stocks or gas oils that contain varying proportions of highly condensed aromatic structures (e.g., naphthalene and phenanthrene) usually crack, in the absence of hydrogen, to yield intractable residues and coke.

An essential difference between *pyrolysis* (thermal decomposition, usually in the absence of any added agent) and *hydrogenolysis* (thermal decomposition in the presence of hydrogen and a catalyst or a hydrogen-donating solvent) of petroleum is that in pyrolysis a certain amount of polymerized heavier products such as cracked residuum and coke is always formed along with the light products such as gas and gasoline. During hydrogenolysis (*destructive hydrogenation*), condensation to coke may be partly or even entirely prevented so that only light products are formed. The prevention of coke formation usually results in an increased distillate (e.g., gasoline) yield. The condensed type of molecule, such as naphthalene or phenanthrene, is one that is closely associated with the formation of coke, but in an atmosphere of hydrogen and in contact with catalysts these condensed molecules are converted into lower molecular weight saturated compounds that boil within the gasoline range.

The mechanism of hydrocracking is basically similar to that of catalytic cracking but with concurrent hydrogenation. The catalyst assists in the production of carbonium ions via olefin intermediates, and these intermediates are quickly hydrogenated under the high hydrogen partial pressures employed in hydrocracking. The rapid hydrogenation prevents adsorption of olefins on the catalyst and hence prevents their subsequent dehydrogenation, which ultimately leads to coke formation so that long on-stream times can be obtained without the necessity of catalyst regeneration.

One of the most important reactions in hydrocracking is the partial hydrogenation of polycyclic aromatics followed by rupture of the saturated rings to form substituted monocyclic aromatics. The side chains may then be split off to give isoparaffins. It is desirable to avoid excessive hydrogenation activity of the catalyst so that the monocyclic aromatics become hydrogenated to naphthenes; furthermore, repeated hydrogenation leads to loss in octane number, which increases the catalytic reforming required to process the hydrocracked naphtha.

Side chains of three or four carbon atoms are easily removed from an aromatic ring during catalytic cracking, but the reaction of aromatic rings with shorter side chains appears to be quite different. For example, hydrocracking single-ring aromatics containing four or more methyl groups produces largely isobutane and benzene. It may be that successive isomerization of the feed molecule adsorbed on the catalyst occurs until a four-carbon side chain is formed, which then breaks off to yield isobutane and benzene. Overall, coke formation is very low in hydrocracking, because the secondary reactions and the formation of the precursors to coke are suppressed as the hydrogen pressure is increased.

When applied to residua, the hydrocracking process can be used for processes such as (1) fuel oil desulfurization and (2) conversion of residuum to lower boiling distillates.

The products from hydrocracking are composed of either saturated or aromatic compounds; no olefins are found. In making gasoline, the lower paraffins formed have high octane numbers; for example, the five- and six-carbon fractions have leaded research octane numbers of 99–100. The remaining gasoline has excellent properties as a feed to catalytic reforming, producing a highly aromatic gasoline that is capable of having a high octane number. Both types of gasoline are suitable for premium grade gasoline. Another attractive feature of hydrocracking is the low yield of gaseous components such as methane, ethane, and propane, which are less desirable than gasoline. When making jet fuel, more hydrogenation activity of the catalysts is used, because jet fuel contains more saturates than gasoline.

Like many refinery processes, the problems encountered in hydrocracking heavy feedstocks can be directly equated to the amount of complex, higher boiling constituents that may require pretreatment (Speight and Moschopedis, 1979; Reynolds and Beret, 1989; Gray, 1994; Speight, 1999, 2000). Processing these feedstocks is not merely a matter of applying know-how derived from refining *conventional* crude oils; it also requires a knowledge of composition (Chapter 3). The materials are complex not only in terms of the carbon number and boiling point ranges (Fig. 17.6) but also because a range of model compounds fall into this envelope (Fig. 17.7), and very little is known about the materials' properties. It is also established that the majority of the higher molecular weight materials produce coke with some liquids, but the majority of the lower molecular weight constituents produce liquids with some coke. Hydrocracking is applicable to both of these groups (Fig. 17.8).

The physical and chemical composition of a feedstock plays a large part not only in determining the nature of the products that arise from refining operations but also in determining the precise manner by which a particular feedstock should be processed (Speight, 1986). Furthermore, it is apparent that the conversion of heavy oils and residua requires new lines of thought to develop suitable processing scenarios (Celestinos et al., 1975). Indeed, the use of thermal (*carbon rejection*) processes and of hydrothermal (*hydro-*

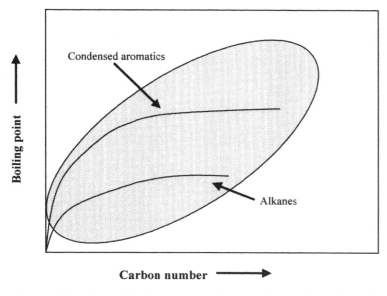

Figure 17.6 General boiling point–carbon number envelope for crude oil.

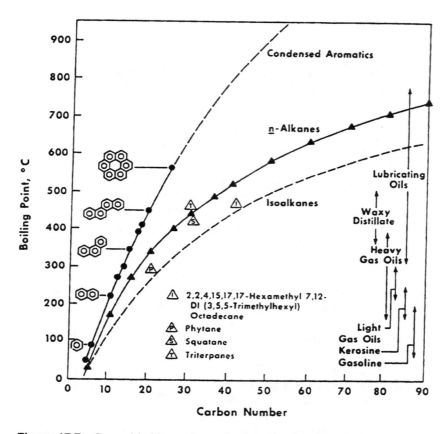

Figure 17.7 General boiling point–carbon number envelope for pure compounds showing crude oil distillation fractions.

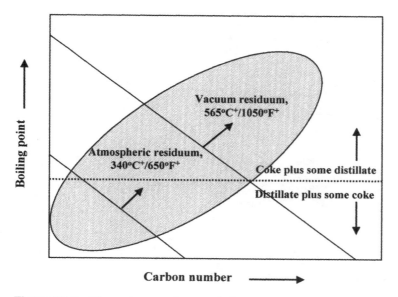

Figure 17.8 Thermal processing trends for crude oil fractions.

492

gen addition) processes, which were inherent in the refineries designed to process lighter feedstocks, have been a particular cause for concern. This has brought about the evolution of processing schemes that accommodate the heavier feedstocks (Chapter 18) (Corbett, 1989; Benton et al., 1986; Boening et al., 1987; Wilson, 1985; Khan and Patmore, 1998). As a point of reference, an example of the former option is the delayed coking process in which the feedstock is converted to overhead with the concurrent deposition of coke, such as the one used by Suncor, Inc., at their oil sands plant (Chapter 14) (Speight, 1990).

The hydrogen addition concept is illustrated by the hydrocracking process in which hydrogen is used in an attempt to *stabilize* the reactive fragments produced during the cracking, thereby decreasing their potential for recombination to heavier products and ultimately to coke. The choice of processing schemes for a given hydrocracking application depends upon the nature of the feedstock as well as the product requirements (Aalund, 1981; Nasution, 1986; Suchanek and Moore, 1986; Murphy and Treese, 1979). The process can be simply illustrated as a single-stage or as a two-stage operation (Fig. 17.9).

The single-stage process can be used to produce gasoline but is more often used to produce middle distillate from heavy vacuum gas oils. The two-stage process was developed primarily to produce high yields of gasoline from straight-run gas oil, and the first stage may actually be a purification step to remove sulfur-containing (as well as nitrogen-containing) organic materials. In terms of sulfur removal, it appears that non-asphaltenic sulfur may be removed before the more refractory asphaltene sulfur (Fig. 17.10), thereby requiring thorough desulfurization. This is a good reason for processes to use an extinction–recycling technique to maximize desulfurization and the yields of the desired product. Significant conversion of heavy feedstocks can be accomplished by hydrocracking at high severity (Howell et al., 1985). For some applications, the products boiling up to 340°C (650°F) can be blended to give the desired final product.

Hydrocracking is similar to catalytic cracking, with hydrogenation superimposed and with the reactions taking place either simultaneously or sequentially. Hydrocracking was initially used to upgrade low value distillate feedstocks such as cycle oils (highly aromatic products from a catalytic cracker that usually are not recycled to extinction for economic

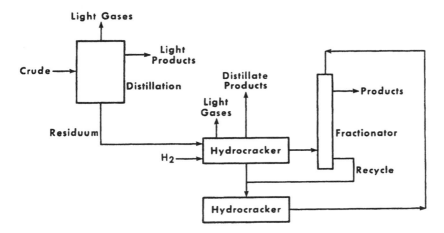

Figure 17.9 A single-stage and two-stage hydrocracking unit.

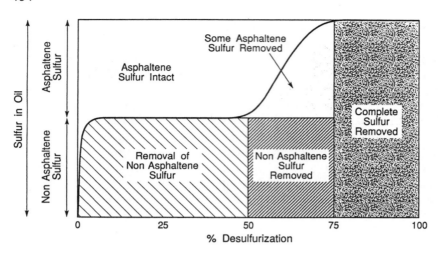

Figure 17.10 Trends for sulfur removal from crude oil.

reasons), thermal and coker gas oils, and heavy-cracked and straight-run naphtha. These feedstocks are difficult to process by either catalytic cracking or reforming, because they are usually characterized by a high polycyclic aromatic content and/or by high concentrations of the two principal catalyst poisons, sulfur and nitrogen compounds (Bland and Davidson, 1967).

Whole families of catalysts are required, depending on feed available and the desired product slate or product character, and the number of process stages is also important to catalyst choice. Generally, one of three options is exercised by the refinery. Thus, depending on the feedstock being processed and the type of plant design employed (single-stage or two-stage), flexibility can be provided to vary product distribution among the principal end products.

Fundamentally, the trend toward lower API gravity feedstocks is related to a decrease in the hydrogen/carbon atomic ratio of crude oils (Fig. 17.11) because of the higher content of residuum. This can be overcome by upgrading methods that increase this ratio by adding hydrogen, rejecting carbon, or using a combination of the two methods.

Though several technologies exist to upgrade heavy feedstocks (Chapter 18), selection of the optimum process units is very much dependent on each refiner's needs and goals, with the market pull being the prime motivator. Furthermore, processing options to dig deeper into the barrel by converting more of the higher boiling materials to distillable products should be not only cost-effective and reliable but also flexible.

Hydrocracking adds that flexibility and offers the refiner a process that can handle varying feeds and operate under diverse process conditions. The use of various types of catalysts (Fig. 17.12) can modify the product slate produced. Reactor design and number of processing stages both play a role in this flexibility.

Finally, a word about conversion measures for upgrading processes. Such measures are necessary for *any conversion process* but more particularly for hydrocracking processes, where hydrogen management is an integral part of process design.

The objective of any upgrading process is to convert heavy feedstock into marketable products by reducing their heteroatom (nitrogen, oxygen, and sulfur) and metal contents,

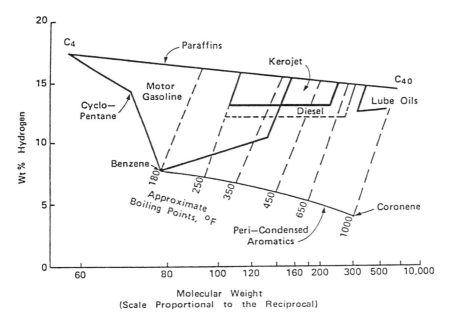

Figure 17.11 Graphical representation of hydrogen content and molecular weight.

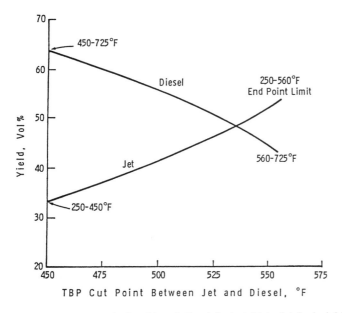

Figure 17.12 Relationship of diesel fuel yield to jet fuel yield.

modifying their asphaltenic structures (reducing coke precursors), and converting the high molecular weight polar molecules into lower molecular weight and lower boiling hydrocarbon products. Upgrading processes are evaluated on the basis of liquid yield (i.e., naphtha, distillate, and gas oil), heteroatom removal efficiency, feedstock conversion (FC), carbon mobilization (CM), and hydrogen utilization (HU), along with other process characteristics. The definitions of RC, CM, and HU are

$$FC = (feedstocks_{IN} - feedstock_{OUT})/feedstock_{IN} \times 100$$

$$CM = carbon_{LIQUIDS}/carbon_{FEEDSTOCK} \times 100$$

$$HU = hydrogen_{LIQUIDS}/hydrogen_{FEEDSTOCK} \times 100$$

High carbon mobilization (CM < 100%) and high hydrogen utilization (HU) correspond to high feedstock conversion (RC) processes that involve hydrogen addition, such as hydrocracking. Because hydrogen is added, hydrogen utilization can be greater than 100%. These tasks can be achieved by using thermal and/or catalytic processes. Low carbon mobilization and low hydrogen utilization correspond to low feedstock conversion such as coking (carbon rejection) processes (Chapter 14). Maximum efficiency from an upgrading process can be attained by maximizing the liquid yield and its quality by minimizing the gas (C_1–C_4) yield. Under these operating conditions hydrogen consumption would be the most efficient, i.e., hydrogen is consumed to increase the liquid yield and its quality (Towler et al., 1996). Several process optimization models can be formulated if the reaction kinetics is known.

2. PROCESSES AND PROCESS DESIGN

The most common type of hydrocracking process is a two-stage operation (Fig. 17.9). This flow scheme has been very popular in the United States because it maximizes the yield of transportation fuels and has the flexibility to produce gasoline, naphtha, jet fuel, or diesel fuel to meet seasonal swings in product demand.

This type of hydrocracker consists of two reactor stages together with a product distillation section. The choice of catalyst in each reaction stage depends on the product slate required and the character of the feedstock. In general, however, the first-stage catalyst is designed to remove nitrogen and heavy aromatics from raw petroleum stocks. The second-stage catalyst carries out a selective hydrocracking reaction on the cleaner oil produced in the first stage.

The two reactor stages have similar process flow schemes. The oil feed is combined with a preheated mixture of makeup hydrogen and hydrogen-rich recycle gas and heated to reactor inlet temperature via a feed–effluent exchanger and a reactor charge heater. The reactor charge heater design philosophy is based on many years of safe operation with such two-phase furnaces. The feed–effluent exchangers take advantage of special high pressure exchanger design features to give leak-free end closures. From the charge heater, the partially vaporized feed enters the top of the reactor. The catalyst is loaded into separate beds in the reactor with facilities between the beds for quenching the reaction mix and ensuring good flow distribution through the catalyst.

The reactor effluent is cooled through a variety of heat exchangers including the feed–effluent exchanger and one or more air coolers. Deaerated condensate is injected into the first-stage reactor effluent before the final air cooler in order to remove ammonia and some of the hydrogen sulfide. This prevents solid ammonium bisulfide from depositing in the system. A body of expertise in the field of materials selection for hydrocracker cooling trains is quite important for proper design.

The reactor effluent leaving the air cooler is separated into hydrogen-rich recycle gas, a sour water stream, and a hydrocarbon liquid stream in the high pressure separator. The sour water effluent stream is often then sent to a plant for ammonia recovery and for purification so water can be recycled back to the hydrocracker. The hydrocarbon-rich stream is pressure reduced and fed to the distillation section after light products are flashed off in a low pressure separator.

The hydrogen-rich gas stream from the high pressure separator is recycled back to the reactor feed by using a recycle compressor. Sometimes with sour feeds the first-stage recycle gas is scrubbed with an amine system to remove hydrogen. If the feed sulfur level is high, this option can improve the performance of the catalyst and result in less costly materials of construction.

The distillation section consists of a hydrogen sulfide (H₂S) stripper and a recycle splitter. This latter column separates the product into the desired cuts. The column bottoms stream is recycled back to the second-stage feed. The recycle cut point is changed depending on the light products needed. It can be as low as 160°C (320°F) if naphtha production is maximized (for aromatics) or as high as 380°C (720°F) if a low pour point diesel is needed. Between these two extremes a recycle cut point of 260–285°C (500–550°F) results in high yields of high smoke point, low freeze point jet fuel.

A *single-stage once-through* (SSOT) unit resembles the first stage of the two-stage plant. This type of hydrocracker usually requires the least capital investment. The feedstock is not completely converted to lighter products. For this application the refiner must have a demand for a highly refined heavy oil. In many refining situations, such an oil product can be used as lube oil plant feed, as FCC plant feed, in low sulfur oil blends, or as ethylene plant feed. It also lends itself to stepwise construction of a future two-stage hydrocracker for full feed conversion.

A *single-stage recycle* (SSREC) unit converts heavy feedstock completely into light products with a flow scheme resembling the second stage of the two-stage plant. Such a unit maximizes the yield of naphtha, jet fuel, or diesel depending on the recycle cut point used in the distillation section. This type of unit is more economical than the more complex two-stage unit when plant design capacity is less than about 10,000–15,000 bbl/day. Commercial Single-stage recycle plants have operated to produce low pour point diesel fuel from waxy Middle East vacuum gas oils. Recent emphasis has been placed on the upgrading of lighter gas oils into jet fuels.

Building on the theme of one- or two-stage hydrocracking, the *once-through partial conversion* (OTPC) concept evolved. This concept offers the means to convert heavy feedstocks into high quality gasoline, jet fuel, and diesel products by a partial conversion operation. The advantage is lower initial capital investment and also lower utilities consumption than a plant designed for total conversion. Because total conversion of the higher molecular weight compounds in the feedstock is not required, once-through hydrocracking can be carried out at lower temperatures and in most cases at lower hydrogen partial pressures than in recycle hydrocracking, where total conversion of the feedstock is normally an objective.

Proper selection of the types of catalysts employed can even permit partial conversion of heavy gas oil feeds to diesel and lighter products at the low hydrogen partial pressures for which gas oil hydrotreaters are normally designed. This so-called *mild hydrocracking* has been attracting a great deal of interest from refiners who have existing hydrotreaters

Table 17.2 Feedstocks and Respective Products

Feed	Products
Straight-run gas oil	Liquefied petroleum gas (LPG)
Vacuum gas oil	Motor gasolines
Fluid catalytic cracking oils and decant oil	Reformer feeds
Coker gas oil	Aviation turbine fuels
Thermally cracked stock	Diesel fuels
Solvent deasphalted residual oil	Heating oils
Straight-run naphtha	Solvents and thinners
Cracked naphtha	Lube oils
	Petrochemical feedstocks
	Ethylene feed pretreatment process (FPP)

and wish to increase their refinery's conversion of fuel oil into lower boiling, higher value products (Table 17.2).

Recycle hydrocracking plants are designed to operate at hydrogen partial pressures of about 1200–2300 psi (8274–15,858 kPa) depending on the type of feed being processed. Hydrogen partial pressure is set in the design in part depending on required catalyst cycle length but also to enable the catalyst to convert high molecular weight polynuclear aromatic and naphthene compounds that must be hydrogenated before they can be cracked. Hydrogen partial pressure also affects properties of the hydrocracked products that depend on hydrogen uptake, such as jet fuel aromatics content and smoke point and diesel cetane number. In general, the higher the feed endpoint, the higher the required hydrogen partial pressure necessary to achieve satisfactory performance of the plant.

Once-through partial conversion hydrocracking of a feedstock may be carried out at hydrogen partial pressures significantly lower than those required for recycle total conversion hydrocracking. The potential higher catalyst deactivation rates experienced at lower hydrogen partial pressures can be offset by using higher activity catalysts and designing the plant for lower catalyst space velocities. Catalyst deactivation is also reduced by the elimination of the recycle stream. The lower capital cost resulting from the reduction in plant operating pressure is much more significant than the increase resulting from the possible additional catalyst requirement and larger volume reactors.

Additional capital cost savings from once-through hydrocracking result from the reduced overall required hydraulic capacity of the plant for a given fresh feed rate as a result of the elimination of a recycle oil stream. Hydraulic capacity at the same fresh feed rate is 30–40% lower for a once-through plant than for one designed for recycle.

Utilities savings for a once-through versus recycle operation arise from lower pumping and compression costs as a result of the lower design pressure possible and also the lower hydrogen consumption. Additional savings are realized as a result of the lower oil and gas circulation rates required, because it isn't necessary to recycle oil from the fractionator bottoms.

Lower capital investment and operating costs are obvious advantages of once-through hydrocracking compared to a recycle design. This type of operation may be adaptable for use in an existing gas oil hydrotreater or atmospheric resid desulfurization plant. The change from hydrotreating to hydrocracking service will require some modifications and capital expenditure, but in most cases these changes will be minimal.

The fact that unconverted oil is produced by the plant is not necessarily a disadvantage. The unconverted oil produced by once-through hydrocracking is a high quality, low sulfur, low nitrogen material that is an excellent feedstock for a fluid catalytic cracking unit or ethylene pyrolysis furnace or a source of high viscosity index lube oil base stock. The properties of the oil are a function of the degree of conversion and other plant operating conditions.

One disadvantage of once-through hydrocracking compared to a recycle operation is a somewhat reduced flexibility for varying the ratio of gasoline to middle distillate that is produced. A greater quantity of naphtha can be produced by increasing conversion, and jet fuel plus diesel yield can also be increased in this manner. But selectivity for higher boiling products is also a function of conversion. Selectivity decreases as once-through conversion increases. If conversion is increased too much, the yield of desired product will decrease, accompanied by an increase in the production of light ends and gas. Higher yields of gasoline or jet fuel plus diesel are possible from a recycle than from a once-through operation.

Middle distillate products made by once-through hydrocracking are generally higher in aromatics content of poorer burning quality than those produced by recycle hydrocracking. However, the quality is generally better than that of products of catalytic cracking or straight-run distillation. Middle distillate product quality improves as the degree of conversion increases and as hydrogen partial pressure is increased.

The hydrocracking process employs high activity catalysts that produce a significant yield of light products. Catalyst selectivity to middle distillate is a function of both the conversion level and operating temperature, with values in excess of 90% being reported in commercial operation. In addition to the increased hydrocracking activity of the catalyst, percent desulfurization and denitrogenation at start-of-run conditions are also substantially increased. End of cycle is reached when product sulfur has risen to the level achieved in conventional vacuum gas oil hydrodesulfurization processes.

An important consideration, however, is that commercial hydrocracking units are often limited by design constraints of an existing vacuum gas oil hydrotreating unit. Thus, the proper choice of catalyst(s) is critical when searching for optimum performance. Typical commercial distillate hydrocracking (DHC) catalysts contain both the hydrogenation (metal) and cracking (acid sites) functions required for service in existing desulfurization units.

The processes that follow are listed in alphabetical order, with no other preference in mind.

2.1 CANMET Process

The CANMET process is a hydrocracking process for heavy oils, atmospheric residua, and vacuum residua (Table 17.3) (Pruden, 1978; Waugh, 1983; Pruden et al., 1993). This technology such as that developed by the CANMET process shows promise for some applications. The scheme is a high conversion, high demetallization residuum hydrocracking process that uses an additive to inhibit coke formation and achieves the conversion of high boiling point hydrocarbons into lighter products. Initially developed to upgrade tar sands bitumen and heavy oils of Alberta, an ongoing program of development has broadened the technology to processing offshore heavy oils and the bottom of the barrel from

Table 17.3 CANMET Process Feedstock and Product Data

Feedstock[a]	
API gravity	4.4
Sulfur, wt%	5.1
Nitrogen, wt%	0.6
Asphaltenes, wt%	15.5
Carbon residue, wt%	20.6
Metals, ppm	
Ni	80
V	170
Residuum (> 525°C; > 975°F), wt%	
Products,[b] wt%	
Naphtha (C_5–204°C; 400°F)	19.8
Nitrogen, wt%	0.1
Sulfur, wt%	0.6
Distillate (204–343°C; 400–650°F)	33.5
Nitrogen, wt%	0.4
Sulfur, wt%	1.8
Vacuum gas oil (343–534°C; 650–975°F)	28.5
Nitrogen, wt%	0.6
Sulfur, wt%	2.3
Residuum (> 534°C; > 975°F)	4.5
Nitrogen, wt%	1.6
Sulfur, wt%	3.1

[a] Cold Lake (Canada) heavy oil vacuum residuum.
[b] Residuum: 93.5% by weight.

so-called conventional crude oils. A 5000 bbl/day demonstration plant was successful at PetroCanada's Montreal East refinery.

The process does not use a catalyst but employs a low cost additive to inhibit coke formation and allow high conversion of heavy feedstocks (such as heavy oil and bitumen) into lower boiling products using a single reactor. The process is unaffected by high levels of feed contaminants such as sulfur, nitrogen, and metals. Conversion of over 90% of the $525°C^+$ ($975°F^+$) fraction into distillates has been attained.

In the process (Fig. 17.13), the feedstock and recycle hydrogen gas are heated to reactor temperature in separate heaters. A small portion of the recycle gas stream and the required amount of additive are routed through the oil heater to prevent coking in the heater tubes. The outlet streams from both heaters are fed to the bottom of the reactor.

The vertical reactor vessel is free of internal equipment and operates in a three-phase mode. The solid additive particles are suspended in the primary liquid hydrocarbon phase, through which the hydrogen and product gases flow rapidly in bubble form. The reactor exit stream is quenched with cold recycle hydrogen prior to teaching the high pressure separator. The heavy liquids are further reduced in pressure as they pass to a hot medium pressure separator and from there to fractionation. The spent additive leaves with the heavy fraction and remains in the unconverted vacuum residue.

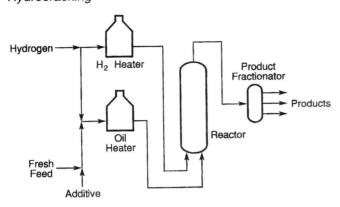

Figure 17.13 The CANMET process.

The vapor stream from the hot high pressure separator is cooled stepwise to produce middle distillate and naphtha that are sent to fractionation. High pressure purge of low-boiling hydrocarbon gases is minimized by a sponge oil circulation system. Product naphtha will be hydrotreated and reformed, light gas oil will be hydrotreated and sent to the distillate pool, the heavy gas oil will be processed in the fluid catalytic cracker.

The additive, prepared from iron sulfate $[Fe_2(SO_4)_3]$, is used to promote hydrogenation and effectively eliminate coke formation. The effectiveness of the dual-role additive permits the use of operating temperatures that give high conversion in a single-stage reactor. The process maximizes the use of reactor volume and provides a thermally stable operation with no possibility of temperature runaway.

The process also offers the attractive option of reducing the coke yield by slurrying the feedstock with less than 10 ppm of catalyst (molybdenum naphthenate) and sending the slurry to a hydroconversion zone to produce low-boiling products (Kriz and Ternan, 1994).

2.2 Gulf HDS Process

The Gulf HDS process is a regenerative fixed-bed process to upgrade residua by catalytic hydrogenation to refined heavy fuel oils or to high quality catalytic charge stocks (Fig. 17.14). Desulfurization and quality improvement are the primary purposes of the process, but if the operating conditions and catalysts are varied, light distillates can be produced and the viscosity of heavy material can be lowered. Long on-stream cycles are maintained by reducing random hydrocracking reactions to a minimum, and whole crude oils or residua may serve as feedstocks.

This process is suitable for the desulfurization of high sulfur residua (atmospheric and vacuum) to produce low sulfur fuel oils or catalytic cracking feedstocks. In addition, the process can be used, through alternative designs, to upgrade high sulfur crude oils or bitumen that are unsuited for the more conventional refining techniques.

The process has three basic variations—the type II unit, the type III unit, and the type IV unit—with the degree of desulfurization and process severity increasing from type I to type IV. Thus, liquid products from type III and IV units can be used directly as catalytic cracker feedstocks and perform similarly to virgin gas oil fractions, whereas liquid products from the type II unit usually need to be vacuum-flashed to provide a feedstock suitable for a catalytic cracker.

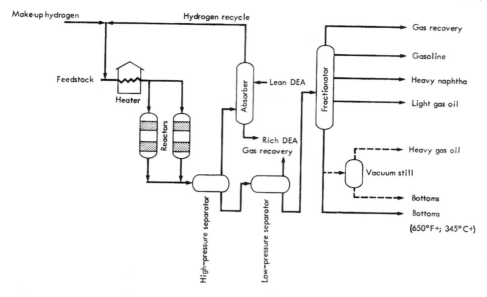

Figure 17.14 The Gulf resid hydrodesulfurization process.

Fresh, filtered feedstock is heated together with hydrogen and recycle gas and charged to the downflow reactor from which the liquid product goes to fractionation after flashing to produce the various product streams. Each process type is basically similar to its predecessor (Fig. 17.14) but will differ in the number of reactors. For example, modifications necessary to convert the type II process to type III process consist of the addition of a reactor and related equipment, whereas the type III process can be modified to a type IV process by the addition of a third reactor section. Types III and IV are especially pertinent to the problem of desulfurizing heavy oils and residua because they have the capability of producing extremely low sulfur liquids from high sulfur residua (Table 17.4).

The catalyst is a metallic compound supported on pelleted alumina and may be regenerated in situ with air and steam or flue gas through a temperature cycle of 400–650°C (750–1200°F). On-stream cycles of 4–5 months can be obtained at desulfurization levels of 65–75%, and catalyst life may be as long as 2 years.

2.3 H-G Hydrocracking Process

The H-G hydrocracking process may be designed with either a single-stage or a two-stage reactor system for conversion of light and heavy gas oils to lower boiling fractions (Fig. 17.15). The feedstock is mixed with recycle gas oil, makeup hydrogen, and hydrogen-rich recycle gas and then heated and charged to the reactor. The reactor effluent is cooled and sent to a high pressure separator, where hydrogen-rich gas is flashed off, scrubbed, and then recycled to the reactor. Separator liquid passes to a stabilizer for removal of butanes and lighter products, and the bottoms are taken to a fractionator for separation; any unconverted material is recycled to the reactor.

Table 17.4 Gulf Resid HDS Process Feedstock and Product Data

Properties	South Louisiana	West Texas	Light Arabian	Kuwait
°API	22.4	19.3	18.5	16.6
Sulfur, wt%	0.46	2.2	2.93	3.8
Nitrogen, wt%	0.16	0.20	0.16	0.21
Carbon residue (Ramsbottom), wt%	1.76	5.5	6.79	8.3
Nickel, ppm	3.2	7.1	7.3	15
Vanadium, ppm	1.7	14.0	27.0	45
Yield on crude, vol%	48	44.1	41	53
Hydrodesulfurization type	II	II	IV	III
Desulfurization, %	87	85	95.9	92
°API	24.0	23.8	25.1	25.7
Sulfur, wt%	0.05	0.33	0.12	0.28
Nitrogen, wt%	0.10	0.14	0.09	0.14
Carbon residue (Ramsbottom), wt%	1.4	2.21	2.28	2.83
Nickel, ppm	0.1	1.1	0.5	1.1
Vanadium, ppm	0.1	0.7	0.3	0.8
Demetallization, %	97.1	87.5	95.3	94.6

2.4 H-Oil Process

The H-Oil process (*Hydrocarbon Processing*, 1998, p. 86) is a catalytic hydrogenation technique that uses a one-, two-, or three-stage ebullated-bed reactor in which, during the reaction, considerable hydrocracking takes place (Fig. 17.16). The process is used to upgrade heavy sulfur-containing crude oils and residua to low sulfur distillates, thereby reducing fuel oil yield (Table 17.5). A modification of H-Oil called Hy-C cracking converts heavy distillates to middle distillates and kerosene.

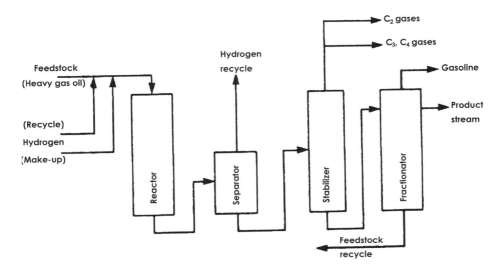

Figure 17.15 The H-G hydrocracking process.

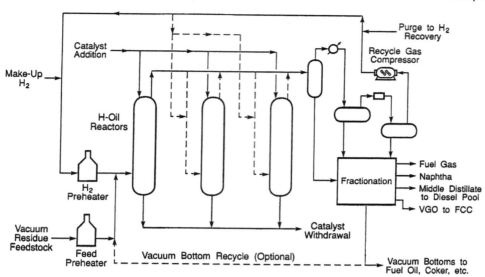

Figure 17.16 The H-Oil catalytic cracking process.

Oil and hydrogen are fed upward through the reactors as a liquid–gas mixture at a velocity such that catalyst is in continuous motion. A catalyst of small particle size can be used, giving efficient contact among gas, liquid, and solid with good mass and heat transfer. Part of the reactor effluent is recycled back through the reactors for temperature control and to maintain the requisite liquid velocity. The entire bed is held within a narrow temperature range, which provides essentially an isothermal operation with an exothermic process. Because of the movement of catalyst particles in the liquid–gas medium, deposition of tar and coke is minimized and fine solids entrained in the feed do not lead to reactor plugging. The catalyst can also be added to and withdrawn from the reactor without destroying the continuity of the process.

The reactor effluent is cooled by exchange and separates into vapor and liquid. After scrubbing in a lean oil absorber, hydrogen is recycled and the liquid product is either stored directly or fractionated before storage and blending.

2.5 IFP Hydrocracking Process

The IFP hydrocracking process features a dual catalyst system. The first catalyst is a promoted nickel-molybdenum amorphous catalyst that acts to remove sulfur and nitrogen and hydrogenate aromatic rings. The second catalyst is a zeolite that finishes the hydrogenation and promotes the hydrocracking reaction.

In the two-stage process (Fig. 17.17), feedstock and hydrogen are heated and sent to the first reaction stage, where conversion to products occurs (Table 17.6) (Kamiya, 1991, p. 85). The reactor effluent phases are cooled and separated, and the hydrogen-rich gas is compressed and recycled. The liquid leaving the separator is fractionated, the middle distillates and lower boiling streams are sent to storage, and the high-boiling stream is transferred to the second reactor section and then recycled back to the separator section.

Table 17.5 H-Oil Process Feedstock and Product Data

	Unspecified	Arabian Medium vacuum residuum[a] 65% conv	90% conv	Unspecified[b] 40% desulf	80% desulf	Athabasca bitumen
Feedstock						
Gravity, °API	8.8	4.9	4.9	14.0	14.0	8.3
Sulfur, wt%	2.0	5.4	5.4	4.0	4.0	4.9
Nitrogen, wt%	0.9					0.5
Carbon residue, wt%	14.6			9.2	9.2	
Metals, ppm		128.0	128.0			
Ni	80.0					
V	170.0					
Residuum (> 525°C, > 975°F), wt%	71.7			50.0	50.0	50.3
Products, wt%						
Naphtha (C$_5$–204°C; 400°F)	33.0	17.6	23.8	1.8	5.0	16.0
Sulfur, wt%	< 0.1				0.1	1.0
Distillate (204–343°C; 400–650°F)	40.0	22.1	36.5			43.0
Sulfur, wt%	0.4					2.0
Vacuum gas oil (343–534°C; 650–975°F)	24.0	34.0	37.1	12.8	15.0	26.4
Sulfur, wt%	1.0			0.7	0.3	3.5
Residuum (> 534°C; > 975°F)	11.0	33.2	9.5	86.6	81.0	16.0
Sulfur, wt%	2.0			2.5	1.0	5.7

[a] conv = conversion.
[b] desulf = desulfurization.

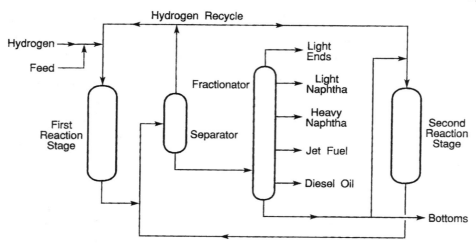

Figure 17.17 The IFP hydrocracking process.

In the single-stage process, the first reactor effluent is sent directly to the second reactor, followed by the separation and fractionation steps. The fractionator bottoms are recycled to the second reactor or sold.

2.6 Isocracking Process

The isocracking process has been applied commercially in the full range of process flow schemes: single-stage, once-through liquid; single-stage, partial recycle of heavy oil; single-stage extinction recycle of oil (100% conversion); and two-stage extinction recycle of oil (Bridge, 1997; *Hydrocarbon Processing*, 1998, p. 84). The preferred flow scheme will depend on the feed properties, the processing objectives, and, to some extent, the specified feed rate.

The process (Fig. 17.18) uses multibed reactors. In most applications, a number of catalysts are used in a reactor. The catalysts are a mixture of hydrous oxides (for cracking) and heavy metal sulfides (for hydrogenation). The catalysts are used in a layered system to optimize the processing of the feedstock, which undergoes changes in its properties along the reaction pathway. In most commercial isocracking units, all of the fractionator bot-

Table 17.6 IFP Hydrocracking Process Feedstock and Product Data

Feedstock: Vacuum gas oil, 350–570°C (660–1060°F)	
API Gravity, °API	20.6
Sulfur, wt%	2.3
Nitrogen, wt%	1.0
Carbon residue, wt%	0.5
Products	
Light naphtha, wt%	6.0
Sulfur, ppm	< 1
Distillate (heavy naphtha, jet fuel, diesel fuel), wt%	91.0
Sulfur, ppm	< 50

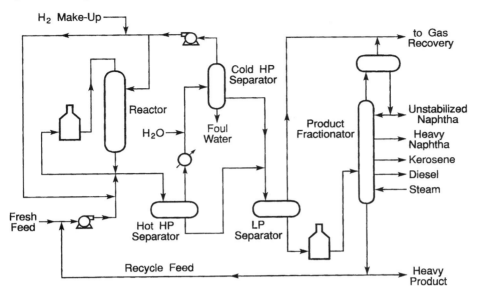

Figure 17.18 The isocracking process.

toms is recycled or all of it is drawn as heavy product, depending on whether the low-boiling or high-boiling products are of greater value. If the low-boiling distillate products (naphtha or naphtha/kerosene) are the most valuable products, the higher boiling point distillates (such as diesel) can be recycled to the reactor for conversion rather than drawn as a product (Table 17.7) (Kamiya, 1991, p. 83). Product distribution depends upon the mode of operation (Table 17.8).

Heavy feedstocks have been used in the process, and the product yield is very much dependent upon the catalyst and the process parameters (Bridge, 1997).

Table 17.7 Isocracking Process Feedstock and Product Data

Feedstock: Vacuum gas oil, 360–530°C (680–985°F)	
API Gravity, °API	22.6
Sulfur, wt%	2.2
Nitrogen, wt%	0.6
Carbon residue, wt%	0.3
Metals	
Ni	0.1
V	0.3
Products	
Naphtha (C_5–124°C; C_5–255°F), wt%	15.9
Sulfur, ppm	< 2
Distillate (124–295°C; 255–565°F), wt%	51.6
Sulfur, ppm	< 5
Heavy distillate (295–375°C; 565–705°F), wt%	42.3
Sulfur, ppm	< 5

Table 17.8 Isocracking Process Desulfurization Data

Operation	Conventional desulfurization	Severe desulfurization	Mild isocracking
% HDS	90.0	99.8	99.6
Yield, LV (%)			
Naphtha	0.2	1.5	3.5
Light isomate	17.2	30.8	37.1
Heavy isomate	84.0	70.0	62.5
Feed			
Gravity, °API	22.6	22.6	23.0
Sulfur, wt%	2.67	2.67	2.57
Nitrogen, ppm	720	720	617
Ni + V, ppm	0.2	0.2	—
Distillation, ASTM, °F	579–993	579–993	552–1031
Light isomate			
Gravity, °API	30.9	37.8	34.0
Sulfur, wt%	0.07	0.002	0.005
Nitrogen, ppm	18	20	20
Pour point, °F	18	14	18
Cetane index	51.5	53.0	53.5
Distillation, ASTM, °F	433–648	298–658	311-683
Heavy isomate			
Gravity, °API	27.1	29.2	30.7
Sulfur, wt%	0.26	0.009	0.013
Nitrogen, ppm	400	60	47
Viscosity, cSt at 122°F	26.2	19.8	17.2
Distillation, ASTM, °F	689–990	691–977	613–1026

2.7 LC-Fining Process

The LC-Fining process is a hydrocracking process capable of desulfurizing, demetallizing, and upgrading a wide spectrum of heavy feedstocks (Table 17.9) by means of an expanded-bed reactor (van Driesen et al., 1979; Fornoff, 1982; Bishop, 1990; Kamiya, 1991, p. 61; Reich et al., 1993; *Hydrocarbon Processing*, 1998, p. 82).

Operating with the expanded bed allows the processing of heavy feedstocks such as atmospheric residua, vacuum residua, and oil sand bitumen. The catalyst in the reactor behaves like fluid, which enables the catalyst to be added to and withdrawn from the reactor during operation. The reactor conditions are near isothermal, because the heat of reaction is absorbed by the cold fresh feed immediately owing to thorough mixing in the reactors.

In the process (Fig. 17.19), the feedstock and hydrogen are heated separately and then pass upward in the hydrocracking reactor through an expanded bed of catalyst (Fig. 17.20). Reactor products flow to the high pressure–high temperature separator. Vapor effluent from the separator is let down in pressure and then goes to the heat exchange and thence to a section for the removal of condensable products and purification.

Liquid is let down in pressure and passes to the recycle stripper. This is a very important part of the high conversion process. The liquid recycle is carried out in the proper boiling range for return to the reactor. In this way the concentration of bottoms in

Table 17.9 LC-Fining Process Feedstock and Product Data

	Kuwait atmospheric residuum	Gach Saran vacuum residuum	Arabian heavy vacuum residuum	Unspecified	AL/AH,[a] atmospheric residuum	AL/AH,[a] vacuum residuum	AL/AH,[a] vacuum residuum	AL/AH,[a] vacuum residuum	Athabasca bitumen
Feedstock									
Conversion, %	15.0			90.0	45.0	60.0	75.0	95.0	
Gravity, °API	.1	6.1	7.5	4.4	12.4	4.7	4.7	4.7	9.1
Sulfur, wt%		3.5	4.9	3.6	3.9	5.0	5.0	5.0	5.5
Nitrogen, wt%				0.5					0.4
Carbon residue, wt%				22.2					
Metals									
Ni				126.0	18.0	39.0	39.0	39.0	
V				80.0	65.0	142.0	142.0	142.0	
Products[b]									
Naphtha (C$_5$–205°C; C$_5$–400°F)	2.5	9.7	14.3	17.4	7.0	12.6	18.3	23.9	11.9
Sulfur, wt%									1.1
Distillate (205–345°C; 400–650°F)	22.7	14.1	26.5	27.2	15.2	30.6	42.7	64.8	37.7
Sulfur, wt%				0.4					0.7
Heavy distillate (345–525°C; 650–975°F)	34.7	24.1	31.1	41.4	53.0	21.5	19.3	11.9	30
Sulfur, wt%				1.1					1.1
Residuum (> 525°C; > 975°F)	35.5	47.5	21.3	7.7	24.5	40.0	25.0	5.0	12.9
Sulfur, wt%				3.0					3.4
Carbon residue, wt%				50.0					

[a] AL/AH: Arabian light crude oil blended with Arabian heavy crude oil.
[b] Distillation ranges may vary by several degrees because of different distillation protocols.

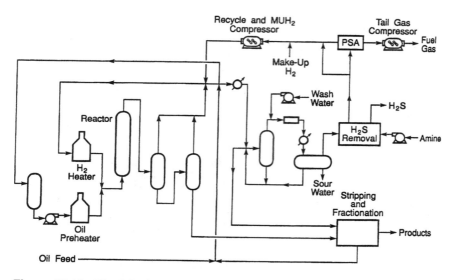

Figure 17.19 The LC-Fining process.

the reactor, and therefore the distribution of products, can be controlled. After the stripping, the recycle liquid is then pumped through the coke precursor removal step, where high molecular weight constituents are removed. The clean liquid recycle then passes to the suction drum of the feed pump. The product from the top of the recycle stripper goes to fractionation, and any heavy oil product is directed from the stripper bottoms to the pump discharge.

The residence time in the reactor is adjusted to provide the desired conversion levels. Catalyst particles are continuously withdrawn from the reactor, regenerated, and recycled back into the reactor, which provides the flexibility to process a wide range of heavy feedstocks such as atmospheric and vacuum tower bottoms, coal-derived liquids, and bitumen. An internal liquid recycle is provided with a pump to expand the catalyst bed continuously. As a result of expanded bed operating mode, small pressure drops and

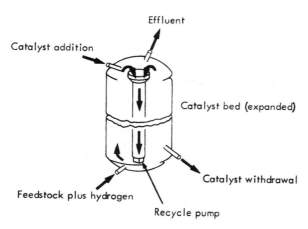

Figure 17.20 The LC-Fining expanded-bed reactor.

isothermal operating conditions are accomplished. Small-diameter extruded catalyst particles as small as 0.8 mm (1/32 in.) can be used in this reactor.

The process offers the means by which conversion of bitumen to synthetic crude oil can be achieved, and is an attractive method of bitumen conversion. Indeed, the process plays the part of the primary conversion process from which liquid products would accrue, these products would then pass to a secondary upgrading (hydrotreating) process to yield a synthetic crude oil.

2.8 MAKfining HDC Process

The MAKfining HDC process uses a multiple catalyst system in multibed reactors that include quench and redistribution system internals (Hunter et al., 1997; *Hydrocarbon Processing*, 1998, p. 86).

In the process (Fig. 17.21), the feedstock and recycle gas are preheated and brought into contact with the catalyst in a downflow fixed-bed reactor. The reactor effluent is sent to high and low temperature separators. Product recovery is a stripper–fractionator arrangement. Typical operating conditions in the reactors are 370–425°C (700–800°F) (single-pass) and 370–425°C (700–800°F) (recycle) with pressures of 1000–2000 psi (6895–13,790 kPa) (single-pass) and 1500–3000 psi (10,342–20,684 kPa) (recycle). Product yields depend upon the extent of the conversion (Table 17.10).

2.9 Microcat-RC Process

The Microcat-RC process (also referred to as the M-Coke process) is a catalytic hydroconversion process operating at relatively moderate pressures and temperatures (Table 17.11) (Bearden and Aldridge, 1981; Bauman et al., 1993). The novel catalyst particles, containing a metal sulfide in a carbonaceous matrix formed within the process, are uniformly dispersed throughout the feed. Because of their ultrasmall size (10^{-4} in. diameter) there are typically several orders of magnitude more of these catalyst particles per cubic centimeter of oil than is possible in other types of hydroconversion reactors using con-

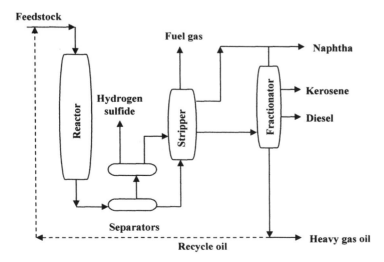

Figure 17.21 The MAKfining process.

Table 17.10   MAKfining Process Feedstock and Product Data[a]

	AL/AH vacuum gas oil	AL/AH vacuum gas oil	AL/AH light cycle oil
Feedstock			
Gravity, °API	20.2	20.2	19.0
Sulfur, wt%	2.9	2.9	1.0
Nitrogen, wt%	0.9	0.9	0.6
Products, vol%			
Naphtha	12.9	22.6	54.0
Kerosene	14.1	24.5	
Diesel	31.8	32.5	54.3
Light gas oil	50.0	30.0	
Conversion, %	50.0	70.0	50.0

[a] AL/AH: Arabian light crude oil blended with Arabian heavy crude oil.

ventional catalyst particles. This results in smaller distances between particles and less time for a reactant molecule or intermediate to find an active catalyst site. Because of their physical structure, microcatalysts suffer none of the pore-plugging problems that plague conventional catalysts.

In the Microcat-RC process (Fig. 17.22), fresh vacuum residuum, microcatalyst, and hydrogen are fed to the hydroconversion reactor. Effluent is sent to a flash separation zone to recover hydrogen, gases, and liquid products, including naphtha, distillate, and gas oil. The residuum from the flash step is then fed to a vacuum distillation tower to obtain a $565°C^-$ ($1050°F^-$) product oil and a $565°C^+$ ($1050°F^+$) bottoms fraction that contains unconverted feed, microcatalyst, and essentially all of the feed metals.

Hydrotreating facilities may be integrated with the hydroconversion section or built on a stand-alone basis, depending on product quality objectives and owner preference.

Table 17.11   Microcat Process Feedstock and Product Data

Feedstock: Cold Lake heavy oil vacuum residuum	
Gravity, °API	4.4
Sulfur, wt%	
Nitrogen, wt%	
Metals (Ni + V), ppm	480.0
Carbon residue, wt%	24.4
Products, vol%	
Naphtha (C_5–177°C; C_5–350°F)	17.2
Distillate (177–343°C; 350–650°F)	63.6
Gas oil (343–566°C; 650–1050°F)	21.9
Residuum (> 566°C; > 1050°F)	2.1

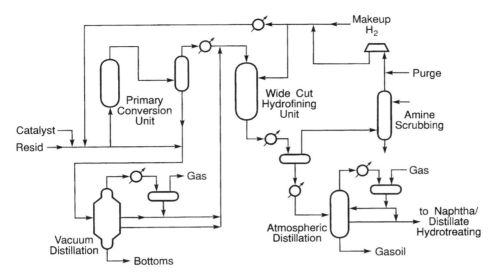

Figure 17.22 The Microcat-RC process.

2.10 Mild Hydrocracking Process

The *mild hydrocracking process* uses operating conditions similar to those of a vacuum gas oil (VGO) desulfurizer to convert a vacuum gas oil to significant yields of lighter products (Table 17.12). Consequently, the flow scheme for a mild hydrocracking unit is virtually identical to that of a vacuum gas oil desulfurizer (Figs. 17.23–17.25).

For example, in a simplified process for vacuum gas oil desulfurization (Fig. 17.23), the vacuum gas oil feedstock is mixed with hydrogen makeup gas and preheated against reactor effluent. Further preheating to reaction temperature is accomplished in a fired heater. The hot feed is mixed with recycle gas before entering the reactor. The temperature rises across the reactor owing to the exothermic heat of reaction. Catalyst bed temperatures are usually controlled by using multiple catalyst beds and

Table 17.12 Vacuum Gas Oil Hydrodesulfurization and Mild Hydrocracking Process Data

	VGO HDS	MHC
Maximum pressure, bar (psi, abs)	75 (1090)	75 (1090)
Maximum temperature, °C (°F)	430 (806)	430 (806)
Makeup gas flow, $N \cdot m^3$/T (SCFB)	90 (500)	150 (835)
Hydrogen content, mol%	90	90
Recycle gas flow, $N \cdot m^3$/T (SCFB)	500 (2780)	500 (2780)
Hydrogen content, mol%	80	80
H_sS content, mol%	0	0
LHSV, h^{-1}	0.5	0.5
Catalyst	S-424, dense loaded	MHC-1
Specification	0.3 wt% S in 370°C$^+$ (698°F$^+$) product	0.3 wt% S at EOR conditions
Cycle length, months	11	11

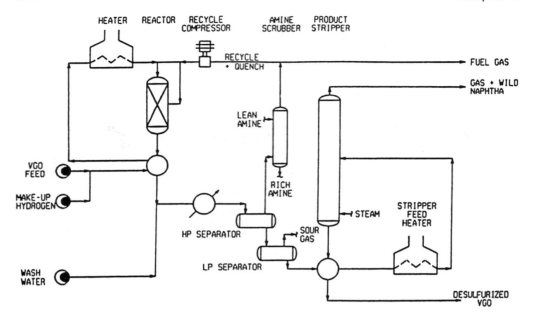

Figure 17.23 Vacuum gas oil desulfurization.

by introducing recycle gas as an interbed quench medium. Reactor effluent is cooled against incoming feed and air or water before entering the high pressure separator. Vapors from this separator are scrubbed to remove hydrogen sulfide (H₂S) before compression back to the reactor as recycle and quench. A small portion of these gases is purged to fuel gas to prevent buildup of light ends. Liquid from the high pressure separator is flashed into the low pressure separator. Sour flash vapors are purged from the unit. Liquid is preheated against stripper bottoms and in a feed heater before steam stripping in a stabilizer tower. Water wash facilities are provided upstream of the last reactor effluent cooler to remove ammonium salts produced by denitrogenation of the vacuum gas oil feedstock.

Variation of this process (Figs. 17.24 and 17.25) leads to the hot separator design. The process flow scheme is identical to that described above up to the reactor outlet. After initial reactor effluent cooling against incoming vacuum gas oil feed and makeup hydrogen, a hot separator is installed. Hot liquid is routed directly to the product stabilizer. Hot vapors are further cooled by air and/or water before entering the cold separator. This arrangement reduces the stabilizer feed preheat duty and the effluent cooling duty by routing hot liquid directly to the stripper tower.

In summary, the so-called *mild hydrocracking process* is a simple form of hydrocracking. The hydrotreaters designed for vacuum gas oil desulfurization and catalytic cracker feed pretreatment are converted to once-through hydrocracking units, and because existing units are being used the hydrocracking is often carried out under nonideal hydrocracking conditions.

The conditions for mild hydrocracking are typical of many low pressure desulfurization units that for hydrocracking units, in general, are marginal in pressure and hydrogen/

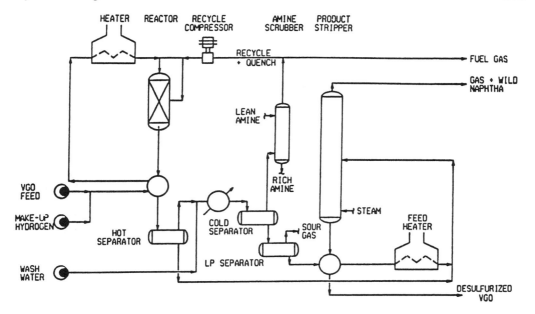

Figure 17.24 Vacuum gas oil desulfurization.

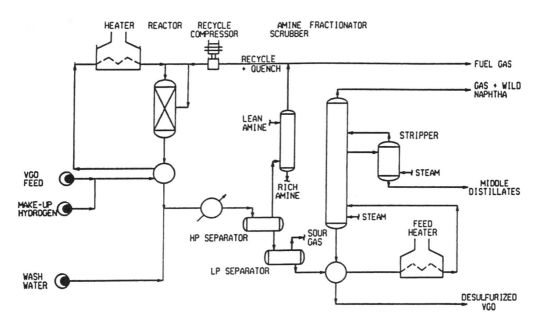

Figure 17.25 Vacuum gas oil desulfurization with mild hydrocracking.

oil ratio capabilities. For hydrocracking, in order to obtain satisfactory run lengths (approximately 11 months), reduction in feed rate or addition of an extra reactor may be necessary. In most cases, because the product slate will be lighter than for normal desulfurization service only, changes in the fractionation system may be necessary. When these limitations can be tolerated, the product value from mild hydrocracking can be greatly enhanced in comparison with that from desulfurization.

2.11 MRH Process

The MRH process is a hydrocracking process designed to upgrade heavy feedstocks containing large amount of metals and asphaltene, such as vacuum residua and bitumen, and to produce mainly middle distillates (Sue, 1989; Kamiya, 1991, p. 65). The reactor is designed to maintain a mixed three-phase slurry of feedstock, fine powder catalyst, and hydrogen and to promote effective contact.

In this process (Fig. 17.26), a slurry consisting of heavy oil feedstock and fine powder catalyst is preheated in a furnace and fed into the reactor vessel. Hydrogen is introduced from the bottom of the reactor and flows upward through the reaction mixture, maintaining the catalyst suspension in the reaction mixture. Cracking, desulfurization, and demetallization reactions take place via thermal and catalytic reactions. In the upper section of the reactor, vapor is disengaged from the slurry and hydrogen and other gases are removed in a high pressure separator. The liquid condensed from the overhead vapor is distilled and then flows out to the secondary treatment facilities.

From the lower section of the reactor, bottom slurry oil (SLO) that contains catalyst, uncracked residuum, and a small amount of vacuum gas oil fraction are withdrawn. Vacuum gas oil is recovered in the slurry separation section, and the remaining catalyst and coke are fed to the regenerator.

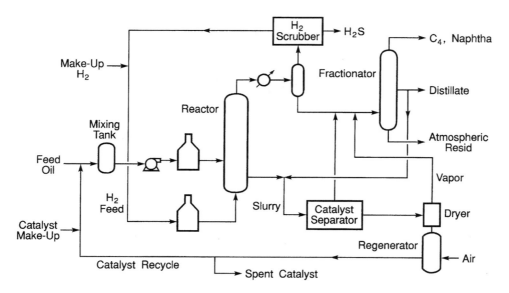

Figure 17.26 The MRH process.

Table 17.13 MRH Process Feedstock and Product Data

	Arabian heavy vacuum residuum	Athabasca (Canada) bitumen
Feedstock		
Gravity, °API	5.9	10.2
Sulfur, wt%	5.1	4.3
Nitrogen, wt%	0.3	0.4
C_7 Asphaltenes, wt%	11.4	8.1
Metals, ppm		
Nickel	41.0	85.0
Vanadium	127.0	182.0
Carbon residue, wt%	21.7	13.3
Distillation profile, vol%		
Naphtha	0.0	2.2
Kerosene	0.0	5.3
Light gas oil	0.0	12.1
Vacuum gas oil	4.0	31.8
Vacuum residuum	96.0	48.6
(Atmospheric residuum)	(100.0)	(80.4)
Products, wt%		
Naphtha	13.0	12.0
Sulfur	0.2	1.1
Nitrogen	0.03	0.05
Kerosene	6.0	11.0
Sulfur	1.0	1.2
Nitrogen	0.06	0.08
Light gas oil	17.0	29.0
Sulfur	2.5	2.2
Nitrogen	0.06	0.11
Atmospheric residuum	55.0	41.0
Sulfur	3.8	3.8
Nitrogen	0.36	0.57

Product distribution focuses on middle distillates (Table 17.13; Fig. 17.27), with a residuum processing unit inserted into a refinery just downstream from the vacuum distillation unit.

2.12 RCD Unibon (BOC) Process

The RCD Unibon (BOC) process is used to upgrade vacuum residua (Table 17.14) (Kamiya, 1991, p. 67; Thompson, 1997; *Hydrocarbon Processing*, 1998).

There are several possible flow scheme variations for the process. It can operate as an independent unit or be used in conjunction with a thermal conversion unit (Fig. 17.28). In the latter configuration, hydrogen and a vacuum residuum are introduced separately to the heater and mixed at the entrance to the reactor. To avoid thermal reactions and premature

(a): Arabian Heavy Crude Oil Vacuum Residuum

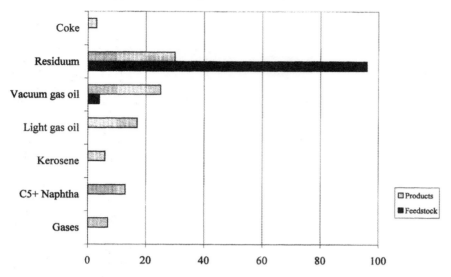

(b): Athabasca bitumen

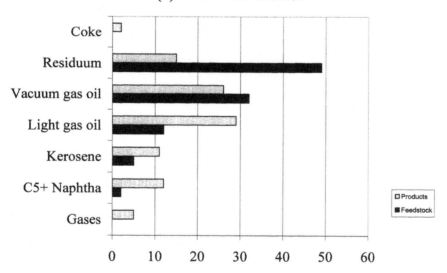

Figure 17.27 Feedstock character and MRH process product distribution.

coking of the catalyst, temperatures are carefully controlled and conversion is limited to approximately 70% of the total projected conversion. The removal of sulfur, heptane-insoluble materials, and metals is accomplished in the reactor. The effluent from the reactor is directed to the hot separator. The overhead vapor phase is cooled and condensed, and the hydrogen separated therefrom is recycled to the reactor.

Table 17.14 MRH Process Feedstock and Product Data

	Unspecified atmospheric residuum	Arabian light, vacuum residuum	Arabian light, vacuum residuum[a]
Feedstock			
Gravity, °API	16.4	7.2	7.2
Sulfur, wt%	3.5	4.0	4.0
Nitrogen, wt%	0.2	0.3	0.3
C_7 Asphaltenes, wt%	2.4		
Carbon residue, wt%	9.5		
Products, wt%			
Naphtha (C_5–180°C; C_5–355°F)	1.0	5.1	9.1
Sulfur		0.04	0.09
Nitrogen		0.01	0.01
Fuel oil (> 190°C; > 375°F)	100.00		
Sulfur	0.3		
Nitrogen	0.1		
Carbon residue	3.8		
Light gas oil (180–343°C; 355–650°F)		6.5	13.4
Sulfur		0.12	0.3
Nitrogen		0.11	0.1
Vacuum gas oil (343–566°C; 650–1050°F)		31.5	37.9
Sulfur		0.4	0.6
Nitrogen		0.2	0.2
Vacuum residuum (> 566°C; > 1050°F)		53.0	34.4
Sulfur		1.0	1.1
Nitrogen		0.3	

[a] With thermal conversion.

Liquid product goes to the thermal conversion heater, where the remaining conversion of nonvolatile materials occurs. The heater effluent is flashed, and the overhead vapors are cooled, condensed, and routed to the cold flash drum. The bottoms liquid stream then goes to the vacuum column, where the gas oils are recovered for further processing and the residuals are blended into the heavy fuel oil pool.

2.13 Residfining Process

Residfining is a catalytic fixed-bed process for the desulfurization and demetallization of residua (Table 17.15) (Kamiya, 1991, p. 69; *Hydrocarbon Processing*, 1998). The process can also be used to pretreat residua to suitably low contaminant levels prior to catalytic cracking.

In this process (Fig. 17.29), liquid feed to the unit is filtered, pumped to pressure, preheated, and combined with treat gas prior to entering the reactors. A small guard reactor is typically employed to prevent plugging or fouling of the main reactors. The guard is periodically removed while the main reactors are kept on-line. The temperature rise associated with the exothermic reactions is controlled by utilizing either a gas or liquid quench. A train of separators is employed to separate the gas and liquid products. The

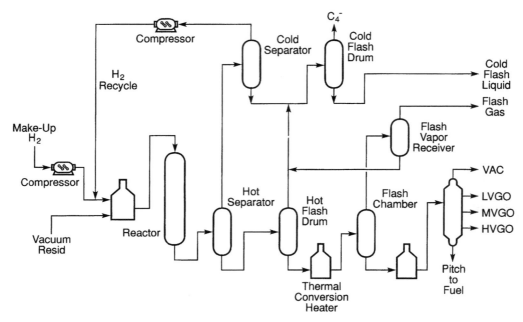

Figure 17.28 The Unibon BOC process.

recycle gas is scrubbed to remove ammonia and hydrogen sulfide. It is then combined with fresh makeup hydrogen before being reheated and recombined with fresh feed. The liquid product is sent to a fractionator, where the product is fractionated.

2.14 Residue Hydroconversion Process

The residue hydroconversion (RHC) process (Fig. 17.30) is a high pressure fixed-bed trickle-flow hydrocatalytic process (Table 17.16) (Kamiya, 1991, p. 71). The feedstock can be desalted atmospheric or vacuum residue.

The reactors are of multibed design with interbed cooling, and the multicatalyst system can be tailored to the nature of the feedstock and the target conversion. For residua with high metals content, a hydrodemetallization catalyst is used in the front-end reactor(s) that excels in its high metal uptake capacity and good activities for metal removal, asphaltene conversion, and residue cracking. Downstream of the demetallization stage, one or more hydroconversion stages, with optimized combination of the catalysts hydrogenation function and texture, are used to achieve desired catalyst stability and activities for denitrogenation, desulfurization, and heavy hydrocarbon cracking.

2.15 Unicracking Process

The Unicracking process is a fixed-bed catalytic process that employs a high activity catalyst that has a high tolerance for sulfur and nitrogen compounds and that can be regenerated (Reno, 1997). The design is based upon a single-stage or two-stage system with provisions to recycle to extinction (Fig. 17.31) (Kamiya, 1991, p. 79).

Table 17.15 Residfining Process Feedstock and Product Data

	Gach Saran atmospheric residuum	Arabian heavy atmospheric residuum	Arabian light atmospheric residuum	AL/AH[a] vacuum residuum
Feedstock				
API gravity	15.0	12.3	14.3	4.7
Sulfur, wt%	2.5	4.2	3.5	5.3
Nitrogen, wt%			0.2	0.4
C_7 Asphaltenes, wt%				
Carbon residue, wt%			9.8	24.6
Metals, ppm				
Nickel			6	50.0
Vanadium			38	170.0
Products, wt%				
Naphtha (C_5–205°C; C_5–400°F)	3.4	6.0		
Naphtha (C_5–220°C; C_5–430°F)			1.9	
Residuum (> 205°C; > 400°F)	98.1	96.4		
Light gas oil (220–345°C; 430–650°F)			11.2	
Distillate (C_5–345°C; C_5–650°F)				13.7
Vacuum gas oil				31.0
Sulfur, wt%				< 0.1
Carbon residue, wt%				0.3
Residuum (> 345°C; > 650°F)			82.2	
Sulfur, wt%			0.1	
Carbon residue, wt%			3.2	
Residuum (> 565°C; > 1050°F)				50.0
Sulfur, wt%				0.8
Carbon residue, wt%				15.7

[a] Arabian light/Arabian heavy blend.

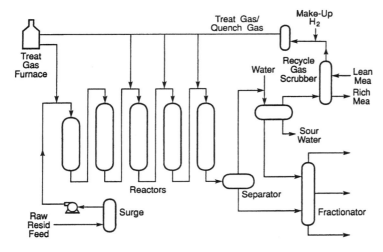

Figure 17.29 The residfining process.

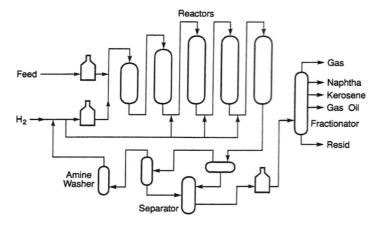

Figure 17.30 The residue hydroconversion (RHC) process.

A two-stage reactor system receives untreated feed, makeup hydrogen, and a recycle gas at the first stage, in which gasoline conversion may be as high as 60 vol%. The reactor effluent is separated into recycle gas, liquid product, and unconverted oil. The second-stage oil may be either once-through or recycle cracking; feed to the second stage is a mixture of unconverted first-stage oil and second-stage recycle.

The process operates satisfactorily for a variety of feedstocks that vary in sulfur content from about 1.0% by weight to about 5% by weight. The rate of desulfurization is dependent on the sulfur content of the feedstock (Tables 17.17 and 17.18), as are catalyst life, product sulfur, and hydrogen consumption (Speight, 2000 and references cited therein).

The high efficiency of the process is due to the excellent distribution of the feedstock and hydrogen that occurs in the reactor, where a proprietary liquid distribution system is employed (Fig. 17.32). In addition, the process catalyst (also proprietary) was designed for the desulfurization of residua and is not merely an upgraded gas oil hydrotreating catalyst as is often used in various processes. It is a cobalt-molybdena-alumina catalyst with a

Table 17.16 RHC Process Feedstock and Product Data

Feedstock: Unspecified residuum	
API Gravity	24.4
Sulfur, wt%	0.1
Nitrogen, wt%	0.1
Carbon residue, wt%	0.3
Products, vol%	
Naphtha	4.5
Light gas oil	19.4
Vacuum gas oil	77.10
Sulfur, wt%	0.01
Carbon residue, wt%	0.20

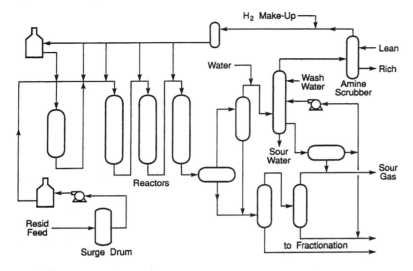

Figure 17.31 The Unicracking process.

controlled pore structure that permits a high degree of desulfurization and at the same time minimizes any coking tendencies.

2.16 Veba Combi-Cracking Process

The Veba Combi-Cracking (VCC) process is a thermal hydrocracking/hydrogenation process for converting residua and other heavy feedstocks (Table 17.19) (Niemann et al., 1988; Kamiya, 1991, p. 81; Wenzel and Kretsmar, 1993; *Hydrocarbon Processing*, 1998, p. 88). The process is based on the Bergius–Pier technology that was used for coal hydrogenation in Germany up to 1945. The heavy feedstock is hydrogenated (hydrocracked) using a commercial catalyst and a liquid-phase hydrogenation reactor operating at 440–485°C (825–905°F) and 2175–4350 psi (14,996–29,993 kPa). The product obtained from the reactor is fed into the hot separator, which operates at temperatures slightly below the reactor temperature. The liquid and solid materials are fed into a vacuum distillation column, and the gaseous products are fed into a gas-phase hydrogenation reactor operating at an identical pressure (Graeser and Niemann, 1982, 1983). This high temperature, high pressure coupling of the reactor products with further hydrogenation provides a specific process economics.

In the Veba Combi-Cracking process (Fig. 17.33), the residua feed is slurried with a small amount of finely powdered additive and mixed with hydrogen and recycle gas prior to preheating. The feed mixture is routed to the liquid-phase reactors. The reactors are operated in an upflow mode and arranged in series. In a once-through operation-conversion rates of > 95% are achieved. Substantial conversion of asphaltenes, desulfurization, and denitrogenation take place at high levels of residue conversion. Temperature is controlled by a recycle gas quench system.

The flow from the liquid-phase hydrogenation reactors is routed to a hot separator, where gases and vaporized products are separated from unconverted material. A vacuum flash recovers distillates in the hot separator bottom product.

Table 17.17 Unicracking Process Feedstock and Product Data

	Alaska North Slope atmospheric residuum	Gach Saran atmospheric residuum	Kuwait atmospheric residuum	Kuwait atmospheric residuum	California atmospheric residuum
Feedstock					
API gravity	15.2	16.3	16.7	14.4	9.9
Sulfur, wt%	1.7	2.4	3.8	4.2	4.5
Nitrogen, wt%	0.4	0.4	0.2	0.2	0.4
Metals (Ni + V), ppm	44.0	220.0	46.0	66.0	213.0
Carbon residue, wt%	8.4	8.5	8.5	10.0	13.6
Products, vol%					
Naphtha (< 185°C; < 365°F)	0.8	1.8	1.1		
Naphtha (C$_5$–205°C; C$_5$–400°F)				2.1	4.2
Light gas oil				11.1	18.0
Residuum (> 185°C; > 365°F)	100.5	100.4	100.4		
API gravity	19.8	22.8	24.4		
Sulfur, wt%	0.3	0.3	0.3		
Nitrogen, wt%	0.3	0.3	0.1		
Metals (Ni + V), ppm	14.0	55.0	15.0		
Carbon residue, wt%	4.0	4.0	3.0		
Residuum (> 345°C; > 650°F)					
API gravity				89.5	81.6
Sulfur, wt%				22.7	21.6
Nitrogen, wt%				< 0.3	< 0.3
Metals (Ni + V), ppm				< 0.2	< 0.2
				< 25	< 5

Table 17.18 Unicracking Desulfurization Rate Data

| Crude source | Resid properties, 650°F⁺ | | | | | |
	Crude (vol%)	Gravity (°API)	Sulfur (wt%)	Ni + V (ppm)	Conradson carbon (wt%)	Relative desulfurization rate
Murban	35.7	22.8	1.5	4	3.7	18.0
North Slope	52.6	15.1	1.7	45	8.9	7.75
Agha Jari (Iranian light)	42.7	17.5	2.4	98	7.7	7.25
Venezuelan (Leona)	55.0	18.9	2.1	187	8.6	7.0
Arabian light	44.7	17.9	2.9	28	7.3	6.5
L.A. Basin (Torrey Canyon)	48.4	8.4	2.0	236	11.7	4.5
West Texas Sour	41.6	15.5	3.4	29	9.0	4.5
Gach Saran (Iranian heavy)	46.7	14.6	2.2	258	9.9	4.25
Sassan	44.7	16.0	3.4	38	8.5	4.25
Kirkuk	39.3	17.6	3.7	84	8.4	4.0
Kuwait	50.5	15.5	3.9	51	8.8	3.5
Safaniyah (Arabian heavy)	53.1	12.3	4.2	123	13.3	1.5
Khafji	53.0	11.9	4.3	75	15.6	1.25
Ratawi	62.0	10.2	5.0	57	14.1	1.0

The hot separator top product, together with recovered distillates and straight-run distillates, enters the gas-phase hydrogenation reactor. The gas-phase hydrogenation reactor operates at the same pressure as the liquid-phase hydrogenation reactor and contains a fixed bed of commercial hydrotreating catalyst. The operation temperature (340–420°C) is controlled by a hydrogen quench. The system operates in a trickle flow mode. The separation of the synthetic crude from associated gases is performed in a cold separator system. The synthetic crude may be sent to a stabilization and fractionation unit as required. The gases are sent to a lean oil scrubbing system for contaminant removal and are recycled to the liquid phase hydrogenation system of the Veba Combi-Cracking process.

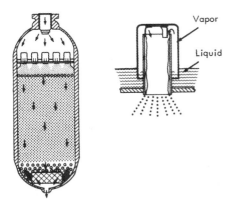

Figure 17.32 Unicracking process reactor.

Table 17.19 VCC Process Feedstock and Product Data

Feedstock: Arabian heavy vacuum residuum[a]	
API gravity	3.4
Sulfur, wt%	5.5
C_7 Asphaltenes, wt%	13.5
Metals (Ni + V), ppm	230.0
Carbon residue, wt%	8.4
Products, vol%	
Naphtha ($< 170°C$; $< 340°F$)	26.9
Middle distillate ($170–370°C$; $340–700°F$)	36.5
Gas oil ($> 370°C$; $700°F$)	19.9

[a] $> 550°C$ ($> 1025°F$); conversion 95%.

3. CATALYSTS

The increasing importance of hydrodesulfurization (HDS) and hydrodenitrogenation (HDN) in petroleum processing for the production of clean-burning fuels has led to a surge of research into the chemistry and engineering of heteroatom removal, with sulfur removal being the most prominent focus. Most of the earlier works focused on (1) catalyst characterization by physical methods, (2) low pressure reaction studies of model compounds having relatively high reactivity, (3) process development, or (4) cobalt-molybdenum (Co-Mo) catalysts, nickel-molybdenum catalysts (Ni-Mo), or nickel-tungsten (Ni-W) supported on alumina, often doped by fluorine or phosphorus.

The need to develop catalysts that can carry out deep hydrodesulfurization and deep hydrodenitrogenation has become even more pressing in view of recent environmental regulations limiting the amount of sulfur and nitrogen emissions. The development of a new generation of catalysts to achieve this objective of low nitrogen and sulfur levels in the processing of various feedstocks presents an interesting challenge for catalyst development.

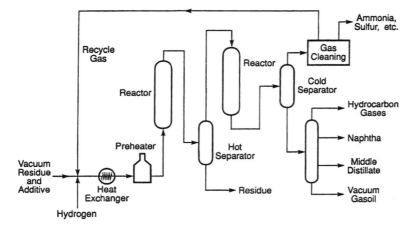

Figure 17.33 The Veba Combi-Cracking (VCC) process.

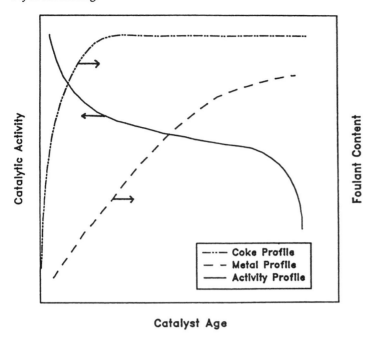

Catalyst Age

Figure 17.34 Schematic of time-dependent catalyst activity.

The deposition of coke and metals onto the catalyst diminishes the cracking activity of hydrocracking catalysts (Fig. 17.34). Basic nitrogen plays a major role because of the susceptibility of such compounds to the catalyst and their predisposition to form coke (Speight, 1999, 2000). However, zeolite catalysts can operate in the presence of substantial concentrations of ammonia, in marked contrast to silica-alumina catalysts, which are strongly poisoned by ammonia. Similarly, sulfur-containing compounds in a feedstock adversely affect the noble metal hydrogenation component of hydrocracking catalysts. These compounds are hydrocracked to hydrogen sulfide, which converts the noble metal to the sulfide form. The extent of this conversion is a function of the hydrogen and hydrogen sulfide partial pressures.

Removal of sulfur from the feedstock results in a gradual increase in catalyst activity, returning to almost the original activity level. As with ammonia, the concentration of the hydrogen sulfide can be used to control precisely the activity of the catalyst. Zeolite catalysts loaded with non-noble metals have an inherently different response to sulfur impurities, because a minimum level of hydrogen sulfide is required to maintain the nickel-molybdenum and nickel-tungsten in the sulfide state.

Hydrodenitrogenation is more difficult to accomplish than hydrodesulfurization, but the relatively smaller amounts of nitrogen-containing compounds in conventional crude oil (Chapter 3) renders this of little concern to refiners. However, the trend to heavier feedstocks in refinery operations, which are richer in nitrogen than the conventional feedstocks, has increased the awareness of refiners of the presence of nitrogen compounds in crude feedstocks. For the most part, hydrodesulfurization catalyst technology has been used to accomplish hydrodenitrogenation (Tøpsoe and Clausen, 1984), although such catalysts are not ideally suited to nitrogen removal (Katzer and Sivasubramanian,

1979). However, in recent years, the limitations of hydrodesulfurization catalysts when applied to hydrodenitrogenation have been recognized, and there are reports of attempts to manufacture catalysts more specific to nitrogen removal (Ho, 1988).

Hydrotreating processes are chemically very simple, because they essentially involve removal of sulfur and nitrogen as hydrogen sulfide and ammonia, respectively:

$$-S- + H_2 \rightarrow H_2S$$
$$2N\equiv + 3H_2 \rightarrow 2NH_3$$

However, nitrogen is the most difficult contaminant to remove from feedstocks, and processing conditions are usually dictated by the requirements for nitrogen removal.

In general, any catalyst capable of participating in hydrogenation reactions can be used for hydrodesulfurization. The sulfides of hydrogenating metals are particularly used for hydrodesulfurization, and catalysts containing cobalt, molybdenum, nickel, and tungsten are widely used on a commercial basis (Kellett et al., 1980).

Hydrotreating catalysts are usually cobalt-molybdenum catalysts, and under the conditions whereby nitrogen removal is accomplished, desulfurization usually occurs as well as oxygen removal. Indeed, it is generally recognized that the fullest activity of the hydrotreating catalyst is not reached until some interaction with the sulfur (from the feedstock) has occurred, with part of the catalyst metals converted to the sulfides. Too much interaction may, of course, lead to catalyst deactivation.

The reactions of hydrocracking require a dual-function catalyst with high cracking and hydrogenation activities. The catalyst base, e.g., an acid-treated clay, usually supplies the cracking function or alumina or silica-alumina that is used to support the hydrogenation function supplied by metals such as nickel, tungsten, platinum, and palladium. These highly acid catalysts are very sensitive to nitrogen compounds in the feed, which break down the conditions of reaction to give ammonia and neutralize the acid sites. Because many heavy gas oils contain substantial amounts of nitrogen (up to approximately 2500 ppm), a purification stage is frequently required. Denitrogenation and desulfurization can be carried out using cobalt-molybdenum or nickel-cobalt-molybdenum on alumina or silica-alumina.

Hydrocracking catalysts typically contain separate hydrogenation and cracking functions. Palladium sulfide and promoted group VI sulfides (nickel-molybdenum or nickel-tungsten) provide the hydrogenation function. These active compositions saturate aromatics in the feed, saturate olefins formed in the cracking, and protect the catalysts from poisoning by coke. Zeolites or amorphous silica-alumina provide the cracking functions. The zeolites are usually type Y (faujasite) ion exchanged to replace sodium with hydrogen and make up 25–50% of the catalysts. Pentasils (silicalite or ZSM-5) may be included in dewaxing catalysts.

Hydrocracking catalysts, such as nickel (5% by weight) on silica-alumina, work best on feedstocks that have been hydrofined to low nitrogen and sulfur levels. The nickel catalyst then operates well at 350–370°C (660–700°F) and a pressure of about 1500 psi to give good conversion of feed to lower boiling liquid fractions with minimum saturation of single-ring aromatics and a high isoparaffin to *n*-paraffin ratio in the lower molecular weight paraffins.

The poisoning effect of nitrogen can be offset to a certain degree by operation at a higher temperature. However, the higher temperature tends to increase the production of material in the methane (CH_4) to butane (C_4H_{10}) range and to decrease the operating stability of the catalyst so that it requires more frequent regeneration. Catalysts containing platinum or palladium (approximately 0.5 wt%) on a zeolite base appear to be somewhat less sensitive to nitrogen than nickel catalysts, and successful operation has been achieved with feedstocks containing 40 ppm nitrogen. This catalyst is also more tolerant of sulfur in the feed, which acts as a temporary poison, the catalyst recovering its activity when the sulfur content of the feed is reduced.

On such catalysts as nickel or tungsten sulfide on silica-alumina, isomerization does not appear to play any part in the reaction, because uncracked normal paraffins from the feedstock tend to retain their normal structure. Extensive splitting produces large amounts of low molecular weight (C_3–C_6) paraffins, and it appears that a primary reaction of paraffins is catalytic cracking followed by hydrogenation to form isoparaffins. With catalysts of higher hydrogenation activity, such as platinum on silica-alumina, direct isomerization occurs. The product distribution is also different, and the ratio of low to intermediate molecular weight paraffins in the breakdown product is reduced.

In addition to the chemical nature of the catalyst, the physical structure of the catalyst is also important in determining the hydrogenation and cracking capabilities, particularly for heavy feedstocks (Kobayashi et al., 1987; Fischer and Angevine, 1986; Kang et al., 1988; van Zijll Langhout et al., 1980). When gas oils and residua are used, the feedstock is present as liquids under the conditions of the reaction. Additional feedstock and the hydrogen must diffuse through this liquid before reaction can take place at the interior surfaces of the catalyst particle.

At high temperatures, reaction rates can be much higher than diffusion rates and concentration gradients can develop within the catalyst particle. Therefore, the choice of catalyst porosity is an important parameter. When feedstocks are to be hydrocracked to liquefied petroleum gas and gasoline, pore diffusion effects are usually absent. Catalysts of high surface area (about $300\,\mathrm{m^2/g}$) and low to moderate porosity (from 12 Å pore diameter with crystalline acidic components to 50 Å or more with amorphous materials) are used. With reactions involving high molecular weight feedstocks, pore diffusion can exert a large influence, and catalysts with pore diameters greater than 80 Å are necessary for more efficient conversion (Scott and Bridge, 1971).

Catalyst operating temperature can influence reaction selectivity because the activation energy is much lower for hydrotreating reactions than for hydrocracking reactions. Therefore, raising the temperature in a residuum hydrotreater increases the extent of hydrocracking relative to hydrotreating, which also increases hydrogen consumption (Bridge et al., 1975, 1981).

Aromatic hydrogenation in petroleum refining may be carried out over supported metal or metal sulfide catalysts depending on the sulfur and nitrogen levels in the feedstock. For hydrorefining of feedstocks that contain appreciable concentrations of sulfur and nitrogen, sulfided nickel-molybdenum (Ni-Mo), nickel-tungsten (Ni-W), or cobalt-molybdenum (Co-Mo) on alumina (γ-Al_2O_3) catalysts are generally used, whereas supported noble metal catalysts have been used for sulfur- and nitrogen-free feedstocks. Catalysts containing noble metals on Y-zeolites have been reported to be more sulfur tolerant than those on other supports (Jacobs, 1986). Within the series of cobalt- or nickel-promoted group VI metal (Mo or W) sulfides supported on γ-Al_2O_3, the ranking for hydrogenation is

Ni-W > Ni-Mo > Co-Mo > Co-W

and nickel-tungsten (Ni-W) and nickel-molybdenum (Ni-Mo) on Al_2O_3 catalysts are widely used to reduce levels of sulfur, nitrogen, and aromatics in petroleum fractions by hydrotreating.

Molybdenum sulfide (MoS_2), usually supported on alumina, is widely used in petroleum processes for hydrogenation reactions. It has a layered structure and can be made much more active by addition of cobalt or nickel. When promoted with cobalt sulfide (CoS), making what is called a *cobalt-moly* catalyst, it is widely used in hydrodesulfurization (HDS) processes. The nickel sulfide (NiS)-promoted version is used for hydrodenitrogenation (HDN) as well as hydrodesulfurization (HDS). The closely related tungsten compound (WS_2) is used in commercial hydrocracking catalysts. Other sulfides [iron sulfide (FeS), chromium sulfide (Cr_2S_3), and vanadium sulfide (V_2S_5)] are also effective and are used in some catalysts. A valuable alternative to the base metal sulfides is palladium sulfide (PdS). Although it is expensive, palladium sulfide forms the basis for several very active catalysts.

Clays have been used as cracking catalysts, particularly for heavy feedstocks (Chapter 15), and their use has also been explored for the demetallization and upgrading of heavy crude oil (Rosa-Brussin, 1995). The results indicated that the catalyst prepared was mainly active toward demetallization and conversion of the heaviest fractions of crude oils.

The choice of hydrogenation catalyst depends on what the catalyst designer wishes to accomplish. In catalysts to make gasoline, for instance, vigorous cracking is needed to convert a large fraction of the feed to the kinds of molecules that will make a good gasoline blending stock. For this vigorous cracking, a vigorous hydrogenation component is needed. Because palladium is the most active catalyst for this, the extra expense is warranted. On the other hand, many refiners wish only to make acceptable diesel, a less demanding application. For this, the less expensive molybdenum sulfides are adequate.

The cracking reaction results from the attack of a strong acid on a paraffinic chain to form a carbonium ion (carbocation, e.g., R^+) (Dolbear, 1998). These are two fundamental types of strong acids, Brønsted and Lewis acids. *Brønsted acids* are the familiar proton-containing acids; *Lewis acids* are a broader class including inorganic and organic species formed by positively charged centers. Both kinds have been identified on the surfaces of catalysts; sometimes both kinds of sites occur on the same catalyst. The mixture of Brønsted and Lewis acids sometimes depends on the level of water in the system.

Examples of Brønsted acids are the familiar proton-containing species such as sulfuric acid (H_2SO_4). Acidity is provided by the very active hydrogen ion (H^+), which has a very high positive charge density. It seeks out centers of negative charge such as the π electrons in aromatic centers. Such reactions are familiar to organic chemistry students, who are taught that bromination of aromatics takes place by attack of the bromonium ion (Br^+) on such a ring system. The proton in strong acid systems behaves in much the same way, adding to the π electrons and then migrating to a site of high electron density on one of the carbon atoms.

However, species other than proton donors behave as acids. These acids all have high positive charge densities. Examples are aluminum chloride ($AlCl_3$) and the bromonium ion (Br^+). Such strong positive species have become known as Lewis acids. This class obviously includes proton acids, but the latter are usually designated Brønsted acids.

In reactions with hydrocarbons, both Lewis and Brønsted acids can catalyze cracking reactions. For example, the proton in Brønsted acids can add to an olefinic double bond to form a carbocation. Similarly, a Lewis acid can abstract a hydride from the corresponding paraffin to generate the same intermediate (Dolbear, 1998). Although these reactions are written to show identical intermediates, in real catalytic systems the intermediates would be different. This is because the carbocations would probably be adsorbed on different surface sites of the two kinds of catalysts.

Zeolites and amorphous silica-alumina provide the cracking function in hydrocracking catalysts. They have similar chemistry at the molecular level, but the crystalline structure of the zeolites provides higher activities and controlled selectivity not found in the amorphous materials.

Chemists outside the catalyst field are often surprised that a solid can have strong acid properties. In fact, many solid materials have acid strength matching that of concentrated sulfuric acid. Some specific examples are

1. Amorphous silica-alumina (SiO_2/Al_2O_3)
2. Zeolites
3. Activated (acid-leached) clays
4. Aluminum chloride ($AlCl_3$) and many related metal chlorides
5. Amorphous silica-magnesia compounds (SiO_2/MgO)
6. Chloride-promoted alumina ($Al_2O_3 \cdot Cl$)
7. Phosphoric acid supported on silica gel (H_3PO_4/SiO_2)

Each of these is applied in one or more commercial catalysts in the petroleum refining industry. For commercial hydrocracking catalysts, only zeolites and amorphous silica-alumina are used commercially.

In 1756 Baron Axel F. Cronstedt, a Swedish mineralogist, made the observation that when certain minerals were heated sufficiently, they bubbled as if they were boiling. He called the substances *zeolites* (from the Greek *zeo*, to boil, and *lithos*, stone). They are now known to consist primarily of silicon, aluminum, and oxygen and to host an assortment of other elements. In addition, zeolites are highly porous crystals veined with submicroscopic channels. The channels contain water (hence the bubbling at high temperatures), which can be eliminated by heating (combined with other treatments) without altering the crystal structure (Occelli and Robson, 1989).

Typical naturally occurring zeolites include *analcite* (also called *analcime*), $Na(AlSi_2O_6)$, and *faujasite*, $Na_2Ca(AlO_2)_2(SiO_2)_4 \cdot H_2O$, which are the structural analogs of the synthetic *zeolite X* and *zeolite Y*. Sodalite ($Na_8[(Al_2O_2)_6(SiO_2)_6]Cl_2$) contains the truncated octahedral structural unit known as the *sodalite cage* that is found in several zeolites. The corners of the faces of the cage are defined by either four or six Al/Si atoms, which are joined together through oxygen atoms. The zeolite structure is generated by joining sodalite cages through the Si/Al rings, so enclosing a cavity or *supercage* bounded by a cube of eight sodalite cages and readily accessible through the faces of that cube (channels or pores). Joining sodalite cages together through the six Si/Al faces generates the structural frameworks of faujasite, zeolite X, and zeolite Y. In zeolites, the effective width of the pores is usually controlled by the nature of the cation (M^+ or M^{2+}).

Natural zeolites form hydrothermally (e.g., by the action of hot water on volcanic ash or lava), and synthetic zeolites can be made by mixing solutions of aluminates and silicates and maintaining the resulting gel at temperatures of 100°C (212°F) or higher for appropriate periods (Swaddle, 1997). *Zeolite A* can form at temperatures below 100°C (212°F), but most zeolite syntheses require hydrothermal conditions (typically 150°F/300°F at the appropriate pressure). The reaction mechanism appears to involve dissolution of the gel and precipitation as the crystalline zeolite, and the identity of the zeolite produced depends on the composition of the solution. Aqueous alkali metal hydroxide solutions favor zeolites with relatively high aluminum contents, whereas the presence of organic molecules such as amines or alcohols favors highly siliceous zeolites such as *silicalite* or ZSM-5. Various tetraalkylammonium cations favor the formation of specific zeolite structures and are known as *template* ions, although it should not be supposed that the channels and cages form simply by the wrapping of aluminosilicate fragments around suitably shaped cations.

Zeolite catalysts have also found use in the refining industry during the last two decades. Like the silica-alumina catalysts, zeolites also consist of a framework of tetrahedra usually with a silicon or aluminum atom at the center. The geometrical characteristics of the zeolites are responsible for their special properties, which are particularly attractive to the refining industry (DeCroocq, 1984). Specific zeolite catalysts have shown up to 10,000 times more activity than the so-called conventional catalysts in specific cracking tests. The mordenite-type catalysts are particularly worthy of mention, because they have shown up to 200 times greater activity for hexane cracking in the temperature range 360–400°C (680–750°F).

Other zeolite catalysts have also shown remarkable adaptability to the refining industry. For example, the resistance to deactivation of the type Y zeolite catalysts containing either noble or non-noble metals is remarkable, and catalyst life of up to 7 years has been obtained commercially in processing heavy gas oils in the Unicracking-JHC processes. Operating life depends on the nature of the feedstock, the severity of the operation, and the nature and extent of operational upsets. Gradual catalyst deactivation in commercial use is counteracted by incrementally raising the operating temperature to maintain the required conversion per pass. The more active a catalyst, the lower is the temperature required. When processing for gasoline, lower operating temperatures have the additional advantage that less of the feedstock is converted to isobutane.

Any given zeolite is distinguished from other zeolites by structural differences in their unit cell, which are tetrahedral structures arranged in various combinations. Oxygen atoms establish the four vertices of each tetrahedron, which are bound to, and enclose, either a silicon (Si) or an aluminum (Al) atom. The vertex oxygen atoms are each shared by two tetrahedra, so every silicon or aluminum atom within the tetrahedral cage is bound to four neighboring caged atoms through an intervening oxygen. The number of aluminum atoms in a unit cell is always smaller than, or at most equal to, the number of silicon atoms, because two aluminum atoms never share the same oxygen.

The aluminum is actually in the ionic form and can readily accommodate electrons donated from three of the bound oxygen atoms. The electron donated by the fourth oxygen imparts a negative, or anionic, charge to the aluminum atom. This negative charge is balanced by a cation from the alkali metal or alkaline earth groups of the periodic table. Such cations are commonly sodium, potassium, calcium, or magnesium. These cations play a major role in many zeolite functions and help to attract polar molecules such as water. However, the cations are not part of the zeolite framework and can be exchanged for other cations without any effect on crystal structure.

Zeolites provide the cracking function in many hydrocracking catalysts, as they do in fluid catalytic cracking catalysts. The zeolites are crystalline aluminosilicates, and in almost all commercial catalysts today, the zeolite used is faujasite. Pentasil zeolites, including silicalite and ZSM-5, are also used in some catalysts for their ability to crack long chain paraffins selectively.

Typical levels are 25–50 wt% zeolite in the catalysts, with the remainder being the hydrogenation component and a silica (SiO_2) or alumina (Al_2O_3) binder. Exact recipes are guarded as trade secrets.

Crystalline zeolite compounds provide a broad family of solid acid catalysts. The chemistry and structures of these solids are beyond the scope of this book. What is important here is that the zeolites are not acidic as crystallized. They must be converted to acidic forms by ion-exchange processes. In the process of this conversion, the chemistry of the crystalline structure is often changed. This complication provides tools for controlling the catalytic properties, and much work has been done on understanding and applying these reactions as a way to make catalysts with higher activities and more desirable selectivity.

As an example, the zeolite faujasite crystallizes with the composition $SiO_2(NaAlO_2)_x \cdot (H_2O)_y$. The ratio of silicon to aluminum, expressed here by the subscript x, can be varied in the crystallization from 1 to greater than 10. What does not vary is the total number of silicon and aluminum atoms per unit cell, 192. For legal purposes, to define certain composition of matter patents, zeolites with a ratio of 1–1.5 are called type X; those with ratios greater than 1.5 are type Y.

Both silicon and aluminum in zeolites are found in tetrahedral oxide sites. The four oxides are shared with another silicon or aluminum (except that two aluminum ions are never found in adjacent, linked tetrahedra). Silicon with a 4+ charge balances exactly half of the charge of the oxide ions it is linked to; because all of the oxygens are shared, silicon balances all of the charge around it and is electrically neutral. Aluminum, with a charge of 3+, leaves one charge unsatisfied; sodium neutralizes this charge.

The sodium, as expected from its chemistry, is not linked to the oxides by covalent bonds as the silicon and aluminum are. The attraction is simply *ionic*, and sodium can be replaced by other cations by ion-exchange processes. In extensive but rarely published experiments carried out by catalyst designers, virtually every metallic and organic cation has been exchanged into zeolites.

The most important ion exchanged for sodium is the proton. In the hydrogen ion form, faujasite zeolites are very strong acids, with strengths approaching that of oleum. Unfortunately, direct exchange using a mineral acid such as hydrochloric acid is not practical. The acid tends to attack the silica-alumina network, in the same way that strong acids attack clays in the activation processes developed by Houdry. The technique adopted to avoid this problem is indirect exchange, beginning with exchange of an ammonium ion for the sodium. When heated to a few hundred degrees, the ammonium decomposes, forming gaseous ammonia and leaving behind a proton:

$$R^-NH_4^+ \rightarrow R^-H^+ + NH_3 \uparrow$$

This step is accompanied by a variety of solid-state reactions that can change the zeolite structure in subtle but important ways.

Zeolites provided a breakthrough that allowed catalytic hydrocracking to become commercially important, and continued advances in the manufacture of amorphous silica-alumina made these materials competitive in certain kinds of applications. This was important because patents controlled by Unocal and Exxon dominated the application of zeolites in this area. Developments in amorphous catalysts by Chevron and UOP allowed them to compete actively in this area.

Typical catalysts of this type contain 60–80 wt% silica-alumina, with the remainder being the hydrogenation component. The compositions of these catalysts are closely held secrets. Over the years, broad ranges of silica/alumina molar ratios have been used in various cracking applications, but silica is almost always in excess for high acidity and stability. A typical level might be 25 wt% alumina.

Amorphous silica-alumina is made by a variety of precipitation techniques. The whole class of materials traces its beginnings to silica gel technology, in which sodium silicate is acidified to precipitate the hydrous silica-alumina sulfate; sulfuric acid is used as some or all of the acid for this precipitation, and a mixed gel is formed. The properties of this gel, including acidity and porosity, can be varied by changing the recipe—concentrations, order of addition, pH, temperature, aging time, and the like. The gels are isolated by filtration and are washed to remove sodium and other ions.

Careful control of the precipitation allows the pore size distributions of amorphous materials to be controlled rather well, but the distributions are still much broader than those in the zeolites. This limits the activity and selectivity. One effect of the reduced activity has been that these materials have been applied only in making middle distillates: diesel and turbine fuels. At higher process severities, the poor selectivity results in production of unacceptable amounts of methane (CH_4) to butane (C_4H_{10}) hydrocarbons.

Hydrocarbons, especially aromatic hydrocarbons, can react in the presence of strong acids to form coke, a complex polynuclear aromatic material that is low in hydrogen. Coke can deposit on the surface of a catalyst, blocking access to the active sites and reducing the activity of the catalyst. Coke poisoning is a major problem in fluid catalytic cracking catalysts, where coked catalysts are circulated to a fluidized-bed combustor to be regenerated. In hydrocracking, coke deposition is virtually eliminated by the catalyst's hydrogenation function.

However, the product referred to as *coke* is not a single material. The first products deposited are tarry deposits that can, with time and temperature, continue to polymerize. Acid catalyzes these polymerizations. The stable product would be graphite, with very large aromatic sheets and no hydrogen. This product forms only with aging at very high temperatures, far more severe than that found in a hydrocracker. The graphitic material is both more thermodynamically stable and less kinetically reactive. This kinetic stability results from the lack of easily hydrogenated functional groups.

In a well-designed hydrocracking system, the hydrogenation function adds hydrogen to the tarry deposits. This reduces the concentration of coke precursors on the surface. There is, however, a slow accumulation of coke that reduces activity over a 1–2 year period. Refiners respond to this slow reduction in activity by raising the average temperature of the catalyst bed to maintain conversions. Eventually, however, an upper limit to the allowable temperature is reached and the catalyst must be removed and regenerated.

Catalysts carrying coke deposits can be regenerated by burning off the accumulated coke. This is done by service in rotary or similar kilns rather than by leaving catalysts in the hydrocracking reactor, where the reactions could damage the metals in the walls. Removing the catalysts also allows inspection and repair of the complex and expensive reactor internals, discussed below. Regeneration of a large catalyst charge can take weeks or months, so refiners may own two catalyst loads, one in the reactor, and one regenerated and ready for reload.

The thermal reactions also convert the metal sulfide hydrogenation functions to oxides and may result in agglomeration. Excellent progress has been made since the 1970s in regenerating hydrocracking catalysts; similar regeneration of hydrotreating catalysts is widely practiced.

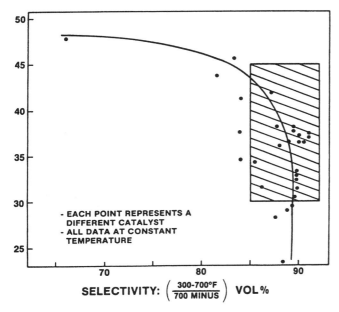

Figure 17.35   Variation of catalyst selectivity with conversion

After combustion to remove the carbonaceous deposits, the catalysts are treated to disperse active metals. Vendor documents claim more than 95% recovery of activity and selectivity in these regenerations. Catalysts can undergo successive cycles of use and regeneration, providing long functional life with these expensive materials.

Hydrocracking allows refiners the potential to balance fuel oil supply and demand by adding vacuum gas oil cracking capacity. Thus, hydrocracking is of particular value to (1) refineries with no existing vacuum gas oil cracking capacity, (2) refineries with a larger supply of vacuum gas oil than their vacuum gas oil conversion capacity, (3) refineries where the addition of vacuum residuum conversion capacity has resulted in production of additional crackable stocks boiling in the vacuum gas oil range (e.g., coker gas oil), and (4) refineries that have one of the two types of vacuum gas oil conversion units but could benefit from adding the second type. In some cases a refiner might add both gas oil cracking and residuum conversion capacity simultaneously.

Those refiners who do choose gas oil cracking as part of their strategy for balancing residual fuel oil supply and demand must decide whether to select a hydrocracker or a fluid catalytic cracking unit. Although the two processes have been compared vigorously over the years, neither process has evolved to be the universal choice for gas oil cracking. Both processes have their advantages and disadvantages, and process selection can be properly made only after careful consideration of many case-specific factors. Among the most important factors are (1) the product slate required, (2) the amount of flexibility required to vary the product slate, (3) the product quality (specifications) required, and (4) the need to integrate the new facilities in a logical and cost-effective way with any existing facilities.

As illustrated above for various forms of more conventional hydrocracking, the type of catalyst used can influence the product slate obtained (Fig. 17.35). For example, for a mild hydrocracking operation at constant temperature, the selectivity of the catalyst varies from about 65% to about 90% by volume. Indeed, several catalytic systems have now been developed with a group of catalysts specifically for mild hydrocracking operations. Depending on the type of catalyst, they may be run as a single catalyst or in conjunction with a hydrotreating catalyst.

REFERENCES

Aalund, L. R. 1981. Oil Gas J. 79(13):70; 79(37):69.

Bauman, R. F., Aldridge, C. L., Bearden, R., Jr., Mayer, F. X., Stuntz, G. F., Dowdle, L. D., and Fiffron, E. 1993. Preprints. Oil Sands—Our Petroleum Future. Alberta Research Council, Edmonton, AB, Canada, p. 269.

Bearden, R., and Aldridge, C. L. 1981. Energy Progr. 1:44.

Bishop, W. 1990. Proceedings. Symposium on Heavy Oil: Upgrading to Refining. Canadian Society for Chemical Engineers, p. 14.

Bland, W. F., and Davidson, R. L. 1967. Petroleum Processing Handbook. McGraw-Hill, New York.

Boening, L. G., McDaniel, N. K., Petersen, R. D., and van Driesen, R. P. 1987. Hydrocarbon Process. 66(9):59.

Bridge, A. G. 1997. In: Handbook of Petroleum Refining Processes. 2nd ed. R. A. Meyers, ed. McGraw-Hill, New York, Chapter 7.2.

Bridge, A. G., Reed, E. M., and Scott, J. W. 1975. Paper presented at the API Midyear Meeting, May.

Bridge, A. G., Gould, G. D., and Berkman, J. F. 1981. Oil Gas J. 79(3):85.

Celestinos, J. A., Zermeno, R. G., Van Driesen, R. P., and Wysocki, E. D. 1975. Oil Gas J. 73(48):127.

Corbett, R. A. 1989. Oil Gas J. 87(26):33.

DeCroocq, D. 1984. Catalytic Cracking of Heavy Petroleum Hydrocarbons. Editions Technip, Paris.

Dolbear, G. E. 1998. In: Petroleum Chemistry and Refining. J. G. Speight, ed. Taylor & Francis, Washington, DC, Chapter 7.

Fischer, R. H., and Angevine, P. V. 1986. Appl. Catal. 27:275.

Fornoff, L. L. 1982. Proc. Second Int. Conf. on the Future of Heavy Crude and Tar Sands, Caracas, Venezuela.

Graeser, U., and Niemann, K. 1982. Oil Gas J. 80(12):121.

Graeser, U., and Niemann, K. 1983. Preprints. Am. Chem. Soc. Div. Petrol. Chem. 28(3):675.

Gray, M. R. 1994. Upgrading Petroleum Residues and Heavy Oils. Marcel Dekker, New York.

Ho, T. C. 1988. Catal. Rev. Sci. Eng. 30:117.

Howell, R. L., Hung, C., Gibson, K. R., and Chen, H. C. 1985. Oil Gas J. 83(30):121.

Hunter, M. G., Pasppal, D. A., and Pesek, C. L. 1997. In: Handbook of Petroleum Refining Processes. 2nd ed. R. A. Meyers, ed. McGraw-Hill, New York, Chapter 7.1.

Hydrocarbon Processing. 1998. Refining processes '96. 77(11):53.

Jacobs, P. A. 1986. In: Metal Clusters in Catalysis. Studies in Surface Science and Catalysis, Vol. 29. B. C. Gates, ed. Elsevier, Amsterdam p. 357.

Kamiya, Y., ed. 1991. Heavy Oil Processing Handbook. Research Association for Residual Oil Processing (RAROP). Ministry of International Trade and Industry, Tokyo, Japan.

Kang, B. C., Wu, S. T., Tsai, H. H., and Wu, J. C. 1988. Appl. Catal. 45:221.

Katzer, J. R., and Sivasubramanian, R. 1979. Catal. Rev. Sci. Eng. 20:155.

Kellett, T. F., Trevino, C. A., and Sartor, A. F. 1980. Oil Gas J. 78(18):244.

Khan, M. R., and Patmore, D. J. 1998. In: Petroleum Chemistry and Refining. J. G. Speight, ed. Taylor & Francis, Washington, DC, Chapter 6.

Kobayashi, S., Kushiyama S., Aizawa, R., Koinuma, Y., Inoue, K., Shmizu, Y., and Egi, K. 1987. Ind. Eng. Chem. Res. 26:2241, 2245.

Kriz, J. F., and Ternan, M. 1994. U.S. Patent 5,296,130. March 22.

Murphy, J. R., and Treese, S. A. 1979. Oil Gas J. 77(26):135.

Nasution, A. S. 1986. Preprints. Div. Petrol. Chem. Am. Chem. Soc. 31(3):722.

Niemann, K., Kretschmar, K., Rupp, M., and Merz, L. 1988. Proceedings 4th UNITAR/UNDP Int. Conf. on Heavy Crude and Tar Sand. Edmonton, AB, Canada. 5:225.

Occelli, M. L., and Robson, H. E. 1989. Zeolite Synthesis. ACS Symp. Ser. No. 398. Am. Chem. Soc., Washington, DC.

Pruden, B. B. 1978. Can. J. Chem. Eng. 56:277.

Pruden, B. B., Muir, G., and Skripek, M. 1993. Preprints. Oil Sands—Our Petroleum Future. Alberta Research Council, Edmonton, AB, Canada, p. 277.

Radler, M. 1999. Oil Gas J. 97(51):445.

Reich, A., Bishop, W., and Veljkovic, M. 1993. Preprints. Oil Sands—Our Petroleum Future. Alberta Research Council, Edmonton, AB, Canada, p. 216.

Reynolds, J. G., and Beret, S. 1989. Fuel Sci. Technol. Int. 7:165.

Rosa-Brussin, M.F. 1995. Catal. Rev. Sci. Eng. 37(1):1.

Scherzer, J., and Gruia, A. J. 1996. Hydrocracking Science and Technology. Marcel Dekker, New York.

Scott, J. W., and Bridge, A. G. 1971. In: Origin and Refining of Petroleum. H. G. McGrath and M. E. Charles, eds. Adv. Chem. Ser. 103. Am. Chem. Soc., Washington, DC, p. 113.

Speight, J. G. 1986. Annu. Rev. Energy 11:253.

Speight, J. G. 1990. In: Fuel Science and Technology Handbook. J. G. Speight, ed. Marcel Dekker, New York, Chapters 12–16.

Speight, J. G. 1999. The Chemistry and Technology of Petroleum. 3rd ed. Marcel Dekker, New York.

Speight, J. G. 2000. The Desulfurization of Heavy Oils and Residua. 2nd ed. Marcel Dekker, New York.

Speight, J. G., and Moschopedis, S. E. 1979. Fuel Process. Technol. 2:295.

Suchanek, A. J., and Moore, A. S. 1986. Oil Gas J. 84(31):36.

Swaddle, T. W. 1997. Inorganic Chemistry. Academic Press, New York.

Thompson, G. J. 1997. In: Handbook of Petroleum Refining Processes. R. A. Meyers, ed. McGraw-Hill, New York, Chapter 8.4.

Tøpsoe, H., and Clausen, B. S. 1984. Catal. Rev. Sci. Eng. 26:395.

Towler, G. P., Mann, R., Serriere, A. J. L., and Gabaude, C. M. D. 1996. Ind. Eng. Chem. Res. 35:278.

van Driesen, R. P., Caspers, J., Campbell, A. R., Lunin, G. 1979. Hydrocarbon Process. 58(5):107.

van Zijll Langhout, W. C., Ouwerkerk, C., and Pronk, K. M. A. 1980. Oil Gas J. 78(48):120.

Waugh, R. J. 1983. Annual Meeting, National Petroleum Refiners Association, San Francisco, CA.

Wenzel, F., and Kretsmar, K. 1993. Preprints. Oil Sands—Our Petroleum Future. Alberta Research Council, Edmonton, AB, Canada, p. 248.

Wilson, J. 1985. Energy Proc. 5:61.

18

The Next Generation Process

1. INTRODUCTION

Petroleum refining is now in a significant transition period as the industry moves into the twenty-first century; the demand for petroleum and petroleum products has shown a sharp growth in recent decades. At the same time, crude oils available to refineries have generally decreased in API gravity (Swain, 1991, 1993, 1998). Thus, the essential step required of refineries is the upgrading of heavy feedstocks, particularly residua (McKetta, 1992; Dickenson et al., 1997).

To satisfy the changing pattern of product demand, significant investments in refining conversion processes will be necessary to profitably utilize heavy feedstocks. The most efficient and economical solution to this problem will depend to a large extent on individual refinery situations. However, the most promising technologies will likely involve the conversion of vacuum residua. Technologies are needed that will take the feedstock beyond current limits (Fig. 18.1) and at the same time reduce the amount of coke and other nonessential products. Such a goal may require the use of two or more technologies in series rather than an attempt to develop a whole new one-stop conversion technology (Fig. 18.2). Such is the nature of the refinery.

Refineries can have any configuration or a conjunction of several, but the refinery of the future will of necessity be required to be a *conversion refinery*. A conversion refinery incorporates all the basic building blocks found in both the *topping refinery* and the *hydroskimming refinery*, but it also features gas oil conversion plants such as catalytic cracking and hydrocracking units, olefin conversion plants such as alkylation or polymerization units, and, frequently, coking units for sharply reducing or eliminating the production of residual fuels. Conversion refineries currently produce as much as two-thirds of their output as unleaded gasoline, with the balance distributed between liquefied petroleum gas (LPG), high quality jet fuel, low sulfur diesel fuel, and coke. Many such refineries also incorporate solvent extraction processes for manufacturing lubricants and petrochemical units with which to recover high purity propylene, benzene, toluene, and xylenes for further processing into polymers.

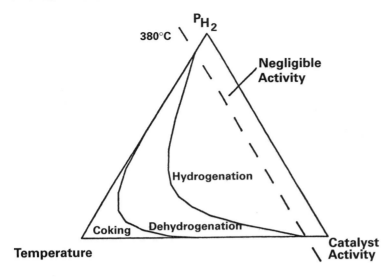

Figure 18.1 Process limits of current technologies. Regions of operation for thermal and catalytic processes are shown in terms of hydrogen pressure, temperature, and catalyst activity.

The manner in which refineries convert heavy feedstocks into low-boiling high value products has become a major focus of operations, with new concepts evolving into new processes (Khan and Patmore, 1998; Speight, 1999, 2000). Even though they may not be classed as conversion processes per se, pretreatment processes for removing asphaltene, metals, sulfur, and nitrogen constituents are also important and can play an important role.

Upgrading heavy oils and residua began with the introduction of desulfurization processes (Speight, 1984, 1999, 2000). In the early days the goal was desulfurization, but in later years the processes were adapted to a 10–30% partial conversion operation intended to achieve desulfurization while simultaneously obtaining low-boiling fractions by increasing severity in operating conditions. Thus, refinery evolution has seen the introduction of a variety of residuum cracking processes based on thermal cracking, catalytic cracking, and hydroconversion. Those processes differ from one another in cracking method, cracked product patterns, and product properties and are employed in refineries according to their respective features.

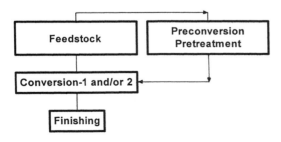

Figure 18.2 Potential processing routes for future refineries.

Using a refinery operation such as the one shown schematically in Fig. 18.3, new processes for the conversion of residua and heavy oils will be used perhaps not in place of but in conjunction with visbreaking and coking options, with some degree of hydro-processing as a primary conversion step. In addition, other processes may replace or, more likely, augment the deasphalting units in many refineries. An exception, which may become the rule, is the upgrading of bitumen from tar sands (Speight, 1990, 2000). The bitumen is subjected to either delayed coking or fluid coking as the primary upgrading step (Fig. 18.4). After primary upgrading, the product streams are hydrotreated and combined to form a synthetic crude oil that is shipped to a conventional refinery for further processing.

Conceivably, a heavy feedstock could be upgraded in the same manner and, depend-ing upon the upgrading facility, upgraded further for sales. However, this is not to be construed as meaning that bitumen upgrading will always involve a coking step as the primary upgrading step. Other options, including some presented elsewhere (Chapter 16) and later in this chapter could well become predominant methods for bitumen upgrading in the future.

Previous chapters (Chapters 14,15,16, and 17) have dealt with the processes used in refineries that were developed initially for application to conventional feedstocks and then to residua and other heavy feedstocks (Fig. 18.3). In this chapter we deal with those processes that are relative latecomers to refinery scenarios insofar as they have come into being and evolved during the last three decades and were developed (and some have been installed) to address the refining of the heavy feedstocks and, where necessary, hydrogen management (Kamiya, 1991; Shih and Oballa, 1991; Dettman et al., 1993; Bridge, 1997; Khan and Patmore, 1998; *Hydrocarbon Processing*, 1998; Speight, 1999, 2000).

As a brief introduction, technologies for upgrading heavy crude oils such as heavy oil, bitumen, and residua can be broadly divided into carbon rejection processes, hydrogen addition processes, and separation processes.

1. *Carbon rejection processes*: Visbreaking, steam cracking, fluid catalytic cracking, coking, and flash pyrolysis; deasphalting may also be considered a carbon rejec-tion process insofar as the rejected material is relatively (feedstock basis) high carbon and low hydrogen.
2. *Hydrogen addition processes*: Catalytic hydroconversion (hydrocracking) using active hydrodesulfurization catalysts, fixed-bed catalytic hydroconversion, ebul-lated catalytic bed hydroconversion, thermal slurry hydroconversion (hydrocrack-ing), hydrovisbreaking, hydropyrolysis, and donor solvent processes.
3. *Separation processes*: Distillation, deasphalting, and extraction.

Carbon rejection processes bring about the redistribution of hydrogen among the various components, resulting in fractions with increased hydrogen/carbon (H/C) atomic ratios and fractions with lower H/C atomic ratios. On the other hand, *hydrogen addition* processes involve the reaction of heavy crude oils with an external source of hydrogen and result in an overall increase in the H/C ratio. *Separation processes* involve either the discharge (as insoluble material) of a (relatively) high carbon, low hydrogen fraction or the extraction (as soluble material) of a low carbon, high hydrogen fraction from the feedstock through the use of a solvent (Speight, 1999, 2000, and references cited therein).

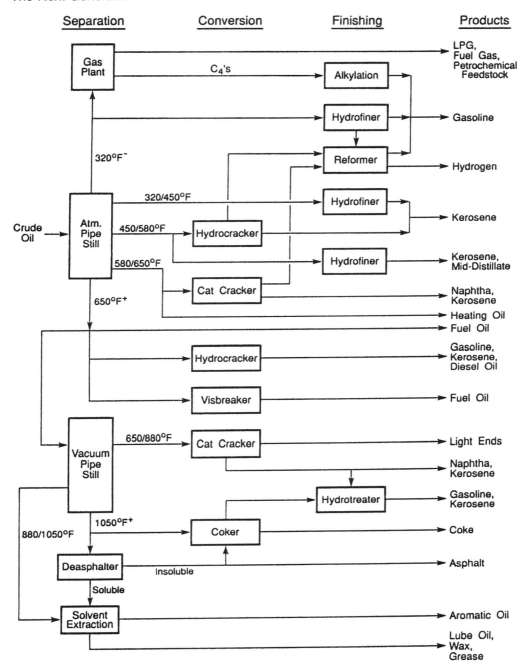

Figure 18.3 Schematic of a refinery showing the relative positions of the conversion units.

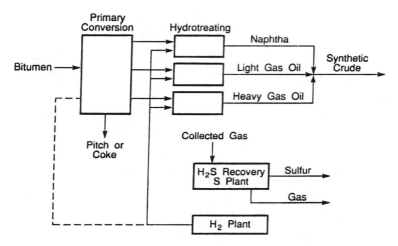

Figure 18.4 A current upgrading facility for tar sand bitumen.

In the not too distant past, and even now, the mature and well-established processes such as visbreaking, delayed coking, fluid coking, flexicoking, propane deasphalting, and butane deasphalting were deemed adequate for upgrading heavy feedstocks.

More options are now being sought to increase process efficiency in terms of the yields of desired products. Thus, it is the purpose of this chapter to present an outline of these options, including some processes that have been on-stream in prior decades. It is the novelty of the concept that leads to the inclusion of such a process (*Hydrocarbon Processing*, 1996, 1998; Meyers, 1997; Speight, 1999, 2000 and references cited therein). Upgrading will be discussed in terms of the type of technology, with occasional reference to the chemistry of the process.

2. THERMAL (CARBON REJECTION) PROCESSES

Thermal cracking processes offer attractive methods of feedstock conversion at low operating pressure without requiring expensive catalysts. Currently, the most widely operated residuum conversion processes are visbreaking, delayed coking, and fluid coking, and these are still attractive processes for refineries from an economic point of view (Chapter 1) (Dickenson et al., 1997).

2.1 Asphalt Coking Technology Process

The asphalt coking technology (ASCOT) process is a residual oil upgrading process that integrates the delayed coking process and the deep solvent deasphalting process (low energy deasphalting, LEDA) (page 579) (Bonilla, 1985; Bonilla and Elliot, 1987; Kamiya, 1991, p. 3; *Hydrocarbon Processing*, 1996). The product yields can be varied according to the desired distribution of products by emphasizing one or both parts of the process (Table 18.1).

In the process, the vacuum residuum is brought to the desired extraction temperature [50–230°C (120–445°F) at 300–500 psig (2060–3430 kPa)] and then sent to the extractor, where solvent (straight-run naphtha, coker naphtha; solvent-to-oil ratio = 4 : 1 to 13 : 1)

Table 18.1 Feedstock and Product Data for the ASCOT Process

	ASCOT	Delayed coking	LEDA
Feedstock			
API gravity	2.8	2.8	2.8
Sulfur, wt%	4.2	4.2	4.2
Nitrogen, wt%	1.0	1.0	1.0
Carbon residue, wt%	22.3	22.3	22.3
Products			
Naphtha, vol%	7.7	19.4	
API gravity	54.7		
Sulfur, wt%	1.1		
Nitrogen, wt%	0.1		
Gas oil, vol%	69.9	51.8	
API gravity	13.4		
Sulfur, wt%	3.4		
Nitrogen, wt%	0.5		
Coke, wt%	25.0	32.5	
Sulfur, wt%	5.8	5.7	
Nitrogen, wt%	2.7	2.6	
Nickel, ppm	774.0	609.0	
Vanadium, ppm	2656.0	2083.0	
Deasphalted oil, vol%			50.0
Asphalt, vol%			50.0
Sulfur, wt%			5.0
Nitrogen, wt%			1.4
Nickel, ppm			365.0
Vanadium, ppm			1250.0

flows upward, extracting soluble material from the downflowing feedstock. The solvent-deasphalted phase leaves the top of the extractor and flows to the solvent recovery system, where the solvent is separated from the deasphalted oil and recycled to the extractor. The deasphalted oil is sent to the delayed coker [heater outlet temperature 480–510°C (900–950°F), at 15–35 psig (105–240 kPa) and a recycle ratio of 0.05–0.25 on fresh feedstock], where it is combined with the heavy coker gas oil from the coker fractionator and sent to the heavy coker gas oil stripper, where low-boiling hydrocarbons are stripped off and returned to the fractionator.

The stripped deasphalted oil–heavy coker gas oil mixture is removed from the bottom of the stripper and used to provide heat to the naphtha stabilizer-reboiler before being sent to battery limits as a cracking stock. The raffinate phase containing the asphalt and some solvent flows at a controlled rate from the bottom of the extractor and is charged directly to the coking section.

The solvent contained in the asphalt and deasphalted oil is condensed in the fractionator overhead condensers, where it can be recovered and used as lean oil for propane/butane recovery in the absorber, eliminating the need for lean oil recirculation from the naphtha stabilizer. The solvent introduced in the coker heater and coke drums results in a significant reduction in the partial pressure of asphalt feed, compared with a regular delayed coking unit. The low asphalt partial pressure results in low coke yield and high liquid yield in the coking reaction.

2.2 Cherry-P (Comprehensive Heavy Ends Reforming Refinery) Process

The Cherry-P process is a process for the conversion of heavy crude oil or residuum into distillate and a cracked residuum (Table 18.2) (Ueda, 1976, 1978; Kamiya, 1991, p. 5).

In the Cherry-P process (Fig. 18.5) the feedstock is mixed with coal powder in a slurry mixing vessel, heated in the furnace, and fed to the reactor, where it undergoes thermal cracking reactions for several (3–5) hours at a temperature higher than 400–430°C (750–805°F) and under pressure (140–280 psig; 980–1960 kPa). Gas and distillate from the reactor are sent to a fractionator, and the cracked residuum is extracted out of the system after low-boiling fractions have been distilled off by the flash drum and vacuum flasher to adjust its softening point.

The distillates produced by this process are generally lower in olefin hydrocarbon content than those produced by the other thermal cracking processes, comparatively easy to desulfurize in hydrotreating units, and compatible with straight-run distillates.

2.3 Deep Thermal Conversion Process

The deep thermal conversion (DTC) process (Fig. 18.6) offers a bridge between visbreaking and coking and provides maximum distillate yields by applying deep thermal conversion to vacuum residua followed by vacuum flashing of the products (*Hydrocarbon Processing*, 1998, p. 69).

Table 18.2 Feedstock and Product Data for the Cherry-P Process

Feedstock: Iranian heavy vacuum residuum	
API gravity	6.4
Sulfur, wt%	3.1
Carbon residue, wt%	21.3
Products	
Naphtha, vol%	9.3
API gravity	58.9
Sulfur, wt%	0.6
Kerosene, vol%	9.2
API gravity	40.4
Sulfur, wt%	0.8
Gas oil, vol%	17.9
API gravity	32.5
Sulfur, wt%	1.4
Vacuum gas oil, vol%	7.7
API gravity	20.5
Sulfur, wt%	2.5
Residuum, wt%	49.8
Sulfur, wt%	3.9
Carbon residue, wt%	42.2

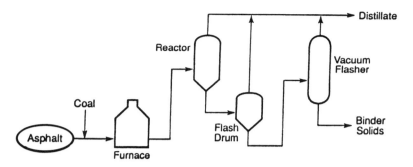

Figure 18.5 The Cherry-P process.

In the DTC process, the heated vacuum residuum is charged to the heater and from there to the soaker, where conversion occurs. The products are then led to an atmospheric fractionator to produce gases, naphtha, kerosene, and gas oil (Table 18.3). The fractionator residuum is sent to a vacuum flasher that recovers additional gas oil and distillate. The next steps for the coke are dependent on its potential use, and it may be isolated as liquid coke (pitch, cracked residuum) or solid coke.

2.4 ET-II Process

The ET-II process is a thermal cracking process for the production of distillates and cracked residuum for use as a metallurgical coke. It is designed to accommodate feedstocks such as heavy oils, atmospheric residua, and vacuum residua (Kuwahara, 1987;

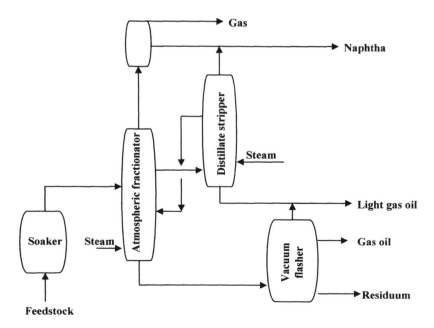

Figure 18.6 The DTC process.

Table 18.3 Feedstock and Product Data for the DTC Process

Feedstock: Middle East vacuum residuum	
API gravity	—[a]
Sulfur, wt%	—[a]
Carbon residue, wt%	—[a]
Products	
Naphtha, vol%	8.0
Gas oil, vol%	18.1
Vacuum gas oil, vol%	22.5
Vacuum residuum, wt%	47.4

[a] No information available.

Kamiya, 1991, p. 9). The distillate (referred to in the process as *cracked oil*) is suitable as a feedstock for hydrocracker and fluid catalytic cracking. The basic technology of the ET-II process is derived from that of the original Eureka process (Section 2.5).

In the ET-II process (Fig. 18.7), the feedstock is heated to 350°C (660°F) by passage through the preheater and fed into the bottom of the fractionator, where it is mixed with recycle oil, the high-boiling fraction of the cracked oil. The weight percent ratio of recycle oil to feedstock is within the range of 0.1–0.3. The feedstock mixed with recycle oil is then pumped out and fed into the cracking heater, where the temperature is raised to approximately 490–495°C (915–925°F), and the outflow is fed to the stirred-tank reactor, where it is subjected to further thermal cracking. Both cracking and condensation reactions take place in this reactor.

The heat required for the cracking reaction is brought in by the effluent itself from the cracking heater as well as by the superheated steam, which is heated in the convection

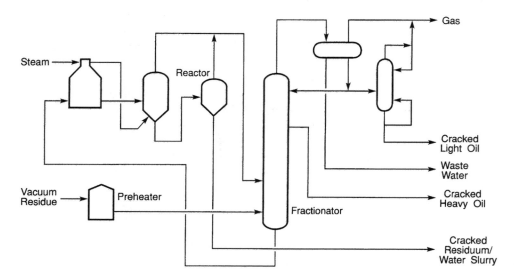

Figure 18.7 The ET-II process.

section of the cracking heater and blown into the reactor bottom. The superheated steam reduces the partial pressure of the hydrocarbons in the reactor and accelerates the stripping of volatile components from the cracked residuum. This residual product is discharged through a transfer pump and transferred to a cooling drum, where the thermal cracking reaction is terminated by quenching with a water spray, after which it is sent to the pitch–water slurry preparation unit.

The cracked oil and gas products together with steam from the top of the reactor are introduced into the fractionator, where the oil is separated into three fractions: *cracked light oil*, *cracked heavy oil*, and *cracked residuum (pitch)* (Table 18.4).

2.5 Eureka Process

The Eureka process is a thermal cracking process to produce a cracked oil and aromatic residuum from heavy residual materials (Aiba et al., 1981; Kamiya, 1991, p. 11; Chen, 1993).

In the Eureka process (Fig. 18.8), the feedstock, usually a vacuum residuum (Table 18.5), is fed to the preheater and then enters the bottom of the fractionator, where it is mixed with the recycle oil. The mixture is then fed to the reactor system, which consists of a pair of reactors operating alternately. In the reactor, thermal cracking reactions occur in the presence of superheated steam that is injected to strip the cracked products out of the

Table 18.4 Feedstock and Product data for the ET-II Process

Feedstock: Mix of Bachaquero, Khafji, and Maya vacuum residua	
API gravity	6.7
Sulfur, wt%	4.1
C_7 Asphaltenes, wt%	10.0
Carbon residue, wt%	22.4
Nickel, ppm	67.0
Vanadium, ppm	343.0
Products	
Light oil (< 315°C; < 600°F), wt%	28.6
API gravity	41.3
Sulfur, wt%	1.7
Nitrogen, wt%	< 0.1
Vacuum gas oil (315–540°C; 600–1000°F), vol%	32.3
API gravity	18.3
Sulfur, wt%	3.2
Nitrogen, wt%	0.2
Carbon residue, wt%	1.0
C_7 Asphaltenes, wt%	< 0.1
Pitch, wt%	32.8
Sulfur, wt%	6.3
Nitrogen, wt%	1.6
Carbon residue, wt%	60.0[a]
Nickel, ppm	204.0
Vanadium, ppm	1045.0

[a] Estimate.

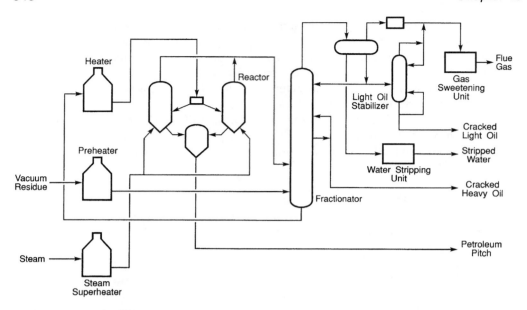

Figure 18.8   The Eureka process.

Table 18.5   Feedstock and Product Data for the
Eureka Process

Feedstock: Middle East vacuum residuum mix[a]	
API gravity	7.6
Sulfur, wt%	3.9
C_7 Asphaltenes, wt%	5.7
Carbon residue, wt%	20.0
Nickel, ppm	136.0
Vanadium, ppm	202.0
Products	
Light oil (C_5–240°C; C_5–465°F), wt%	14.9
API gravity	53.0
Sulfur, wt%	1.1
Nitrogen, wt%	< 0.1
Gas oil (240–540°C; 465–1000°F), wt%	50.7
API gravity	21.3
Sulfur, wt%	2.7
Nitrogen, wt%	0.3
Pitch (> 540°C; > 1000°F), wt%	29.6
Sulfur, wt%	5.7
Nitrogen, wt%	1.2
Nickel, ppm	487.0
Vanadium, ppm	688.0

[a] > 500°C (> 930°F), not defined by name.

reactor and supply part of the heat required for cracking. At the end of the reaction, the bottom product is quenched.

The oil and gas products (and steam) pass from the top of the reactor to the lower section of the fractionator, where a small amount of entrained material is removed by a wash operation. The upper section is an ordinary fractionator, where the heavier fraction of cracked oil is drawn as a sidestream.

The original Eureka process uses two batch reactors, whereas the newer ET II and HSC processes (Sections 2.4 and 2.7) both employ continuous reactors.

2.6 Fluid Thermal Cracking Process

The fluid thermal cracking (FTC) process is a heavy oil and residuum upgrading process in which the feedstock is thermally cracked to produce distillate and coke, which is gasified to fuel gas (Table 18.6) (Miyauchi et al., 1981; Miyauchi and Ikeda, 1988; Kamiya, 1991, p. 17).

The feedstock, mixed with recycle stock from the fractionator and injected into the cracker, is immediately absorbed into the pores of the particles by capillary force and is subjected to thermal cracking (Fig. 18.9). In consequence, the surface of the noncatalytic particles is kept dry and good fluidity is maintained, allowing a good yield of, and selectivity for, middle distillate products. Hydrogen-containing gas from the fractionator is used for the fluidization in the cracker.

Excessive coke caused by the metals accumulated on the particle is suppressed under the presence of hydrogen. The particles with deposited coke from the cracker are sent to the gasifier, where the coke is gasified and converted into carbon monoxide (CO), hydrogen (H_2), carbon dioxide (CO_2), and hydrogen sulfide (H_2S) with steam and air. Regenerated hot particles are returned to the cracker.

Table 18.6 Feedstock and Product Data for the FTC Process

Feedstock: Bachaquero, Khafji, and Maya vacuum residua mix	
API gravity	6.4
Sulfur, wt%	4.5
Carbon residue, wt%	21.9
Nickel + vanadium, ppm	500.0
Products	
Naphtha (C_5–150°C; C_5–300°F), wt%	14.9
API gravity	57.1
Sulfur, wt%	0.4
Middle distillate (150–310°C; 300–590°F), wt%	34.4
API gravity	39.5
Sulfur, wt%	1.4
Heavy distillate (310–525°C; 590–975°F), wt%	21.8
API gravity	18.7
Sulfur, wt%	3.1
Nitrogen, wt%	0.2
Coke	19.0[a]

[a] Estimate.

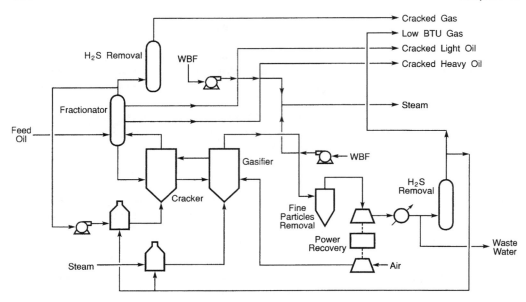

Figure 18.9 The FTC process.

2.7 High Conversion Soaker Cracking Process

The high conversion soaker cracking (HSC) process (Fig. 18.10) is a cracking process to achieve moderate conversion, higher than visbreaking but lower than coking, to various products (Table 18.7) (Watari et al., 1987; Washimi, 1989; Kamiya, 1991, p. 19; Washimi and Hamamura, 1993). The process features a lower yield of gas and a higher yield of distillate than other thermal cracking processes. The process can be used to convert a wide

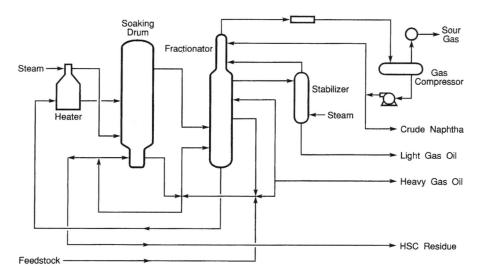

Figure 18.10 The HSC process.

Table 18.7 Feedstock and Product Data for the HSC Process

	Iranian heavy vacuum residuum	Maya vacuum residuum
Feedstock		
API gravity	5.7	2.2
Sulfur, wt%	4.8	5.1
Nitrogen, wt%	0.6	0.8
C_7 Asphaltenes, wt%	11.3	19.0
Carbon residue, wt%	22.6	28.7
Nickel, ppm	69.0	121.0
Vanadium, ppm	205.0	649.0
Products		
Naphtha (C_5–200°C, C_5–390°F), wt%	6.3	3.8
API gravity	54.4	51.5
Sulfur, wt%	1.1	1.1
Light gas oil (200–350°C; 390–660°F), wt%	15.0	12.3
API gravity	30.2	29.5
Sulfur, wt%	2.6	3.1
Nitrogen, wt%	0.1	< 0.1
Heavy gas oil (350–620°C; 660–970°F), wt%	32.2	18.6
API gravity	16.4	16.5
Sulfur, wt%	3.5	3.7
Nitrogen, wt%	0.3	0.3
Vacuum residue (> 520°C; > 970°F), wt%	43.2	62.0
Sulfur, wt%	5.8	5.6
Nitrogen, wt%	1.0	1.2
Carbon residue, wt%	49.2	49.2
Nickel, ppm	148.0	148.0
Vanadium, ppm	453.0	453.0

range of feedstocks with high sulfur and heavy metals, including heavy oils, oil sand bitumen, residua, and visbroken residua. As a note of interest, the HSC process employs continuous reactors whereas the original Eureka process uses two batch reactors.

The preheated feedstock enters the bottom of the fractionator, where it is mixed with the recycle oil. The mixture is pumped up to the charge heater and fed to the soaking drum (at approximately atmospheric pressure, steam injection at the top and bottom), where sufficient residence time is provided to complete the thermal cracking. In the soaking drum, the feedstock and some product flows downward, passing through a number of perforated plates, while steam with cracked gas and distillate vapors flow countercurrently through the perforated plates.

The volatile products from the soaking drum enter the fractionator, where the distillates are fractionated into the desired product oil streams, including a heavy gas oil fraction. The cracked gas product is compressed and used as refinery fuel gas after sweetening. After hydrotreating, the cracked oil product is used as feedstock for fluid catalytic cracking or hydrocracking. The residuum is suitable for use as boiler fuel, as road asphalt, as binder for the coking industry, and as a feedstock for partial oxidation.

2.8 Tervahl Process

In the Tervahl T process (LePage et al., 1987; Kamiya, 1991, p. 25) (Fig. 18.11a), the feedstock is heated to the desired temperature using the coil heater and heat recovered in the stabilization section and is held for a specified residence time in the soaking drum. The soaking drum effluent is quenched and sent to a conventional stabilizer or fractionator, where the products are separated into the desired streams. The gas produced from the process is used for fuel.

In the Tervahl H process (Fig. 18.11b), the feedstock and hydrogen-rich stream are heated using heat recovery techniques and a fired heater and held in the soaking drum as in the Tervahl T process. The gas and oil from the soaking drum effluent are mixed with recycle hydrogen and separated in the hot separator, where the gas is cooled; then it is passed through a separator, and recycled to the heater and soaking drum effluent. The liquids from the hot and cold separators are sent to the stabilizer section, where purge gas and synthetic crude are separated (Table 18.8). The gas is used as fuel, and the synthetic crude can now be transported or stored.

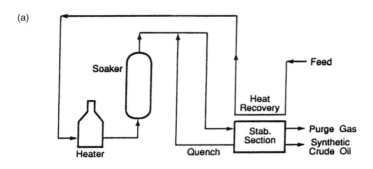

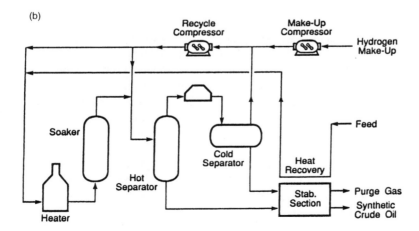

Figure 18.11   (a) The Tervahl T process; (b) the Tervahl H process.

Table 18.8 Feedstock and Product Data for the Tervahl T and
H Processes

Feedstock: Boscan heavy crude oil		
API gravity		10.5
Distillate (< 500°C; < 930°F), wt%		35.5
	Process	
	Tervahl T	Tervahl H
Product		
API gravity	11.7	14.8
Distillate (< 500°C; < 930°F), wt%	52.5	55.3

3. CATALYTIC CRACKING (CARBON REJECTION) PROCESSES

The *fluid catalytic cracking process*, using vacuum gas oil as the feedstock, was introduced into refineries in the 1930s. In recent years, because of a trend toward the use of low-boiling products, most refineries perform the operation by partially blending residua into vacuum gas oil. However, conventional fluid catalytic cracking processes have limits in residuum processing, so residuum fluid catalytic cracking processes have lately been employed one after another. Because the residuum fluid catalytic cracking process enables efficient gasoline production directly from residua, it will play the most important role as a residuum cracking process, along with a residuum hydrotreating process (Table 18.9; Fig. 18.12) (Reynolds et al., 1992).

Another role of the residuum fluid catalytic cracking process is to generate high quality gasoline blending stock and petrochemical feedstock. Olefins (propene, butenes, and pentenes) serve as feed for alkylating processes, for polymer gasoline, and as additives for reformulated gasoline.

The processes described in this section are the evolutionary offspring of the fluid catalytic cracking and residuum catalytic cracking processes. Some of these newer pro-

Table 18.9 Feedstock and Product Data for the Fluid Catalytic Cracking
Process With and Without Feedstock Hydrotreating

	No pretreatment	With hydrotreatment
Feedstock (> 370°C; > 700°F)		
API gravity	15.1	20.1
Sulfur, wt%	3.3	0.5
Nitrogen, wt%	0.2	0.1
Carbon residue, wt%	8.9	4.9
Nickel + vanadium, ppm	51.0	7.0
Products		
Naphtha (C_5–221°C; C_5–430°F), vol%	50.6	58.0
Light cycle (221–360°C; 430–680°F), vol%	21.4	18.2
Residuum (> 360°C; > 680°F), wt%	9.7	7.2
Coke, wt%	10.3	7.0

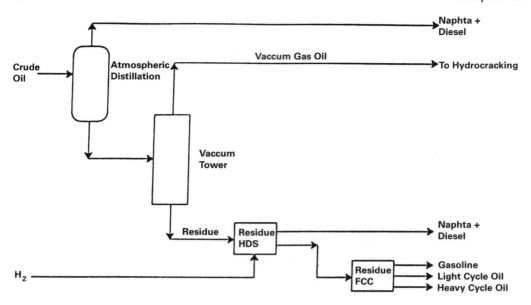

Figure 18.12 Schematic for residuum hydroconversion with fluid catalytic cracking.

cesses use catalysts with different silica/alumina ratios as acid support of metals such as molybdenum, cobalt, nickel, and tungsten. The first catalyst used to remove metals from oils was the conventional hydrodesulfurization (HDS) catalyst. Diverse natural minerals are also used as raw materials for elaborating catalysts addressed to the upgrading of heavy fractions. Among these minerals are clays; manganese nodules; bauxite activated with vanadium, nickel, chromium, iron, and cobalt, as well as iron laterites and sepiolites; and mineral nickel and transition metal sulfides supported on silica and alumina. Other kinds of catalysts, such as vanadium sulfide, are generated in situ, possibly in colloidal states.

3.1 Asphalt Residua Treating Process

The asphalt residua treating (ART) process is a process for increasing the production of high value distillates (Table 18.10) (Logwinuk and Caldwell, 1983; Green and Center, 1985; Gussow and Kramer, 1990; Kamiya, 1991, p. 29; Bartholic et al., 1992; *Hydrocarbon Processing*, 1996). The process is a combination of selective vaporization and fluid decarbonization and demetallization. This process removes more than 95% of the metals, all the asphaltenes, and 30–50% of the sulfur and nitrogen from residual oil but preserves the hydrogen content of the feedstock. It also enables the subsequent conversion step in residual oil processing to be accomplished in conventional downstream catalytic processing units.

The ART process is capable of handling whole crude oils, atmospheric residua, vacuum residua, and tar sand bitumen. It is designed to minimize molecular conversion of the noncontaminant hydrocarbon portion of the feedstock, thereby maximizing the hydrogen content of the liquid products. It accomplishes this while at the same time removing or reducing the constituents of the feedstock that would poison conventional catalysts contained in downstream conversion units. End products are therefore suitable

Table 18.10 Feedstock and Product Data for the ART Process

	Mixed residua[a]
Feedstock	
API gravity	14.9
Sulfur, wt%	4.1
Nitrogen, wt%	0.3
C_5 Asphaltenes	12.4
Carbon residue, wt%	15.8
Nickel, ppm	52.0
Vanadium, ppm	264.0
Products	
Naphtha, vol%	62.1
No. 2 fuel oil, vol%	35.5
No. 6 fuel oil, vol%	2.3

[a] Arabian light vacuum residuum plus Arabian heavy vacuum residuum.

for processing in any of the conventional downstream catalytic refinery processes such as fluid catalytic cracking, hydrotreating, and hydrocracking.

In the ART process (Fig. 18.13), the preheated feedstock (which may be whole crude, atmospheric residuum, vacuum residuum, or bitumen) is injected into a stream of fluidized hot catalyst (trade name: ArtCat). The unit configuration is similar to that of a riser fluid catalytic cracking unit where complete mixing of the feedstock with the catalyst is achieved in the contactor, which is operated within a pressure–temperature envelope to ensure selective vaporization. The vapor and the contactor effluent are quickly and efficiently separated from each other, and entrained hydrocarbons are stripped from the contaminant (containing spent solid) in the stripping section. The contactor vapor effluent and vapor from the stripping section are combined and are rapidly quenched in a quench drum to minimize product degradation. The cooled products are then transported to a conventional fractionator that is similar to that found in a fluid catalytic cracking unit. Spent solid from the stripping section is transported to the combustor bottom zone for carbon burn-off.

Contact of the feedstock (residuum) with the fluidizable catalyst in a short residence time contactor causes the lower boiling components of the feedstock to vaporize, and asphaltene constituents (high molecular weight compounds) are cracked to yield lower boiling compounds and coke. The metals present, as well as some of the sulfur and the nitrogen compounds in the nonvolatile constituents, are retained on the catalyst. At the exit of the contacting zone, the oil vapors are separated from the catalyst and are rapidly quenched to minimize thermal cracking of the products. The catalyst, which holds metals, sulfur, nitrogen, and coke, is transferred to the regenerator, where the combustible portion is oxidized and removed. Regenerated contact material, bearing metals but very little coke, exits the regenerator and passes to the contactor for further removal of contaminants from the charge stock.

The metals level on the catalyst in the system is controlled by the addition of fresh catalyst and the removal of spent material. A high metals level can be maintained without

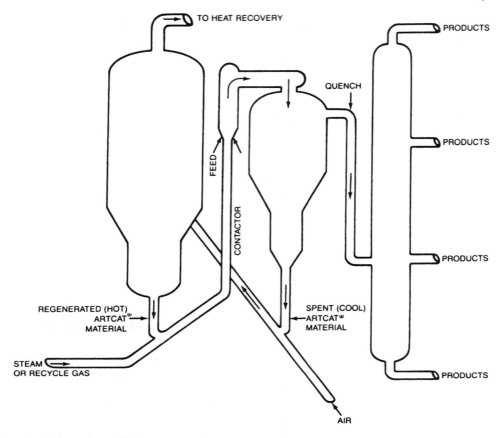

Figure 18.13   The ART process.

detrimentally affecting its performance. Because the contact material is essentially inert, very little molecular conversion of the light gas oil and lighter fractions takes place. The molecular conversion that does occur is due to the disproportionation of the higher molecular weight compounds present in the feedstock.

In the combustor, coke is burned from the spent solid, which is then separated from combustion gas in the surge vessel. The surge vessel circulates regenerated catalyst streams to the contactor inlet for feed vaporization and to the combustor bottom zone for pre-mixing.

The components of the combustion gases include carbon dioxide (CO_2), nitrogen (N_2), oxygen (O_2), sulfur oxides (SO_x), and nitrogen oxides (NO_x) that are released from the catalyst with the combustion of the coke in the combustor. The concentration of sulfur oxides in the combustion gas requires treatment for their removal.

3.2 Heavy Oil Treating Process

The heavy oil treating (HOT) process is a catalytic cracking process for upgrading heavy feedstocks such as topped crude oils, vacuum residua, and solvent-deasphalted bottoms using a fluidized bed of iron ore particles (Table 18.11) (Ozaki, 1982; Kamiya, 1991, p. 35).

Table 18.11 Feedstock and Product Data for the HOT Process

Feedstock: Arabian light vacuum residuum	
API gravity	7.1
Sulfur, wt%	4.2
Carbon residue, wt%	21.6
Products	
Light naphtha (C_5–180°C; C_5–355°F), vol%	15.2
Heavy naphtha (180–230°C; 355–445°F), vol%	8.2
Light gas oil (230–360°C; 445–680°F), vol%	2.3
Heavy gas oil (360–510°C; 680–950°F, vol%	28.2

The main section of the process (Fig. 18.14) consists of three fluidized reactors, and separate reactions take place in each reactor (*cracker*, *regenerator*, and *desulfurizer*):

$$Fe_3O_4 + asphaltenes \rightarrow coke/Fe_3O_4 + oil + gas \quad \text{(in the cracker)}$$

$$3FeO + H_2O \rightarrow Fe_3O_4 + H_2 \quad \text{(in the cracker)}$$

$$Coke/Fe_3O_4 + O_2 \rightarrow 3FeO + CO + CO_2 \quad \text{(in the regenerator)}$$

$$FeO + SO_2 + 3CO \rightarrow FeS + 3CO_2 \quad \text{(in the regenerator)}$$

$$3FeS + 5O_2 \rightarrow Fe_3O_4 + 3SO_2 \quad \text{(in the desulfurizer)}$$

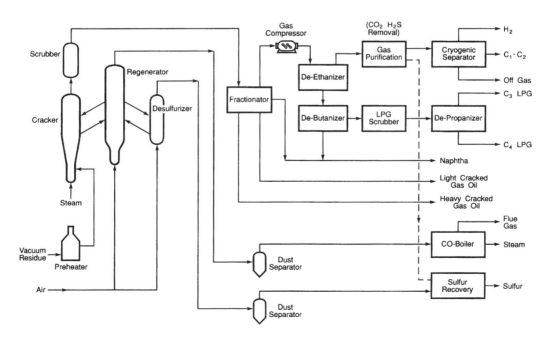

Figure 18.14 The HOT process.

In the *cracker*, heavy oil cracking and the steam–iron reaction take place simultaneously under conditions similar to those used in thermal cracking. Any unconverted feedstock is recycled to the cracker from the bottom of the scrubber. The scrubber effluent is separated into hydrogen gas, liquefied petroleum gas (LPG), and liquid products that can be upgraded by conventional technologies to priority products.

In the *regenerator*, coke deposited on the catalyst is partially burned to form carbon monoxide in order to reduce iron tetroxide and act as a heat supply. In the *desulfurizer*, sulfur in the solid catalyst is removed and recovered as molten sulfur in the final recovery stage.

3.3 R2R Process

The R2R process is a fluid catalytic cracking process for conversion of heavy feedstocks (Table 18.12) (Kamiya, 1991, p. 37).

In the R2R process (Fig. 18.15), the feedstock is vaporized upon contacting hot regenerated catalyst at the base of the riser and lifts the catalyst into the reactor vessel separation chamber, where rapid disengagement of the hydrocarbon vapors from the catalyst is accomplished by both a special solids separator and cyclones. The bulk of the cracking reactions take place at the moment of contact and continue as the catalyst and hydrocarbons travel up the riser. The reaction products, along with a minute amount of entrained catalyst, then flow to the fractionation column. The stripped spent catalyst, deactivated with coke, flows into regenerator 1.

Partially regenerated catalyst is pneumatically transferred via an air riser to regenerator 2, where the remaining carbon is completely burned in a drier atmosphere. This regenerator is designed to minimize catalyst inventory and residence time at high temperature while optimizing the coke burning rate. Flue gases pass through external cyclones to a waste heat recovery system. Regenerated catalyst flows into a withdrawal well and after stabilization is charged back to the oil riser.

3.4 Reduced Crude Oil Conversion Process

In the reduced crude oil conversion (RCC) process (Fig. 18.16), the clean regenerated catalyst enters the bottom of the reactor riser, where it contacts low-boiling hydrocarbon *lift gas* that accelerates the catalyst up the riser prior to feed injection (*Hydrocarbon Processing*, 1996; Kamiya, 1991, p. 39). At the top of the lift gas zone, the feed is injected through a series of nozzles located around the circumference of the reactor riser.

The catalyst–oil disengaging system is designed to separate the catalyst from the reaction products and then rapidly remove the reaction products from the reactor vessel. Spent catalyst from the reaction zone is first steam-stripped to remove adsorbed hydrocarbon and then routed to the regenerator. In the regenerator all of the carbonaceous deposits are removed from the catalyst by combustion, restoring the catalyst to an active state with a very low carbon content. The catalyst is then returned to the bottom of the reactor riser at a controlled rate to achieve the desired conversion and selectivity to the primary products (Table 18.13).

3.5 Residual Fluid Catalytic Cracking Process

The residual fluid catalytic cracking (HOC) process is a version of the fluid catalytic cracking process that has been adapted to conversion of a wide range of feedstocks that

Table 18.12 Feedstock and Product Data for the R2R Process

	Atmospheric residuum + vacuum gas oil[a]	Atmospheric residuum[a]	Atmospheric residuum[a]	Hydrotreated atmospheric residuum[a]	Hydrotreated atmospheric residuum
Feedstock					
API gravity	28.4	26.4	22.6	20.7	20.7
Carbon residue, wt%	0.2	6.4	5.5	4.8	4.8
Nickel + vanadium, ppm	1.0	22.0	34.0	20.0	20.0
Products					
Naphtha, vol%	63.4	59.5	58.0	60.4	49.6
Distillate, vol%	16.6	14.1	16.3	17.7	29.7
Heavy gas oil/residuum, vol%	3.9	6.7	9.8	6.8	12.5
Coke, wt%	4.6	7.4	6.9	7.2	6.7

[a] Processed for maximum gasoline production.

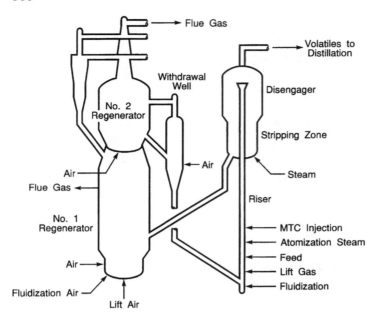

Figure 18.15 The R2R process.

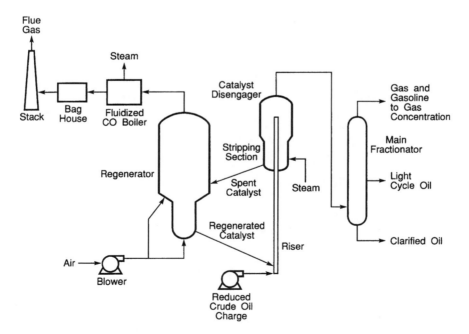

Figure 18.16 The RCC process.

Table 18.13 Feedstock and Product Data for the RCC Process

	Residuum[a]	Residuum[a]	Residuum[a]
Feedstock			
API gravity	22.8	21.3	19.2
Sulfur, wt%			
Nitrogen, wt%			
Carbon residue, wt%	0.2	6.4	5.5
Nickel, ppm	1.0	22.0	34.0
Vanadium, ppm			
Products			
Gasoline (C$_5$–221°C; C$_5$–430°F), vol%	59.1	56.6	55.6
Light cycle oil (221–322°C; 430–610°F), vol%	16.3	15.4	16.3
Gas oil (> 322°C; > 610°F), vol%	6.2	9.0	9.6
Coke, wt%	8.4	9.1	10.8

[a] Unspecified.

contain high amounts of metal and asphaltenes (Table 18.14) (Finneran, 1974; Murphy and Treese, 1979; Johnson, 1982; Kamiya, 1991, p. 33; Feldman et al., 1992). It provides a flexible upgrading that can be the answer to many bottom-of-the-barrel processor problems caused by heavy crude oils containing high amounts of metal, sulfur, or asphaltenes.

In the HOC process (Fig. 18.17), the flow of solids is essentially vertical. Regenerator bed temperatures are limited to around 730°C (1300°F), and feed introduction systems are designed for efficient mixing of oil and catalyst and rapid quenching of catalyst temperature to the equilibrium mix temperature. The reaction system is an external vertical riser, which provides very low contact times and terminates in riser cyclones for rapid separation of catalyst and vapors. A two-stage stripper is used to remove hydrocarbons from the catalyst. Hot catalyst flows at low velocity in dense phase through the catalyst cooler and returns to the regenerator. Regenerated catalyst flows to the bottom of the riser to meet the feed.

The coke deposited on the catalyst is burned off in the regenerator along with the coke formed during the cracking of the gas oil fraction. The large amounts of coke produced in cracking residua can cause extreme temperatures and excessive catalyst deactivation. Steam coils located within the regenerator bed and/or external catalyst coolers remove the excess heat produced by the high coke yields.

Table 18.14 Feedstock Categories for the HOC Process

Category		Property	
Number	Descriptor	Carbon residue (wt%)	Metals, Ni + V (ppm)
1	Good	< 5	< 10
2	Medium	> 5, < 10	> 10, < 30
3	Poor	> 10, < 20	> 30, < 150
4	Bad	> 20	> 150

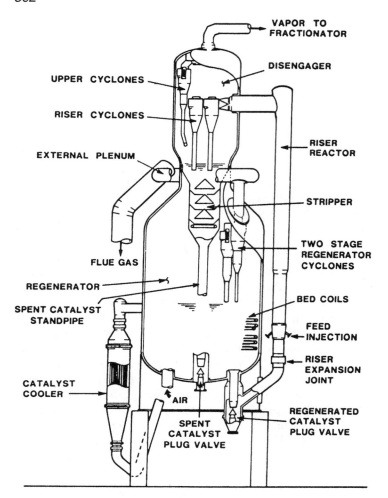

Figure 18.17 The HOC process.

The process is a version of the fluid catalytic cracking process that has been adapted to the conversion of residua having a wide range of properties. Depending on quality and product objectives, feedstocks with vanadium-plus-nickel content of up to 30 ppm and carbon residue of up 10% by weight can be processed without feed pretreatment (Table 18.15). When the metals content and carbon residua are in excess of 30 ppm and 10 wt%, respectively, the feedstock may require pretreatment. If the feedstock contains high proportions of metals, control of the metals on the catalyst requires excessive amounts of catalyst withdrawal and fresh catalyst addition. This problem can be addressed by feedstock pretreatment, such as hydrodesulfurization (Table 18.16).

Catalysts for heavy oil cracking units are required to be tolerant of the constituents of the residua. Some have greater zeolite content or have pore structures that prevent the trapping of large molecules and avoid coke production. Poisons such as sodium and vanadium accelerate the deactivation rate of catalyst, and high amounts of sodium are usually avoided by double desalting of the crude. Means of avoiding deactivation due to vanadium

Table 18.15 Feedstock (Medium Quality) and Product Data for the HOC Process

Feedstock property		
Gravity, °API	21.3	
Sulfur, wt%	1.5	
Conradson carbon, wt%	5.0	
Metals, ppm		
Nickel	15	
Vanadium	7	
Conversion, vol% of fresh feed	76	
Product yield		
Gasoline, vol%	57.8	
Butanes, butylene	12.0	
Isobutane		3.1
n-Butane		0.9
Butylenes		8.0
Propane–propylene	7.0	
Propane		1.5
Propylene		5.5
Light cycle oil	15.0	
Decanted oil	9.0	
Total		100.8
Coke, wt%	11.0	

Table 18.16 Effect of Hydrotreating on HOC Products

	Untreated atmospheric tower bottoms		Hydrodesulfurized atmospheric tower bottoms	
Feedstock				
Gravity, °API	16.1		23.3	
Carbon residue content	8.6		4.2	
Aniline point, °F	161		202	
Sulfur, %	3.5		0.3	
Metals, ppm				
Vanadium	51		9	
Nickel	16		6	
HOC yields at constant conversion of 77%				
Gasoline, C_5–430°F TBP, vol%	52.6		65.0	
C_4–C_4, vol%	15.5		16.0	
C_3–C_3, vol%	11.2		8.0	
Catalytic gas oil, vol%	23.0		23.0	
Total, vol%		99.3		112.0
Coke, wt%	12.4		8.0	
C_2 and lighter, wt%	4.2		1.4	

have included designs for low temperatures that result in higher equilibrium activities and lower catalyst costs. Up to 10,000 ppm of nickel plus vanadium on the catalyst and temperatures up to 730°C (of 1350°F) with an equilibrium activity of about 65–70 are ideal for heavy oil cracking units. The catalyst makeup rate to achieve this metals concentration can be determined by a simple metals balance; for a feedstock of 30 ppm nickel plus vanadium, a makeup rate of metals-free catalyst equal to 1 lb/bbl would be required.

3.6 Shell FCC Process

In the Shell FCC process (Fig. 18.18), the preheated feedstock (vacuum gas oil, atmospheric residuum) is mixed with the hot regenerated catalyst (Table 18.17) (Khouw, 1990; Kamiya, 1991, p. 41). After reaction in a riser, volatile materials and catalyst are separated, then the spent catalyst is immediately stripped of entrained and adsorbed hydrocarbons in a very effective multistage stripper.

The stripped catalyst gravitates through a short standpipe into a single-vessel, simple, reliable, and efficient catalyst regenerator. Regenerative flue gas passes via a cyclone–swirl tube combination to a power recovery turbine. From the expander turbine the heat in the flue gas is further recovered in a waste heat boiler.

Depending on the environmental conservation requirements, a deNO$_x$ing, deSO$_x$ing, and particulate emission control device can be included in the flue gas train.

It has been claimed (Sato et al., 1992) that a hydrogenation pretreatment of bitumen prior to fluid catalytic cracking (or for that matter any catalytic cracking process) can enhance the yield of naphtha. It has been suggested that mild hydrotreating be carried out upstream of a fluid catalytic cracking unit to increase the yield and quality of distillate products. This is in keeping with earlier work (Speight and Moschopedis, 1979) in which mild hydrotreating of bitumen was reported to produce low sulfur liquids that would be amenable to further catalytic processing.

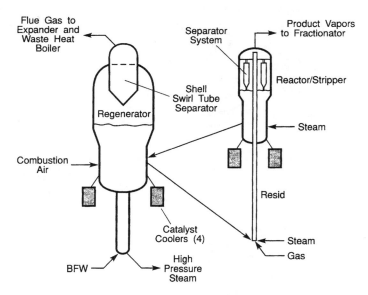

Figure 18.18 The Shell fluid catalytic cracking process.

Table 18.17 Feedstock and Product Data for the Shell Fluid Catalytic Cracking Process

	Residuum[a]	Residuum[a]
Feedstock		
API gravity	18.2	13.4
Sulfur, wt%	1.1	1.3
Carbon residue, wt%	1.2	4.7
Products		
Gasoline (C_5–221°C; C_5–430°F), wt%	49.5	46.2
Light cycle oil (221–370°C; 430–700°F), wt%	20.1	19.1
Heavy cycle oil (> 370°C; > 700°F), wt%	5.9	10.8
Coke, wt%	5.9	7.6

[a] Unspecified.

3.7 S&W Fluid Catalytic Cracking Process

In the S&W fluid catalytic cracking process (Fig. 18.19), the heavy feedstock is injected into a stabilized, upward-flowing catalyst stream, whereupon the feedstock–steam–catalyst mixture travels up the riser and is separated by a high efficiency inertial separator. The product vapor goes overhead to the main fractionator for separation of various distillate products (Table 18.18) (Long, 1987; Kamiya, 1991, p. 43).

The spent catalyst is immediately stripped in a staged, baffled stripper to minimize hydrocarbon carryover to the regenerator system. The first regenerator (650–700°C; 1200–1290°F) burns 50–70% of the coke in an incomplete carbon monoxide combustion mode

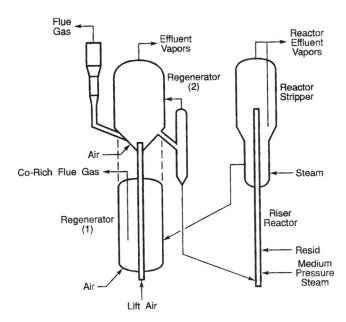

Figure 18.19 The S&W fluid catalytic cracking process.

Table 18.18 Feedstock and Product Data for the S&W Fluid Catalytic Cracking Process

	Residuum[a]	Residuum[a]
Feedstock		
API gravity	24.1	22.3
Sulfur, wt%	0.8	1.0
Carbon residue, wt%	4.4	6.5
Products		
Naphtha, vol%	61.5	60.2
Light cycle oil, vol%	16.6	17.5
Heavy cycle oil, vol%	5.6	6.6
Coke, wt%	7.1	7.8
Conversion, vol%	77.7	75.9

[a]Unspecified.

running countercurrently. This relatively mild, partial regeneration step minimizes the significant contribution of hydrothermal catalyst deactivation. The remaining coke is burned in the second regenerator ($\sim 775°C$; $\sim 1425°F$) with an extremely low steam content. Hot clean catalyst enters a withdrawal well that stabilizes its fluid qualities prior to being returned to the reaction system.

4. HYDROGEN ADDITION PROCESSES

Residuum hydrotreating processes have three definite roles: (1) desulfurization to supply low sulfur fuel oils, (2) pretreatment of feed residua for residuum fluid catalytic cracking processes, and (3) hydrocracking to produce feedstocks for fluid catalytic cracking processes. The hydrotreating processes are becoming more popular as pretreating processes whose main goal is to reduce the sulfur, metal, and asphaltene contents of residua and other heavy feedstocks to a desired level and at the same time maintain hydrogen consumption at acceptable levels that are dictated, for example, by the degree of desulfurization and/or hydrocracking (Table 18.19) (Maples, 2000; Speight, 2000). Hydrocracking process options for residua are also seeing an increase in use but have been considered in the past as *hydrogen sinks* (being somewhat uncontrollable and therefore wasteful in hydrogen use) and are therefore seemingly less popular that the hydrotreating processes. Hopefully this attitude is changing as more hydroconversion processes are being introduced for heavy feed conversion.

Thus, the major goal of *residuum hydroconversion* is cracking of residua with desulfurization, metal removal, denitrogenation, and asphaltene conversion. The residuum hydroconversion process produces kerosene and gas oil and enables the production of feedstocks for hydrocracking, fluid catalytic cracking, and petrochemical applications.

4.1 Asphaltenic Bottoms Cracking Process

The alsphaltenic bottoms cracking (ABC) process can be used for hydrodemetallization, asphaltene cracking, and moderate hydrodesulfurization and also affords resistance to

Table 18.19 Hydrogen Consumption for Heavy Feedstocks

Feedstock	Gravity (°API)	Sulfur (wt%)	Carbon residue (Conradson) (wt%)	Nitrogen (wt%)	Hydrogen (scf/bbl)	Desulfurization (%)
Venezuela, atmospheric	15.3–17.2	2.1–2.2	9.9–10.4	—	425–730	48–86
Venezuela, vacuum	4.5–7.5	2.9–3.2	20.5–21.4	—	825–950	47–60
Boscan (whole crude)	10.4	5.6	—	0.52	1100	70
Tia Juana, vacuum	7.8	2.5	21.4[a]	0.52	490–770	39–67
Bachaquero, vacuum	5.8	3.7	23.1[a]	0.56	1080–1260	59–73
West Texas, atmospheric	17.7–17.9	2.2–2.5	8.4	—	520–670	80–86
West Texas, vacuum	10.0–13.8	2.3–3.2	12.2–14.8	—	675–1200	55–92
Khafji, atmospheric	15.1–15.7	4.0–4.1	11.0–12.2	—	725–800	71–88
Khafji, vacuum	5.0	5.4	21.0	—	1000–1100	68–78
Arabian light, vacuum	8.5	3.8	—	—	435–1180	53–92
Kuwait, atmospheric	15.7–17.2	3.7–4.0	8.6–9.5	0.20–0.23	470–815	63–92
Kuwait, vacuum	5.5–8.0	5.1–5.5	16.0	—	290–1200	30–95

[a] Ramsbottom carbon residue.

coke fouling and metal deposition using such feedstocks as vacuum residua, thermally cracked residua, solvent-deasphalted bottoms, and bitumen with fixed catalyst beds (Table 18.20) (Takeuchi, 1982; Kamiya, 1991, p. 45).

The process can be combined with

1. Solvent deasphalting for complete or partial conversion of the residuum
2. Hydrodesulfurization to promote the conversion of residua, to treat feedstock with high metals, and to increase catalyst life
3. Hydrovisbreaking to attain high conversion of residua with favorable product stability

In the ABC process (Fig. 18.20), the feedstock is pumped up to the reaction pressure and mixed with hydrogen. The mixture is heated to the reaction temperature in the charge heater after a heat exchange and fed to the reactor. In the reactor, hydrodemetallization and subsequent asphaltene cracking with moderate hydrodesulfurization take place simultaneously under conditions similar to those of residuum hydrodesulfurization. The reactor effluent gas is cooled, cleaned, and recycled to the reactor section, while the separated

Table 18.20 Feedstock and Product Data for the ABC Process

	California atmospheric residuum	Arabian light vacuum residuum	Arabian light vacuum residuum	Arabian heavy vacuum residuum	Cero Negro vacuum residuum
Feedstock					
API gravity	6.1	5.5	7.0	5.1	1.7
Sulfur, wt%	6.5	4.4	4.0	5.3	4.3
Carbon residue, wt%	17.5	24.6	20.8	23.3	23.6
C_7 Asphaltenes	16.2	8.9	7.0	13.1	19.8
Nickel, ppm	150.0	30.0	223.0	52.0	150.0
Vanadium, ppm	380.0	90.0	76.0	150.0	640.0
Products					
Naphtha, vol%	12.6	9.3	6.5	7.7	15.1
API gravity	56.7	58.7	57.2	57.2	54.7
Distillate, vol%	23.0	20.1	16.0	19.8	21.3
API gravity	34.6*	33.2	34.2*	34.2*	32.5*
Vacuum gas oil, vol%	38.4	33.1	34.3	38.1	32.8
API gravity	23.1	20.3	24.7	21.6	15.4
Sulfur, wt%	0.4	0.4	0.2	1.7	0.5
Vacuum residuum, vol%	29.2	40.8	46.2	37.9	34.7
API gravity	5.4	7.5	10.6	7.8	< 0.0
Sulfur, wt%	2.4	1.2	0.6	1.7	2.2
Carbon residue, wt%	26.0	28.6	13.6	26.5	13.6
C_7 Asphaltenes, wt%		12.0			
Nickel, ppm	100.0		9.0	45.0	117.0
Vanadium, ppm	180.0		11.0	75.0	371.0
Conversion, %	60.0		55.0	60.0	60.0

a Estimated.

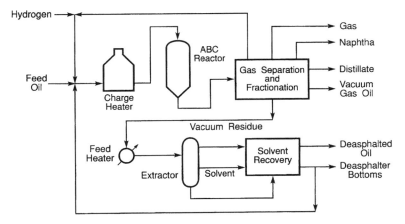

Figure 18.20 The ABC process.

liquid is distilled into distillate fractions and vacuum residuum, which is further separated into deasphalted oil and asphalt using butane or pentane.

In the case of the ABC–hydrodesulfurization catalyst combination, the ABC catalyst is placed upstream of the hydrodesulfurization catalyst (Fig. 18.21a) and can be operated at a higher temperature than the hydrodesulfurization catalyst under conventional residuum hydrodesulfurization conditions. In the VisABC process (Fig. 18.21b), a soaking

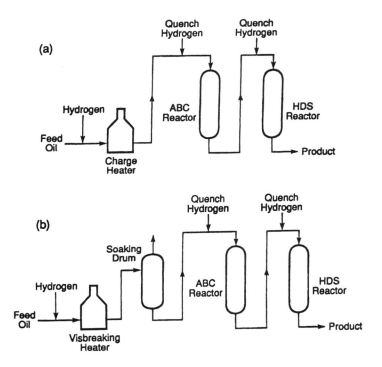

Figure 18.21 ABC process options.

drum is provided after the heater when necessary. Hydrovisbroken oil is first stabilized by the ABC catalyst through hydrogenation of coke precursors and then desulfurized by the HDS catalyst.

4.2 Hydrovisbreaking (HYCAR) Process

By way of recall, visbreaking is a fairly simple process that consists of a large furnace that heats the feedstock to the range of 450–500°C (840–930°F) at an operating pressure of about 140 psi (965 kPa). The residence time in the furnace is carefully limited to prevent much of the reaction from taking place and clogging the furnace tubes. The heated feed is then charged to a coil or a reaction chamber that is maintained at a pressure high enough to permit cracking of the large molecules but restrict coke formation. From the reaction area the process fluid is cooled (quenched) to inhibit further cracking and is then charged to a distillation column for separation into components.

Visbreaking units typically convert about 15% of the feedstock to naphtha and diesel oils and produce a lower viscosity residual fuel. Thermal cracking units provide more severe processing and often convert as much as 50–60% of the incoming feed to naphtha and light diesel oils.

Hydrovisbreaking, a noncatalytic process, is conducted under conditions similar to those of visbreaking and involves treatment with hydrogen under mild conditions (Kamiya, 1991, p. 57). The presence of hydrogen leads to more stable products (lower flocculation threshold) than can be obtained with straight visbreaking, which means that higher conversions can be achieved, producing a lower viscosity product.

The HYCAR process (Fig. 18.22) is composed fundamentally of three parts: (1) visbreaking, (2) hydrodemetallization, and (3) hydrocracking. In the visbreaking section, the heavy feedstock (e.g., vacuum residuum or bitumen) is subjected to moderate thermal cracking while no coke formation is induced. The visbroken oil is fed to the demetallization reactor in the presence of catalysts, which provides sufficient pore volume for the diffusion and adsorption of high molecular weight constituents. The product from this second stage proceeds to the hydrocracking reactor, where desulfurization and denitrogenation take place along with hydrocracking.

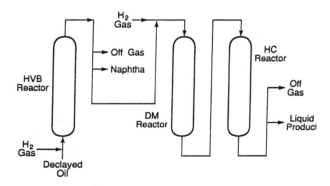

Figure 18.22 The HYCAR process.

4.3 Hyvahl F Process

The Hyvahl F process is used to hydrotreat atmospheric and vacuum residua to convert the feedstock to naphtha and middle distillates (Table 18.21) (*Hydrocarbon Processing*, 1996; Kamiya, 1991, p. 59).

The main features of this process are its dual catalyst system and its fixed-bed swing-reactor concept. The first catalyst has a high capacity for metals (to 100 wt% of new catalyst) and is used for both hydrodemetallization (HDM) and most of the conversion. This catalyst is resistant to fouling, coking, and plugging by asphaltenes and shields the second catalyst from them. Protected from metal poisons and the deposition of coke-like products, the highly active second catalyst can carry out its deep hydrodesulfurization (HDS) and refining functions. Both catalyst systems use fixed beds that are more efficient than moving beds and are not subject to attrition problems.

The swing-reactor design reserves two of the HDM reactors for use as guard reactors; one of them can be removed from service for catalyst reconditioning and put on standby while the rest of the unit continues to operate. More than 50% of the metals is removed from the feed in the guard reactors.

In the Hyvahl F process (Fig. 18.23), the preheated feedstock enters one of the two guard reactors, where a large proportion of the nickel and vanadium is adsorbed and hydroconversion of the high molecular weight constituents commences. Meanwhile the second guard reactor catalyst undergoes a reconditioning process and then is put on standby. From the guard reactors, the feedstock flows through a series of hydrodemetallization reactors that continue the metals removal and the conversion of heavy ends.

The next processing stage, hydrodesulfurization, is where most of the sulfur, some of the nitrogen, and the residual metals are removed. A limited amount of conversion also takes place. From the final reactor, the gas phase is separated, hydrogen is recirculated to the reaction section, and the liquid products are sent to a conventional fractionation section for separation into naphtha, middle distillates, and heavier streams.

Table 18.21   Feedstock and Product Data for the Hyvahl F Process

	Hyvahl F (once-through)	Hyvahl F plus R2R
Feedstock: Iranian crude oil (topped) ($> 360°C; > 680°F$)		
API gravity	15.2	
Sulfur, wt%	2.6	
Nitrogen, wt%	0.4	
Carbon residue, wt%	9.4	
C_7 Asphaltenes	2.9	
Nickel + vanadium, ppm	191.0	
Products		
Naphtha, wt%	4.0	48.0
Distillate/gas oil/vacuum gas oil, wt%	24.5	17.5
Vacuum residuum, wt%	67.5	8.4
Coke		6.4

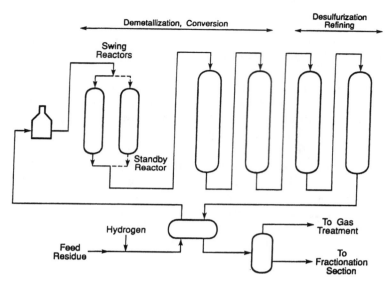

Figure 18.23   The Hyvahl F process.

5. SOLVENT PROCESSES

Solvent deasphalting processes have not realized their maximum potential. With ongoing improvements in energy efficiency, such processes would display their effects in combination with other processes. Solvent deasphalting allows removal of sulfur and nitrogen compounds as well as metallic constituents by balancing yield with the desired feedstock properties (Fig. 18.24) (Ditman, 1973).

5.1   Deasphalting Process

Petroleum processing normally involves separation into various fractions that require further processing in order to produce marketable products. The initial separation process is distillation (Chapter 13), in which crude oil is separated into fractions of increasingly higher boiling range fractions. Because petroleum fractions are subject to thermal degradation, there is a limit to the temperatures that can be used in simple separation processes. The crude cannot be subjected to temperatures much above 395°C (740°F), irrespective of the residence time, without encountering some thermal cracking. Therefore, to separate the higher molecular weight and higher boiling fractions from the crude, special processing steps must be used.

Thus, although a crude oil might be subjected to atmospheric distillation and vacuum distillation, there may still be some valuable oils left in the vacuum residuum. These valuable oils are recovered by solvent extraction, and the first application of solvent extraction in refining was the recovery of heavy lube oil base stocks by propane (C_3H_8) deasphalting. To recover more oil from vacuum-reduced crude, mainly for catalytic cracking feedstocks, higher molecular weight solvents such as butane (C_4H_{10}) and even pentane (C_5H_{12}) have been employed.

The deasphalting process is a mature process, but as refinery operations evolve it is necessary to include a description of the process here so that the new processes might be

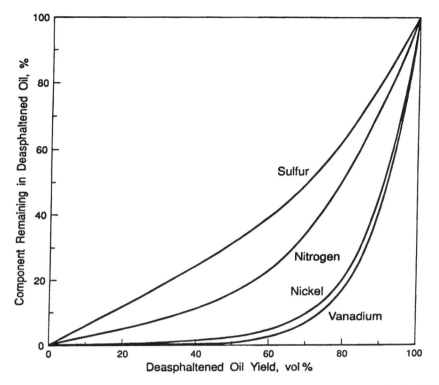

Figure 18.24 Relationship of deasphalted oil properties to deasphalted oil yield.

compared with new options that also provide for deasphalting various feedstocks. Indeed, several of these options, such as the ROSE process, have been on-stream for several years and are included here for the same reason. Thus, this section provides a one-stop discussion of solvent recovery processes and their integration into refinery operations.

The separation of residua into oil and asphalt fractions was first performed on a production scale by mixing the vacuum residuum with propane (or mixtures of normally gaseous hydrocarbons) and continuously decanting the resulting phases in a suitable vessel. Temperature was maintained within about 55°C (100°F) of the critical temperature of the solvent, at a level that would regulate the yield and properties of the deasphalted oil in solution and reject the heavier undesirable components as asphalt.

Currently, deasphalting and delayed coking are used frequently for residuum conversion. The high demand for petroleum coke, mainly for use in the aluminum industry, has made delayed coking a major residuum conversion process. However, many crude oils will not produce coke that meets the sulfur and metals specifications for aluminum electrodes, and coke gas oils are less desirable feedstocks for fluid catalytic cracking than virgin gas oils. In comparison, the solvent deasphalting process can apply to most vacuum residua. The deasphalted oil is an acceptable feedstock for both fluid catalytic cracking and, in some cases, hydrocracking. Because it is relatively less expensive to desulfurize the deasphalted oil than the heavy vacuum residuum, the solvent deasphalting process offers a more economical route for disposing of vacuum residuum from high sulfur crude.

However, the question of disposal of the asphalt remains. Use as a road asphalt is common; use as a refinery fuel is less common because expensive stack gas cleanup facilities may be required when asphalt is used as fuel.

In the deasphalting process (Fig. 18.25), the feedstock is mixed with dilution solvent from the solvent accumulator and then cooled to the desired temperature before it enters the extraction tower. Because of its high viscosity, the charge oil cannot be cooled easily to the required temperature, nor will it mix readily with solvent in the extraction tower. By adding a relatively small portion of solvent upstream of the charge cooler (insufficient to cause phase separation), the viscosity problem is avoided.

The feedstock, with a small amount of solvent, enters the extraction tower at a point about two-thirds up the column. The solvent is pumped from the accumulator, is cooled, and enters near the bottom of the tower. The extraction tower is a multistage contactor normally equipped with baffle trays, and the heavy oil flows downward while the light solvent flows upward. As the extraction progresses, the desired oil goes to the solvent and the asphalt separates and moves toward the bottom.

As the extracted oil and solvent rise in the tower, the temperature is increased to control the quality of the product by providing adequate reflux for optimum separation. Separation of oil from asphalt is controlled by maintaining a temperature gradient across the extraction tower and by varying the solvent/oil ratio. The tower top temperature is regulated by adjusting the feed inlet temperature and the steam flow to the heating coils in

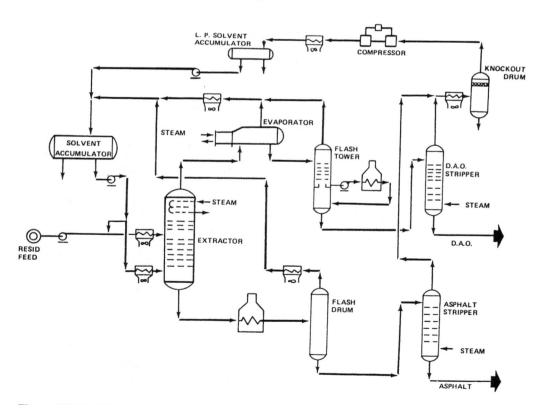

Figure 18.25 The deasphalting process.

the top of the tower. The temperature at the bottom of the tower is maintained at the desired level by the temperature of the entering solvent. The deasphalted oil–solvent mixture flows from the top of the tower under controlled pressure to a kettle-type evaporator heated by low pressure steam. The vaporized solvent flows through the condenser into the solvent accumulator.

The liquid phase flows from the bottom of the evaporator, to the deasphalted oil flash tower, where it is reboiled by means of a fired heater. In the flash tower, most of the remaining solvent is vaporized and flows overhead, joining the solvent from the low pressure steam evaporator. The deasphalted oil, with a relatively minor amount of solvent, flows from the bottom of the flash tower to a steam stripper operating at essentially atmospheric pressure. Superheated steam is introduced into the lower portion of the tower. The remaining solvent is stripped out and flows overhead with the steam through a condenser into the compressor suction drum, where the water drops out. The water flows from the bottom of the drum to appropriate disposal.

The asphalt–solvent mixture is advanced by pressure from the extraction tower bottom under flow control to the asphalt heater and on to the asphalt flash drum, where the vaporized solvent is separated from the asphalt. The drum operates essentially at the solvent condensing pressure so that the overhead vapors flow directly through the condenser into the solvent accumulator.

Hot asphalt with a small quantity of solvent flows from the asphalt flash drum bottom, under level control, to the asphalt stripper, which is operated at near atmospheric pressure. Superheated steam is introduced into the bottom of the stripper. The steam and solvent vapors pass overhead, join the deasphalted oil stripper, and flow through the condenser into the compressor suction drum. The asphalt is pumped from the bottom of the stripper, under level control, to storage.

The yield of deasphalted oil varies with the feedstock (Table 18.22), but the deasphalted oil does make less coke and more distillate than the feedstock. The metals content of the deasphalted oil is relatively low, and the nitrogen and sulfur contents in the deasphalted oil are also related to the deasphalted oil yield (Fig. 18.24).

The process parameters for a deasphalting unit must be selected with care according to the nature of the feedstock and the desired final products.

The choice of solvent is vital to the flexibility and performance of the unit. The solvent must be suitable not only for the extraction of the desired oil fraction but also for control of the yield and/or quality of the deasphalted oil at temperatures that are within the operating limits. If the temperature is too high (i.e., close to the critical temperature of the solvent), the operation becomes unreliable in terms of product yields and character. If the temperature is too low, the feedstock may be too viscous and have an adverse effect on contact with the solvent in the tower.

Liquid propane is by far the most selective solvent among the light hydrocarbons used for deasphalting. At temperatures ranging from 38°C–65°C (100–150°F), most hydrocarbons are soluble in propane, whereas asphaltic and resinous compounds are not; this allows the latter compounds to be rejected, resulting in a drastic reduction (relative to the feedstock) of the nitrogen and metals content in the deasphalted oil. Although the deasphalted oil from propane deasphalting has the best quality, the yield is usually less than that of deasphalted oil produced using a higher molecular weight (higher boiling) solvent.

The propane/oil ratios required vary from 6:1 to 10:1 by volume, with the ratio occasionally being as high as 13:1. Since the critical temperature of propane is 97°C

Table 18.22 Feedstock and Product Data for the Deasphalting Process

Crude source	Arab	West Texas	California	Canadian	Kuwait	Kuwait
Feedstock						
Crude, vol%	23.0	29.2	20.0	16.0	22.2	32.3
Gravity, °API	6.8	12.0	6.3	9.6	5.6	8.1
Conradson carbon, wt%	15.0	12.1	22.2	18.9	24.0	19.7
SUS at 210°F	75,000	526	9600	1740	14,200	3270
Metals, wppm						
Ni	73.6	16.0	139	46.6	29.9	29.7
V	365.0	27.6	136	30.9	110.0	89
Cu + Fe	15.5	14.8	94	40.7	13.7	7.5
Deasphalted oil						
Vol% feed	49.8	66.0	52.8	67.8	45.6	54.8
Gravity, °API	18.1	19.6	18.3	17.8	16.2	17.1
Conradson carbon, wt%	5.9	2.2	5.3	5.4	4.5	5.4
SUS at 210°F	615	113	251	250	490	656
Metals, wppm						
Ni	3.5	1.0	8.1	3.9	0.9	0.6
V	12.4	1.3	2.3	1.4	0.7	4.0
Cu + Fe	0.2	0.8	3.5	0.2	0.8	0.8
Asphalt						
Vol% feed	50.2	34.0	47.2	32.2	54.4	45.2
Gravity, °API	−1.3	−0.9	−5.1	−5.1	−1.3	−2.0

(206°F), this limits the extraction temperature to about 82°C (180°F). Therefore, propane alone may not be suitable for high viscosity feedstocks because of its relatively low operating temperature.

Isobutane and *n*-butane are more suitable for deasphalting high viscosity feedstocks, because their critical temperatures are higher [134°C (273°F), and 152°C (306°F), respectively] than that of propane. Higher extraction temperatures can be used to reduce the viscosity of the heavy feed and to increase the rate of transfer of oil to solvent.

Although *n*-pentane is less selective for the removal of metals and carbon residue, it can increase the yield of deasphalted oil from a heavy feed by a factor of 2–3 over propane (Speight, 1999, 2000). However, if the metals and carbon residue content of the pentane-deasphalted oil is too high (defined by the ensuing process), the deasphalted oil may be unsuitable as a cracking feedstock. In certain cases, the nature of the cracking catalyst may dictate that the pentane-deasphalted oil be blended with vacuum gas oil that, after further treatment such as hydrodesulfurization, produces a good cracking feedstock.

Solvent composition is an important process variable for deasphalting units. The use of a single solvent may (depending on the nature of the solvent) limit the range of feedstocks that can be processed in a deasphalting unit. When a deasphalting unit is required to handle a variety of feedstocks and/or produce various yields of deasphalted oil (as is the case in these days of variable feedstock quality), a dual solvent may be the only option to provide the desired flexibility. For example, a mixture of propane and *n*-butane might be suitable for feedstocks that vary from vacuum residua to both the heavy residua and heavy gas oils that contain asphaltic materials. Adjusting the solvent composition allows the most desirable product quantity and quality within the range of temperature control.

Besides the solvent composition, the *solvent/oil ratio* also plays an important role in a deasphalting operation. Solvent/oil ratios vary considerably and are governed by feedstock characteristics and desired product qualities, and for each individual feedstock there is a minimum operable solvent/oil ratio. Increasing the solvent/oil ratio almost invariably results in improving the quality of the deasphalted oil at a given yield, but other factors must also be taken into consideration, and (generalities aside) each plant and feedstock will have an optimum ratio.

The main consideration in the selection of the *operating temperature* is its effect on the yield of deasphalted oil. For practical applications, the lower limits of operable temperature are set by the viscosity of the oil-rich phase.

When the operating temperature is near the critical temperature of the solvent, control of the extraction tower becomes difficult, because the rate of change of solubility with temperature becomes very large at conditions close to the critical point of the solvent. Such changes in solubility cause large amounts of oil to transfer between the solvent-rich and oil-rich phases, which in turn causes flooding and/or uncontrollable changes in product quality. To mitigate such effects, the upper limits of operable temperatures must lie below the critical temperature of the solvent in order to ensure good control of the product quality and to maintain a stable condition in the extraction tower.

The *temperature gradient* across the extraction tower influences the sharpness of separation of the deasphalted oil and the asphalt because of internal reflux that occurs when the cooler oil–solvent solution in the lower section of the tower attempts to carry a large portion of oil to the top of the tower. When the oil–solvent solution reaches the steam-heated, higher temperature area near the top of the tower, some oil of higher molecular weight in the solvent solution is rejected because the oil is less soluble in solvent at the higher temperature. The heavier oil rejected from the solution at the top of the tower attempts to flow downward and causes internal reflux. In fact, generally, the greater the temperature difference between the top and the bottom of the tower, the greater will be the internal reflux and the better will be the quality of the deasphalted oil. However, too much internal reflux can cause tower flooding and jeopardize the process.

The *process pressure* is usually not considered to be an operating variable, because it must be higher than the vapor pressure of the solvent mixture at the tower operating temperature to maintain the solvent in the liquid phase. The tower pressure is usually subject to change only when there is a need to change the solvent composition or the process temperature.

Proper *contact* and *distribution of the oil and solvent* in the *tower* are essential to the efficient operation of any deasphalting unit. In early units, mixer-settlers were used as contactors, but they proved to be less efficient than the countercurrent contacting devices. Packed towers are difficult to operate in this process because of the large differences in viscosity and density between the asphalt phase and the solvent-rich phase.

The *extraction tower* for solvent deasphalting consists of two contacting zones: (1) a rectifying zone above the oil feed and (2) a stripping zone below the oil feed. The rectifying zone contains some elements designed to promote contact and to avoid *channeling*. Steam-heated coils are provided to raise the temperature sufficiently to induce an oil-rich reflux in the top section of the tower. The stripping zone has disengaging spaces at the top and bottom and consists of contacting elements between the oil inlet and the solvent inlet.

A *countercurrent tower* with static baffles is widely used in solvent deasphalting service. The baffles consist of fixed elements formed of expanded metal gratings in groups of two or more to provide maximum change of direction without limiting capacity. The *rotating disk contactor* has also been employed. It consists of disks connected to a rotating shaft that are used in place of the static baffles in the tower. The rotating element is driven by a variable-speed drive at either the top or bottom of the column, and operating flexibility is provided by controlling the speed of the rotating element and thus the amount of mixing in the contactor.

In the deasphalting process, the solvent is recovered for circulation, and the efficient operability of a deasphalting unit is dependent on the design of the *solvent recovery system.*

The solvent may be separated from the deasphalted oil in several ways such as conventional evaporation or the use of a flash tower. Irrespective of the method of solvent recovery from the deasphalted oil, it is usually most efficient to recover the solvent at a temperature close to the extraction temperature. If a higher temperature is used for solvent recovery, heat is wasted in the form of high vapor temperature, and, conversely, if a lower temperature is used, the solvent must be reheated, thereby requiring additional energy input. The solvent recovery pressure should be low enough to maintain a smooth flow under pressure from the extraction tower.

The asphalt solution from the bottom of the extraction tower usually contains less than an equal volume of solvent. A fired heater is used to maintain the temperature of the asphalt solution well above the foaming level and to keep the asphalt phase in a fluid state. A flash drum is used to separate the solvent vapor from asphalt, with the design being such as to prevent carryover of asphalt into the solvent outlet line and to avoid fouling the downstream solvent condenser. The solvent recovery system from asphalt is not usually subject to the same degree of variation as the solvent recovery system for the deasphalted oil, and operation at constant temperature and pressure with a separate solvent condenser and accumulator is possible.

Asphalt from different crude oils varies considerably, but its viscosity is often too high for fuel oil, although in some cases the different asphalts can be blended with refinery cutter stocks to make No. 6 fuel oil. When the sulfur content of the original residuum is high, even the blended fuel oil will not be able to meet the sulfur specification of fuel oil unless stack gas cleanup is available.

The deasphalted oil and solvent asphalt are not finished products; they require further processing or blending, depending on the final use. The manufacture of lubricating oil is one possibility, and the deasphalted oil may also be used as a *catalytic cracking feedstock* or it may be desulfurized. These last two options are perhaps more pertinent to the present text and future refinery operations.

Briefly, catalytic cracking or hydrodesulfurization of atmospheric and vacuum residua from high sulfur, high metal crude oil is theoretically the best way to enhance their value. However, the concentrations of sulfur (in the asphaltene fraction) in the residua can severely limit the performance of cracking catalysts and hydrodesulfurization catalysts (Speight, 2000). Both processes generally require tolerant catalysts as well as (in the case of hydrodesulfurization) high hydrogen pressure, low space velocity, and high hydrogen recycle ratio.

For both processes, the advantage of using the deasphalting process to remove the troublesome compounds becomes obvious. The deasphalted oil, with no asphaltene constituents and low metal content, is easier to process than the residua. Indeed, in the hydrodesulfurization process, the deasphalted oil may consume only 65% as much hydrogen as is required for direct hydrodesulfurization of topped crude oil.

As always, the use of the deasphalter reject (solvent asphalt) remains an issue. It can be used (apart from its use for various types of asphalt) as feed to a partial oxidation unit to make a hydrogen-rich gas for use in hydrodesulfurization and hydrocracking processes. Alternatively, the asphalt may be treated in a visbreaker to reduce its viscosity, thereby minimizing the need for cutter stock to be blended with the solvent asphalt for making fuel oil. Hydrovisbreaking also offers an option of converting the asphalt to feedstocks for other conversion processes.

5.2 Deep Solvent Deasphalting Process

The deep solvent deasphalting process is an application of the low energy deasphalting (LEDA) process, which is used to extract high quality lubricating oil bright stock or prepare catalytic cracking feeds, hydrocracking feeds, hydrodesulfurizer feeds, and asphalt from vacuum residua (Table 18.23) (Kamiya, 1991, p. 91; *Hydrocarbon Processing*, 1998, p. 67).

The LEDA process (Fig. 18.26) uses a low-boiling hydrocarbon solvent specifically formulated to ensure the most economical deasphalting design for each operation. For example, a propane solvent may be specified for a low deasphalted oil yield operation, whereas a solvent containing hydrocarbons such as hexane may be used to obtain a high deasphalted oil yield from a vacuum residuum. The deep deasphalting process can be integrated with a delayed coking operation (ASCOT process; Section 2.1); in this case the solvent can be a low-boiling naphtha.

Low energy deasphalting operations are usually carried out in a rotating disk contactor (RDC) that provides more extraction stages than a mixer-settler or baffle-type column. Although not essential to the process, the rotating disk contactor provides higher quality deasphalted oil at the same yield or higher yields at the same quality.

The low energy solvent deasphalting process selectively extracts the more paraffinic components from vacuum residua while rejecting the condensed ring aromatics. As

Table 18.23 Feedstock and Product Data for the LEDA Process

	Residuum[a]	Residuum[a]
Feedstock		
API gravity	6.5	6.5
Sulfur, wt%	3.0	3.0
Carbon residue, wt%	21.8	21.8
Nickel, ppm	46.0	46.0
Vanadium, ppm	125.0	125.0
Product		
Deasphalted oil, vol%	53.0	65.0
API gravity	17.6	15.1
Sulfur, wt%	1.9	2.2
Carbon residue, wt%	3.5	6.2
Nickel, ppm	1.8	4.5
Vanadium, ppm	3.4	10.3

[a] Unspecified.

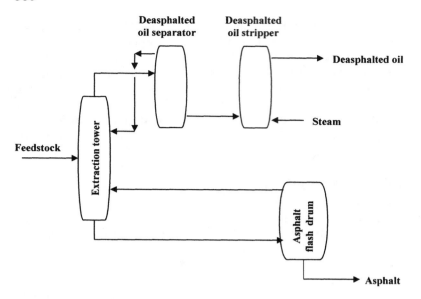

Figure 18.26 The LEDA process.

expected, deasphalted oil yields vary as a function of solvent type and quantity and feed-stock properties (Chapter 19).

   In the deep solvent deasphalting process (Fig. 18.27), vacuum residuum feed is combined with a small quantity of solvent to reduce its viscosity, then cooled to a specific extraction temperature before entering the rotating disk contactor. Recovered solvent from the high pressure and low pressure solvent receivers is combined, adjusted to a specific temperature by the solvent heater-cooler, and injected into the bottom section of the rotating disk contactor. Solvent flows upward, extracting the paraffinic hydrocarbons from the vacuum residuum, which is flowing downward through the rotating disk contactor.

   Steam coils at the top of the tower maintain the specified temperature gradient across the rotating disk contactor. The higher temperature in the top section of the rotating disk contactor results in separation of the less soluble heavier material from the deasphalted oil mix and provides internal reflux, which improves the separation. The deasphalted oil mix leaves the top of the rotating disk contactor tower. It flows to an evaporator, where it is heated to vaporize a portion of the solvent. It then flows into the high pressure flash tower, and high pressure solvent vapors are taken overhead.

   The deasphalted oil mix from the bottom of this tower flows to the heat exchanger, where additional solvent is vaporized from the deasphalted oil mix by condensing high pressure flash. The high pressure solvent, totally condensed, flows to the high pressure solvent receiver. Partially vaporized, the deasphalted oil mix flows from the heat exchanger to the low pressure flash tower, where low pressure solvent vapor is taken overhead, condensed, and collected in the low pressure solvent receiver. The deasphalted oil mix flows down the low pressure flash tower to the reboiler, where it is heated, and then to the deasphalted oil stripper, where the remaining solvent is stripped overhead with super-

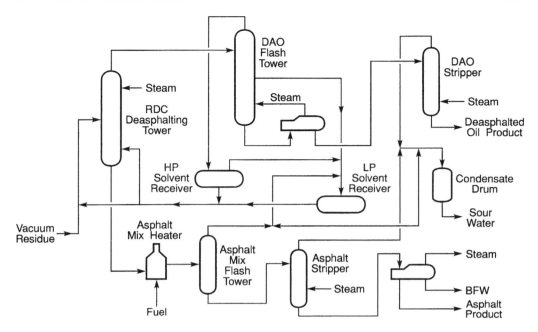

Figure 18.27 The deep solvent deasphalting process.

heated steam. The deasphalted oil product is pumped from the stripper bottom, then cooled, if required, before flowing to battery limits.

The raffinate phase containing asphalt and a small amount of solvent flows from the bottom of the rotating disk contactor to the asphalt mix heater. The hot two-phase asphalt mix from the heater is flashed in the asphalt mix flash tower, where solvent vapor is taken overhead, condensed, and collected in the low pressure solvent receiver. The remaining asphalt mix flows to the asphalt stripper, where the remaining solvent is stripped overhead with superheated steam. The asphalt stripper overhead vapors are combined with the overhead from the deasphalted oil stripper, condensed, and collected in the stripper drum. The asphalt product is pumped from the stripper and is cooled by generating low pressure steam.

5.3 Demex Process

The Demex process is a solvent extraction demetallizing process that separates high metal vacuum residuum into demetallized oil of relatively low metal content and asphaltene of high metal content (Table 18.24) (Salazar, 1986; Kamiya, 1991, p. 93; Houde, 1997). The asphaltene and condensed aromatic contents of the demetallized oil are very low. The demetallized oil is a desirable feedstock for fixed-bed hydrodesulfurization, and in cases where the metals content and carbon residue are sufficiently low, it is a desirable feedstock for fluid catalytic cracking and hydrocracking units.

The Demex process is an extension of the propane deasphalting process. It employs a less selective solvent to recover not only the high quality oils but also higher molecular weight aromatics and other processable constituents present in the feedstock.

Table 18.24 Feedstock and Product Data for the Demex Process

	Vacuum residuum[a]	Vacuum residuum[a]	Arabian light vacuum residuum	Arabian light vacuum residuum	Arabian light vacuum residuum	Arabian heavy vacuum residuum	Arabian heavy vacuum residuum
Feedstock							
API gravity	7.2	7.2	6.9	6.9	6.9	3.0	3.0
Sulfur, wt%	4.0	4.0	4.0	4.0	4.0	6.0	6.0
Nitrogen, wt%	0.3	0.3	0.3	0.3	0.3	0.5	0.5
Carbon residue, wt%	20.8	20.8	20.8	20.8	20.8	27.7	27.7
Nickel, ppm			23.0	23.0	23.0	64.0	64.0
Vanadium, ppm			75.0	75.0	75.0	205.0	205.0
Nickel + vanadium, ppm	98.0	98.0					
C_6 Asphaltenes, wt%	10.0	10.0	10.0	10.0	10.0	15.0	15.0
C_7 Asphaltenes, wt%							
Products							
Demetallized oil, vol%	56.0	78.0	40.0	60.0	78.0	30.0	55.0
API gravity	16.0	12.0	18.9	15.3	12.0	16.3	12.0
Sulfur, wt%	2.7	3.3	2.3	2.8	3.3	3.5	4.3
Nitrogen, wt%	0.1	0.2	0.1	0.2	0.2	0.1	0.2
Carbon residue, wt%	5.6	10.7	2.9	6.4	10.7	4.8	10.1
Nickel + vanadium, ppm	6.0	19.0	2.5	7.2	19.0	16.0	38.0
C_6 Asphaltenes, wt%	< 0.1	< 0.1					
Asphalt, vol%	44.0	22.0					
API gravity	< 0.0	< 0.0	< 0.0	< 0.0	< 0.0	< 0.0	< 0.0
Sulfur, wt%	5.4	6.3	5.0	5.5	6.3	6.9	7.8
Nickel + vanadium, ppm	201.0	341.0	154.0	216.0	341.0	364.0	515.0

[a] Unspecified.

Furthermore, the Demex process requires much less solvent circulation to achieve its objectives, thus reducing the utility costs and unit size significantly.

The Demex process selectively rejects asphaltenes, metals, and high molecular weight aromatics from vacuum residua. The resulting demetallized oil can then be combined with vacuum gas oil to give a greater availability of acceptable feed to subsequent conversion units.

In the Demex process (Fig. 18.28), the vacuum residuum feedstock mixed with Demex solvent recycling from the second stage is fed to the first-stage extractor. The pressure is kept high enough to maintain the solvent in the liquid phase. The temperature is controlled by the degree of cooling of the recycle solvent. The solvent rate is set near the minimum required to ensure the desired separation.

Asphaltenes are rejected in the first stage. Some resins are also rejected to maintain sufficient fluidity of the asphaltene for efficient solvent recovery. The asphaltene is heated and steam-stripped to remove solvent.

The first-stage overhead is heated by exchange with hot solvent. The increase in temperature decreases the solubility of resins and high molecular weight aromatics. These precipitate in the second-stage extractor. The bottom stream of the second-stage extractor is recycled to the first stage. A portion of this stream can also be drawn as a separate product.

The overhead from the second stage is heated by exchange with hot solvent. The fired heater further raises the temperature of the solvent–demetallized oil mixture to a point above the critical temperature of the solvent. This causes the demetallized oil to separate. It is then flashed and steam-stripped to remove all traces of solvent. The vapor streams from the demetallized oil and asphalt strippers are condensed, dewatered, and pumped up to process pressure for recycle. The bulk of the solvent goes overhead in the supercritical separator. This hot solvent stream is then effectively used for process heat exchange. The

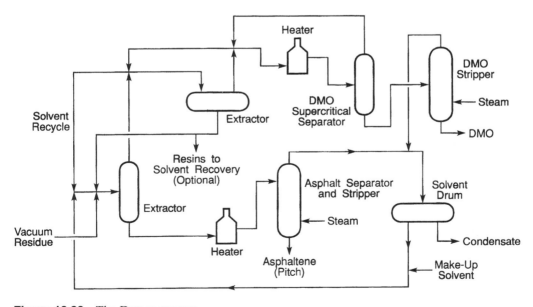

Figure 18.28 The Demex process.

subcritical solvent recovery techniques, including multiple effect systems, allow much less heat recovery. Most of the low grade heat in the solvent vapors from the subcritical flash vaporization must be released to the atmosphere, requiring additional heat input to the process.

5.4 MDS Process

The MDS process is a technical improvement of the solvent deasphalting process that is particularly effective for upgrading heavy crude oils (Table 18.25) (Kashiwara, 1980; Kamiya, 1991, p. 95). Combined with hydrodesulfurization, the process is fully applicable to feed preparation for fluid catalytic cracking and hydrocracking. The process is capable of using a variety of feedstocks, including atmospheric and vacuum residua derived from various crude oils, oil sand, and visbreaker nonvolatile products.

In the MDS process (Fig. 18.29), the feed and the solvent are mixed and fed to the deasphalting tower. Deasphalting extraction proceeds in the upper half of the tower. After the removal of the asphalt, the mixture of deasphalted oil and solvent flows out of the tower through the tower top. Asphalt flows downward to come in contact with a counter-current of rising solvent. The contact eliminates oil from the asphalt, and the asphalt then accumulates on the bottom.

Deasphalted oil containing solvent is heated through a heating furnace and fed to the deasphalted oil flash tower, where most of the solvent is separated under pressure. Deasphalted oil, still containing a small amount of solvent, is again heated and fed to the stripper, where the remaining solvent is completely removed.

Table 18.25 Feedstock and Product Data for the MDS Process

	Iranian heavy atmospheric residuum	Kuwait atmospheric residuum	Khafji vacuum residuum
Feedstock			
API gravity	17.0	16.4	5.2
Sulfur, wt%	2.7	3.7	5.2
Carbon residue, wt%	9.1	9.4	21.9
Nickel, ppm	40.0	14.0	49.0
Vanadium, ppm	130.0	48.0	140.0
Products			
Deasphalted oil, vol%	93.4	93.8	72.4
API gravity	19.0	16.4	11.3
Sulfur, wt%	2.4	3.7	4.3
Carbon residue, wt%	5.9	9.4	10.9
Nickel, ppm	18.0	14.0	6.0
Vanadium, ppm	53.0	48.0	28.0
Asphalt, vol%	6.6	6.2	27.6
API gravity	< 0.0	< 0.0	< 0.0
Sulfur, wt%	5.4	7.2	7.3
Carbon residue, wt%			49.3
Nickel, ppm	320.0	113.0	150.0
Vanadium, ppm	1010.0	425.0	400.0

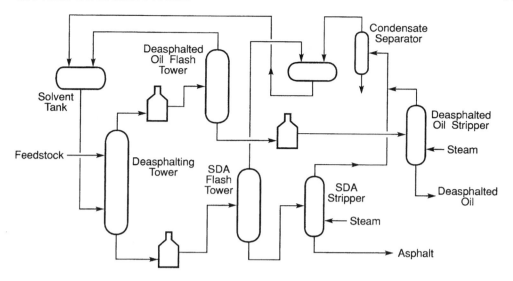

Figure 18.29 The MDS process.

Asphalt is withdrawn from the bottom of the extractor. Because this asphalt contains a small amount of solvent, it is heated through a furnace and fed to the flash tower to remove most of the solvent. Asphalt is then sent to the asphalt stripper, where the remaining portion of solvent is completely removed.

Solvent recovered from the deasphalted oil and asphalt flash towers is cooled and condensed into liquid and sent to a solvent tank. The solvent vapor leaving both strippers is cooled to remove water and compressed for condensation. The condensed solvent is then sent to the solvent tank for further recycling.

5.5 Residuum Oil Supercritical Extraction Process

The residuum oil supercritical extraction (ROSE) process is a solvent deasphalting process with minimum energy consumption using a supercritical solvent recovery system and is of value in obtaining oils for further processing (Gearhart, 1980; Kamiya, 1991, p. 97; Low et al., 1995, *Hydrocarbon Processing*, 1996; Northrup and Sloan, 1996).

In the ROSE process (Fig. 18.30), the residuum is mixed with several volumes of a low-boiling hydrocarbon solvent and passed into the asphaltene separator vessel. Asphaltenes rejected by the solvent are separated from the bottom of the vessel and are further processed by heating and steam stripping to remove a small quantity of dissolved solvent. The solvent-free asphaltenes are sent to another section of the refinery for further processing.

The main flow, solvent and extracted oil, passes overhead from the asphaltene separator through a heat exchanger and heater into the oil separator, where the extracted oil is separated without solvent vaporization. The solvent, after heat exchange, is recycled to the process. The small amount of solvent contained in the oil is removed by steam stripping, and the resulting vaporized solvent from the strippers is condensed and returned to the process. Product oil is cooled by heat exchange before being pumped to storage or further processing.

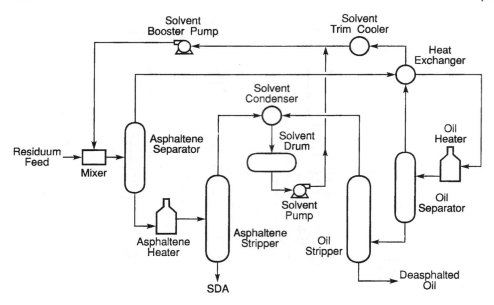

Figure 18.30   The ROSE process.

Table 18.26   Feedstock and Product Data for the
Solvahl Process

Feedstock: Arabian light vacuum residuum	
API gravity	9.6
Sulfur, wt%	4.1
Nitrogen, wt%	0.3
C_7 Asphaltenes, wt%	4.2
Carbon residue, wt%	16.4
Nickel, ppm	19.0
Vanadium, ppm	61.0
Products	
C_4-Deasphalted oil, wt%	70.1
API gravity	16.0
Sulfur, wt%	3.3
Nitrogen, wt%	0.2
C_7 Asphaltenes, wt%	< 0.1
Carbon residue, wt%	5.3
Nickel, ppm	2.0
Vanadium, ppm	3.0
C_5-Deasphalted oil, wt%	85.5
API gravity	13.8
Sulfur, wt%	3.7
Nitrogen, wt%	0.2
C_7 Asphaltenes, wt%	< 0.1
Carbon residue, wt%	7.9
Nickel, ppm	7.0
Vanadium, ppm	16.0

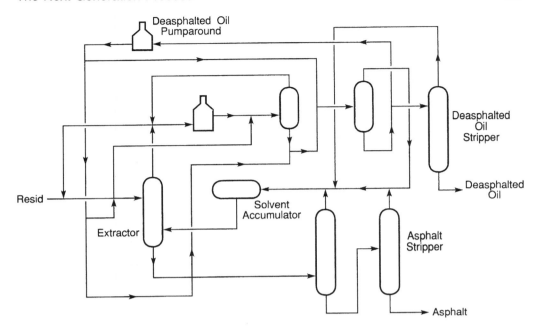

Figure 18.31   The Solvahl process.

5.6 Solvahl Process

The Solvahl process is a solvent deasphalting process for application to vacuum residua (Table 18.26) (Kamiya, 1991, p. 9).

The process (Fig. 18.31) was developed to give maximum yields of deasphalted oil while eliminating asphaltenes and reducing the metals content to a level compatible with the reliable operation of downstream units (Peries et al., 1995; *Hydrocarbon Processing*, 1996).

REFERENCES

Aiba, T., Kaji, H., Suzuki, T., and Wakamatsu, T. 1981. Chem. Eng. Progr., February, p. 37.

Bartholic, D. B., Center, A. M., Christian, B. R., and Suchanek, A. J. 1992. In: Petroleum Processing Handbook. J. J. McKetta, ed. Marcel Dekker, New York, p. 108.

Bonilla, J. 1985. Energy Progr. December, p. 5.

Bonilla, J., and Elliott, J. D. 1987. U.S. Patent 4,686,027, Aug. 11.

Bridge, A. G. 1997. In: Handbook of Petroleum Refining Processes. 2nd edition. R. A. Meyers, ed. McGraw-Hill, New York, Chapter 14.1.

Chen, R. 1993. Preprints. Oil Sands—Our Petroleum Future. Alberta Oil Sands Technology and Research Authority, Edmonton, AB, Canada, p. 287.

Dettman, H. D., Patmore, D. J., and Ng, S. H. 1993. Preprints. Oil Sands—Our Petroleum Future. Alberta Oil Sands Technology and Research Authority, Edmonton, AB, Canada, p. 365.

Dickenson, R. L., Biasca, F. E., Schulman, B. L., and Johnson, H. E. 1997. Hydrocarbon Process. 76(2):57.

Ditman, J. G. 1973. Hydrocarbon Process. 52(5):110.

Feldman, J. A., Lutter, B. E., and Hair, R. L. 1992. In: Petroleum Processing Handbook. J. J. McKetta, ed. Marcel Dekker, New York, p. 480.

Finneran, J. A. 1974. Oil Gas J. Jan. 14, No. 2, p. 72.

Gearhart, J. A. 1980. Hydrocarbon Process. 59(5):150.

Green, P. M., and Center, A. M. 1985. Proc. 3rd UNITAR/UNDP Int. Conf. on Heavy Crude and Tar Sands, Long Beach, CA, p. 1219.

Gussow, S., and Kramer, R. 1990. Annual Meeting, National Petroleum Refiners Association, March.

Houde, E. J. 1997. In: Handbook of Petroleum Refining Processes. 2nd ed. R. A. Meyers, ed. McGraw-Hill, New York, Chapter 10.4.

Hydrocarbon Processing. 1996. 75(11):89 et seq.

Hydrocarbon Processing. 1998. 77(11):53 et seq.

Johnson, T. E. 1982. Oil Gas J. March 22.

Kamiya, Y., ed. 1991. Heavy Oil Processing Handbook. Research Association for Residual Oil Processing (RAROP). Ministry of Trade and International Industry, Tokyo, Japan

Kashiwara, H. 1980. Kagaku Kogaku 7:44.

Khan, M. R., and Patmore, D. J. 1998. In: Petroleum Chemistry and Refining. J. G. Speight, ed. Taylor & Francis, Washington, DC, Chapter 6.

Khouw, F. H. H. 1990. Annual Meeting, Natl. Petroleum Refiners Assoc., March.

Kuwahara, I. 1987. Kagaku Kogaku 51:1.

LePage, J. F., Morel, F., Trassard, A. M. and Bousquet, J. 1987. Preprints Div. Fuel Chem. 32:470.

Logwinuk, A. K., and Caldwell, D. L. 1983. Annual Meeting, Natl. Petroleum Refiners Assoc., March.

Long, S. 1987. Annual Meeting, Natl. Petroleum Refiners Assoc., San Antonio, TX, March.

Low, J. Y., Hood, R. L., and Lynch, K. Z. 1995. Preprints Div. Petrol. Chem. Am. Chem. Soc. 40:780.

Maples, R. E. 2000. Petroleum Refinery Process Economics. 2nd ed. PennWell Corp., Tulsa, OK.

McKetta, J. J., ed. 1992. Petroleum Processing Handbook. Marcel Dekker, New York.

Meyers, R. A., ed., 1997. Petroleum Refining Processes. 2nd ed. McGraw-Hill, New York.

Miyauchi, T., and Ikeda, Y. 1998. U.S. Patent 4,722,378.

Miyauchi, T., Furusaki, S., and Morooka, Y. 1981. Adv. Chem. Eng. Chapter 11.

Murphy, J. R., and Treese, S. A. 1979. Oil Gas J, June 25, p. 135.

Northrup, A. H., and Sloan, H. D. 1996. Annual Meeting. Natl. Petroleum Refiners Assoc., Houston, TX, Paper AM-96-55.

Ozaki, 1982. Proc. 32nd Annu. Conf. Can. Soc. Chem. Eng., Vancouver, BC, Canada.

Peries, J. P., Billon, A., Hennico, A., Morrison, E., and Morel, F., 1995. Proc. 6th UNITAR Int. Conf. on Heavy Crude and Tar Sand 2:229.

Reynolds, B. E., Brown, E. C., and Silverman, M. A. 1992. Hydrocarbon Process. 71(4):43.

Salazar, J. R. 1986. In: Handbook of Petroleum Refining Processes. R. A. Meyers, ed. McGraw-Hill, New York, Chapter 8.5.

Sato, Y., Yamamoto, Y., Kamo, T., and Miki, K. 1992. Energy Fuels 6:821.

Shih, S. S., and Oballa, M. C., eds. 1991. Tar Sand Upgrading Technology. Symp. Ser. No. 282. Am. Inst. Chem. Eng., New York.

Speight, J. G. 1984. In: Catalysis on the Energy Scene. S. Kaliaguine and A. Mahay, eds. Elsevier, Amsterdam.

Speight, J. G. 1990. In: Fuel Science and Technology Handbook. J. G. Speight, ed. Marcel Dekker, New York, Chapters 12–16.

Speight, J. G. 1999. The Chemistry and Technology of Petroleum. 3rd ed. Marcel Dekker, New York.

Speight, J. G. 2000. The Desulfurization of Heavy Oils and Residua. 2nd ed. Marcel Dekker, New York.

Speight, J. G., and Moschopedis, S. E. 1979. Fuel Process. Technol. 2:295.

Swain, E. J. 1991. Oil Gas J. 89(36):59.

Swain, E. J. 1993. Oil Gas J. 91(9):62.

Swain, E. J. 1998. Oil Gas J. 96(40):43.

Takeuchi, C. 1982. Proc. 2nd Int. Conf. on the Future of Heavy Crude and Tar Sands, Caracas, Venezuela.

Ueda, K. 1976. J. Jpn. Petrol. Inst. 19(5):417.

Ueda, H. 1978. J. Fuel Soc. Jpn. 57:963.

Washimi, K. 1989. Hydrocarbon Process. 68(9):69.

Washimi, K., and Hamamura, M. 1993. Preprints. Oil Sands—Our Petroleum Future. Alberta Oil Sands Technology and Research Authority, Edmonton, AB, Canada, p. 283.

Watari, R., Shoji, Y., Ishikawa, T., Hirotani, H., and Takeuchi, T. 1987. Annual Meeting, Natl. Petroleum Refiners Assoc., San Antonio, TX, Paper AM-87-43.

19

Product Improvement

1. INTRODUCTION

As already noted in Chapter 1, petroleum and its derivatives have been used by humans for centuries (Henry, 1873; Abraham, 1945; Forbes, 1958a, 1958b, 1959; James and Thorpe, 1994), but the petroleum industry is essentially a modern industry, and petroleum is perhaps the most important raw material consumed in modern society. It provides not only raw materials for the ubiquitous plastics and other products but also fuel for energy, industry, heating, and transportation.

From a chemical standpoint petroleum is an extremely complex mixture of hydrocarbon compounds, usually with minor amounts of nitrogen-, oxygen-, and sulfur-containing compounds as well as trace amounts of metal-containing compounds. In addition, the properties of petroleum vary widely (Speight, 1999, 2000) and are not conducive to its use in its raw state. A variety of processing steps are required to convert petroleum from its raw state to products that are usable in modern society.

Petroleum is the source of a wide variety of products (Table 19.1). The properties and character of these products are not covered here, because they have been discussed in detail elsewhere (Speight, 1999). However, the processes that are used to *clean* these products and enable them to meet desired specifications are extremely important in refinery operations and are therefore covered in this chapter.

The fuel products (liquefied petroleum gas, naphtha, gasoline, kerosene, diesel fuel, and fuel oil) that are derived from petroleum constitute more than half of the world's total supply of energy. Gasoline, kerosene, and diesel oil provide fuel for automobiles, tractors, trucks, aircraft, and ships. Fuel oil and natural gas are used to heat homes and commercial buildings as well as to generate electricity. Petroleum products are basic materials in the manufacture of synthetic fibers for clothing and in plastics, paints, fertilizers, insecticides, soaps, and synthetic rubber. The uses of petroleum as a source of raw material in manufacturing are central to the functioning of modern industry.

For the purposes of this chapter, products from petroleum can be divided into three major classes: (1) products that are of natural origin; (2) products that are manufactured; and (3) petrochemical products. The materials included in categories 1 and 2 are relevant here because of their uses, either individually or as part of a blending stock for product

Table 19.1 Brief Summation of Petroleum Products and Their Uses

Petroleum fraction	Refinery product	Consumer product
Gases	Distillation gases	Fuel gas
	Propane (C_3H_8)	Liquefied petroleum gas (LPG)
	Butane (C_4H_{10})	Liquefied petroleum gas (LPG)
Light naphtha	Gasoline	Automobile fuel, engine fuel
	Aviation fuel	Jet-B (naphtha type)
	Solvents	Rubber solvent, lacquer diluent
Heavy naphtha	Gasoline	Automobile fuel, engine fuel
	Aviation fuel	Jet-B (naphtha type)
	Solvents	Varnish solvent, dyer's naphtha, cleaner's naphtha
Kerosene	Aviation fuel	Jet-A (kerosene type)
	Lamp fuel	Illuminating (lamp) oil
	No. 1 fuel oil	Kerosene (range oil, stove oil)
	Diesel fuel	Motor and engine fuel
	Refined oil	Spray oil, insecticides, fungicides
Light gas oil	Diesel fuel	Motor and engine fuel
	No. 2 fuel oil	Home heating oil, domestic heating oil
	Wax	Grease, soap
		Medicinal oil
Heavy gas oil	No. 4 fuel oil	Commercial heating oil, light industrial heating oil
	No. 5 fuel oil	Heavy industrial heating oil
	Bright stock	Lubricants
Residuum	No. 6 fuel oil	Bunker C oil (marine engines)
	Atmospheric residuum	Asphalt and asphaltic oils
	Vacuum residuum	Road asphalt
Coke	Coke	Burner fuel
	Refinery fuel	Hydrogen

manufacture. Straight-run constituents of petroleum (i.e., constituents distilled without change) are used in products. Manufactured materials are produced by a variety of processes and are also used in product streams. Category 3 materials (petrochemical products) are not discussed here.

The production of liquid product streams by distillation (Chapter 13) or by thermal cracking processes (Chapter 14) or by catalytic cracking processes (Chapter 15) is only the first of a series of steps that lead to the production of marketable liquid products. Several other unit processes are involved in the production of a final product (Fig. 19.1; not all such processes are shown). Such processes may be generally termed *finishing processes*, *product improvement processes*, or *secondary processes*, because they are not used directly on the crude petroleum but are used on primary product streams that have been produced from the crude petroleum (Bland and Davidson, 1967; Hobson and Pohl, 1973).

The term "product improvement" as used in this chapter includes processes such as reforming processes in which the molecular structure of the feedstock is reorganized. An example is the conversion (*reforming, molecular rearrangement*) of *n*-hexane to cyclohexane or of cyclohexane to benzene. These processes reform or rearrange one particular molecular type to another, thereby changing the properties of the product relative to those

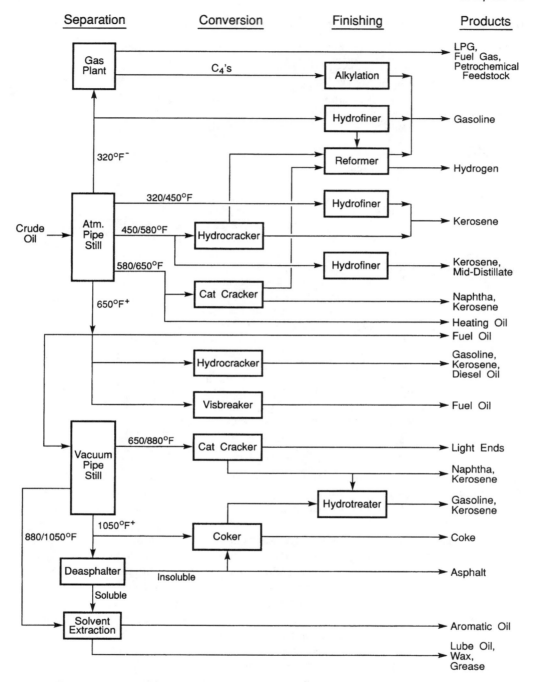

Figure 19.1 Schematic of a petroleum refinery.

of the feedstock. Such processes are conducive to expansion of the utility of petroleum products and to sales.

2. DESULFURIZATION AND HETEROATOM REMOVAL

Product improvement is the treatment of petroleum products to ensure that they meet utility and performance specifications. Product improvement usually involves changes in molecular shape (reforming and isomerization) or change in molecular size (alkylation and polymerization). However, other processes that might be classified as product improvement processes are also an integral part of refinery operations. These are the processes that remove sulfur and other heteroatomic species, including metals. Such processes were described in Chapter 18 with the focus mostly on the higher boiling feedstocks. In the present context the focus is on the lower boiling products that require treatment before being used as feedstocks in any product improvement process.

Most refinery liquid streams contain some sulfur (as well as other heteroatoms depending on the boiling range of the stream) in various chemical forms, even if they come from sweet crude oil. Removal of such heteroatoms by distillation is extremely difficult, if not impossible, and chemical methods of removal are necessary; hydrotreating is one such method.

Thus, the removal of heteroatoms from petroleum products is often achieved by washing or by hydrotreating (Speight, 1999, 2000). These procedures are an integral part of refinery operations and are not often recognized as also being product improvement processes, whether or not the heteroatoms are removed during primary processing or after primary processing. Thus, with the inception of hydrogenation as a process by which petroleum could be converted into lower boiling products, it was also recognized that hydrogenation would be effective for product improvement by the removal of nitrogen, oxygen, and sulfur compounds from the feedstock (Fig. 19.2).

In fact, in the early days of refinery operations, it was the reforming processes that provided substantial quantities of *by-product* hydrogen for the hydrotreating processes. The need for such commercial operations has become more acute because of a shift in supply trends that has increased the amount of high-heteroatom crude oils employed as refinery feedstocks.

There are several valid reasons for removing heteroatoms from petroleum fractions:

1. Reduction, or elimination, of corrosion during refining, handling, or use of the various products
2. Production of products having an acceptable odor and specification
3. Increasing the performance (and stability) of gasoline
4. Decreasing smoke formation in kerosene
5. Reduction of heteroatom content in fuel oil to a level that improves burning characteristics and is environmentally acceptable

Extraction and chemical treatment of various petroleum fractions are still used as means of removing certain heteroatomic constituents from petroleum products, but hydroprocessing is generally applicable to the removal of all types of heteroatomic compounds (Speight, 2000).

Heteroatom removal, as practiced in various refineries, can take several forms such as concentration in refinery products (e.g., coke), hydroprocessing, or chemical removal (acid treating and caustic treating, i.e., sweetening or finishing processes) (Speight, 1999, 2000).

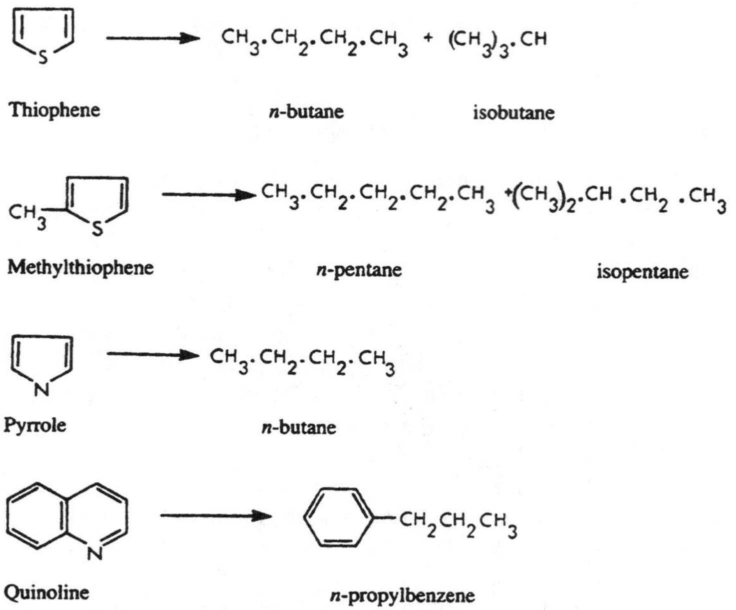
CH3.CH2.CH2.CH3 + (CH3)3.CH
Thiophene
n-butane
isobutane
CH3
S
CH3.CH2.CH2.CH2.CH3 +(CH3)2.CH .CH2 .CH3
Methylthiophene
n-pentane
isopentane
N
CH3.CH2.CH2.CH3
Pyrrole
n-butane
N
CH2CH2CH3
Quinoline
n-propylbenzene

Figure 19.2 Heteroatom removal by hydrotreating.

Nevertheless, the removal of heteroatoms from petroleum feedstocks is almost universally accomplished by the catalytic reaction of hydrogen with the feedstock constituents. Certain other refinery processes are adaptable to residua and heavy oils and may be effective for reducing, but not necessarily for completely removing, the heteroatomic content.

Propane deasphalting (Chapter 18), or modifications of it, is most efficient in improving cracking feedstocks because the asphaltic (heteroatom-containing or coke-forming) constituents are lower in the deasphalted oil than in the original residuum or heavy oil. It has been claimed that installation of a deasphalting unit just prior to a catalytic cracking (hydrocracking) unit or heavy oil conversion unit may actually improve the yield and quality of the products from the cracker, thereby reducing the amount of sulfur to be removed prior to reforming.

It is usually claimed that *visbreaking* (Chapter 14) has little or no effect on the heteroatomic sulfur compounds in a residuum and that its sole purpose is to reduce the viscosity of the feedstock. However, visbreaking is capable of producing product streams that have reduced heteroatomic content compared to the feedstock (Speight, 2000). The products are more amenable to subsequent hydroprocessing for further heteroatom removal. The heteroatoms tend to concentrate in the sediment, and the liquid products can then be blended back with the original feedstock or with the residuum (from the visbreaking) to produce an overall low sulfur stream.

Coking processes (Chapter 14) are different from the other thermal processes found in a refinery insofar as the reaction times may be longer and the reactions are usually allowed to proceed to completion (in contrast to, say, visbreaking, where the reactions are terminated by quenching with gas oil). The coke obtained from the coking processes is usually used as a fuel for the process, although marketing for specialty uses, such as electrode manufacture, increases its value. But, because of the tendency of the process to concentrate the feedstock heteroatoms in the coke, the coker feedstock may have to be chosen carefully to produce a coke of sufficiently low heteroatomic content for a specialty use.

Catalytic cracking (Chapter 15) also makes it possible to process heavy feedstocks to bring about some degree of heteroatom removal. The effectiveness, and degree, of heteroatom removal depends upon the heteroatomic content of the feedstock. However, generally, to optimize the use of a catalytic cracking unit, feedstocks should be treated to remove excess asphaltic material and metals by processes such as visbreaking, coking, or deasphalting (Chapter 18).

Hydrocracking (Chapter 17) is probably the most versatile of petroleum refining processes because of its applicability to a wide range of feedstocks. Chemically, hydrocracking can be regarded as a combination of cracking, hydrogenation, and isomerization. It is also a treating operation, because elements such as sulfur, nitrogen, and oxygen are almost completely eliminated during the process. Hydrocracking employs a catalyst and an environment of hydrogen at pressures on the order of 800–2500 psi (5515–17,237 kPa).

However, what is more important here are the processes that will be employed to ensure that a product meets specifications or that it is suitable for (for example) reforming processes and that the levels of sulfur (and other heteroatoms) will not unduly poison the catalyst.

2.1 Hydrotreating

In the hydrotreating process (Chapter 16), the stream is mixed with hydrogen and heated to approximately 260–425°C (500–800°F) (*Hydrocarbon Processing*, 1998, pp. 89, 90, 92, 94–96; Bridge, 1997; Kennedy, 1997; Thompson, 1997). The oil–hydrogen stream is then charged to a reactor filled with a pelletized catalyst (Fig. 19.3), where several reactions occur:

1. The hydrogen combines with the sulfur (or other heteroatoms—nitrogen and oxygen) to form the hydrogenated analogs (H_2S, NH_3, and H_2O).
2. Metals are entrained in the oil deposit on the catalyst.

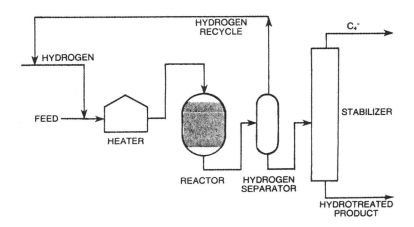

Figure 19.3 The hydrotreating process.

3. Some of the olefins and aromatics become saturated with hydrogen.
4. As the contaminants are removed from the hydrocarbon skeleton, dealkylation can occur, forming methane, ethane, propane, and butanes.

The most popular catalysts contain cobalt and molybdenum oxides on alumina and are resistant to poisoning from contaminants and easier to regenerate. The stream coming from the reactor is sent to a flash unit, where the low-boiling hydrocarbons, propane and lighter, and the unused ones pass through overhead.

The importance of hydrotreating has been gradually increasing for several years not only because of environmental concerns related to fuel specifications but also because pretreating units immediately upstream of catalytic reformers, catalytic crackers, and hydrocracking units protect the increasingly sensitive catalysts from fouling by sulfur, nitrogen, and metals. In addition, hydrotreating will change the bulk composition of liquids that are destined for use as fuels, reducing the olefins and aromatics while increasing the paraffins and naphthenes.

Naphtha hydrotreating that can reduce the sulfur content from several thousand ppm to less than 50 ppm and in some cases less than 10 ppm (*Hydrocarbon Processing*, 1998, pp. 89, 92, 94–96) is often employed to prepare a feedstock for catalytic reforming. The naphtha hydrotreater consists of a feedstock heater, the reactor, high pressure and low pressure separators, a recycle compressor, and a treated naphtha splitter. In addition, when a highly unsaturated stock such as coker naphtha is being fed, a separate additional reactor may precede the main reactor (see Fig. 16.1 of Chapter 16). The additional reactor is used to selectively saturate (under milder conditions than those in the main reactor) any unsaturated hydrocarbons in order to prevent runaway temperature increases due to the highly exothermic reactions that occur when such compounds are present. A hydrogen sulfide stripper is often placed between the separators. If the naphtha comes from storage where there is no inert gas blanket, a stripper will be needed ahead of the heater to remove oxygen.

Jet fuel hydrotreating and *distillates hydrotreating* (*Hydrocarbon Processing*, 1998, pp. 94–96) usually need to be treated to meet specifications. In addition, hydrotreating cracked light gas oil reduces the amount of aromatic compounds and raises the cetane number of the diesel fuel product. Feedstocks for catalytic cracking units and for reforming units need to be hydrotreated to protect the catalyst and improve conversion rates. In fact, feedstocks for catalytic reforming units are almost always hydrotreated, and hydrotreaters in front of catalytic cracking units serve to (1) protect the catalyst, (2) produce cleaner products, and (3) convert some of the aromatics in the feedstock, which results in better conversion rates.

2.2 Hydrogen Sulfide Removal

For the main part, and not to ignore the other heteroatoms, hydrotreating creates hydrogen sulfide gas that must be disposed of. The usual process is a two-stage operation involving (1) the removal of the hydrogen sulfide stream from the hydrocarbon stream and (2) conversion of the hydrogen sulfide to elemental sulfur.

Removal of the hydrogen sulfide from the hydrocarbon stream can be accomplished by a number of different chemical processes of which the most widely used is solvent extraction using diethanolamine (DEA) (Fig. 19.4). A mixture of diethanolamine and water trickles down a contactor containing trays or packing that distribute the liquid, while the gas stream containing the hydrogen sulfide enters at the base of the contactor.

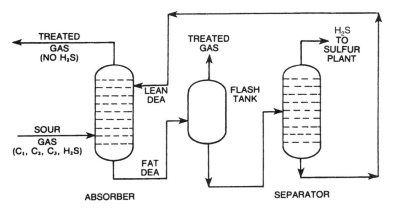

Figure 19.4 Hydrogen sulfide removal using anethanolamine.

The diethanolamine selectively absorbs the hydrogen sulfide, after which the hydrogen sulfide–rich (fat) diethanolamine is fractionated to separate the hydrogen sulfide, which is sent to a sulfur recovery plant, and the stripped (lean) diethanolamine is recycled (Speight, 1993, 1999, 2000 and references cited therein).

The conversion of hydrogen sulfide to sulfur is accomplished by means of the Claus process (Fig. 19.5), and although there are variations on the process, most are two-stage, split-stream processes. In the first stage, part of the hydrogen sulfide stream is burned in a furnace in a limited supply of oxygen to produce not only sulfur dioxide and water but also sulfur:

$$2H_2S + 2O_2 \rightarrow SO_2 + H_2O + S$$

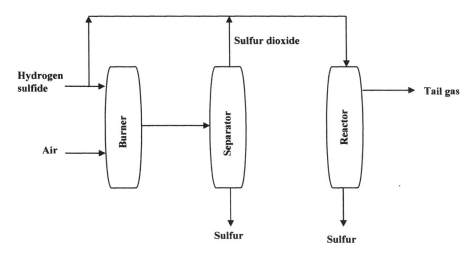

Figure 19.5 The Claus process.

The remainder of the hydrogen sulfide is mixed with the combustion products and passed over a catalyst. The hydrogen sulfide reacts with the sulfur dioxide to form sulfur and water:

$$2H_2S + SO_2 \rightarrow 2H_2O + 3S$$

Thus, Claus plants convert about 90–93% of the hydrogen sulfide to sulfur, and removal of the remaining sulfur can be achieved by the use of more advanced processes, such as the Stretford process (Speight, 1993, 1999, 2000, and other processes cited therein).

3. REFORMING

When the demand for higher octane gasoline developed during the 1930s, attention was directed to ways and means of improving the octane number of fractions within the boiling range of gasoline. Straight-run gasoline, for example, frequently had a low octane number, and any process that would improve the octane number would aid in meeting the demand for higher quality (higher octane number) gasoline. Such a process, called *thermal reforming*, was developed and used widely but to a much lesser extent than thermal cracking.

3.1 Thermal Reforming

Thermal reforming was a natural development from thermal cracking, because reforming is also a decomposition reaction to heat. Cracking converts heavier oils into naphtha or gasoline; reforming converts or reforms naphtha or gasoline into higher octane naphtha or gasoline. The equipment for thermal reforming is essentially the same as for thermal cracking, but higher temperatures are used (Nelson, 1958; Speight, 1999 and references cited therein). However, except where older units are still on-stream, thermal reforming has been largely supplanted by *catalytic reforming*, which is more effective. Nevertheless, because of the existence of these older units, some comment and description is warranted here.

Upgrading by reforming is essentially a treatment to improve a gasoline octane number. It can be accomplished in part by an increase in the volatility (reduction in molecular size) but chiefly by the conversion of *n*-paraffins to isoparaffins, olefins, and aromatics and the conversion of naphthenes to aromatics. The nature of the final product is, of course, influenced by the source (and composition) of the feedstock. In thermal reforming, the reactions resemble the reactions that occur during gas oil cracking; molecular size is reduced, and olefins and some aromatics are produced.

Thermal reforming of gasoline is a severe, vapor-phase thermal conversion process conducted under pressure. Its purpose is to increase the octane number of the gasoline charge stock, but this octane number increase is not obtained without a reduction in gasoline yield. A recycle stream is not required and is not normally used. Significant quantities of light hydrocarbon gases, pentanes and lighter compounds, are produced. These gases contain sizable quantities of light olefins that are useful as alkylation, polymerization, or petrochemical feedstocks. In thermal reforming the conditions are sufficiently severe to cause some polymerization of the light olefins and condensation to polynuclear compounds. These compounds form a stream with a higher boiling range than the gasoline charge. This heavy stream is removed and can be used as a fuel oil component.

Initially designed to upgrade the octane of low octane straight-run gasoline, thermal reforming has been extended to a variety of gasoline stocks, e.g., catalytic cracked naphtha, thermal cracked naphtha, catalytic reformed naphtha, and paraffinic raffinate. Originally, the primary product was a higher octane gasoline, and light olefinic gases and a fuel oil polymer were by-products. Thermal reforming capacity has declined rapidly from about 500,000 bbl/day in the early 1950s to less than 25,000 bbl/day today. This rapid reduction in capacity can be attributed entirely to the development of catalytic reforming.

Thermal reforming of gasoline stocks requires more severe operating conditions than thermal cracking of gas oils because the gasoline boiling range hydrocarbons are more difficult to crack thermally. Furnace outlet temperatures are on the order of 525–540°C (975–1000°F) with operating pressures ranging from 200–1000 psi (1380–6895 kPa). The desired conversion can be achieved in single-pass operation without significant heater coking because gasoline stocks have low coking tendencies. The high conversions obtained at the high temperatures also yield a high octane product. Thermal reforming can be applied to many types of gasoline stocks to increase their octane rating, but the degree of octane improvement is dependent upon the octane number of the charge stock. Generally, the lower the octane number of the charge stock, the greater the octane number increase that can be achieved.

The primary octane improving reactions in thermal reforming are similar to some of the reactions that occur during thermal cracking: (1) the cracking of paraffins into lower molecular weight paraffins and olefin chains and (2) the dehydrogenation of naphthenes to aromatics (Fig. 19.6). Other preferred reactions are (1) the conversion of paraffins to isoparaffins; (2) the conversion of paraffins to naphthenes, with the production of hydrogen; and (3) the conversion of naphthenes to aromatics, with the production of hydrogen (Fig. 19.6). The elimination and conversion of the low octane normal paraffins and the production of gasoline boiling range, high octane aromatics greatly improve the octane number of the gasoline stock being processed. The polymerization of the light olefins and subsequent condensation to polynuclear compounds to form a stream heavier than the gasoline charge is also significant at the high temperatures and moderate pressures used in thermal reforming.

Thermal reforming is considered to be a first-order reaction. Although the cracking severity is much greater than for visbreaking or thermal cracking, the charge stock is lighter and thus more refractory, which compensates for the increased severity. Again it was necessary to slightly alter the form of the first-order equation.

$$K = \frac{1}{t} \ln\left(\frac{100}{X_2}\right)$$

where K = first-order reaction rate constant, s^{-1}; t = residence time at thermal reforming conditions, s (based on charge liquid volume); $X_2 = C_5$–205°C (C_5–400°F) gasoline yield, vol%. This allows the derivation of a series of reaction velocity constants for thermal reforming for various types of feedstocks (Fig. 19.7) (Turpin, 1995). Although thermal reforming is accomplished in the vapor phase, the residence time used in the first-order equation is based on the liquid volume of the feedstock. This was done to simplify calculations and to maintain uniform calculation procedures among the thermal reforming, thermal cracking, and visbreaking discussions. Thus the residence time can be easily calculated by dividing the reactor volume above 480°C (900°F) by the charge rate. The actual residence time of the vaporized charge stock is much shorter than the pseudo

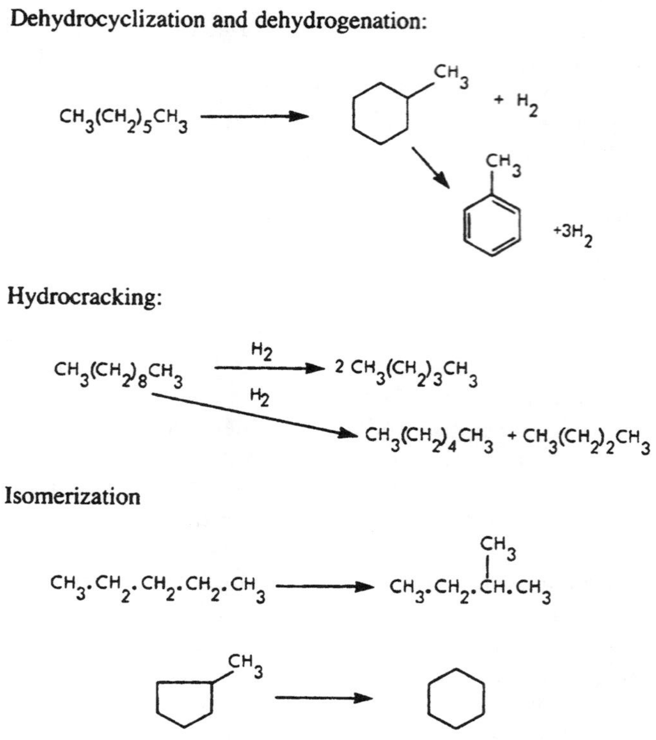

Dehydrocyclization and dehydrogenation:

Hydrocracking:

Isomerization

Figure 19.6   Reactions that occur during reforming.

residence time used in the first-order equation. If required, the actual vapor residence time can be calculated by standard procedures.

In carrying out thermal reforming (Fig. 19.8), the feedstock, such as a 205°C (400°F) endpoint naphtha, is heated to 510–595°C (950–1100°F) in a furnace much the same as a cracking furnace, with pressures from 400–1000 psi (2758–6895 kPa). As the heated naphtha leaves the furnace, it is cooled or quenched by the addition of cold naphtha. The quenched, reformed material then enters a fractional distillation tower, where any heavy products are separated. The remainder of the reformed material leaves the top of the tower to be separated into gases and reformate. The higher octane number of the product (*reformate*) is due primarily to the cracking of longer chain paraffins into higher octane olefins.

The products of thermal reforming are gases, gasoline, and residual oil, the latter formed in very small amounts (about 1%). The amount and the quality of the reformate are very dependent on the temperature. A general rule is the higher the reforming temperature, the higher the octane number of the product but the lower the yield of reformate. By using catalysts, as in the catalytic reforming processes, higher yields of much higher octane gasoline can be obtained at a given temperature.

As noted, thermal reforming is less effective than catalytic processes and has been largely supplanted, other than where older units are still on-stream, by catalytic reforming.

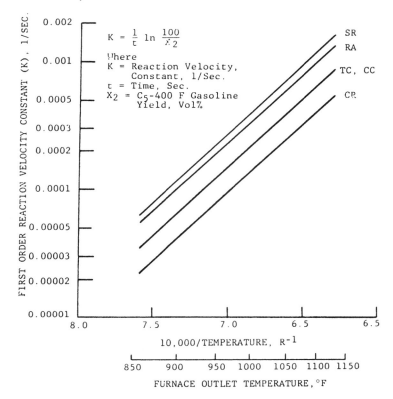

Figure 19.7 First-order rate constants for thermal reforming.

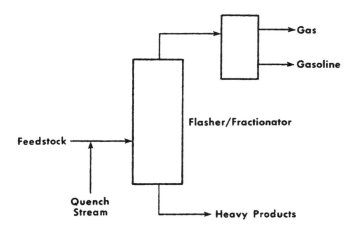

Figure 19.8 The thermal reforming process.

3.2 Catalytic Reforming

Like thermal reforming, *catalytic reforming* converts low octane naphtha or gasoline into high octane naphtha or gasoline (reformate) through the conversion of feedstock constituents to products of varying octane numbers (Table 19.2) (Fig. 19.9) (Aitani, 1995; Kelly et al., 1997; *Hydrocarbon Processing*, 1998, p. 60). Although thermal reforming can produce reformate with a research octane number of 65–80 depending on the yield, catalytic reforming produces reformate with octane numbers of the order of 90–95.

The importance of catalytic reforming is reflected in capacity on a worldwide and national (U.S.) basis. Thus, on a worldwide basis, catalytic reforming capacity is approximately 11,050,000 bbl/day of a total refining capacity of approximately 81,500,000 bbl/day. In the United States the analogous data are approximately 3,500,000 bbl/day capacity for catalytic refining out of a national refining capacity of approximately 16,500,000 bbl/day. Catalytic reforming was commercially nonexistent in the United States before 1940 and is really a process of the 1950s that showed phenomenal growth in the period 1953–1959 (Bland and Davidson, 1967; Riediger, 1971; Sivasanker and Ratnasamy, 1995).

Catalytic reforming is conducted in the presence of hydrogen over hydrogenation–dehydrogenation catalysts, which may be supported on alumina or silica-alumina. Depending on the catalyst, a definite sequence of reactions takes place, involving structural changes in the charge stock. Furthermore, this process has rendered thermal reforming somewhat obsolete (Schwarzenbek, 1971). Thermal reforming may still exist in modern refineries as an integral add-on part of a catalytic reforming process such as, Iso-Plus Houdriforming (Section 3.2.2).

3.2.1 Process Types

Reforming processes are generally classified into three types: semiregenerative processes, cyclical (fully regenerative) processes, and continuous regenerative (moving-bed) processes. Currently, most reformers are equipped with continuous catalyst regeneration

Table 19.2   Octane Numbers of Various Hydrocarbons

Hydrocarbon homologs		Octane number, clear	
		Motor	Research
C_7 hydrocarbons			
n-Paraffin	C_7H_{16} (*n*-heptane)	0.0	0.0
Naphthene	C_7H_{14} (cycloheptane)	40.2	38.8
	C_7H_{14} (methylcyclohexane)	71.1	74.8
Aromatic	C_7H_8 (toluene)	103.5	120.1
C_8 hydrocarbons			
n-Paraffin	C_8H_{18} (*n*-octane)	-15^a	-19^a
Naphthene	$C_{18}H_{16}$ (cyclooctane)	58.2	71.0
	C_8H_{16} (ethylcyclohexane)	40.8	45.6
Aromatic	C_8H_{10} (ethylbenzene)	97.9	107.4
	C_8H_{10} (*o*-xylene)	100.0	120^a
	C_8H_{10} (*m*-xylene)	115.0	117.5
	C_8H_{10} (*p*-xylene)	109.6	116.4

[a] Blending value at 20 vol% in 60 octane number reference fuel.

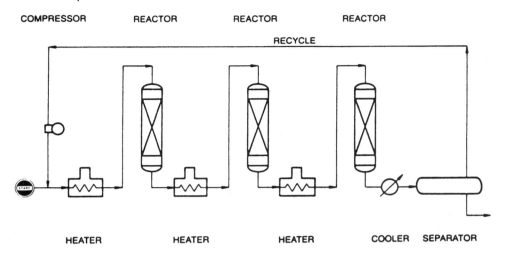

Figure 19.9 General schematic of the catalytic reforming process.

options. The bed type (fixed bed, moving bed, or fluid bed) also plays a role in determining the type of process.

The *semiregenerative reforming process* (Fig. 19.10) is characterized by continuous operation over long periods with decreasing catalyst activity. The reformers are shut down periodically as a result of coke deposition to regenerate the catalyst in situ (Aitani, 1995). Regeneration is carried out at low pressure [approximately 116 psi (800 kPa)] with air as the source of oxygen. The development of bimetallic and multimetallic reforming catalysts with the ability to tolerate high coke levels has allowed the semi-

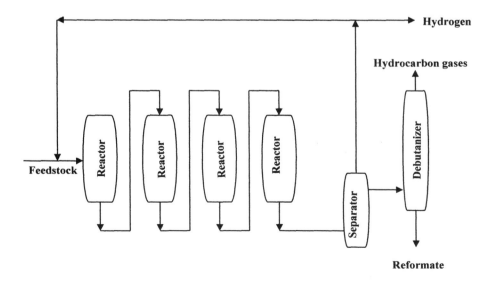

Figure 19.10 General schematic of the semiregenerative catalytic reforming process.

regenerative units to operate at 203–247 psi (1400–1700 kPa) with similar cycle lengths obtained at higher pressures. The semiregenerative process is a conventional reforming process that operates continuously over a period of up to 1 year. As the catalytic activity decreases, the yield of aromatics and the purity of the by-product hydrogen drop because of increased hydrocracking.

Semiregenerative reformers are usually built with three or four catalyst beds in series, with another reactor added to some units to allow an increase in either severity or throughput while maintaining the same cycle length. Conversion is maintained at a constant level by raising the reactor temperature as catalyst activity declines. Sometimes, when the capacity of a semiregenerative reformer is expanded, two existing reactors are placed in parallel and a new, usually smaller, reactor is added. Frequently, the parallel reactors are placed in the terminal position. When evaluating unit performance, these reactors are treated as though they were a single reactor of equivalent volume (Aitani, 1995). The catalyst can be regenerated in situ at the end of an operating cycle, and the catalyst inventory can be regenerated five to ten times before its activity falls below the economic minimum, whereupon it is removed and replaced (Aitani, 1995).

The *fixed-bed catalytic reforming* process can be classified by catalyst type: (1) cyclical regenerative with non-precious metal oxide catalysts and (2) cyclical regenerative with platinum-alumina catalysts. Both types use swing reactors (Fig. 19.11) to regenerate a portion of the catalyst while the remainder stays on-stream.

The *cyclical reforming process* typically uses five or six fixed catalyst beds, similar to the semiregenerative process, with a swing reactor as a spare reactor (Fig. 19.12) (Aitani, 1995). This reactor can substitute for any of the regular reactors in a train while the catalyst in the regular reactor is being regenerated. Thus, only one reactor at a time has to be taken out of operation for regeneration, while the process continues.

Usually the reactors are of the same size; in this case the catalyst in the early stages is less utilized and will therefore be regenerated at much longer intervals than catalyst in the later stages. The cyclical process may be operated at a low pressure for a wide boiling

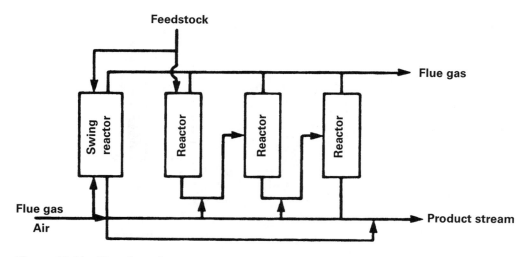

Figure 19.11   Use of a swing reactor in fixed-bed catalytic reforming.

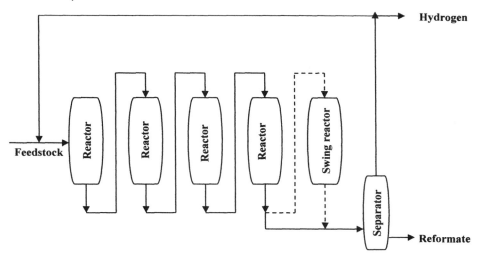

Figure 19.12 General schematic of the cyclical catalytic reforming process.

range feedstock and a low hydrogen-to-feed ratio. Coke laydown rates at these low pressures and high octane severity are so high that the catalyst in individual reactors becomes exhausted in time intervals of less than 1 week to 1 month.

The cyclical regenerative process using a platinum catalyst is basically a low pressure process (250–350 psi; 1724–2413 kPa), which gives higher gasoline yields as well as higher octane products from a given naphtha charge and better hydrogen yields because of more dehydrogenation and fewer hydrocracking reactions. The coke yield, with attendant catalyst deactivation, increases rapidly at low pressures.

The process design of the cyclical reforming process takes advantage of low unit pressures to gain a higher C_{5+} reformates yield and hydrogen production. The overall catalyst activity, conversion, and hydrogen purity vary much less with time than in the semiregenerative process. However, in the cyclical reforming process, all reactors alternate frequently between a reducing atmosphere during normal operation and an oxidizing atmosphere during regeneration. This switching technique requires that all the reactors be of the same maximum size to make switches between them possible.

The *continuous reforming (moving-bed) process* (Fig. 19.13) is characterized by high catalyst activity with reduced catalyst requirements, uniform reformate of higher aromatic content, and high hydrogen purity (Aitani, 1995). The process can achieve and surpass reforming severities as applied in the cyclical process but avoids the drawbacks of the cyclical process. In this process, small quantities of catalyst are continuously withdrawn from an operating reactor, transported to a regeneration unit, regenerated, and returned to the reactor system. In the most common moving-bed design, all the reactors are stacked on top of one other. The fourth (last) reactor may be set beside the other stacked reactors. The reactor system has a common catalyst bed that moves as a column of particles from top to bottom of the reactor section.

Coked catalyst is withdrawn from the last reactor and sent to the regeneration reactor, where the catalyst is regenerated on a continuous basis. However, the final step of the regeneration, i.e., reduction of the oxidized platinum and second metal, takes place in the

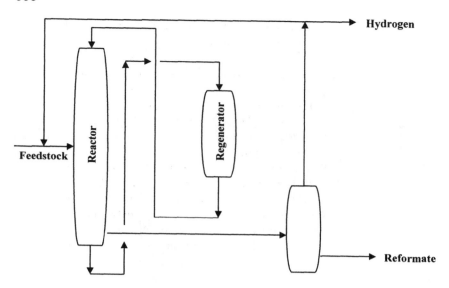

Figure 19.13 The continuous (moving-bed) reforming process (see also Fig. 19.19).

top of the first reactor or at the bottom of the regeneration train. Fresh or regenerated catalyst is added to the top of the first reactor to maintain a constant quantity of catalyst. Catalyst transport through the reactors and the regenerator is by gravity flow, whereas the transport of catalyst from the last reactor to the top of the regenerator and back to the first reactor is by the gas lift method (Aitani, 1995). The catalyst circulation rate is controlled to prevent any decline in reformate yield or hydrogen production over time on-stream.

In another design, the individual reactors are placed separately, as in the semiregenerative process, with modifications for moving the catalyst from the bottom of one reactor to the top of the next reactor in line. The regenerated catalyst is added to the first reactor, and the spent catalyst is withdrawn from the last reactor and transported back to the regenerator. The continuous reforming process is capable of operation at low pressures and high severity by managing the rapid coke deposition on the catalyst at an acceptable level. Additional benefits include elimination of downtime for catalyst regeneration and the steady production of hydrogen of constant purity. Operating pressures are in the 50–247 psi (350–1700 kPa) range, and design reformate octane number is in the range of 95–108.

Like thermal reforming, the primary octane-improving reactions in thermal reforming are similar to some of the reactions that occur during thermal cracking: (1) the cracking of paraffins into lower molecular weight paraffins and olefin chains and (2) the dehydrogenation of naphthenes to aromatics (Fig. 19.6) together with (3) the conversion of paraffins to isoparaffins, (4) the conversion of paraffins to naphthenes, with the production of hydrogen, and (5) the conversion of naphthenes to aromatics, with the production of hydrogen (Fig. 19.6) (Parera and Fígoli, 1995a, 1995b; Paál, 1995.

The elimination and conversion of the low octane normal paraffins and the production of gasoline boiling range high octane aromatics greatly improve the octane number of the gasoline stock being processed. The polymerization of the light olefins and subsequent condensation to polynuclear compounds to form a stream heavier than the gasoline charge are also significant at the high temperatures and moderate pressures used in thermal reforming. Whatever the products of catalytic reforming, and they may vary depending on the feedstock and processes parameters (Table 19.3), the important reactions are the conversion of paraffins to isoparaffins and naphthenes and the conversion of naphthenes to aromatic compounds.

Although subsequent olefin reactions occur in thermal reforming, the product contains appreciable amounts of unstable unsaturated compounds. In the presence of catalysts and hydrogen (available from dehydrogenation reactions), hydrocracking of paraffins yields two lower paraffins. Olefins that do not undergo dehydrocyclization are also produced. The olefins are hydrogenated with or without isomerization, so the end product contains only traces of olefins.

The addition of a hydrogenation–dehydrogenation catalyst to the system yields a dual-function catalyst complex. Hydrogen reactions—hydrogenation, dehydrogenation, dehydrocyclization, and hydrocracking—take place on the one catalyst, and cracking, isomerization, and olefin polymerization take place on the acid catalyst sites.

Under the high hydrogen partial pressure conditions used in catalytic reforming, sulfur compounds are readily converted into hydrogen sulfide, which, unless removed, builds up to a high concentration in the recycle gas. Hydrogen sulfide is a reversible poison for platinum and causes a decrease in the catalyst dehydrogenation and dehydrocyclization activities (Beltramini, 1995). In the first catalytic reformers the hydrogen sulfide was removed from the gas cycle stream by absorption in, for example, diethanolamine. Sulfur is generally removed from the feedstock by use of conventional desulfurization over a cobalt-molybdenum catalyst. An additional benefit of desulfurization of the feed to a level of < 5 ppm sulfur is the elimination of hydrogen sulfide (H_2S) corrosion problems in the heaters and reactors.

Organic nitrogen compounds are converted into ammonia under reforming conditions, and this neutralizes acid sites on the catalyst and thus represses the activity for isomerization, hydrocracking, and dehydrocyclization reactions. Straight-run materials do not usually present serious problems with regard to nitrogen, but feeds such as coker naphtha may contain around 50 ppm nitrogen, and removal of this quantity may require high pressure hydrogenation (800–1000 psi; 5515–6895 kPa) over nickel-cobalt-molybdenum on an alumina catalyst.

Catalytic reformer feeds are saturated (i.e., not olefinic) materials. In the majority of cases the feed may be a straight-run naphtha, but other by-product low octane naphtha (e.g., coker naphtha) can be processed after treatment to remove olefins and other contaminants. Hydrocarbon naphtha that contains substantial quantities of naphthenes is also a suitable feed.

The yield of gasoline of a given octane number and at given operating conditions depends on the hydrocarbon types in the feed. For example, high naphthene stocks, which readily give aromatic gasoline, are the easiest to reform and give the highest gasoline yields. Paraffinic stocks, however, which depend on the more difficult isomerization, dehydrocyclization, and hydrocracking reactions, require more severe conditions and give lower gasoline yields than the naphthenic stocks. The endpoint of the feed is usually limited to about 190°C (375°F), partially because of increased coke deposition on the catalyst as the endpoint increases (Marécot and Barbier, 1995). Limiting the feed endpoint avoids redistillation of the product to meet the gasoline end point specification of 205°C (400°F) maximum.

Table 19.3 Effect of Process Parameters on Product Yields

Feed	Reaction	Product	Desired rate	To get desired rate	
				Pressure	Temp.
Paraffins	Isomerization	Isoparaffins	Increase	Increase	Increase
	Dehydrocyclization	Naphthenes	Increase	Decrease	Increase
	Hydrocracking	Lower molecular weight paraffins	Decrease	Decrease	Decrease
Naphthenes	Dehydrogenation	Aromatics	Increase	Decrease	Increase
	Isomerization	Isoparaffins	Increase	Increase	Increase
	Hydrocracking	Lower molecular weight naphthenes	Decrease	Decrease	Increase
Aromatics	Hydrodealkylation	Lower molecular weight aromatics	Decrease	Decrease	Decrease

Dehydrogenation is a main chemical reaction in catalytic reforming, and hydrogen gas is consequently produced in quantity. The hydrogen is recycled through the reactors where the reforming takes place to provide the atmosphere necessary for the chemical reactions and also to prevent the carbon from being deposited on the catalyst, thus extending its operating life. An excess of hydrogen above whatever is consumed in the process is produced, and as a result catalytic reforming processes are unique in that they are the only petroleum refinery processes to produce hydrogen as a by-product.

Catalytic reforming is usually carried out by feeding a mixture of naphtha, after pretreating with hydrogen if necessary for catalyst protection and process control (Lovink, 1995), and hydrogen to a furnace, where the mixture is heated to the desired temperatures (450–520°C; 840–965°F) and then passed through fixed-bed catalytic reactors at hydrogen pressures of 100–1000 psi (689–6895 kPa). The operating conditions that promote each of the chemical reactions (Fig. 19.6) are different, as measured by pressure, temperature, and residence time. For that reason, catalytic reformers typically have three reactors, each performing a different function. The reactors operate at 480–525°C (900–975 °F) and 200–500 psi (1380–3450 kPa) and are characteristically spherical in shape. Reheaters are located between adjoining reactors to compensate for the endothermic reactions taking place. Sometimes as many as four or five are kept on-stream in series while one or more are being regenerated. The on-stream cycle of any one reactor may vary from several hours to many days, depending on the feedstock and reaction conditions.

The naphtha feed is pressurized, heated, and charged to the first reactor, where it trickles through the catalyst and out the bottom of the reactor. This process is repeated in the next two reactors, and the product issuing from the last catalytic reactor is cooled and sent to a high pressure separator, where the hydrogen-rich gas is split into two streams. One stream goes to recycle, and the remaining portion represents excess hydrogen available for other uses. The excess hydrogen is vented from the unit and is used in hydrotreating, as a fuel, or for the manufacture of chemicals (e.g., ammonia). The liquid product from the bottom of the separator is sent to a *stabilizer* or debutanizer, where the bottom product is called the *reformate;* butanes and lighter gases go overhead to the gas plant. The liquid reformate is used directly in gasoline or extracted for aromatic blending stocks for aviation gasoline.

As the process proceeds, carbonaceous deposits (coke) appear on the catalyst, causing a decline in its performance, which is reflected in a reduction in the octane number of the reformate and a lower reformate yield per barrel of feed. In the early catalytic reforming processes, units were shut down for 3 weeks to one month to regenerate the catalyst. By the 1960s, an extra (fourth) reactor was added so that one reactor could be taken off-line at a time for catalyst regeneration. Because the catalyst in each reactor could be regenerated as frequently as desired, the regeneration could be much milder and take as little as 30 h. A catalytic reforming unit could be on-stream for as much as 36 months without a shutdown. Thus, in the newer reforming units, three reactors are in operation at any one time, with the fourth in a regeneration mode. Regeneration is accomplished by blowing hot air into the reactor to remove the carbonaceous deposit from the catalyst by forming carbon monoxide and carbon dioxide. A small amount of chlorine in the hot air will also remove some of the metal deposits, and with the off-line time for a reactor being only about 30 h the catalyst is kept fresh virtually all the time.

Despite the continuous regeneration, over a long period of time the activity of the catalyst will decay. The high temperatures required for regeneration cause the catalyst's pores to collapse; some metals such as vanadium or nickel deposit on the catalyst.

Consequently, every 2–3 years the entire reformer may require shutdown so the aged catalyst can be replaced by fresh catalyst. The platinum and palladium in the catalyst are the expensive parts, and usually refiners either own or lease these precious metals, returning them to the processors for refurbishing (Rosso and El Guindy, 1995; Chaudhuri, 2000).

Catalytic reforming units require various temperatures, pressures, residence times, feedstock quality, and feedstock cut points to strike a balance between the volume of the reformate produced and its quality due to reaction variance. As the octane number increases, the percent of volume reformate decreases with an increase in the yield of butanes and lower molecular weight gases (Fig. 19.14). That happens because more reforming of the heavier molecules inevitably leads to side chains breaking off or to free carbon atoms forming methane or ethane. The operation of the catalytic reformer must be tuned very closely to the gasoline blending operations and the gasoline component yields of the other processing units.

3.2.2 Commercial Processes

The *Catforming process*, a fixed-bed process in which the catalyst is composed of platinum, alumina, and silica-alumina, permits relatively high space velocities and results in very high hydrogen purity. Regeneration to prolong catalyst life is practiced on a block-out basis with a dilute air in-stream mixture.

The *Houdriforming process* is a process in which the catalyst may be regenerated, if necessary, on a block-out basis. It is used to upgrade naphtha to aviation blending stock, aromatics, and high octane gasoline in the range of 80–100 research octane number (RON) (Table 19.4). The process operates in a conventional semiregenerative mode with four reactors in series for benzene–toluene–xylene production, compared with three reactors

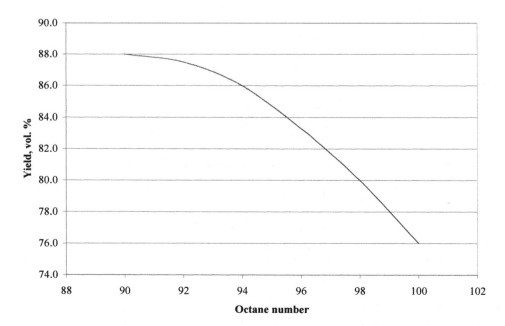

Figure 19.14 Relationship of reformate octane number to yield.

Table 19.4 Feedstock and Product Data for the
Houdriforming Process

Feedstock	
API gravity	52.6
Boiling range	92–192°C; 197–377°F
Composition, vol%	
Paraffins	53
Naphthenes	38
Aromatics	9
Product	
Research octane number	100
Composition, vol%	
Paraffins	21
Naphthenes	2
Aromatics	77

for gasoline. The catalyst used is usually Pt/Al_2O_3, but it may be bimetallic. A small *guard case* hydrogenation pretreater can be used to prevent catalyst poisons in the naphtha feedstock from reaching the catalyst in the reforming reactors. The guard case is filled with the usual reforming catalyst but operated at a lower temperature. The guard case can use the same Houdry catalyst as the Houdriformer reactors for high-sulfur feedstocks. Lead and copper salts are also removed under the mild conditions of the guard case operation.

At moderate severity, the process may be operated continuously for either high octane gasoline or aromatics without provision for catalyst regeneration. However, operation at high severity requires frequent in situ catalyst regeneration. Typical operating conditions are a temperature of 480–540°C (900–1000°F) and a pressure in the range 145–390 psi (1000–2700 kPa).

A modification of the Houdriforming process, the *Iso-Plus Houdriforming* process is a combination process using a conventional Houdriformer operated at moderate severity in conjunction with one of three possible alternatives:

1. Conventional catalytic reforming plus aromatic extraction and separate catalytic reforming of the aromatic raffinate (Fig. 19.15a)
2. Conventional catalytic reforming plus aromatic extraction and recycling of the aromatic raffinate aligned to the reforming state (Fig. 19.15b)
3. Conventional catalytic reforming followed by thermal reforming of the Houdriformer product and catalytic polymerization of the C_3 and C_4 olefins from thermal reforming (Fig. 19.15c)

A typical feedstock for this type of unit is naphtha. The use of a Houdry guard case permits charging stocks of relatively high sulfur content.

The *hydroforming process* (Fig. 19.16) made use of molybdena-alumina (MoO_2-Al_2O_3) catalyst pellets arranged in mixed beds; hence the process is known as fixed-bed hydroforming. The hydroformer had four reaction vessels or catalyst cases, two of which were regenerated; the other two were on the process cycle. Naphtha feed was preheated to 400–540°C (900–1000°F) and passed in series through the two catalyst cases under a pressure of

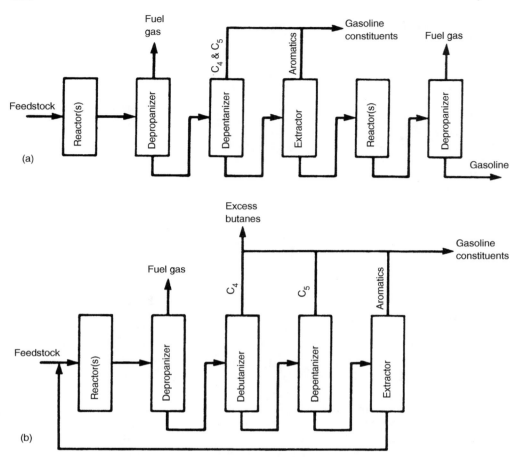

Figure 19.15 The Iso-Plus Houdriforming process with (a) catalytic reforming of the aromatic raffinate, (b) raffinate recycle, and (c) thermal reforming.

150–300 psi (1034–2068 kPa). Gas containing 70% hydrogen produced by the process was passed through the catalyst cases with the naphtha. The material leaving the final catalyst case entered a four-tower system where fractional distillation separated hydrogen-rich gas, a product (reformate) suitable for motor gasoline and an aromatic polymer boiling above 205°C (400°F).

After 4–16 h on process cycle, the catalyst was regenerated. This was done by burning carbon deposits from the catalyst at a temperature of 565°C (1050°F) by blowing air diluted with flue gas through the catalyst. The air also reoxidized the reduced catalyst (9% molybdenum oxide on activated alumina pellets) and removed sulfur from the catalyst.

The *hyperforming process* is a moving-bed reforming process (Fig. 19.17) that uses catalyst pellets of cobalt molybdate with a silica-stabilized alumina base. In operation, the catalyst moves downward through the reactor by gravity flow and is returned to the top by means of a solids-conveying technique (hyperflow), which moves the catalyst at low velo-

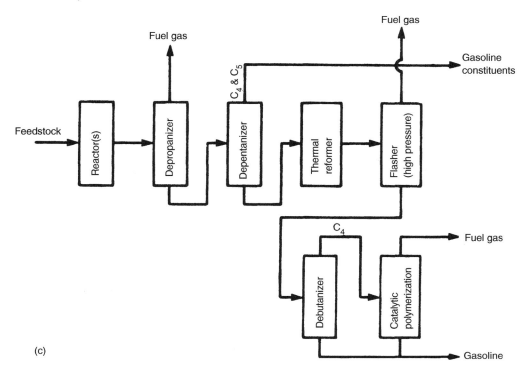

(c)

cities and with minimum attrition loss. Feedstock (naphtha vapor) and recycle gas flow upward, countercurrent to the catalyst, and regeneration of catalyst is accomplished in either an external vertical lift line or a separate vessel.

Hyperforming naphtha (65–230°C; 150–450°F) can result in improvement of the motor fuel component; in addition, sulfur and nitrogen are removed. Light gas oil stocks can also be charged to remove sulfur and nitrogen under mild hydrogenation conditions for the production of premium diesel fuels and middle distillates. Operating conditions in the reactor are 425–480°C (800–895°F) and 400 psi (2758 kPa), the higher temperature being employed for a straight-run naphtha feedstock. Catalyst regeneration takes place at 510°C (950°F) and 415 psi (2861 kPa).

The *octanizing* process (*Hydrocarbon Processing*, 1998, p. 61) is used to upgrade naphtha to high octane reformate as well as benzene, toluene, xylene (BTX), and liquefied petroleum gas. Two process designs are available; in one the catalyst is regenerated at the end of each cycle, whereas the other uses continuous catalyst regeneration that is made possible by the use of moving-bed reactors (Fig. 19.18). Using these two concepts, the *Dualforming* (conventional) and *Dualforming +* (regenerative) processes have been developed in which the main feature is the addition of the high severity regenerator for the platinum-based or bimetallic or multimetallic catalysts.

The *platforming process* (Table 19.5) (Fig. 19.19) (Weiszmann, 1986a; Dachos et al., 1997; *Hydrocarbon Processing*, 1998, p. 61) is a semiregenerative or continuous process for gasoline manufacture. The first step is the preparation of the naphtha feedstock by dis-

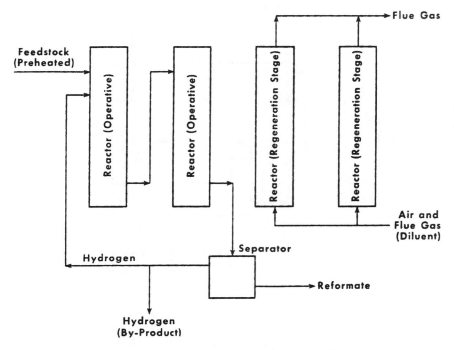

Figure 19.16 The Hydroforming process.

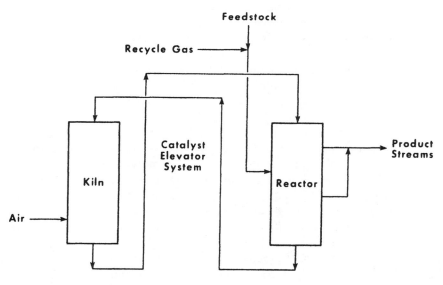

Figure 19.17 The Hyperforming process.

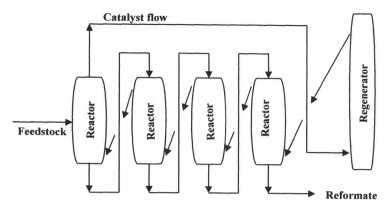

Figure 19.18   The Octanizing-Dualforming process.

tillation to separate a fraction boiling in the 120–205°C (250–400°F) range. Because sulfur adversely affects the platinum catalyst, the naphtha fraction may be treated to remove sulfur compounds. Otherwise, the hydrogen-rich gas produced by the process, which is cycled through the catalyst cases, must be scrubbed free of its hydrogen sulfide content.

The prepared naphtha feed is heated to 455–540°C (850–1000°F) and passed into a series of three catalyst cases under a pressure of 200–1000 psi (1379–6895 kPa). Further heat is added to the naphtha between each of the catalyst cases in the series. The material from the final case is fractionated into hydrogen-rich gas and reformates. The catalyst is composed of 1/8 in. (3.5 mm) pellets of alumina containing chlorine and about 0.5% platinum. Each pound of catalyst reforms up to 100 bbl of naphtha before losing its

Table 19.5   Feedstock and Product Data for the Platforming Process

Feedstock			
API gravity	59.0		
Boiling range	92–112°C; 197–233°F		
Composition, vol%			
Paraffins	69		
Naphthenes	21		
Aromatics	10		
Process operation	Semiregenerative	Continuous	
Pressure			
psi	295	123	49
kPa	2040	850	340
Product			
Research octane number	100	100	100
Composition			
C_{5+}, vol%	70	78	82
Aromatics, vol%	46	55	58
Hydrogen, wt%	2	3	4

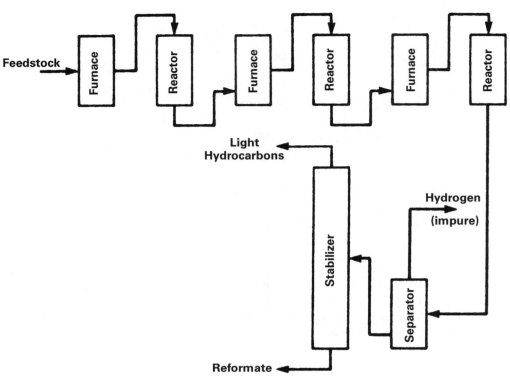

Figure 19.19 The platforming process.

activity. It is possible to regenerate the catalyst, but it is more usual to replace the spent catalyst with new catalyst.

The *powerforming process* (Table 19.6) (Fig. 19.20) is a cyclical or regenerative process that is based on frequent regeneration by carbon burn-off and permits continuous operation. Reforming takes place in four or five reactors, and regeneration is carried out in a swing reactor. Thus the plant need not be shut down to regenerate a catalyst reactor. The cyclical process ensures a continuous supply of hydrogen gas for hydrorefining operations and tends to produce a greater yield of higher octane reformate. The choice between the semiregenerative process and the cyclical process depends on the size of plant required, type of feedstock available, and the octane number needed in the product.

Rexforming is a combination process that uses platforming and aromatic extraction processes (Fig. 19.21) in which low octane raffinate is recycled to the platformer. Operating temperatures may be as much as 27°C (50°F) lower than in conventional platforming, and higher space velocities are used. A balance is struck between hydrocyclization and hydrocracking, thus avoiding the formation of excessive coke and gas. The glycol solvent in the aromatic extraction section is designed to extract low boiling high octane isoparaffins as well as aromatics.

The *Rheniforming process* (Table 19.7) (Fig. 19.22) is used to convert naphthas to high octane gasoline or aromatics plant feedstock (Aitani, 1995). Rheniforming is a semiregenerative process that comprises a sulfur-adsorbing unit, three radial flow reactors in series, a

Table 19.6 Feedstock and Product Data for the
Powerforming Process

Feedstock		
API gravity	57.2	
Composition, vol%		
Paraffins	57	
Naphthenes	30	
Aromatics	13	
Process operation	Semiregenerative	Cyclical
Product		
Research octane number	99	101
Composition		
C_1–C_4, vol%	13	11
C_{5+}, vol%	79	79
Hydrogen, wt%	2	3

separator, and a stabilizer. The sulfur control adsorbing unit which reduces sulfur to 0.2 ppm in the reformer feed, characterizes the process. The high resistance to fouling of the catalyst system increases the yields of aromatic naphtha product and hydrogen due to the long cycle lengths, which reach 6 months or more.

The *selectoforming* process (Fig. 19.23) uses a fixed-bed reactor operating under a hydrogen partial pressure. Typical operating conditions depend on the process configuration but are in the ranges 315–450°C (600–900°F) and 200–600 psi (1379–4137 kPa). The catalyst used in the *selectoforming* process is non-noble metal with a low potassium content. As with the large-pore hydrocracking catalysts, the cracking activity increases with decreasing alkali metal content.

There are two configurations of the selectoforming process that are being used commercially. The first selectoformer was designed as a separate system (Fig. 19.23) and

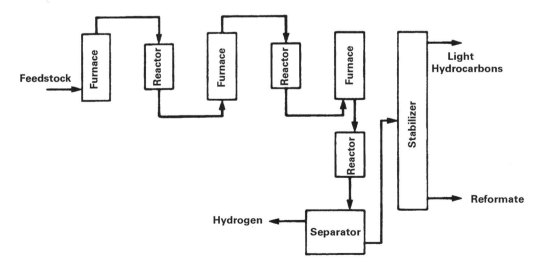

Figure 19.20 The powerforming process.

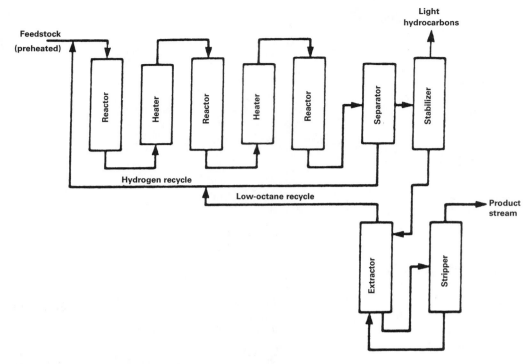

Figure 19.21   The Rexforming process.

Table 19.7   Feedstock and Product Data for the Rheniforming Process

Feedstock			
Pretreatment	Hydrotreating	Hydrotreating	Hydrocracking
Type	Paraffinic	Paraffinic	Naphthenic
Boiling range			
°C	92–117	92–117	92–201
°F	197–242	197–242	197–395
Composition, vol%			
Paraffins	69	69	33
Naphthenes	23	23	55
Aromatics	8	8	12
Process pressure			
psi	87	197	197
kPa	600	1360	1360
Product			
Research octane number	98	99	100
Composition			
C_{5+}, vol%	80	74	85
Paraffins, vol%	32	31	28
Naphthenes, vol%	1	1	3
Aromatics, vol%	67	68	70
Hydrogen, wt%	3	3	3

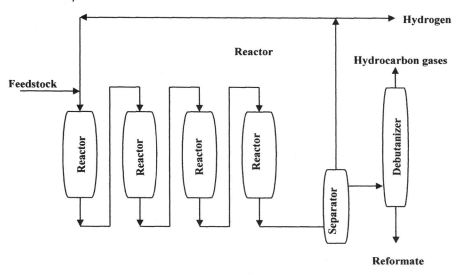

Figure 19.22 The Rheniforming process.

integrated with the reformer only to the extent of having a common hydrogen system. The reformer naphtha is mixed with hydrogen and passed into the reactor, which contains the shape-selective catalyst. The reactor effluent is cooled and separated into hydrogen, liquid petroleum, gas, and high octane gasoline. The removal of *n*-paraffins reduces the vapor pressure of the reformate because these paraffins are in higher concentration in the front end of the feed. The separate selectoforming system has the additional flexibility of being able to process other refinery streams.

The second process modification is the terminal reactor system. In this system, the shape-selective catalysts replace all or part of the reforming catalyst in the last reforming reactor. Although this configuration is more flexible, the high reforming operating temperature causes butane and propane cracking and consequently decreases the liquid pet-

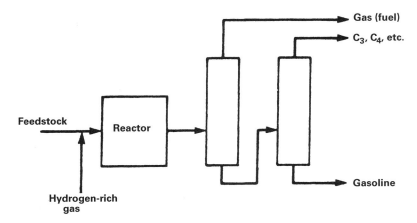

Figure 19.23 The selectoforming process.

roleum gas yield and generates higher production of ethane and methane. The life of a selectoforming catalyst used in a terminal system is between 2 and 3 years, and regeneration only partially restores fresh catalytic activity.

The Thermofor catalytic reforming (TCR) process is also a moving-bed process that uses a synthetic bead coprecipitated chromia (CrO_2) and alumina (Al_2O_3) catalyst. Catalyst/naphtha ratios have little effect on product yield or quality when varied over a wide range. The catalyst flow downward through the reactor and the naphtha–recycle gas feed enters the center of the reactor. The catalyst is transported from the base of the reactor to the top of the regenerator by bucket-type elevators.

In catalytic reforming processes that use a fluidized solids catalyst bed, continuous regeneration with a separate or integrated reactor is practiced to maintain catalyst activity by removing coke and sulfur. Cracked or virgin naphtha is charged with hydrogen-rich recycle gas to the reactor. A molybdena (Mo_2O_3, 10.0%) on alumina catalyst, which is not materially affected by normal amounts of arsenic, iron, nitrogen, or sulfur, is used. Operating conditions in the reactor are about 480–545°C (900–950°F), 200–300 psi (1379–2068 kPa). The fluidized-bed operation with its attendant excellent temperature control prevents over- and underreforming operations, resulting in more selectivity in the conditions needed for optimum yield of the desired product.

The *Ultraforming process* is used to upgrade low octane naphthas to high octane blending stocks and aromatics (Table 19.8) (Aitani, 1995). The process is a fixed-bed cyclical system with a swing reactor incorporated in the reaction section that is usually specified for aromatic (BTX) production. The system can be adapted to semiregenerative operation with the conventional three radial-flow reactors in series. The process uses low precious metal catalysts that permit frequent regeneration and high severity operations at low pressures with a service life of up to 4 years for cyclical operation versus 8 years for semiregenerative operation (Aitani, 1995). The swing reactor in cyclical operation replaces any reactor while the catalyst bed in this reactor is being regenerated. Normally, the reactors are all the same size; however, the first reactor is loaded with half the usual amount of catalyst.

Table 19.8 Feedstock and Product Data for the Ultraforming Process

Feedstock			
Boiling range			
°C	101–187	77–182	107–192
°F	215–368	170–359	224–377
Composition, vol%			
Paraffins	68	52	25
Naphthenes	19	35	36
Aromatics	13	13	39
Product			
Research octane number	99	103	106
Composition			
C_1–C_4 wt%	13	12	10
C_{5+}, vol%	78	78	84
Aromatics, vol%	68	78	84

Ultraforming units may be designed to produce high purity xylene and toluene, which can be separated by straight distillation before the extraction step. The benzene fraction can be recovered by extractive distillation. High yields of C_{5+} reformate and hydrogen have been reported for the Ultraforming process (Table 19.8).

3.3 Dehydrogenation

Dehydrogenation, a subset of the reactions that occur during reforming (Fig. 19.6), is a class of chemical reactions by means of which less saturated and more reactive compounds can be produced. There are many important conversion processes in which hydrogen is directly or indirectly removed, because, in principle, any compound containing hydrogen atoms can be dehydrogenated (Gregor, 1997). Other than the dehydrogenation reactions that occur during reforming, the dehydrogenation processes of industrial importance (which are often carried out separately from reforming processes per se) are

$$CH_3CH_3 \rightarrow CH_2 {=} CH_2$$
Ethane   Ethylene

$$CH_3CH_2CH_3 \rightarrow CH_3CH_2 {=} CH_2$$
   Propane   Propylene

$$CH_3CH_3CH_3CH_3 \rightarrow CH_3CH_2CH {=} CH_2 + CH_3CH {=} CHCH_3 \rightarrow CH_2 {=} CHCH {=} CH_2$$
   n-Butane   *n*-Butenes   Butadiene

Also included are the processes that convert higher molecular weight paraffins to olefins, cyclohexanes to aromatics, and ethylbenzene to styrene:

$$C_6H_5CH_2CH_3 \rightarrow C_6H_5CH {=} CH_2$$

In general, dehydrogenation reactions are difficult reactions and require high temperatures for favorable equilibria as well as for adequate reaction velocities. Active catalysts are usually necessary, and because permissible hydrogen partial pressures are inadequate to prevent coke deposition, periodic catalyst regeneration is necessary.

The endothermic heat of hydrocarbon dehydrogenation (Table 19.9) may be supplied through the walls of tubes (2–6 in. i.d.), by preheating the feedstock, by adding hot diluents, by reheaters between stages, or by heat stored in periodically regenerated fixed

Table 19.9 Dehydrogenation Reaction Energy for Various Hydrocarbons (527°C, 980°F, 800 K)

Reaction	ΔH_{800} [cal/(g · mol)]
Ethane → ethylene + H_2	34,300
Propane → propylene + H_2	30,900
Butane → 1-butene + H_2	31,300
Butane → *tert*-2-butene + H_2	28,500
n-Dodecane → 1-dodecene + H_2	31,200
1-Butene → 1, 3-butadiene + H_2	28,400
tert-2-Butene → 1, 3-butadiene + H_2	31,100
Methylcyclohexane → toluene + $3H_2$	51,500
Ethylbenzene → styrene + H_2	29,700

or fluidized solid catalyst beds. Usually, fairly large temperature gradients will have to be tolerated, either from wall to center of tube, from inlet to outlet of bed, or from start to finish of a processing cycle between regenerations. The ideal profile of a constant temperature (or even a rising temperature) is seldom achieved in practice. In oxidative dehydrogenation the complementary problem of temperature rise because of the exothermic nature of the reaction is encountered.

Other requirements of dehydrogenation reactions are the needs for rapid heating and quenching to prevent side reactions, the need for low pressure drops through catalyst beds, and the selection of reactor materials that can withstand the operating conditions.

Selection of operating conditions for a dehydrogenation reaction often requires a compromise. For example, the reactor *temperature* must be high enough for a favorable equilibrium and for a good reaction rate but not so high as to cause excessive cracking or catalyst deactivation. The rate of dehydrogenation reaction diminishes as conversion increases, not only because equilibrium is approached more closely, but also because in many cases reaction products act as inhibitors. The ideal temperature profile in a reactor would probably show an increase with distance, but practically attainable profiles normally are either flat or show a decline. Large adiabatic beds in which the decline is steep are often used.

The reactor *pressure* should be as low as possible (usually close to atmospheric pressure) without excessive recycle costs or equipment size. However, reduced pressures have been used in the Houdry butane dehydrogenation process, and the catalyst bed is often designed for a low pressure drop.

Rapid *feedstock preheating* is desirable to minimize cracking and is usually achieved by mixing prewarmed feedstock with superheated diluent just as the two streams enter the reactor. Rapid cooling or quenching at the exit of the reactor is necessary to prevent condensation reactions of the olefinic products. Materials of construction must be resistant to attack by hydrogen and capable of prolonged operation at high temperature and should not catalyze the conversion of hydrocarbons to carbon. Alloy steels containing chromium are usually favored, and steel containing nickel is also used, but they can cause trouble from carbon formation. If steam is not present, traces of sulfur compounds may be needed to avoid carbonization, because both steam and sulfur compounds act to keep metal walls in a passive condition.

These dehydrogenation reactions are generally applied to the production of low-boiling or light olefins, which may then be suitable for the production of isomerate, alkylate, polymer gasoline, or petrochemical intermediates. However, the STAR process is specific to the dehydrogenation and dehydrocyclization of paraffins (Hutson and McCarthy, 1986a), and the PACOL process is specific to the production of monoolefins from paraffins (Pujadó, 1997).

3.4 Catalysts

Reforming consists of two types of chemical reactions that are catalyzed by two different types of catalysts: (1) isomerization of straight-chain paraffins and isomerization (simultaneously with hydrogenation) of olefins to produce branched-chain paraffins and (2) dehydrogenation–hydrogenation of paraffins to produce aromatics and of olefins to produce paraffins.

The most important aspect of the catalytic reforming process is the catalyst, which is composed of alumina, silica, platinum, and sometimes palladium. Platinum and palladium are the key ingredients that cause paraffins to cyclize to naphthenes by forming a carbon–carbon bond and losing hydrogen (Fig. 19.6).

However, although several reactions are desirable, the composition of a reforming catalyst is dictated by the composition of the feedstock and the desired reformate. The most common catalysts are molybdena-alumina (MoO_2-Al_2O_3), chromia-alumina (Cr_2O_3-Al_2O_3), and platinum (Pt) on a silica-alumina (SiO_2-Al_2O_3) or alumina (Al_2O_3) base. The non-platinum catalysts are widely used in regenerative processes for feeds containing, for example, sulfur, which poisons platinum catalysts, although pretreatment processes (e.g., hydrodesulfurization) may permit platinum catalysts to be employed (Boitiaux et al., 1995; Murthy et al., 1995; Sie, 1995).

The purpose of platinum on the catalyst is to promote dehydrogenation and hydrogenation reactions, i.e., the production of aromatics, participation in hydrocracking, and rapid hydrogenation of carbon-forming precursors. For the catalyst to have an activity for the isomerization of both paraffins and naphthenes—the initial cracking step of hydrocracking—and to participate in paraffin dehydrocyclization, it must have an acid activity. The balance between these two activities is most important in a reforming catalyst.

In the production of aromatics from cyclic saturated materials (naphthenes), it is important that hydrocracking be minimized to avoid loss of the desired product. Thus, the catalytic activity must be moderated relative to the case of gasoline production from a paraffinic feed, where dehydrocyclization and hydrocracking play an important part.

The acid activity can be obtained by means of halogens (usually fluorine or chlorine up to about 1% by weight in catalyst) or silica incorporated in the alumina base. The platinum content of the catalyst is normally in the range of 0.3–0.8 wt%. At higher levels there is some tendency to effect demethylation and naphthene ring opening, which is undesirable; at lower levels the catalysts tend to be less resistant to poisons.

Most processes have a means of regenerating the catalyst as needed. The time between periods of regeneration—which varies with the process, the severity of the reforming reactions, and the impurities of the feedstock—ranges from a few hours to several months. Several processes employ a nonregenerative catalyst that can be used for a year or more, after which it is returned to the catalyst manufacturer for reprocessing. Processes that employ moving beds of catalysts continuously regenerate the catalyst in separate regenerators.

The processes using bauxite (*Cycloversion process*) and clay (*Isoforming process*) differ from other catalytic reforming processes in that hydrogen is not formed and hence none is recycled through the reactors. Because hydrogen does not take part in the reforming reactions there is no limit to the amount of olefin that may be present in the feedstock. The Cycloversion process is also used as a catalytic cracking process and as a desulfurization process. The Isoforming process causes only a moderate increase in octane number.

4. ISOMERIZATION

Catalytic reforming processes provide high octane constituents in the heavier gasoline fraction, but the *n*-paraffin components of the lighter gasoline fraction, especially butane (C_4) to hexane (C_6), have poor octane ratings. The conversion of these *n*-paraffins to their isomers (*isomerization*) yields gasoline components of high octane ratings in this lower boiling range (*Hydrocarbon Processing*, 1998, pp. 98–104). Conversion is achieved in the presence of a catalyst (aluminum chloride activated with hydrochloric acid or a noble metal or zeolite catalyst):

$$CH_3CH_2CH_2CH_3 \rightarrow CH_3CH(CH_3)CH_3$$

      n-Butane   Isobutane

$$CH_3CH_2CH_2CH_2CH_3 \rightarrow CH_3CH_2CH(CH_3)CH_3$$

       n-Pentane   Isopentane

It is essential to inhibit side reactions such as cracking and olefin formation. Various isomerization processes have been developed that increase the octane numbers of light naphtha from, say, 70 or less to more than 80. In a typical process, naphtha is passed over an aluminum chloride catalyst at 120°C (250°F) and at a pressure of about 800 psi (5515 kPa) to produce the isomerate.

Isomerization, another innovation specific to recent times, found initial commercial applications during World War II for making high octane aviation gasoline components and additional feed for alkylation units. The lowered alkylate demands in the post-World War II period caused a shutdown of the majority of the butane isomerization units. In recent years the greater demand for high octane motor fuel has resulted in the installation of new butane isomerization units.

4.1 Process Types

The earliest important process was the formation of isobutane, which is required as an alkylation feed. The isomerization may take place in the vapor phase with the activated catalyst supported on a solid phase, or in the liquid phase with a dissolved catalyst. A pure butane feed is mixed with hydrogen (to inhibit olefin formation) and passed to the reactor at 110–170°C (230–340°F) and 200–300 psi (1379–2068 kPa). The product is cooled and the hydrogen separated; the cracked gases are then removed in a stabilizer column. The stabilizer bottom product is passed to a superfractionator, and the *n*-butane and iso-butane are separated. With pentanes, the equilibrium is favorable at higher temperatures, and operating conditions of 240–500°C (465–930°F) and 300–1000 psi (2068–6895 kPa) may be used.

Isomerization is presently used to provide additional feedstock for alkylation units or high octane fractions for gasoline blending. Straight-chain paraffins (*n*-butane, *n*-pentane, and *n*-hexane) are converted to the respective iso-compounds by continuous catalytic (aluminum chloride and noble metals) processes. Natural gasoline or light straight-run gasoline can provide feed by first fractionating as a preparatory step. High volumetric yields (> 95%) and 40–60% conversion per pass are characteristic of the isomerization reaction.

Nonregenerable aluminum chloride catalyst has been, and still is, employed with various carriers in a fixed bed or liquid contactor. Platinum or other metal catalyst processes use a fixed-bed operation and can be regenerable or nonregenerable (Figs. 19.24 and 19.25). The reaction conditions vary widely; depending on the particular process and feedstock, the ranges are 40–480°C (100–900°F) and 150–1000 psi (1034–6895 kPa); residence time in the reactor is 10–40 min.

4.2 Commercial Processes

The *Butamer* process (Fig. 19.26) is designed to convert *n*-butane to isobutane under mild operating conditions (Rosati, 1986; Cusher, 1997a). A platinum catalyst on a support is used in a fixed-bed reactor system. The use of reformer offgas can readily satisfy the low hydrogen requirement. The operation can be designed for once-through or recycle operation and is normally tied in with alkylation unit deisobutanizer operations to provide additional feed.

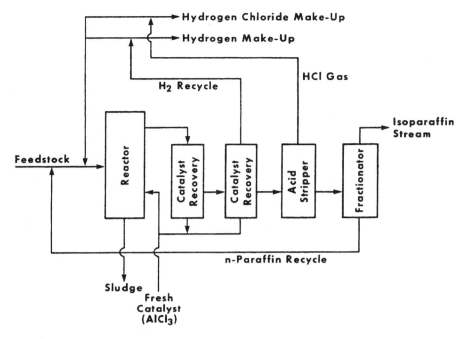

Figure 19.24 Isomerization using aluminum chloride.

Butane feed is mixed with hydrogen, heated, and charged to the reactor at moderate pressure. The effluent is cooled before light gas separation and stabilization. The resultant butane mixture is then charged to a deisobutanizer to separate a recycle stream from the isobutane product.

The *Butomerate* process is specially designed to isomerize *n*-butane to produce additional alkylation feedstock. The catalyst contains a small amount of non-noble hydrogenation metal on a high surface area support. The process operates with hydrogen recycle

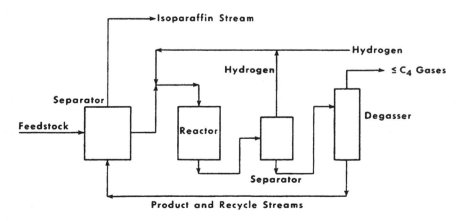

Figure 19.25 Isomerization using noble metal catalysts.

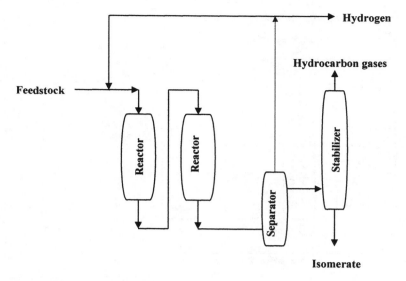

Figure 19.26   The Butamer process.

to eliminate coke deposition on the catalyst, but the isomerization reaction can continue for extended periods in the absence of hydrogen.

The feedstock should be dry and comparatively free of sulfur and water. The feed is heated, mixed with hydrogen, and conveyed to the reactor. Operating conditions range from 150°C to 260°C (300–500°F) and from 150–450 psi (1034–3103 kPa). The effluent is cooled and flashed, and the liquid product is stripped of light material.

The *Hysomer* process (Fig. 19.27) uses hydrotreated feedstocks containing pentane(s) and hexane(s) without further pretreatment. Operating conditions are approximately

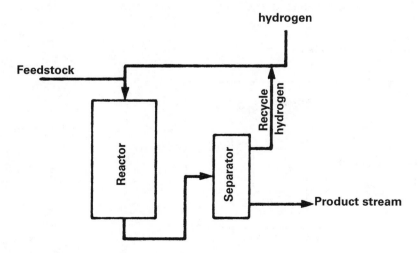

Figure 19.27   The Hysomer process.

290°C (550°F) and 400–450 psi (2758–3103 kPa) hydrogen; the catalyst (zeolite) life is approximately 2 years (Cusher, 1986).

The influence of sulfur on the catalyst activity is minimal, and the catalyst can tolerate a permanent sulfur level of 10 ppm in the feedstock, but concentrations up to 35 ppm are not harmful. The process can operate at a water level of 50 ppm, and feedstocks having saturated water contents can be processed without a deleterious effect on either catalyst stability or conversion. A minimum quantity of water is essential to the activity of zeolite catalysts in this application.

The *Iso-Kel* process is a fixed-bed, vapor-phase isomerization process that employs a precious metal catalyst and hydrogen. A wide variety of feedstocks, including natural gasoline, pentane, and/or hexane cuts, can be processed. Operating conditions include reactor temperatures and pressures from 345°C–455°C (650–550°F) and from 350–600 psi (2413–4137 kPa).

The *Isomate* process is a nonregenerative pentane and hexane or naphtha (C_6) isomerization process using an aluminum chloride–hydrocarbon complex catalyst with anhydrous hydrochloric acid as a promoter. Hydrogen partial pressure is maintained to suppress undesirable reactions (cracking and disproportionation) and retain catalyst activity. The feed is saturated with anhydrous hydrogen chloride in an absorber, then heated and combined with hydrogen and charged to the reactor [115°C (240°F), and 700 psi (4826 kPa)]. Catalyst is added to the reactor separately, and the reaction takes place in the liquid phase. The product is washed (caustic and water), acid-stripped, and stabilized before going to storage.

The *Isomerate* process is a continuous isomerization process designed to convert pentanes and hexanes into highly branched isomers; a dual-function catalyst is used in a fixed-bed reactor system. Operating conditions are mild: 400°C (750°F) and less than 750 psi (5171 kPa). Hydrogen is added to the feed along with recycle gas, and the usual operation includes fractionation facilities to allow the recycling of *n*-paraffins almost to extinction.

The *Penex* process (Fig. 19.28) is a nonregenerative pentane(s) and/or hexane(s) isomerization process (Weiszmann, 1986b; Cusher, 1997b; *Hydrocarbon Processing*, 1998, p. 104). The reaction takes place in the presence of hydrogen with a platinum catalyst, and the reactor conditions are selected so that catalyst life is long and regeneration is not required. The reactor temperatures range from 260°C–480°C (500–900°F) and pressures from 300–1000 psi (2068–6895 kPa).

The *Penex* process may be applied to many feedstocks by varying the fractionating system. Mixed feeds may be split into pentane and hexane fractions and the respective iso-fractions separated from each. The system can also be operated in conjunction with reforming of the naphtha ($> C_7$) fraction.

The *Pentafining* process is a regenerable pentane isomerization process that uses platinum catalyst on a silica-alumina support. A number of process combinations are possible. For example, with natural gasoline and hydrogen as starting materials, the feed is depentanized, and heavy material goes to a low pressure reformer. The pentane stream is split, and the *n*-fraction is combined with recycle and makeup hydrogen and charged to the reactor [425–480°C (800–900°F, and 300–700 psi (2068–4826 kPa)]. Hydrogen is removed from the effluent, which is degassed and fractionated to separate *n*- and *iso* (95% purity) cuts. The catalyst is regenerated at 260–540°C (500–1000°F) using a steam–air mixture.

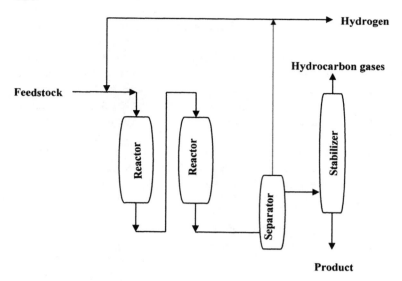

Figure 19.28   The Penex process.

4.3 Catalysts

During World War II aluminum chloride was the catalyst used to isomerize butane, pentane, and hexane. Since then, supported metal catalysts have been developed for use in high temperature processes that operate in the range of 370–480°C (700–900°F) at 300–750 psi (2068–5171 kPa). Aluminum chloride plus hydrogen chloride is universally used for the low temperature processes. However, aluminum chloride is volatile at commercial reaction temperatures and is somewhat soluble in hydrocarbons, and techniques must be employed to prevent its migration from the reactor. This catalyst is nonregenerable and is used in either a fixed-bed or liquid contactor.

5. ALKYLATION

The combination of olefins with paraffins to form higher *iso*paraffins is termed alkylation:

$$(CH_3)_3CH + CH_2{=}CH_2 \rightarrow (CH_3)_3CHCH_2CH_3$$

Because olefins are reactive (hence unstable) and are responsible for exhaust pollutants, their conversion to high octane *iso*paraffins is desirable when possible. In refinery practice, only *iso*butane is alkylated by reaction with *iso*butene or *n*-butene, and *iso*octane is the product. Although alkylation is possible without catalysts, commercial processes use sulfuric acid (Lerner, 1997) or hydrogen fluoride (Scheckler and Shah, 1997) as catalysts when the reactions can take place at low temperatures, minimizing undesirable side reactions such as the polymerization of olefins.

Alkylate is composed of a mixture of *iso*paraffins that have octane numbers that vary with the olefins from which they were made. Butylenes produce the highest octane numbers, propylene the lowest, and pentylenes the intermediate. All alkylates, however, have high octane numbers (> 87) and are particularly valuable because of them.

Alkylation developments in petroleum processing in the late 1930s and during World War II were directed to the production of high octane blending stock for aviation gasoline. The sulfuric acid process was introduced in 1938, and hydrogen fluoride alkylation was introduced in 1942. Rapid commercialization took place during the war to supply military needs, but many of these plants were shut down at the end of the war.

In the mid-1950s aviation gasoline demand started to decline, and motor gasoline quality requirements rose sharply. Whenever practical, refiners shifted the use of alkylate to premium motor fuel. The alkylate endpoint was increased for this service, and total alkylate was often used without rerunning. To help improve the economics of the alkylation process and also the sensitivity of the premium gasoline pool, additional olefins were gradually added to alkylation feed. New plants were built to alkylate propylene and the butylenes (butanes) produced in the refinery rather than the butane–butylene stream formerly used. More recently *n*-butane isomerization has been used to produce additional *iso*butane for alkylation feed.

5.1 Process Types

The alkylation reaction as practiced in petroleum refining is the union, through the agency of a catalyst, of an olefin [ethylene, $CH_2 = CH_2$; propylene, $CH_3CH = CH_2$; butene (also called butylene), $CH_3CH_2CH = CH_2$; and amylene, $CH_3CH_2CH_2CH = CH_2$] with *iso*butane [$(CH_3)_3CH$] to yield high octane branched-chain hydrocarbons in the gasoline boiling range. Olefin feedstock is derived from the gas make of a catalytic cracker; *iso*butane is recovered from refinery gases or produced by the catalytic isomerization of butane.

In thermal catalytic alkylation, ethylene or propylene is combined with *iso*butane at 50–280°C (125–450°F) and 300–1000 psi (2068–6895 kPa) in the presence of a metal halide catalyst such as aluminum chloride. Conditions are less stringent in catalytic alkylation; olefins (C_3, C_4, and C_5) are combined with *iso* butane in the presence of an acid catalyst (sulfuric or hydrofluoric) at low temperatures (1–40°C; 30–105°F) and pressures from atmospheric (15 psi) to 150 psi (101–1034 kPa).

Zeolite catalysts are also used for alkylation processes. For example, cumene (*iso*propylbenzene) is produced by the alkylation of benzene by propylene (Wallace and Gimpel, 1997). The cumene can then be used for the production of phenol and acetone by means of oxidation processes.

Prior to the alkylation process, it is often necessary to ensure the purity of the butane stream to the alkylation reactor. This can be achieved by hydrotreating in which the cracked product (C_4) stream from a fluid catalytic cracking unit is washed with caustic soda to remove sulfur compounds before being sent to the hydrotreating unit. This unit may comprise a pair of reactors, with only one in service at any specific time while the other reactor is inactive due to catalyst regeneration (Fig. 19.29) (Schneider, 2000).

5.2 Commercial Processes

Sulfuric acid alkylation has been practiced for a considerable time using the cascade sulfuric acid alkylation process to react olefins with *iso*butane to produce high octane aviation or motor fuel blending stock (Fig. 19.30) (Lerner, 1997; *Hydrocarbon Processing*, 1998, pp. 55, 56). A variant of the sulfuric acid alkylation technology, the *effluent refrigeration alkylation* process (Fig. 19.31) is also in operation.

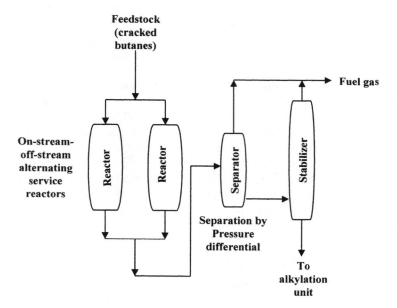

Figure 19.29 Schematic of a dual-reactor C$_4$ hydrotreater.

In the *cascade sulfuric acid alkylation* process, a low temperature process (Fig. 19.30), the olefin feed is split into equal streams and charged to the individual reaction zones of the cascade reactor. *Iso*butane-rich recycle and refrigerant streams are introduced in the front of the reactor and pass through the reaction zones. The olefin is contacted with the *iso*butane and acid in the reaction zones, which operate at 2–7°C (35–45°F) and 5–15 psi (34–103 kPa), after which vapors are withdrawn from the top of the reactor, compressed, and condensed. Part of this stream is sent to a depropanizer to control propane concentration in the unit.

The depropanizer bottoms and the remainder of the stream are combined and returned to the reactor. Spent acid is withdrawn from the bottom of the settling zone; hydrocarbons spill over a baffle into a special withdrawal section and are hot water washed with caustic addition for pH control before being successively depropanized, de*iso*butanized, and debutanized. Alkylate can then be taken directly to motor fuel blending or be rerun to produce aviation grade blending stock.

In the *effluent refrigeration alkylation* process (Fig. 19.31), olefins and an isobutane-rich stream with a recycle stream of sulfuric acid are charged to the contactor, where the mixture is circulated at high velocity, producing a large interfacial area between the hydrocarbons and the acid. The temperature in the reactor is maintained constant throughout the procedure. Reactor products pass through a flash drum and a deisobutanizer. The overhead from the deisobutanizer is isobutane, which is recycled to the contactor.

In the sulfuric acid process, alkylation usually works best at temperatures on the order of 4°C (40°F) and at an isobutane/olefin ratio of 5 : 1 to 15 : 1. Lower temperatures cause an increase in the viscosity of the acid, thereby preventing complete reactions. Higher temperatures promote reactions that produce unwanted side products that lower the

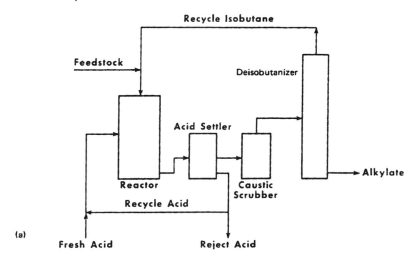

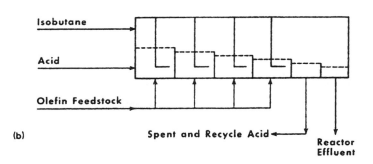

Figure 19.30 Sulfuric acid alkylation process using (a) a conventional reactor and (b) a cascade reactor.

quality of the alkylate. Acid strength must also be maintained by refortification or regeneration, usually on-site.

The *hydrogen fluoride alkylation* process (Fig. 19.32) uses regenerable hydrofluoric acid as a catalyst to unite olefins with *iso*butane to produce high octane blending stock (Fig. 19.33) (Shah, 1986; Hutson and McCarthy, 1986b; Sheckler and Shah, 1997; *Hydrocarbon Processing*, 1998, pp. 55, 56). The dried charge is intimately contacted in the reactor with acid at 20–140°C (70–100°F) and a high (15:1) *iso*butane/olefin ratio. The mixture is separated in a settler, and acid is returned to the reactor, but an acid sidestream must be continuously regenerated to 88% purity by fractionation to remove acid-soluble oils. The hydrocarbon fraction from the settler is de*iso*butanized, and alkylate is run to storage.

The hydrofluoric acid alkylation process operates at temperatures somewhat higher than the sulfuric acid alkylation processes. Reactor temperatures are maintained in the range 27–38°C (80–100°F), i.e., at temperatures that are attainable with cooling water.

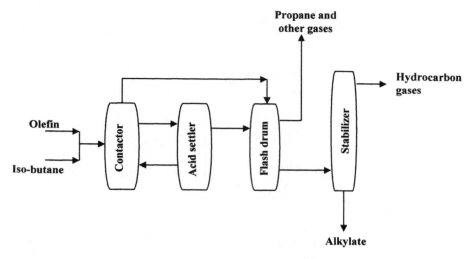

Figure 19.31   The effluent refrigeration alkylation process.

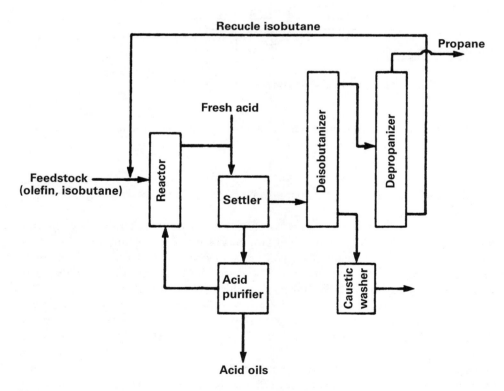

Figure 19.32   Hydrogen fluoride alkylation process.

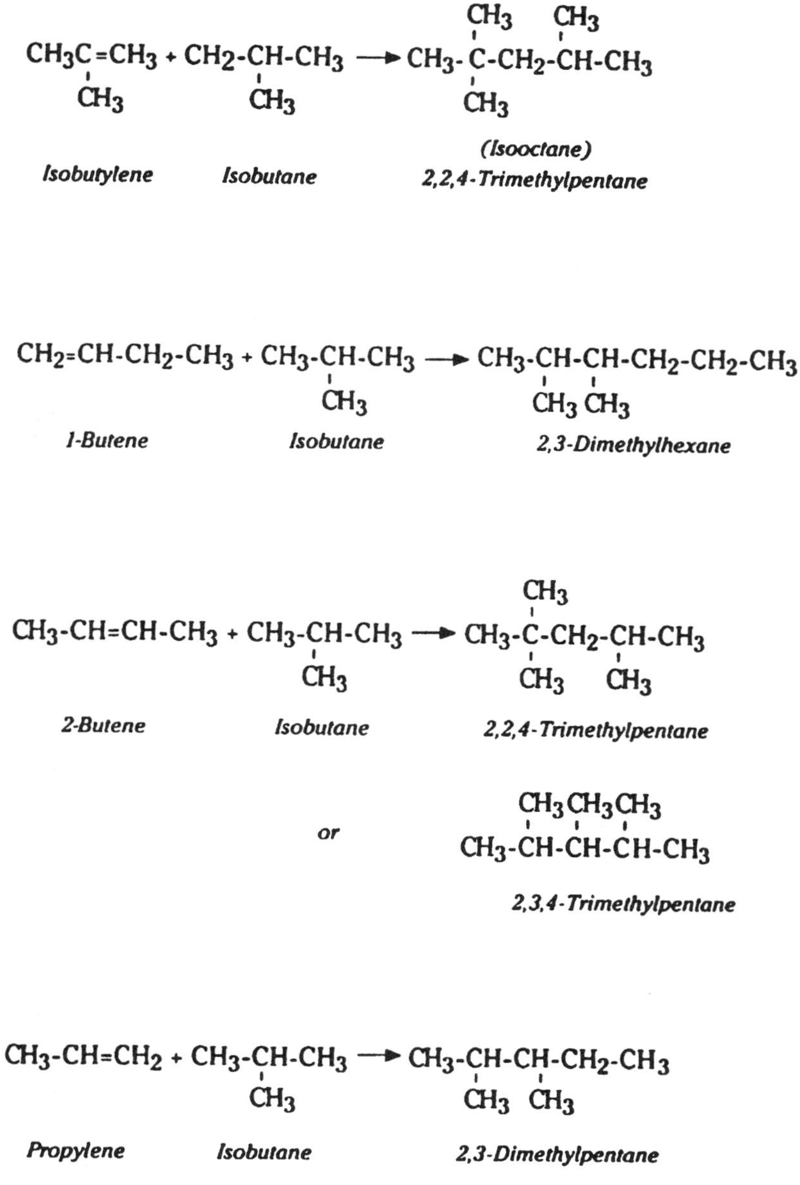

$$CH_3C=CH_3 + CH_2-CH-CH_3 \longrightarrow CH_3-\overset{\overset{\displaystyle CH_3}{|}}{C}-CH_2-\overset{\overset{\displaystyle CH_3}{|}}{CH}-CH_3$$
$$\quad\;\; \overset{|}{CH_3} \qquad\quad\; \overset{|}{CH_3} \qquad\qquad\quad\; \overset{|}{CH_3}$$

Isobutylene   *Isobutane*   *(Isooctane)*
2,2,4-Trimethylpentane

$$CH_2=CH-CH_2-CH_3 + CH_3-\overset{\overset{\displaystyle CH_3}{|}}{CH}-CH_3 \longrightarrow CH_3-\overset{\overset{\displaystyle CH_3}{|}}{CH}-\overset{\overset{\displaystyle CH_3}{|}}{CH}-CH_2-CH_2-CH_3$$

1-Butene   *Isobutane*   *2,3-Dimethylhexane*

$$CH_3-CH=CH-CH_3 + CH_3-\overset{\overset{\displaystyle CH_3}{|}}{CH}-CH_3 \longrightarrow CH_3-\overset{\overset{\displaystyle CH_3}{|}}{C}-CH_2-\overset{\overset{\displaystyle CH_3}{|}}{CH}-CH_3$$

2-Butene   *Isobutane*   *2,2,4-Trimethylpentane*

or

$$CH_3-\overset{\overset{\displaystyle CH_3}{|}}{CH}-\overset{\overset{\displaystyle CH_3}{|}}{CH}-\overset{\overset{\displaystyle CH_3}{|}}{CH}-CH_3$$

2,3,4-Trimethylpentane

$$CH_3-CH=CH_2 + CH_3-\overset{\overset{\displaystyle CH_3}{|}}{CH}-CH_3 \longrightarrow CH_3-\overset{\overset{\displaystyle CH_3}{|}}{CH}-\overset{\overset{\displaystyle CH_3}{|}}{CH}-CH_2-CH_3$$

Propylene   *Isobutane*   *2,3-Dimethylpentane*

Figure 19.33 Alkylation chemistry.

5.3 Catalysts

Sulfuric acid, hydrogen fluoride, and aluminum chloride are the only catalysts used commercially. Sulfuric acid is used with propylene and higher boiling feeds but not with ethylene, because the latter reacts to form ethyl hydrogen sulfate, and a suitable catalyst contains a minimum of 85% acidity, as determined by titration. The acid is pumped through the reactor and forms an air emulsion with reactants; the emulsion is maintained at 50% acid. The rate of deactivation varies with the feed and *iso*butane charge rate. Butene feeds cause less acid consumption than propylene feeds.

Aluminum chloride is not widely used as an alkylation catalyst, but when it is, hydrogen chloride is used as a promoter and water is injected to activate the catalyst. The catalyst is in the form of an aluminum chloride–hydrocarbon complex, and the aluminum chloride concentration is 63–84%.

Hydrogen fluoride is used for alkylation of higher boiling olefins. The advantage of hydrogen fluoride is that it is more readily separated and recovered from the resulting product. The usual concentration is 85–92% titratable acid, with about 1.5% water.

6. POLYMERIZATION

In the usual industrial sense, polymerization is a process in which a substance of low molecular weight is transformed into one of the same composition but of higher molecular weight while maintaining the atomic arrangement present in the basic molecule. It has also been described as the successive addition of one molecule to another by means of a functional group, such as the one present in an aliphatic olefin. In the petroleum industry, polymerization is the process by which olefin gases are converted to liquid condensation products that may be suitable for gasoline (hence polymer gasoline) or other liquid fuels.

The feedstock usually consists of propylene (propene, $CH_3CH=CH_2$) and butylenes (butenes, various isomers of C_4H_8) from cracking processes or may even be selective olefins for dimer, trimer, or tetramer production:

$$CH_2=CH_2 \quad -(CH_2=CH_2)_2- \quad -(CH_2=CH_2)_3- \quad -(CH_2=CH_2)_4-$$
Olefin   Dimer   Trimer   Tetramer

This type of reaction is actually a copolymerization reaction in which the molecular size of the product is limited. Polymerization in the true sense of the word is usually prevented, and all attempts are made to terminate the reaction at the dimer or trimer (three monomers joined together) stage. The four- to twelve-carbon compounds that are required as the constituents of liquid fuels are the prime products. However, in the petrochemical section of a refinery, polymerization, which results in the production of polyethylene, for example, is allowed to proceed until products having the required high molecular weight have been produced.

6.1 Process Types

Polymerization is a process that can claim to be the earliest to employ catalysts on a commercial scale. Catalytic polymerization came into use in the 1930s and was one of the first catalytic processes to be used in the petroleum industry as a means of producing saturated hydrocarbons for motor fuel (York et al., 1997).

Polymerization can be accomplished thermally or in the presence of a catalyst at lower temperatures (York et al., 1997). Thermal polymerization is considered to be not as effective as catalytic polymerization but has the advantage that it can be used to polymerize saturated materials that cannot be induced to react by catalysts. The process consists essentially of vapor-phase cracking of, say, propane and butane followed by prolonged periods at high temperature (510–590°C; 950–1100°F) for the reactions to proceed to near completion.

On the other hand, olefins can be conveniently polymerized by means of an acid catalyst. Thus, the treated olefin-rich feedstream is contacted with a catalyst (sulfuric acid, copper pyrophosphate, or phosphoric acid) at 150–220°C (300–425°F) and 150–1200 psi (1034–8274 kPa), depending on feedstock and product requirements. The reaction is exothermic, and temperature is usually controlled by heat exchange. Stabilization and/or fractionation systems separate saturated and unreacted gases from the product. In both thermal and catalytic polymerization processes, the feedstock is usually pretreated to remove sulfur and nitrogen compounds.

6.2 Commercial Processes

Thermal polymerization converts butanes and lighter gases into liquid condensation products. Olefins are produced by thermal decomposition and polymerized by heat and pressure (Fig. 19.34). Thus liquid feed under a pressure of 1200–2000 psi (8274–13,790 kPa) is pumped to a furnace heated to 510–595°C (950–1100°F), from which the various streams are separated by fractionation.

Thermal polymerization is considered to be not as effective as catalytic polymerization but has the advantage that it can be used to polymerize saturated materials that cannot be induced to react by catalysts. The process consists essentially of vapor-phase cracking of, say, propane and butane followed by prolonged periods at high temperature (510–595°C; 950–1100°F) for the reactions to proceed to near completion.

Olefins can be conveniently polymerized by means of an acid catalyst such as *solid phosphoric acid* (Tajbl, 1986; York et al., 1997). Thus, the treated olefin-rich feed stream is contacted with a catalyst (sulfuric acid, copper pyrophosphate, or phosphoric acid fixed in a silica matrix) at 150–220°C (300–425°F) and 150–1200 psi (1034–8274 kPa), depending on the feedstock and the desired product(s). The reaction is exothermic, and temperature is usually controlled by heat exchange. Stabilization and/or fractionation systems separate saturated and unreacted gases from the product. In both thermal and catalytic polymer-

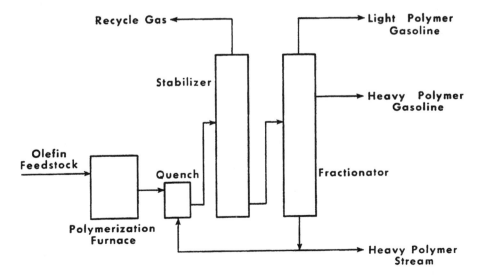

Figure 19.34 Thermal polymerization process.

ization processes the feedstock is usually pretreated to remove sulfur and nitrogen compounds.

This process (Fig. 19.35) converts propylenes and/or butylenes to high octane gasoline or petrochemical polymers. The catalyst, pelleted kieselguhr (diatomaceous earth) impregnated with phosphoric acid, is used in either a chamber or a tubular reactor. The exothermic reaction temperature is controlled by using saturates (separated from the effluent as recycle to the feed) as a quench liquid between the catalyst chamber beds. Tubular reactors are temperature-controlled by water or oil circulating around the catalyst tubes.

Reaction temperatures and pressures are 175–225°C (350–435°F) and 400–1200 psi (2758–8274 kPa). Olefins and aromatics may be united by alkylation for special applications at 205–315°C (400–600°F) and 400–900 psi (2758–6205 kPa), and a rerun column is required in addition to the usual fractionating column.

Bulk acid polymerization is a process used to produce high octane polymer gasoline from all types of light olefin feed, and the olefin concentration can be as high as 95%. Liquid phosphoric acid is used as the catalyst.

The olefin feed is washed (with caustic and water) and then contacted thoroughly by liquid phosphoric acid in a small reactor. The effluent stream and the acid are separated in a settler, and acid is returned to the reactor through a cooler. Gasoline is first stabilized and washed with caustic before storage. The heat of reaction is removed by circulation through an exchanger before contact with the olefin feed, and catalyst activity is maintained by continuous addition of fresh acid and withdrawal of spent acid.

6.3 Catalysts

Phosphates are the principal catalysts for polymerization. The commercially used catalysts are liquid phosphoric acid, phosphoric acid on diatomaceous earth, copper pyrophosphate pellets, and phosphoric acid film on quartz. The latter is the least active but the most used and easiest to regenerate simply by washing and recoating; the serious disadvantage is that residue must occasionally be burned off the support. The process using liquid phosphoric acid catalyst is far more responsive to attempts to raise production by increasing temperature than the other processes.

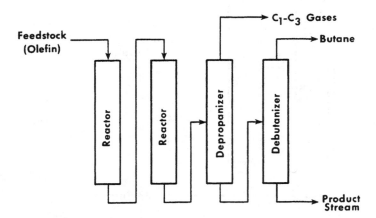

Figure 19.35   The catalytic polymerization process

REFERENCES

Abraham, H. 1945. Asphalts and Allied Substances. Van Nostrand, New York.

Aitani, A. M. 1995. In: Catalytic Naphtha Reforming. G. J. Antos, A. M. Aitani, and J. M. Parera, eds. Marcel Dekker, New York, Chapter 13.

Beltramini, J. N. 1995. In: Catalytic Naphtha Reforming. G. J. Antos, A. M. Aitani, and J. M. Parera, eds. Marcel Dekker, New York, Chapter 10.

Bland, W. F., and Davidson, R. L. 1967. Petroleum Processing Handbook. McGraw-Hill, New York.

Boitiaux, J. P., Devès, J. M., Didillon, B., and Marcilly, C. R. 1995. In: Catalytic Naphtha Reforming. G. J. Antos, A. M. Aitani, and J. M. Parera, eds. Marcel Dekker, New York, Chapter 4.

Bridge, A. J. 1997. In: Handbook of Petroleum Refining Processes. 2nd ed. R. A. Meyers, ed. McGraw-Hill, New York, Chapter 14.1.

Chaudhuri, P. M. 2000. Oil Gas J. 98(44):46.

Cusher, N. A. 1986. In: Handbook of Petroleum Refining Processes. R. A. Meyers, ed. McGraw-Hill, New York, Chapter 5.5.

Cusher, N. A. 1997a. In: Handbook of Petroleum Refining Processes. 2nd ed. R. A. Meyers, ed. McGraw-Hill, New York, Chapter 9.2.

Cusher, N. A. 1997b. In: Handbook of Petroleum Refining Processes. 2nd ed. R. A. Meyers, ed. McGraw-Hill, New York, Chapter 9.3.

Dachos, N., Kelly, A., Felch, D., and Reis, E. 1997. In: Handbook of Petroleum Refining Processes. 2nd ed. R. A. Meyers, ed. McGraw-Hill, New York, Chapter 4.1.

Forbes, R. J. 1958a. A History of Technology, Oxford Univ. Press, Oxford, England.

Forbes, R. J. 1958b. Studies in Early Petroleum Chemistry. E. J. Brill, Leiden, The Netherlands.

Forbes, R. J. 1959. More Studies in Early Petroleum Chemistry. E. J. Brill, Leiden, The Netherlands.

Gregor, J. 1997. In: Handbook of Petroleum Refining Processes. 2nd ed. R. A. Meyers, ed. McGraw-Hill, New York, Chapter 5.1.

Henry, J. T. 1873. The Early and Later History of Petroleum. Vols. I and II. APRP Co., Philadelphia, PA.

Hobson, G. D., and Pohl, W. 1973. Modern Petroleum Technology. Applied Science, Barking, Essex, England.

Hutson, T., Jr., and McCarthy, W. C. 1986a. In: Handbook of Petroleum Refining Processes. R. A. Meyers, ed. McGraw-Hill, New York, Chapter 4.3.

Hutson, T., Jr., and McCarthy, W. C. 1986b. In: Handbook of Petroleum Refining Processes. R. A. Meyers, ed. McGraw-Hill, New York, Chapter 1.2.

Hydrocarbon Processing. 1998. 77(11):53 et seq.

James, P., and Thorpe, N. 1994. Ancient Inventions. Ballantine Books, New York.

Kelly, A., Felch, D., and Reis, E. 1997. In: Handbook of Petroleum Refining Processes. 2nd ed. R. A. Meyers, ed. McGraw-Hill, New York, Chapter 4.1.

Kennedy, J. E. 1997. In: Handbook of Petroleum Refining Processes. 2nd ed. R. A. Meyers, ed. McGraw-Hill, New York, Chapter 8.3.

Lerner, H. 1997. In: Handbook of Petroleum Refining Processes. 2nd ed. R. A. Meyers, ed. McGraw-Hill, New York, Chapter 1.1.

Lovink, H. J. 1995. In: Catalytic Naphtha Reforming. G. J. Antos, A. M. Aitani, and J. M. Parera, eds. Marcel Dekker, New York, Chapter 8.

Marécot, P., and Barbier, J. 1995. In: Catalytic Naphtha Reforming. G. J. Antos, A. M. Aitani, and J. M. Parera, eds. Marcel Dekker, New York, Chapter 10.

Murthy, K. R., Sharma, N., and George, N. 1995. In: Catalytic Naphtha Reforming. G. J. Antos, A. M. Aitani, and J. M. Parera, eds. Marcel Dekker, New York, Chapter 7.

Nelson, W. L. 1958. Petroleum Refinery Engineering. McGraw-Hill, New York.

Paál, Z. 1995. In: Catalytic Naphtha Reforming. G. J. Antos, A. M. Aitani, and J. M. Parera, eds. Marcel Dekker, New York, Chapter 2.

Parera, J. M., and Fígoli, N. S. 1995a. In: Catalytic Naphtha Reforming. G. J. Antos, A. M. Aitani, and J. M. Parera, eds. Marcel Dekker, New York, Chapter 1.

Parera, J. M., and Fígoli, N. S. 1995b. In: Catalytic Naphtha Reforming. G. J. Antos, A. M. Aitani, and J. M. Parera, eds. Marcel Dekker, New York, Chapter 3.

Pujadó, P. R. 1997. In: Handbook of Petroleum Refining Processes. 2nd ed. R. A. Meyers, ed. McGraw-Hill, New York, Chapter 5.2.

Riediger, B. 1971. The Refining of Petroleum. Springer-Verlag, Heidelberg.

Rosati, D. 1986. In: Handbook of Petroleum Refining Processes. R. A. Meyers, ed. McGraw-Hill, New York, Chapter 5.4.

Rosso, J. P., and El Guindy, M. I. 1995. In: Catalytic Naphtha Reforming. G. J. Antos, A. M. Aitani, and J. M. Parera, eds. Marcel Dekker, New York, Chapter 12.

Scheckler, J. C., and Shah, B. R. 1997. In: Handbook of Petroleum Refining Processes. 2nd ed. R. A. Meyers, ed. McGraw-Hill, New York, Chapter 1.4.

Schneider, D. F. 2000. Hydrocarbon Process. 79(10):39.

Schwarzenbek, E. F. 1971. In: Origin and Refining of Petroleum. ACS Ser. No. 103. H. G. McGrath and M. E. Charles, eds. Am. Chem. Soc., Washington, DC.

Shah, B. R. 1986. In: Handbook of Petroleum Refining Processes. R. A. Meyers, ed. McGraw-Hill, New York, Chapter 1.1.

Sie, S. T. 1995. In: Catalytic Naphtha Reforming. G. J. Antos, A. M. Aitani, and J. M. Parera, eds. Marcel Dekker, New York, Chapter 6.

Sivasanker, S., and Ratnasamy, P. 1995. In: Catalytic Naphtha Reforming. G. J. Antos, A. M. Aitani, and J. M. Parera, eds. Marcel Dekker, New York, Chapter 15.

Speight, J. G. 1993. Gas Processing: Environmental Aspects and Methods. Butterworth-Heineman, Oxford, England.

Speight, J. G. 1999. The Chemistry and Technology of Petroleum. 3rd ed. Marcel Dekker, New York.

Speight, J. G. 2000. The Desulfurization of Heavy Oils and Residua. 2nd ed. Marcel Dekker, New York.

Tajbl, D. G. 1986. In: Handbook of Petroleum Refining Processes. R. A. Meyers, ed. McGraw-Hill, New York, Chapter 3.1.

Thompson, G. J. 1997. In: Handbook of Petroleum Refining Processes. 2nd ed. R. A. Meyers, ed. McGraw-Hill, New York, Chapter 8.4.

Turpin, L. E. 1995. In: Catalytic Naphtha Reforming. G. J. Antos, A. M. Aitani, and J. M. Parera, eds. Marcel Dekker, New York, Chapter 2.

Wallace, J. W., and Gimpel, H. E. 1997. In: Handbook of Petroleum Refining Processes. 2nd ed. R. A. Meyers, ed. McGraw-Hill, New York, Chapter 1.2.

Weiszmann, J. A. 1986a. In: Handbook of Petroleum Refining Processes. R. A. Meyers, ed. McGraw-Hill, New York, Chapter 3.1.

Weiszmann, J. A. 1986b. In: Handbook of Petroleum Refining Processes. R. A. Meyers, ed. McGraw-Hill, New York, Chapter 5.5.

York, D., Scheckler, J. C., and Tajbl, D. G. 1997. In: Handbook of Petroleum Refining Processes. 2nd ed. R. A. Meyers, ed. McGraw-Hill, New York, Chapter 1.3.

20
Hydrogen Production

1. INTRODUCTION

In the previous chapters there have been several references and/or acknowledgments of a very important property of petroleum and petroleum products—their hydrogen content or the use of hydrogen during refining in conversion processes and in finishing processes (Table 20.1) (Fig. 20.1). As hydrogen use has become more widespread in refineries, hydrogen production has moved from the status of a high tech specialty operation to an integral feature of most refineries. This was made necessary by the increase in hydrotreating and hydrocracking, including the treatment of progressively heavier feedstocks (Chapter 18).

In fact, the use of hydrogen in thermal processes was perhaps the single most significant advance in refining technology during the twentieth century (Bridge et al., 1981; Scherzer and Gruia, 1996; Bridge, 1997; Dolbear, 1998). The continued increase in hydrogen demand over the last several decades is a result of the conversion of petroleum to match changes in product slate and the supply of heavy, high sulfur oil and in order to make lower boiling, cleaner, and more salable products. There are also many reasons other than product quality for using hydrogen at relevant stages of the refining process (Table 20.2).

As an example of the popularity or necessity of using of hydrotreating (hydrogen addition), data (Radler, 1999) show that of a total worldwide daily refinery capacity of approximately 81,500,000 bbl of oil, approximately 4,000,000 bbl/day is dedicated to catalytic hydrocracking, 8,500,000 bbl/day is dedicated to catalytic hydrorefining, and approximately 28,100,000 bbl/day is dedicated to catalytic hydrotreating. On a national (U.S.) basis, the corresponding data are approximately 1,400,000 bbl/day hydrocracking, approximately 1,780,000 bbl/day hydrorefining, and approximately 1,900,000 bbl/day catalytic hydrotreating out of a daily total refining capacity of approximately 16,500,000 bbl/day.

Thus, the production of hydrogen for refining purposes requires a major effort by refiners. In fact, the trend to increasing the number of hydrogenation (hydrocracking and/ or hydrotreating) processes in refineries coupled with the need to process the heavier oils, which require substantial quantities of hydrogen for upgrading because of the increased

Table 20.1 Process Characteristics for Hydrogen-Dependent Processes

Feedstock	Process characteristics								Products
	Hydro-cracking	Aromatics removal	Sulfur removal	Nitrogen removal	Metals removal	Coke mitigation	n-Paraffins removal	Olefins removal	
Naphtha	✓		✓	✓					Reformer feedstock
								✓	Liquefied petroleum gas (LPG)
Gas oil									
Atmospheric		✓							Diesel fuel
		✓							Jet fuel
		✓					✓		Petrochemical feedstock
	✓								Naphtha
Vacuum	✓	✓	✓	✓	✓				Catalytic cracker feedstock
	✓	✓	✓						Diesel fuel
	✓	✓	✓						Kerosene
	✓		✓						Jet fuel
	✓								Naphtha
									LPG
	✓	✓							Lubricating oil
Residuum			✓	✓	✓	✓			Catalytic cracker feeder
			✓	✓	✓	✓			Coker feedstock
	✓								Diesel fuel (others)

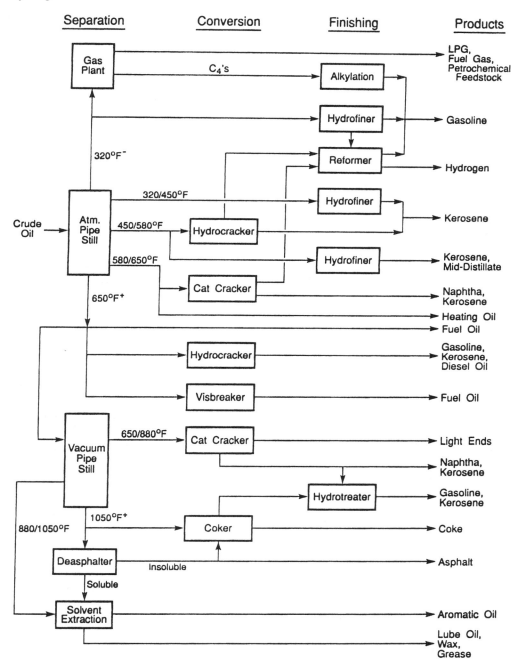

Figure 20.1 Schematic of a petroleum refinery showing processes where hydrogen is used.

Table 20.2 Effects of Hydrogen Use in Refinery Processes

Reaction[a]	Feedstock	Purpose
HDS	Catalytic reformer feedstocks	Reduce catalyst poisoning
	Diesel fuel	Meet environmental specifications
	Distillate fuel oil	Meet environmental specifications
	Hydrocracker feedstocks	Reduce catalyst poisoning
	Coker feedstocks	Reduce sulfur content of coke
HDN	Lubricating oil	Improve stability
	Catalytic cracking feedstocks	Reduce catalyst poisoning
	Hydrocracker feedstocks	Reduce catalyst poisoning
HDM	Catalytic cracking feedstocks	Avoid metals deposition
		Avoid coke buildup
		Avoid catalyst destruction
	Hydrocracker feedstocks	Avoid metals deposition
		Avoid coke buildup
		Avoid catalyst destruction
CRR	Catalytic cracker feedstocks	Reduce coke buildup on catalyst
	Residua	Reduce coke yield
	Heavy oils	Reduce coke yield

[a] HDS, Hydrodesulfurization; HDN, hydrodenitrogenation; HDM, hydrodemetallization; CRR, carbon residue reduction.

use of hydrogen in hydrocracking processes, has resulted in vastly increased demands for this gas. The hydrogen demands can be estimated to a very rough approximation using API gravity and the extent of reaction, particularly the hydrodesulfurization reaction (Fig. 20.2) (Speight, 2000). Accurate estimation requires equivalent process parameters and a thorough understanding of the nature of each process. Thus, as hydrogen production increases, a better understanding of the capabilities and requirements of a hydrogen plant becomes ever more important to overall refinery operations as a means of making the best use of hydrogen supplies in the refinery.

An early use of hydrogen in refineries was in naphtha hydrotreating as feed pre-treatment for catalytic reforming (which in turn produced hydrogen as a by-product). As environmental regulations tightened, the technology matured and heavier streams were hydrotreated. Thus, in the early refineries the hydrogen for hydroprocesses was provided as a result of catalytic reforming processes in which dehydrogenation (Chapter 19) is a major chemical reaction and, as a consequence, hydrogen gas is produced. The light ends from the catalytic reformer contain a high ratio of hydrogen to methane, so the stream is deethanized and/or depropanized to get a high concentration of hydrogen in the stream.

The hydrogen is recycled through the reactors where the reforming takes place to provide the atmosphere necessary for the chemical reactions and also to prevent the carbon from being deposited on the catalyst, thus extending its operating life. An excess of hydrogen above whatever is consumed in the process is produced, and as a result catalytic reforming processes are unique in that they are the only petroleum refinery processes to produce hydrogen as a by-product. However, as refineries and refinery feed-stocks evolved during 1960s–1990s, the demand for hydrogen increased, and reforming

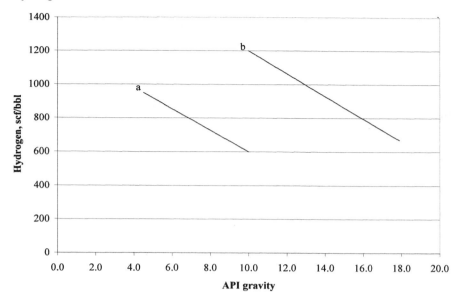

Figure 20.2 Approximate relationship of hydrogen requirements to API gravity for (a) 40% desulfurization and (b) 90% desulfurization.

processes are no longer capable of providing the quantities of hydrogen necessary for feedstock hydrogenation. Within the refinery, other processes are used as sources of hydrogen. Thus, the recovery of hydrogen from the by-products of the coking units, visbreaker units, and catalytic cracking units is also practiced in some refineries.

In coking units and visbreaker units, heavy feedstocks are converted to petroleum coke, oil, light hydrocarbons (benzene, naphtha, liquefied petroleum gas), and gas (Chapter 14). Depending on the process, hydrogen is present in a wide range of concentrations. Because petroleum coking processes need gas for heating purposes, adsorption processes are best suited to recover the hydrogen because they feature a very clean hydrogen product and an offgas suitable as fuel.

Catalytic cracking is the most important process step for the production of light products from gas oil and increasingly from vacuum gas oil and heavy feedstocks (Chapter 15). In catalytic cracking the molecular mass of the main fraction of the feed is lowered while another part is converted to coke that is deposited on the hot catalyst. The catalyst is regenerated in one or two stages by burning the coke off with air, which also provides the energy for the endothermic cracking process. In the process, paraffins and naphthenes are cracked to olefins and to alkanes with shorter chain lengths, mono aromatic compounds are dealkylated without ring cleavage, and diaromatics and polyaromatics are dealkylated and converted to coke. Hydrogen is formed in the last type of reaction, whereas the first two reactions produce light hydrocarbons and therefore require hydrogen. Thus, a catalytic cracker can be operated in such a manner that enough hydrogen is formed for subsequent processes.

In reforming processes, naphtha fractions are reformed to improve the quality of gasoline (Speight, 1999, 2000). The most important reaction occurring during this process is the dehydrogenation of naphthenes to aromatics. This reaction is endothermic and is favored by low pressures, and the reaction temperature lies in the range of 300–450°C (570–840°F). The reaction is performed on platinum catalysts, with other metals, e.g., rhenium, as promoters.

The chemical nature of the crude oil used as the refinery feedstock has always played the major role in determining the hydrogen requirements of any particular refinery. For example, the lighter, more paraffinic crude oils will require somewhat less hydrogen for upgrading to, say, a gasoline product than a heavier more asphaltic crude oil (Speight, 2000). It follows that the hydrodesulfurization of heavy oils and residua (which, by definition, is a hydrogen-dependent process) needs substantial amounts of hydrogen.

In general, considerable variation exists from one refinery to another in the balance between hydrogen produced and hydrogen consumed in the refining operations. However, what is more pertinent to the present text is the excessive amounts of hydrogen that are required for hydroprocessing operations—whether these be hydrocracking or the somewhat milder hydrotreating processes. For effective hydroprocessing, a substantial hydrogen partial pressure must be maintained in the reactor and in order to meet this requirement, an excess of hydrogen above what is actually consumed by the process must be fed to the reactor. Part of the hydrogen requirement is met by recycling a stream of hydrogen-rich gas. However, the need still remains to generate hydrogen as makeup material to accommodate the process consumption of 500–3000 scf/bbl depending upon whether the heavy feedstock is being subjected to a predominantly hydrotreating (hydrodesulfurization) or a predominantly hydrocracking process.

Hydrogen is generated in a refinery by the catalytic reforming process, but there may not always be the need to have a catalytic reformer as part of the refinery sequence. Nevertheless, assuming that a catalytic reformer is part of the refinery sequence, the hydrogen production from the reformer usually falls well below the amount required for hydroprocessing purposes. For example, in a 100,000 bbl/day hydrocracking refinery, assuming intensive reforming of hydrocracked gasoline, the hydrogen requirements of the refinery may still fall some 500–900 scf/bbl of crude charge below that necessary for the hydrocracking sequences.

The trend to greater numbers of hydrogenation (*hydrocracking* and/or *hydrotreating*) processes in refineries (Dolbear, 1998) coupled with the need to process the heavier oils, which require substantial quantities of hydrogen for upgrading, has resulted in a vast increase in the demand for this gas. Part of the hydrogen requirements can be satisfied by hydrogen recovery from catalytic reformer product gases, but other external sources are required. Most of the external hydrogen is manufactured either by steam–methane reforming or by oxidation processes. However, other processes, such as steam–methanol interaction or ammonia dissociation, may also be used as sources of hydrogen. Electrolysis of water produces high purity hydrogen, but the power costs may be prohibitive.

Several processes are available for the production of the additional hydrogen that is necessary for the various heavy feedstock hydroprocessing sequences that have been outlined elsewhere (Chapters 17 and 18), and it is the purpose of the present chapter to present a general description of these processes. In general, most of the external hydrogen is manufactured by steam–methane reforming or by oxidation processes. Other processes such as ammonia dissociation, steam–methanol interaction, or electrolysis are also available for hydrogen production, but economic factors and feedstock availability assist in the choice between processing alternatives.

The gasification of residua and coke to produce hydrogen and/or power may become an attractive option for refiners (Campbell, 1997; Dickenson et al., 1997; Fleshman, 1997).

The processes described in this chapter are those gasification processes that are often referred to the *garbage disposal units* of the refinery. Hydrogen is produced for use in other parts of the refinery as well as for energy, and it is often produced from process by-products that may not be of any use elsewhere. Such by-products might be the highly aromatic, heteroatomic, and metal-containing rejects from a deasphalting unit or from a mild thermal (hydro)cracking process. However attractive this may seem, there will be the need to incorporate a gas-cleaning operation to remove any environmentally objectionable components from the hydrogen gas.

2. PROCESS VARIABLES

Hydrogen has historically been produced during catalytic reforming processes as a by-product of the production of the aromatic compounds used in gasoline and in solvents. As reforming processes changed from fixed-bed to cyclical to continuous regeneration, process pressures dropped and hydrogen production per barrel of reformate tended to increase. However, hydrogen production as a by-product is not always adequate to the needs of the refinery, and other processes are necessary. Thus, hydrogen production by steam reforming or by partial oxidation of residua has also been used, particularly where heavy oil is available. Steam reforming is the dominant method for hydrogen production and is usually combined with pressure swing adsorption (PSA) to purify the hydrogen to greater than 99% by volume.

2.1 Feedstocks

The most common, and perhaps the best, feedstocks for steam reforming are low-boiling saturated hydrocarbons that have a low sulfur content, including natural gas, refinery gas, liquefied petroleum gas (LPG), and low-boiling naphtha.

Natural gas is the most common feedstock for hydrogen production because it meets all the requirements for reformer feedstock. Natural gas typically contains more than 90% methane and ethane and only a few percent being propane and higher boiling hydrocarbons (Table 20.3) (Speight, 1993). Natural gas may (most likely will) contain traces of carbon dioxide with some nitrogen and other impurities.

Purification of natural gas before reforming is usually relatively straightforward (Speight, 1993). Traces of sulfur must be removed to avoid poisoning the reformer catalyst; zinc oxide treatment in combination with hydrogenation is usually adequate.

Light *refinery gas*, containing a substantial amount of hydrogen, can be an attractive steam reformer feedstock because it is produced as a by-product. Processing of refinery gas will depend on its composition, particularly the levels of olefins, propane, and heavier hydrocarbons.

Olefins, which can cause problems by forming coke in the reformer, are converted to saturated compounds in the hydrogenator. Higher boiling hydrocarbons in refinery gas can also form coke, either on the primary reformer catalyst or in the preheater. If there is more than a few percent of C_3 and higher compounds, a promoted reformer catalyst should be considered in order to avoid carbon deposits.

When the hydrogen content of the refinery gas is greater than 50% by volume, the gas should first be considered for hydrogen recovery, using a membrane or pressure swing adsorption unit. The tail gas or reject gas, which will still contain a substantial amount of hydrogen, can then be used as steam reformer feedstock.

Table 20.3 Composition of Natural Gas

Category	Component	Amount (%)
Paraffinic	Methane (CH_4)	70–98
	Ethane (C_2H_6)	1–10
	Propane (C_3H_8)	Trace–5
	Butane (C_4H_{10})	Trace–2
	Pentane (C_5H_{12})	Trace–1
	Hexane (C_6H_{14})	Trace–0.5
	Heptane and higher (C_{7+})	None–trace
Cyclic	Cyclopropane (C_3H_6)	Traces
	Cyclohexane (C_6H_{12})	Traces
Aromatic	Benzene (C_6H_6), others	Traces
Nonhydrocarbon	Nitrogen (N_2)	Trace–15
	Carbon dioxide (CO_2)	Trace–1
	Hydrogen sulfide (H_2S)	Trace occasionally
	Helium (He)	Trace–5
	Other sulfur and nitrogen compounds	Trace occasionally
	Water (H_2O)	Trace–5

Refinery gas from different sources varies in suitability as hydrogen plant feed. Catalytic reformer offgas (Table 20.4) (Chapter 19), for example, is saturated and very low in sulfur and often has a high hydrogen content. The process gases from a coking unit (Chapter 14) or from a fluid catalytic cracking unit (Chapter 15) are much less desirable because of their unsaturated constituents. In addition to olefins, these gases contain substantial amounts of sulfur that must be removed before the gas is used as feedstock. These gases are also generally unsuitable for direct hydrogen recovery, because the hydrogen content is usually too low. Hydrotreater offgas lies in the middle of the range. It is saturated, so it is readily used as hydrogen plant feed. Its content of hydrogen and heavier hydrocarbons depends to a large extent on the upstream pressure. Sulfur removal will generally be required.

The feedstock purification process (Fig. 20.3) uses three different refinery gas streams to produce hydrogen. First, high pressure hydrocracker purge gas is purified in a membrane unit that produces hydrogen at medium pressure and is combined with medium

Table 20.4 Composition of Catalytic Reformer Product Gas

Constituent	Vol%
Hydrogen	75–85
Methane	5–10
Ethane	5–10
Ethylene	0
Propane	5–10
propylene	0
Butane	< 5
Butylenes	0
Pentane plus	< 2

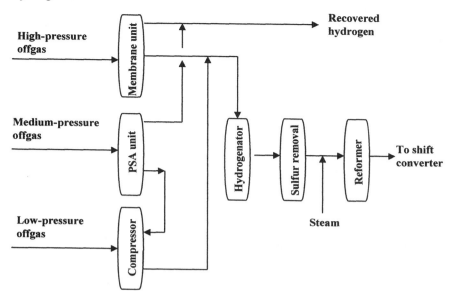

Figure 20.3 Feedstock handling and purification.

pressure offgas that is first purified in a pressure swing adsorption unit. Finally, low pressure offgas is compressed, mixed with reject gases from the membrane and pressure swing adsorption units, and used as steam reformer feed.

Liquid feedstocks, either liquefied petroleum gas or naphtha, can also provide backup feed if there is a risk of natural gas curtailments. The feed handling system needs to include a surge drum, feed pump, and vaporizer (usually steam-heated), followed by further heating before desulfurization. The sulfur in liquid feedstocks occurs as mercaptans, thiophenes, or higher boiling compounds. These compounds are stable and will not be removed by zinc oxide; therefore, a hydrogenation unit will be required. In addition, as with refinery gas, olefins must be hydrogenated if they are present.

The reformer will generally use a potash-promoted catalyst to avoid coke buildup from cracking of the heavier feed. If liquefied petroleum gas is to be used only occasionally, it is often possible to use a methane-type catalyst at a higher steam/carbon ratio to avoid coking. Naphtha will require a promoted catalyst unless a preformer is used.

2.2   Process Chemistry

Before the feedstock is introduced into a process, it should be purified according to a strict purification protocol. Prolonging catalyst life in hydrogen production processes is attributable to effective feedstock purification, particularly sulfur removal. A typical natural gas or other light hydrocarbon feedstock contains traces of hydrogen sulfide and organic sulfur.

To remove sulfur compounds, it is necessary to hydrogenate the feedstock to convert the organic sulfur to hydrogen sulfide, which is then reacted with zinc oxide (ZnO) at approximately 370°C (700°F). This results in the optimal use of the zinc oxide as well as ensuring complete hydrogenation. Thus, assuming assiduous feedstock purification and removal of all of the objectionable contaminants, the chemistry of hydrogen production can be defined.

In *steam reforming*, low-boiling hydrocarbons such as methane are reacted with steam to form hydrogen:

$$CH_4 + H_2O \rightarrow CO + 3H_2 \qquad \Delta H = 97,400 \text{ Btu/lb}$$

where ΔH is the heat of reaction. A more general form of the equation that shows the chemical balance for higher boiling hydrocarbons is

$$C_nH_n + nH_2O \rightarrow nCO\left(n + \frac{m}{2}\right)H_2$$

The reaction is typically carried out at approximately 815°C (1500°F) over a nickel catalyst packed into the tubes of a reforming furnace. The high temperature also causes the hydrocarbon feedstock to undergo a series of cracking reactions, plus the reaction of carbon with steam:

$$CH_4 \rightarrow 2H_2 + C$$

$$C + H_2O \rightarrow CO + H_2$$

Carbon is produced on the catalyst at the same time that hydrocarbon is reformed to hydrogen and carbon monoxide. With natural gas or similar feedstock, reforming predominates, and the carbon can be removed by reaction with steam as fast as it is formed. When higher boiling feedstocks are used, the carbon is not removed fast enough and builds up, thereby requiring that the catalyst be regenerated or replaced. Carbon buildup on the catalyst (when high-boiling feedstocks are employed) can be avoided by the addition of alkali compounds, such as potash, to the catalyst, thereby encouraging or promoting the carbon–steam reaction.

However, even with an alkali-promoted catalyst, feedstock cracking limits the process to hydrocarbons with a boiling point below 180°C (350°F). Natural gas, propane, butane, and light naphtha are most suitable. Prereforming, a process that uses an adiabatic catalyst bed operating at a lower temperature, can be used as a pretreatment to allow heavier feedstocks to be used with lower potential for carbon deposition (coke formation) on the catalyst.

After reforming, the carbon monoxide in the gas is reacted with steam to form additional hydrogen (the *water–gas shift* reaction):

$$CO + H_2O \rightarrow CO_2 + H_2 \qquad \Delta H = -16,500 \text{ Btu/lb}$$

This leaves a mixture consisting primarily of hydrogen and carbon monoxide that is removed by conversion to methane:

$$CO + 3H_2O \rightarrow CH_4 + H_2O$$

$$CO_2 + 4H_2 \rightarrow CH_4 + 2H_2O$$

The critical variables for steam reforming processes are (1) temperature, (2) pressure, and (3) the steam/hydrocarbon ratio. Steam reforming is an equilibrium reaction, and conversion of the hydrocarbon feedstock is favored by high temperature, which in turn requires higher fuel use. Because of the volume increase in the reaction, conversion is also favored by low pressure, which conflicts with the need to supply the hydrogen at high pressure. In practice, materials of construction limit temperature and pressure.

On the other hand, and in contrast to reforming, shift conversion is favored by low temperature. The gas from the reformer is reacted over iron oxide catalyst at 315–370°C (600–700°F), with the lower limit being dictated by the activity of the catalyst at low temperature.

Hydrogen can also be produced by *partial oxidation* (POX) of hydrocarbons in which the hydrocarbon is oxidized in a limited or controlled supply of oxygen:

$$2CH_4 + O_2 \rightarrow CO + 4H_2 \qquad \Delta H = -10,195 \text{ Btu/lb}$$

The shift reaction also occurs, and a mixture of carbon monoxide and carbon dioxide is produced in addition to hydrogen. The catalyst tube materials do not limit the reaction temperatures in partial oxidation processes, and higher temperatures may be used that enhance the conversion of methane to hydrogen. Indeed, much of the design and operation of hydrogen plants involves protecting the reforming catalyst and the catalyst tubes because of the extreme temperatures and the sensitivity of the catalyst. In fact, minor variations in feedstock composition or operating conditions can have significant effects on the life of the catalyst or the reformer itself. This is particularly true of changes in the molecular weight of the feed gas or poor distribution of heat to the catalyst tubes.

Because the high temperature takes the place of a catalyst, partial oxidation is not limited to the lower boiling feedstocks that are required for steam reforming. Partial oxidation processes (Fig. 20.4) were first considered for hydrogen production because of expected shortages of lower boiling feedstocks and the need to have available a disposal method for higher boiling, high sulfur streams such as asphalt or petroleum coke.

Catalytic partial oxidation, also known as autothermal reforming, reacts oxygen with a light feedstock, passing the resulting hot mixture over a reforming catalyst. The use of a catalyst allows the use of lower temperatures than in noncatalytic partial oxidation and reduces the oxygen demand.

The feedstock requirements for catalytic partial oxidation processes are similar to the feedstock requirements for steam reforming, and light hydrocarbons from refinery gas to naphtha are preferred. The oxygen substitutes for much of the steam in preventing coking, and a lower steam/carbon ratio is required. In addition, because a large excess of steam is not required, catalytic partial oxidation produces more carbon monoxide and less hydrogen than steam reforming. Thus, the process is more suited to situations where carbon monoxide is the more desirable product such as in synthesis gas.

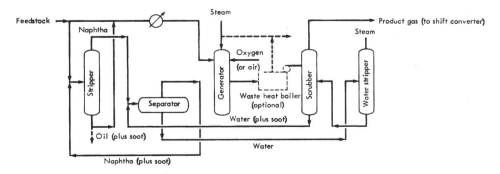

Figure 20.4 Partial oxidation process.

3. COMMERCIAL PROCESSES

3.1 Heavy Residue Gasification and Combined Cycle Power Generation

Heavy residua are gasified using the Texaco gasification process (partial oxidation process), and the produced gas is purified to clean fuel gas. Then electric power is generated by applying a combined cycle system. As an example, solvent deasphalter residuum (Fig. 20.1) is gasified by the partial oxidation method under pressure of about 570 psi (3923 kPa) and at a temperature of 1300–1500°C (2370–2730°F). The gas generated at high temperature flows into the specially designed waste heat boiler, in which the hot gas is cooled and high pressure saturated steam is generated. The gas from the waste heat boiler is then heat exchanged with the fuel gas and flows to the carbon scrubber, where unreacted carbon particles are removed from it by water scrubbing.

The gas from the carbon scrubber is further cooled by the fuel gas and boiler feed water and led into the sulfur compound removal section, where hydrogen sulfide (H_2S) and carbonyl sulfide (COS) are removed from the gas to obtain clean fuel gas. This clean fuel gas is heated with the hot gas generated in the gasifier and finally supplied to the gas turbine at a temperature of 250–300°C (480–570°F).

The exhaust gas from the gas turbine, which has a temperature of about 550–600°C (1020–1110°F), flows into the heat recovery steam generator, which consists of five heat exchange elements. The first element is a superheater in which the combined stream of the high pressure saturated steam generated in the waste heat boiler and in the second element (high pressure steam evaporator) is superheated. The third element is an economizer; the fourth element is a low pressure steam evaporator; and the final or fifth element is a deaerator heater. The offgas from the heat recovery steam generator, which has a temperature of about 130°C, is emitted into the air via a stack.

To decrease the nitrogen oxide (NO_x) content in the flue gas, two methods can be applied. The first method is to inject water into the gas turbine combustor. The second is to selectively reduce the nitrogen oxide content by injecting ammonia gas in the presence of de-NO_x catalyst, which is packed in the proper position in the heat recovery steam generator. The latter is more effective than the former to lower the nitrogen oxide emissions to the air.

3.2 Hybrid Gasification Process

In the hybrid gasification process, a slurry of coal and residual oil is injected into the gasifier, where it is pyrolyzed in the upper part of the reactor to produce gas and chars. The chars produced are then partially oxidized to ash. The ash is removed continuously from the bottom of the reactor.

In this process, coal and vacuum residua are mixed together into a slurry to produce clean fuel gas. The slurry fed into the pressurized gasifier is thermally cracked at a temperature of 850–950°C (1560–1740°F) and is converted into gas, tar, and char. The mixture of oxygen and steam in the lower zone of the gasifier gasifies the char. The gas leaving the gasifier is quenched to a temperature of 450°C (840°F), in the fluidized bed heat exchanger and is then scrubbed to remove tar, dust, and steam at around 200°C (390°F).

The coal and residual oil slurry is gasified in the fluidized bed gasifier. The charged slurry is converted to gas and char by thermal cracking reactions in the upper zone of the fluidized bed. The produced char is further gasified with steam and oxygen, which enter the gasifier just below the fluidizing gas distributor. Ash is discharged from the gasifier, indirectly cooled with steam, and then discharged into the ash hopper. It is burned in an incinerator to produce process steam. Coke deposited on the silica sand is regenerated by the incinerator.

3.3 Hydrocarbon Gasification

The gasification of hydrocarbons to produce hydrogen is a continuous, noncatalytic process (Fig. 20.5) that involves partial oxidation of the hydrocarbon. Air or oxygen (with steam or carbon dioxide) is used as the oxidant at 1095–1480°C (2000–2700°F). Any carbon produced (2–3 wt% of the feedstock) during the process is removed as a slurry in a carbon separator and pelleted for use either as a fuel or as raw material for carbon-based products.

3.4 Hypro Process

The Hypro process is a continuous catalytic method (Fig. 20.6) for hydrogen manufacture from natural gas or from refinery effluent gases. The process is designed to convert natural gas,

$$CH_4 \rightarrow C + 2H_2$$

and recover hydrogen by phase separation to yield hydrogen of about 93% purity. The principal contaminant is methane.

3.5 Shell Gasification (Partial Oxidation) Process

The Shell gasification process is a flexible process for generating synthesis gas, principally hydrogen and carbon monoxide, for the ultimate production of high purity, high pressure hydrogen, ammonia, methanol, fuel gas, town gas, or reducing gas by reaction of gaseous or liquid hydrocarbons with oxygen, air, or oxygen-enriched air.

The most important step in converting heavy residua to industrial gas is the partial oxidation of the oil using oxygen with the addition of steam. The gasification process takes

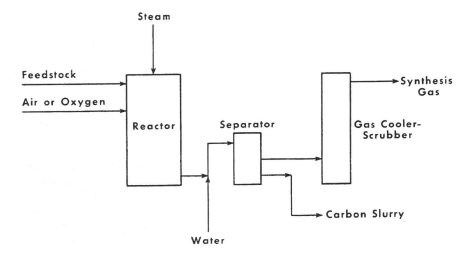

Figure 20.5 Hydrocarbon gasification.

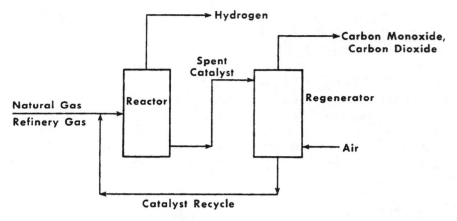

Figure 20.6   The Hypro process.

place in an empty, refractory-lined reactor at temperatures of about 1400°C (2550°F) and pressures between 29 and 1140 psi (196–7845 kPa). The chemical reactions in the gasification reactor proceed without catalyst to produce gas containing carbon amounting to some 0.5–2% by weight, based on the feedstock. The carbon is removed from the gas with water, extracted in most cases with feed oil from the water, and returned to the feed oil. The reformed gas at high temperature is utilized in a waste heat boiler for generating steam. The steam is generated at 850–1565 psi (5884–10,787 kPa). Some of this steam is used as process steam and for oxygen and oil preheating. The surplus steam is used for energy production and heating purposes.

3.6   Steam–Methane Reforming

Steam–methane reforming is a continuous catalytic process that has been employed for hydrogen production over a period of several decades. The major reaction is the formation of carbon monoxide and hydrogen from methane and steam,

$$CH_4 + H_2O \rightarrow CO + 3H_2$$

but higher molecular weight feedstocks may also yield hydrogen:

$$C_3H_8 + 3H_2O \rightarrow 3CO + 7H_2$$

That is,

$$C_nH_m + nH_2O \rightarrow nCO + \left(\frac{m}{2} + n\right)H_2$$

In the actual process (Fig. 20.7), the feedstock is first desulfurized by passage through activated carbon, which may be preceded by caustic and water washes. The desulfurized material is then mixed with steam and passed over a nickel-based catalyst [730–845°C (1350–1550°F) and 400 psi (2758 kPa)]. Effluent gases are cooled by the addition of steam or condensate to about 370°C (700°F), at which point carbon monoxide reacts with steam in the presence of iron oxide in a shift converter to produce carbon dioxide and hydrogen:

$$CO + H_2O \rightarrow CO_2 + H_2$$

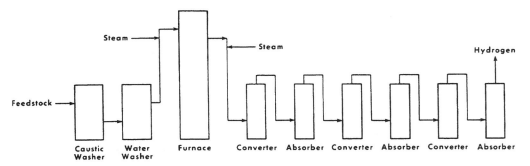

Figure 20.7 Steam–methane reforming process.

The carbon dioxide is removed by amine washing; the hydrogen is usually a high purity (> 99%) material.

Because the presence of any carbon monoxide or carbon dioxide in the hydrogen stream can interfere with the chemistry of the catalytic application, a third stage is used to convert these gases to methane:

$$CO + 3H_2 \rightarrow CH_4 + H_2O$$

$$CO_2 + 4H_2 \rightarrow CH_4 + 2H_2O$$

For many refiners, sulfur-free natural gas (CH_4) is not always available to produce hydrogen by this process. In that case, higher boiling hydrocarbons (such as propane, butane, or naphtha) may be used as the feedstock.

3.7 Steam–Naphtha Reforming

Steam–naphtha reforming is a continuous process (Fig. 20.8) for the production of hydrogen from liquid hydrocarbons and is, in fact, similar to steam–methane reforming. A variety of naphtha types in the gasoline boiling range may be employed, including feeds containing up to 35% aromatics. Thus, following pretreatment to remove sulfur compounds, the feedstock is mixed with steam and taken to the reforming furnace [675–815°C (1250–1500°F), 300 psi (2068 kPa)], where hydrogen is produced.

3.8 Synthesis Gas Generation

The synthesis gas generation process is a noncatalytic process for producing synthesis gas (principally hydrogen and carbon monoxide) for the ultimate production of high purity hydrogen from gaseous or liquid hydrocarbons.

In this process (Fig. 20.9), a controlled mixture of preheated feedstock and oxygen is fed to the top of the generator, where carbon monoxide and hydrogen emerge as the products. Soot produced in this part of the operation is removed in a water scrubber from the product gas stream and is then extracted from the resulting carbon–water slurry with naphtha and transferred to a fuel oil fraction. The oil–soot mixture is burned in a boiler or recycled to the generator to extinction to eliminate carbon production as part of the process.

The soot-free synthesis gas is then charged to a shift converter where the carbon monoxide reacts with steam to form additional hydrogen and carbon dioxide at the

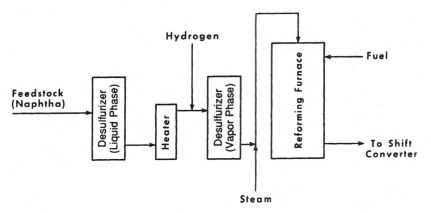

Figure 20.8 Steam–naphtha reforming.

stoichiometric rate of 1 mol of hydrogen for every mole of carbon monoxide charged to the converter.

The reactor temperatures vary from 1095°C to 1490°C (2000–2700°F), while pressures can vary from approximately atmospheric pressure to approximately 2000 psi (13,790 kPa). The process has the capability of producing high purity hydrogen, although the extent of the purification procedure depends upon the use to which the hydrogen is to be put. For example, carbon dioxide can be removed by scrubbing with various alkaline reagents, and carbon monoxide can be removed by washing with liquid nitrogen; if nitro-

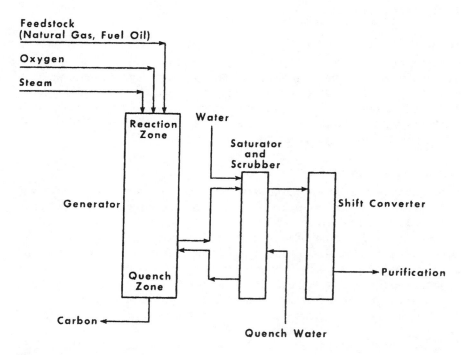

Figure 20.9 Synthesis gas generation.

gen is undesirable in the product, the carbon monoxide should be removed by washing with copper–amine solutions.

This particular partial oxidation technique has also been applied to a whole range of liquid feedstocks for hydrogen production (Table 20.5). Serious consideration is being given to hydrogen production by the partial oxidation of solid feedstocks such as petroleum coke (from both delayed and fluid-bed reactors), lignite, and coal as well as petroleum residua.

The chemistry of the process, using naphthalene as an example, may be simply represented as the selective removal of carbon from the hydrocarbon feedstock and further conversion of a portion of this carbon to hydrogen:

$$C_{10}H_8 + 5O_2 \rightarrow 10CO + 4H_2$$

$$10CO + 10H_2O \rightarrow 10CO_2 + 10H_2$$

Although these reactions may be represented very simply by using equations of this type, the reactions can be complex and result in carbon deposition on parts of the equipment, thereby requiring careful inspection of the reactor.

3.9 Texaco Gasification (Partial Oxidation) Process

The Texaco gasification process is a partial oxidation gasification process for generating synthesis gas, principally hydrogen and carbon monoxide. The characteristic of the Texaco gasification process is that feedstock is injected together with carbon dioxide, steam, or water into the gasifier. Therefore, solvent-deasphalted residua or petroleum coke rejected from any coking method can be used as feedstock. The gas produced by this gasification process can be used for the production of high purity, high pressure hydrogen, ammonia, and methanol. The heat recovered from the high temperature gas is used for the generation of steam in the waste heat boiler. Alternatively, the less expensive quench-type configuration is preferred when high pressure steam is not needed or when a high degree of shift is needed in the downstream CO converter.

In the Texaco process, the feedstock, together with the feedstock–carbon slurry recovered in the carbon recovery section, is pressurized to a given pressure, mixed with high pressure steam, and then blown into the gas generator through the burner together with oxygen.

The gasification reaction is a partial oxidation of hydrocarbons to carbon monoxide and hydrogen:

$$C_xH_{2y} + \frac{x}{2O_2} \rightarrow xCO + yH_2$$

$$C_xH_{2y} + xH_2O \rightarrow xCO + (x + y)H_2$$

The gasification reaction is instantly completed, thus producing gas that consists mainly of H_2 and CO ($H_2 + CO > 90\%$). The high temperature gas leaving the reaction chamber of the gas generator enters the quenching chamber linked to the bottom of the gas generator and is quenched to 200–260°C (390–500°F) with water.

Table 20.5 Synthesis Gas Generation by Partial Oxidation of Various Feedstocks

Feedstock	Natural gas	Propane	654°API naphtha	9.6°API fuel oil	9.7°API fuel oil	−11.4°API coal tar
Feedstock composition, wt%						
Carbon	73.40	81.69	83.8	87.2	87.2	88.1
Hydrogen	22.76	18.31	16.2	9.9	9.9	5.7
Oxygen	−0.76	—	—	0.8	0.8	4.4
Nitrogen	3.08	—	—	0.7	0.7	0.9
Sulfur	—	—	—	1.4	1.4	0.8
Ash	—	—	—	—	—	0.1
Gross heating value, Btu/lb	22,630	21,662	20,300	18,200	18,200	15,690
Composition of product gas, mol						
Hydrogen	61.1	54.0	51.2	45.9	45.8	38.9
Carbon monoxide	35.0	43.7	45.3	48.5	47.5	54.3
Carbon dioxide	2.6	2.1	2.7	4.6	5.7	5.7
Nitrogen	1.0	0.1	0.1	0.7	0.2	0.8
Methane	0.3	0.1	0.7	0.2	0.5	0.1
Hydrogen sulfide	—	—	—	0.1	0.3	0.2

4. CATALYSTS

Hydrogen plants are among the heaviest users of catalysts in the refinery. Catalytic operations include hydrogenation, steam reforming, shift conversion, and methanation.

4.1 Reforming Catalysts

The reforming catalyst is usually supplied as nickel oxide that, during start-up, is heated in a stream of inert gas, then steam. When the catalyst is near the normal operating temperature, hydrogen or a light hydrocarbon is added to reduce the nickel oxide to metallic nickel.

The high temperatures [up to 870°C (1600°F)], and the nature of the reforming reaction require that the reforming catalyst be used inside the radiant tubes of a reforming furnace. The active agent in the reforming catalyst is nickel, and normally the reaction is controlled both by diffusion and by heat transfer. Catalyst life is limited as much by physical breakdown as by deactivation.

Sulfur is the main catalyst poison, and the catalyst poisoning is theoretically reversible, with the catalyst being restored to nearly full activity by steaming. However, in practice the deactivation may cause the catalyst to overheat and coke, to such an extent that it must be replaced. Reforming catalysts are also sensitive to poisoning by heavy metals, although these are rarely present in low-boiling hydrocarbon feedstocks or in naphtha feedstocks.

Coking deposition on the reforming catalyst and ensuing loss of catalyst activity is the most characteristic issue that must be assessed and mitigated.

Although light methane-rich streams such as natural gas or light refinery gas are the most common feeds to hydrogen plants, there is often a requirement for a variety of reasons to process a variety of higher boiling feedstocks such as liquefied petroleum gas and naphtha. Feedstock variations may also be inadvertent due, for example, to changes in refinery offgas composition from another unit or because of variations in naphtha composition because of variance in feedstock delivered to the naphtha unit.

Thus, in using higher boiling feedstocks in a hydrogen plant, coke deposition on the reformer catalyst becomes a major issue. Coking is most likely to occur in the reformer unit when both temperature and hydrocarbon content are high. In this region, hydrocarbons crack and form coke faster than the coke is removed by reaction with steam or hydrogen, and when catalyst deactivation occurs there is a simultaneous increase in temperature with a concomitant increase in coke formation and deposition. In other zones, where the hydrocarbon/hydrogen ratio is lower, there is less risk of coking.

Coking depends to a large extent on the balance between catalyst activity and heat input, with the more active catalysts producing higher yields of hydrogen at lower temperature, thereby reducing the risk of coking. A uniform input of heat is important in this region of the reformer, because any catalyst voids or variations in catalyst activity can produce localized hot spots, leading to coke formation and/or reformer failure.

Coke formation results in hot spots in the reformer, which increases pressure drop and reduces feedstock (methane) conversion, leading eventually to reformer failure. Coking may be partially mitigated by increasing the steam/feedstock ratio to change the reaction conditions, but the most effective solution may be to replace the reformer catalyst with one designed for higher boiling feedstocks.

A *standard* steam–methane reforming catalyst uses nickel on an α-alumina ceramic carrier that is acidic in nature. Promotion of hydrocarbon cracking with such a catalyst leads to coke formation from higher boiling feedstocks. Some catalyst formulations use a magnesia-alumina ($MgO\text{-}Al_2O_3$) support that is less acidic than α-alumina and that reduces cracking on the support and allows higher boiling feedstocks (such as liquefied petroleum gas) to be used.

Further resistance to coking can be achieved by adding an alkali promoter, typically some form of potash (KOH), to the catalyst. Besides reducing the acidity of the carrier, the promoter catalyzes the reaction of steam and carbon. Although carbon continues to be formed, it is removed faster than it can build up. This approach can be used with naphtha feedstocks with boiling points of up to approximately 180°C (350°F). Under the conditions in a reformer, potash is volatile, and it is incorporated into the catalyst as a more complex compound that slowly hydrolyzes to release potassium hydroxide (KOH). Alkali-promoted catalyst allows the use of a wide range of feedstocks, but, in addition to possible potash migration, which can be minimized by proper design and operation, the catalyst is also somewhat less active than conventional catalyst.

Another option to reduce coking in steam reformers is to use a *prereformer* in which a fixed bed of catalyst is used that operates at a lower temperature upstream of the fired reformer (Fig. 20.10). Inlet temperatures are selected such that there is minimal risk of coking and the gas leaving the prereformer contains only steam, hydrogen, carbon monoxide, carbon dioxide, and methane. This allows a standard methane catalyst to be used in the fired reformer, and this approach has been used with feedstocks up to light kerosene. Because the gas leaving the prereformer poses a reduced risk of coking, it can compensate to some extent for variations in catalyst activity and heat flux in the primary reformer.

4.2 Shift Conversion Catalysts

The second important reaction in a steam reforming plant is the shift conversion reaction,

$$CO + H_2O \rightarrow CO_2 + H_2$$

Two basic types of shift catalysts are used in steam reforming plants: iron–chrome high temperature shift catalysts, and copper–zinc low temperature shift catalysts.

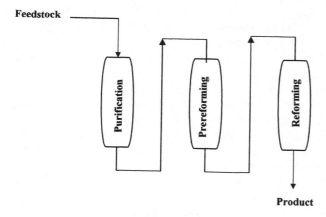

Figure 20.10 Schematic of a prereformer.

High temperature shift catalysts operate in the range of 315–430°C (600–800°F) and consist primarily of magnetite (Fe₃O₄) with three-valent chromium oxide (Cr₂O₃) added as a stabilizer. The catalyst is usually supplied in the form of ferric oxide (Fe₂O₃) and six-valent chromium oxide (CrO₃) and is reduced by the hydrogen and carbon monoxide in the shift feed gas as part of the start-up procedure to produce the catalyst in the desired form. However, caution is necessary because if the steam/carbon ratio of the feedstock is too low and the reducing environment too strong, the catalyst can be reduced further to metallic iron. Metallic iron is a catalyst for Fischer-Tropsch reactions, and hydrocarbons will be produced.

Low temperature shift catalysts operate at temperatures on the order of 205–230°C (400–450°F). Because of the lower temperature, the reaction equilibrium is more controllable and smaller amounts of carbon monoxide are produced. The low temperature shift catalyst is used primarily in wet scrubbing plants that use methanation for final purification. Pressure swing adsorption plants do not generally use a low temperature because any unconverted carbon monoxide is recovered as reformer fuel. Low-temperature shift catalysts are sensitive to poisoning by sulfur and are sensitive to water (liquid), which can cause softening of the catalyst followed by crusting or plugging.

The catalyst is supplied as copper oxide (CuO) on a zinc oxide (ZnO) carrier, and the copper must be reduced by heating in a stream of inert gas with measured quantities of hydrogen. The reduction of the copper oxide is strongly exothermic and must be closely monitored.

4.3 Methanation Catalysts

In wet scrubbing plants, the final hydrogen purification procedure is methanation in which the carbon monoxide and carbon dioxide are converted to methane:

$$CO + 3H_2O \rightarrow CH_4 + H_2O$$

$$CO_2 + 4H_2 \rightarrow CH_4 + 2H_2O$$

The active agent is nickel on an alumina carrier.

The catalyst has a long life, because it operates under ideal conditions and is not exposed to poisons. The main source of deactivation is plugging from carryover of carbon dioxide from removal solutions.

The most severe hazard arises from high levels of carbon monoxide or carbon dioxide, which can result from breakdown of the carbon dioxide removal equipment or from exchanger tube leaks, which quench the shift reaction. The results of breakthrough can be severe, because the methanation reaction produces a temperature rise of 70°C (125°F) for each 1% of carbon monoxide or a temperature rise of 33°C (60°F) for each 1% of carbon dioxide. Although the normal operating temperature during methanation is approximately 315°C (600°F), it is possible to reach 700°C (1300°F) in cases of major breakthrough.

5. HYDROGEN PURIFICATION

Various processes are available to purify the hydrogen stream, but because the product streams are available in a wide variety of compositions, flows, and pressures, the best method of purification will vary. In addition, several other factors must also be taken into consideration in the selection of a purification method:

1. Hydrogen recovery
2. Product purity
3. Pressure profile
4. Reliability

Cost, an equally important parameter, is not considered here, because the emphasis is on the technical aspects of the purification process.

5.1 Wet Scrubbing

Wet scrubbing systems, particularly amine or potassium carbonate systems, are used to remove acidic gases such as hydrogen sulfide and carbon dioxide (Speight, 1993). Most systems depend on chemical reaction and can be designed for a wide range of pressures and capacities. They were once widely used to remove carbon dioxide in steam reforming plants but have generally been replaced by pressure swing adsorption units except where carbon monoxide is to be recovered. Wet scrubbing is still used to remove hydrogen sulfide and carbon dioxide in partial oxidation plants.

Wet scrubbing systems remove only acidic gases or heavy hydrocarbons; they do not remove methane or other hydrocarbon gases and therefore have little influence on product purity. Therefore, wet scrubbing systems are most often used as a pretreatment step or where a hydrogen-rich stream is to be desulfurized for use as fuel gas.

5.2 Pressure Swing Adsorption Units

Many hydrogen plants that formerly used a wet scrubbing process (Fig. 20.11) for hydrogen purification are now using the *pressure swing adsorption* (PSA) concept (Fig. 20.12) for purification. The pressure swing adsorption process is a cyclical process that uses beds of solid adsorbent to remove impurities from the gas and generally produces a higher purity

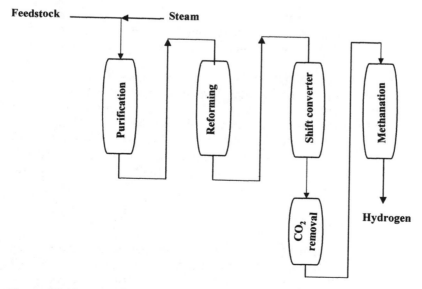

Figure 20.11 Purification by wet scrubbing.

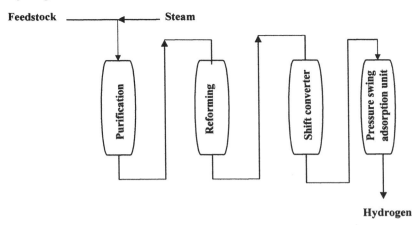

Figure 20.12 Purification by pressure swing adsorption.

hydrogen (99.9 vol% compared to less than 97 vol% for wet scrubbing). The purified hydrogen passes through the adsorbent beds with only a tiny fraction absorbed. The beds are regenerated by depressurization followed by purging at low pressure.

When the beds are depressurized, a waste gas (or *tail gas*) stream is produced that consists of the impurities from the feed (carbon monoxide, carbon dioxide, methane, and nitrogen) plus some hydrogen. This stream is burned in the reformer as fuel, and reformer operating conditions in a pressure swing adsorption plant are set so that the tail gas provides no more than about 85% of the reformer fuel. This gives good burner control because the tail gas is more difficult to burn than regular fuel gas and the high content of carbon monoxide can interfere with the stability of the flame. As the reformer operating temperature is increased, the reforming equilibrium shifts, resulting in more hydrogen and less methane in the reformer outlet and hence less methane in the tail gas.

Pressure swing adsorption is generally the purification method of choice for steam reforming units because of its production of high purity hydrogen and is also used for purification of refinery offgases, where it competes with membrane systems.

5.3 Membrane Systems

Membrane systems separate gases by taking advantage of the differences in their rates of diffusion through the membranes. Gases that diffuse faster (including hydrogen) become the permeate stream and are available at low pressure, whereas the more slowly diffusing gases become the nonpermeate and leave the unit at a pressure close to that of the feedstock at the entry point. Membrane systems contain no moving parts or switch valves and have potentially very high reliability. The major threat is from components in the gas (such as aromatics) that attack the membranes or from liquids, which plug them.

Membranes are fabricated in relatively small modules; for larger capacity more modules are added. Cost is therefore virtually linear with capacity, making the modules more competitive at lower capacities. The design of membrane systems involves a trade-off between pressure drop (or diffusion rate) and surface area as well as between product purity and recovery. As the surface area is increased, the recovery of fast-diffusing components increases; however, more of the slow components are recovered, which lowers the purity.

5.4 Cryogenic Separation

Cryogenic separation units operate by cooling the gas and condensing some or all of the constituents of the gas stream. Depending on the product purity required, separation may involve flashing or distillation. Cryogenic units offer the advantage of being able to separate a variety of products from a single feedstream. One specific example is the separation of light olefins from a hydrogen stream.

Hydrogen recovery is in the range of 95%, with purity above 98% obtainable.

REFERENCES

Bridge, A. G. 1997. In: Handbook of Petroleum Refining Processes. 2nd ed. R. A. Meyers, ed. McGraw-Hill, New York, Chapter 14.1.

Bridge, A. G., Gould, G. D., and Berkman, J. F. 1981. Oil Gas J. 79(3):85.

Campbell, W. M. 1997. In: Handbook of Petroleum Refining Processes. 2nd ed. R. A. Meyers, ed. McGraw-Hill, New York, Chapter 6.1.

Dickenson, R. L., Biasca, F. E., Schulman, B. L., and Johnson, H. E. 1997. Hydrocarbon Process. 76(2):57.

Dolbear, G. E. 1998. In: Petroleum Chemistry and Refining. J. G. Speight, ed. Taylor & Francis, Washington, DC, Chapter 7.

Fleshman, J. D. 1997. In: Handbook of Petroleum Refining Processes. 2nd ed. R. A. Meyers, ed. McGraw-Hill, New York, Chapter 6.2.

Radler, M. 1999. Oil Gas J. 97(51):445 et seq.

Scherzer, J., and Gruia, A. J. 1996. Hydrocracking Science and Technology. Marcel Dekker, New York.

Speight, J. G. 1993. Gas Processing: Environmental Aspects and Methods. Butterworth-Heineman, Oxford, England.

Speight, J. G. 1999. The Chemistry and Technology of Petroleum. 3rd ed. Marcel Dekker, New York.

Speight, J. G. 2000. The Desulfurization of Heavy Oils and Residua. 2nd ed. Marcel Dekker, New York.

Glossary

Absorber   See **Absorption tower**.

Absorption gasoline   Gasoline extracted from natural gas or refinery gas by contacting the absorbed gas with an oil and subsequently distilling the gasoline from the higher boiling components.

Absorption oil   Oil used to separate the heavier components from a vapor mixture by absorption of the heavier components during intimate contacting of the oil and vapor; used to recover natural gasoline from wet gas.

Absorption plant   A plant for recovering the condensable portion of natural or refinery gas by absorbing the higher boiling hydrocarbons in an absorption oil followed by separation and fractionation of the absorbed material.

Absorption tower   A tower or column that promotes contact between a rising gas and a falling liquid so that part of the gas will be dissolved in the liquid.

Acetone–benzol process   A dewaxing process in which acetone and benzol (benzene or aromatic naphtha) are used as solvents.

Acid catalyst   A catalyst having acidic character. Alumina (*q.v.*) is an example of such a catalyst.

Acidity   The capacity of an acid to neutralize a base such as a hydroxyl ion (OH^-).

Acidizing   A technique for improving the permeability (*q.v.*) of a reservoir by injecting acid.

Acid number   A measure of the reactivity of petroleum with a caustic solution; expressed as milligrams of potassium hydroxide that are neutralized by 1 g of petroleum.

Acid sludge   The residue left after treating petroleum oil with sulfuric acid for the removal of impurities; a black, viscous substance containing the spent acid and impurities.

Acid treating   A process in which unfinished petroleum products, such as gasoline, kerosene, and lubricating oil stocks, are contacted with sulfuric acid to improve their color, odor, and other properties.

Adapted from J. G. Speight, The Chemistry and Technology of Petroleum. 3rd ed. Marcel Dekker, New York, 1999.

Additive   A material added to another (usually in small amounts) to enhance desirable properties or to suppress undesirable properties.

Adsorption   Transfer of a substance from a solution to the surface of a solid, resulting in relatively high concentration of the substance at the place of contact. *See also* **Chromatographic adsorption**.

Adsorption gasoline   Natural gasoline (*q.v.*) obtained by adsorption from wet gas.

Afterburn   The combustion of carbon monoxide (CO) to carbon dioxide (CO_2), usually in the cyclones of a catalyst regenerator.

Air-blown asphalt   Asphalt produced by blowing air through residua at elevated temperatures.

Airlift Thermofor catalytic cracking   A moving-bed continuous catalytic process for conversion of heavy gas oils into lighter products; a stream of air moves the catalyst.

Air sweetening   A process in which air or oxygen is used to oxidize lead mercaptides to disulfides instead of using elemental sulfur.

Alicyclic hydrocarbon   A compound containing only carbon and hydrogen that has a cyclic structure (e.g., cyclohexane); also collectively called naphthenes.

Aliphatic hydrocarbon   A compound containing only carbon and hydrogen that has an open-chain structure (e.g., ethane, butane, octane, butene) or a cyclic structure (e.g., cyclohexane).

Alkalinity   The capacity of a base to neutralize the hydrogen ion (H^+).

Alkali treatment   *See* **Caustic wash**.

Alkali wash   *See* **Caustic wash**.

Alkylate   The product of an alkylation (*q.v.*) process.

Alkylate bottoms   Residua from fractionation of alkylate; the alkylate product that boils at a higher temperature than aviation gasoline; sometimes called heavy alkylate or alkylate polymer.

Alkylation   In the petroleum industry, a process by which an olefin (e.g., ethylene) is combined with a branched-chain hydrocarbon (e.g., isobutane); alkylation may be accomplished as a thermal or catalytic reaction.

Alphascission   The rupture of the aromatic carbon–aliphatic carbon bond that joins an alkyl group to an aromatic ring.

Alumina   Aluminum oxide (Al_2O_3) used in separation methods as an adsorbent and in refining as a catalyst.

American Society for Testing and Materials (ASTM)   The official organization in the United States for designing standard tests for petroleum and other industrial products.

Aniline point   The temperature, usually expressed in degrees Fahrenheit, above which equal volumes of a petroleum product are completely miscible; a qualitative indication of the relative proportions of paraffins in a petroleum product that are miscible with aniline only at higher temperatures. A high aniline point indicates low aromatics.

Antiknock   Resistance to detonation or pinging in spark ignition engines.

Antiknock agent   A chemical compound such as tetraethyllead that, when added in small amount to the fuel charge of an internal combustion engine, tends to lessen knocking.

API gravity   A measure of the *lightness* or *heaviness* of petroleum that is related to density and specific gravity.

$$°API = 141.5/(sp\ gr\ at\ 60°F) - 131.5$$

Apparent bulk density   The density of a catalyst as measured, usually loosely compacted in a container.

Apparent viscosity   The viscosity of a fluid, or several fluids flowing simultaneously, measured in a porous medium (rock) and subject to both viscosity and permeability effects; also called **effective viscosity**.

Aromatic hydrocarbon   A hydrocarbon characterized by the presence of an aromatic ring or condensed aromatic rings; benzene and substituted benzene, naphthalene and substituted naphthalene, phenanthrene and substituted phenanthrene, as well as the higher condensed ring systems; compounds that are distinct from those of aliphatic compounds (*q.v.*) or alicyclic compounds (*q.v.*).

Aromatization   The conversion of nonaromatic hydrocarbons to aromatic hydrocarbons by (1) rearrangement of aliphatic (noncyclic) hydrocarbons (*q.v.*) into aromatic ring structures or (2) dehydrogenation of alicyclic hydrocarbons (naphthenes).

Arosorb process   A process for the separation of aromatics from nonaromatics by adsorption on a gel, from which they are recovered by desorption.

Asphalt   The nonvolatile product obtained by distillation and treatment of an asphaltic crude oil; a manufactured product.

Asphaltene (asphaltenes)   The brown to black powdery material produced by treatment of petroleum, petroleum residua, or bituminous materials with a low-boiling liquid hydrocarbon such as pentane or heptane; soluble in benzene (and other aromatic solvents), carbon disulfide, and chloroform (or other chlorinated hydrocarbon solvents).

Asphaltite   A variety of naturally occurring, dark brown to black, solid, nonvolatile bituminous material that is differentiated from bitumen primarily by a high content of asphaltene.

Asphaltoid   A brown to black, solid bituminous material differentiated from asphaltite by infusibility and low solubility in carbon disulfide.

Asphaltum   *See* **Asphalt**.

Atmospheric residuum   A residuum (*q.v.*) obtained by distillation of a crude oil under atmospheric pressure that boils above 350°C (660°F).

Atmospheric equivalent boiling point (AEBP)   A mathematical method of estimating the boiling point at atmospheric pressure of nonvolatile fractions of petroleum.

Attapulgus clay   *See* **Fuller's earth**.

Autofining   A catalytic process for desulfurizing distillates.

Average particle size   The weighted average particle diameter of a catalyst.

Aviation gasoline   Any of the special grades of gasoline suitable for use in certain airplane engines.

Aviation turbine fuel   *See* **Jet fuel**.

Backmixing   The phenomenon observed when a catalyst travels at a slower rate in the riser pipe than the vapors.

Baghouse   A filter system for the removal of particulate matter from gas streams; so called because of the similarity of the filters to coal bags.

Bari-Sol process   A dewaxing process that employs a mixture of ethylene dichloride and benzol as the solvent.

Barrel   The unit of measurement of liquids in the petroleum industry; equivalent to 42 U.S. standard gallons or 33.6 Imperial gallons; abbreviated bbl.

Basic nitrogen   Nitrogen (in petroleum) that occurs in pyridine form

Basic sediment and water (bs&w, bsw)   The material that collects in the bottom of storage tanks; usually composed of oil, water, and foreign matter; also called **bottoms, bottom settlings**.

Battery   A series of stills or other refinery equipment operated as a unit.

Baumé gravity   The specific gravity of liquids expressed as degrees on the Baumé scale (°Bé). For liquids lighter than water,

$$\text{Sp gr at } 60°F = 140/(130 + °Bé)$$

and for liquids heavier than water,

$$\text{Sp gr at } 60°F = 145/(145 - °Bé)$$

Bauxite   Mineral matter used as a treating agent; hydrated aluminum oxide formed by the chemical weathering of igneous rocks.

bbl   *See* **Barrel**.

Bell cap   A hemispherical or triangular cover placed over the riser in a (distillation) tower to direct the vapors through the liquid layer on the tray. *See also* **Bubble cap**.

Bender process   A chemical treating process using lead sulfide catalyst for sweetening light distillates that converts mercaptans to disulfides by oxidation.

Bentonite   Montmorillonite (a magnesium-aluminum silicate) used as a treating agent.

Benzene   A colorless aromatic liquid hydrocarbon (C_6H_6).

Benzin   A refined light naphtha used for extraction purposes.

Benzine   An obsolete term for light petroleum distillates covering the gasoline and naphtha range. *See* **Ligroine**.

Benzol   A general term for commercial or technical (not necessarily pure) benzene; also a term used for aromatic naphtha.

Beta scission   The rupture of a carbon–carbon bond two bonds removed from an aromatic ring.

Billion 1×10^9.

Bitumen   A semisolid to solid hydrocarbonaceous material found filling the pores and crevices of sandstone, limestone, or argillaceous sediments.

Bituminous   Containing bitumen or constituting the source of bitumen.

Bituminous rock   *See* **Bituminous sand**.

Bituminous sand   A formation in which the bituminous material (*see* **Bitumen**) is found as a filling in veins and fissures in fractured rocks or impregnating relatively shallow sand, sandstone, and limestone strata; a sandstone reservoir that is impregnated with a heavy, viscous black petroleum-like material that cannot be retrieved through a well by conventional production techniques, including enhanced oil recovery techniques.

Black acid(s)   A mixture of the sulfonates found in acid sludge that are insoluble in naphtha, benzene, and carbon tetrachloride; very soluble in water; but insoluble in 30% sulfuric acid. In the dry, oil-free state, the sodium soaps are black powders.

Black oil   Any dark-colored oil; a term now often applied to heavy oil (*q.v.*).

Black soap   *See* **Black acid**.

Black strap   The black material (mainly lead sulfide) formed in the treatment of sour light oils with doctor solution (*q.v.*) and found at the interface between the oil and the solution.

Blown asphalt   The asphalt prepared by air blowing a residuum (*q.v.*) or an asphalt (*q.v.*).

Boiling range   The range of temperature, usually determined at atmospheric pressure in standard laboratory apparatus, over which the distillation of an oil commences, proceeds, and finishes.

Bottled gas   Usually butane or propane, or butane–propane mixtures, liquefied and stored under pressure for domestic use. *See also* **Liquefied petroleum gas**.

Bottoms   The liquid that collects in the bottom of a vessel (tower bottoms, tank bottoms) during distillation; also the deposit or sediment formed during storage of petroleum or a petroleum product. *See also* **Residuum** *and* **Basic sediment and water**.

Bright stock   Refined, high viscosity lubricating oils usually made from residual stocks by processes such as a combination of acid treatment and solvent extraction with dewaxing or clay finishing.

British thermal unit   *See* **Btu**.

Bromine number   The number of grams of bromine absorbed by 100 g of oil, which indicates the percentage of double bonds in the material.

Brown acid   Oil-soluble petroleum sulfonates found in acid sludge that can be recovered by extraction with naphtha solvent. Brown acid sulfonates are somewhat similar to mahogany sulfonates but are more water soluble. In the dry, oil-free state, the sodium soaps are light-colored powders.

Brown soap   *See* **Brown acid**.

BS&W   *See* **Basic sediment and water**.

Brønsted acid   A chemical species that can act as a source of protons.

Brønsted base   A chemical species that can accept protons.

Btu (British thermal unit)   The energy required to raise the temperature of 1 lb of water one degree Fahrenheit.

Bubble cap   An inverted cup with a notched or slotted periphery to disperse the vapor in small bubbles beneath the surface of the liquid on the bubble plate in a distillation tower.

Bubble plate   A tray in a distillation tower.

Bubble point   The temperature at which incipient vaporization of a liquid in a liquid mixture occurs, corresponding with the equilibrium point of 0% vaporization or 100% condensation.

Bubble tower   A fractionating tower so constructed that the vapors rising pass up through layers of condensate on a series of plates or trays (*see* **Bubble plate**). The vapor passes from one plate to the next above by bubbling under one or more caps (*see* **Bubble cap**) and out through the liquid on the plate, where the less volatile portions of vapor condense in bubbling through the liquid on the plate, overflow to the next lower plate, and ultimately flow back into the reboiler, thereby effecting fractionation.

Bubble tray   A circular perforated plate having the internal diameter of a bubble tower (*q.v.*). Bubble trays are set at specified distances in a tower to collect the various fractions produced during distillation.

Bumping   The knocking against the walls of a still that occurs during distillation of petroleum or a petroleum product that (usually) contains water.

Bunker C oil   *See* **No. 6 fuel oil**.

Burner fuel oil   Any petroleum liquid suitable for combustion.

Burning oil   An illuminating oil, such as kerosene (kerosine), suitable for burning in a wick lamp.

Burning point   *See* **Fire point**.

Burning quality index An empirical numerical indication of the likely burning performance of a furnace or heater oil; derived from the distillation profile (*q.v.*) and the API gravity (*q.v.*) and generally recognizing the factors of paraffinicity and volatility.

Burton process An older thermal cracking process in which oil was cracked in a pressure still and any condensation of the products of cracking also took place under pressure.

Butane dehydrogenation A process for removing hydrogen from butane to produce butenes and, on occasion, butadiene.

Butane vapor-phase isomerization A process for isomerizing *n*-butane to isobutane using aluminum chloride catalyst on a granular alumina support and with hydrogen chloride as a promoter.

C_1, C_2, C_3, C_4, C_5 fractions A common way of representing fractions containing a preponderance of hydrocarbons having 1, 2, 3, 4, or 5 carbon atoms, respectively, and without reference to hydrocarbon type.

Capillary forces Interfacial forces between immiscible fluid phases, resulting in pressure differences between the two phases.

Capillary number (N_c) The ratio of viscous forces to capillary forces; equal to viscosity times velocity divided by interfacial tension.

Carbene The pentane- or heptane-insoluble material that is insoluble in benzene or toluene but is soluble in carbon disulfide (or pyridine).

Carboid The pentane- or heptane-insoluble material that is insoluble in benzene or toluene and is also insoluble in carbon disulfide (or pyridine).

Carbon-forming propensity *See* **Carbon residue**.

Carbonization The conversion of an organic compound into char or coke by heat in the substantial absence of air; often used in reference to the destructive distillation (with simultaneous removal of distillate) of coal.

Carbon rejection An upgrading process in which coke is produced, e.g., coking.

Carbon residue The amount of carbonaceous residue remaining after thermal decomposition of petroleum, a petroleum fraction, or a petroleum product in a limited amount of air; also called the *coke-* or *carbon-forming propensity*; often prefixed by the terms Conradson or Ramsbottom in reference to the inventors of the respective tests.

Cascade tray A fractionating device consisting of a series of parallel troughs arranged in stairstep fashion in which liquid from the tray above enters the uppermost trough, liquid thrown from this trough by vapor rising from the tray below impinges against a plate and a perforated baffle, and liquid passing through the baffle enters the next longer of the troughs.

Casinghead gas Natural gas that issues from the casinghead (the mouth or opening) of an oil well.

Casinghead gasoline The liquid hydrocarbon product extracted from casinghead gas (*q.v.*) by one of three methods: compression, absorption, or refrigeration. *See also* **Natural gasoline**.

Catalyst A chemical agent that when added to a reaction (process) will enhance the conversion of a feedstock without being consumed in the process.

Catalyst selectivity The relative activity of a catalyst with respect to a particular compound in a mixture, or the relative rate in competing reactions of a single reactant.

Catalyst stripping The introduction of steam, at a point where spent catalyst leaves the reactor, in order to strip (i.e., remove) deposits retained on the catalyst.

Catalytic activity   The ratio of the space velocity of the catalyst under test to the space velocity required for the standard catalyst to give the same conversion as the catalyst being tested; usually multiplied by 100 before being reported.

Catalytic cracking   The conversion of high-boiling feedstocks into lower boiling products by means of a catalyst that may be used in a fixed bed (q.v.) or fluid bed (q.v.).

Cat cracking   *See* **Catalytic cracking**.

Catalytic reforming   Rearranging hydrocarbon molecules in a gasoline boiling range feedstock to produce other hydrocarbons having a higher antiknock quality; isomerization of paraffins, cyclization of paraffins to naphthenes (*q.v.*), dehydrocyclization of paraffins to aromatics (*q.v.*).

Catforming   A process for reforming naphtha using a platinum-silica-alumina catalyst that permits relatively high space velocities and results in the production of high purity hydrogen.

Caustic wash   The process of treating a product with a solution of caustic soda to remove minor impurities; often used in reference to the solution itself.

Cetane index   An approximation of the cetane number (*q.v.*) calculated from the density (*q.v.*) and mid-boiling point temperature (*q.v.*). *See also* **Diesel index**.

Cetane number   A number indicating the ignition quality of diesel fuel. A high cetane number represents a short ignition delay time; the ignition quality of diesel fuel can also be estimated from the formula

Diesel index = aniline point ($^\circ$F) $\times$ API gravity $\times$ 100

Characterization factor   The UOP characterization factor K, defined as the ratio of the cube root of the molal average boiling point, T_B, in degrees Rankine ($^\circ$R $= ^\circ$F $+ 460$), to the specific gravity at 60°F$/60^\circ$F:

$$K = T_B)^{1/3}/(\text{sp gr})$$

K ranges from 12.5 for paraffinic stocks to 10.0 for the highly aromatic stocks. Also called the Watson characterization factor.

Cheesebox still   An early type of vertical cylindrical still designed with a vapor dome.

Chemical octane number   The octane number added to gasoline by refinery processes or by the use of octane number (*q.v.*) improvers such as tetraethyllead.

Chlorex process   A process for extracting lubricating oil stocks in which the solvent used is Chlorex (β,β-dichlorodiethyl ether).

Chromatographic adsorption   Selective adsorption on materials such as activated carbon, alumina, or silica gel. Liquid or gaseous mixtures of hydrocarbons are passed through the adsorbent in a stream of diluent, and certain components are preferentially adsorbed.

Chromatography   A method of separation based on selective adsorption. *See also* **Chromatographic adsorption**.

Clarified oil   The heavy oil that has been taken from the bottom of a fractionator in a catalytic cracking process and from which residual catalyst has been removed.

Clarifier   Equipment for removing the color or cloudiness of an oil or water by separating the foreign material through mechanical or chemical means; may involve centrifugal action, filtration, heating, or treatment with acid or alkali.

Clay   Silicate minerals that also usually contain aluminum and have particle sizes less than $0.002\,\mu$m. Used in separation methods as an adsorbent and in refining as a catalyst.

Clay contact process   *See* **Contact filtration**.

Clay refining   A treating process in which vaporized gasoline or other light petroleum product is passed through a bed of granular clay such as fuller's earth (*q.v.*).

Clay regeneration   A process in which spent coarse-grained adsorbent clays from percolation processes are cleaned for reuse by deoiling with naphtha, steaming out the excess naphtha, and then roasting in a stream of air to remove carbonaceous matter.

Clay treating   *See* **Gray clay treating**.

Clay wash   A light oil, such as kerosene (kerosine) or naphtha, used to clean fuller's earth after it has been used in a filter.

Cloud point   The temperature at which paraffin wax or other solid substances begin to crystallize or separate from the solution, imparting a cloudy appearance to the oil when the oil is chilled under prescribed conditions.

Coal   An organic rock.

Coal tar   The specific name for the tar (*q.v.*) produced from coal.

Coal tar pitch   The specific name for the pitch (*q.v.*) produced from coal.

Coke   A gray to black solid carbonaceous material produced from petroleum during thermal processing; characterized by having a high carbon content (95% + by weight) and a honeycomb type of appearance; insoluble in organic solvents.

Coke drum   A vessel in which coke is formed

Coke number   A number used, particularly in Great Britain, to report the results of the Ramsbottom carbon residue test (*q.v.*), which is also referred to as a coke test.

Coker   The processing unit in which coking takes place.

Coking   A process for the thermal conversion of petroleum in which gaseous, liquid, and solid (coke) products are formed.

Cold pressing   The process of separating wax from oil by first chilling it (to help form wax crystals) and then filtering it under pressure in a plate-and-frame press.

Cold settling   Processing for the removal of wax from high viscosity stocks, wherein a naphtha solution of the waxy oil is chilled and the wax crystallizes out of the solution.

Color stability   The resistance of a petroleum product to color change due to light, aging, etc.

Combustible liquid   A liquid with a flash point in excess of 37.8°C (100°F) but below 93.3°C (200°F).

Composition   The general chemical makeup of petroleum.

Composition map   A means of illustrating the chemical makeup of petroleum using chemical and/or physical property data.

Con carbon   *See* **Carbon residue**.

Conradson carbon residue   *See* **Carbon residue**.

Contact filtration   A process in which finely divided adsorbent clay is used to remove color bodies from petroleum products.

Continuous contact coking   A thermal conversion process in which petroleum-wetted coke particles move downward into the reactor in which cracking, coking, and drying take place to produce coke, gas, gasoline, and gas oil.

Continuous contact filtration   A process to finish lubricants, waxes, or special oils after acid treating, solvent extraction, or distillation.

Conventional recovery   Primary and/or secondary recovery.

Conversion   The thermal treatment of petroleum that results in the formation of new products by the alteration of the original constituents.

Conversion factor The percentage of feedstock converted to light ends, gasoline, other liquid fuels, and coke.

Copper sweetening Processes involving the oxidation of mercaptans to disulfides by oxygen in the presence of cupric chloride.

Cracked residua Residua that have been subjected to temperatures above 350°C (660°F) during the distillation process.

Cracking The thermal processes by which the constituents of petroleum are converted to lower molecular weight products.

Cracking activity *See* **Catalytic activity**.

Cracking coil Equipment used for cracking heavy petroleum products consisting of a coil of heavy pipe running through a furnace so that the oil passing through it is subjected to high temperature.

Cracking still The combined equipment—furnace, reaction chamber, fractionator—for the thermal conversion of heavier feedstocks to lighter products.

Cracking temperature The temperature (350°C; 660°F) at which the rate of thermal decomposition of petroleum constituents becomes significant.

Crude assay A procedure for determining the general distillation characteristics (e.g., distillation profile, *q.v.*) and other quality information of crude oil.

Crude oil *See* **Petroleum**.

Crude scale wax The wax product from the first sweating of the slack wax.

Crude still Distillation (*q.v.*) equipment in which crude oil is separated into various products.

Cumene A colorless liquid [$C_6H_5CH(CH_3)_2$] used as an aviation gasoline blending component and as an intermediate in the manufacture of chemicals.

Cut point The boiling temperature division between distillation fractions of petroleum.

Cutback The term applied to the products from blending heavier feedstocks or products with lighter oils to bring the heavier materials to the desired specifications.

Cutting oil An oil used to lubricate and cool metal-cutting tools; also called *cutting fluid, cutting lubricant*.

Cycle stock The product taken from some later stage of a process and recharged (recycled) to the process at some earlier stage.

Cyclic steams injection The alternating injection of steam and production of oil with condensed steam from the same well or wells.

Cyclization The process by which an open-chain hydrocarbon structure is converted to a ring structure, e.g., hexane to benzene.

Cyclone A device for extracting dust from industrial waste gases. It is in the form of an inverted cone into which the contaminated gas enters tangentially from the top; the gas is propelled down a helical pathway, and the dust particles are deposited by means of centrifugal force onto the wall of the scrubber.

Deactivation Reduction in catalyst activity by the deposition of contaminants (e.g., coke, metals) during a process.

Dealkylation The removal of an alkyl group from aromatic compounds.

Deasphaltened oil The fraction of petroleum remaining after the asphaltenes have been removed.

Deasphaltening Removal of a solid powdery asphaltene fraction from petroleum by the addition of the low-boiling liquid hydrocarbons such as *n*-pentane or *n*-heptane under ambient conditions.

Deasphalting The removal of the asphalt fraction (tacky, semisolid) from petroleum (as occurs in a refinery asphalt plant) by the addition of liquid propane or liquid butane under pressure.

Debutanization Distillation to separate butane and lighter components from higher boiling components.

Decant oil The highest boiling product from a catalytic cracker; also referred to as *slurry oil, clarified oil*, or *bottoms*.

Decarbonizing A thermal conversion process designed to maximize coker gas oil production and minimize coke and gasoline yields; operated at lower temperatures and pressures than delayed coking (*q.v.*).

Decoking Removal of petroleum coke from equipment such as coking drums. Hydraulic decoking uses high velocity water streams.

Decolorizing Removal of suspended, colloidal, and dissolved impurities from liquid petroleum products by filtering, adsorption, chemical treatment, distillation, bleaching, etc.

Deethanization Distillation to separate ethane and lighter components from propane and higher boiling components; also called *deethanation*.

Dehydrating agents Substances capable of removing water (drying, *q.v.*) or the elements of water from another substance.

Dehydrocyclization Any process by which both dehydrogenation and cyclization reactions occur.

Dehydrogenation The removal of hydrogen from a chemical compound; for example, the removal of two hydrogen atoms from butane to make butene(s) as well as the removal of additional hydrogen to produce butadiene.

Delayed coking A coking process in which thermal reactions are allowed to proceed to completion to produce gaseous, liquid, and solid (coke) products.

Demethanization The process of distillation in which methane is separated from the higher boiling components; also called *demethanation*.

Density The mass (or weight) of a unit volume of any substance at a specified temperature. *See also* **Specific gravity**.

Deoiling Reduction in quantity of liquid oil entrained in solid wax by draining (sweating) or by a selective solvent. *See also* **MEK deoiling**.

Depentanizer A fractionating column for the removal of pentane and lighter fractions from a mixture of hydrocarbons.

Depropanization Distillation in which lighter components are separated from butanes and higher boiling material; also called *depropanation*.

Desalting Removal of mineral salts (mostly chlorides) from crude oils.

Desorption The reverse process of adsorption whereby adsorbed matter is removed from the adsorbent; also used as the reverse of absorption (*q.v.*).

Desulfurization The removal of sulfur or sulfur compounds from a feedstock.

Detergent oil A lubricating oil possessing special sludge-dispersing properties for use in internal combustion engines.

Dewaxing *See* **Solvent dewaxing**.

Devolatilized fuel Smokeless fuel; coke that has been reheated to remove all of the volatile material.

Diesel cycle A repeated succession of operations representing the idealized working behavior of the fluids in a diesel engine.

Diesel fuel   Fuel used for internal combustion in diesel engines; usually the fraction that distills after kerosene.

Diesel index   An approximation of the cetane number (*q.v.*) of diesel fuel (*q.v.*) calculated from the density (*q.v.*) and aniline point (*q.v.*).

Diesel knock   The result of a delayed period of ignition and the accumulation of diesel fuel in the engine.

Distillation   A process for separating liquids with different boiling points.

Distillation curve   *See* **Distillation profile**.

Distillation loss   The difference, in a laboratory distillation, between the volume of liquid originally introduced into the distilling flask and the sum of the residue and condensate recovered.

Distillation profile   The distillation characteristics of petroleum or a petroleum products showing the temperature and the percent distilled.

Distillation range   The difference between the temperatures at the initial boiling point and at the endpoint, as obtained by the distillation test.

Doctor solution   A solution of sodium plumbite used to treat gasoline or other light petroleum distillates to remove mercaptan sulfur. *See also* **Doctor test**.

Doctor sweetening   A process for sweetening gasoline, solvents, and kerosene by converting mercaptans to disulfides using sodium plumbite and sulfur.

Doctor test   A test used for the detection of compounds in light petroleum distillates that react with sodium plumbite. *See also* **Doctor solution**.

Domestic heating oil   *See* **No. 2 fuel oil**.

Donor solvent process   A conversion process in which hydrogen donor solvent is used in place of or to augment hydrogen.

Downcomer   A means of conveying liquid from one tray to the next below in a bubble tray column (*q.v.*).

Dropping point   The temperature at which grease passes from a semisolid to a liquid state under prescribed conditions.

Drying   Removal of a solvent or water from a chemical substance; also referred to as the removal of solvent from a liquid or suspension.

Dry gas   A gas that does not contain fractions that may easily condense under normal atmospheric conditions.

Dry point   The temperature at which the last drop of petroleum fluid evaporates in a distillation test.

Dualayer distillate process   A process for removing mercaptans and oxygenated compounds from distillate fuel oils and similar products by using a combination of treatment with concentrated caustic solution and electrical precipitation of the impurities.

Dualayer gasoline process   A process for extracting mercaptans and other objectionable acidic compounds from petroleum distillates. *See also* **Dualayer solution**.

Dualayer solution   A solution that consists of concentrated potassium or sodium hydroxide containing a solubilizer. *See also* **Dualayer gasoline process**.

Dubbs cracking   A continuous, liquid-phase thermal cracking process no longer used.

Ebullated bed   A process in which the catalyst bed is in a suspended state in the reactor by means of a feedstock recirculation pump that pumps the feedstock upward at sufficient speed to expand the catalyst bed at approximately 35% above the settled level.

Edeleanu process   A process for refining oils at low temperature with liquid sulfur dioxide (SO_2) or with liquid sulfur dioxide and benzene; applicable to the recovery of aromatic concentrates from naphthas and heavier petroleum distillates.

Effective viscosity   *See* **Apparent viscosity**.

Electric desalting   A continuous process to remove inorganic salts and other impurities from crude oil by settling out in an electrostatic field.

Electrical precipitation   A process using an electric field to improve the separation of hydrocarbon reagent dispersions. May be used in chemical treating processes on a wide variety of refinery stocks.

Electrofining   A process for contacting a light hydrocarbon stream with a treating agent (acid, caustic, doctor, etc.), then assisting the action of separation of the chemical phase from the hydrocarbon phase by an electrostatic field.

Electrolytic mercaptan process   A process in which aqueous caustic solution is used to extract mercaptans from refinery streams.

Electrostatic precipitators   Devices used to trap fine dust particles (usually in the size range 30–60 μm) that operate on the principle of imparting an electric charge to particles in an incoming airstream, which are then collected on an oppositely charged plate across a high voltage field.

Engler distillation   A standard test for determining the volatility characteristics of a gasoline by measuring the percent distilled at various specified temperatures.

Enhanced oil recovery   Petroleum recovery following recovery by conventional (i.e., primary and/or secondary) methods (*q.v.*).

Entrained bed   A bed of solid particles suspended in a fluid (liquid or gas) at such a rate that some of the solid is carried over (entrained) by the fluid.

Ethanol   *See* **Ethyl alcohol**.

Ethyl alcohol   An inflammable organic compound (C_2H_5OH) formed during fermentation of sugars; used as an intoxicant and as a fuel. Also called *ethanol* or *grain alcohol*.

Evaporation   A process for concentrating nonvolatile solids in a solution by boiling off the liquid portion of the waste stream.

Expanding clays   Clays that expand or swell on contact with water, e.g., montmorillonite.

Explosive limits   The limits of percentage composition of mixtures of gases and air within which an explosion takes place when the mixture is ignited.

Extractive distillation   The separation of different components of mixtures that have similar vapor pressures by flowing a relatively high boiling solvent that is selective for one of the components in the feed down a distillation column as the distillation proceeds; the selective solvent scrubs the soluble component from the vapor.

Fabric filters   Filters made from fabric materials and used for removing particulate matter from gas streams. *See* **Baghouse**.

Fat oil   The bottom or enriched oil drawn from the absorber as opposed to lean oil (*q.v.*).

Faujasite   A naturally occurring silica-alumina (SiO_2-Al_2O_3) mineral.

FCC   Fluid catalytic cracking (*q.v.*).

Feedstock   Petroleum as it is fed to the refinery; a refinery product that is used as the raw material for another process; the term is also generally applied to raw materials used in other industrial processes.

Filtration   The use of an impassable barrier to collect solids but allow liquids to pass.

Fire point   The lowest temperature at which, under specified conditions in standardized apparatus, a petroleum product vaporizes sufficiently rapidly to form above its surface an air–vapor mixture that burns continuously when ignited by a small flame.

Fischer-Tropsch process   A process for synthesizing hydrocarbons and oxygenated chemicals from a mixture of hydrogen and carbon monoxide.

Fixed bed   A stationary bed (of catalyst) to accomplish a process. *See* **Fluid bed**.

Flammability range   The range of temperature over which a chemical is flammable.

Flammable   Capable of burning readily.

Flammable liquid   A liquid with a flash point below 37.8°C (100°F).

Flammable solid   A solid that can ignite from friction or from heat remaining from its manufacture or may cause a serious hazard if ignited.

Flash point   The lowest temperature to which the product must be heated under specified conditions to give off sufficient vapor to form a mixture with air that can be ignited momentarily by a flame.

Flexicoking   A modification of the fluid coking process that includes a gasifier adjoining the burner/regenerator to convert excess coke to a clean fuel gas.

Floc point   The temperature at which wax or solids separate as a definite floc.

Flue gas   Gas from the combustion of fuel, the heating value of which has been substantially spent and which is therefore discarded to the flue or stack.

Fluidbed   A bed of catalyst that is agitated by an upward-passing gas in such a manner that the catalyst particles simulate the movement of a fluid and that has the characteristics associated with a true liquid. Also called *fluidized bed*. Cf. **Fixed bed**.

Fluid catalytic cracking   Cracking in the presence of a fluidized bed of catalyst.

Fluid coking   A continuous fluidized solids process that cracks feed thermally over heated coke particles in a reactor vessel to gas, liquid products, and coke.

Fly ash   Particulate matter produced from mineral matter in coal that is converted during combustion to finely divided inorganic material and emerges from the combustor in the gases.

Foots oil   The oil sweated out of slack wax. Named from the fact that the oil goes to the foot, or bottom, of the pan during the sweating operation.

Fossil fuel   A gaseous, liquid, or solid fuel material formed in the ground by chemical and physical changes (diagenesis, *q.v.*) in plant and animal residues over geological time; natural gas, petroleum, coal, and oil shale.

Fractional composition   The composition of petroleum as determined by fractionation (separation) methods.

Fractional distillation   The separation of the components of a liquid mixture by vaporizing and collecting the fractions, or cuts, that condense in different temperature ranges.

Fractionating column   A column arranged to separate various fractions of petroleum by a single distillation; may be tapped at different points along its length to separate various fractions in the order of their boiling points.

Fractionation   The separation of petroleum into the constituent fractions using solvent or adsorbent methods; chemical agents such as sulfuric acid may also be used.

Frasch process   A process formerly used for removing sulfur by distilling oil in the presence of copper oxide.

Fuel oil   A distillate product that covers a wide range of properties. Also called **heating oil**. *See also* **No. 1 to No. 4 Fuel oils**.

Fuller's earth   A clay that has high adsorptive capacity for removing color from oils. Attapulgus clay is a widely used fuller's earth.

Functional group   The portion of a molecule that is characteristic of a family of compounds and determines the properties of these compounds.

Furfural extraction   A single-solvent process in which furfural is used to remove aromatic, naphthenic, olefinic, and unstable hydrocarbons from a lubricating oil charge stock.

Furnace oil   A distillate fuel primarily intended for use in domestic heating equipment.

Gas cap   A part of a hydrocarbon reservoir at the top that will produce only gas.

Gasohol   A term for motor vehicle fuel comprising 80–90% unleaded gasoline and 10–20% ethanol. *See also* **Ethyl alcohol**.

Gas oil   A petroleum distillate with a viscosity and boiling range between those of kerosene and lubricating oil.

Gas/oil ratio   Ratio of the number of cubic feet of gas measured at atmospheric (standard) conditions to barrels of produced oil measured at stocktank conditions.

Gasoline   Fuel for the internal combustion engine that is commonly, but improperly, referred to simply as *gas*.

Gasoline blending   Combining the components that make up motor gasoline.

Gas reversion   A combination of thermal cracking or reforming of naphtha with thermal polymerization or alkylation of hydrocarbon gases carried out in the same reaction zone.

Gilsonite   An asphaltite (*q.v.*) that is > 90% bitumen.

Glance pitch   An asphaltite (*q.v.*).

Glycol-amine gas treating   A continuous, regenerative process to simultaneously dehydrate and remove acid gases from natural gas or refinery gas.

Grahamite   An asphaltite.

Gray clay treating   A fixed-bed (*q.v.*), usually fuller's earth (*q.v.*), vapor-phase treating process to selectively polymerize unsaturated gum-forming constituents (diolefins) in thermally cracked gasoline.

Grain alcohol   *See* **Ethyl alcohol**.

Gravity drainage   The movement of oil in a reservoir that results from the force of gravity.

Gravity segregation   Partial separation of fluids in a reservoir caused by the gravity force acting on differences in density.

Greenhouse effect   Warming of the earth due to entrapment of the sun's energy by the atmosphere.

Greenhouse gases   Gases that contribute to the greenhouse effect (*q.v.*)

Guard bed   A bed of disposal adsorbent used to protect process catalysts from contamination by feedstock constituents.

Gulf HDS process   A fixed-bed process for the catalytic hydrocracking of heavy stocks to lower boiling distillates with accompanying desulfurization.

Gulfining   A catalytic hydrogen treating process for cracked and straight-run distillates and fuel oils to reduce sulfur content; improve carbon residue, color, and general stability; and effect a slight increase in gravity.

Gum   An insoluble tacky semisolid material formed as a result of the storage instability and/or thermal instability of petroleum and petroleum products.

Heating oil   *See* **Fuel oil**.

Heavy ends   The highest boiling portion of a petroleum fraction. *See also* **Light ends**.

Heavy fuel oil   Fuel oil having a high density and viscosity; generally residual fuel oil such as No. 5 and No. 6 fuel oil (*q.v.*).

Heavy oil Petroleum having a gravity of less than 20°API.

Heavy petroleum *See* **Heavy oil**.

Heteroatom compounds Chemical compounds that contain nitrogen and/or oxygen and/or sulfur and/or metals bound within their molecular structure(s).

HF alkylation An alkylation process whereby olefins (C_3, C_4, C_5) are combined with isobutane in the presence of hydrofluoric acid catalyst.

Hortonsphere A spherical pressure-type tank used to store volatile liquids that prevents the excessive evaporation loss that occurs when such products are placed in conventional storage tanks.

Hot filtration test A test for the stability of a petroleum product.

Hot spot An area of a vessel or line wall appreciably above normal operating temperature, usually as a result of the deterioration of an internal insulating liner that exposes the line or vessel shell to the temperature of its contents.

Houdresid catalytic cracking A continuous moving-bed process for catalytically cracking reduced crude oil to produce high octane gasoline and light distillate fuels.

Houdriflow catalytic cracking A continuous moving-bed catalytic cracking process employing an integrated single vessel for the reactor and regenerator kiln.

Houdriforming A continuous catalytic reforming process for producing aromatic concentrates and high octane gasoline from low octane straight naphthas.

Houdry butane dehydrogenation A catalytic process for dehydrogenating light hydrocarbons to their corresponding mono- or diolefins.

Houdry fixed-bed catalytic cracking A cyclical regenerable process for cracking of distillates.

Houdry hydrocracking A catalytic process combining cracking and desulfurization in the presence of hydrogen.

Hydraulic fracturing The opening of fractures in a reservoir by high pressure, high volume injection of liquids through an injection well.

Hydrocarbon compounds Chemical compounds containing only carbon and hydrogen.

Hydrocarbon resources Resources such as petroleum and natural gas that can produce naturally occurring hydrocarbons without the application of conversion processes.

Hydrocarbon-producing resource A resource such as coal and oil shale (kerogen) that produces derived hydrocarbons by the application of conversion processes; the hydrocarbons so produced are not naturally occurring materials.

Hydroconversion A term often applied to hydrocracking (*q.v.*).

Hydrocracking A catalytic high pressure, high temperature process for the conversion of petroleum feedstocks in the presence of fresh and recycled hydrogen; carbon–carbon bonds are cleaved in addition to the removal of heteroatomic species.

Hydrocracking catalyst A catalyst used for hydrocracking that typically contains separate hydrogenation and cracking functions.

Hydrodenitrogenation The removal of nitrogen by hydrotreating (*q.v.*).

Hydrodesulfurization The removal of sulfur by hydrotreating (*q.v.*).

Hydrofining A fixed-bed catalytic process to desulfurize and hydrogenate a wide range of charge stocks from gases through waxes.

Hydroforming A process in which naphthas are passed over a catalyst at elevated temperatures and moderate pressures, in the presence of added hydrogen or hydrogen-containing gases, to form high octane motor fuel or aromatics.

Hydrogen addition An upgrading process in the presence of hydrogen, e.g. hydrocracking. *See* **Hydrogenation**.

Hydrogen blistering Blistering of steel caused by trapped molecular hydrogen formed as atomic hydrogen during corrosion of steel by hydrogen sulfide.

Hydrogenation The chemical addition of hydrogen to a material. In nondestructive hydrogenation, hydrogen is added to a molecule only if, and where, unsaturation with respect to hydrogen exists.

Hydrogen transfer The transfer of inherent hydrogen within the feedstock constituents and products during processing.

Hydroprocessing A term often equally applied to hydrotreating (*q.v.*) and to hydrocracking (*q.v.*); also often collectively applied to both.

Hydrotreating The removal of heteroatomic (nitrogen, oxygen, and sulfur) species by treatment of a feedstock or product at relatively low temperatures in the presence of hydrogen.

Hydrovisbreaking A noncatalytic process, conducted under similar conditions to visbreaking, which involves treatment with hydrogen to reduce the viscosity of the feedstock and produce more stable products than is possible with visbreaking.

Hydropyrolysis A short residence time high temperature process using hydrogen.

Hyperforming A catalytic hydrogenation process for improving the octane number of naphthas through removal of sulfur and nitrogen compounds.

Hypochlorite sweetening The oxidation of mercaptans in a sour stock by agitation with aqueous, alkaline hypochlorite solution; used where avoidance of free sulfur addition is desired because of stringent copper strip requirements and minimum expense is not the primary object.

Ignitability Characteristic of liquids whose vapors are likely to ignite in the presence of an ignition source; also a characteristic of nonliquids that may catch fire from friction or contact with water and that burn vigorously.

Illuminating oil Oil used for lighting purposes.

Immiscible The property of two or more fluids that do not have complete mutual solubility and coexist as separate phases.

Inhibitor A substance, the presence of which, in small amounts, in a petroleum product prevents or retards undesirable chemical changes from taking place in the product or in the condition of the equipment in which the product is used.

Inhibitor sweetening A treating process to sweeten gasoline of low mercaptan content, using a phenylenediamine type of inhibitor, air, and caustic.

Initial boiling point The recorded temperature when the first drop of liquid falls from the end of the condenser.

Initial vapor pressure The vapor pressure of a liquid of a specified temperature and *zero percent evaporated*.

Instability The inability of a petroleum product to exist for periods of time without change.

Incompatibility The *immiscibility* of petroleum products and also of different crude oils; often reflected in the formation of a separate phase after mixing and/or storage.

In situ combustion Combustion of oil in the reservoir, sustained by continuous air injection carried out to displace unburned oil toward producing wells.

Iodine number   A measure of the iodine absorption by oil under standard conditions; used to indicate the quantity of unsaturated compounds present. Also called *iodine value*.

Ion exchange   A means of removing cations or anions from solution onto a solid resin.

Isocracking   A hydrocracking process for conversion of hydrocarbons that operates at relatively low temperatures and pressures in the presence of hydrogen and a catalyst to produce more valuable, lower boiling products.

Isoforming   A process in which olefinic naphtha is contacted with an alumina catalyst at high temperature and low pressure to produce isomers of higher octane number.

Iso-Kel process   A fixed-bed, vapor-phase isomerization process using a precious metal catalyst and external hydrogen.

Isomate process   A continuous, nonregenerative process for isomerizing C_5–C_8 normal paraffinic hydrocarbons, using a mixed aluminum chloride–hydrocarbon catalyst with anhydrous hydrochloric acid as a promoter.

Isomerate process   A fixed-bed isomerization process to convert pentane, heptane, and heptane to high octane blending stocks.

Isomerization   The conversion of a *normal* (straight-chain) paraffin hydrocarbon into an isoparaffin (branched-chain) hydrocarbon having the same atomic composition.

Iso-Plus Houdriforming   A combination process using a conventional Houdriformer operated at moderate severity, in conjunction with one of three possible alternatives including the use of an aromatic recovery unit or a thermal reformer. *See also* **Houdriforming**.

Jet fuel   Fuel meeting the required properties for use in jet engines and aircraft turbine engines.

Kaolinite   A clay mineral formed by hydrothermal activity at the time of rock formation or by chemical weathering of rocks with high feldspar content; usually associated with intrusive granite rocks with high feldspar content.

Kata-condensed aromatic compounds   Compounds based on linear condensed aromatic hydrocarbon systems, e.g., anthracene and naphthacene (tetracene).

Kerogen   A complex carbonaceous (organic) material that occurs in sedimentary rocks and shales; generally insoluble in common organic solvents.

Kerosene   A fraction of petroleum that was initially sought as an illuminant in lamps; a precursor to diesel fuel. Also spelled *kerosine*.

K-factor   *See* **Characterization factor**.

Kinematic viscosity   The ratio of viscosity (*q.v.*) to density, both measured at the same temperature.

Knock   The noise associated with self-ignition of a portion of the fuel–air mixture ahead of the advancing flame front. *See also* **Antiknock, Antiknock agent**.

Lamp burning   A test of burning oils in which the oil is burned in a standard lamp under specified conditions in order to observe the steadiness of the flame, the degree of encrustation of the wick, and the rate of consumption of the kerosene.

Lamp oil   *See* **Kerosene**.

Leaded gasoline   Gasoline containing tetraethyllead or other organometallic lead antiknock compounds.

Lean gas   The residual gas from the absorber after the condensable gasoline has been removed from the *wet gas* (*q.v.*).

Lean oil   Absorption oil from which gasoline fractions have been removed; oil leaving the stripper in a natural gasoline plant.

Lewis acid A chemical species that can accept an electron pair from a base.

Lewis base A chemical species that can donate an electron pair.

Light ends The lower boiling components of a mixture of hydrocarbons. *See also* **Heavy ends**; **Light hydrocarbons**.

Light hydrocarbons Hydrocarbons with molecular weights less than that of heptane (C_7H_{16}).

Light oil The products distilled or processed from crude oil up to, but not including, the first lubricating oil distillate.

Light petroleum Petroleum having an API gravity greater than 20°.

Ligroine (ligroin) A saturated petroleum naphtha boiling in the range of 20–135°C (68–275°F) and suitable for general use as a solvent. Also called *benzine* or *petroleum ether*.

Linde copper sweetening A process for treating gasoline and distillates with a slurry of clay and cupric chloride.

Liquid petrolatum *See* **White oil**.

Liquefied petroleum gas Propane, butane, or mixtures thereof, gaseous at atmospheric temperature and pressure, held in the liquid state by pressure to facilitate storage, transport, and handling.

Liquid sulfur dioxide–benzene process A mixed-solvent process for treating lubricating oil stocks to improve viscosity index; also used for dewaxing.

Lithology The geological characteristics of the reservoir rock.

Live steam Steam coming directly from a boiler before being utilized for power or heat.

Liver The intermediate layer of dark-colored, oily material, insoluble in weak acid and in oil, that is formed when acid sludge is hydrolyzed.

Lube *See* **Lubricating oil**.

Lube cut A fraction of crude oil of suitable boiling range and viscosity to yield lubricating oil when completely refined. Also referred to as *lube oil distillates* or *lube stock*.

Lubricating oil A fluid lubricant used to reduce friction between bearing surfaces.

Mahogany acids Oil-soluble sulfonic acids formed by the action of sulfuric acid on petroleum distillates. They may be converted to their sodium soaps (mahogany soaps) and extracted from the oil with alcohol for use in the manufacture of soluble oils, rust preventives, and special greases. The calcium and barium soaps of these acids are used as detergent additives in motor oils. *See also* **Brown acids** *and* **Sulfonic acids**.

Maltenes That fraction of petroleum that is soluble in, for example, pentane or heptane; deasphaltened oil (*q.v.*); also the term arbitrarily assigned to the pentane-soluble portion of petroleum that is relatively high boiling (> 300°C, 760 mm). *See also* **Petrolenes**.

Marine engine oil Oil used as a crankcase oil in marine engines.

Marine gasoline Fuel for motors in marine service.

Marine sediment The organic biomass from which petroleum is derived.

Marsh An area of spongy waterlogged ground with large numbers of surface water pools. Marshes usually result from (1) an impermeable underlying bedrock; (2) surface deposits of glacial boulder clay; (3) a basin-like topography from which natural drainage is poor; (4) very heavy rainfall in conjunction with a correspondingly low evaporation rate; (5) low-lying land, particularly at estuarine sites at or below sea level.

Mayonnaise Low temperature sludge; a black, brown, or gray deposit having a soft, mayonnaise-like consistency.

Methanol *See* **Methyl alcohol**.

Medicinal oil Highly refined, colorless, tasteless, and odorless petroleum oil used as a medicine in the nature of an internal lubricant; sometimes called *liquid paraffin*.

MEK (Methyl ethyl ketone) A colorless liquid ($CH_3COCH_2CH_3$) used as a solvent, as a chemical intermediate, and in the manufacture of lacquers, celluloid, and varnish removers.

MEK deoiling A wax-deoiling process in which the solvent is generally a mixture of methyl ethyl ketone and toluene.

MEK dewaxing A continuous solvent dewaxing process in which the solvent is generally a mixture of methyl ethyl ketone and toluene.

Mercapsol process A regenerative process for extracting mercaptans, utilizing aqueous sodium (or potassium) hydroxide containing mixed cresols as solubility promoters.

Mercaptans Organic compounds having the general formula R—SH.

Methyl alcohol A colorless, volatile, inflammable, and poisonous alcohol (CH_3OH) traditionally formed by destructive distillation of wood or, more recently, as a result of synthetic distillation in chemical plants. Also called *methanol* or *wood alcohol*.

Methyl *t*-butyl ether (MTBE) An ether added to gasoline to improve its octane rating and to decrease gaseous emissions. *See also* **Oxygenate**.

Methyl ethyl ketone *See* **MEK**.

Mica A complex aluminum silicate mineral that is transparent, tough, flexible, and elastic.

Micelle The structural entity in which asphaltenes are dispersed in petroleum.

Microcarbon residue The carbon residue determined using a thermogravimetric method. *See also* **Carbon residue**.

Microcrystalline wax Wax extracted from certain petroleum residua and having a finer and less apparent crystalline structure than paraffin wax.

Mid-boiling point The temperature at which approximately 50% of a material has distilled under specific conditions.

Middle distillate Distillate boiling between the kerosene and lubricating oil fractions.

Mineral oil A term for petroleum introduced in the nineteenth century as a means of differentiating petroleum (rock oil) from whale oil, which at the time was the predominant illuminant for oil lamps.

Minerals Naturally occurring inorganic solids with well-defined crystalline structures.

Mineral seal oil A distillate fraction boiling between kerosene and gas oil.

Mineral wax Yellow to dark brown solid substances that occur naturally and are composed largely of paraffins. Usually found associated with considerable mineral matter, as a filling in veins and fissures, or as an interstitial material in porous rocks.

Mitigation Identification, evaluation, and cessation of potential impacts of a process product or by-product.

Mixed-phase cracking The thermal decomposition of higher boiling hydrocarbons to gasoline components.

Modified naphtha insolubles (MNI) An insoluble fraction obtained by adding naphtha to petroleum; usually adding paraffinic constituents modify the naphtha. The fraction might be equated to asphaltenes if the naphtha is equivalent to *n*-heptane, but usually it is not.

Molecular sieve A synthetic zeolite mineral having pores of uniform size. It is capable of separating molecules on the basis of their size, structure, or both, by absorption or sieving.

Motor octane method A test for determining the knock rating of fuels for use in spark ignition engines. *See also* **Research octane method**.

Moving-bed catalytic cracking A cracking process in which the catalyst is continuously cycled between the reactor and the regenerator.

MTBE *See* **Methyl *t*-butyl ether**.

Naft Pre-Christian era (Greek) term for naphtha (*q.v.*).

Napalm A thickened gasoline used as an incendiary medium that adheres to the surface it strikes.

Naphtha A generic term applied to refined, partly refined, or unrefined petroleum products and liquid products of natural gas, the majority of which distill below 240°C (464°F); the volatile fraction of petroleum that is used as a solvent or as a precursor to gasoline.

Naphthenes Cycloparaffins.

Native asphalt *See* **Bitumen**.

Natural asphalt *See* **Bitumen**.

Natural gas The naturally occurring gaseous constituents that are found in many petroleum reservoirs; also found in reservoirs in which it is the sole occupant.

Natural gas liquids (NGL) The hydrocarbon liquids that condense during the processing of hydrocarbon gases that are produced from oil or gas reservoir. *See also* **Natural gasoline**.

Natural gasoline A mixture of liquid hydrocarbons extracted from natural gas (*q.v.*) suitable for blending with refinery gasoline.

Natural gasoline plant A plant for the extraction of fluid hydrocarbon, such as gasoline and liquefied petroleum gas, from natural gas.

Neutralization A process for reducing the acidity or alkalinity of a waste stream by mixing acids and bases to produce a neutral solution. Also known as *pH adjustment*.

Neutral oil A distillate lubricating oil with a viscosity usually not above 200 s at 100°F.

Neutralization number The weight, in milligrams, of potassium hydroxide needed to neutralize the acid in 1 g of oil; an indication of the acidity of an oil.

Non-Newtonian fluid A fluid that exhibits a change of viscosity with flow rate.

No. 1 Fuel oil An oil very similar to kerosene (*q.v.*) and used in burners where vaporization before burning is usually required and a clean flame is specified.

No. 2 Fuel oil An oil with properties similar to those of diesel fuel and heavy jet fuel; used in burners where complete vaporization is not required before burning. Also called *domestic heating oil*.

No. 4 Fuel oil A light industrial heating oil used where preheating is not required for handling or burning. There are two grades of No. 4 fuel oil, differing in safety (flash point) and flow (viscosity) properties.

No. 5 Fuel oil A heavy industrial fuel oil that requires preheating before burning.

No. 6 Fuel oil A heavy fuel oil known as *Bunker C oil* when it is used to fuel ocean going vessels. Preheating is always required for burning this oil.

Octane barrel yield A measure used to evaluate fluid catalytic cracking processes; defined as (RON + MON)/2 times the gasoline yield, where RON is the research octane number and MON is the motor octane number.

Octane number A number indicating the antiknock characteristics of gasoline.

Oils That portion of the maltenes (*q.v.*) that is not adsorbed by a surface-active material such as clay or alumina.

Oil sand *See* **Tar sand**.

Oil shale A fine-grained impervious sedimentary rock that contains an organic material called kerogen.

Organic sedimentary rocks Rocks containing organic material such as residues of plant and animal remains.

Overhead That portion of the feedstock that is vaporized and removed during distillation.

Oxidation A process that can be used for the treatment of a variety of inorganic and organic substances.

Oxidized asphalt *See* **Air-blown asphalt**.

Ozokerite (ozocerite) A naturally occurring wax; when refined also known as *ceresin*.

Oxygenate An oxygen-containing compound that is blended into gasoline to improve its octane number and to decrease gaseous emissions.

Pale oil A lubricating or process oil refined until its color, by transmitted light, is straw-colored to pale yellow.

Paraffinum liquidum *See* **Liquid petrolatum**.

Paraffin wax The colorless, translucent, highly crystalline material obtained from the light lubricating fractions of paraffinic crude oils (wax distillates).

Particle density The density of solid particles.

Particulate matter Particles in the atmosphere or in a gas stream that may be organic or inorganic and originate from a wide variety of sources and processes.

Particle size distribution The particle size distribution (of a catalyst sample) expressed as a percent of the whole.

Penex process A continuous, nonregenerative process for isomerization of C_5 and/or C_6 fractions in the presence of hydrogen (from reforming) and a platinum catalyst.

Pentafining A pentane isomerization process using a regenerable platinum catalyst on a silica-alumina support and requiring outside hydrogen.

Pepper sludge The fine particles of sludge produced in acid treating that may remain in suspension.

Pericondensed aromatic compounds Compounds based on angular condensed aromatic hydrocarbon systems, e.g., phenanthrene, chrysene, picene, etc..

Permeability The ease of flow of water through rock.

Petrol A term commonly used in some countries for *gasoline*.

Petrolatum A semisolid product, ranging from white to yellow in color, produced during refining of residual stocks. *See also* **Petroleum jelly**.

Petrolenes The term applied to that part of the pentane-soluble or heptane-soluble material that is low boiling [$< 300°C$ ($< 570°F$), 760 mm] and can be distilled without thermal decomposition. *See also* **Maltenes**.

Petroleum Crude oil. A naturally occurring mixture of gaseous, liquid, and solid hydrocarbon compounds usually found trapped deep underground beneath impermeable cap rock and above a lower dome of sedimentary rock such as shale. Most petroleum reservoirs occur in sedimentary rocks of marine, deltaic, or estuarine origin.

Petroleum asphalt *See* **Asphalt**.

Petroleum ether *See* **Ligroine**.

Petroleum jelly A translucent, yellowish to amber or white hydrocarbon substance (m.p. 38–54°C) having almost no odor or taste, derived from petroleum and used principally in medicine and pharmaceuticals as a protective dressing and as a substitute for fats in ointments and cosmetics; also used in many types of polishes and in lubricating greases, rust preventives, and modeling clay; obtained by dewaxing heavy lubricating oil stocks.

Petroleum refinery   *See* **Refinery.**

Petroleum refining   A complex sequence of events that result in the production of a variety of products.

Petroporphyrins   *See* **Porphyrins.**

Phase separation   The formation of a separate phase that is usually the prelude to coke formation during a thermal process; the formation of a separate phase as a result of the instability or incompatibility of petroleum and petroleum products.

pH adjustment   Neutralization.

Phosphoric acid polymerization   A process using a phosphoric acid catalyst to convert propene, butene, or both, to gasoline or petrochemical polymers.

Pipe still   A still in which heat is applied to the oil while it is being pumped through a coil or pipe arranged in a suitable firebox.

Pipestill gas   The most volatile fraction that contains most of the gases that are generally dissolved in the crude. Also known as *pipestill light ends.*

Pipestill light ends   *See* **Pipestill gas.**

Pitch   The nonvolatile, brown to black, semisolid to solid viscous product from the destructive distillation of many bituminous or other organic materials, especially coal.

Platforming   A reforming process using a platinum-containing catalyst on an alumina base.

PNA   Polynuclear aromatic compound (*q.v.*).

Polar aromatics   Resins; the constituents of petroleum that are predominantly aromatic in character and contain polar (nitrogen, oxygen, and sulfur) functions in their molecular structure(s).

Polyforming   A process charging both C_3 and C_4 gases with naphtha or gas oil under thermal conditions to produce gasoline.

Polymer gasoline   The product of polymerization of gaseous hydrocarbons to hydrocarbons boiling in the gasoline range.

Polymerization   The combination of two olefin molecules to form a higher molecular weight paraffin.

Polynuclear aromatic compound   An aromatic compound having two or more fused benzene rings, e.g., naphthalene, phenanthrene.

Polysulfide treating   A chemical treatment used to remove elemental sulfur from refinery liquids by contacting them with a nonregenerable solution of sodium polysulfide.

PONA analysis   A method of analysis for paraffins (P), olefins (O), naphthenes (N), and aromatics (A).

Pore diameter   The average pore size of a solid material such as a catalyst.

Pore space   A small hole in reservoir rock that contains fluid or fluids. 4 in. cube of reservoir rock may contain millions of interconnected pore spaces.

Pore volume   Total volume of all pores and fractures in a reservoir or part of a reservoir; also applied to catalyst samples.

Porosity   The percentage of rock volume available to contain water or other fluid.

Porphyrins   Organometallic constituents of petroleum that contain vanadium or nickel; the degradation products of chlorophyll that became included in the protopetroleum.

Possible reserves   Petroleum reserves about which there is a greater degree of uncertainty than that associated with potential reserves but about which there is some information.

Potential reserves   A measure of petroleum reserves based upon geological information about the types of sediments where such resources are likely to occur and considered to represent an educated guess.

Pour point   The lowest temperature at which oil will pour or flow when it is chilled without disturbance under definite conditions.

Powerforming   A fixed-bed naphtha-reforming process using a regenerable platinum catalyst.

Precipitation number   The number of milliliters of precipitate formed when 10 mL of lubricating oil is mixed with 90 mL of petroleum naphtha of a definite quality and centrifuged under definitely prescribed conditions.

Primary oil recovery   Oil recovery utilizing only naturally occurring forces.

Primary structure   The chemical arrangement of atoms in a molecule.

Probable reserves   Mineral reserves that are nearly certain but about which a slight doubt exists.

Propane asphalt   *See* **Solvent asphalt**.

Propane deasphalting   Solvent deasphalting using propane as the solvent.

Propane decarbonizing   A solvent extraction process used to recover catalytic cracking feed from heavy fuel residues.

Propane dewaxing   A process for dewaxing lubricating oils in which propane serves as solvent.

Propane fractionation   A continuous extraction process employing liquid propane as the solvent; a variant of propane deasphalting (*q.v.*).

Protopetroleum   A generic term used to indicate the initial product formed by changes that have occurred to the precursors of petroleum.

Proved reserves   Mineral reserves that have been positively identified as recoverable with current technology.

Pyrobitumen   *See* **Asphaltoid**.

Pyrolysis   Exposure of a feedstock to high temperatures in an oxygen-poor environment.

Pyrophoric   Describing substances that catch fire spontaneously in air without an ignition source.

Quadrillion 1×10^{15}.

Quench   The sudden cooling of hot material discharging from a thermal reactor.

Raffinate   That portion of the oil that remains undissolved in a solvent refining process.

Ramsbottom carbon residue   *See* **Carbon residue**.

Raw materials   Minerals extracted from the earth prior to any refining or treating.

Recycle ratio   The volume of recycle stock per volume of fresh feed; often expressed as the volume of recycle divided by the total charge.

Recycle stock   The portion of a feedstock that has passed through a refining process and is recirculated through the process.

Recycling   The use or reuse of chemical waste as an effective substitute for a commercial product or as an ingredient or feedstock in an industrial process.

Reduced crude A residual product remaining after the removal, by distillation or other means, of an appreciable quantity of the more volatile components of crude oil.

Refinery A series of integrated unit processes by which petroleum can be converted to a slate of useful (salable) products.

Refinery gas A gas (or a gaseous mixture) produced as a result of refining operations.

Refining The process(es) by which petroleum is distilled and/or converted by application of physical and chemical processes to form a variety of products.

Reformate The liquid product of a reforming process.

Reformed gasoline Gasoline made by a reforming process.

Reforming The conversion of hydrocarbons with low octane numbers (*q.v.*) into hydrocarbons having higher octane numbers; e.g., the conversion of an *n*-paraffin into an isoparaffin.

Reformulated gasoline (RFG) Gasoline designed to mitigate smog production and to improve air quality by limiting the emission levels of certain chemical compounds such as benzene and other aromatic derivatives; often contains oxygenates (*q.v.*).

Reid vapor pressure A measure of the volatility of liquid fuels, especially gasoline.

Regeneration The reactivation of a catalyst by burning off the coke deposits.

Regenerator A reactor for catalyst reactivation.

Renewable energy sources Solar, wind, and other non-fossil fuel energy sources.

Rerunning The distillation of an oil that has already been distilled.

Research octane method (ROM) A test for determining the knock rating, in terms of octane numbers, of fuels for use in spark ignition engines. *See also* **Motor octane method**.

Reserves Well-identified resources that can be profitably extracted and utilized with existing technology.

Reservoir A domain where a petroleum may reside for an indeterminate time.

Residual asphalt *See* **Straight-run asphalt**.

Residual fuel oil Fuel oil obtained by blending the residual product(s) from various refining processes with suitable diluent(s) (usually middle distillates) to obtain the required fuel oil grades.

Residual oil *See* **Residuum**.

Residuum (resid; *pl.,* **residua)** The residue obtained from petroleum after nondestructive distillation has removed all the volatile materials from crude oil, e.g., an atmospheric ($345°C$; $650°F^+$) residuum.

Resins That portion of the maltenes (*q.v.*) that is adsorbed by a surface-active material such as clay or alumina; the fraction of deasphaltened oil that is insoluble in liquid propane but soluble in *n*-heptane.

Resource The total amount of a commodity (usually a mineral but can include non-minerals such as water and petroleum) that has been estimated to be ultimately available.

Rexforming A process combining platforming (*q.v.*) with aromatics extraction, wherein low octane raffinate is recycled to the platformer.

Rich oil Absorption oil containing dissolved natural gasoline fractions.

Riser The part of the bubble-plate assembly that channels the vapor and causes it to flow downward to escape through the liquid; also the vertical pipe where fluid catalytic cracking reactions occur.

Rock asphalt   Bitumen that occurs in formations that have a limiting ratio of bitumen to rock matrix.

Run-of-the-river reservoirs   Reservoirs with a large rate of flow-through compared to their volume.

Sand   A coarse granular mineral mainly comprising quartz grains that is derived from the chemical and physical weathering of rocks rich in quartz, notably sandstone and granite.

Sandstone   A sedimentary rock formed by compaction and cementation of sand grains; can be classified according to the mineral composition of the sand and cement.

SARA separation   A method of fractionation by which petroleum is separated into saturates, aromatics, resins, and asphaltene fractions.

Saturates   Paraffins and cycloparaffins (naphthenes).

Saybolt Furol viscosity   The time, in seconds (*Saybolt Furol seconds*; *SFS*), for 60 mL of fluid to flow through a capillary tube in a Saybolt Furol viscometer at specified temperatures between 70 and 210°F. The method is appropriate for high viscosity oils such as transmission, gear, and heavy fuel oils.

Saybolt Universal viscosity   The time, in seconds (*Saybolt Universal seconds*; *SUS*), for 60 mL of fluid to flow through a capillary tube in a Saybolt Universal viscometer at a given temperature.

Scale wax   The paraffin derived by removing the greater part of the oil from slack wax by sweating or solvent deoiling.

Scrubbing   Purifying a gas by washing with water or chemical; less frequently, the removal of entrained materials.

Secondary structure   The ordering of the atoms of a molecule in space relative to each other.

Sediment   An insoluble solid formed as a result of the storage instability and/or the thermal instability of petroleum and petroleum products.

Sedimentary strata   Natural layers of the earth that typically consist of mixtures of clay, silt, sand, organic matter, and various minerals; formed by or from deposits of sediments, especially from sand grains or silts transported from their source and deposited in water, such as sandstone and shale; or from calcareous remains of organisms, such as limestone.

Selective solvent   A solvent that, at certain temperatures and ratios, will preferentially dissolve more of one component of a mixture than of another and thereby permit partial separation.

Separation process   An upgrading process in which the constituents of petroleum are separated, usually without thermal decomposition, e.g., distillation and deasphalting.

Separator-Nobel dewaxing   A solvent (tricholoethylene) dewaxing process.

Shell fluid catalytic cracking   A two-stage fluid catalytic cracking process in which the catalyst is regenerated.

Shell still   A formerly used still in which the oil was charged into a closed cylindrical shell and the heat required for distillation was applied to the outside of the bottom from a firebox.

Sidestream   A liquid stream taken from any one of the intermediate plates of a bubble tower.

Sidestream stripper   A device used to perform further distillation on a liquid stream from any one of the plates of a bubble tower, usually by the use of steam.

Slack wax   The soft, oily crude wax obtained from the pressing of paraffin distillate or wax distillate.

Slime   A name used for petroleum in ancient texts.

Sludge   A semisolid to solid product that results from the storage instability and/or the thermal instability of petroleum and petroleum products.

Slurry hydroconversion process   A process in which the feedstock is contacted with hydrogen under pressure in the presence of a catalytic coke-inhibiting additive.

Slurry phase reactors   Tanks into which wastes, nutrients, and microorganisms are placed.

Smoke point   A measure of the burning cleanliness of jet fuel and kerosene.

Sodium hydroxide treatment   *See* **Caustic wash**.

Sodium plumbite   A solution prepared from a mixture of sodium hydroxide, lead oxide, and distilled water; used in making the doctor test for light oils such as gasoline and kerosene.

Solubility parameter   A measure of the solvent power and polarity of a solvent.

Solutizer–steam regenerative process   A chemical treating process for extracting mercaptans from gasoline or naphtha, using solutizers (potassium isobutyrate, potassium alkyl phenolate) in strong potassium hydroxide solution.

Solvent asphalt   The asphalt (*q.v.*) produced by solvent extraction of residua (*q.v.*) or by light hydrocarbon (propane) treatment of a residuum (*q.v.*) or an asphaltic crude oil.

Solvent deasphalting   A process for removing asphaltic and resinous materials from reduced crude oils, lubricating oil stocks, gas oils, or middle distillates through the extraction or precipitant action of low molecular weight hydrocarbon solvents. *See also* **Propane deasphalting**.

Solvent decarbonizing   *See* **Propane decarbonizing**.

Solvent deresining   *See* **Solvent deasphalting**.

Solvent dewaxing   A process for removing wax from oils by means of solvents, usually by chilling a mixture of solvent and waxy oil, filtering or centrifuging the wax that precipitates, and recovering the solvent.

Solvent extraction   A process for separating liquids by mixing the stream with a solvent that is immiscible with part of the waste but that will extract certain components of the waste stream.

Solvent naphtha   A refined naphtha of restricted boiling range used as a solvent. *Also* called *petroleum naphtha* or *petroleum spirits*.

Solvent refining   *See* **Solvent extraction**.

Sonic log   An oil well log based on the time required for sound to travel through rock, useful in determining porosity.

Sour crude oil   Crude oil containing an abnormally large amount of sulfur compounds. *See also* **Sweet crude oil**.

Spontaneous ignition   Ignition of a fuel, such as coal, under normal atmospheric conditions; usually induced by climatic conditions.

Specific gravity   The mass (or weight) of a unit volume of any substance at a specified temperature compared to the mass of an equal volume of pure water at a standard temperature. *See also* **Density**.

Spent catalyst   Catalyst that has lost much of its activity due to the deposition of coke and metals.

Stabilization   The removal of volatile constituents from a higher boiling fraction or product (*see* **Stripping**); the production of a product that to all intents and purposes does not undergo any further reaction when exposed to the air.

Stabilizer   A fractionating tower for removing light hydrocarbons from an oil to reduce vapor pressure; particularly applied to gasoline.

Standpipe   The pipe by which catalyst is conveyed between the reactor and the regenerator.

Steam cracking   A conversion process in which the feedstock is treated with superheated steam.

Steam distillation   Distillation in which vaporization of the volatile constituents is effected at a lower temperature by introduction of steam (*open steam*) directly into the charge.

Steam drive injection (steam injection)   The continuous injection of steam into one set of wells (injection wells) or other injection source to effect oil displacement toward and production from a second set of wells (production wells). Steam stimulation of production wells is *direct steam stimulation*, whereas driving steam injection to increase production from other wells is *indirect steam stimulation*.

Storage stability   The ability of a liquid to remain in storage over extended periods of time without appreciable deterioration as measured by gum formation and the depositions of insoluble material (sediment).

Straight-run asphalt   The asphalt (*q.v.*) produced by the distillation of asphaltic crude oil.

Straight-run products   Products obtained from a distillation unit and used without further treatment.

Strata   Layers including the solid iron-rich inner core, molten outer core, mantle, and crust of the earth.

Straw oil   Paraffin oil of pale straw color used for many process applications.

Stripping   A means of separating volatile components from less volatile ones in a liquid mixture by the partitioning the more volatile materials to a gas phase of air or steam (*see* **stabilization**).

Sulfonic acids   Acids obtained by contacting petroleum or a petroleum product with strong sulfuric acid.

Sulfuric acid alkylation   An alkylation process in which olefins (C_3, C_4, and C_5) combine with isobutane in the presence of a catalyst (sulfuric acid) to form branched-chain hydrocarbons; used especially in gasoline blending stock.

Suspensoid catalytic cracking   A nonregenerative cracking process in which cracking stock is mixed with a slurry of catalyst (usually clay) and cycle oil and passed through the coils of a heater.

Sweated wax   A crude wax freed from oil by having been passed through a sweater.

Sweating   The separation of paraffin oil and low-melting wax from paraffin wax.

Sweet crude oil   Crude oil containing little sulfur *See also* **Sour crude oil**.

Sweetening   The process that improves the odor and color of petroleum products by oxidizing or removing the sulfur-containing and unsaturated compounds.

Synthetic crude oil (syncrude)   A hydrocarbon product produced by the conversion of coal, oil shale, or tar sand bitumen that resembles conventional crude oil. Can be refined in a petroleum refinery *See* **Refinery**.

Tar   The volatile, brown to black, oily, viscous product from the destructive distillation of many bituminous or other organic materials, especially coal; a name used for petroleum in ancient texts.

Tar sand   *See* **Bituminous sand**.

Tertiary structure   The three-dimensional structure of a molecule.

Tetraethyllead (TEL)   An organic compound of lead, $Pb(CH_3)_4$, that when added in small amounts, increases the antiknock quality of gasoline.

Thermal coke   The carbonaceous residue formed as a result of a noncatalytic thermal process; the Conradson carbon residue; the Ramsbottom carbon residue.

Thermal cracking   A process that decomposes, rearranges, or combines hydrocarbon molecules by the application of heat without the aid of catalysts.

Thermal polymerization   A thermal process to convert light hydrocarbon gases into liquid fuels.

Thermal process   Any refining process that utilizes heat without the aid of a catalyst.

Thermal reforming   A process using heat (but no catalyst) to effect molecular rearrangement of low octane naphtha into gasoline of higher antiknock quality.

Thermal stability   The ability of a liquid to withstand relatively high temperatures for short periods of time without the formation of carbonaceous deposits (sediment or coke).

Thermofor catalytic cracking   A continuous, moving-bed catalytic cracking process.

Thermofor catalytic reforming   A reforming process in which the synthetic, bead-type catalyst of coprecipitated chromia (Cr_2O_3) and alumina (Al_2O_3) flows down through the reactor concurrently with the feedstock.

Thermofor continuous percolation   A continuous clay-treating process to stabilize and decolorize lubricants or waxes.

Topped crude   Petroleum that has had volatile constituents removed up to a certain temperature, e.g., $250°C^+$ ($480°F^+$) topped crude; not always the same as a residuum (*q.v.*).

Topping   The distillation of crude oil to remove light fractions only.

Tower   Equipment for increasing the degree of separation obtained during the distillation of oil in a still.

Trace element   An element that occurs at very low levels in a given system.

Traps   Sediments in which oil and gas accumulate and from which further migration is prevented.

Treatment   Any method, technique, or process that changes the physical and/or chemical character of petroleum.

Trickle hydrodesulfurization   A fixed-bed process for desulfurizing middle distillates.

Trillion 1×10^{12}.

True boiling point (True boiling range)   The boiling point (boiling range) of a crude oil fraction or a crude oil product under standard conditions of temperature and pressure.

Tube-and-tank cracking   An older liquid-phase thermal cracking process.

Ultimate analysis   Elemental composition.

Ultrafining   A fixed-bed catalytic hydrogenation process to desulfurize naphthas and upgrade distillates by essentially removing sulfur, nitrogen, and other materials.

Ultraforming   A low pressure naphtha-reforming process employing on-stream regeneration of a platinum-on-alumina catalyst and producing high yields of hydrogen and high octane number reformate.

Unassociated molecular weight   The molecular weight of asphaltenes in a nonassociating (polar) solvent such as dichlorobenzene, pyridine, or nitrobenzene.

Unifining   A fixed-bed catalytic process to desulfurize and hydrogenate refinery distillates.

Unisol process   A chemical process for extracting mercaptan sulfur and certain nitrogen compounds from sour gasoline or distillates using regenerable aqueous solutions of sodium or potassium hydroxide containing methanol.

Universal viscosity   *See* **Saybolt Universal viscosity**.

Unstable   Usually refers to a petroleum product that has fairly volatile constituents present or to the presence of olefins and other unsaturated constituents.

UOP alkylation   A process using hydrofluoric acid (which can be regenerated) as a catalyst to unite olefins with isobutane.

UOP copper sweetening   A fixed-bed process for sweetening gasoline by converting mercaptans to disulfides by contact with ammonium chloride and copper sulfate in a bed.

UOP fluid catalytic cracking   A fluid process of using a reactor-over-regenerator design.

Upgrading   The conversion of petroleum to value-added salable products.

Urea dewaxing   A continuous dewaxing process for producing low pour point oils, using urea, which forms a solid complex (adduct) with the straight-chain wax paraffins in the stock. The complex is readily separated by filtration.

Vacuum distillation   Distillation (*q.v.*) under reduced pressure.

Vacuum residuum   A residuum (*q.v.*) obtained by distillation of a crude oil under vacuum (reduced pressure); the portion of petroleum that boils above a selected temperature such as 510°C (950°F) or 565°C (1050°F).

Vapor-phase cracking   A high temperature, low pressure conversion process.

Vapor-phase hydrodesulfurization   A fixed-bed process for desulfurization and hydrogenation of naphtha.

Visbreaking   A process for reducing the viscosity of heavy feedstocks by controlled thermal decomposition.

Viscosity   A measure of the ability of a liquid to flow or a measure of its resistance to flow; the force required to move a plane surface of area $1\,m^2$ over another parallel plane surface 1 m away at a rate of 1 m/s when both surfaces are immersed in the fluid.

VGC (viscosity-gravity constant)   An index of the chemical composition of crude oil defined by the general relation between specific gravity (sp gr) at 60°F and Saybolt Universal viscosity (SUV) at 100°F:

$$a = 10 \times sp\ gr - \frac{1.0752 \log (SUV - 38)}{10 \times sp\ gr} - \log (SUV - 38)$$

The constant, *a*, is low for the paraffinic crude oils and high for the naphthenic crude oils.

VI (Viscosity index)   An arbitrary scale used to show the magnitude of viscosity changes in lubricating oils with changes in temperature.

Viscosity-gravity constant   *See* **VGC**.

Viscosity index   *See* **VI**.

Watson characterization factor   *See* **Characterization factor**.

Wax   *See* **Mineral wax** *and* **Paraffin wax**.

Wax distillate   A neutral distillate containing a high percentage of crystallizable paraffin wax, obtained on the distillation of paraffin or mixed-base crude and the reduction of neutral lubricating stocks.

Wax fractionation   A continuous process for producing waxes of low oil content from wax concentrates. *See also* **MEK deoiling**.

Wax manufacturing   A process for producing oil-free waxes.

Weathered crude oil   Crude oil that, due to natural causes during storage and handling, has lost an appreciable quantity of its more volatile components; also crude oil that has taken up oxygen.

Wet gas   Gas containing a relatively high proportion of hydrocarbons that are recoverable as liquids. *See also* **Lean gas**.

Wet scrubbers   Devices in which a countercurrent liquid spray is used to remove impurities and particulate matter from a gas stream.

White oil   A generic term applied to highly refined, colorless hydrocarbon oils of low volatility; covers a wide range of viscosity.

Wood alcohol   *See* **Methyl alcohol.**

Zeolite   A crystalline aluminosilicate used as a catalyst and having a particular chemical and physical structure.

Conversion Factors and SI Units

CONVERSION FACTORS

1 acre $= 43,560$ sq ft
1 acre-foot $= 7758.0$ bbl
1 atmosphere (atm) $= 760$ mmHg $= 14.696$ psi $= 29.91$ in.Hg
1 atmosphere (atm) $= 1.0133$ bars $= 33.899$ ftH$_2$O
1 barrel (bbl) (oil) $= 42$ gal $= 5.6146$ cu ft
1 barrel (bbl) (water) $= 350$ lb at $60°$F
1 bbl/day $= 1.84$ cm^3/s
1 British thermal unit (Btu) $= 778.26$ ft $\cdot$ lb
1 centipoise (cP) $\times 2.42 =$ lb mass/(ft $\cdot$ h), viscosity
1 cP $0.000672 =$ lb mass/(ft s), viscosity
1 cuf $= 28,317$ cm^3 $= 7.4805$ gal
Density of water at $60°$F $= 0.999$ g/cm^3 $= 62.367$ lb/cu ft $= 8.337$ lb/gal
1 gal $= 231$ cu in. $= 3,785.4$ cm^3 $= 0.13368$ cu ft
1 horsepower-hour (hp $\cdot$ h) $= 0.7457$ kWh $= 2544.5$ Btu
1 h $= 550$ ft $\cdot$ lb/s $= 745.7$ watts (W)
1 in. $= 2.54$ cm
1 m $= 100$ cm $= 1000$ mm $= 10^6$ μm $= 10^{10}$ angstroms (Å)
1 ounce (oz) $= 28.35$ g
1 lb $= 453.59$ g $= 7000$ grains (gr)
1 square mile $= 640$ acres

SI METRIC CONVERSION FACTORS (E = exponent; i.e., E + 03 = 10^3)

acre-feet $\times$ 1.233482 E + 03 = meters cubed
barrels $\times$ 1.589873 E − 01 = meters cubed
centipoise $\times$ 1.000000 E − 03 = pascal-seconds
darcy $\times$ 9.869233 E − 01 = micrometers squared

feet $\times$ 3.048000 E $-$ 01 $=$ meters
pounds/acre-foot $\times$ 3.677332 E $-$ 04 $=$ kilograms/meters cubed
pounds/square inch $\times$ 6.894757 E $+$ 00 $=$ kilopascals
dyn/cm $\times$ 1.000000 E $+$ 00 $=$ mN/m
parts per million $\times$ 1.000000 E $+$ 00 $=$ milligrams/kilograms

Index